(Generalized) Fuzzy Matrices and Relations

(Generalized) Fuzzy Matrices and Relations

Herbert Toth

World Scientific

NEW JERSEY • LONDON • SINGAPORE • BEIJING • SHANGHAI • HONG KONG • TAIPEI • CHENNAI • TOKYO

Published by

World Scientific Publishing Co. Pte. Ltd.
5 Toh Tuck Link, Singapore 596224
USA office: 27 Warren Street, Suite 401-402, Hackensack, NJ 07601
UK office: 57 Shelton Street, Covent Garden, London WC2H 9HE

Library of Congress Cataloging-in-Publication Data
Names: Toth, Herbert, author.
Title: (Generalized) fuzzy matrices and relations / Herbert Toth.
Other titles: Fuzzy matrices and relations
Description: New Jersey : World Scientific, [2025] | Includes bibliographical references and index.
Identifiers: LCCN 2024033297 | ISBN 9789811298356 (hardcover) |
ISBN 9789811298363 (ebook for institutions) | ISBN 9789811298370 (ebook for individuals)
Subjects: LCSH: Fuzzy sets. | Matrices.
Classification: LCC QA248.5 .T68 2025 | DDC 512.9/434--dc23/eng/20250107
LC record available at https://lccn.loc.gov/2024033297

British Library Cataloguing-in-Publication Data
A catalogue record for this book is available from the British Library.

For any available supplementary material, please visit
https://www.worldscientific.com/worldscibooks/10.1142/13989#t=suppl

Desk Editors: Soundararajan Raghuraman/Rok Ting Tan

Typeset by Stallion Press
Email: enquiries@stallionpress.com

CONTENTS

PREFACE

My intention with this book is to provide an *overview* on the main concepts and results around fuzzy matrices and fuzzy relations. Concerning the book's title you may ask

(1) why it starts with '*(Generalized)*': This is, because there is a clear emphasis on underlying structures more general than the two-valued Boolean algebra $\{0, 1\}$ or the real unit interval $[0, 1]$ (such as various kinds of semirings, inclines and residuated lattices). The parentheses indicate that these two special classical cases are, of course, also considered.

(2) why both '*Matrices*' and '*Relations*' do appear in it: The answer is that although matrices and relations usually may be identified from a syntactical point of view (by a suitable isomorphism), semantically different concepts are associated to these two notions as you can see from the text.
I have tried to be consequent in using that one of these two terms which is semantically adequate in a given context. But I am sure that I did not always succeed, so some naming inconsistencies may have slipped in. However, this should not be an obstacle for understanding the material presented.

There are two ways how this volume can be regarded: (i) as a companion to my *Relations: Concrete, Abstract, and Applied - An Introduction* (World Scientific, 2020) with a clear emphasis on the many-valued concepts, or (ii) as a stand-alone volume that can be read independently from it. But since the topics of the two books are overlapping, the current one borrows from the former where appropriate. Hence there are some repetitions as preliminary or reference material.

As my former book did, also this one does not really present new results, but tries to give a rather self-contained compilation of topics selected from the huge amount of existing knowledge attained over the last five decades in a hopefully somewhat systematic arrangement. So, of course, you will find repetitions of stuff that has already been presented elsewhere in books or papers. Though very many researchers have been active, you can (and reading through this text hopefully will) see later on that almost always there are dominant contributors to the various topics.

The material in this book will not only be presented from a purely technical, or *syntactic*, but also from a *semantic*, and sometimes from a historical viewpoint if appropriate: This way the reader will get an impression of how things have evolved within fuzzy set theory and what has been the original motivation behind, but also how the formal notions have become more and more abstract and advanced in the course of time.

There is a considerable variety of ways to extend definitions of properties of classical binary relations. As you can see from the considerations in this book, there is a big step from two-valued all-or-nothing concepts to many-valued graded ones, drastically changing both the syntactical and semantical points of view.

A first fundamental task is to decide between working with fuzzy relations on fuzzy sets or on ordinary sets, with the latter one having evolved as the standard framework. Another fundamental choice is the one between working with 'crisp' properties of fuzzy relations or with 'fuzzified' (graded) ones.

Therefore, dealing with fuzzy relations, one has to choose from the possibilities available for generalization: The considerations in this book will take place in the context of fuzzy relations on ordinary sets with crisp properties. There is one exception: In Chapter 13 on fuzzy graphs, following the mainstream, we consider fuzzy relations on fuzzy sets or, alternatively, fuzzy graphs with fuzzy node sets.

Only few basic mathematical and logical preliminaries will be collected in Chapter 1. Further notions and concepts will be introduced in due course at the places where they are first needed. The term "*fuzzy*" is used in a generic, i.e. not only in the narrow original sense of Zadeh's seminal paper of 1965: It will denote general many-valued structures and concepts, but of course including the usual two-valued ones as special cases.

As a final remark let me say that, although this book is a one-person endeavor, I nonetheless hope that I succeeded in keeping the number of all kind of errors in the text within a tolerable extent.

Vienna, May 2024 *Herbert Toth*

ACKNOWLEDGMENTS

General Thanks

As in my former book: My sincere thanks go to all the many pioneers, in both mathematics and technology; to the famous and well known, but also to the nameless and unknown ones!

Without their visions, and without the concepts they have invented, the tools they have programmed, the machines they have built, and the data they have collected, writing this book would have been absolutely impossible for me.

Special Thanks

First of all, I am very grateful to Dr. Furusawa for helping me with the proof of Lemma 14.13(5) for which I have not been clever enough to find it myself.

My thanks also go to the anonymous reviewers for spending their time to gain a positive opinion of this text.

And, of course, I would like to express my sincere thanks to all the people from WSPC involved in the publishing process, and to Ms. Rok Ting Tan and Mr. Soundara Rajan in particular, for their always smooth and friendly cooperation and great support.

I will also be grateful to any reader who will be so kind to email me hints on remaining typos and other errors to **herbert.toth@inode.at**.

INDEX OF NOTATIONS

PART 1

Preliminaries and Classical Two-valued Concepts

CHAPTER 1

SOME MATHEMATICAL PREREQUISITES

We assume that the reader is familiar with the most elementary concepts of set theory and first order logic and their standard notations. But in order to achieve the goal of getting this book self-contained to a reasonable degree, some mathematical prerequisites are arranged in Chapter 1, usually without further comments or proofs. They provide the meta-level infrastructure which we need for the considerations about our object-level topics, i.e. fuzzy matrices and relations. So this chapter includes some theorems of both propositional and predicate logic as well as some basic general results on algebraic structures together with examples that will become essential later in the book. We will also have to include some basic facts on relations and functions.

Most of the concepts in this chapter will not be introduced in their full algebraic generality, but only to the extent that is necessary for the understanding of subsequent material. Readers interested in more details than those presented in this text, may consult one of the books mentioned in the chapter's references for further information.

Important note: *If not obvious from their meaning, as for e.g. the set of natural numbers* $\mathbb{N}$*, or if not explicitly stated otherwise, the sets mentioned in this book are assumed to be finite. However, for sake of mathematical generality, completeness assumptions are usually added where appropriate.*

1.1 Sets and their Operations

The reader is assumed to be familiar with the basic operations on sets and their properties. For finite sets S, $|S|$ denotes their **cardinality**. There is a special and uniquely determined set $\emptyset$ containing no element, whence it is called the **empty set**; it may formally be defined as $\emptyset := \{x : x \neq x\}$. Thus, $|\emptyset| = 0$, but $|\{\emptyset\}| = |\{\{1,\{a,b,c\}\}\}| = 1$, $|\{1,\{a,b,c\}\}| = 2$, and $|\{a,b,c\}| = 3$. Another simple example is $D_{12} = \{1,2,3,4,6,12\}$ containing the number 12 and all its divisors. The set $C_{12} = \{\{1,2\},\{1,3\},\{1,2,4\}, \{1,2,3,6\},\{1,2,3,4,6,12\}\}$ provides additional information about the divisors of the divisors of 12; it has only sets as elements. Sets with this property will be called **collections** and denoted by calligraphic letters like $\mathscr{C}, \mathscr{D}$.

Definition 1.1.

(a) *Two non-empty sets S and T are called* **disjoint** *if* $S \cap T = \emptyset$.
(b) $\mathscr{C}$ *is called a collection of* **pairwise disjoint** *sets if* $\forall A, B \in \mathscr{C}[A \neq B \rightarrow A \cap B = \emptyset]$.
(c) *The collection* $\mathscr{P} = \{B : B \subseteq A\}$ *of all subsets of some set A is called the* **power set** *of A, denoted by either $\mathcal{P}(A)$ or $\mathbf{2}^A$.*
(d) *A collection $\mathscr{A} \subseteq \mathcal{P}(A)$ is called a* **system over** *A.*
(e) *A system $\mathscr{A} = \{A_i : i \in I\} \neq \emptyset$ is called a* **partial partition** *of A if $A_i \neq \emptyset$ for every $i \in I$, and A_i and A_j are pairwise disjoint sets.*

Since $\{\emptyset, A\} \subseteq \mathcal{P}(A)$ we always have $\mathcal{P}(A) \neq \emptyset$. How the power set is related to subset, intersection, and union is described in

Proposition 1.2. *For all sets A, B we have:*

(1) $A \subseteq B \leftrightarrow \mathcal{P}(A) \subseteq \mathcal{P}(B)$.
(2) $\mathcal{P}(A) \cap \mathcal{P}(B) = \mathcal{P}(A \cap B)$.
(3) $\mathcal{P}(A) \cup \mathcal{P}(B) \subseteq \mathcal{P}(A \cup B)$.

Two important kinds of systems over a set A are introduced in

Definition 1.3.

(a) *A system $\mathscr{A} = \{A_i, i \in I\} \neq \emptyset$ over a set $A \neq \emptyset$ is called a* **covering** *of A iff*

$$(1) \quad \forall i \in I[A_i \neq \emptyset], \qquad \text{and} \qquad (2) \quad \bigcup_{i \in I} A_i = A.$$

(b) *A system $\mathscr{A} = \{A_i, i \in I\} \neq \emptyset$ over a set $A \neq \emptyset$ is called a* **partition** *of A iff it is a partial partition that is also a covering of A. A set $A_i \in \mathscr{A}$ is called a* **block** *of the partition. The set of the partitions of A is denoted by Π_A. If $\pi \in \Pi_A$ then we denote the block of a in π by $[a]_\pi$.*

Definition 1.4. *Let A be a set, and $\pi, \pi' \in \Pi_A$ be two partitions of A. Then π is called a* **refinement** *of (or* **finer** *than) π', denoted by $\pi \leq \pi'$, if every block of π' is a union of blocks of π, i.e. π is a further fragmentation of π'.*

There are two special partitions in Π_A: The partition of A consisting of all singletons: $\alpha_A = \{\{a\} | a \in A\}$; and the partition that contains A itself: $\omega_A = \{A\}$. Obviously, we have $\alpha_A \leq \pi \leq \omega_A$ for all $\pi \in \Pi_A$.

1.1.1 *Cartesian Products, Relations, and Functions*

Definition 1.5. *A set $A := \{\{x\}, \{x, y\}\}$ is called an* **ordered pair** *with first component x and second component y and will be denoted by $\langle x, y \rangle$.*

The following theorem implies that for an ordered pair its first and second component are uniquely determined. This justifies the definition given above.

Theorem 1.6. *Let $A = \langle x, y\rangle$ and $B = \langle u, v\rangle$; then $A = B$ if, and only if, $x = u$ and $y = v$.*

Proof. We prove the non-trivial direction by distinction in two cases:
Case 1. $x = y$ yields $A = \{\{x\}\}$, whence B must also consist of a single set, i.e. $\{u, v\} = \{u\}$. This implies $u = v$, and from our assumption $A = B$ we get $\{x\} = \{u\}$, and finally $x = u$ and trivially also $y = v$.
Case 2. If $x \neq y$ then neither A nor B are singletons themselves (i.e. $|A| > 1$), but each of them contains exactly one singleton; thus we again get $\{x\} = \{u\}$. Since $A = B$ we must also have $\{x, y\} = \{u, v\}$. Since $v \in \{x, y\}$ and $v \neq u = x$ we get $v = y$ as desired. □

Definition 1.7.

(a) *The* **Cartesian product** *of sets $A_1, \ldots, A_n$ is defined as*

$$\prod_{i\in I} A_i := A_1 \times \ldots, \times A_n := \{\langle a_1, \ldots, a_n\rangle : a_i \in A_i\}.$$

The vectors $\langle a_1, \ldots, a_n\rangle$ are called n-tuples. In case $A_i = A_j$ for all i, j, the product is written as A^n.

(b) *The mapping $\pi_i : \prod_{i\in I} A_i \longrightarrow A_i, i \in I$, defined by $\pi_i(\langle a_1, \ldots, a_n\rangle) = a_i$, is called the* **$i$-th projection** *of $\prod_{i\in I} A_i$.*

Def. 1.7 describes a concept in a general set-theoretical formalism. By taking $n = 2$ we get ordered pairs as described in Def. 1.5 above: If R designates a concept that applies to pairs of elements from a certain set S, we usually speak of "*the binary relation R on S*". It is also common practice to tacitly identify the property R with the relation it defines, so that $R(x, y)$ means the same as $\langle x, y\rangle \in R$. And it will be binary relations that are almost exclusively in the focus of this book.

The Cartesian product is (left and right) distributive over union and intersection of sets. To keep notation simple, we will usually restrict our considerations to the case $n = 2$.

Theorem 1.8. *Let $* \in \{\cup, \cap\}$; then for any sets R, S, and T the following holds:*

(1) $(R * S) \times T = (R \times T) * (S \times T)$
(2) $T \times (R * S) = (T \times R) * (T \times S)$

Proof. As an example we give a proof for $(R \cap S) \times T = (R \times T) \cap (S \times T)$: From $\langle x, y\rangle \in (R \cap S) \times T$ we get $x \in R \cap S$ and $y \in T$. Thus $\langle x, y\rangle \in R \times T$ and $\langle x, y\rangle \in S \times T$, whence $\langle x, y\rangle \in (R \times T) \cap (S \times T)$. The other way round, $\langle x, y\rangle \in (R \times T) \cap (S \times T)$ implies $x \in R$, $x \in S$, and $y \in T$, whence $\langle x, y\rangle \in (R \cap S) \times T$. □

Proposition 1.9. *For all sets S_1, S_2, T_1, T_2 we have: If $S_1 \subseteq S_2$ and $T_1 \subseteq T_2$, then $S_1 \times T_1 \subseteq S_2 \times T_2$.*

Relations are so omnipresent in our daily life that we rarely become aware of them as a mathematical concept. Who reasons about what follows from "*She is a relative of mine*"? E.g. that I am also a relative of her, and that her relatives are also my relatives, some of them probably not as near as she is. Or that a cooler is part of your car, but, quite naturally, your car is not part of its cooler. Readers can surely find a lot more of such everyday examples. So it seems justified to assert that without relations we would not be able to manage our daily routine and life as a whole (and, of course, one could not do mathematics). This means that without a rudimentary knowledge of relations on the meta-language level we cannot introduce and analyze the notion of a relation and its properties on the object level. As a consequence, we have to include the definitions of some basic notions around the concepts of relation and function to our preliminaries, and we will now do so without many comments.

Functions are a special kind of relations, and we will return to them later in Section 2.8.1. We will then notice that function composition fg will correspond to relation composition $g \circ f$. Though both can be taken as the standard notation, do always assure which is the one used in your current text, and whether it is about functions or relations: *left* properties of functions will become *right* properties of relations.

Definition 1.10. *Let X and Y denote two sets.*

(a) *$R \subseteq X \times Y$ is called a (**binary**) **relation** from X to Y.*

(b) *A binary relation R on X, i.e. $R \subseteq X \times X$, is called*

- **reflexive** *iff $\langle x, x\rangle \in R$, for all $x \in X$;*
- **symmetric** *iff $\langle x, y\rangle \in R \rightarrow \langle y, x\rangle \in R$, for all $x, y \in X$;*
- **transitive** *iff $\langle x, y\rangle \in R \wedge \langle y, z\rangle \in R \rightarrow \langle x, z\rangle \in R$, for all $x, y, z \in X$.*

(c) *$R \subseteq X \times X$, is called **equivalence relation** if it is reflexive, symmetric, and transitive. The set $x/R = [x]_R := \{y \in X : \langle x, y\rangle \in R\}$ is called the **equivalence class** of x in R, and the set of set of all equivalence classes is called the **quotient set** of X modulo R, denoted as X/R.*

Definition 1.11. *Let X and Y denote two sets.*

(a) *$f \subseteq X \times Y$ is called a (**total**) **function** or **mapping** from X into Y iff*

$$|f \cap (\{x\} \times Y)| = 1 \text{ for all } x \in X,$$

*and is denoted by $f : X \longrightarrow Y$. $f(x)$ is the second component of the unique element in $f \cap (\{x\} \times Y)$. The set of all functions from X into Y is denoted by Y^X. f is called a **partial function** iff $|f \cap (\{x\} \times Y)| \leq 1$ for all $x \in X$.*

(b) *A function f is called*

- **injective** *(or **one-to-one**) iff $|f \cap (X \times \{y\})| \leq 1$ for all $y \in Y$;*
- **surjective** *(or **onto**) iff $|f \cap (X \times \{y\})| \geq 1$ for all $y \in Y$;*
- **bijective** *iff $|f \cap (X \times \{y\})| = 1$ for all $y \in Y$.*

(c) *The **domain** of a function $f : X \longrightarrow Y$ is*

$$\mathsf{dom}(f) = \{x \in X : \exists y \in Y[f(x) = y]\} \subseteq X,$$

*and the **range** of f is*

$$\mathsf{rng}(f) = \{y \in Y : \exists x \in X[f(x) = y]\} \subseteq Y.$$

(d) *The* **composition** *of two functions* $f: X \longrightarrow Y$ *and* $g: Y \longrightarrow Z$*, denoted as* $g(f(.))$ *or simply as* $gf(.)$*, is the function* $h: X \longrightarrow Z$*, where* $h(x) = g(f(x))$ *is the uniquely determined element* $z \in Z$ *such that* $f(x) = y$ *and* $g(y) = z$*.*

A more familiar, but less concise, characterization of the notions defined in item *(b)* above is as follows: A function $f: X \longrightarrow Y$ is called **injective** if

$$\forall x_1, x_2 \in X\,[f(x_1) = f(x_2) \rightarrow x_1 = x_2], \tag{1.1}$$

i.e. elements of X with the same image must be equal. The contrapositive is equally useful:

$$\forall x_1, x_2 \in X\,[x_1 \neq x_2 \rightarrow f(x_1) \neq f(x_2)], \tag{1.2}$$

i.e. different elements of X yield different images. f is called **surjective** if

$$\forall y \in Y\, \exists x \in X\,[f(x) = y], \tag{1.3}$$

i.e. each element of Y has a pre-image. By definition, a function is **bijective**, if it is injective and surjective, i.e.

$$\forall y \in Y\, \exists! x \in X\,[f(x) = y],$$

where $\exists!$ means "*there is exactly one*".

If $S \subseteq X$ then the **restriction of** f **to** S, denoted by $f|_S$, is a function $f|_S: S \longrightarrow Y$ defined by $f|_S(x) = f(x)$, for all $x \in S$. id_X denotes the bijection $id_X: X \longrightarrow X$ with $id_X(x) = x$ and is called the **identity function** on X.

A function $f: X \longrightarrow Y$ gives rise to two induced functions between the powersets of X and Y. Since one of them is also denoted by f, the mapping $f: X \longrightarrow Y$ will be written as $f(x) = y$ which is different from $f[S] = T$ for the induced map $f: \mathcal{P}(X) \longrightarrow \mathcal{P}(Y)$. The latter one being defined as

$$f: \mathcal{P}(X) \longrightarrow \mathcal{P}(Y),\ f[S] = \{f(x) : x \in S\} \subseteq Y \text{ for all } S \subseteq X,$$

where $f[S]$ is called **image of** S **under** f. The **pre-image of** T **under** f is

$$f^{-1}: \mathcal{P}(Y) \longrightarrow \mathcal{P}(X),\ f^{-1}[T] = \{x : f(x) \in T\} \subseteq X, \text{ for all } T \subseteq Y.$$

f **maps** S **into** T if $f[S] \subseteq T$ and **maps** S **onto** T if $f[S] = T$.

1.1.1.1 *Some Basic Function Properties*

In this section we list some important and often used properties of functions. The next proposition reveals the surprising fact that from the two induced functions introduced above the pre-image function is "better behaved" with respect to the usual set operations, since it gives equality for both union and intersection.

Proposition 1.12. *Suppose* $f: X \longrightarrow Y$ *is a function, and* $S, T \subseteq X$ *and* $U, V \subseteq Y$*. Then*

(1) $f[S \cup T] = f[S] \cup f[T]$
(2) $f[S \cap T] \subseteq f[S] \cap f[T]$
(3) $f^{-1}[U \cup V] = f^{-1}[U] \cup f^{-1}[V]$
(4) $f^{-1}[U \cap V] = f^{-1}[U] \cap f^{-1}[V]$

Proof. (2): $b \in f[S \cap T] \to \exists a \in S \cap T\,[b = f(a)]$. For this a we have $a \in S \wedge a \in T$, so $b \in f[S] \wedge b \in f[T]$.

(4): For $x \in X$ we have, $x \in f^{-1}[U \cap V] \leftrightarrow f(x) \in U \cap V \leftrightarrow f(x) \in U \wedge f(x) \in V \leftrightarrow x \in f^{-1}[U] \wedge x \in f^{-1}[V]$. □

Proposition 1.13.

(1) *Let $f : X \longrightarrow Y$ be a mapping. Then f is surjective iff there exists a mapping $g : Y \longrightarrow X$, called a* **right inverse** *of f, such that $fg = id_Y$.*
(2) *Let $f : X \longrightarrow Y$ be a surjective mapping, and let $g, h : Y \longrightarrow Z$ be two mappings from Y into Z. If $gf = hf$ then $g = h$. Thus, surjective mappings are* **right cancellative.**
(3) *Let $f : X \longrightarrow Y$ be a mapping. Then f is injective iff there exists a mapping $g : \mathrm{rng}(f) \longrightarrow X$, called a* **left inverse** *of f, such that $gf = id_X$.*
(4) *Let $f : X \longrightarrow Y$ be a injective mapping, and let $g, h : Y \longrightarrow Z$ be two mappings from Y into Z. If $fg = fh$ then $g = h$. Thus, injective mappings are* **left cancellative.**

Proof. (1): If there exists a mapping g such that $f(g(y)) = y$ for all $y \in Y$, then f clearly is surjective. Conversely, suppose f to be surjective; then, for each $y \in Y$, select $g(y)$ in the non-empty set $f^{-1}(y)$. This defines a mapping $g : Y \longrightarrow X$ such that $f(g(y)) = y$ for all $y \in Y$.

(2): By (1) there exists a mapping $\tilde{f} : Y \longrightarrow X$ such that $f\tilde{f} = id_Y$. Therefore, $gf = hf$ implies $gf\tilde{f} = hf\tilde{f}$, whence $g = h$. □

Proposition 1.14. *For functions $f : X \longrightarrow Y$ and $g : Y \longrightarrow Z$ the following holds:*

(1) *If f and g are injective then also $gf : X \longrightarrow Z$.*
(2) *If f and g are surjective then also $gf : X \longrightarrow Z$.*

Proof. (1): Since g is injective, $g(f(a)) = g(f(a'))$ implies $f(a) = f(a')$, and since f is injective we get $a = a'$. So gf is injective.

(2): Since g is surjective, $\forall c \in Z \exists b \in Y\,[g(b) = c]$, and similarly $\forall b \in Y \exists a \in X\,[f(a) = b]$. Together this implies $\forall c \in Z \exists a \in X\,[g(f(a)) = c]$. So gf is surjective. □

Proposition 1.15. *For functions $f : X \longrightarrow Y$ and $g : Y \longrightarrow X$ the following holds:*

(1) *If gf is injective then also f.*
(2) *If gf is surjective then also g.*

Proof. (1): Assume $x, x' \in X$ and $f(x) = f(x')$. Then $g(f(x)) = g(f(x'))$, and since gf is injective we get $x = x'$, whence f is injective too. □

Proposition 1.16. *Let $f : X \longrightarrow Y$ be a mapping. Then the following statements are equivalent:*

(1) *f is bijective;*
(2) *there exists a (unique and bijective) mapping $g: Y \longrightarrow X$ that is both a right and left inverse of f;*
(3) *f has both a right and left inverse.*

Proof. (1) → (2): Let f be bijective; then for each $y \in Y$ there is a unique $x \in X$ such that $f(x) = y$. By Prop. 1.13(1) there exists a function $g: Y \longrightarrow X$ such that $fg = id_Y$, defined by $g(y) = x$, where x is the unique element such that $f(x) = y$. This mapping g also satisfies $gf = id_X$, whence it is also a left inverse of f.

Moreover, g is bijective by Prop. 1.15, since fg is injective and gf is surjective. Finally assume that both g_1 and g_2 are two right (hence, by what we have just shown, also left) inverses of f; then $g_1 = g_1 id_Y = g_1(fg_2) = (g_1 f)g_2 = id_X g_2 = g_2$. This unique mapping g is called the **inverse** of f, and is denoted by f^{-1}. (In the proof we have tacitly used that function composition is associative; this is shown in Th. 2.22(3) below for general relations.)

(2) → (3): Trivial.

(3) → (1): If f has both a right and left inverse, say r and l respectively, then $fr = id_Y$ and $lf = id_X$. So f is bijective by Prop. 1.15. □

Instead of proving that $f: X \longrightarrow Y$ is a bijection by directly showing that it is both one-to-one and onto, Prop. 1.16 provides an alternative way: If we can define a function $g: Y \longrightarrow X$ which is both a left and right inverse of f, then f is bijective and $g = f^{-1}$.

1.2 Basics from Predicate Logic

In this section we will list some basic facts and theorems from both propositional (also: sentential) and first order (also: predicate) logic without much further discussion or justification. They will be used later on in order to facilitate the comprehension of some proofs which will be formulated in a free mixture of the usual "colloquial mathematical speech" and a pure first order logic style. In the following we assume acquaintance with the most basic concepts and notions of logic only as e.g. axioms, connectives, quantifiers, well formed formulas (wffs), inference rules; the rest should become obvious from the text.

We will not focus on mathematical elegance of axiom systems and the like; that is, we avoid 'technical' details which are not necessary for our current needs. Instead we try to provide a toolbox that helps mathematicians or computer scientists master the problems in their daily proof work. The theorems and rules given below should appear in any standard introduction to logic like e.g. Copi et al. [7] or Hurley [13].

Typically, the development of a formal calculus for classical propositional logic, say $\mathscr{L}$, starts with the definition of what counts as a *formula* in $\mathscr{L}$, followed by the proof theory (around the notions of *rule of inference, proof,* and *theorem*) and the semantics of $\mathscr{L}$ (*validity* and *tautology*). There is a set of wffs, called the set $A_{\mathscr{L}}$ of *axioms of* $\mathscr{L}$ which are taken as given, and a finite set of *inference rules* which tell us when a conclusion can be derived from a set of premises.

A *proof* in $\mathscr{L}$ is a sequence Φ_i such that, for each i, $\Phi_i \in A_{\mathscr{L}}$ or Φ_i is a direct consequence of some of the preceding wffs in the sequence by application of one of the rules of inference of $\mathscr{L}$. A *theorem* of $\mathscr{L}$ is a wff Φ such that it is the last wff in a proof in $\mathscr{L}$. In a second step the system for $\mathscr{L}$ becomes extended with further axioms and rules to properly deal with quantification.

In our situation, viz. as users of predicate logic, we do not need to take care what are the axioms of a certain logical calculus, and what are theorems derived from them; we also need not differentiate between given and derived rules of inference. We know that we simply can trust the work of the many logicians and their results, and are, therefore, only interested in what are legal steps in the search for a proof of a certain statement.

(A typographical note: $A \equiv B$ means the same as $A \leftrightarrow B$, but it gives better readability of longer formulas, especially if implications are involved.)

Theorem 1.17. (Some useful laws from propositional logic)

(1)	$A \wedge B \rightarrow A$	$A \wedge B \rightarrow B$
(2)	$A \rightarrow A \vee B$	$B \rightarrow A \vee B$
(3)	$\mathsf{true} \wedge A \equiv A$	$\mathsf{false} \vee A \equiv A$
(4)	$A \rightarrow \mathsf{true} \equiv \mathsf{true}$	$\mathsf{true} \rightarrow A \equiv A$
(5)	$A \rightarrow \mathsf{false} \equiv \neg A$	$\mathsf{false} \rightarrow A \equiv \mathsf{true}$
(6)	$\neg A \equiv (A \rightarrow \neg A)$	$A \equiv (\neg A \rightarrow A)$
(7)	$A \rightarrow B \equiv \neg A \vee B$	
(8)	$A \rightarrow B \equiv \neg B \rightarrow \neg A$	
(9)	$(A \rightarrow B) \wedge (B \rightarrow C) \rightarrow (A \rightarrow C)$	
(10)	$A \wedge (A \rightarrow B) \equiv A$	
(11)	$A \wedge B \rightarrow C \equiv A \rightarrow (B \rightarrow C)$	
(12)	$A \rightarrow (B \rightarrow C) \equiv B \rightarrow (A \rightarrow C)$	
(13)	$(A \rightarrow B) \wedge (A \rightarrow C) \equiv A \rightarrow (B \wedge C)$	
(14)	$(A \rightarrow C) \wedge (B \rightarrow C) \equiv (A \vee B) \rightarrow C$	
(15)	$B \rightarrow (A \rightarrow B)$	
(16)	$(A \rightarrow C) \rightarrow (A \wedge B \rightarrow C)$	
(17)	$(A \rightarrow C) \wedge (B \rightarrow D) \rightarrow [(A \wedge B) \rightarrow (C \wedge D)]$	

Do not forget all the laws concerning commutativity and distributivity of conjunction $\wedge$ and disjunction $\vee$ not mentioned above, as well as the laws concerning negation. All of them should be familiar from the analogies to the operations in set theory.

Before we can continue with some theorems from first order logic (FOL), a few basic notions must be introduced and illustrated by examples. Reading mathematical texts you are likely to encounter strings of the form $\forall x$ and $\exists x$ (to be read as "*for all x*" and "*there is an x*") called *prefixes* as part of some logical formula. Formulas in which $\forall$ and $\exists$ do not occur are termed *quantifier-free* or *open*.

Intuitively, the formulas $(a)\ \forall x \exists y[x+y=0]$ and $(b)\ \exists y[x+y=0]$ are different in that in every context with a given meaning for + and 0, the former is either true or false, whereas the latter is not: in formula (b) the variable x is waiting to be assigned a value before something can be said about its truth. One says that all variables in (a) are bound, while (b) contains the free variable x. The *free variables* of a formula φ are those variables occurring in φ that are not quantified, and the *bound variables* of φ are the ones that do have a quantifier referring to them. Formulas without free variables are called *sentences*, or *closed* formulas. For a fully-fledged formal exposition of the topics of free and bound occurrences of variables, substitution etc. see e.g. [9, 18].

In the following, let P denote a predicate with at most x and y occurring as free variables, Q a predicate with at most x as a free variable, and A a predicate with no free occurrence of variable x.

Theorem 1.18. (Laws of Quantifier (In)Dependence)

(1) $\forall x \forall y P(x,y) \equiv \forall y \forall x P(x,y)$
(2) $\exists x \exists y P(x,y) \equiv \exists y \exists x P(x,y)$
(3) $\exists x \forall y P(x,y) \rightarrow \forall y \exists x P(x,y)$

Concerning the last implication it is not too hard to give some examples showing the converse direction to fail: From "*To each natural number y there exists another one, x, which is greater than y.*" we cannot conclude "*There is a natural number x that is greater than all other natural numbers y.*". Likewise, from "*Every child has a mother.*" and "*To every event there is a cause.*" we cannot infer "*There is a mother of all children.*" or "*There is one cause for all events.*".

Theorem 1.19. (Duality Laws)

(1) $\forall x Q(x) \equiv \neg\exists x[\neg Q(x)]$ $\qquad \forall x \neg Q(x) \equiv \neg\exists x Q(x)$
(2) $\neg\forall x Q(x) \equiv \exists x[\neg Q(x)]$ $\qquad \neg\forall x \neg Q(x) \equiv \exists x Q(x)$
(3) $\forall x[P(x) \rightarrow Q(x)] \equiv \neg\exists x[P(x) \wedge \neg Q(x)]$
(4) $\forall x[P(x) \rightarrow \neg Q(x)] \equiv \neg\exists x[P(x) \wedge Q(x)]$
(5) $\forall x[\neg P(x) \rightarrow Q(x)] \equiv \neg\exists x[\neg P(x) \wedge \neg Q(x)]$

Theorem 1.20. (Laws of Quantifier Movement)

(1) (a) $\exists x[A \wedge Q(x)] \equiv A \wedge \exists x Q(x)$
(b) $\forall x[A \wedge Q(x)] \equiv A \wedge \forall x Q(x)$
(2) (a) $\exists x[A \vee Q(x)] \equiv A \vee \exists x Q(x)$
(b) $\forall x[A \vee Q(x)] \equiv A \vee \forall x Q(x)$
(3) (a) $\forall x[Q(x) \rightarrow A] \equiv \exists x Q(x) \rightarrow A$
(b) $\exists x[Q(x) \rightarrow A] \equiv \forall x Q(x) \rightarrow A$
(4) (a) $\exists x[A \rightarrow Q(x)] \equiv A \rightarrow \exists x Q(x)$
(b) $\forall x[A \rightarrow Q(x)] \equiv A \rightarrow \forall x Q(x)$

Theorem 1.21. (Laws of Quantifier Distribution)

(1) (a) $\exists x[P(x) \vee Q(x)] \equiv \exists x P(x) \vee \exists x Q(x)$
(b) $\forall x[P(x) \wedge Q(x)] \equiv \forall x P(x) \wedge \forall x Q(x)$
(2) (a) $\forall x P(x) \vee \forall x Q(x) \rightarrow \forall x[P(x) \vee Q(x)]$
(b) $\exists x[P(x) \wedge Q(x)] \rightarrow \exists x P(x) \wedge \exists x Q(x)$
(3) (a) $\forall x P(x) \wedge \exists x Q(x) \rightarrow \exists x[P(x) \wedge Q(x)]$
(b) $\forall x[P(x) \vee Q(x)] \rightarrow \forall x P(x) \vee \exists x Q(x)$
(4) (a) $\forall x[P(x) \rightarrow Q(x)] \rightarrow (\forall x P(x) \rightarrow \forall x Q(x))$
(b) $\forall x[P(x) \leftrightarrow Q(x)] \rightarrow (\forall x P(x) \leftrightarrow \forall x Q(x))$
(5) $\exists x[P(x) \rightarrow Q(x)] \equiv (\forall x P(x) \rightarrow \exists x Q(x))$

All the logical theorems shown above tell us how to manipulate existential and universal quantifiers if they are already there. However, none of them allows us to prove the following kind of everyday arguments: "All cats are mammals, Luna is a cat, so Luna is a mammal", in symbols $\forall x[C(x) \rightarrow M(x)], C(L) \vdash M(L)$. But if the universal quantifier could be eliminated, then we could use *modus ponens* $p,\ p \rightarrow q \vdash q$ to get the desired result.

This example indicates that we have to think about the questions of where and when quantifiers can be added to or eliminated from a formula. And as you will see immediately, this is a more subtle thing than listing a few theorems. Our presentation will be rather informal but must, of course, be sufficiently precise, so it may become a little bit longer than many would probably expect. It is based upon [7, Section 10.5], [13, Section 8.2], and the detailed discussion in [23, Chapter 6]. For the technical formalities we refer to one of the books mentioned at the end of this chapter. We will follow the usual way of presentations as derivation rules.

Theorem 1.22. (Rules for Quantifier Instantiation and Generalization)
In FOL the following are valid inference rules for quantifiers, provided that the respective restrictions are fulfilled:

(1) **Universal instantiation**, (UI): $\forall x[P(x)] \vdash P(b)$,
where b is an individual constant symbol of the underlying language used.
(2) **Universal generalization**, (UG): $P(a) \vdash \forall x[P(x)]$,
where a stands for "any individual arbitrarily chosen" from some domain S, and x is not free in P.
(3) **Existential instantiation**, (EI): $\exists x[P(x)] \vdash P(b)$,
where b is any individual constant symbol that does not occur somewhere earlier in the derivation.
(4) **Existential generalization**, (EG): $P(b) \vdash \exists x[P(x)]$,
where b is is any individual constant symbol.

Instead of *universal/existential instantiation* also the names *universal/existential elimination* can be found, and instead of *universal/existential generalization* the names *universal/existential introduction*. Sometimes these rules are formulated as axiom schemes, e.g. for (UI) as $\forall x[P(x)] \Rightarrow P(t/x)$ for some term t where $P(t/x)$ is the result of substituting t for all free occurrences of x in P.

The *rule (UI)* means: If we have shown somewhere in a derivation that a property P holds for each member of some class C of individuals, then we can conclude that P is true for an arbitrary particular individual of that class. Applications of (UI) are obviously and intuitively valid quite generally.

However, for the *rule (UG)* one must take care to ensure that it is truth-preserving: (UG) allows to generalize universally from saying something about some particular individual element of the domain under consideration to saying something about *every* element of this domain. In doing so we must be cautious: the crucial point is what has been formulated above as "*any individual arbitrarily chosen*", i.e. all we should know about that individual, say a, is that it is an element of the domain and nothing else; we *must not* associate any particular property with it. And this must hold for the all derivation steps preceding the one in which we plan to use (UG). In the practice of proof-construction we can always look back at previous lines of the proof, checking carefully that there is no line which contains a formula telling us something special about that named individual. Assume we have ... $\vdash x = 0$ then $\forall y[x = 0]$ is a valid, but $\forall x[x = 0]$ is an invalid application of (UG).

(EG), in contrast to (UG), allows us to make inferences whose validity is perfectly obvious, whence there is no need to impose any sort of special restriction upon its use: If we know that $P(a)$, with a any individual constant, then we clearly can infer that there is an individual in the domain with property P, i.e. $\exists x[P(x)]$.

Concerning the last rule, viz. existential instantiation (EI), we know that the existential quantification of a propositional function $P(x)$ is true if and only if there is at least one individual a in the domain such that $P(a)$ holds. If an individual constant appears nowhere earlier in the derivation, we may use it to denote an individual that has property P.

(To tell the truth, things are still more complicated, but the comments just given should give you a feeling about the kind of problems one has to consider. (i) Assume you have been able to derive $\forall x[P(x)]$ with $P(x) := \exists y[x \neq y]$; then it is, of course, invalid to conclude $P(y/x) = \exists y[y \neq y]$ by usage of (UI). Such situations need additional restrictions, for which you may consult e.g. [25, 4.1.28 Definition (Substitution of Terms into Variables)]. (ii) In case you really have to work with more complex formal derivations you cannot save the labor of thoroughly studying one of the books on logic mentioned at the end of the chapter.)

1.3 Classical Algebraic Structures

The notion of an *operation on a set* is needed for the definition of various algebraic structures. Binary operations on a set S are rules for combining two elements of S to get another element of the same set. They occur in a lot of contexts and are a rather familiar thing: Union and intersection of two subsets of a set S are binary operations on $\mathcal{P}(S)$; other examples are the addition of two reals, the multiplication of matrices, etc.

Definition 1.23. *Let $S \neq \emptyset$; a* **binary operation** *on S is a function $f : S \times S \longrightarrow S$ mapping the ordered pair $\langle x, y\rangle$ onto $f(x, y)$.*

Commonly used notations for the image $f(a, b) \in S$ are e.g. $a + b, a \cdot b$, or also $a * b$ or $a \circ b$ as more neutral and generic versions, and sometimes only ab. We will not be very careful on which sign to use: you simply will have to live with diverse notations if you look through various books and papers.

Definition 1.24. *Let S be a set and $\cdot$ be a binary operation on S; $\cdot$ is called*

(a) **associative** *if $(a \cdot b) \cdot c = a \cdot (b \cdot c)$ for all $a, b, c \in S$;*
(b) **commutative** *if $a \cdot b = b \cdot a$ for all $a, b \in S$;*
(c) **idempotent** *if $a \cdot a = a$ for all $a \in S$.*

Set union and intersection are both associative, commutative, and idempotent operations. The addition of real numbers is associative and commutative on $\mathbb{R}$.

A binary operation $\cdot$ on a finite set $S = \{x_1, \ldots, x_n\}$ can conveniently be defined by using an $n \times n$ matrix, called the **Cayley table**[1] for the operation $\cdot$. It is customary to write elements x_i in front of the i-th line (yielding the 0-column), and x_j at the top of the j-th column (yielding the 0-row). Following this convention for the set $\{x_1, x_2, \ldots, x_n\}$ and the binary operation $\cdot$ on it, we might define $\cdot$ as e.g. in Table 1.1. If x occupies i-th position in the 0-column and y occupies j-th position in the 0-row, then we get $\cdot : S \times S \longrightarrow S, \langle x, y\rangle \mapsto x \cdot y$, the element at the (i, j)-th position in the table.

TABLE 1.1: A Caley table example.

$\cdot$	x_1	x_2	$\cdots$	x_j	$\cdots$	x_n
x_1	$x_1 \cdot x_1$	$x_1 \cdot x_2$	$\cdots$	$x_1 \cdot x_j$	$\cdots$	$x_1 \cdot x_n$
x_2	$x_2 \cdot x_1$	$x_2 \cdot x_2$	$\cdots$	$x_2 \cdot x_j$	$\cdots$	$x_2 \cdot x_n$
$\vdots$	$\vdots$	$\vdots$		$\vdots$		$\vdots$
x_i	$x_i \cdot x_1$	$x_i \cdot x_2$	$\cdots$	$x_i \cdot x_j$	$\cdots$	$x_i \cdot x_n$
$\vdots$	$\vdots$	$\vdots$		$\vdots$		$\vdots$
x_n	$x_n \cdot x_1$	$x_n \cdot x_2$	$\cdots$	$x_n \cdot x_j$	$\cdots$	$x_n \cdot x_n$

Definition 1.25. *Let S be a set and $\cdot$ be a binary operation on S.*

(a) *$l \in S$ is called a* **left identity** *of S if $l \cdot a = a$ for all $a \in S$.*
(b) *$r \in S$ is called a* **right identity** *of S if $a \cdot r = a$ for all $a \in S$.*
(c) *$e \in S$ is called an* **identity** *of S if it is both a left and a right identity.*
(d) *$l \in S$ is called a* **left inverse** *for a, and a* **left invertible**, *if $l \cdot a = e$*

[1] ARTHUR CAYLEY(1821-1895), a British mathematician, was the first to define the concept of a group in the modern way, i.e. as a set with a binary operation satisfying certain laws.

(e) *$r \in S$ is called a* **right inverse** *for a, and a* **right invertible**, *if $a \cdot r = e$.*

(f) *$b \in S$ is called an* **inverse** *for a, and a* **invertible**, *if it is both a left and a right inverse for a.*

(g) *$l \in S$ is called* **left absorbing** *for a if $l \cdot a = l$*

(h) *$r \in S$ is called* **right absorbing** *for a if $a \cdot r = r$.*

(i) *$b \in S$ is called* **absorbing** *if it is both left and a right absorbing.*

Like identities and inverses, also absorbing elements are quite common: The most well known example comes from elementary algebra, where any number multiplied by zero equals zero, which thus is an absorbing element. The empty relation is absorbing in the set of binary relations over a set S, since $R \circ \emptyset = \emptyset$; and the empty set is with respect to intersection: $S \cap \emptyset = \emptyset$. For $(\mathcal{P}(S), \cup)$ S is absorbing since $M \cup S = S$, for all $M \in \mathcal{P}(S)$.

Definition 1.26. *Let S be a set with a binary operation $\cdot$, and ρ be a relation on S. ρ is called* **compatible with** $\cdot$ *if $x\rho y$ implies $(z \cdot x)\rho(z \cdot y)$ and $(x \cdot z)\rho(y \cdot z)$, for all $x, y, z \in S$.*

1.3.1 *A Simple Example: Semigroups*

Definition 1.27. *Let S be a set and $\circ$ be a binary operation on S.*

(a) *$(S, \circ)$ is called* **groupoid**.

(b) *A* **semigroup** *is a groupoid with $\circ$ an associative operation. A semigroup is called* **Abelian** *(or* **commutative***) if $\circ$ is a commutative operation.*

Proposition 1.28. *If a semigroup S has a left identity l and a right identity r, then $l = r$ is a neutral element. There is only one neutral element in S.*

Proof. By definition of a left identity we have $lr = r$, and also $lr = l$ since r is right identity. Therefore, $l = r$ is a neutral element, say e. Now suppose, there exists another neutral element e'; then e' is both left and right identity and therefore $e = ee' = e'$. □

This proposition shows that a semigroup has at most one identity element which we will usually denote by 1, e or I, dependent on the context. As a consequence, exactly one of the following statements must hold for a groupoid S:

(1) S has no left and no right identity element.
(2) S has one or more left, but no right identity elements.
(3) S has one or more right, but no left identity elements.
(4) S has a unique two-sided, and no other right or left identity element.

Proposition 1.29. *If, in a semigroup S, x has a left inverse l and a right inverse r, then $l = r$ is an inverse x' for x. x' is the unique inverse for x in S.*

Proof. We obviuosly have $l = le = l(xr) = (lx)r = er = r$. Assume that both y and z are inverses of x; then y is a left inverse, and z is a right inverse, whence $y = z$ as we have just shown. □

1.3.2 *Algebras and Homomorphisms*

Semigroups are but a simple example of a general concept which will be introduced in this section. Let A be a nonempty set and $n \in \mathbb{N}$ a natural number, $n \geq 0$. Then A^n denotes the set of all n-tuples of elements of A; we set $A^0 = \{\emptyset\}$. A function $f : A^n \longrightarrow A$ is called an **n-ary operation on** A, and n is called the **arity** or **rank of an operation** on f. Operations of arity 0,1, and 2 are called *nullary, unary, and binary operations*, respectively. A nullary operation on A is a mapping $f : \{\emptyset\} \longrightarrow A$, which is completely determined by $f(\emptyset) \in A$, the image of the only element $\emptyset \in A^0$. Nullary operations may be identified with elements $a \in A$, and are often called **constants:** $\emptyset \mapsto f(\emptyset) = a_f$.

Definition 1.30.

(a) *A* **type** *(also* **similarity type** *or* **signature***) is a pair* (Ω, θ), *with* $\Omega = \{f_i | i \in I\}$ *a set of operation symbols and a mapping* $\theta : I \longrightarrow \mathbb{N}$ *indicating their ranks. Usually we will speak of* θ *as the type, leaving* Ω *implicit. Two algebras are said to be* **similar** *if and only if they have the same type.*

(b) *An* **algebra of type** θ *(also* Ω**-algebra***) is a pair* $\mathfrak{A} = (A, \Omega^A)$ *such that (1) A is a non-void set, and (2)* $\Omega^A = \{f_i^A | f_i \in \Omega\}$ *is a set of operations on A, and* $f_i^A \in \Omega^A$ *has arity* $\theta(i)$ *for every* $i \in I$.

If the type θ is clear from the context we will shortly speak of an algebra. Let $\mathfrak{A} = (A, \Omega^A)$ be an algebra; A is called the **carrier** of $\mathfrak{A}$, and $\mathfrak{A}$ is called a **finite algebra** if its carrier is a finite, and **trivial** (or **degenerate**) if it is a singleton set. If the indexing set I is finite, we say that the type θ is finite and that $\mathfrak{A}$ is an **algebra of finite type.** For finite types the indexing set I we will usually be assumed to have the form $\{0, 1, \ldots, n-1\}$ and θ will be written as $\langle \theta(0), \theta(1), \ldots, \theta(n-1) \rangle \in \mathbb{N}^n$. Algebras of finite type are usually denoted in a simplified manner as $(A, f_1, f_2, \ldots, f_k)$, with the operations f_i usually ordered by decreasing rank or by structural aspects.

Roughly speaking, an algebra is to be thought of as a set with certain operations defined on it; thus, an algebra can also be said to be an **operational structure.** But note that, given any operation f on a set A with arity $n-1$, there is a naturally corresponding n-ary relation ρ on A consisting of those n-tuples $a_1, a_2, \ldots, a_n$ for which $f(a_1, a_2, \ldots, a_{n-1}) = a_n$.

The concept of an algebra as defined above includes most of the algebraic structures encountered in practice as you can see from the examples given later on. But there are some important exceptions like an ordered group, which is a structure that needs a relation for its description in addition to the usual group operations. For a convenient handling one needs to have relations as well as operations, or relations alone, since operations may be obtained as a special case as we have just argued. For sake of convenience we vote for the first possibility. The corresponding algebraic concept is that of a *relational structure*, which generalizes the notion of an algebra:

Definition 1.31. *Let* $\Omega = \{f_i | i \in I\}$ *and* $\Sigma = \{r_j | j \in J\}$ *be disjoint sets of operation and relation symbols, respectively. A* **type** *is a mapping* $\theta : I \cup J \longrightarrow \mathbb{N}$. *A* **relational structure of**

type θ *is a triple* $\mathfrak{R} = (A, \Omega^A, \Sigma^A)$ *such that*

(1) *A is a non-void set,*
(2) *$\Omega^A = \{f_i^A | f_i \in \Omega\}$ is a set of operations on A, and $f_i^A \in \Omega^A$ has arity $\theta(i)$ for every $i \in I$, and*
(3) *$\Sigma^A = \{r_i^A | r_i \in \Sigma\}$ is a set of relations on A such that for every relation $r_j^A \in \Sigma^A$ we have $r_j^A \subseteq A^{\theta(j)}$.*

Depending on which of Ω^A or Σ^A, if any, is empty, we can distinguish three types of algebraic structures: we speak of *algebras* or *operational structures* if $\Sigma^A = \emptyset$, and of *relational structures* if $\Sigma^A \neq \emptyset$. In the latter case we say that $\mathfrak{R} = (A, \Omega^A, \Sigma^A)$ is a **hybrid structure** if neither $\Omega^A = \emptyset$ nor $\Sigma^A = \emptyset$. If the indexing sets I and J are finite, we say that the type θ is finite and that $\mathfrak{R}$ is an **relational structure of finite type**. For finite types the indexing sets I and J we will usually be assumed to have the forms $\{0, 1, \ldots, n-1\}$ and $\{n, n+1, \ldots, n+k-1\}$, respectively. In this case θ will be written as $\langle \theta(0), \theta(1), \ldots, \theta(n-1); \theta(n), \ldots, \theta(n+k-1) \rangle \in \mathbb{N}^{n+k}$. Algebras of finite type are usually denoted in a simplified manner as $(A, f_1, f_2, \ldots, f_n; r_1, r_2, \ldots, r_k)$, with the f_i and r_j ordered by decreasing rank.

Definition 1.32. *Let $\mathfrak{A} = (A, \Omega^A)$ and $\mathfrak{B} = (B, \Omega^B)$ be two similar algebras. A function $h : A \longrightarrow B$ is called a* **homomorphism** *from $\mathfrak{A}$ to $\mathfrak{B}$, and is denoted by $h : \mathfrak{A} \longrightarrow \mathfrak{B}$, if*

$$h(f_i^A(a_1, a_2, \ldots, a_n)) = f_i^B(h(a_1), \ldots, h(a_n))$$

for any $n \in \mathbb{N}$, $i \in I$, and $(a_1, \ldots, a_n) \in A^n$.

Lemma 1.33. [4, Theorem 6.5] *Let $\mathfrak{A}$, $\mathfrak{B}$, and $\mathfrak{C}$ be algebras; furthermore, suppose that $g : \mathfrak{A} \longrightarrow \mathfrak{B}$ and $h : \mathfrak{B} \longrightarrow \mathfrak{C}$ are homomorphisms. Then the composition $hg : \mathfrak{A} \longrightarrow \mathfrak{C}$ is a homomorphism too.*

A homomorphism $h : \mathfrak{A} \longrightarrow \mathfrak{B}$ which is bijective is called an **isomorphism**, and $\mathfrak{A}$ and $\mathfrak{B}$ are said to be *isomorphic* algebras, denoted as $\mathfrak{A} \cong \mathfrak{B}$. In mathematics, the notion of an isomorphism between two structures is fundamental, since it allows to prove that objects of different nature do obey identical laws, i.e. that they have the same properties. In other words, two isomorphic structures may be regarded identical up to the naming of their elements.

There are some other special cases of homomorphisms: a homomorphism is called a **monomorphism** if it is injective, and an **epimorphism** if it is surjective. A homomorphism $h : \mathfrak{A} \longrightarrow \mathfrak{A}$ is called **endomorphism**, and an **automorphism** if h is an isomorphism. Given two Ω-algebras $\mathfrak{A}$ and $\mathfrak{B}$, a monomorphism $h : \mathfrak{A} \longrightarrow \mathfrak{B}$, $\mathfrak{A}$ is called an **embedding** of $\mathfrak{A}$ into $\mathfrak{B}$. If there is an epimorphism from $\mathfrak{A}$ to $\mathfrak{B}$, then $\mathfrak{B}$ is called a **homomorphic image** of $\mathfrak{A}$.

Usually, compared to Def. 1.32, we can simplify notation because it is clear from the context, if the operation belongs to $\mathfrak{A}$ or to $\mathfrak{B}$. Look for the following

Example: Let h be a homomorphism between two groups $\mathfrak{A} = (A, \cdot, ^{-1}, e)$ and $\mathfrak{B} = (B, \cdot, ^{-1}, e)$, i.e. a function $h : A \longrightarrow B$ that satisfies:

(1) $h(e) = e$

(2) $h(x_1 \cdot x_2) = h(x_1) \cdot h(x_2)$, and

(3) $h(x_1^{-1}) = (h(x_1))^{-1}$.

(Note the simplification from $\cdot^A$ and $\cdot^B$ to $\cdot$; analogous for e^A and ${}^{-1A}$.) Now, choose $x_2 = e$; then $h(x_1 \cdot x_2) = h(x_1) \cdot h(x_2)$ gives $h(x_1) = h(x_1) \cdot h(e)$. Multiplying both sides from the left with $h(x_1)^{-1}$ yields $h(x_1)^{-1} \cdot h(x_1) = e = h(x_1)^{-1} \cdot h(x_1) \cdot h(e) = h(e)$, thus $e = h(e)$. Choosing $x_2 = x_1^{-1}$ we get $e = h(e) = h(x_1) \cdot h(x_1^{-1})$, which implies $h(x_1^{-1}) = (h(x_1))^{-1}$. Thus, in case of groups, homomorphisms can be characterized by imposing the condition on the binary operation alone.

1.3.3 *Some more Examples of Algebraic Structures*

With the new concepts at hand, taking a short look back to Section 1.3.1 we see that a *groupoid* on a set A is an algebra of type $\langle 2 \rangle$, $\mathfrak{A} = (A, \cdot)$. Groupoids with an associative operation are called *semigroups*, and *monoids* are semigroups with an identity, i.e. algebras $\mathfrak{A} = (A, \cdot, e)$ of type $\langle 2, 0 \rangle$. Some familiar kinds of algebras, repeating semigroup definition for sake of completeness, are presented in

Definition 1.34.

(a) *A* **semigroup** *is an algebra* $(A, \cdot)$ *of type* $\langle 2 \rangle$ *such that for all* $x, y, z \in A$

(G1) $x \cdot (y \cdot z) = (x \cdot y) \cdot z$ *(associativity).*

(b) *A* **monoid** *is a semigroup with identity* e, *i.e. an algebra* $(A, \cdot, e)$ *of type* $\langle 2, 0 \rangle$ *such that for all* $x \in A$

(G2) $x \cdot e = e \cdot x = x$ *(identity, neutral element).*

(c) *A* **group** *is an algebra* $(A, \cdot, {}^{-1}, e)$ *of type* $\langle 2, 1, 0 \rangle$ *where* $(A, \cdot, e)$ *is a monoid in which for every* $x \in A$ *there is an element* x^{-1} *such that:*

(G3) $x \cdot x^{-1} = x^{-1} \cdot x = e$ *(inverse).*

(d) *A semigroup is called* **Abelian** *(or* **commutative***) if for all* $x, y \in A$

(G4) $x \cdot y = y \cdot x$ *(commutativity).*

(e) *An element* $e \in A$ *is called* **idempotent** *if* $e^2 = ee = e$. *The subset of idempotents of a semigroup* A *is often denoted* E_A. A *is called an* **idempotent semigroup** *or* **band** *if* $E_A = A$, *and an* **orthodox semigroup** *if* E_A *is a subsemigroup of* A.

Examples: Let $\mathbb{N}^+$ denote the set of positive integers; then both $(\mathbb{N}^+, +)$ and $(\mathbb{N}^+, \cdot)$ are commutative semigroups. The set $\mathbb{N}$ of nonnegative integers is a commutative monoid for both addition and multiplication, with 0, respectively 1, as neutral element. Also $(\mathcal{P}(A), \cup)$ is a commutative monoid with $\emptyset$ as neutral element. And the set $\mathbb{Z}$ of integers is a commutative group for addition, with 0 as identity. In the foregoing examples the semigroup operation is denoted as addition '+' or as multiplication '·'; in the first case, an identitiy is called the **zero**, and in the second case the **one** or **multiplicative unit** of the semigroup.

Note that any semigroup can be extended to a monoid by what is called "the adjunction of an identity element to S" ([6, p. 4]): Let S be any semigroup, and let 1 be a symbol not in S. In $S \cup \{1\}$ we define multiplication by $11 = 1$ and $1a = a1 = a$ for every $a \in S$. Then $S \cup \{1\}$ becomes a monoid. It is common to use the following notation:

$$S^1 = \begin{cases} S & \text{if } S \text{ has an identity element,} \\ S \cup \{1\} & \text{otherwise.} \end{cases}$$

Definition 1.35.

(a) A **ring** *is an algebra* $(R, +, \cdot, -, 0, 1)$ *of type* $\langle 2, 2, 1, 0, 0\rangle$ *such that*

(R1) $(R, +, -, 0)$ *is an Abelian group,*
(R2) $(R, \cdot, 1)$ *is a monoid with neutral element* 1,
(R3) $x \cdot (y + z) = (x \cdot y) + (x \cdot z)$,
$(x + y) \cdot z = (x \cdot z) + (y \cdot z)$, *for all* $x, y, z \in R$.

(b) A **division ring** *is an algebra* $(R, +, \cdot, -, ^{-1}, 0, 1)$ *of type* $\langle 2, 2, 1, 1, 0, 0\rangle$ *such that*

(DR1) $(R, +, -, 0)$ *is an Abelian group,*
(DR2) $(R_0, \cdot, ^{-1}, 1)$ *is a group, with* $R_0 := R - \{0\}$.
(DR3) $x \cdot (y + z) = (x \cdot y) + (x \cdot z)$,
$(x + y) \cdot z = (x \cdot z) + (y \cdot z)$, *for all* $x, y, z \in R$.

(c) A ring is called **commutative** *if* $xy = yx$, *for all* $x, y \in R$. *A* **field** *is a commutative division ring.*

The prototype example for a ring is $(\mathbb{Z}, +, \cdot, -, 0)$, where $- : \mathbb{Z} \longrightarrow \mathbb{Z}$, $z \mapsto -z$. Another one is the set of all $n \times n$ matrices over an arbitrary ring R which form a ring, usually referred to as the "*full ring of n-by-n matrices*". The rational, real and complex numbers form fields, usually denoted by $\mathbb{Q}$, R and $\mathbb{C}$.

1.3.4 *Partial Orders and Lattices*

This section is devoted to the presentation of basic notions and results around the concept of an order relation. In this chapter, providing the mathematical preliminaries, this has to be done on the metalanguage level. As you will see later on, the stuff included here is indispensable for the further development of the subject matter, especially in part III. Moreover, the contents of this section are usually regarded to belong to lattice and order theory rather than to a theory of relations. You only have to look at e.g. the books by Blyth [2] or Schröder [21] which are recommended for further reading.

Definition 1.36.

(a) A **partially ordered set** *(or* **poset***) is a relational structure* $\mathfrak{P} = (A, \leq)$ *of type* $\langle 2\rangle$ *such that for all* $a, b, c \in A$:

(PO1) $a \leq a$ *(reflexive)*
(PO2) $a \leq b \wedge b \leq a \rightarrow a = b$ *(antisymmetric)*
(PO3) $a \leq b \wedge b \leq c \rightarrow a \leq c$ *(transitive)*

$\leq$ is called a **partial order** *on A.*
Two elements $a, b \in A$ are called **comparable** *if $a \leq b$ or $b \leq a$, denoted by $a \asymp b$, and* **incomparable** *otherwise, denoted by $a \not\asymp b$.*

(b) *A functin (mapping) $f : A \longrightarrow A$ is said to be*

- **isotone** *iff* $\forall x, y \in A[x \leq y \rightarrow f(x) \leq f(y)]$;
- **antitone** *iff* $\forall x, y \in A[x \leq y \rightarrow f(x) \geq f(y)]$.

(c) *A poset A is called a* **chain** *(or* **linearly ordered set***) iff $\forall a, b \in A[a \asymp b]$. A subset $B \subseteq A$ of a poset A with pairwise incomparable elements, i.e. $\forall a, b \in B[a \not\asymp b]$, is called an* **antichain**.

(A warning: Terminology in literature is not consistent. The unspecified term '*order*' is sometimes used for our notion of a partial order and sometimes for the notion of a total order.) For poset examples see Section 2.8.4.

Some more notions need to be mentioned in connection with partial orders: Let $(A, \leq)$ be a poset. We shall use $x < y$ as abbreviation for $x \leq y \wedge x \neq y$. Note that the relation $<$ thus defined is transitive but is not an order since it fails to be reflexive. Nonetheless, $<$ is called a **strict partial order**. We say that y **covers** x in $(A, \leq)$, denoted as $x \prec y$, if $x < y$ and $x \leq z \leq y \rightarrow z = x \vee z = y$, i.e. there is no z in between x and y.

Definition 1.37. *Let $(A, \leq)$ be a poset, and $B \subseteq A$.*

(a) *$a \in A$ is called a* **lower bound** *of B in A if $\forall b \in B[a \leq b]$, and an* **upper bound** *of B in A if $\forall b \in B[b \leq a]$.*

(b) *B is said to be* **bounded from below** *if there exists a lower bound, and* **bounded from above** *if there exists a upper bound of B in A. If B is bounded from both below and above, it is called* **bounded**.

(c) *If for every two-element set $\{a, b\} \subseteq B$ there exists a lower (respectively upper) bound of $\{a, b\}$, then $(A, \leq)$ is called* **directed from below** *(respectively* **directed from above***). $(A, \leq)$ is* **directed** *if it is directed from below and from above.*

(d) *The* **infimum**, *or* **greatest lower bound** *(g.l.b.), of B (when it exists) is a lower bound $a \in A$ with $a \geq x$ for all lower bounds x of B; it is denoted by $\inf B$. Dually, if it exists, the* **supremum**, *or* **least upper bound** *(l.u.b.) of B is an upper bound $a \in A$ such that $a \leq x$ for all upper bounds x of B; it is denoted by $\sup B$.*

(e) *If $a = \inf B \in B$ then a is called the* **least element** *or* **minimum** *of B; dually, $a \in B$ is the* **greatest element** *or* **maximum** *of B if $\forall x \in B[a \geq x]$.*

(f) *An element $m \in A$ is said to be* **minimal** *if there is no $a \in A$ with $a < m$, i.e. $\forall a \in A[a \leq m \rightarrow a = m]$, and* **maximal** *if $\forall a \in A[m \leq a \rightarrow a = m]$.*

Obviously, a maximum, if it exists, is uniquely determined: Suppose that both a and b are a maximum of A, then both $a \leq b$ and $b \leq a$, so $a = b$ by the antisymmetry of $\leq$. An analogous result holds for the minimum. It is customary to denote maximum and minimum by 1 and 0, respectively.

Having defined the notion of a poset, our next step is to introduce semilattices and lattices. Both of them have two forms of existence which, however, stand in close relationship as will be shown below: The first form arises from a pure structural view as a special algebra, the other as a special kind of poset as described in the two definitions below.

Definition 1.38.

(a) *An algebra* $(A,\cdot)$ *with a binary operation* $\cdot$ *on* A *is called* **semilattice** *iff* $\cdot$ *is idempotent, commutative, and associative.*

(b) *An algebraic structure* $(L,+,\cdot)$ *of type* $\langle 2,2\rangle$ *is called* **lattice** *iff*

(AL1) *both* $(L,+)$ *and* $(L,\cdot)$ *are semilattices,*

(AL2) *the* **absorption laws** $a+(a\cdot b)=a=a\cdot(a+b)$ *hold.*

Definition 1.39.

(a) *A poset* $(L,\leq)$ *in which* $\inf\{a,b\}$ *exists for all* $a,b\in L$ *is called* **meet semilattice**; *this element is denoted by* $a\wedge b$ *or* $a\cdot b$ *and called the meet of* a *and* b.

(b) *A poset* $(L,\leq)$ *in which* $\sup\{a,b\}$ *exists for all* $a,b\in L$ *is called* **join semilattice**; *this element is denoted by* $a\vee b$ *or* $a+b$ *and called the join of* a *and* b.

(c) *A* **lattice** *is a poset* $(L,\leq)$ *which is both a meet and join semilattice, i.e. an algebra* $(L,\vee,\wedge)$ *of type* $\langle 2,2\rangle$.

(d) *A* **distributive lattice** *is a lattice which satisfies*

(D1) $x\wedge(y\vee z)=(x\wedge y)\vee(x\wedge z)$

(D2) $x\vee(y\wedge z)=(x\vee y)\wedge(x\vee z)$

(e) L *is called* **complete lattice** *if every* $S\subseteq L$ *has both a supremum and an infimum.*

(f) *A complete lattice in which arbitrary joins and arbitrary meets distribute over each other is called* **completely distributive.**

In its simplest form complete distributivity means that

$$x\wedge\Big(\bigvee_{i\in I}y_i\Big)=\bigvee_{i\in I}(x\wedge y_i)\qquad\text{and}\qquad x\vee\Big(\bigwedge_{i\in I}y_i\Big)=\bigwedge_{i\in I}(x\vee y_i),$$

and in the most general version (see e.g. [1, p. 119] or [19, p. 140])

$$\bigwedge_{C}\Big(\bigvee_{A_\gamma}x_{\gamma,\alpha}\Big)=\bigvee_{\Phi}\Big(\bigwedge_{C}x_{\gamma,f(\gamma)}\Big)\qquad\text{and}\qquad\bigvee_{C}\Big(\bigwedge_{A_\gamma}x_{\gamma,\alpha}\Big)=\bigwedge_{\Phi}\Big(\bigvee_{C}x_{\gamma,f(\gamma)}\Big),$$

for any nonvoid familiy of index sets $A_\gamma,\gamma\in C$, with Φ being the set of all functions f with domain C and $f(\gamma)\in A_\gamma$.

Note that by taking $S=L$ in (e), we see that every complete lattice has a top and a bottom element since clearly $\bigwedge L=0$ and $\bigvee L=1$, with the special cases $\bigvee\emptyset=0$ and $\bigwedge\emptyset=1$. Complete lattices thus form a special class of bounded lattices.

A familiar example of a meet semilattice is $(\mathcal{P}(X),\cap)$, or more generally any collection of subsets of X closed under intersection, and the prototype example for a (distributive) lattice is $(\mathcal{P}(X),\cup,\cap)$.

Between semilattices and partial orders there exist important relationships, which are presented below.

Theorem 1.40.

(1) *Let* $(S,\cdot)$ *be a semilattice; the relation* $\leq$ *on* S*, defined by* $x \leq y$ *iff* $x \cdot y = x$ *for* $x, y \in S$*, is a partial order on* S*.*
(2) *Moreover, in* $(S,\leq)$ $\inf\{s,t\}$ *exists for all* $s, t \in S$ *and* $s \cdot t = \inf\{s,t\}$*.*

Proof. (1): Since $\cdot$ is idempotent we get $x = x \cdot x$, thus $x \leq x$. And if $x \leq y \wedge y \leq x$, then $x \cdot y = x \wedge y \cdot x = y$, whence $x = y$ since $\cdot$ is commutative. If $x \leq y \wedge y \leq z$ then $x \cdot y = x \wedge y \cdot z = y$, so we have $x \cdot z = (x \cdot y) \cdot z = x \cdot (y \cdot z) = x \cdot y = x$, i.e. $x \leq z$.

(2) Let $s, t \in S$. First we show that $s \cdot t \leq s$ and $s \cdot t \leq t$: We have $(s \cdot t) \cdot s = s \cdot (s \cdot t) = (s \cdot s) \cdot t = s \cdot t$ and $(s \cdot t) \cdot t = s \cdot (t \cdot t) = s \cdot t$. Thus $s \cdot t$ is a lower bound of $\{s,t\}$. Now suppose that b is an arbitrary lower bound of $\{s,t\}$, i.e. $b \cdot s = b$ and $b \cdot t = b$. We have $b \cdot (s \cdot t) = (b \cdot s) \cdot t = b \cdot t = b$, whence $b \leq s \cdot t$. □

Proposition 1.41.

(1) *Let* $(S,\leq)$ *be a partially ordered set such that* $\inf\{s,t\}$ *exists for all* $s, t \in S$*. If* $\cdot$ *is the operation defined by* $s \cdot t = \inf\{s,t\}$ *for* $s, t \in S$*, then* $(S,\cdot)$ *is a semilattice.*
(2) *Let* $(S,\leq)$ *be a partially ordered set such that* $\sup\{s,t\}$ *exists for all* $s, t \in S$*. If* $+$ *is the operation defined by* $s + t = \sup\{s,t\}$ *for* $s, t \in S$*, then* $(S,+)$ *is a semilattice.*

Since the infimum and supremum are uniquely determined if they exist, we can define two pairs of functions $\wedge, \vee : S \times S \to S$ or $\cdot, + : S \times S \to S$ such that for all $a, b \in S$: $\wedge(a,b) = \cdot(a,b) = \inf\{a,b\}$ and $\vee(a,b) = +(a,b) = \sup\{a,b\}$. Using an infix notation for these functions we arrive at the familiar forms $a \wedge b$ and $a \vee b$, or $a \cdot b$ and $a + b$, respectively.

(There are several reasons for introducing two different notations for the same thing: You can find both of them in existing literature and should therefore get accustomed to both; the $\cdot/+$-combination is the one commonly used in the context of relation algebras. And sometimes there is the practical need to typographically distinguish between the logical and '$\wedge$' and the order theoretical meet '$\wedge$'. Nonetheless, both notations will still be used in a generic sense also, but it should always be clear from the context which is the case. If appropriate for better readability, we will alternatively write & for the logical and, and $\|$ for the logical or in the context of lattices and orders.)

Proposition 1.42. *A meet semilattice* $(S,\wedge)$ *is a commutative and idempotent semigroup, hence a semilattice i.e. the following identities hold:*

(SL1) $x \wedge (y \wedge z) = (x \wedge y) \wedge z$ *(associative).*
(SL2) $x \wedge x = x$ *(idempotent)*
(SL3) $x \wedge y = y \wedge x$ *(commutative)*

Proof. We give only some parts of the proofs, since they are easy applications of the definitions and properties of logical operations:

(SL1): Let $p = a \wedge (b \wedge c)$ & $q = (a \wedge b) \wedge c$; then by definition of $\wedge$

$$p \leq a \;\&\; p \leq b \wedge c, \text{ so } p \leq a \;\&\; (p \leq b \;\&\; p \leq c),$$

and by the associativity of logical conjunction we get

$$(p \leq a \;\&\; p \leq b) \;\&\; p \leq c, \text{ i.e. } p \leq a \wedge b \;\&\; p \leq c, \text{ whence } p \leq (a \wedge b) \wedge c = q.$$

A similar argument shows $q \leq p$.

(SL2): Clearly $a \wedge a = \inf\{a, a\} = a$.

(SL3): Let $p = a \wedge b$ & $q = b \wedge a$; then by definition of $\wedge$ we have $p \leq a$ & $p \leq b$, whence $p \leq b$ & $p \leq a$, i.e. $p \leq b$ & $a = q$. A similar argument shows $q \leq p$. □

Theorem 1.43. *Let $(L, \vee, \wedge)$ be a lattice according to* Def. 1.39(c)*; then it is also a lattice according to* Def. 1.38(b).

Proof. From Prop. 1.42 we get that $(L, \wedge)$ and $(L, \vee)$ are semilattices according to Def. 1.38(a). So we have to show that the absorption laws $a \vee (a \wedge b) = a = a \wedge (a \vee b)$ hold. By definition of $\wedge$ we have $a \wedge (a \vee b) \leq a$. On the other hand, $a \leq a$ and $a \leq a \vee b$, whence $a \leq a \wedge (a \vee b)$. □

Proposition 1.44. *In any lattice $(L, +, \cdot)$ the following holds:*

(L1) $a + (a \cdot b) = a \;\wedge\; a \cdot (a + b) = a.$ *(absorption)*

(L2) $a \leq b \;\leftrightarrow\; a \cdot b = a \;\leftrightarrow\; a + b = b.$

(L3) $a \cdot 0 = 0,\; a + 0 = a,\; a \cdot 1 = a$, *and* $a + 1 = 1$.

(L4) $a \leq c \;\&\; b \leq d \;\rightarrow\; a \cdot b \leq c \cdot d \;\&\; a + b \leq c + d.$

Proof. (L1): By Th. 1.43.

(L2): If $a \leq b$ then a fulfills exactly the constraints defining $a \cdot b$: $a \leq a \wedge a \leq b$, and $c \leq a \wedge c \leq b \rightarrow c \leq a$; thus $a \cdot b = a$. Conversely, $a = a \cdot b \leq b \rightarrow a \leq b$. For the second equivalence we have: $a + b = b$ and (L1) imply $a \cdot b = a$, and assuming $a \cdot b = a$ we get $b = b + (b \cdot a) = b + (a \cdot b) = b + a$ by (L1).

(L3): Apply (L2).

(L4): By (L2) we get $a + c = c$ & $b + d = d$ & $a \cdot c = a$ & $b \cdot d = b$; then

$$(a \cdot b)(c \cdot d) = (a \cdot c)(b \cdot d) = a \cdot b, \quad \text{and}$$

$$(a + b) + (c + d) = (a + c) + (b + d) = c + d.$$

□

Proposition 1.45. *In any lattice $(L, \wedge, \vee)$ the following distributive inequalities hold for all $a, b, c \in L$:*

(L5) $a \wedge (b \vee c) \geq (a \wedge b) \vee (a \wedge c)$

(L5') $a \vee (b \wedge c) \leq (a \vee b) \wedge (a \vee c)$.

Proof. (L5): Clearly $a \wedge b \leq a$ and $a \wedge b \leq b \leq b \vee c$; hence $a \wedge b \leq a \wedge (b \vee c)$. We also have $a \wedge c \leq a$ and $a \wedge c \leq c \leq b \vee c$; hence $a \wedge c \leq a \wedge (b \vee c)$. Thus, $a \wedge (b \vee c)$ is an upper bound of $a \wedge b$ and $a \wedge c$, and we have shown (L5). (L5') follows from (L5) by duality. □

Proposition 1.46. *In any lattice $(L, \wedge, \vee)$ the following holds for all $a, b, c \in L$:*

(L6) $a \wedge (b \vee c) = (a \wedge b) \vee (a \wedge c)$ *iff*
(L6') $a \vee (b \wedge c) = (a \vee b) \wedge (a \vee c)$.

Proof. We show that (L6) implies (L6'): For any $a, b, c \in L$ we have

$$\begin{aligned}
(a \vee b) \wedge (a \vee c) &= [(a \vee b) \wedge a] \vee [(a \vee b) \wedge c] && \text{by (L6)} \\
&= a \vee [c \wedge (a \vee b)] && \text{by absorption, commutativity} \\
&= a \vee [(c \wedge a) \vee (c \wedge b)] && \text{by (L6)} \\
&= [a \vee (c \wedge a)] \vee (c \wedge b) && \text{by associativity} \\
&= a \vee (b \wedge c) && \text{by absorption, commutativity.}
\end{aligned}$$

□

Definition 1.47. *Let L be lattice.*

(a) A **bounded lattice** *is a structure $(L, \vee, \wedge, 0, 1)$ of type $\langle 2, 2, 0, 0\rangle$ such that 0 is the least and 1 the greatest element in L, i.e. $\forall x \in L[0 \le x \le 1]$.*
(b) Let L be bounded lattice. $x \in L$ is called complemented *if there is an element $y \in L$ such that $x \wedge y = 0$ and $x \vee y = 1$. Every element y with this property is called a* **complement** *of x. The set of complemented elements in L is denoted by C_L. L is called* **complemented** *if $C_L = L$, i.e. every $x \in L$ is complemented.*
(c) An **interval** *of a lattice L is a subset of L of the form $[a, b] = \{x : a \le x \le b\}$, $a, b \in L$ with $a \le b$. Let $a \in [b, c]$; x is a* **relative complement** *of a in $[b, c]$ if $a \wedge x = b$ and $a \vee x = c$. L is* **relatively complemented** *if every interval $[x, y]$ of L is complemented.*

Note that if y is a complement of x then x is also a complement of y. An element may have several complements, a single one, or none.

Proposition 1.48. *For a bounded distributive lattice $(L, \vee, \wedge, 0, 1)$ the following holds:*

(1) Every $x \in L$ has at most one complement denoted by $\overline{x}$.
(2) $\{0, 1\} \subseteq C_L$.
(3) $\{x, y\} \subseteq C_L \;\to\; \{x \wedge y, x \vee y, \overline{x}\} \subseteq C_L$: We have $\overline{x \wedge y} = \overline{x} \vee \overline{y}$, $\overline{x \vee y} = \overline{x} \wedge \overline{y}$, and $\overline{\overline{x}} = x$.

Proof. (1) Let both y and y' be complements of x. Then

$$\begin{aligned}
y &= y \wedge 1 = y \wedge (x \vee y') = (y \wedge x) \vee (y \wedge y') \\
&= 0 \vee (y \wedge y') = (x \wedge y') \vee (y \wedge y') = (x \vee y) \wedge y' = 1 \wedge y' \\
&= y'.
\end{aligned}$$

(3) Using distributivity we get

$$\begin{aligned}
(x \wedge y) \wedge (\overline{x} \vee \overline{y}) &= (x \wedge y \wedge \overline{x}) \vee (x \wedge y \wedge \overline{y}) \\
&= (0 \wedge y) \vee (x \wedge 0) = (0 \vee 0) = 0, \text{ and}
\end{aligned}$$

$$\begin{aligned}
(x \wedge y) \vee (\overline{x} \vee \overline{y}) &= (x \vee \overline{x} \vee \overline{y}) \wedge (y \vee \overline{x} \vee \overline{y}) \\
&= (1 \vee \overline{y}) \wedge (1 \vee \overline{x}) = (1 \wedge 1) = 1.
\end{aligned}$$

This implies $\overline{x \wedge y} = \overline{x} \vee \overline{y}$; the second equation follows by duality. And since $\overline{x} \wedge x = 0$ and $\overline{x} \vee x = 1$, x is the complement of $\overline{x}$, i.e. $\overline{\overline{x}} = x$. □

Proposition 1.49. *In a bounded distributive lattice* $(L, \vee, \wedge, 0, 1)$ *the following properties hold for every* $x \in C_L$ *and* $y \in L$:

(1) $x \wedge (\overline{x} \vee y) = x \wedge y$,
(2) $x \vee (\overline{x} \wedge y) = x \vee y$,
(3) $x \geq y \;\leftrightarrow\; \overline{x} \wedge y = 0$.
(4) $x \leq y \;\leftrightarrow\; \overline{x} \vee y = 1$.

Proof. (1) $x \wedge (\overline{x} \vee y) = (x \wedge \overline{x}) \vee (x \wedge y) = x \wedge y$.
(3) $x \geq y \rightarrow 0 = \overline{x} \wedge x \geq \overline{x} \wedge y$. Conversely, $\overline{x} \wedge y = 0 \rightarrow y = (x \vee \overline{x}) \wedge y = (x \wedge y) \vee (\overline{x} \wedge y) = x \wedge y$. □

1.3.5 *Boolean Algebras*

Boolean algebras will play a major role later on in part IV dealing with the abstract algebraic aspects of relations: *Relation algebras* are Boolean algebras with additional operators and properties. And the *binary Boolean algebra* (see below) is the special case all generalizations of its valuation space $\{0, 1\}$ must be consistent with. So there is need to provide some relevant concepts and results (see e.g. [10, 15]).

Definition 1.50.

(a) *An algebra* $(L, \vee, \wedge, \eta, 0, 1)$ *of type* $\langle 2, 2, 1, 0, 0\rangle$ *is called a* **De Morgan algebra** *if* $(L, \vee, \wedge, 0, 1)$ *is a bounded distributive lattice and* η *fulfills*

(DM1) $\eta(\eta(x)) = x$,
(DM2) $\eta(x \wedge y) = \eta x \vee \eta y$,
(DM3) $\eta(x \vee y) = \eta x \wedge \eta y$.

(b) *A* **Boolean algebra** *is an algebra* $\mathfrak{B} = (B, \vee, \wedge, \overline{}, 0, 1)$ *such that*

(B1) $\mathfrak{B}$ *is a De Morgan algebra, and*
(B2) $L = C_L$, *i.e. every element is complemented.*

Examples: The most common example for a Boolean algebra, and the motivating source for the algebraic abstraction, is of course $(\mathcal{P}(X), \cup, \cap, \overline{}_X, \emptyset, X)$, with $\overline{}_X$ denoting the set theoretical complement of $A \subseteq X$ with respect to X. Another well-known example is $\mathbf{2} = (\{0, 1\}, \vee, \wedge, \neg, 0, 1)$,with $\vee, \wedge$ denoting max, min, $\neg 1 = 0$ and $\neg 0 = 1$. $\mathbf{2}$ is called the **binary Boolean algebra** , and it is the simplest non-trivial Boolean algebra, i.e. having more than one element. By Prop. 1.48, the set C_L of complemented elements in a bounded distributive lattice is a Boolean algebra.

Another example of a Boolean algebra is a tuple $\mathscr{F} = \langle F, \emptyset, X, \cup, \overline{}\rangle$, where $\emptyset \subset F \subseteq \mathbf{2}^X$ such that $S, T \in F \rightarrow (S \cup T) \in F$ and $\overline{S} \in F$, called a **field of sets**. A field of sets is thus closed under finite unions and under complementation relative to X. It also follows that $\emptyset, X \in F$ and that F is also closed under finite intersections.

The notion of a Boolean algebra has been introduced above by building it on top of more simpler structures by gradually extending the set of requirements. For sake of convenience, it makes sense to list them as a set of axioms that is larger than necessary, but fine for doing calculations. A much more parsimonious characterization is given by the requirements (R1) - (R3) of Def. 14.1.

Definition 1.51. *A* **Boolean algebra** *is an algebra* $(B, \vee, \wedge, \overline{\ }, 0, 1)$ *of type* $\langle 2,2,1,0,0\rangle$ *such that for all* $a, b, c \in B$

(a) $\overline{0} = 1, \quad \overline{1} = 0,$
(b) $a \wedge 0 = 0, \quad a \vee 1 = 1,$
(c) $a \wedge 1 = a, \quad a \vee 0 = a,$
(d) $a \wedge \overline{a} = 0, \quad a \vee \overline{a} = 1,$
(e) $\overline{\overline{a}} = a$
(f) $a \wedge a = a, \quad a \vee a = a,$
(g) $\overline{(a \wedge b)} = \overline{a} \vee \overline{b}, \quad \overline{(a \vee b)} = \overline{a} \wedge \overline{b},$
(h) $a \wedge b = b \wedge a \quad a \vee b = b \vee a$
(i) $(a \wedge b) \wedge c = a \wedge (b \wedge c) \quad (a \vee b) \vee c = a \vee (b \vee c)$
(j) $a \wedge (b \vee c) = (a \wedge b) \vee (a \wedge c), \quad a \vee (b \wedge c) = (a \vee b) \wedge (a \vee c).$

Proposition 1.52. *The following identities are satisfied in every Boolean algebra* $\mathfrak{B} = (B, +, \cdot, \overline{\ }, 0, 1)$ *for arbitrary elements* $x, y \in B$:

(1) $\overline{\overline{x}} = x.$
(2) $\overline{x+y} = \overline{x} \cdot \overline{y}$ *and* $\overline{x \cdot y} = \overline{x} + \overline{y}.$ *(De Morgan Laws)*
(3) $\overline{1} = 0$, *and* $\overline{0} = 1.$
(4) $x \cdot y = \overline{\overline{x} + \overline{y}}.$
(5) $x = \overline{\overline{x} + y} + \overline{\overline{x} + \overline{y}}.$ *(Huntington's axiom)*

Proof. (1)-(3): By Prop. 1.48.
(4): From Prop. 1.48 we get $\overline{x \cdot y} = \overline{x} + \overline{y}$ and $x \cdot y = \overline{\overline{x \cdot y}} = \overline{\overline{x} + \overline{y}}$.
(5): $\overline{\overline{x} + y} + \overline{\overline{x} + \overline{y}} = x \cdot \overline{y} + x \cdot y = x \cdot (y + \overline{y}) = x$, using (4) and distributivity. □

Proposition 1.53. *Let* $\mathfrak{B}$ *be a Boolean algebra; then the following holds:*

(1) *The following statements are equivalent for all* $x, y \in B$:
   (a) $x \leq y,$
   (b) $\overline{y} \leq \overline{x},$
   (c) $x + y = y,$
   (d) $x \cdot y = x,$
   (e) $\overline{x} + y = 1,$
   (f) $x \cdot \overline{y} = 0.$
(2) $\forall x, y, z \in B[x \cdot y \leq z \leftrightarrow y \leq \overline{x} + z].$

Proof. (1) (a), (c) and (d) are equivalent by Prop. 1.44(L2). Using Prop. 1.49(3) we get (f) equivalent to (b): $x\cdot\overline{y}=0 \leftrightarrow \overline{x}\geq\overline{y} \leftrightarrow \overline{y}\leq\overline{x}$; and (f) equivalent to (a): $0=x\cdot\overline{y}=\overline{y}\cdot x \leftrightarrow y\geq x \leftrightarrow x\leq y$. Finally, by Prop. 1.49(4) (a) is equivalent to (e).

(2)

$$
\begin{array}{lll}
 & x\cdot y\leq z & \\
\leftrightarrow & x\cdot y\cdot\overline{z}=0 & \text{by Prop. 1.49(3)}\\
\leftrightarrow & y\cdot(x\cdot\overline{z})=0 & \text{by associativity of }\cdot\\
\leftrightarrow & y\cdot\overline{(\overline{x}+\overline{\overline{z}})}=0 & \text{by Prop. 1.52(4)}\\
\leftrightarrow & y\cdot\overline{(\overline{x}+z)}=0 & \text{by Prop. 1.52(1)}\\
\leftrightarrow & y\leq\overline{x}+z & \text{by Prop. 1.49(3)}
\end{array}
$$

□

1.4 More on Semigroups

1.4.1 *Various Kinds of Semigroups*

Before going on with semigroups you may want to recall what has been said about commutative and idempotent semigroups in Def. 1.34.

Definition 1.54. *Let S be a semigroup.*

(a) An element $a\in S$ is called **regular** *if $\exists x\in S[axa=a]$. If every element of S is regular, then S is called a* **regular semigroup**.

(b) Two elements a and b of S are said to be **inverse elements** *of each other if $aba=a$ and $bab=b$. If, for every $a\in S$, there is a unique inverse $b\in S$, denoted by a^{-1}, then S is called an* **inverse semigroup** .

(c) An algebra $(A,\cdot,^*)$ of type $\langle 2,1\rangle$ such that $(A,\cdot)$ is a semigroup and such that $\forall a,b\in A[(a^*)^*=a\wedge(a\cdot b)^*=b^*\cdot a^*]$, is called **involution semigroup**.

(d) An element $z\in S$ is called a **left zero** *of S if $\forall a\in S[za=z]$ and a* **right zero** *of S if $\forall a\in S[az=z]$. An element of $z\in S$ is called a* **zero element** *of S, denoted by 0, if it is both a left and a right zero element, i.e. $\forall a\in S[0a=a0=0]$. A semigroup (with zero) in which the product of any two elements is zero is called* **zero semigroup**.

(Note that in the terminology of Def. 2.85, the x occurring in item (a) is called a *generalized inverse*, and b of item (b) a *pseudoinverse* of a.)

Example: Unlike a group, a semigroup S can have a zero element which is then unique. Consider the following set $B\subseteq\mathbb{B}_2$:

$$B:=\left\{\begin{pmatrix}0&0\\0&0\end{pmatrix},\begin{pmatrix}1&0\\0&0\end{pmatrix},\begin{pmatrix}0&0\\0&1\end{pmatrix}\right\}. \tag{1.4}$$

B forms a zero semigroup under Boolean matrix multiplication (see Def. 2.12).

By defining a product in a set S by $\forall x,y\in S[xy=x]$, S becomes a semigroup, in which each element $x\in S$ is a right identity and at the same time a left zero. Moreover, $E_S=S$. Note that, contrary to a semigroup, a group has exactly one idempotent: its identity.

Proposition 1.55. *Let S be a semigroup.*

(1) *S is regular, i.e. $\exists x \in S[axa = a]$, iff every element a has at least one inverse b, in the sense that $aba = a$ and $bab = b$.*
(2) *If $aba = a$ then $e = ab$ and $f = ba$ are idempotent, and $ea = a = af$.*
(3) *If e, f, ef and fe are idempotent elements of S, then ef and fe are inverses of each other.*
(4) *An idempotent e is an inverse of itself.*

Proof. (1): If S is regular then $a(xax)a = axa(xa) = axa = a$ and $(xax)a(xax) = x(axa)(xax) = xa(xax) = x(axa)x = xax$; so xax is an inverse for a. Conversely, for every $a \in S$, there is a b with $aba = a$; so S is regular.

(2): We have $e^2 = (ab)(ab) = (aba)b = ab = e$.

(3): $(ef)(fe)(ef) = ef^2e^2f = (ef)^2 = ef$, and by symmetry $(fe)(ef)(fe) = fe$. □

Definition 1.56. *Let $(S,\cdot;\leq)$ be an algebraic structure of type $\langle 2;2\rangle$. S is called a* **partially ordered semigroup** *(also* **posemigroup***) if*

(POSG1) $(S,\cdot)$ *is a semigroup,*
(POSG2) $(S,\leq)$ *is a partially ordered set, and*
(POSG3) $\forall a,b,c \in S[a \leq b \to ac \leq bc \wedge ca \leq cb]$

If the partial order on S is total then S is called an **ordered semigroup.**

Examples:

- $(\mathbb{N}_0,\cdot)$, ordered by divisibility, i.e. $m \leq n$ iff $m|n$ iff $\exists k \in \mathbb{N}_0\,[mk = n]$, is a partially ordered monoid, **pomonoid** for short, with top element 0 and bottom element 1, since for all $n \in \mathbb{N}_0$ we have $1|n$ and $n|0$. Reflexivity and transitivity of $|$ are obvious, and for antisymmetry we have: $mk = n \wedge nl = m$ implies $nlk = n$, so $lk = 1$. Thus $l = k = 1$ and $m = n$.
- For a semigroup S $(\mathbf{2}^S,\cdot;\subseteq)$ is a posemigroup under product $XY = \{xy : x \in X, y \in Y\}$.
- The additive groups $(\mathbb{Z},+)$, $(\mathbb{Q},+)$, and $(\mathbb{R},+)$ are all ordered Abelian groups under the natural ordering. Note that $(\mathbb{Z},+)$ is the usual shorthand for $(\mathbb{Z},+,-,0)$ (where $-$ denotes the unary operator $- : z \mapsto -z$), so the 'official' notation would be $(\mathbb{Z},+,-,0;\leq)$.
- In an ordered group G the subset $P_G = \{x : x \geq 1_G\}$, called the *positive cone* of G, is a partially ordered monoid.

Definition 1.57. ([13, Section 1.7],[5, Definition 1.1]) *Let $(S,\cdot,1)$ be a monoid.*

(a) *If for $p \in S$ there is a $q \in S$ such that $pq = qp = 1$, then p will be called a* **left inverse** *of q and q a* **right inverse** *of p.*
(b) *A* **right [left] unit** *in S is an element of S having a right [left] inverse in S. Thus, if $pq = 1$, p is a right and q a left unit. By a* **unit** *in S we mean an element of S having both a right and a left inverse in S.*

(c) *A nonzero noninvertible element p is called a* **prime element** *of S if p cannot be expressed as a product of two noninvertible elements of S.*
(d) *p is called* **factorizable** *in S if p is not a prime element of S.*

1.4.2 *Indices of Semigroups*

Let S be a semigroup, and $\emptyset \neq A \subseteq S$. A is called a **subsemigroup** of S if it forms a semigroup under the same operation as S. Let $\{A_i : i \in I \neq \emptyset\}$ be a family of subsemigroups of S. Then, if it is non-empty, the intersection $\bigcap_{i\in I} A_i$ is again a subsemigroup of S. For every $\emptyset \neq A \subseteq S$ there is at least one subsemigroup of S containing it, namely S itself; thus the intersection of all semigroups $B \supseteq A$ is a subsemigroup of S containing A, denoted by $\langle A \rangle$. Note that $\langle A \rangle$ is defined by two properties: (a) $A \subseteq \langle A \rangle$, and (b) $\langle A \rangle \subseteq B$ for any subsemigroup B of S containing A.

$\langle A \rangle$ consists of all $s \in S$ that can be expressed as finite products $a_1 a_2 \ldots a_k$ with $a_i \in A, i \in \underline{k}$. If $\langle A \rangle = S$, then A is said to be a **set of generators** for S, or to **generate** S. Of special interest is the case where $A = \{a\}$; then we write $\langle a \rangle = \{a, a^2, a^3, \ldots\}$. $\langle a \rangle$ is called the **cyclic** subsemigroup of S generated by a.

For a semigroup S, $|S|$ is called the **order** of S, and $|\langle a \rangle|$ is called the **order of the element** $a \in S$. A semigroup is called **periodic** if all its elements are of finite order. Thus, a finite semigroup is necessarily periodic.

Definition 1.58. *Let S be a semigroup, $a \in S$, and assume that $a^m = a^{m+d}$ for some $m, d \geq 1$. The least such positive integers m and d are called the* **index** *(of convergence) and the* **period** *of a, denoted by $k = k(a) \geq 1$ and $p = p(a) \geq 1$, respectively. If $p = 1$, a is said to* **converge**, *and otherwise, to* **oscillate** *with finite period p.*

The following is a well known result from semigroup theory (see e.g. [6, THEOREM 1.9] or [21, p. 10]):

Theorem 1.59. *Let S be a semigroup and $a \in S$. Then either*

(1) *all the powers of a are distinct, and $\langle a \rangle$ isomorphic to $(\mathbb{N}, +)$, i.e. a has (countably) infinite order; or*
(2) *there exist the index k and the period p of a such that the following holds*
    (a) *$a^k = a^{k+mp}$, for each $m \geq 1$;*
    (b) *for all $u, v \in \mathbb{N}_0$, $a^{k+u} = a^{k+v}$ iff $u \equiv v \pmod p$;*
    (c) *$\langle a \rangle = \{a, a^2, \ldots, a^{k+p-1}\}$, i.e. a has finite order;*
    (d) *$K_a := \{a^k, \ldots, a^{k+p-1}\}$ is a cyclic subgroup of $\langle a \rangle$ of order p.*

Proof. (1): If there are no repetitions in the list $a, a^2, a^3, \ldots$, i.e. $a^m = a^n \rightarrow m = n$, then all the powers are distinct and, obviously, $(\langle a \rangle, \cdot)$ is isomorphic to $(\mathbb{N}, +)$, since multiplication is carried out according to the rule of addition of exponents.

(2): *(a):* Let k be the index and p be the period of a; then by successive substitution of a^k by a^{k+p} we get $a^k = a^{k+p} = a^k a^p = a^{k+p} a^p = a^{k+2p} = \ldots$, and finally $a^k = a^{k+mp}$, for each $m \geq 1$.

(b): Without loss of generality we let $i < j$. If $i, j < k$, then clearly $i \equiv j \pmod p$ $\leftrightarrow$ $i = j$ $\leftrightarrow$ $a^i = a^j$. Assume $i, j \geq k$ and $i \equiv j \pmod p$; then $j = i + mp$ with $m > 0$; so $a^j = a^{k+mp+(i-k)} = a^{k+i-k} = a^i$. Conversely, if $a^i = a^j$, then $i \geq k$ by definition of k. If $j - i = mp + r$ with $0 \leq r < p$ and $m \geq 0$, then

$$a^{k+r} = a^{k+mp+r} = a^{k+j-i} = a^{k+ip+j-i} = a^j\, a^{k+ip-i} = a^i\, a^{k+ip-i} = a^{k+ip} = a^k$$

By definition of the period p, $r \equiv 0 \pmod p$, whence $i \equiv j \pmod p$.

(c): From the minimality of k and p it follows that $k + p$ is the least positive integer such that a^{k+p} equals some lower power of a. Thus $a, a^2, \ldots, a^{k+p-1}$ are all distinct. It is a elementary fact from number theory that any $n \geq 1$ can be written as $n = mp + r$ with $m \geq 0$ and $0 \leq r < p$. Thus from $a^{k+n} = a^{k+mp+r} = a^{k+r}$ we see that, for each $s > k$, $a^s \in K_a := \{a^k, \ldots, a^{k+p-1}\}$ whence $\langle a \rangle = \{a, a^2, \ldots, a^{k+p-1}\}$.

(d): K_a is a cyclic subgroup of $\langle a \rangle$: For $a^{k+u}, a^{k+v} \in K_a$, choosing s such that $s \equiv v - u - k \pmod p$ and $0 \leq s \leq p - 1$, we have $a^{k+u}\, a^{k+s} = a^{k+v}$. Further note that $k, k+1, \ldots, k+p-1$ form a complete set of incongruent residues modulo p. Then there exists an l such that $0 \leq l < p$ and $k + l \equiv 1 \pmod p$, Thus for each $m \in \mathbb{N}$ $m(k+l) \equiv m \pmod p$, whence the powers $(a^{k+l})^m$ exhaust K_a. And for $0 \leq q < p$ with $k + q \equiv 0$ (mod p) a^{k+q} is idempotent, since $a^{k+q}\, a^{k+q} = a^{k+q}\, a^{mp} = a^{k+mp+q} = a^{k+q}$, and hence the identity of K_a. Thus K_a is a cyclic group of order p, generated by the element a^{k+l}. □

Proposition 1.60. *Every element of a periodic semigroup has a power which is idempotent. Hence, in every periodic (and in particular, in every finite) semigroup there is at least one idempotent.*

Proof. If S is a periodic semigroup then $\langle a \rangle$ is finite; thus, for some k, a^k is the identity of the group K_a from Th. 1.59(2)(d). □

So, in addition to the two integer numbers that have been introduced in Def. 1.58, viz. the index $k = k(A)$ and the period $p = p(A)$, we can associate the following two numbers to any relation or matrix $A \in \mathbb{B}_n$ (see e.g. [42, p. 634]):

- With $r := r(A)$ we denote the least integer $r \geq 1$ such that A^r is idempotent, i.e. $A^{2r} = A^r$; and
- with $t := t(A)$ we denote the least integer $t \geq 1$ such that A^t is transitive, i.e. $A^{2t} \leq A^t$.

Clearly, $t(A) \leq r(A)$.

Corollary 1.61. *Let S be a semigroup and $a \in S$ with index k and period p. Then $a^m = a^{m+d}$ if and only if $k \geq m$ and $p|d$.*

Proof. Assume $a^m = a^{m+d}$. By definition of index and period we have $k \leq m$ and $p \leq d$. Let $d = qp + r$, $0 \leq r < p$ and $q \geq 1$. Then $a^k = a^{k+p}$ implies $a^m = a^{m+p}$, whence $a^k = a^{k+p} = a^{k+2p} = \cdots = a^{k+qp}$. Thus, $a^k = a^{k+d} = a^{k+qp+r} = a^{k+qp}\, a^r = a^k\, a^r = a^{k+r}$. But since $0 \leq r < p$, we get $r = 0$ by the minimality of p; therefore $p|d$. The converse direction is obvious. □

1.5 Residuation

We now turn to one of the central notions that we need in the area of (generalized) fuzzy relations: In this section we will first consider *residuation* in the algebraic context of mappings between ordered sets, and later on how such mappings can be used in algebraic structures, e.g. semigroups. Much more can be found in the books of Blyth and Janowitz [2,3].

As is well known, if $(A,\leq)$ is a poset, then $(A,\geq)$, with $\geq := \leq\breve{}$ is also a poset, called the **dual poset** of A. So, defining $(A,\leq)$, $x \leq y$ iff $y \geq x$, as is usually done for partial orders, $\leq$ and $\geq$ become dual partial orders on A. Concepts defined or valid for a poset $(A,\leq)$ have always counterparts for the dual poset $(A,\geq)$: e.g. lower and upper bound, infimum and supremum, or maximum and minimum. Furthermore, if a statement S holds for a poset $(A,\leq)$, then its dual holds in $(A,\geq)$. With the exception of few examples, in the following we will restrict our considerations to that of the two variants which we will directly need later on.

1.5.1 *Residuated Mappings*

Definition 1.62. ([3, p. 5]) *Let $(A,\leq)$ be a poset.*

(a) A subset $\emptyset \neq D \subseteq A$ is called a **down-set** *(or* **ideal***) of A if for $y \in A$*

$$x \in D \;\&\; y \leq x \;\rightarrow\; y \in D.$$

The set $x^{\downarrow} = \{y \in A : y \leq x\}$ is called the **principal down-set** *(or* **principal ideal***) generated by x.*

(b) Dually, we define an **up-set** *(or* **filter***) of A as a subset $\emptyset \neq F \subseteq A$ such that for $y \in A$*

$$x \in F \;\&\; y \geq x \;\rightarrow\; y \in F.$$

The set $x^{\uparrow} = \{y \in A : y \geq x\}$ is called the **principal up-set** *(or* **principal filter***) generated by x.*

We continue with a characterization of isotone mappings in

Proposition 1.63. ([3, THEOREM 2.1]) *Let $f : A \longrightarrow B$ be any mapping between posets $(A,\leq)$ and $(B,\leq)$; then the following conditions are equivalent:*

(1) f is isotone;

(2) the pre-image under f of every principal ideal of B is either empty or is an ideal of A;

(3) the pre-image under f of every principal filter of B is either empty or is a filter of A.

Proof. (1) → (2): Suppose f is isotone, and let $y \in B$ and $D = f^{-1}[y^{\downarrow}]$. If $x \in D$, then for each $z \in A$ with $z \leq x$ we have $f(z) \leq f(x) \leq y$. Thus $z \in D$, whence D is an ideal of A. (2) → (1): Note that for each $y \in A$ trivially $f(y) \leq f(y)$, so $y \in f^{-1}[f(y)^{\downarrow}]$ which, by assumption is an ideal of A. Thus, $x \leq y \to x \in f^{-1}[f(y)^{\downarrow}] \to f(x) \leq f(y)$, i.e. f is isotone. □

In the next definition, we introduce a kind of mappings that will soon turn out to strengthen the conditions equivalent to the isotonicity of f given in the previous lemma: the pre-images of principal ideals (filters) will again be principal ideals (filters). To show this property we follow the presentation in [3, p. 6ff] and [2, Section 1.3].

Definition 1.64. *Let $(A, \leq)$ and $(B, \leq)$ be posets, and f be an isotone mapping $f: A \longrightarrow B$. f is said to be* **residuated** *if there is an isotone mapping $g: B \longrightarrow A$ such that*

$$gf \geq \mathrm{id}_A \quad \text{and} \quad fg \leq \mathrm{id}_B, \tag{1.5}$$

or, as it is often formulated equivalently,

$$f(x) \leq y \quad \text{iff} \quad x \leq g(y), \tag{1.6}$$

for all $x \in A$ and $y \in B$.

As shown below in Prop. 1.65(1), the mapping $g: B \longrightarrow A$ mentioned above is unique and is called the **residual of** f, denoted as $f^{\sharp}$ and given by $f^{\sharp}(y) = \max\{x \in A : f(x) \leq y\}$. The pair (f, g) is also called a **residuated pair**, and condition (1.6) is called the **residuation property**.

Proposition 1.65. *Let $f: A \longrightarrow B$ be a residuated mapping. Then the following holds:*

(1) *An isotone mapping $g: B \longrightarrow A$ such that $gf \geq \mathrm{id}_A$ and $fg \leq \mathrm{id}_B$ is unique.*
(2) *Residuated mappings preserve existing suprema.*
(3) *$ff^{\sharp}f = f$ and $f^{\sharp}ff^{\sharp} = f^{\sharp}$.*
(4) *Let $g: B \longrightarrow C$ be another residuated mapping; then also $fg: A \longrightarrow C$ is residuated and $(fg)^{\sharp} = f^{\sharp}g^{\sharp}$.*

Proof. (1): Assume that there are two mappings g and g' satisfying (1.5). Then $g = \mathrm{id}_A g \leq (g'f)g = g'(fg) \leq g'\mathrm{id}_B = g'$. Similarly we also have $g' \leq g$, whence $g = g'$.

(2): Let $x = \bigvee_{i \in I} x_i$ exist in A, then, since f is isotone, $f(x) \geq f(x_i)$ for every $i \in I$. Using (1.5) a few times we get: If $y \geq f(x_i)$ for all $i \in I$ then $g(y) \geq g(f(x_i)) \geq x_i$ for each $i \in I$. So $g(y) \geq \bigvee_{i \in I} x_i = x$, and $y \geq f(g(y)) \geq f(x)$. Thus we have shown that $\bigvee_{i \in I} f(x_i)$ exists and that $f(x) = f(\bigvee_{i \in I} x_i)$.

(3): Since $f$is isotone, it follows from Def. 1.64 that $ff^{\sharp}f \geq f\mathrm{id}_A = f$, and that $ff^{\sharp}f \leq \mathrm{id}_B f = f$, from which the first equality follows. The second is established similarly.

(4): Both fg and gf are isotone, and by the monotonicity of f and g together with $f^{\sharp}f \geq \mathrm{id}_A$, $ff^{\sharp} \leq \mathrm{id}_B$, $g^{\sharp}g \geq \mathrm{id}_B$, $gg^{\sharp} \leq \mathrm{id}_C$ we get

$$(f^{\sharp}g^{\sharp})(gf) \geq f^{\sharp}\mathrm{id}_B f = f^{\sharp}f \geq \mathrm{id}_A,$$
$$(gf)(f^{\sharp}g^{\sharp}) \leq g\,\mathrm{id}_B\, g^{\sharp} = gg^{\sharp} \geq \mathrm{id}_C.$$

Thus, using the uniqueness of residuals shown in (1), $(gf)^{\sharp}$ exists and is $f^{\sharp}g^{\sharp}$. □

Without proof we state the following characterization of residuated mappings which is a result analogous to that of Prop. 1.63:

Proposition 1.66. ([24, Proposition 7.10])
Let $(A, \leq)$ and $(B, \leq)$ be posets and $f : A \longrightarrow B$ be any mapping; then the following conditions are equivalent:

(1) *f is residuated;*
(2) *there exists a function $g : B \longrightarrow A$ such that $f(x) \leq y \leftrightarrow x \leq g(y)$;*
(3) *the pre-image of every principal ideal of B is a principal ideal of A, i.e.* $\forall y \in B \exists x \in A[f^{-1}[y^{\downarrow}] = x^{\downarrow}]$;
(4) *f is isotone and for every $y \in B$ the set $\{x \in A : f(x) \leq y\}$ has a maximal element.*

1.5.2 *Residuated Semigroups*

Definition 1.67. *Let $(S, \cdot)$ be a semigroup.*

(a) *Let $a \in S$. The* **left** *(resp.* **right***)* **translation** *on S determined by a is defined as $\lambda_a(x) = a \cdot x$ (resp. $\rho_a(x) = x \cdot a$) for all $x \in S$.*
(b) *An ordered semigroup $(S, \cdot; \leq)$ is* **left (right) residuated** *if each right (left) translation is a residuated mapping, and* **residuated** *if it is both left and right residuated.*
(c) *Let $(S, \cdot; \leq)$ be a residuated semigroup and $a, b \in S$; then*
$$a.{\cdot}b := \lambda_a^{\sharp}(b) = \max\{x \in S : ax \leq b\}.$$
is the called **right residual** *of b by a. In an analogous way the* **left residual** *of b by a is*
$$a{\cdot}.b := \rho_a^{\sharp}(b) = \max\{x \in S : xa \leq b\}.$$

Of course, in Abelian semigroups we need not make a distinction between right and left residuals.

The definition given above is according to [2, Section 11.2] and [14, p. 186]. Unfortunately, notation is not only different in these two sources, but in the existing literature in general, which makes things cumbersome for the reader. We hope the variant used is sufficiently manageable: As done in [14] a is always on the left of b in the notations used, contrary to the notation in e.g. [2]. You may think of the *right* residual as what remains of b after "dividing" it on the *left* by a: in *left* translations $\lambda_a(x) = a \cdot x$ the indeterminate or variable is the *right* factor in the product.

So you have to observe the 'directional cross relationship' between the residuation of mappings and semigroup, and to be careful to not confuse them: If a *left* translation of S, say $\lambda_a : S \longrightarrow S : x \mapsto ax$ is residuated then, according to Prop. 1.66 (4), the set $\{x \in S : ax \leq b\}$ has a maximal element $\lambda_a^{\sharp}(b)$, the *right* residual $a.{\cdot}b$ of b by a. Therefore, $a.{\cdot}b$ is characterized by the properties

(1) $a(a.{\cdot}b) \leq b$,
(2) $ac \leq b \rightarrow c \leq a.{\cdot}b$.

Analogously, we have for the left residual

(3) $(a{\cdot}.b)a \leq b$,
(4) $ca \leq b \rightarrow c \leq a{\cdot}.b$.

Later on we will learn about other residuated algebraic structures, like semirings and lattices, and we will see that they are all based on a residuation semigroup forming an essential constituent.

Additional Reading

For set theory you may wish to have a look at one of the introductory, but rigorous texts by Devlin [8], Monk [17], or Suppes [22].

Readers interested in more details and results, or in the foundations of mathematical logic might want to consult e.g. one of the following books: Hinman [12], Mendelson [16], Rautenberg [18], or Tourlakis [25].

Concerning general algebraic concepts, in addition to [4] a good source for information are the opening chapters of Grätzer [11] or Wechler [26].

References

[1] Birkhoff, G.: *Lattice Theory,* 3rd ed., American Mathematical Society, Providence, Rhode Island (1973).

[2] Blyth, T.S.: *Lattices and Ordered Algebraic Structures,* Springer-Verlag, London (2005).

[3] Blyth, T.S., Janowitz, M.F.: *Residuation Theory,* Pergamon Press, Oxford (1972).

[4] Burris, S., Sankappanavar, H.P.: *A Course in Universal Algebra,* The Millenium Edition, Update 2012, available online at `https://www.math.uwaterloo.ca/~snburris/htdocs/UALG/univ-algebra2012.pdf` (2012).

[5] Cho, H.H.: Prime Boolean matrices and factorizations, Linear Algebra and its Applications **190**, 87-98 (1993).

[6] Clifford, A.H., Preston, G.B.: *The Algebraic Theory of Semigroups, Vol. 1,* 2nd ed., reprinted with corrections, American Mathematical Society, Providence, Rhode Island (1977).

[7] Copi, I.M., Cohen, C., McMahon, K.: *Introduction to Logic,* 14th ed., Pearson Education Limited, Harlow, Essex (2014).

[8] Devlin, K.: *The Joy of Sets,* 2nd ed., Springer, New York (1993).

[9] Ebbinghaus, H.-D., Flum, J., Thomas, W.: *Mathematical Logic,* 2nd ed., Springer Science+Business Media, New York (1994).

[10] Givant, S., Halmos, P.: *Introduction to Boolean Algebras,* Springer Science+Business Media (2009).

[11] Grätzer, G.: *Universal Algebra,* 2nd ed., Springer Science+Business Media (2008).

[12] Hinman, P.G.: *Fundamentals of Mathematical Logic,* A K Peters, Wellesley, Massachusetts (2005).

[13] Hurley, P.J.: *A Concise Introduction to Logic,* 12th ed., Wadsworth, Belmont (2015).

[14] Ignjatović, J., Ćirić, M.: Moore-Penrose equations in involutive residuated semigroups and involutive quantales, Filomat **31**, 183-196 (2017).

[15] Maddux, R.D.: Relation-algebraic semantics, Theoretical Computer Science **160**, 1-85 (1996).

[16] Mendelson, E.: *Introduction to Mathematical Logic*, 5th ed., Taylor & Francis Group (2010).

[17] Monk, J.D.: *Introduction to Set Theory*, McGraw-Hill, New York (1969).

[18] Rautenberg, W.: *A Concise Introduction to Mathematical Logic*, 3rd ed., Springer Science+Business Media, New York (2010).

[19] Roman, S.: *Lattices and Ordered Sets*, Springer Science+Business Media, New York (2008).

[20] Simovici, D.A., Djeraba, C.: *Mathematical Tools for Data Mining*, 2nd ed., Springer, London (2014).

[21] Schröder, B.S.W.: *Ordered Sets - An Introduction*, Springer Science+Business Media, New York (2003).

[22] Suppes, P.: *Axiomatic Set Theory*, Dover Publications, New York (1972).

[23] Tomassi, P.: *Logic*, Routledge, London and New York (1999).

[24] Toth, H.: *Relations: Concrete, Abstract and Applied - An Introduction*, World Scientific Publishing, Singapore (2020).

[25] Tourlakis, G.: *Mathematical Logic*, John Wiley & Sons, Inc., New Jersey (2008).

[26] Wechler, W.: *Universal Algebra for Computer Scientists*, Springer-Verlag, Berlin Heidelberg (1992).

CHAPTER 2

RELATIONS

This and the following chapter are devoted to the most special case of fuzzy relations and matrices, viz. the classical two-valued binary relations and their representing matrices. We will call the former simply '*relations*' and the latter '*Boolean matrices*', to be distinguished from matrices over general Boolean algebras, i.e. ones with more than only two elements.

These two kinds of mathematical concepts have now been investigated for a considerable number of decades and have found a lot of applications in various fields during that time, as e.g. in linear algebra, combinatorics, and graph theory, and also in computer science and in the analysis of switching circuits. So both as a guideline for many-valued generalizations and because they are, of course, interesting of their own and are the type of relations one meets first in the course of mathematical education, a detailed exposition is provided.

As already mentioned in the preface, matrices and relations usually may be identified from a syntactical point of view. From a semantical point of view, however, different concepts and application areas are associated to these two notions. To make this statement more explicit, we start with a closer look on the interrelationships between relations and Boolean matrices in Section 2.2. Since it is a common habit, we will henceforth use the simple notation $\mathbb{B}_n$ for $\mathbb{B}_{n\times n}$, and also for the set of relations $\mathcal{B}_X$ over a set X, with $|X| = n$.

It is a well established result that $\mathbb{B}_n$ with Boolean matrix multiplication (recall that for Boolean addition we have 1+1=1) and $(\mathcal{B}_X, \cdot)$, with $|X| = n$, are isomorphic semigroups. Therefore, we will write $\rho \in \mathbb{B}_n$ for relations, as well as $A \in \mathbb{B}_n$ for Boolean matrices, but will otherwise use small Greek letters for relations and uppercase letters for matrices to indicate where concepts and notations come from, and which is the context and vocabulary we are just using.

2.1 Some Basic Facts on Relations

Relations and their properties have been investigated for more than 100 years and are the basis on which all of the generalizations presented later in this book are built upon. It is thus worthwhile, and also a matter of convenience, to recall basic properties of relations and their composition for later use. For sake of brevity, only sample proofs are provided (but enough of them to get a feeling for the techniques to be used).

We start off with the general case: Let $X_1, X_2, \ldots, X_n$ be sets; an **n-ary relation** ρ is a subset of the Cartesian product of the X_i, i.e. $\rho \subseteq X_1 \times X_2 \times \cdots \times X_n$. The special case $n = 2$ most often occurs in practice and theory, and we make official our habit of speech: as mentioned above, the term "*relation*" always means "*binary relation*".

Definition 2.1.

(a) *A* (**binary**) **relation** ρ *is a set of ordered pairs.*
(b) *If* $\rho \subseteq S \times T$ *for two sets* S *and* T*, then* ρ *is said to be a* **relation from** S **to** T.
(c) *A relation from* S *to* S *is called a* **relation on** S.
(d) *If* $\langle x, y\rangle \in \rho$ *we say that* x *is in relation* ρ *to* y *and usually write this as* $x\rho y$*. For* $\langle x, y\rangle \notin \rho$ *we also use* $\neg x\rho y$*, or* $x\overline{\rho}y$. $\overline{\rho}$ *is called the* **complement of** ρ.
(e) *The* **restriction of a relation** $\rho \subseteq S \times T$ *to a subset* $A \subseteq S$ *is the relation* $\rho|_A \subseteq S \times T$ *defined by* $\rho|_A = \rho \cap (A \times T) = \{\langle x, y\rangle | \langle x, y\rangle \in \rho \wedge x \in A\}$.

Two special relations are the **empty relation** $\emptyset$ and the **universal relation** $S \times T$, also denoted as $U_{S\times T}$, and as U_S if $S = T$. The **identity relation** on a set S is $I_S := \{\langle x, x\rangle : x \in S\}$. Relations *on* a set are also called **homogeneous**, and those between two different sets **heterogeneous.**

Examples of relations are such diverse sets as $\{\langle Adam, Eve\rangle, \langle Bonny, Clyde\rangle\}$ and $\{\langle x, y\rangle : x, y \in S \wedge \exists z \in \mathbb{R}_{\geq 0}(y = x + z)\}$, the latter one representing "*less than or equal*" on S, with $\mathbb{R}_{\geq 0}$ denoting the non-negative reals. A relation between different kinds of objects is "*capital city of*" which is a relation from cities to countries.

From Def. 2.1 we get $\mathcal{P}(S \times T)$ as the set of all relations from S to T. Following [3], an element $\rho \in \mathcal{P}(S \times T)$ is said to be a relation of type $S \times T$, and to emphasize our focus on binary relations we will write $\mathcal{B}_{S\times T}$ for this set of relations, and $\mathcal{B}_S$ if $S = T$. Now, $\mathcal{B}_{S\times T}$ is defined to be a power set, viz. that of $S \times T$, which means that it bears already the following lattice theoretical structure:

(1) Existence of the null relation: $\emptyset \in \mathcal{B}_{S\times T}$;
(2) Existence of the full (or universal) relation: $U_{S\times T} \in \mathcal{B}_{S\times T}$;
(3) Three set theoretical operations are defined as follows:

 (a) $\cup : \mathcal{B}_{S\times T} \times \mathcal{B}_{S\times T} \mapsto \mathcal{B}_{S\times T}$ (Union)
 (b) $\cap : \mathcal{B}_{S\times T} \times \mathcal{B}_{S\times T} \mapsto \mathcal{B}_{S\times T}$ (Intersection)
 (c) $\overline{} : \mathcal{B}_{S\times T} \mapsto \mathcal{B}_{S\times T}$ (Complement)

Theorem 2.2. *The structure* $(\mathcal{B}_{S\times T}, \cup, \cap, \overline{})$ *is a complete atomic Boolean algebra with least element* $\emptyset$ *and greatest element* $U_{S\times T}$*, with its set of atoms given by* $\mathsf{At}(\mathcal{B}_{S\times T}) = \{\{\langle x,y\rangle\} : x \in S, y \in T\}$.

Definition 2.3. *Let* $\rho \subseteq S \times T$ *be a relation from* S *to* T*. Then*

(a) $\mathsf{dom}(\rho) = \{x \in S : \exists y \in T[x\rho y]\}$ *is called the* **domain** *of* ρ;
(b) $\mathsf{rng}(\rho) = \{y \in T : \exists x \in S[x\rho y]\}$ *is called the* **range** *of* ρ;
(c) $\mathsf{fld}(\rho) = \mathsf{dom}(\rho) \cup \mathsf{rng}(\rho)$ *is called the* **field** *of* ρ.

Definition 2.4. *Let* $\rho \subseteq S \times T$, $A \subseteq S$ *and* $B \subseteq T$.

(a) The **image of** A **under** ρ *is the set* $\rho[A] = \{y \in T : \exists x \in A[x\rho y]\}$.
(b) The **pre-image of** B **under** ρ *(also* **inverse image of** B **under** ρ*)) is the set* $\rho^{-1}(B) = \{x \in S : \exists y \in B[x\rho y]\}$.

We continue with the definitions of converse and composition, the two basic operations on relations, and present some of their fundamental properties.

Definition 2.5. *The relation* $\rho^{\smile} = \{\langle y,x\rangle : \langle x,y\rangle \in \rho\}$ *is called the* **converse** *of* ρ.

It has turned out convenient to use the following abbreviations: Let $\rho \subseteq A \times B$. The **afterset** of $a \in A$ is $a\rho = \rho[\{a\}] = \{b \in B : a\rho b\}$, and the **foreset** of $b \in B$ is $\rho b = \rho^{\smile}[\{b\}] = \{a \in A : a\rho b\}$. Due to their meaning in Boolean matrix representations (see Def. 2.12) aftersets (foresets) are also called *rows* (*columns*) of a relation. Sometimes we will also use the following generalized notation: we write e.g. $A\rho := \bigcup_{x\in A} x\rho = \rho[A]$. Note that our notation from Def. 2.4 follows the common habit: Of course $\rho^{-1}(B) = \rho^{\smile}[B]$. Using all these notations one gets, and can also find in the pertinent literature:

$$\mathsf{rng}(\rho) = \rho[A] = A\rho \quad \text{and} \quad \mathsf{dom}(\rho) = \rho^{-1}(B) = \rho B. \tag{2.1}$$

Definition 2.6. *Let* ρ *and* σ *be two relations. The* **product** *or* **composition** *of* ρ *and* σ *is the relation* $\rho\sigma = \{\langle x,z\rangle : \exists y[x\rho y \wedge y\sigma z]\} = \{\langle x,z\rangle : \exists y[x\rho y\sigma z]\}$.

For $\langle x,z\rangle \in \rho\sigma$ we may also write $x(\rho\sigma)z$; occasionally $\rho\cdot\sigma$ or $\rho\circ\sigma$ will be used for typographical reasons, e.g. in $x\,\mathsf{childOf}\cdot\mathsf{sonOf}\,y$, saying that x is a grandchild of y, where we know that the father of x connects him or her and y, whose gender we also do not know. And sometimes, e.g. in historical contexts or when we will consider (abstract) relation algebras later on, we will adhere to the commonly used $R;S$ or $x;y$ instead of $\rho\sigma$ to be nearer to the notation of the original literature.

Proposition 2.7. *Let* $\rho,\sigma \subseteq A \times B$ *and* $\tau \subseteq B \times C$*; then the following holds:*

(1) $x\overline{\rho} = \overline{x\rho}$ *and* $\overline{\rho}y = \overline{\rho y}$.
(2) $x(\rho\cup\sigma) = x\rho \cup x\sigma$ *and* $(\rho\cup\sigma)y = \rho y \cup \sigma y$.
(3) $x(\rho\cap\sigma) = x\rho \cap x\sigma$ *and* $(\rho\cap\sigma)y = \rho y \cap \sigma y$.
(4) $x(\rho\tau) = (x\rho)\tau$.

Proof. (4): $y \in x(\rho\tau) \leftrightarrow \exists z[x\rho z\tau y] \leftrightarrow \exists z[z \in x\rho \wedge z \in \tau y] \leftrightarrow y \in (x\rho)\tau$. □

To keep notation simple in complex cases, we fix the following priority for operations: negation, conversion, composition, intersection, and union. We keep, however, freedom in inserting parentheses and the product operator. Thus, $(\overline{\rho})^{\smile}$ is the same as $\overline{\rho}^{\smile}$, and the short version of $(\rho\cdot\sigma)\cap(\sigma\cdot\tau)$ is $\rho\sigma\cap\sigma\tau$.

Some further remarks concerning notation are advisable at this place: Whenever we write a relational expression using operations like e.g. $\cap, \cup, \circ$, we assume the relations involved to be of appropriate type, without stating this explicitly. We will also refrain from always stating all technical details if they become obvious from the context. This simplifies formulas and makes them concise and clear.

Thus, the assertion $UU = U$ is written as a concise and parsimonious (but maybe somewhat sloppy) shorthand for a whole set of equations like $U_S U_S = U_S, \ldots, U_{S\times T}U_{T\times V} = U_{S\times V}$. Take a look at the Tarski rule in Prop. 2.26: $\rho \neq \emptyset \;\rightarrow\; U\rho U = U$. Here it is e.g. assumed that the domain of the first and third universal relation are the same, and that the range of the first coincides with the domain of ρ. Similar remarks hold for the identity relations I which look alike with respect to comprising only diagonal elements, but which may differ in size for different sets S and T. But for a relation $\rho \subseteq S\times T$we always have $I_S\rho = \rho = \rho I_T$. Henceforth we will, with some exceptions, use the generic shorthand notation for identity and universal relations.

Proposition 2.8. *For any relations ρ and σ we have*

(1) $\mathsf{dom}(\rho^{\smile}) = \mathsf{rng}(\rho)$.
(2) $\mathsf{rng}(\rho^{\smile}) = \mathsf{dom}(\rho)$.
(3) *If $\rho \subseteq S\times T$, then $\rho^{\smile} \subseteq T\times S$.*
(4) $\rho^{\smile\smile} = \rho$.
(5) $\overline{\rho}^{\smile} = \overline{\rho^{\smile}}$.
(6) $\rho \subseteq \sigma$ *iff* $\rho^{\smile} \subseteq \sigma^{\smile}$.
(7) $(\rho\cap\sigma)^{\smile} = \rho^{\smile}\cap\sigma^{\smile}$.
(8) $(\rho\cup\sigma)^{\smile} = \rho^{\smile}\cup\sigma^{\smile}$.
(9) $(\rho-\sigma)^{\smile} = \rho^{\smile}-\sigma^{\smile}$.
(10) $x\rho = \rho^{\smile}x$ *and* $\rho y = y\rho^{\smile}$.

Proof. We give two example proofs, both easy applications of the definitions:
(5): $x(\overline{\rho})^{\smile}y \leftrightarrow y\overline{\rho}x \leftrightarrow \neg y\rho x \leftrightarrow \neg x\rho^{\smile}y \leftrightarrow x\overline{\rho^{\smile}}y$.
(7): $x(\rho\cap\sigma)^{\smile}y \leftrightarrow y(\rho\cap\sigma)x \leftrightarrow y\rho x \wedge y\sigma x \leftrightarrow x\rho^{\smile}y \wedge x\sigma^{\smile}y \leftrightarrow x(\rho^{\smile}\cap\sigma^{\smile})y$. □

Theorem 2.9. *Let ρ and σ be relations. The following holds:*

(1) $\mathsf{dom}(\rho\cup\sigma) = \mathsf{dom}(\rho)\cup\mathsf{dom}(\sigma)$ *and* $\mathsf{rng}(\rho\cup\sigma) = \mathsf{rng}(\rho)\cup\mathsf{rng}(\sigma)$.
(2) $\mathsf{dom}(\rho\cap\sigma) \subseteq \mathsf{dom}(\rho)\cap\mathsf{dom}(\sigma)$ *and* $\mathsf{rng}(\rho\cap\sigma) \subseteq \mathsf{rng}(\rho)\cap\mathsf{rng}(\sigma)$.
(3) $\mathsf{dom}(\rho)-\mathsf{dom}(\sigma) \subseteq \mathsf{dom}(\rho-\sigma)$ *and* $\mathsf{rng}(\rho)-\mathsf{rng}(\sigma) \subseteq \mathsf{rng}(\rho-\sigma)$.
(4) *If $\rho \subseteq \sigma$ then* $\mathsf{dom}(\rho) \subseteq \mathsf{dom}(\sigma)$ *and* $\mathsf{rng}(\rho) \subseteq \mathsf{rng}(\sigma)$.
(5) $\mathsf{dom}(\emptyset) = \emptyset = \mathsf{rng}(\emptyset)$.
(6) $\mathsf{dom}(I_S) = S = \mathsf{rng}(I_S)$.

(7) $\mathsf{dom}(\rho\sigma) \subseteq \mathsf{dom}(\rho)$ *and* $\mathsf{rng}(\rho\sigma) \subseteq \mathsf{rng}(\sigma)$.
(8) $(\rho\sigma)[A] = \sigma[\rho[A]]$.
(9) $\mathsf{dom}(\rho\sigma) = \rho^{\smile}[\mathsf{dom}(\sigma)]$ *and* $\mathsf{rng}(\rho\sigma) = \sigma[\mathsf{rng}(\rho)]$.
(10) $\mathsf{dom}(\rho\sigma) = [\mathsf{rng}(\rho) \cap \mathsf{dom}(\sigma)]\rho^{\smile}$.
(11) $\mathsf{rng}(\rho\sigma) = [\mathsf{rng}(\rho) \cap \mathsf{dom}(\sigma)]\sigma$.

Proof. (7): $\mathsf{dom}(\rho\sigma) = \{x : \exists z[x\rho\sigma z]\} = \{x : \exists y, z[x\rho y\sigma z]\} = \{x : \exists y[x\rho y \wedge \exists z[y\sigma z]]\} \subseteq \{x : \exists y[x\rho y]\} = \mathsf{dom}(\rho)$.

(8): $y \in (\rho\sigma)[A] \to \exists a \in A[a(\rho\sigma)y]$; let $a\rho x \wedge x\sigma y$; then, by definition, $x \in \rho[A]$, whence $y \in \sigma[\rho[A]]$. Conversely, assume $z \in \sigma[\rho[A]]$, and let $t \in \rho[A]$ such that $t\sigma z$. Then choose $b \in A$ with $b\rho t$; we get $b(\rho\sigma)z$, i.e. $z \in (\rho\sigma)[A]$.

(9): Let $x \in \mathsf{dom}(\rho\sigma)$, and choose a z with $x \in (\rho\sigma)z$. Then $\exists y[y \in \mathsf{dom}(\sigma) \wedge x\rho y\sigma z]$; thus $y\rho^{\smile}x$, whence $x \in \rho^{\smile}[\mathsf{dom}(\sigma)]$. To demonstrate the other direction, suppose $x \in \rho^{\smile}[\mathsf{dom}(\sigma)]$. Then choose a $y \in \mathsf{dom}(\sigma)$ with $y\rho^{\smile}x$; thus $x\rho y$, and taking z such that $y\sigma z$ we get $x(\rho\sigma)z$, so that finally $x \in \mathsf{dom}(\rho\sigma)$.

(10): Assume that $x \in \mathsf{dom}(\rho\sigma)$. Then there exist $y, z \in X$ such that $x\rho z\sigma y$. Clearly, $z \in \mathsf{rng}(\rho) \cap \mathsf{dom}(\sigma)$ and $z\rho^{\smile}x$. Hence $x \in z\rho^{\smile} \subseteq [\mathsf{rng}(\rho) \cap \mathsf{dom}(\sigma)]\rho^{\smile}$. Conversely, suppose that $x \in [\mathsf{rng}(\rho) \cap \mathsf{dom}(\sigma)]\rho^{\smile}$. Then there exists a $z \in \mathsf{rng}(\rho) \cap \mathsf{dom}(\sigma)$ such that $x \in z\rho^{\smile}$, i.e. such that $x\rho z$. Since $z \in \mathsf{dom}(\sigma)$, there exists a y with $z\sigma y$. Therefore, $x\rho\sigma y$ and so $x \in \mathsf{dom}(\rho\sigma)$. Thus $[\mathsf{rng}(\rho) \cap \mathsf{dom}(\sigma)]\rho^{\smile} \subseteq \mathsf{dom}(\rho\sigma)$. □

It should be mentioned that there exists a second pair of related concepts also bearing the names *domain* and *range* (or *codomain*) of a relation which are themselves relations again and not sets (see [4]).

Definition 2.10. *Let $\rho \subseteq X \times Y$ be a (possibly heterogeneous) relation.*

(a) $\rho_{\mathsf{dom}} = I \cap \rho\rho^{\smile}$ *is called the* **domain** *of* ρ.
(b) $\rho_{\mathsf{rng}} = I \cap \rho^{\smile}\rho$ *is called the* **range** *of* ρ.

By definition, we have that both ρ_{dom} and ρ_{rng} are partial identities, i.e. relations included in the identity relation. ρ_{dom} obviously identifies those elements in X which are in relation ρ to at least one element of Y. Some straightforward consequences of the definition are the following:

Proposition 2.11. *For the domain and range of a relation ρ the following holds:*

(1) $\rho_{\mathsf{dom}} = \{\langle x, x\rangle : \exists y \in Y[x\rho y]\} = I \cap \rho U$.
(2) $\rho_{\mathsf{rng}} = \{\langle y, y\rangle : \exists x \in X[x\rho y]\} = I \cap U\rho$.
(3) $\pi_1(\rho_{\mathsf{dom}}) = \pi_2(\rho_{\mathsf{dom}}) = \{x \in X : \exists y \in Y[x\rho y]\} = \mathsf{dom}(\rho)$.
(4) $\pi_1(\rho_{\mathsf{rng}}) = \pi_2(\rho_{\mathsf{rng}}) = \{y \in Y : \exists x \in X[x\rho y]\} = \mathsf{rng}(\rho)$.
(5) $\rho_{\mathsf{rng}} = \rho^{\smile}{}_{\mathsf{dom}}$.
(6) $\rho_{\mathsf{dom}}\rho = \rho = \rho\rho_{\mathsf{rng}}$.

Proof. (1): Clearly, $\rho_{\mathsf{dom}} \subseteq I$. Thus, $x(I \cap \rho\rho^{\smile})x \leftrightarrow \exists y[x\rho y \wedge y\rho^{\smile} x] \leftrightarrow \exists y[x\rho y \wedge x\rho y] \leftrightarrow \exists y[x\rho y] \leftrightarrow \exists y[x\rho y \wedge yUx] \leftrightarrow x(I \cap \rho U)x$.

(6): By Prop. 2.25. □

2.2 Representing Relations

A convenient way to represent relations, particularly relations between finite sets, is by using Boolean matrices. We do already know from Section 1.3.5 that $\mathbf{2} = (\{0,1\}, \vee, \wedge, 1-.,0,1)$ is a Boolean algebra.

Definition 2.12.

(a) *A $m \times n$ matrix over* **2** *is called a* **Boolean matrix** *(also* **binary matrix***). The set of all $m \times n$ Boolean matrices is denoted by $\mathbb{B}_{m\times n}$; we let $\mathbb{B}_n := \mathbb{B}_{n\times n}$.*

(b) *Let $\rho \subseteq A \times B$ be a relation from $A = \{a_1, a_2, \ldots, a_m\}$ to $B = \{b_1, b_2, \ldots, b_n\}$. $M_\rho = (m_{ij}) \in \mathbb{B}_{m,n}$ is the* **representation of** ρ *in case $m_{ij} = 1$ iff $a_i \rho b_j$.*

(c) *Let $M = (m_{ij}) \in \mathbb{B}_{m\times p}$ and $P = (p_{ij}) \in \mathbb{B}_{p\times n}$. The* **product of Boolean matrices** *M and P, denoted by MP is the matrix $Q = (q_{ij}) \in \mathbb{B}_{m\times n}$ with*

$$q_{ij} = \bigvee_{1\le k\le p} (m_{ik} \wedge p_{kj}). \tag{2.2}$$

Proposition 2.13. *Let $\rho \subseteq A \times B$ and $\sigma \subseteq B \times C$ be relations from A to B, and from B to C, respectively, with $A = \{a_1, a_2, \ldots, a_m\}$, $B = \{b_1, b_2, \ldots, b_p\}$, and $C = \{c_1, c_2, \ldots, c_n\}$. Furthermore, let $R = (r_{ij}) \in \mathbb{B}_{m\times p}$ and $S = (s_{ij}) \in \mathbb{B}_{p\times n}$ be representations of ρ and σ. Then $P = RS \in \mathbb{B}_{m\times n}$ is a representation of $\rho\sigma$.*

Proof. We have to show that $a_i \rho\sigma c_j$ if and only if $p_{ij} = 1$ (see Def. 2.6). Assuming $p_{ij} = 1$, we get from

$$p_{ij} = \bigvee_{1\le k\le p} (r_{ik} \wedge s_{kj}) = 1$$

that at least one of the expressions $r_{ik} \wedge s_{kj}$ must be equal to 1, for $1 \le i \le m$ and $1 \le j \le n$. Thus, $\exists l[r_{il} = 1 \wedge s_{lj} = 1]$, which implies $a_i \rho b_l$ and $b_l \sigma c_j$ for a $b_l \in B$, and therefore $a_i \rho\sigma c_j$. The proof of the other direction is just as easy. □

In a similar way we can use Boolean matrices for the representation of relation union and intersection by defining $M \vee P$ and $M \wedge P$ for two matrices $M, P \in \mathbb{B}_{m\times n}$ by

$$(m \vee p)_{ij} = m_{ij} \vee p_{ij} \text{ and } (m \wedge p)_{ij} = m_{ij} \wedge p_{ij} \tag{2.3}$$

Then, by assuming R and S to be representations of the relations ρ and σ as above in Prop. 2.13, it can easily be seen that $R \vee S$ and $R \wedge S$ are representations of $\rho \cup \sigma$ and $\rho \cap \sigma$. Moreover, the operations $\wedge$ and $\vee$ are associative, commutative, and idempotent, and also distributivity holds:

$$M(P \vee Q) = MP \vee MQ,$$
$$(M \vee P)Q = MQ \vee PQ.$$

Another often used way to visualize relations are directed graphs. Viewed informally, a *directed graph* consists of a set of *vertices*, depicted as points, and of a set of *arcs* represented by arrows joining one vertex to another. The position of the vertices in a drawing of the directed graph is irrelevant: only which vertices are joined by arcs is significant. For some details in the context of fuzzy graphs see Chapter 13.

2.3 Relation Properties and some of their Interrelationships

The number of genuine relation properties that can be found in book and journal literature is considerably larger then the one presented in the definition below. Looking at Def. 2.14 below you will notice:

- that a few common synonyms have been included;
- that, in rare cases, alternative equivalent definitions are given;
- that, usually, the sets underlying the quantifiers are not mentioned. (This serves to simplify the list of definitions without damaging the presentation: the characteristic features of the property to be defined remain clearly visible. Take as an example for a generally formulated definition the following long version: $\rho \in \mathcal{B}_{S\times T}$ is called *functional* if $\forall x \in S \forall y, z \in T[x\rho y \wedge x\rho z \rightarrow y = z]$.)

Definition 2.14. *The definitions given below are to be read as: A relation* ρ *is* "Name" *iff the related* "FOL Description" *holds.*

Name	**FOL Description**
antisymmetric (1)	$\forall x, y[x\rho y \wedge y\rho x \rightarrow x = y]$
antisymmetric (2)	$\forall x, y[x\rho y \wedge x \neq y \rightarrow \neg y\rho x]$
asymmetric	$\forall x, y[x\rho y \rightarrow \neg y\rho x]$
bijective	*injective* $\wedge$ *surjective*
circular	$\forall x, y, z[x\rho y \wedge y\rho z \rightarrow z\rho x]$
complete	$\forall x, y[x \neq y \rightarrow x\rho y \vee y\rho x]$
connected	see *complete*
coreflexive	$\forall x, y[x\rho y \rightarrow x = y]$
Euclidean	$\forall x, y, z[y\rho x \wedge y\rho z \rightarrow x\rho z]$
functional	$\forall x, y, z[x\rho y \wedge x\rho z \rightarrow y = z]$
injective	$\forall x, y, z[x\rho z \wedge y\rho z \rightarrow x = y]$
irreflexive	$\forall x[\neg x\rho x]$
left Euclidean	$\forall x, y, z[x\rho y \wedge z\rho y \rightarrow x\rho z]$
locally reflexive	$\forall x\big[\exists y[x\rho y \vee y\rho x] \rightarrow x\rho x\big]$
many-to-one	see *functional*
mapping	*total* $\wedge$ *functional*

one-to-many	see *injective*
onto	$\forall y \exists x[x\rho y]$
reflexive	$\forall x[x\rho x]$
serial	see *total*
strongly complete	$\forall x, y[x\rho y \vee y\rho x]$
strongly connected	see *strongly complete*
a subidentity	see *coreflexive*
surjective	see *onto*
symmetric	$\forall x, y[x\rho y \to y\rho x]$
total	$\forall x \exists y[x\rho y]$
totally connected	see *strongly complete*
transitive	$\forall x, y, z[x\rho y \wedge y\rho z \to x\rho z]$
univalent	see *functional*

The proposition below lists interrelationships for some of the more prominent properties relations may have. In doing so we will keep formulations as concise as possible, bypassing grammatical rules for sake of brevity.

Proposition 2.15. *Let ρ be a binary relation on some set S. The following relationships hold for properties of ρ:*

(1) *Euclidean $\wedge$ surjective $\to$ reflexive*
(2) *Euclidean $\wedge$ reflexive $\to$ symmetric*
(3) *Euclidean $\wedge$ symmetric $\to$ transitive $\wedge$ circular*
(4) *ρ symmetric $\wedge$ ρ transitive $\to$* $\mathsf{dom}(\rho) = \mathsf{rng}(\rho) = \mathsf{fld}(\rho)$
(5) *ρ symmetric $\wedge$ ρ transitive $\to$ ρ reflexive on* $\mathsf{fld}(\rho)$
(6) *symmetric $\wedge$ transitive $\to$ Euclidean*
(7) *reflexive $\wedge$ symmetric $\wedge$ transitive $\leftrightarrow$ symmetric $\wedge$ transitive $\wedge$ total $\leftrightarrow$ reflexive $\wedge$ Euclidean*
(8) *strongly complete $\leftrightarrow$ complete $\wedge$ reflexive*
(9) *reflexive $\wedge$ circular $\to$ equivalence*

Proof. (1): $x\rho y \to y\rho x \to x\rho x$.

(4): ([27, p. 21]) If $x \in \mathsf{dom}(\rho)$ then $\exists y[x\rho y$, whence $\exists y[y\rho x$ by symmetry. So $x \in \mathsf{rng}(\rho)$, i.e. $\mathsf{dom}(\rho) \subseteq \mathsf{rng}(\rho)$. The other direction follows by a similar argument. Therefore, $\mathsf{dom}(\rho) = \mathsf{rng}(\rho)$, yielding the assertion.

(5): ([27, p. 21]) If $x \in \mathsf{fld}(\rho)$ then by (8) $x \in \mathsf{dom}(\rho)$, so that $\exists y[x\rho y$. Since by symmetry $\exists y[y\rho x$ we get $x\rho x$ by transitivity.

(7): The assertion has the logical structure $(A) \leftrightarrow (B) \leftrightarrow (C)$. $(A) \to (B)$ follows from (5); so we need to show $(B) \to (C)$: Let $x \in S$; then $\exists y[x\rho y]$. For each such y we have $y\rho x$ by symmetry, whence $x\rho x$ by transitivity. Since $y\rho x \wedge y\rho z \to x\rho y \wedge y\rho z \to x\rho z, \rho$ is also Euclidean. $(C) \to (A)$: By (2) ρ is symmetric, and by (3) transitive.

(9): Since ρ is circular, $x\rho x \wedge x\rho y \rightarrow y\rho x$, so ρ is symmetric; and $x\rho y \wedge y\rho z \rightarrow z\rho x$, whence $x\rho z$ by symmetry, i.e. ρ is transitive. □

Proposition 2.16. *Some basic relationships involving irreflexivity, antisymmetry, and asymmetry are:*

(1) *irreflexive* $\wedge$ *transitive* $\rightarrow$ *asymmetric.*
(2) *irreflexive* $\wedge$ *transitive* $\rightarrow$ *acyclic.*
(3) *asymmetric* $\leftrightarrow$ *(irreflexive* $\wedge$ *antisymmetric).*
(4) *asymmetric* $\wedge$ *antisymmetric* $\rightarrow$ *transitive.*
(5) *transitive* $\rightarrow$ *(irreflexive* $\leftrightarrow$ *symmetric).*
(6) *asymmetric* $\wedge$ *negatively transitive* $\rightarrow$ *transitive.*
(7) *asymmetric* $\wedge$ *transitive* $\rightarrow$ *negatively transitive.*

Proof. (1): $a\rho b \wedge b\rho a \rightarrow a\rho a$ by the transitivity of ρ, a contradiction to its irreflexivity. So ρ must be asymmetric.

(3): Assume $x\rho x$; then by asymmetry we get $x\rho x \rightarrow \neg x\rho x$ and, from Th. 1.17(6), $\neg x\rho x$. Thus, ρ is irreflexive. By asymmetry we have $x\rho y \rightarrow \neg y\rho x$, whence $x\rho y \wedge x \neq y \rightarrow \neg y\rho x$ by Th. 1.17(16). Thus, ρ is also antisymmetric. To prove the other direction, assume $x\rho y$; then $x \neq y$ by the irreflexivity of ρ, whence $\neg y\rho x$ by antisymmetry. Thus ρ is asymmetric.

(4): Let $x\rho y \wedge y\rho z$; then $y\rho x$ by symmetry, But by the antisymmetry of ρ we get $x\rho y \wedge y\rho x \rightarrow x = y$, so from $y\rho z$ we get $x\rho z$, whence ρ is transitive.

(5): By (3) asymmetric relations are irreflexive. So it remains to show that a transitive and irreflexive relation is asymmetric: Assume that ρ is not asymmetric, i.e. $x\rho y \wedge y\rho x$ for some x, y; then transitivity implies $x\rho x$, a contradiction to irreflexivity.

(6): From $x\rho y \wedge y\rho z$, by definition of negative transitivity, we get $(x\rho z \vee z\rho y) \wedge (y\rho x \vee x\rho z)$. Since $\neg z\rho y$ and $\neg y\rho x$ by asymmetry, $x\rho z$ follows.

(7): From $x\rho y \wedge y\rho z$ we get $y\overline{\rho}x \wedge z\overline{\rho}y$ by asymmetry and $x\rho z$ by transitivity; so $z\overline{\rho}x$ by asymmetry. □

Proposition 2.17. *Properties of converse and complement relations* (see e.g. [29]):

(1) ρ *reflexive* $\rightarrow$ $\rho^{\smile}$ *reflexive* $\wedge$ $\overline{\rho}$ *irreflexive.*
(2) ρ *irreflexive* $\rightarrow$ $\rho^{\smile}$ *irreflexive* $\wedge$ $\overline{\rho}$ *reflexive.*
(3) ρ *symmetric* $\rightarrow$ $\rho^{\smile}$ *symmetric* $\wedge$ $\overline{\rho}$ *symmetric.*
(4) ρ *asymmetric* $\rightarrow$ $\rho^{\smile}$ *asymmetric* $\wedge$ $\overline{\rho}$ *not symmetric.*
(5) ρ *antisymmetric* $\rightarrow$ $\rho^{\smile}$ *antisymmetric.*
(6) ρ *transitive* $\rightarrow$ $\rho^{\smile}$ *transitive.*
(7) ρ *intransitive* $\rightarrow$ $\rho^{\smile}$ *intransitive.*
(8) ρ *connected* $\rightarrow$ $\rho^{\smile}$ *connected.*

Proof. (3): Assuming that $\overline{\rho}$ is not symmetric we get $\exists x, y[x\overline{\rho}y \wedge \neg y\overline{\rho}x]$. Thus $y\rho x$, and the symmetry of ρ gives $x\rho y$, a contradiction to $x\overline{\rho}y$, whence $\overline{\rho}$ must be symmetric.

(6): Assume $z\rho^{\smile}y \wedge y\rho^{\smile}x$; then $x\rho y \wedge y\rho z$, whence $x\rho z$ by transitivity of ρ. Thus, also $z\rho^{\smile}x$. (We anticipate the algebraic version of transitivity from Prop. 2.18, viz. $\rho\rho \subseteq \rho$, to give an illustration of the different character of the proof methods: We have $\rho\rho \subseteq \rho$, whence $(\rho\rho)^{\smile} = \rho^{\smile}\rho^{\smile} \subseteq \rho^{\smile}$ by Prop. 2.8(6) and Th. 2.22(8).)

(7): Assume that $z\rho^{\smile}y \wedge y\rho^{\smile}x$; then we have $x\rho y \wedge y\rho z$, and by intransitivity of ρ we get $\neg x\rho z$. This implies $\neg z\rho^{\smile}x$. □

2.4 Point-free Representation of Relation Properties

As a mathematician you might perhaps wonder about what might have been the reason for using the cumbersome FOL-descriptions, if we can do it much more compact and elegant in many cases, as shown below. In my view, this question has a simple and obvious answer: because we usually think in an element-based, or set theoretical way - or at least start to think so. For transitivity we analyze: if I can reach B from A and C from B, then I can get from A to C, and not $R \circ R \subseteq R$.

We present the algebraic representations using the structure of Def. 2.14, and we will also give some of the proofs. It should be clear from the context if ρ denotes a relation on S or a relation from S to T without overloading the list with this detail. Using the relational, or algebraic, instead of the set-theoretical definitions leads to equational proofs that are more compact and easier to verify. Working with this representation makes up what is usually called the *calculus of relations* and which is the starting point for the axiomatic abstraction that is named *relation algebra* (see Section 14.2). For illustration purposes we include some of the proofs for the equivalence of the corresponding FOL and algebraic description of a property.

Proposition 2.18. *The list of properties mirrors the one given in Def. 2.14, but here the properties are described algebraically, i.e. point-free. The assertions are to be read as: A relation ρ is* "Name" *iff the related* "Algebraic Property" *holds.*

Name	Algebraic Property
antisymmetric	$\rho \cap \rho^{\smile} \subseteq I_S \leftrightarrow \rho^{\smile} \subseteq \overline{\rho} \cup I$
asymmetric	$\rho \cap \rho^{\smile} = \emptyset \leftrightarrow \rho^{\smile} \subseteq \overline{\rho}$
bijective	*injective* $\wedge$ *surjective*
circular	$\rho \cdot \rho \subseteq \rho^{\smile}$
complete	see *connected*
connected	$\overline{I_S} \subseteq \rho \cup \rho^{\smile}$
coreflexive	$\rho \subseteq I$
deterministic	see *functional*
Euclidean	$\rho^{\smile} \cdot \rho \subseteq \rho$
functional	$\rho^{\smile} \cdot \rho \subseteq I_T$
injective	$\rho \cdot \rho^{\smile} \subseteq I_S \leftrightarrow \rho^{\smile}$ is *functional*
irreflexive	$I_S \cap \rho = \emptyset \leftrightarrow I_S \subseteq \overline{\rho}$

left Euclidean	$\rho\cdot\rho^{\smile}\subseteq\rho$
locally reflexive	$(\rho\cup\rho^{\smile})_{\mathsf{dom}}\subseteq\rho$
many-to-one	see *functional*
mapping	$\rho\cdot\overline{I}=\overline{\rho}\ \leftrightarrow\ \rho$ is *functional* and *total*
one-to-many	see *injective*
onto	$I_T\subseteq\rho^{\smile}\cdot\rho$
reflexive	$I_S\subseteq\rho$
serial	see *total*
strongly complete	see *strongly connected*
strongly connected	$\rho\cup\rho^{\smile}=U_{S\times S}$
a subidentity	see *coreflexive*
surjective	$I_T\subseteq\rho^{\smile}\cdot\rho$
symmetric	$\rho^{\smile}=\rho\ \leftrightarrow\ \rho^{\smile}\cap\overline{\rho}=\emptyset$
total	$I_S\subseteq\rho\cdot\rho^{\smile}$
totally connected	see *strongly connected*
transitive	$\rho\cdot\rho\subseteq\rho$
univalent	see *functional*

Proof. ***(connected)***: $\forall x,y[x\neq y\rightarrow x\rho y\vee y\rho x]\iff\overline{I_S}\subseteq\rho\cup\rho^{\smile}$

[→] Assume

$$x\overline{I_S}y\,, \tag{2.4}$$

then

$$\{x,y\}\subseteq\mathsf{fld}(\rho)\wedge x\neq y\,, \tag{2.5}$$

which, by connectedness of ρ, gives

$$x\rho y\vee y\rho x. \tag{2.6}$$

Thus

$$x(\rho\cup\rho^{\smile})y. \tag{2.7}$$

[⇐] We have to demonstrate that (2.6) follows from (2.5) based on the hypothesis

$$\overline{I_S}\subseteq\rho\cup\rho^{\smile}\,. \tag{2.8}$$

But by (2.8) we obviously have (2.4) → (2.7), and, since (2.4) ↔ (2.5) and (2.6) ↔ (2.7), we get the desired result.

(functional): $\forall x,y,z[x\rho y\wedge x\rho z\rightarrow y=z]\iff\rho^{\smile}\cdot\rho\subseteq I_T$

[⇒] $\forall x,y,z[x\rho y\wedge x\rho z\rightarrow y=z]\leftrightarrow\forall y,z\exists x[y\rho^{\smile}x\wedge x\rho z]\rightarrow y=z$ by definition of conversion and Th. 1.20(3)(a).

[⇐] Assume that $\rho^{\smile}\cdot\rho\subseteq I_T$, and let $x\rho^{\smile}y$ and $y\rho z$. This gives $x\rho^{\smile}\cdot\rho z$ and, by assumption, $x=z$.

(injective): $\forall x,y,z[x\rho z\wedge y\rho z\rightarrow x=y]\iff\rho\cdot\rho^{\smile}\subseteq I_S$

[⇒] $x\rho\cdot\rho^{\smile}z\rightarrow\exists y[x\rho y\wedge z\rho y]$. Thus $x=z$ since ρ is assumed to be injective, whence $\langle x,z\rangle=\langle x,x\rangle$, and finally $\rho\cdot\rho^{\smile}\subseteq I_S$.

[⇐] Now assume that $\rho\cdot\rho^{\smile}\subseteq I_S$, and let $x\rho y$ and $y\rho^{\smile}z$. This gives $x\rho\cdot\rho^{\smile}z$ and, by the assumption, $x=z$.

(locally reflexive): $\forall x[\exists y[x\rho y \vee y\rho x] \to x\rho x] \iff (\rho \cup \rho^{\smile})_{\text{dom}} \subseteq \rho$
Use Prop. 2.11 (1).

***(total)**:* $\forall x \exists y[x\rho y] \iff I_S \subseteq \rho \cdot \rho^{\smile}$
[⇒] $\forall x \exists y[x\rho y]$ implies for each $x \in S$ there is a y such that: $x\rho y$, whence $y\rho^{\smile}x$, and thus $x\rho\rho^{\smile}x$ for each $x \in S$. So $I_S \subseteq \rho \cdot \rho^{\smile}$.

$$
\begin{aligned}
[\Leftarrow]\ I_S \subseteq \rho \cdot \rho^{\smile} &\to \forall x \exists y[x\rho y \wedge y\rho^{\smile}x]] \\
&\to \forall x[\exists y[x\rho y] \wedge \exists y[y\rho^{\smile}x]] && \text{by Th. 1.21(2)(b)} \\
&\to \forall x \exists y[x\rho y] \wedge \forall x \exists y[y\rho^{\smile}x]. && \text{by Th. 1.21(1)(b)}
\end{aligned}
$$

***(transitive)**:* $\forall x,y,z[x\rho z \wedge z\rho y \to x\rho y] \iff \rho \cdot \rho \subseteq \rho$

$$
\begin{aligned}
\rho \cdot \rho \subseteq \rho &\leftrightarrow \forall x,z[\exists y[x\rho y \wedge y\rho z] \to x\rho z] \\
&\leftrightarrow \forall x,z[\forall y[\neg x\rho y \vee \neg y\rho z] \vee x\rho z] && \text{by Th. 1.19(1), Th. 1.17(7)} \\
&\leftrightarrow \forall x,y,z[\neg x\rho y \vee \neg y\rho z \vee x\rho z] \\
&\leftrightarrow \forall x,y,z[x\rho y \wedge y\rho z \to x\rho z].
\end{aligned}
$$

□

To make you get some feeling for how to work with this algebraic machinery we insert proofs of two more interrelationships in addition to those given above in Prop. 2.15. (Prop. 2.19(2) shows that the algebraic need not necessarily be shorter or simpler than the point-based proof which is: $x\rho y \to y\rho y \wedge x\rho y \to y\rho x$).

Do also note that we indirectly anticipate the SCHRÖDER rule Th. 2.20(1) in the proof below by using Th. 2.22. But their proofs are pure FOL proofs that do not need any additional assumptions besides theorems from set theory and logic, so we can safely use them without thr danger of introducing a vicious circle in the proof.

Proposition 2.19.

(1) total ∧ left Euclidean → reflexive
(2) left Euclidean ∧ reflexive → symmetric

Proof. (1): $I_S \subseteq \rho \cdot \rho^{\smile} \wedge \rho \cdot \rho^{\smile} \subseteq \rho \to I_S \subseteq \rho$

(2): By assumption we have $\rho \cdot \rho^{\smile} \subseteq \rho$ and $I_S \subseteq \rho$. Thus, by monotonicity of composition, $I_S \cdot \rho^{\smile} \subseteq \rho \cdot \rho^{\smile}$, and by Th. 2.22(1) $I_S \cdot \rho^{\smile} = \rho^{\smile}$, whence $\rho^{\smile} \subseteq \rho \cdot \rho^{\smile} \subseteq \rho$. To demonstrate the other inequality we apply Prop. 2.8(6) and Th. 2.22(8): From $\rho \cdot \rho^{\smile} \subseteq \rho$, we get $(\rho \cdot \rho^{\smile})^{\smile} = \rho \cdot \rho^{\smile} \subseteq \rho^{\smile}$; and $I_S \subseteq \rho$ gives $I_S^{\smile} = I_S \subseteq \rho^{\smile}$. Thus $\rho = \rho \cdot I_S \subseteq \rho \cdot \rho^{\smile} \subseteq \rho^{\smile}$. □

2.5 From the Calculus of Relations

Among the most important results in the calculus of relations are the following two theorems.

Theorem 2.20. *Let ρ, σ, and τ be relations, and, as usual, assume that the constructs are well formed. Then the following holds:*

(1) **Schröder *equivalences:***

(a) $\rho\sigma \subseteq \tau \;\leftrightarrow\; \breve{\rho}\overline{\tau} \subseteq \overline{\sigma}$,

(b) $\rho\sigma \subseteq \tau \;\leftrightarrow\; \overline{\tau}\breve{\sigma} \subseteq \overline{\rho}$.

(c) $\breve{\rho}\overline{\tau} \subseteq \overline{\sigma} \;\leftrightarrow\; \overline{\tau}\breve{\sigma} \subseteq \overline{\rho}$.

(2) ***Modular laws:***

(a) $\rho\sigma \cap \tau \subseteq (\rho \cap \tau\breve{\sigma})\sigma$,

(b) $\rho\sigma \cap \tau \subseteq \rho(\sigma \cap \breve{\rho}\tau)$.

(3) **Dedekind *rule:*** $\rho\sigma \cap \tau \subseteq (\rho \cap \tau\breve{\sigma})(\sigma \cap \breve{\rho}\tau)$.

Proof. (1)(a) *Schröder rule,* first equivalence:

$$\begin{aligned}
\rho\sigma \subseteq \tau &\leftrightarrow \forall x,y[x\rho\sigma y \;\rightarrow\; x\tau y] \\
&\leftrightarrow \forall x,y[\exists z[x\rho z \wedge z\sigma y] \;\rightarrow\; x\tau y] \\
&\leftrightarrow \forall x,y[\neg\exists z[x\rho z \wedge z\sigma y] \;\vee\; x\tau y] \\
&\leftrightarrow \forall x,y,z[\neg z\breve{\rho}x \vee x\tau y \vee \neg z\sigma y] && \text{Th. 1.21(1)} \\
&\leftrightarrow \forall y,z[\neg\exists x[z\breve{\rho}x \wedge \neg x\tau y] \vee \neg z\sigma y] \\
&\leftrightarrow \forall y,z[\exists x[z\breve{\rho}x \wedge \neg x\tau y] \rightarrow \neg z\sigma y] && \text{Th. 1.17(7)} \\
&\leftrightarrow \forall y,z[z\breve{\rho}\overline{\tau}y \;\rightarrow\; \neg z\sigma y] \\
&\leftrightarrow \breve{\rho}\overline{\tau} \subseteq \overline{\sigma}
\end{aligned}$$

(1)(b) *Schröder rule,* second equivalence:

$$\begin{aligned}
\rho\sigma \subseteq \tau &\leftrightarrow \forall x,y[x\rho\sigma y \;\rightarrow\; x\tau y] \\
&\leftrightarrow \forall x,y[\exists z[x\rho z \wedge z\sigma y] \;\rightarrow\; x\tau y] \\
&\leftrightarrow \forall x,y[\neg\exists z[x\rho z \wedge z\sigma y] \;\vee\; x\tau y] && \text{Th. 1.17(7)} \\
&\leftrightarrow \forall x,y,z[x\tau y \vee \neg x\rho z \vee \neg z\sigma y] && \text{Th. 1.19(1)} \\
&\leftrightarrow \forall x,y,z[\neg x\overline{\tau}y \vee \neg y\breve{\sigma}z \vee \neg x\rho z] \\
&\leftrightarrow \forall x,z[\neg\exists y[x\overline{\tau}y \wedge y\breve{\sigma}z] \vee \neg x\rho z] \\
&\leftrightarrow \forall x,z[x\overline{\tau}\breve{\sigma}z \;\rightarrow\; \neg x\rho z] \\
&\leftrightarrow \overline{\tau}\breve{\sigma} \subseteq \overline{\rho}
\end{aligned}$$

(2)(a) *Modular laws,* relational proof:

$$\begin{aligned}
\rho\sigma \cap \tau &= [\rho \cap (\tau\breve{\sigma} \cup \overline{\tau\breve{\sigma}})]\sigma \cap \tau && \text{since } \rho \cap U = \rho \\
&= [(\rho \cap \tau\breve{\sigma})\sigma \cup (\rho \cap \overline{\tau\breve{\sigma}})\sigma] \cap \tau && \text{distributivity, twice} \\
&= [(\rho \cap \tau\breve{\sigma})\sigma \cap \tau] \cup [(\rho \cap \overline{\tau\breve{\sigma}})\sigma \cap \tau] && \text{distributivity} \\
&= [(\rho \cap \tau\breve{\sigma})\sigma \cap \tau] \\
&\subseteq (\rho \cap \tau\breve{\sigma})\sigma
\end{aligned}$$

(The last equation follows, because $(\rho \cap \overline{\tau\breve{\sigma}})\sigma \cap \tau = \emptyset$ as shown below:

$$\begin{aligned}
\tau\breve{\sigma} \subseteq \tau\breve{\sigma} &\leftrightarrow \overline{\tau\breve{\sigma}}\sigma \subseteq \overline{\tau} && \text{Schröder rule} \\
&\rightarrow (\rho \cap \overline{\tau\breve{\sigma}})\sigma \subseteq \overline{\tau} && \text{monotonicity} \\
&\leftrightarrow (\rho \cap \overline{\tau\breve{\sigma}})\sigma \cap \tau = \emptyset && \text{set theory)}
\end{aligned}$$

(3) *Dedekind rule,* relational proof:

$$\begin{aligned}
\rho\sigma \cap \tau &= [\rho\sigma \cap \tau] \cap \tau \\
&\subseteq [(\rho \cap \tau\breve{\sigma})\sigma] \cap \tau && \text{by (2)(a), monotonicity} \\
&= [(\rho \cap \tau\breve{\sigma})\sigma \cap \tau] \cap \tau \\
&\subseteq [(\rho \cap \tau\breve{\sigma})\,((\rho \cap \tau\breve{\sigma})\breve{}\,\tau \cap \sigma)] \cap \tau && \text{by (2)(b), with } \rho := \rho \cap \tau\breve{\sigma}
\end{aligned}$$

$\subseteq [(\rho \cap \tau\sigma^{\smile})(\rho^{\smile}\tau \cap \sigma)] \cap \tau$ monotonicity
$\subseteq (\rho \cap \tau\sigma^{\smile})(\sigma \cap \rho^{\smile}\tau)$

(3) *Dedekind rule*, FOL proof for direct comparison:

$$
\begin{aligned}
x(\rho\sigma \cap \tau)y &\leftrightarrow \exists z[x\rho z \wedge z\sigma y] \wedge x\tau y \\
&\leftrightarrow \exists z[x\rho z \wedge x\tau y \wedge y\sigma^{\smile} z \wedge z\sigma y \wedge z\rho^{\smile} x \wedge x\tau y] && \text{by Th. 1.20(1)} \\
&\rightarrow \exists z[x\rho z \wedge \exists s[x\tau s \wedge s\sigma^{\smile} z] \wedge z\sigma y \wedge \exists t[z\rho^{\smile} t \wedge t\tau y]] && \text{by Th. 1.22(EG)} \\
&\leftrightarrow \exists z[x(\rho \cap \tau\sigma^{\smile})z \wedge z(\sigma \cap \rho^{\smile}\tau)y] \\
&\leftrightarrow x(\rho \cap \tau\sigma^{\smile})(\sigma \cap \rho^{\smile}\tau)y
\end{aligned}
$$

□

Theorem 2.21. ([45, Theorem 3.27]) *For relations ρ, σ and τ the following holds:*

(1) The two modular laws together are equivalent to the DEDEKIND *rule.*
(2) The SCHRÖDER *equivalences are equivalent to the* DEDEKIND *rule.*

Theorem 2.22. *Let ρ, σ, and τ be relations such that the various compositions are well defined. Then the following holds:*

(1) $I_S\rho = \rho = \rho I_T$.
(2) $\emptyset\rho = \emptyset = \rho\emptyset$.
(3) Associativity of relation product: $\rho(\sigma\tau) = (\rho\sigma)\tau$.
(4) Subdistributivity of relation product over intersection:

$$\rho(\sigma \cap \tau) \subseteq (\rho\sigma) \cap (\rho\tau) \quad \textit{and} \quad (\rho \cap \sigma)\tau \subseteq (\rho\tau) \cap (\sigma\tau).$$

(5) Monotonicity of relation product: If $\sigma \subseteq \tau$, then

$$\rho\sigma \subseteq \rho\tau \quad \textit{and} \quad \sigma\rho \subseteq \tau\rho.$$

(6) Distributivity of relation product over union:

$$\rho(\sigma \cup \tau) = (\rho\sigma) \cup (\rho\tau) \quad \textit{and} \quad (\rho \cup \sigma)\tau = (\rho\tau) \cup (\sigma\tau).$$

(7) If $\rho \subseteq \sigma$ and $\tau \subseteq \omega$, then $\rho\tau \subseteq \sigma\omega$.
(8) $(\rho\sigma)^{\smile} = \sigma^{\smile}\rho^{\smile}$.
(9) $\rho\sigma = \bigcup_{y\in Y}(\rho y \times y\sigma)$.

Proof. For illustration purposes we give some proofs, most of them being easy routine applications of FOL.

$$
\begin{aligned}
\text{(3):}\quad x\rho(\sigma\tau)w &\leftrightarrow \exists y[x\rho y \wedge y(\sigma\tau)w] \\
&\leftrightarrow \exists y[x\rho y \wedge \exists z(y\sigma z \wedge z\tau w)] \\
&\leftrightarrow \exists y\exists z[x\rho y \wedge (y\sigma z \wedge z\tau w)] && \text{Th. 1.20(1)(a)} \\
&\leftrightarrow \exists z[\exists y(x\rho y \wedge y\sigma z) \wedge z\tau w)] && \text{Th. 1.18(2)} \\
&\leftrightarrow \exists z[x\rho\sigma z \wedge z\tau w] \\
&\leftrightarrow x(\rho\sigma)\tau w
\end{aligned}
$$

(5): Assume $x\rho y\sigma z$; by hypothesis we have $\sigma \subseteq \tau$, whence $y\tau z$. Thus $x\rho y\tau z$, i.e. $\rho\sigma \subseteq \rho\tau$.

(6): By (5) we get $\rho\sigma \subseteq \rho(\sigma \cup \tau)$ and $\rho\tau \subseteq \rho(\sigma \cup \tau)$, whence $\rho\sigma \cup \rho\tau \subseteq \rho(\sigma \cup \tau)$. For the other direction we start from $\rho\sigma \subseteq \rho\sigma$ and $\rho\tau \subseteq \rho\tau$, from which, by the first equivalence of Th. 2.20(1), we get $\rho\breve{}\,\overline{\rho\sigma} \subseteq \overline{\sigma}$ and $\rho\breve{}\,\overline{\rho\tau} \subseteq \overline{\tau}$. Furthermore, using (4) we have $\rho\breve{}(\overline{\rho\sigma \cup \rho\tau}) = \rho\breve{}(\overline{\rho\sigma} \cap \overline{\rho\tau}) \subseteq \rho\breve{}\,\overline{\rho\sigma} \cap \rho\breve{}\,\overline{\rho\tau} \subseteq \overline{\sigma} \cap \overline{\tau} = \overline{\sigma \cup \tau}$, i.e. $\rho\breve{}(\overline{\rho\sigma} \cap \overline{\rho\tau}) \subseteq \overline{\sigma \cup \tau}$. Again applying Th. 2.20(1) we finally get $\rho(\sigma \cup \tau) \subseteq \rho\sigma \cup \rho\tau$.

(7): By (5) $\rho \subseteq \sigma \wedge \tau \subseteq \omega \;\rightarrow\; \rho\tau \subseteq \rho\omega \wedge \rho\omega \subseteq \sigma\omega$, whence $\rho\tau \subseteq \sigma\omega$.

(8): $x(\rho\sigma)\breve{}y \;\leftrightarrow\; y(\rho\sigma)x \;\leftrightarrow\; \exists z[y\rho z \wedge z\sigma x] \;\leftrightarrow\; \exists z[x\sigma\breve{}z \wedge z\rho\breve{}y] \;\leftrightarrow\; x(\sigma\breve{}\rho\breve{})y$.

(9): $\langle x, z\rangle \in \bigcup_{y \in Y}(\rho y \times y\sigma) \;\leftrightarrow\; \exists y \in Y[x \in \rho y \wedge z \in y\sigma] \;\leftrightarrow\; \exists y \in Y[y \in x\rho \wedge y \in \sigma z] \;\leftrightarrow\; x\rho \cap \sigma z \neq \emptyset \;\leftrightarrow\; \langle x, z\rangle \in \rho\sigma$. □

2.6 Properties of Compound Relations

Proposition 2.23. *Some basic relationships concerning reflexivity, symmetry, and transitivity:*

(1) ρ reflexive $\rightarrow$ $\rho \cup \sigma$ reflexive.
(2) ρ, σ reflexive $\rightarrow$ $\rho \cap \sigma$ reflexive.
(3) ρ, σ irreflexive $\rightarrow$ $\rho \cap \sigma$ irreflexive $\wedge$ $\rho \cup \sigma$ irreflexive.
(4) ρ, σ symmetric $\rightarrow$ $\rho \cap \sigma$ symmetric $\wedge$ $\rho \cup \sigma$ symmetric.
(5) $\rho \cup \rho\breve{}, \rho \cap \rho\breve{}, \rho\rho\breve{}$, and $\rho\breve{}\rho$ are symmetric.
(6) ρ, σ transitive $\rightarrow$ $\rho \cap \sigma$ transitive.

Proof. (1): $I \subseteq \rho \;\rightarrow\; I \subseteq \rho \cup \sigma$.

(2): $I \subseteq \rho \wedge I \subseteq \sigma \;\rightarrow\; I \subseteq \rho \cap \sigma$.

(3): We have $I \cap (\rho \cap \sigma) \subseteq I \cap \rho = \emptyset$, and $\emptyset = (I \cap \rho) \cup (I \cap \sigma) = I \cap (\rho \cup \sigma)$.

(4): $\rho\breve{} = \rho \;\wedge \sigma\breve{} = \sigma \;\rightarrow\; (\rho \cap \sigma)\breve{} = \rho\breve{} \cap \sigma\breve{} = \rho \cap \sigma$, by Prop. 2.8(7).

(5): For union we have: $(\rho \cup \rho\breve{})\breve{} = \rho\breve{}\breve{} \cup \rho\breve{} = \rho \cup \rho\breve{}$; and for composition: $(\rho\rho\breve{})\breve{} = \rho\breve{}\breve{}\rho\breve{} = \rho\rho\breve{}$.

(6): We have $\rho \cap \sigma \subseteq \rho$ and $\rho \cap \sigma \subseteq \sigma$, and by hypothesis $\rho^2 \subseteq \rho$ and $\sigma^2 \subseteq \sigma$. From this it follows that $(\rho \cap \sigma)^2 \subseteq \rho(\rho \cap \sigma) \subseteq \rho^2 \cap \rho\sigma \subseteq \rho^2 \subseteq \rho$ by Th. 2.22(4) and (5). An analogous argument gives $(\rho \cap \sigma)^2 \subseteq \sigma$, whence $(\rho \cap \sigma)^2 \subseteq \rho \cap \sigma$. □

Proposition 2.24.

(1) Let ρ and σ be reflexive; then also $\rho\sigma$ and $\sigma\rho$ are reflexive.
(2) Let ρ and σ be reflexive; then $\rho \cup \sigma \subseteq \rho\sigma \cap \sigma\rho$.
(3) Let ρ be reflexive and transitive; then ρ is idempotent, i.e. $\rho^2 = \rho$.
(4) Let ρ and σ be symmetric, and $\rho\sigma \subseteq \sigma\rho$; then $\rho\sigma = \sigma\rho$.
(5) Let ρ be coreflexive; then ρ is symmetric, transitive, and $\rho^2 = \rho$.
(6) Let ρ and σ be coreflexive relations; then $\rho\sigma = \rho \cap \sigma$.
(7) Let ρ and σ be coreflexive relations; then $\rho\sigma = \sigma\rho$.

Proof. (1): From $I \subseteq \rho \wedge I \subseteq \sigma$ we get $I = II \subseteq \rho\sigma$ by Th. 2.22(7).

(2): From $I \subseteq \rho \wedge I \subseteq \sigma$ we get $\rho = \rho I \subseteq \rho\sigma$ and $\sigma = I\sigma \subseteq \rho\sigma$. The remaining assertions follow in a similar way.

(3): By assumption $I \subseteq \rho \wedge \rho^2 \subseteq \rho$; then $\rho = \rho I \subseteq \rho^2$ by monotonicity of composition.

(4): We have $(\rho\sigma)^{\smile} \subseteq (\sigma\rho)^{\smile}$ by Prop. 2.8(6); furthermore $(\rho\sigma)^{\smile} = \sigma^{\smile}\rho^{\smile} = \sigma\rho$, and $(\sigma\rho)^{\smile} = \rho^{\smile}\sigma^{\smile} = \rho\sigma$, i.e. $\sigma\rho \subseteq \rho\sigma$.

(5): Symmetry is obvious, and transitivity follows by monotonicity: $\rho \subseteq I \rightarrow \rho\rho \subseteq I\rho = \rho$. Now assume that $\rho^2 \subset \rho$, i.e. $\exists x[x\rho x \wedge \neg x\rho^2 x]$; but then it follows that $t\rho t \wedge \neg\exists y[t\rho y \wedge y\rho t]$ for some t, a contradiction, whence $\rho^2 = \rho$.

(6): $\rho\sigma \subseteq \rho I = \rho \wedge \rho\sigma \subseteq I\sigma = \sigma$, whence $\rho\sigma \subseteq \rho \cap \sigma$. On the other hand, $\rho \cap \sigma$ is coreflexive; thus $(\rho \cap \sigma) = (\rho \cap \sigma)^2$ by (5), and $(\rho \cap \sigma)^2 \subseteq \rho\sigma$ by monotonicity.

(7): Follows from (6). □

Some more properties of compound relations are given later on in Section 2.8.1 on relations and functions.

Proposition 2.25. *For an arbitrary relation ρ the following relationships hold:*

(1) $\rho \subseteq \rho\rho^{\smile}\rho$.
(2) $\rho = \rho(\rho^{\smile}\rho \cap I) = \rho \cdot \mathsf{rng}(\rho)$.
(3) $\rho = (\rho\rho^{\smile} \cap I)\rho = \mathsf{dom}(\rho) \cdot \rho$.

Proof. (1) Let ρ be reflexive; thus $I \subseteq \rho \rightarrow I \subseteq \rho^{\smile} \rightarrow \rho = I\rho \subseteq \rho^{\smile}\rho$; and from $I \subseteq \rho \subseteq \rho^{\smile}\rho$ we finally get $\rho = \rho I \subseteq \rho\rho^{\smile}\rho$. And if ρ is coreflexive, then $\rho = \rho^{\smile} = \rho^2$ by Prop. 2.24(5); thus $\rho\rho^{\smile}\rho = \rho^3 = \rho$.

(2): $x\rho y \rightarrow x\rho y \wedge y\rho^{\smile}x \rightarrow y\rho^{\smile}\rho y \rightarrow y(\rho^{\smile}\rho \cap I)y \rightarrow x[\rho(\rho^{\smile}\rho \cap I)]y$, whence $\rho \subseteq \rho(\rho^{\smile}\rho \cap I)$. For the other direction we have $x[\rho(\rho^{\smile}\rho \cap I)]y \rightarrow \exists z[x\rho z \wedge z(\rho^{\smile}\rho)y \wedge zIy]$. So $z = y$, whence $x\rho y \wedge y(\rho^{\smile}\rho \cap I)y$. □

The next proposition is formulated in a way that makes the types of the identity and universal relations explicit.

Proposition 2.26. *Let $\rho \subseteq S \times T$. Then for universal relations U the following relationships hold:*

(1) $U_{S\times T}U_{T\times V} = U_{S\times V}$.
(2) $\rho \subseteq \rho U_T \subseteq U_{S\times T}$ *and* $\rho \subseteq U_S\rho \subseteq U_{S\times T}$.
(3) If $\rho \subseteq S \times S$ *is reflexive, then* $\rho U_S = U_S = U_S\rho$.
(4) $\rho \neq \emptyset \leftrightarrow U_{X\times S}\,\rho\,U_{T\times Y} = U_{X\times Y}$ (**Tarski rule**).

Proof. (1): For homogeneous instantiations U_S we can use Prop. 2.24(3); for the general, i.e. heterogeneous, case the assertion follows from the fact that for each $s \in S, t \in T$, and $v \in V$ we have $\langle s,t\rangle \in U_{S\times T}, \langle t,v\rangle \in U_{T\times V}$. Thus, $\langle s,v\rangle \in U_{S\times V}$ for each $s \in S$ and $v \in V$.

(2): We have $I_T \subset U_T$ and $\rho \subseteq U_{S\times T}$; thus $\rho = \rho I_T \subseteq \rho U_T \subseteq U_{S\times T} U_T = U_{S\times T}$, using monotonicity, and (1) for the last equation.

$I_S \subseteq \rho \rightarrow U_S = I_S U_S \subseteq \rho U_S \subseteq U_S$.

(4): For the aftersets and foresets induced by U we have

$$\forall x \in X[xU_{X\times S} = S] \quad \text{and} \quad \forall y \in Y[T = U_{T\times Y}\, y].$$

To show the non-trivial direction assume $\rho \neq \emptyset$ and $s\rho t$; then we get $xUs\rho tUy$ for all x, y, whence $U_{X\times S}\, \rho\, U_{T\times Y} = U_{X\times Y}$. □

2.7 Powers and Closures of Relations

As it is customary for various kinds of multiplication arising in mathematics, we define the powers of a relation ρ on a set S defined recursively as follows:

$$\rho^0 = I_S \text{ and } \rho^{n+1} = \rho^n \rho \text{ for } n \geq 1.$$

Proposition 2.27. *Let ρ be relation relation on a set S; for powers of ρ we have:*

(1) $\rho^m \rho^n = \rho^{m+n}$ *(for $m, n \geq 0$).*
(2) $(\rho^m)^n = \rho^{m\cdot n}$ *(for $m, n \geq 0$).*

Two unary oprations related to the powers of a relation that are useful for further considerations are the following:

Definition 2.28.

(a) $\rho^+ = \bigcup_{n\geq 1} \rho^n$.
(b) $\rho^* = I_S \cup \rho^+ = \bigcup_{n\geq 0} \rho^n$ *is called the* **star** *of ρ.*

Definition 2.29. *A relation ρ' on a set S is the* **closure of a relation** *ρ on S with respect to a property P if*

(1) $\rho \subseteq \rho'$;
(2) *ρ' has property P;*
(3) *if ρ'' has property P and $\rho \subseteq \rho''$, then $\rho' \subseteq \rho''$.*

We will use $\mathbf{r}(\rho)$, $\mathbf{s}(\rho)$, and $\mathbf{t}(\rho)$ to denote the **reflexive**, **symmetric**, and **transitive closure** of relation ρ, respectively; composite closures will be written with only the innermost parentheses, e.g. $\mathbf{rt}(\rho)$.

Theorem 2.30. *Let ρ be a relation on a set S, then*

(1) $\mathbf{r}(\rho) = \rho \cup I_S$.
(2) $\mathbf{s}(\rho) = \rho \cup \rho^{\smile}$.
(3) $\mathbf{t}(\rho) = \rho^+$.

Proof. (1): Trivially we have (a) $\rho \subseteq \mathbf{r}(\rho)$, and (b) $I_S \subseteq \mathbf{r}(\rho)$, whence $\mathbf{r}(\rho)$ is reflexive. And for ρ' a reflexive relation with $\rho \subseteq \rho'$, we get (c) $I_S \subseteq \rho' \wedge \rho \subseteq \rho' \rightarrow I_S \cup \rho \subseteq \rho'$; thus $\mathbf{r}(\rho) \subseteq \rho'$.

(2): Obviously, (a) $\rho \subseteq \mathbf{s}(\rho)$, and (b) $\mathbf{s}(\rho)$ is symmetric by Prop. 2.23(5). (c) If ρ' is symmetric with $\rho \subseteq \rho'$, then $\rho^{\smile} \subseteq \rho'^{\smile} = \rho'$. So $\rho \subseteq \rho' \wedge \rho^{\smile} \subseteq \rho'$, whence $\rho \cup \rho^{\smile} = \mathbf{s}(\rho) \subseteq \rho'$.

(3): As before (a) $\rho \subseteq \rho^+$. (b) ρ^+ is transitive: $x\rho^+ y \wedge y\rho^+ z \rightarrow \exists j,k \in \mathbb{N}[x\rho^k y \wedge y\rho^j z]$. Thus $x\rho^k\rho^j z$; therefore $x\rho^{j+k}z$ whence $x\rho^+ z$. (c) Let ρ' be transitive with $\rho \subseteq \rho'$. To show that $\rho^+ \subseteq \rho'$ we prove $\rho^n \subseteq \rho'$ for each $n \in \mathbb{N}$: The case $n = 1$ follows from the assumption $\rho \subseteq \rho'$; if $\rho^k \subseteq \rho'$, then $\rho^{k+1} = \rho^k\rho \subseteq \rho'\rho \subseteq \rho'\rho' \subseteq \rho'$ by monotonicity of composition and transitivity of ρ'. Thus $\rho^{k+1} \subseteq \rho'$ for all $k \in \mathbb{N}$, whence $\rho^+ \subseteq \rho'$. So $\rho^+ = \mathbf{t}(\rho)$. □

Proposition 2.31. *For a relation ρ on S the following holds:*

(1) $\rho = \mathbf{r}(\rho)$ iff ρ is reflexive; furthermore, $\mathbf{rr}(\rho) = \mathbf{r}(\rho)$.
(2) $\rho = \mathbf{s}(\rho)$ iff ρ is symmetric; furthermore, $\mathbf{ss}(\rho) = \mathbf{s}(\rho)$.
(3) $\rho = \mathbf{t}(\rho)$ iff ρ is transitive; furthermore, $\mathbf{tt}(\rho) = \mathbf{t}(\rho)$.

Proof. (1): A reflexive relation ρ is clearly the smallest reflexive relation containing itself, thus $\rho = \mathbf{r}(\rho)$. If $\rho = \mathbf{r}(\rho)$, then, since $\mathbf{r}(\rho)$ is reflexive, ρ also is. And clearly $\mathbf{rr}(\rho) = \mathbf{r}(I_S \cup \rho) = I_S \cup I_S \cup \rho = I_S \cup \rho = \mathbf{r}(\rho)$. □

Note that each $\mathbf{o} \in \{\mathbf{r},\mathbf{s},\mathbf{t}\}$ is indeed a closure operator: We have $\rho \subseteq \mathbf{o}\rho$ by Th. 2.30, isotonicity by the properties of set union, and idempotency holds by Prop. 2.31.

Proposition 2.32. ([37], p. 584) *For a relation ρ on S the following holds:*

(1) If ρ is reflexive, then also $\mathbf{s}(\rho)$ and $\mathbf{t}(\rho)$.
(2) If ρ is symmetric, then also $\mathbf{r}(\rho)$ and $\mathbf{t}(\rho)$.
(3) If ρ is transitive, then also $\mathbf{r}(\rho)$.

Proof. (1): $I_S \subseteq \rho \rightarrow I_S \subseteq \mathbf{s}(\rho) \wedge I_S \subseteq \mathbf{t}(\rho)$.

(2): I_S is symmetric, whence $\mathbf{r}(\rho)$ is symmetric by Prop. 2.23(4). Now we demonstrate that ρ^k is symmetric for each $n \in \mathbb{N}$: ρ^1 is symmetric by assumption; if $\rho^k = (\rho^k)^{\smile}$, then $\rho^{k+1} = \rho^k\rho = (\rho^k)^{\smile}\rho^{\smile} = (\rho\rho^k)^{\smile} = (\rho^{k+1})^{\smile}$. So, again by Prop. 2.23(4), it follows that $\mathbf{t}(\rho)$ is symmetric.

(3): $\mathbf{r}(\rho)^2 = (\rho \cup I_S)(\rho \cup I_S) = (\rho \cup I_S)\rho \cup (\rho \cup I_S)I_S = \rho^2 \cup \rho \cup \rho \cup I_S = \rho \cup I_S$, by the transitivity of ρ. (For sake of direct comparison I include the point-based proof given in [37]: Let $x(\rho \cup I_S)y$ and $y(\rho \cup I_S)z$. If $\langle x,y\rangle \in I_S$, then $x = y$ and $\langle x,z\rangle = \langle y,z\rangle \in \rho \cup I_S$. If $\langle y,z\rangle$ is in I_S, then $y = z$ and $\langle x,z\rangle = \langle x,y\rangle \in \rho \cup I_S$. If none of these two cases holds, then both pairs are in ρ and therefore $x(\rho \cup I_S)z$ by the transitivity of ρ.) □

The following examples are taken from [18]: Let $<$ be taken on e.g. the integers, then $\mathbf{r}(<) = \leq$. For the symmetric closure of parentOf we get $\sigma := \mathbf{s}(\text{parentOf}) = \text{parentOf} \cup$

childOf: clearly, $a\sigma b$ implies $b\sigma a$. And $\mathbf{t}(\text{parentOf}) = \text{ancestorOf}$: whenever $\langle a,b\rangle$ and $\langle b,c\rangle$ are in ancestorOf, then also $\langle a,c\rangle$.

Proposition 2.33. *Let ρ and σ be relations on a set S. Some consequences from the definitions of ρ^+ and ρ^* are:*

(1) $\rho \subseteq \rho^+, \qquad \rho \subseteq \rho^*.$
(2) $\rho \subseteq \sigma \;\rightarrow\; \rho^+ \subseteq \sigma^+ \;\wedge\; \rho^* \subseteq \sigma^*.$
(3) $\rho^*\rho^* = \rho^*.$
(4) $\rho^* \subseteq \rho^*\sigma^*, \quad \rho^* \subseteq \sigma^*\rho^*, \quad \sigma^* \subseteq \rho^*\sigma^*, \quad \sigma^* \subseteq \sigma^*\rho^*.$
(5) $\rho^*\sigma^* \subseteq (\rho\cup\sigma)^*.$
(6) $\rho^* = \mathbf{rt}(\rho).$
(7) $\rho^{**} = \rho^*.$
(8) $(\rho^+)^{\smile} = (\rho^{\smile})^+, \qquad (\rho^*)^{\smile} = (\rho^{\smile})^*.$
(9) *Let σ be transitive with $\rho \subseteq \sigma$; then $\mathbf{t}(\rho) \subseteq \sigma$.*

Proof. (3): ρ^+ is transitive, thus $\mathbf{r}(\rho^+) = I_S \cup \rho^+ = \rho^*$ is transitive by Prop. 2.32(3). Idempotency then follows by Prop. 2.24(3).

(4): We have $I \subseteq \sigma^*$, thus $\rho^* = I\rho^* \subseteq \sigma^*\rho^*$.

(5): Applying (2) and (3) we get $\rho,\sigma \subseteq \rho\cup\sigma \;\rightarrow\; \rho^*,\sigma^* \subseteq (\rho\cup\sigma)^* \;\rightarrow\; \rho^*\sigma^* \subseteq (\rho\cup\sigma)^*(\rho\cup\sigma)^* = (\rho\cup\sigma)^*$.

(6): By Th. 2.30 (3), $\mathbf{t}(\rho) = \rho^+$, whence $\mathbf{rt}(\rho^+) = \rho^+ \cup I_S = \rho^*$. Moreover, (a) $\rho \subseteq \rho^*$ by (1); (b) ρ^* is transitive by (3) and reflexive by definition; and (c) If σ is a reflexive and transitive relation with $\rho \subseteq \sigma$, then $\sigma = \mathbf{t}(\sigma) = \sigma^+$ by Prop. 2.31. By (2), $\rho^+ \subseteq \sigma^+$, whence $\rho^* \subseteq \sigma^*$ by monotonicity of union.

(7): We will show $\rho^{**} = \mathbf{rtrt}(\rho) = \mathbf{rt}(\rho)$. By construction $\mathbf{rt}(\rho)$ is reflexive, and by Prop. 2.32(1) also $\mathbf{trt}(\rho)$, whence $\mathbf{trt}(\rho) = \mathbf{rtrt}(\rho)$ by Prop. 2.31(1). And $\mathbf{rt}(\rho)$ is transitive by Prop. 2.32(3), whence $\mathbf{rt}(\rho) = \mathbf{trt}(\rho)$ by Prop. 2.31(3).

(8): $(\rho^+)^{\smile} = \left(\bigcup_{i\geq 1}\rho^i\right)^{\smile} = \bigcup_{i\geq 1}(\rho^i)^{\smile} = \bigcup_{i\geq 1}(\rho^{\smile})^i = (\rho^{\smile})^+$, using Th. 2.22(8).

(9): $I_S\cup\sigma$ is reflexive and transitive, hence $\rho^* \subseteq I_S\cup\sigma$. Thus, $\mathbf{t}(\rho) = \rho^+ = \rho\rho^* \subseteq \rho(I_S\cup\sigma) = \rho\cup\rho\sigma \subseteq \sigma\cup\sigma^2 = \sigma$. □

Proposition 2.34. *For relations ρ and σ on S the following holds:*

(1) $\mathbf{rs}(\rho) = \mathbf{sr}(\rho)$.
(2) $\mathbf{rt}(\rho) = \mathbf{tr}(\rho)$.
(3) $\mathbf{st}(\rho) \subseteq \mathbf{ts}(\rho)$.
(4) $\mathbf{rts}(\rho) = \mathbf{trs}(\rho) = \mathbf{tsr}(\rho)$ *and* $\mathbf{rst}(\rho) = \mathbf{srt}(\rho) = \mathbf{str}(\rho)$.

Proof. (1): $\mathbf{rs}(\rho) = I_S \cup \rho \cup \rho^{\smile} = I_S \cup \rho \cup I_S \cup \rho^{\smile} = \mathbf{s}(I_S \cup \rho) = \mathbf{sr}(\rho)$.

(2): It is not hard to show by induction that $(I_S\cup\rho)^k = \bigcup_{0\leq i\leq k}\rho^i$. Then $\mathbf{tr}(\rho) = \mathbf{t}(I_S\cup\rho) = \bigcup_{1\leq n}(I_S\cup\rho)^n = I_S \cup \bigcup_{1\leq n}\bigcup_{0\leq i\leq n}\rho^i = I_S\cup\bigcup_{1\leq n}\rho^n = I_S\cup\rho^+ = \rho^* = \mathbf{rt}(\rho)$ by Prop. 2.33(6).

(3): We have $\mathbf{st}(\rho) = \mathbf{s}(\rho^+) = \rho^+ \cup (\rho^+)\breve{}$. We show by induction that $\rho^n \cup (\rho^n)\breve{} \subseteq (\rho \cup \rho\breve{})^n$ for all $n \in \mathbb{N}$: For $n = 1$ we trivially have $\rho^1 \cup (\rho^1)\breve{} \subseteq (\rho \cup \rho\breve{})^1$. Now assume $\rho^k \cup (\rho^k)\breve{} \subseteq (\rho \cup \rho\breve{})^k$; then, by monotonicity of composition,

$$\begin{aligned}
&(\rho \cup \rho\breve{})^k(\rho \cup \rho\breve{}) \\
&\quad \supseteq [\rho^k \cup (\rho^k)\breve{}](\rho \cup \rho\breve{}) \\
&\quad = \rho^{k+1} \cup (\rho^k)\breve{}\rho \cup \rho^k\rho\breve{} \cup (\rho^k)\breve{}\rho\breve{} \\
&\quad = \rho^{k+1} \cup (\rho^{k+1})\breve{} \cup \ldots, && \text{by applying Th. 2.22(8) twice to the last term} \\
&\quad \supseteq \rho^{k+1} \cup (\rho^{k+1})\breve{}.
\end{aligned}$$

Thus, $\bigcup_{n\geq 1} \rho^i \cup (\rho^i)\breve{} \subseteq \bigcup_{n\geq 1} (\rho \cup \rho\breve{})^i$, and therefore $\mathbf{st}(\rho) \subseteq \mathbf{ts}(\rho)$.

(4): By (1) we have $\mathbf{trs}(\rho) = \mathbf{tsr}(\rho)$, and by (2) $\mathbf{rts}(\rho) = \mathbf{trs}(\rho)$. □

Proposition 2.35. *For relations ρ and σ on S the following holds:*

(1) $\mathbf{r}(\rho \cup \sigma) = \mathbf{r}(\rho) \cup \mathbf{r}(\sigma)$.
(2) $\mathbf{s}(\rho \cup \sigma) = \mathbf{s}(\rho) \cup \mathbf{s}(\sigma)$.
(3) $\mathbf{t}(\rho_1 \cup \ldots \cup \rho_K) = \bigcup_{1\leq i_k\leq K,\ n\geq 1} \rho_{i_1}\rho_{i_2}\cdots\rho_{i_n}$, *for all* $K \geq 1$.
(4) $\mathbf{r}(\rho \cap \sigma) = \mathbf{r}(\rho) \cap \mathbf{r}(\sigma)$.
(5) $\mathbf{s}(\rho \cap \sigma) \subseteq \mathbf{s}(\rho) \cap \mathbf{s}(\sigma)$.

Proof. (2): $\mathbf{s}(\rho \cup \sigma) = (\rho \cup \sigma) \cup (\rho \cup \sigma)\breve{} = \rho \cup \rho\breve{} \cup \sigma \cup \sigma\breve{} = \mathbf{s}(\rho) \cup \mathbf{s}(\sigma)$.

(3): Let (TC_K) denote the equation of the assertion. We have to show that (TC_K) holds for all $K \geq 1$. (TC_1) holds by Th. 2.30(3). For sake of ease of notation writing $\rho := \rho_1$ and $\sigma := \rho_2$, we show (TC_2), because it will be needed later on in the proof of the general case:

$$\begin{aligned}
\mathbf{t}(\rho \cup \sigma) &= \bigcup_{i\geq 1}(\rho \cup \sigma)^i \\
&= (\rho \cup \sigma) \cup (\rho^2 \cup \sigma\rho \cup \rho\sigma \cup \sigma^2) \cup \\
&\quad (\rho^3 \cup \sigma\rho^2 \cup \rho\sigma\rho \cup \sigma^2\rho \cup \rho^2\sigma \cup \sigma\rho\sigma \cup \sigma^2\rho \cup \rho\sigma^2 \cup \sigma^3) \cup \ldots \\
&= \bigcup_{1\leq i_k\leq 2,\ n\geq 1} \rho_{i_1}\rho_{i_2}\cdots\rho_{i_n}.
\end{aligned}$$

To prove the general case, we use mathematical induction, and assume that (TC_K) holds. Then we have $\mathbf{t}(\rho_1 \cup \ldots \cup \rho_K \cup \rho_{K+1}) = \mathbf{t}((\rho_1 \cup \ldots \cup \rho_K) \cup \rho_{K+1})$. Setting $\rho := (\rho_1 \cup \ldots \cup \rho_K)$ and $\sigma := \rho_{K+1}$ we now apply (TC_2) to get

$$\begin{aligned}
\bigcup_{i\geq 1}(\rho \cup \sigma)^i &= \bigcup_{i\geq 1}\rho^i \cup (\rho\sigma \cup \sigma\rho \cup \sigma^2) \cup \\
&\quad (\sigma\rho^2 \cup \rho\sigma\rho \cup \sigma^2\rho \cup \rho^2\sigma \cup \sigma\rho\sigma \cup \sigma^2\rho \cup \rho\sigma^2 \cup \sigma^3) \cup \ldots \\
&= \bigcup_{1\leq i_k\leq K,\ n\geq 1} \rho_{i_1}\rho_{i_2}\cdots\rho_{i_n} \cup \text{ all products involving } \sigma \\
&= \bigcup_{1\leq i_k\leq K+1,\ n\geq 1} \rho_{i_1}\rho_{i_2}\cdots\rho_{i_n}.
\end{aligned}$$

(5): $\mathbf{s}(\rho \cap \sigma) = (\rho \cap \sigma) \cup (\rho\breve{} \cap \sigma\breve{}) = (\rho \cup \rho\breve{}) \cap (\sigma \cup \rho\breve{}) \cap (\rho \cup \sigma\breve{}) \cap (\sigma \cup \sigma\breve{}) \subseteq \mathbf{s}(\rho) \cap \mathbf{s}(\sigma)$. □

2.8 Some Basic Kinds of Relations

In this section we present the most often used and hence most popular kinds of relations together with some of their basic properties. We begin with functions (in contrast to Chapter 1 now on the object level), and then continue with relations which are of foremost importance in pure and applied mathematics, in computer science, and also, of course, in daily life: partial orders and equivalence relations.

2.8.1 *Functions*

Though we do already know about the concept of a function from Def. 1.11, to avoid cumbersome look-ups we shortly recall the relevant notions in

Definition 2.36. *A relation $\rho \subseteq S \times T$ is called*

(a) a **partial mapping**, *or* **functional**, *if $\forall x \in S, \forall y, z \in T[x\rho y \wedge x\rho z \rightarrow y = z]$.*
(b) **total** *if $\forall x \in S\,\exists y \in T[x\rho y]$, i.e. $\mathrm{dom}(\rho) = S$.*
(c) a **mapping** *or* **function** *if it is functional and total.*
(d) **injective** *or* **one-to-one** *if $\forall x, y \in S, \forall z \in T[x\rho z \wedge y\rho z \rightarrow x = y]$.*
(e) **surjective** *or* **onto** *if $\forall y \in T\,\exists x \in S[x\rho y]$.*
(f) **bijective** *if it is injective and surjective.*

We continue with listing some easy consequences from the definitions above. Some characterizations of functional, injective and surjective relations have already been shown in Prop. 2.18. We will use these results to give some more relationships of that kind in

Proposition 2.37. *Let $\rho \subseteq S \times T$ be a relation.*

(1) ρ is functional $\iff \rho\breve{}\rho \subseteq I_T \iff \rho\overline{I_T} \subseteq \overline{\rho} \iff \rho\breve{}$ is injective.
(2) ρ is total $\iff I_S \subseteq \rho\rho\breve{} \iff \overline{\rho} \subseteq \rho\overline{I_T} \iff U_{S\times T} = \rho U_T$.
(3) ρ is injective $\iff \rho\rho\breve{} \subseteq I_S \iff \overline{I_S}\rho \subseteq \overline{\rho} \iff \rho\breve{}$ is functional.
(4) ρ is surjective $\iff I_T \subseteq \rho\breve{}\rho \iff \overline{\rho} \subseteq \overline{I_T}\rho \iff U_{S\times T} = U_S\rho$
$\iff \rho\breve{}$ is total.
(5) ρ is a mapping $\iff \rho\overline{I_T} = \overline{\rho}$.
(6) If both ρ and $\rho\breve{}$ are mappings, then both are bijective.

Proof. Note that the first equivalences in (1) - (4) are due to Prop. 2.18.

(1): From $\rho\breve{}\rho \subseteq I_T$ we get $\rho\overline{I_T} \subseteq \overline{\rho}$ by applying the SCHRÖDER equivalences Th. 2.20(1)(a). Now assume ρ to be functional, and let $\{\langle y_1, x\rangle, \langle y_2, x\rangle\} \subseteq \rho\breve{}$. Then $\{\langle x, y_1\rangle, \langle x, y_2\rangle\} \subseteq \rho$; thus $y_1 = y_2$, i.e. $\rho\breve{}$ is injective. The proof for the converse direction is similar.

(2): $I_S \subseteq \rho\rho\breve{}$ and $\rho\breve{}U_{S\times X} \subseteq U_{T\times X}$ imply $U_{S\times X} = I_S U_{S\times X} \subseteq \rho\rho\breve{}U_{S\times X} \subseteq \rho U_{T\times X} \subseteq U_{S\times X}$, i.e. $U_{S\times X} = \rho U_{T\times X}$, for any set X, which is more general than we need to show. For the proof of the other direction assume $U_{S\times T} = \rho U_T$. Then $I_S = \rho U_{T\times S} \cap I_S$ and, by applying the Dedekind rule Th. 2.20(3), $\rho U_{T\times S}\cap I_S \subseteq (\rho\cap I_S U_{S\times T})(U_{T\times S}\cap\rho^{\smile} I_S) = \rho\rho^{\smile} I_S = \rho\rho^{\smile}$, so $I_S \subseteq \rho\rho^{\smile}$. (Note that $U_{T\times S}{}^{\smile} = U_{S\times T}$.)

Furthermore, we have $\overline{\rho} \subseteq S\times T$ and $\rho\overline{I_T} \subseteq S\times T$, whence $\overline{\rho}\subseteq\rho\overline{I_T} \iff \rho\cup\rho\overline{I_T} = U_{S\times T}$. But $\rho\cup\rho\overline{I_T} = \rho I_T\cup\rho\overline{I_T} = \rho(I_T\cup\overline{I_T}) = \rho U_T$.

(3): $\rho\rho^{\smile}\subseteq I_S \iff \overline{I_S}\rho\subseteq\overline{\rho}$ holds due to the SCHRÖDER equivalence Th. 2.20(1)(b). By (1) we have $\rho^{\smile}$ is functional iff $\rho^{\smile\smile} = \rho$ is injective.

(4): Similar to (2).

(5): Follows directly from the inequalities shown in (1) and (2).

(6): By hypothesis ρ is functional and total, i.e. $\rho^{\smile}\rho\subseteq I_T$ and $I_S\subseteq\rho\rho^{\smile}$; for $\rho^{\smile}$ we similarly get $\rho\rho^{\smile}\subseteq I_S$ and $I_T\subseteq\rho^{\smile}\rho$. From the last two inequalities ρ is injective and surjective, according to (3) and (4), and from the first two the same follows for $\rho^{\smile}$. □

Corollary 2.38. *Let $\rho\subseteq X\times Y$ and $\sigma\subseteq Y\times Z$ be relations. If both ρ and σ are*

(1) functional,
(2) total,
(3) injective,
(4) surjective,
(5) mappings,
(6) bijective,

then also $\rho\sigma$ and $\sigma\rho$.

Proof. (1): $(\rho\sigma)^{\smile}\rho\sigma = \sigma^{\smile}\rho^{\smile}\rho\sigma\subseteq\sigma^{\smile} I_Y\sigma = \sigma^{\smile}\sigma\subseteq I_Z$.

(2): $I_X\subseteq\rho\rho^{\smile}\subseteq\rho I_Y\rho^{\smile}\subseteq\rho\sigma\sigma^{\smile}\rho^{\smile} = \rho\sigma(\rho\sigma)^{\smile}$.

(3): $(\rho\sigma)(\rho\sigma)^{\smile} = \rho\sigma\sigma^{\smile}\rho^{\smile}\subseteq\rho I_Y\rho^{\smile} = \rho\rho^{\smile}\subseteq I_X$.

(4): $I_Z\subseteq\sigma^{\smile}\sigma\subseteq\sigma^{\smile} I_Y\sigma\subseteq\sigma^{\smile}\rho^{\smile}\rho\sigma = (\rho\sigma)^{\smile}\rho\sigma$.

(5): By (1) and (2).

(6): By (3) and (4). □

Corollary 2.39. *For all relations ρ,σ and τ the following holds:*

(1) τ is functional $\iff \tau(\rho\cap\sigma) = \tau\rho\cap\tau\sigma$.
(2) τ is functional $\iff \rho\sigma\cap\tau = (\rho\cap\tau\sigma^{\smile})\sigma$.
(3) τ is functional $\iff \tau\overline{\rho} = \tau U\cap\overline{\tau\rho}$.
(4) τ is functional $\wedge\ \rho\subseteq\tau\ \wedge\ \rho U\supseteq\tau U \Longrightarrow \rho = \tau$.

Proof. (1) [⇒] By Th. 2.22(4) we have $\tau(\rho\cap\sigma)\subseteq\tau\rho\cap\tau\sigma$. Using the Dedekind rule Th. 2.20(3), monotonicity, and $\tau^{\smile}\tau\subseteq I$ gives $\tau\rho\cap\tau\sigma\subseteq(\tau\cap\tau\sigma\rho^{\smile})(\rho\cap\tau^{\smile}\tau\sigma)\subseteq\tau(\rho\cap\tau^{\smile}\tau\sigma)\subseteq\tau(\rho\cap I\sigma)$.

[⇐] We take $\rho := \overline{I}$ and $\sigma := I$; then $\tau\overline{I}\cap\tau = \tau\overline{I}\cap\tau I = \tau(\overline{I}\cap I) = \tau\emptyset = \emptyset$. But $\tau\overline{I}\cap\tau = \emptyset\rightarrow\tau\overline{I}\subseteq\overline{\tau}$, whence τ is functional by Prop. 2.37(1).

(2) [⇒] $(\rho\cap\tau\sigma^{\smile})\sigma\subseteq\rho\sigma\cap\tau\sigma^{\smile}\sigma\subseteq\rho\sigma\cap\tau\subseteq(\rho\cap\tau\sigma^{\smile})(\sigma\cap\rho^{\smile}\tau)\subseteq(\rho\cap\tau\sigma^{\smile})\sigma$.

[⇐] Taking $\rho := \sigma^{\smile}$ and $\sigma := \overline{I}$ we get $\sigma^{\smile}\sigma \cap \overline{I} = (\sigma^{\smile} \cap \overline{I}\sigma^{\smile})\sigma = (I\sigma^{\smile} \cap \overline{I}\sigma^{\smile})\sigma$. Now $I\sigma^{\smile} \cap \overline{I}\sigma^{\smile} = \{\langle x, y\rangle : xIx \wedge x\sigma^{\smile}y \wedge \neg xIx \wedge x\sigma^{\smile}y\} = \emptyset$. So $\sigma^{\smile}\sigma \cap \overline{I} = \emptyset$, whence $\sigma^{\smile}\sigma \subseteq I$.

(3) [⇒] We have $\rho \subseteq \rho \;\rightarrow\; \tau^{\smile}\tau\rho \subseteq I\rho \;\leftrightarrow\; \tau\overline{\rho} \subseteq \overline{\tau\rho}$ by the functionality of τ and Th. 2.20(1). Since $\tau\overline{\rho} \subseteq \tau U$, we get $\tau\overline{\rho} \subseteq \overline{\tau\rho} \cap \tau U$. For the other inequality, using set theory and distributivity, we conclude as follows: $U = U \;\leftrightarrow\; \tau U \cup \overline{\tau U} = U \;\leftrightarrow\; \tau(\rho \cup \overline{\rho}) \cup \overline{\tau U} = U \;\leftrightarrow\; \tau\rho \cup \tau\overline{\rho} \cup \overline{\tau U} = U \;\leftrightarrow\; \overline{\overline{\tau\rho} \cap \tau U} \cup \tau\overline{\rho} = U \;\leftrightarrow\; \overline{\tau\rho} \cap \tau U \subseteq \tau\overline{\rho}$.
[⇐] Taking $\rho := I$, from the assumption we get $\tau\overline{I} = \tau U \cap \overline{\tau} = \tau(I \cup \overline{I}) \cap \overline{\tau} = (\tau I \cup \tau\overline{I}) \cap \overline{\tau} = \tau\overline{I} \cap \overline{\tau}$. So $\tau\overline{I} \subseteq \overline{\tau}$, whence τ is functional by Prop. 2.37(1).

(4): By Prop. 2.26(2) and our hypothesis we have $\tau \subseteq \tau U \subseteq \rho U$, whence $\tau = \tau U \cap \tau \subseteq \rho U \cap \tau$. Applying the Dedekind rule Th. 2.20(3) this gives

$$\tau \subseteq \rho U \cap \tau \subseteq (\rho \cap \tau U^{\smile})(U \cap \rho^{\smile}\tau) \subseteq \rho\rho^{\smile}\tau \subseteq \rho\tau^{\smile}\tau,$$

with the last inequality following from $\rho \subseteq \tau \;\rightarrow\; \rho^{\smile} \subseteq \tau^{\smile}$. And by the functionality of τ we finally get $\tau \subseteq \rho\tau^{\smile}\tau \subseteq \rho I = \rho$. □

Proposition 2.40. *For all relations ρ and τ the following holds:*

(1) τ is total $\Longrightarrow \overline{\tau\rho} \subseteq \tau\overline{\rho}$.
(2) τ is total $\Longrightarrow (\rho\tau = \emptyset \;\leftrightarrow\; \rho = \emptyset)$.

Proof. Note that for τ total we have $U = \tau U$ by Prop. 2.37.
(1): Then $U = \tau U = \tau(\rho \cup \overline{\rho}) = \tau\rho \cup \tau\overline{\rho}$. So we have $\tau\rho \supseteq \overline{\tau\overline{\rho}}$, whence $\overline{\tau\rho} \subseteq \tau\overline{\rho}$.
(2): Using the SCHRÖDER equivalence Th. 2.20(1)(b) we get, since $U = U^{\smile}$,

$$\rho\tau \subseteq \emptyset \;\leftrightarrow\; U\tau^{\smile} \subseteq \overline{\rho} \;\leftrightarrow\; U^{\smile}\tau^{\smile} \subseteq \overline{\rho} \;\leftrightarrow\; (\tau U)^{\smile} \subseteq \overline{\rho} \;\leftrightarrow\; U \subseteq \overline{\rho} \;\leftrightarrow\; \rho \subseteq \emptyset.$$

□

Corollary 2.41. *Let τ be a mapping; then for all relations ρ and σ the following holds:*

(1) $\tau\overline{\rho} = \overline{\tau\rho}$.
(2) $\rho\tau \subseteq \sigma \iff \rho \subseteq \sigma\tau^{\smile}$.

Proof. (1): From Cor. 2.39(3), since $\tau U = U$ by Prop. 2.37(2).
(2): [⇒] Since τ is total and $U = U^{\smile}$ we get: $U = \tau U = U^{\smile} = U\tau^{\smile} = (\overline{\sigma} \cup \sigma)\tau^{\smile} = \overline{\sigma}\tau^{\smile} \cup \sigma\tau^{\smile} \subseteq \overline{\rho} \cup \sigma\tau^{\smile}$ (for the last inequality we used that , by Th. 2.20(1)(b), $\rho\tau \subseteq \sigma \;\rightarrow\; \overline{\sigma}\tau^{\smile} \subseteq \overline{\rho}$). Thus $U \subseteq \overline{\rho} \cup \sigma\tau^{\smile}$, whence $\rho \cap \overline{\sigma\tau^{\smile}} \subseteq \emptyset$, i.e. $\rho \subseteq \sigma\tau^{\smile}$.
[⇐] $\rho \subseteq \sigma\tau^{\smile} \;\rightarrow\; \rho\tau \subseteq \sigma\tau^{\smile}\tau \subseteq \sigma$, since τ is functional by hypothesis. □

We close this section with the following two remarks:

- Given two relations ρ and σ, then their composition $\rho\sigma$ is commonly denoted as $\sigma\rho$ in case ρ and σ are functions: $x(\rho\sigma) := \sigma(\rho(x))$.
- A relation $\rho \subseteq S \times T$ from S to T can also be viewed as a function

$$f_\rho : S \longrightarrow \mathbf{2}^T \text{ with } f_\rho(x) := x\rho \tag{2.9}$$

for each $x \in S$. Similarly, $\rho^{\smile}$ con be considered as a function

$$f_{\rho^{\smile}} : T \longrightarrow \mathbf{2}^S \text{ with } f_{\rho^{\smile}}(y) := \rho y. \tag{2.10}$$

2.8.2 *Dual Relations*

Definition 2.42. $\rho^d = \{\langle x,y\rangle : \langle y,x\rangle \notin \rho\}$ *is called the* **dual relation** *for* ρ.

Note that by Prop. 2.8(5) we have $\rho^d = \overline{\rho^{\breve{}}} = \overline{\rho}^{\breve{}}$. For dual relations the following easy to prove relationships hold:

Corollary 2.43. *Let* $\rho, \sigma \subseteq X \times Y$ *be binary relations. Then*

(1) $\rho \subseteq \sigma$ *iff* $\sigma^d \subseteq \rho^d$.
(2) $(\rho^d)^d = \rho$.
(3) $(\rho^d)^{\breve{}} = \overline{\rho} = (\rho^{\breve{}})^d$.
(4) $\overline{\rho}^d = \rho^{\breve{}}$.
(5) $\rho\sigma^{\breve{}}\rho \subseteq \rho \;\leftrightarrow\; \sigma \subseteq \overline{\rho\rho^d\rho}$.

Proof. (1): By Prop. 2.8(6), $\rho \subseteq \sigma$ iff $\rho^{\breve{}} \subseteq \sigma^{\breve{}}$. But $\rho^{\breve{}} \subseteq \sigma^{\breve{}}$ iff $\overline{\sigma^{\breve{}}} \subseteq \overline{\rho^{\breve{}}}$.
(2): $(\rho^d)^d = \overline{(\overline{\rho^{\breve{}}})^{\breve{}}} = (\overline{\overline{\rho^{\breve{}}}})^{\breve{}} = \rho^{\breve{}\breve{}} = \rho$.
(3): $(\rho^d)^{\breve{}} = (\overline{\rho^{\breve{}}})^{\breve{}} = \overline{\rho} = \overline{\rho^{\breve{}\breve{}}} = (\rho^{\breve{}})^d$.
(4): By (2) and (3).

$$
\begin{array}{llll}
(5): & & \rho\sigma^{\breve{}}\rho \subseteq \rho & \\
& \leftrightarrow & \sigma^{\breve{}}\rho \subseteq \overline{\rho^{\breve{}}\overline{\rho}} & \text{by (1)} \\
& \leftrightarrow & \sigma^{\breve{}} \subseteq \overline{\rho^{\breve{}}\overline{\rho}\rho^{\breve{}}} & \text{by (1)} \\
& \leftrightarrow & \sigma \subseteq \overline{\rho\rho^d\rho} & \text{by Prop. 2.8(5),(6)}
\end{array}
$$

□

Lemma 2.44. ([9, Lemma 7.10]) *Let* ρ *be a binary relation on* X. *Then the following properties are equivalent:*

(a) $\rho\rho^d$ *is asymmetric.*
(b) $\rho^d\rho$ *is asymmetric.*
(c) $\rho\rho^d$ *is a weak order (i.e. asymmetric and negatively transitive).*
(d) $\rho^d\rho$ *is a weak order.*

Proof. Assuming that $\rho\rho^d$ is asymmetric, but $\rho^d\rho$ is not, implies $x\,\rho^d\rho\,y$ and $y\,\rho^d\rho\,x$ for two elements $x \neq y$. So there are two distinct elements z and t with $z\overline{\rho}x \wedge z\rho y$ from the first, and $t\overline{\rho}y \wedge t\rho x$ from the second relationship, whence we have $t\,\rho\rho^d\,z$. But from $z\rho y$ and $t\overline{\rho}y$ we also get $z\,\rho\rho^d\,t$, a contradiction to the asymmetry of $\rho\rho^d$. In a similar way one can show that the asymmetry of $\rho^d\rho$ implies that of $\rho\rho^d$; thus (a) is equivalent to (b).

Since $\rho^d\rho$ is negatively transitive, (b) implies (c), and (c) trivially implies (b) by the definition of a weak order. So we get that (b) is equivalent to (c). And using that $\rho\rho^d$ is also negatively transitive, (a), and (b), we have that (c) is equivalent to (d). □

2.8.3 *Residuals*

Since we have defined an operation called *product*, a natural question is, if we can also define an operation that gives something like a *quotient* of two relations. To possibly find an answer to this question, the concept of a (left and right, since product of relations is not commutative) **residual** is introduced: Residuals are the greatest solutions χ of the relational inequalities $\rho\chi \subseteq \sigma$ and $\chi\rho \subseteq \sigma$, respectively.

In the expressions $\rho(\sigma/\rho) \subseteq \sigma$ and $(\sigma\backslash\rho)\rho \subseteq \sigma$, the ρ's "cancel out". Moreover, in this scheme of notation the upper end of the / or \ symbol points in the direction indicated by its name, which is the place of the unknown factor χ in the product.

Definition 2.45.

(a) *Let* $\rho \in \mathcal{B}_{C\times A}$ *and* $\sigma \in \mathcal{B}_{C\times B}$ *be two relations; then the* **right residual** *of* σ *by* ρ *is* $\sigma/\rho \in \mathcal{B}_{A\times B}$ *defined as* $\sigma/\rho := \overline{\rho\breve{}\,\overline{\sigma}}$.

(b) *Let* $\rho \in \mathcal{B}_{A\times C}$ *and* $\sigma \in \mathcal{B}_{B\times C}$ *be two relations; then the* **left residual** *of* σ *by* ρ *is* $\sigma\backslash\rho \in \mathcal{B}_{B\times A}$ *defined as* $\sigma\backslash\rho := \overline{\overline{\sigma}\,\rho\breve{}}$.

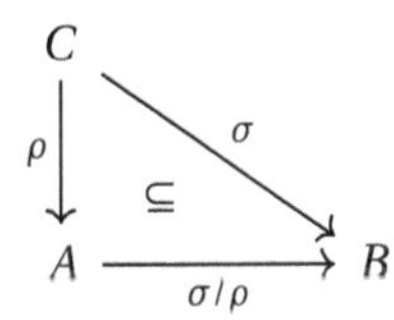

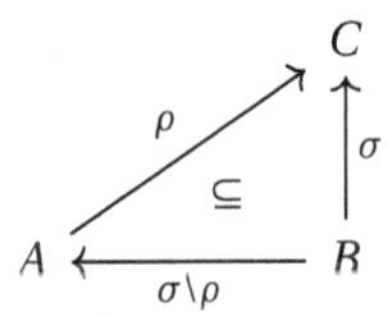

Proposition 2.46. *Left and right residuals have the following properties:*

(1) $\rho\chi \subseteq \sigma \;\leftrightarrow\; \chi \subseteq \sigma/\rho$.
(2) $\chi\rho \subseteq \sigma \;\leftrightarrow\; \chi \subseteq \sigma\backslash\rho$.
(3) $x(\sigma/\rho)y \;\leftrightarrow\; \rho x \subseteq \sigma y$, *i.e.* $\forall z[z\rho x \;\rightarrow\; z\sigma y]$.
(4) $x(\sigma\backslash\rho)y \;\leftrightarrow\; x\rho \subseteq y\sigma$, *i.e.* $\forall z[x\rho z \;\rightarrow\; y\sigma z]$.
(5) $\rho(\sigma/\rho) \subseteq \sigma$, *i.e.* σ/ρ *is the greatest solution* χ *of* $\rho\chi \subseteq \sigma$.
(6) $(\sigma\backslash\rho)\rho \subseteq \sigma$, *i.e.* $\sigma\backslash\rho$ *is the greatest solution* χ *of* $\chi\rho \subseteq \sigma$.
(7) $\rho\tau \subseteq \sigma \;\leftrightarrow\; \tau \subseteq \sigma/\rho \;\leftrightarrow\; \rho \subseteq \sigma\backslash\tau$.
(8) $\rho\tau \subseteq \sigma \;\leftrightarrow\; \rho \subseteq \overline{\overline{\sigma}\tau\breve{}} \;\leftrightarrow\; \tau \subseteq \overline{\rho\breve{}\,\overline{\sigma}}$.

Proof. (1): Using the first equivalence of the Schröder rule, we have $\rho\chi \subseteq \sigma \;\leftrightarrow\; \rho\breve{}\,\overline{\sigma} \subseteq \overline{\chi}$, and $\rho\breve{}\,\overline{\sigma} \subseteq \overline{\chi}$ is equivalent to $\chi \subseteq \overline{\rho\breve{}\,\overline{\sigma}}$.

(2): $\chi\rho \subseteq \sigma \;\leftrightarrow\; \overline{\sigma}\rho\breve{} \subseteq \overline{\chi} \;\leftrightarrow\; \chi \subseteq \overline{\overline{\sigma}\rho\breve{}} = \sigma\backslash\rho$.

(3): $x(\sigma/\rho)y \;\leftrightarrow\; x(\overline{\rho\breve{}\,\overline{\sigma}})y \;\leftrightarrow\; \neg\exists z[x\rho\breve{}z \wedge z\overline{\sigma}y]$ by definition of composition
$\leftrightarrow\; \forall z[\overline{x\rho\breve{}z} \vee z\sigma y]$ by Th. 1.19(1)
$\leftrightarrow\; \forall z[x\rho\breve{}z \;\rightarrow\; z\sigma y]$ by Th. 1.17(7)
$\leftrightarrow\; \forall z[z\rho x \;\rightarrow\; z\sigma y]$
$\leftrightarrow\; \rho x \subseteq \sigma y$.

(5): We have $\rho\breve{}\,\overline{\sigma} \subseteq \rho\breve{}\,\overline{\sigma} \;\leftrightarrow\; \rho\breve{}\,\overline{\sigma} \subseteq \overline{\overline{\rho\breve{}\,\overline{\sigma}}} \;\leftrightarrow\; \rho\breve{}\,\overline{\sigma} \subseteq \overline{\sigma/\rho}$, which is equivalent to $\rho(\sigma/\rho) \subseteq \sigma$ by the Schröder rule, and by (1) σ/ρ is the greatest solution.

(7): By (1) and (2).

(8): Using Th. 2.20(1)(a),(b) we get $\rho\tau \subseteq \sigma \leftrightarrow \rho^{\smile}\overline{\sigma} \subseteq \overline{\tau} \leftrightarrow \tau \subseteq \overline{\rho^{\smile}\overline{\sigma}}$, and similarly Th. 2.20(1)(b) implies $\rho\tau \subseteq \sigma \leftrightarrow \overline{\sigma}\tau^{\smile} \subseteq \overline{\rho} \leftrightarrow \rho \subseteq \overline{\overline{\sigma}\tau^{\smile}}$. □

By (3) of the above proposition, x is (σ/ρ)-related to y iff the x-column of ρ is a subset of the y-column of σ; and by (4) $\sigma\backslash\rho$ is characterized by the fact that the y-row of ρ is a subset of the x-row of σ. Do also note that by (1) and (2) of the above proposition the following holds:

Corollary 2.47. *Let ρ be a transitive relation; then $\rho \subseteq \rho/\rho = \overline{\rho^{\smile}\overline{\rho}}$ and $\rho \subseteq \rho\backslash\rho = \overline{\overline{\rho}\rho^{\smile}}$.*

Proposition 2.48. ([5, Lemma 2.3]) *Let $\rho, \rho' \subseteq Y \times Z$, $\sigma, \sigma' \subseteq X \times Y$, and $\tau, \tau' \subseteq X \times Z$ be relations. Then the following properties hold:*

(1) *If $\tau \subseteq \tau'$ and $\rho' \subseteq \rho$ then $\tau\backslash\rho \subseteq \tau'\backslash\rho'$.*
(2) *If $\tau' \subseteq \tau$ and $\sigma \subseteq \sigma'$ then $\tau/\sigma \subseteq \tau'/\sigma'$.*
(3) *$\rho\backslash I = \rho$, $I/\rho = \rho$, and $\overline{I}\backslash\overline{\rho} = \rho^{\smile}$.*
(4) *$I \subseteq \rho/\rho$ and $I \subseteq \rho\backslash\rho$.*
(5) *Both ρ/ρ and $\rho\backslash\rho$ are transitive.*
(6) *$\rho(\rho/\rho) = \rho$ and $(\rho\backslash\rho)\rho = \rho$.*
(7) *$(\rho\backslash\sigma)(\sigma\backslash\tau) \subseteq \rho\backslash\tau$ and $(\sigma/\rho)(\tau/\sigma) \subseteq \tau/\rho$.*

Proof. (1): By hypothesis $\overline{\tau'} \subseteq \overline{\tau}$ and $\rho'^{\smile} \subseteq \rho^{\smile}$. So $\overline{\tau'}\rho'^{\smile} \subseteq \overline{\tau}\rho^{\smile}$ which implies $\tau\backslash\rho \subseteq \tau'\backslash\rho'$.

(3): Trivially, $\overline{I}\backslash\overline{\rho} = \overline{\overline{I}\,\overline{\rho}^{\smile}} = \overline{\overline{\rho^{\smile}}} = \rho^{\smile}$, using Prop. 2.8(5).

(4): By Prop. 2.46(6) $\rho\backslash\rho$ is the greatest solution χ of $\chi\rho \subseteq \rho$, so $I \subseteq \rho\backslash\rho$.

(5): By Prop. 2.46(5) we have $\rho(\rho/\rho) \subseteq \rho$, whence $\rho(\rho/\rho)^2 \subseteq \rho(\rho/\rho) \subseteq \rho$. And since ρ/ρ is the greatest solution χ of $\rho\chi \subseteq \rho$, $(\rho/\rho)^2 \subseteq \rho/\rho$ follows.

(6): By Prop. 2.46(5) we have $\rho(\rho/\rho) \subseteq \rho$, and by (4) $\rho = \rho I \subseteq \rho(\rho/\rho)$.

(7): Using Prop. 2.46(5),(6) we get $(\rho\backslash\sigma)(\sigma\backslash\tau)\tau \subseteq (\rho\backslash\sigma)\sigma \subseteq \rho$ and $\rho(\sigma/\rho)(\tau/\sigma) \subseteq \sigma(\tau/\sigma) \subseteq \tau$, from which the assertions follow by Prop. 2.46(1),(2). □

There are still more properties for residuals, which exhibit a further regularity of this concept:

Proposition 2.49. *Using conversion and complement, the following interrelationships hold for residuals:*

(1) *$\sigma/\rho = (\sigma^{\smile}\backslash\rho^{\smile})^{\smile}$.*
(2) *$\sigma\backslash\rho = (\sigma^{\smile}/\rho^{\smile})^{\smile}$.*
(3) *$\overline{\sigma}/\overline{\rho} = \rho^{\smile}\backslash\sigma^{\smile}$.*
(4) *$\overline{\rho}/\overline{\sigma} = \rho^{\smile}/\sigma^{\smile}$.*

Proof.

(1): σ/ρ	$= \overline{\rho^{\smile}\overline{\sigma}}$	by definition of right residual
	$= \overline{(\overline{\sigma}^{\smile}\rho)^{\smile}}$	by Th. 2.22(8)

$$
\begin{aligned}
&= \overline{(\overline{\sigma^{\smile}}\,\rho^{\smile\smile})}^{\smile} && \text{using Prop. 2.8(5) twice}\\
&= (\sigma^{\smile}\backslash\rho^{\smile})^{\smile} && \text{by definition of left residual}
\end{aligned}
$$

(3): $\overline{\sigma}/\overline{\rho} \;= \overline{\overline{\rho}^{\smile}\sigma} = \rho^{\smile}\backslash\sigma^{\smile}.$

□

2.8.4 *Order Relations*

We can be sure of the fact that what we commonly call '*order relations*' nowadays played a significant, yet unconscious role in everyday life already very early in the development of mankind: It has always been important to be stronger, to associate with a companion more durable than most of the others, to own a larger area of land, to have available richer stocks for the winter period, to raise a greater number of cattle, to have more money at one's disposal than … . So, without doubt, various kinds of order do not only play a prominent role in pure and applied mathematics, but also in daily life.

Definition 2.50. *Let ρ be a relation on a set S. ρ is called a*

(a) **preorder** *(or* **quasi-order***), if ρ is reflexive and transitive.*
(b) **partial order***, if it is a preorder that is additionally antisymmetric, usually denoted in a generic sense by $\preceq$.*
(c) **strict partial order***, if it is irreflexive and transitive.*
(d) **total order** *if it is a partial order that is additionally complete.*
(e) **linear order** *if it is a strict partial order that is additionally complete.*
(f) If $\preceq$ is a partial order on a set S, then the pair $\langle S, \preceq\rangle$ is called a **partially ordered set** *or* **poset** *for short.*

A partial order is thus characterized by the following relationships: $I \subseteq \rho, \rho^2 \subseteq \rho$, and $\rho \cap \rho^{\smile} \subseteq I$ (see Prop. 2.18); and for a total order $\rho \cup \rho^{\smile} = U$ additionally holds.

We proceed with some elementary properties and examples of pre- and partial orders. Two more advanced results for partial orders will be mentioned later on in Section 10.4 in connection with fuzzy partial orders.

Proposition 2.51. *Let ρ and σ be two preorders on a set A. Then*

(1) $\rho^{\smile}$ is a preorder.
(2) $\rho \cap \sigma$ is a preorder.
(3) $\rho^2 \subseteq \rho^{\smile}\rho$.

Proof. (1): By Prop. 2.17(1) and (6).
(2): By Prop. 2.23.
(3): By (1) $\rho^{\smile}$ is a preorder, so $I \subseteq \rho^{\smile}$. Therefore, $\rho^2 = (I \cap \rho^{\smile})\rho^2 \subseteq \rho^{\smile}\rho^2 \subseteq \rho^{\smile}\rho$ by transitivity of ρ. □

Proposition 2.52. *Let ρ and σ be two preorders on a set S. Then ρ and σ commute iff both $\rho\sigma$ and $\sigma\rho$ are preorders.*

Proof. Assume $\rho\sigma = \sigma\rho$. $\rho\sigma$ is reflexive by Prop. 2.24(1). Furthermore, using Prop. 2.24(3), we have $(\rho\sigma)^2 = \rho\sigma\sigma\rho = \rho\sigma\rho = \rho^2\sigma = \rho\sigma$. Thus $\rho\sigma$ and $\sigma\rho$ are preorders. Conversely, assume both $\rho\sigma$ and $\sigma\rho$ to be preorders, i.e. each of $\rho, \sigma, \rho\sigma$ and $\sigma\rho$ are reflexive, transitive and idempotent. Then, using $I \subseteq \sigma$ and $I \subseteq \rho$, we get $\rho\sigma \subseteq \sigma(\rho\sigma) \subseteq \sigma\rho\sigma\rho = \sigma\rho$. Similarly one shows $\sigma\rho \subseteq \rho\sigma$. □

From Prop. 2.51(2) it follows that, given a set $R \subseteq \mathcal{B}_X$, there is a smallest preorder containing all the relations of R, called the *preorder generated by R*.

Proposition 2.53. *Let ρ and σ be two preorders on a set S. If they commute, the preorder generated by ρ and σ is equal to $\rho\sigma$.*

Proof. Suppose that ρ and σ commute; then by Prop. 2.52 both $\rho\sigma$ and $\sigma\rho$ are preorders. Then, using $I \subseteq \sigma$ and $I \subseteq \rho$, we have $\rho \subseteq \rho\sigma$ and $\sigma \subseteq \rho\sigma$. Now let τ be another preorder with $\rho \subseteq \tau$ and $\sigma \subseteq \tau$. Then $\rho\sigma \subseteq \tau\tau = \tau$, since preorders are idempotent by Prop. 2.24(3). □

As an example of a preorder consider the following (from [19, Section 1.4]): Let S denote the set of all sequences $(a_n)_{n\in\mathbb{N}}$ of real numbers. For $\boldsymbol{a}, \boldsymbol{b} \in S$ we define $\boldsymbol{a} \le \boldsymbol{b}$ if $\boldsymbol{a}$ is a subsequence of $\boldsymbol{b}$. Then $\le$ is reflexive and transitive, and thus a preorder, but not antisymmetric since $(0,1,0,1,\dots) \le (1,0,1,0,\dots) \le (0,1,0,1,\dots)$ without the last two sequences being equal.

Proposition 2.54. *Let ρ and σ be two partial orders on a set A. Then*

(1) *$\rho^\smallsmile$ is a partial order.*
(2) *$\rho \cap \sigma$ is a partial order.*

Proof. (1): By Prop. 2.51 and Prop. 2.17(5).
(2): From the antisymmetry of ρ and σ we get $\rho\cap\rho^\smallsmile \subseteq I$ and $\sigma\cap\sigma^\smallsmile \subseteq I$; thus $\rho\cap\rho^\smallsmile\cap\sigma\cap\sigma^\smallsmile = \rho\cap\sigma\cap(\rho\cap\sigma)^\smallsmile \subseteq I$. □

The strict order $\prec$ associated with a partial order $\preceq$ is the relation defined as $x \prec y := x \preceq y \wedge x \neq y$. Note that if $\prec$ is the strict order associated with $\preceq$, then $\prec$ is irreflexive. Conversely, if $\prec$ is an irreflexive and transitive relation on a set S, then $\preceq$ defined by $x \preceq y := x \prec y \vee x = y$ is a partial order.

As a shorthand we will often refer to $\langle S, \preceq\rangle$ as the poset S, leaving $\preceq$ implicit. And if it is necessary to distinguish it from others, we will write $\preceq_S$ for a partial order on S. If there is already a commonly used sign for a certain partial order, like $\le$ for numbers or $\subseteq$ for sets, we will of course use these signs instead of $\preceq$. A few examples of partial orders are (see [15]):

(1) The subset relation $\subseteq$: For any set S, $\langle \mathbf{2}^S, \subseteq\rangle$ is a partial order.
(2) The natural orders on $\mathbb{N}, \mathbb{Q}$ or $\mathbb{Z}$.
(3) Also $(\mathbb{N}, |)$ is a poset, where $m|n$ means $\exists k \in \mathbb{N}[k \ge 1 \wedge n = km]$.

Besides the above mentioned there are other kinds of order relations which play a crucial role in the study of preferences and measurement theory (see e.g. [45, Chapter 18] for an overview) which we mention below, first of all the Ferrers relation.

2.8.4.1 *Ferrers Relations*

This and the subsequent section can be skipped at first reading. They contain classical two-valued concepts and results which are needed later on in connection to our considerations on fuzzy preferences in Chapter 11.

Definition 2.55. *A relation ρ is said to be a* **Ferrers** *relation if*

$$\forall u,v,x,y[u\rho v \wedge x\rho y \rightarrow u\rho y \vee x\rho v]$$

Note that the condition above is equivalent to $\forall u,v,x,y[u\rho v \wedge \neg x\rho v \wedge x\rho y \rightarrow u\rho y]$. So ρ is a Ferrers relation if, and only if, $\rho\rho^d\rho \subseteq \rho$.

In terms of graphs this means: Given any two arrows, there exists an arrow from one of the starting points leading to the ending point of the other arrow. Visualized in matrix representation the definition means that $\begin{pmatrix}1&0\\0&1\end{pmatrix}$ must not occur as a pattern, i.e. three of the four cells of such a square must be filled. This is equivalent to the property that such a matrix can be transformed into a stair-shaped block form by rearranging its rows and columns. Ferrers relations thus have the interesting property that they can be decomposed in rectangular blocks of either 0's or 1's.

Proposition 2.56.

(1) *Either none or all of $\rho, \rho^\smile$, and $\overline{\rho}$ are of Ferrers type.*
(2) *ρ is of Ferrers type iff $\rho^\smile\overline{\rho}$ is.*
(3) *ρ is of Ferrers type iff $\overline{\rho}\rho^\smile$ is.*

Proof. (1): From Prop. 2.8(6) we get $\rho\rho^d\rho \subseteq \rho$ iff $\rho^\smile\overline{\rho}\rho^\smile \subseteq \rho^\smile$; and since $\overline{\rho} = (\rho^\smile)^d$, by Cor. 2.43(3), $\rho^\smile$ is Ferrers iff ρ is. From the first Schröder equivalence Th. 2.20(1)(a) we get $(\rho\rho^d)\rho \subseteq \rho \leftrightarrow ((\rho^d)^\smile\rho^\smile)\overline{\rho} \subseteq \overline{\rho}$. By Cor. 2.43 we then have $(\rho^d)^\smile = \overline{\rho}$ and $\rho^\smile = \overline{\rho}^d$, which gives the desired result.

(2): For each relation ρ, from $\rho^\smile\overline{\rho} = \rho^\smile\overline{\rho}$ we have $\rho\overline{\rho^\smile\overline{\rho}} \subseteq \rho$ by Th. 2.20(1)(a), and $(\rho^\smile\overline{\rho})^\smile = \rho^d\rho$. Applying these two results and Prop. 2.8(6) we get

$$\rho^\smile\overline{\rho}\cdot\overline{(\rho^\smile\overline{\rho})}^\smile\cdot\rho^\smile\overline{\rho} \subseteq \rho^\smile\overline{\rho} \quad \text{iff}$$
$$\rho^d\rho\cdot\overline{(\rho^\smile\overline{\rho})}\cdot\rho^d\rho \subseteq \rho^d\rho$$

But this last inequality holds since $\rho^d[\rho\cdot\overline{(\rho^\smile\overline{\rho})}]\cdot\rho^d\rho \subseteq \rho^d[\rho]\cdot\rho^d\rho \subseteq \rho^d\rho$ due to the assumption that ρ is of Ferrers type. So the first inequality also holds, i.e. $\rho^\smile\overline{\rho}$ is a Ferrers relation. □

Proposition 2.57.

(1) *A reflexive Ferrers relation is strongly complete.*
(2) *An irreflexive Ferrers relation is asymmetric and transitive.*
(3) *An irreflexive Ferrers relation is negatively transitive.*
(4) *If ρ is a Ferrers relation on a set X then $\rho\rho^d$ is a weak order.*

Proof. (1): $x\rho x \wedge y\rho y \rightarrow x\rho y \vee y\rho x$.

(2): $x\rho y \wedge y\rho z \rightarrow x\rho z \vee y\rho y$, so ρ is transitive. Thus, by Prop. 2.16(1) it is also asymmetric.

(3): By (2) and Prop. 2.16(7).

(4): According to Lemma 2.44 it suffices to show that $\rho\rho^d$ is asymmetric. Assume this to fail; then there exist $x, y \in X$ such that $x\rho\rho^d z$ and $z\rho\rho^d x$. This implies $x\rho y \wedge z\overline{\rho}y \wedge z\rho t \wedge x\overline{\rho}t$ for some $t, y \in X$, contradicting the definition of a Ferrers relation. □

2.8.4.2 *More on Order Relations*

Below we will elaborate on some kinds of relations that play a central role in the context of describing and analyzing preferences. For this we need some new concepts, and we begin by listing the corresponding formal definitions, together with some replications for easier look-up:

Definition 2.58. *Let ρ be a relation on a set X. Then ρ is called*

(a) **negatively transitive** *(1) if $\forall x, y, z[x\rho z \rightarrow x\rho y \vee y\rho z]$.*
(b) **negatively transitive** *(2) if $\forall x, y, z[\neg x\rho y \wedge \neg y\rho z \rightarrow \neg x\rho z]$.*
(c) **semitransitive** *if $\forall x, y, z, t \in X[x\rho y \wedge y\rho z \rightarrow x\rho t \vee t\rho z]$, or $\rho\rho\rho^d \subseteq \rho$.*

Lemma 2.59. *Let ρ be an asymmetric relation on X. Then*

(1) *$\rho \subseteq \rho^d$.*
(2) *Let $\asymp_\rho := \overline{\rho} \cap \rho^d$; then $\rho \cup \asymp_\rho = \rho^d$.*

Proof. (1): $x\rho y \rightarrow \neg y\rho x \rightarrow \overline{x\breve{\rho} y} \rightarrow x\rho^d y$.
(2): $\rho \cup \asymp_\rho = (\rho \cup \overline{\rho}) \cap (\rho \cup \rho^d) = U \cap \rho^d = \rho^d$, using (1). □

Definition 2.60. *Let ρ be a relation on a set X. Then ρ is called*

(a) *a* **strict partial order** *if it is irreflexive and transitive.*
(b) *a* **weak order** *if it is asymmetric and negatively transitive;.*
(c) *an* **interval order** *if it is an irreflexive Ferrers relation;.*
(d) *a* **semiorder** *if it is a semitransitive interval order;.*
(e) *a* **total preorder** *if it is a complete preorder.*

For sake of completeness we list some easy consequences of the above definition in

Corollary 2.61.

(1) An interval order is asymmetric and transitive.
(2) A semiorder is asymmetric and transitive.
(3) Interval orders, and hence also semiorders, are weak orders.
(4) A weak order is acyclic.
(5) A complete weak order is linear.

Proof. (1): By Prop. 2.57(2).
(2): By (1), since semiorders are interval orders.
(3): By (1),(2) and Prop. 2.16(7).
(4): By definition, a weak order is asymmetric and negatively transitive: Thus it is also transitive by Prop. 2.16(6) and irreflexive by Prop. 2.16(3), so finally acyclic by Prop. 2.16(2).
(5): By Prop. 2.16(3),(6). □

The intuition behind the definition of an interval order $\prec$ is that four elements w, x, y, z behave as if they were the end-points of two intervals $[x, y]$ and $[z, w]$ in $(\mathbb{R}, \leq)$, for which clearly the Ferrers property $x \prec y \wedge z \prec w \rightarrow x \prec w \vee z \prec y$ holds. This justifies the name for this kind of order on an arbitrary set.

2.8.5 *Equivalence and Tolerance Relations*

From a formal point of view we now turn to a few basic properties of three variants of symmetric relations, as given in

Definition 2.62. *A symmetric binary relation ρ on a set A is said to be a*

(a) **tolerance relation** *iff ρ is also reflexive.*
(b) **partial equivalence relation** *(short: PER), iff ρ is also transitive.*
(c) **equivalence relation**, *iff ρ is also transitive and reflexive.*

An equivalence relation thus is a tolerance that is also transitive or, equivalently, a preorder that is also symmetric (see Def. 2.50(a)), or a partial equivalence relation that is reflexive too; therefore an equivalence relation is a special case of each of the other three mentioned concepts. Because of their prevalence (and popularity) we begin our exposition with this special case, followed by the presentation of the two generalizations mentioned in Def. 2.62. This way, most of the material on equivalence relations can be read independently from the rest.

Equivalence relations are the most radical notion to model similarity in the sense that they partition a collection of things into pairwise disjoint sub-collections. Note that we do already know two trivial examples for an equivalence relation on a set S: I_S which fragments S into single-element sets, and U_S which does not fragment S at all.

A result that is inseparably connected to equivalence relations concerns their relationship to partitions. The fundamental significance of equivalence relations is that they motivate a rather general principle of abstraction: If we want to ignore some inessential attributes of the objects in a certain collection, then all objects with the same essential attributes can be identified. And analysis of such equivalence classes of objects rather than of the objects themselves often simplifies a problem significantly.

2.8.5.1 *Equivalence Relations*

Before we turn to details, we have to introduce some new notation that has become standard for equivalences:

Definition 2.63.

(a) *Let ρ be an equivalence relation on a set S, then it is customary to denote the aftersets $x\rho$ by $[x]_\rho$, x/ρ, or simply by $[x]$ if no confusion about the equivalence involved can occur: $[x]_\rho = \{y \in S : x\rho y\}$ is called the* **equivalence class** *of x induced by ρ.*

(b) *The set of all equivalence classes is called the* **quotient set** *of S modulo ρ, denoted by S/ρ.*

Proposition 2.64. *Let ρ be an equivalence relation on a set S. Then the following three statements are equivalent:*

(i) $x\rho y$.
(ii) $[x] = [y]$.
(iii) $[x] \cap [y] \neq \emptyset$.

Proof. (1) $\to$ (2): Suppose $x\rho y \wedge z \in [x]$; then $x\rho z$ and, by symmetry, $y\rho x$. Thus, $y\rho z$ by transitivity, whence $z \in [y]$. So $[x] \subseteq [y]$. Conversely, from $z \in [y]$ we get $y\rho z$ and, by transitivity, $x\rho z$. Thus, $[y] \subseteq [x]$.
(2) $\to$ (3): Obvious, since $x \in [x]$.
(3) $\to$ (1): $[x] \cap [y] \neq \emptyset$ implies $\exists z[x\rho z \wedge y\rho z]$. Thus, $z\rho y$, which gives $x\rho y$ by the transitivity of ρ. □

The following two theorems contain the presumably most well known and important results on equivalence relations:

Theorem 2.65. *Let ρ be an equivalence relation on a set S. Then the following holds:*

(1) $[x] \neq \emptyset$.
(2) *For $x, y \in S$, either $[x] = [y]$ or $[x] \cap [y] = \emptyset$.*
(3) $\bigcup\{[x] : x \in S\} = S$.
(4) *S/ρ is a partition of S.*

Proof. (1): Since ρ is reflexive, for each $x \in S : x \in [x]$.
(2): By Prop. 2.64 $[x] = [y] \leftrightarrow [x] \cap [y] \neq \emptyset$.

(3): Since for each $x \in S$ we have $x \in [x]$, we get $S \subseteq \bigcup\{[x] : x \in S\}$. And $[x] \subseteq S$ gives $\bigcup\{[x] : x \in S\} \subseteq S$.
(4): An immediate consequence from (1) - (3). □

The set of all equivalence relations on a set S will be denoted by Eq_S. It is well known that there is a one-to-one correspondence between Π_S, the set of partitions of the set S, and the set of equivalence relations Eq_S.

In what follows we will consider some properties of equivalence relations besides their correspondence to partitions, beginning with

Proposition 2.66. *Let $\rho, \sigma \in \mathsf{Eq}_S$. Then the following holds:*

(1) $\rho \cap \sigma$ is an equivalence relation.
(2) $\rho\rho = \rho$.
(3) $\rho\sigma$ is an equivalence iff $\rho\sigma = \sigma\rho$.
(4) $\rho\rho^{\smile}\rho = \rho$.
(5) $\rho \subseteq \sigma$ implies $\rho\sigma = \sigma$.
(6) If $\rho \cup \sigma$ is an equivalence then $\rho \cup \sigma = \rho\sigma$.
(7) If $\rho \cup \sigma$ is an equivalence then ρ and σ commute.

Proof. (1): By Prop. 2.23.
(2): By Prop. 2.24(3).
(3): If $\rho\sigma$ is an equivalence, then $\rho\sigma = (\rho\sigma)^{\smile} = \sigma^{\smile}\rho^{\smile} = \sigma\rho$, by the symmetry of each of $\rho\sigma, \rho$ and σ. Conversely, assume $\rho\sigma = \sigma\rho$. By Prop. 2.24(1) $\rho\sigma$ is reflexive; by the symmetry of ρ and σ we get $\rho\sigma = \sigma\rho = \sigma^{\smile}\rho^{\smile} = (\rho\sigma)^{\smile}$, whence $\rho\sigma$ is symmetric; and, finally, $\rho\sigma$ is transitive: $\rho\sigma\rho\sigma = \rho\sigma\sigma\rho = \rho\sigma\rho = \rho\rho\sigma = \rho\sigma$ by (2).
(4): Using symmetry and (2) we get $\rho\rho^{\smile}\rho = \rho^2 = \rho$.
(5): Obviously, $\sigma = I\sigma \subseteq \rho\sigma$; by assumption we also have $\rho\sigma \subseteq \sigma\sigma = \sigma$ by monotonicity of composition and (2).
(6): By Prop. 2.24(2) we have $\rho \cup \sigma \subseteq \rho\sigma$. Moreover, $\rho\sigma \subseteq (\rho \cup \sigma)^2 \subseteq \rho \cup \sigma$.
(7): By (6) and (3). □

The natural question about the structure of Eq_S is answered in the first part of

Proposition 2.67. *Let $\rho, \sigma \in \mathsf{Eq}_S$. Then the following holds:*

(1) $(\mathsf{Eq}_S, \subseteq)$ is a complete lattice.
(2) $\rho \vee \sigma = \mathbf{t}(\rho \cup \sigma)$.
(3) $\rho \vee \sigma = \rho \cup \rho\sigma \cup \rho\sigma\rho \cup \ldots \cup \sigma \cup \sigma\rho \cup \sigma\rho\sigma \ldots$.
(4) ρ and σ commute iff $\rho \vee \sigma = \rho\sigma$ iff $\rho\sigma \in \mathsf{Eq}_S$.

Proof. (1): Obviously, for every $S \neq \emptyset$ we have $\mathsf{Eq}_S \neq \emptyset$, since $\{I_S, U_S\} \subseteq \mathsf{Eq}_S$. Furthermore, we have $I_S \subseteq \rho \subseteq U_S$ for each $\rho \in \mathsf{Eq}_S$, whence Eq_S is a bounded poset. We have to show that any non-empty family $\{\rho_j : j \in J\} \subseteq \mathsf{Eq}_S$ has a supremum and infimum. From

Prop. 2.66(1) we get that $\bigwedge\{\rho_j : j \in J\} = \bigcap_{j\in J} \rho_j \in \mathsf{Eq}_S$. The supremum $\bigvee\{\rho_j : j \in J\}$ of a familiy of equivalence relations on S is $\bigcap\{\rho \in \mathsf{Eq}_S : \rho \supseteq \rho_j$ for all $j \in J\}$. But $U_S \in \mathsf{Eq}_S$ and $\rho_j \subseteq U_S$ for each $j \in J$, whence this intersection exists and is an equivalence on S due to what has just been shown. Moreover, it is the least such equivalence and therefore the supremum of $\{\rho_j : j \in J\} \subseteq \mathsf{Eq}_S$.

(2): By Prop. 2.23(1),(4) $\rho \cup \sigma$ is reflexive and symmetric, and by Prop. 2.32 $\mathbf{t}(\rho \cup \sigma)$ is also reflexive and symmetric. So by Def. 2.29 $\mathbf{t}(\rho \cup \sigma)$, is the smallest equivalence relation with $\rho \subseteq \mathbf{t}(\rho \cup \sigma)$ and $\sigma \subseteq \mathbf{t}(\rho \cup \sigma)$.

(3):
$$\begin{aligned}
\rho \vee \sigma &= \textstyle\bigcup_{i\ge 1} (\rho \cup \sigma)^i \\
&= (\rho \cup \sigma) \cup (\rho^2 \cup \sigma\rho \cup \rho\sigma \cup \sigma^2) \cup (\rho \cup \sigma)^3 \ldots \\
&= (\rho \cup \sigma) \cup (\sigma\rho \cup \rho\sigma) \cup (\sigma\rho \cup \rho\sigma)(\rho \cup \sigma) \ldots \\
&= (\rho \cup \sigma) \cup (\sigma\rho \cup \rho\sigma) \cup (\sigma\rho^2 \cup \rho\sigma\rho \cup \sigma\rho\sigma \cup \rho\sigma^2) \ldots \\
&= (\rho \cup \sigma) \cup (\sigma\rho \cup \rho\sigma) \cup (\rho\sigma\rho \cup \sigma\rho\sigma) \ldots \\
&= \rho \cup \rho\sigma \cup \rho\sigma\rho \cup \ldots \cup \sigma \cup \sigma\rho \cup \sigma\rho\sigma \ldots
\end{aligned}$$
(Note that by transitivity $\rho^2 \subseteq \rho$, whence $\rho^2\chi \subseteq \rho\chi$ and thus $\rho^2\chi \cup \rho\chi = \rho\chi$.)

(4): Assume $\rho\sigma = \sigma\rho$; then $\rho \vee \sigma$ reduces to $\rho \cup \rho\sigma \cup \sigma$. But by reflexivity we get $\rho = \rho I_S \subseteq \rho\sigma$ and $\sigma = I_S\sigma \subseteq \rho\sigma$, so $\rho \vee \sigma = \rho\sigma$. By (2), $\rho \vee \sigma \in \mathsf{Eq}_S$; thus $\rho\sigma \in \mathsf{Eq}_S$ and by Prop. 2.66(3) ρ and σ commute. □

2.8.5.2 *Tolerance Relations*

Among the three generalizations of equivalences (recall the other two are *preorders* and *partial equivalence relations*) *tolerances* are presumably the most popular because of their appealing intuitive interpretation: Abandoning transitivity, they offer a less radical notion of similarity than equivalence does and provide a general tool for studying indiscerniblity phenomena in the investigation of resemblance between perceptual objects, occurring with close-by sensations. With this purpose in mind they have been introduced by Poincaré [32] and Carnap [8], who called them *similarity.* The name "*tolerance*" itself is due to [51]; for more details see [30].

A paradigmatic example was given in [32]: Consider a *metric* d[1] on a set X, and define $x\tau y := d(x, y) < \Delta$ for some small positive value Δ. Since $2\Delta > \Delta$, we do not necessarily get $x\tau z$ from $x\tau y$ and $y\tau z$, i.e. τ is, in general, not transitive. So $x\tau_\epsilon y := |x - y| < \epsilon$ for some fixed real $\epsilon > 0$ can be regarded as the "classical" tolerance.

We continue with some basic concepts and properties around tolerances, using as main sources [1,30] as well as some early papers (given in their logical order) of Chajda and his colleagues [10–12]. Before we are going to have a closer look at what makes

[1] A **metric** on a set X is a function $d : X \times X \longrightarrow \mathbb{R}_0$, called *distance*, such that

(1) $d(x, y) \ge 0$ (non-negativity),

(2) $d(x, y) = 0$ iff $x = y$ (identity of indiscernibles),

(3) $d(x, y) = d(y, x)$ (symmetry), and

(4) $d(x, z) \le d(x, y) + d(y, z)$ (triangle inequality).

equivalences so special compared with other kinds of relations, we list some elementary properties of tolerance relations in

Proposition 2.68. *Let ρ and σ be two tolerances on a set S. Then*

(1) $\rho^{\smile} \in \mathsf{Tol}_S$.
(2) $\rho \cap \sigma \in \mathsf{Tol}_S$.
(3) $\rho\sigma \in \mathsf{Tol}_S$ *iff* $\rho\sigma = \sigma\rho$.
(4) $\rho^2 \in \mathsf{Tol}_S$.

Proof. (1): By Prop. 2.17.
(2): By Prop. 2.23.
(3): By Prop. 2.66(3), using the parts for reflexivity and symmetry.
(4): By (3). □

The concept of an equivalence class can be generalized in a natural way for arbitrary relations as done in Def. 2.69 below. Recall that an equivalence relation ρ on a set S defines a partition of S via its aftersets $x\rho, x \in S$. For arbitrary relations the notion of an equivalence class splits into different concepts:

Definition 2.69. ([1, 11]) *Let ρ be an arbitrary, and τ be a tolerance relation on a set S.*

(a) *A* **pre-block** *of ρ is a non-void subset $B \subseteq S$, such that $B \times B \subseteq \rho$. The set of pre-blocks of ρ is denoted by* PBl_ρ.
(b) *A* **block** *of ρ is a maximal pre-block: if $B \subseteq C$ and $C \times C \subseteq \rho$, then $B = C$. The set of blocks of ρ is denoted by Bl_ρ. We say that ρ has blocks provided $\mathsf{Bl}_\rho \neq \emptyset$.*
(c) *The* **tolerance class** *of an element $x \in S$ is the afterset $x\tau$, also denoted as x/τ, and sometimes also called the τ-***neighborhood of** *x. The set of all tolerance classes is called the* **quotient set** *of S modulo τ, denoted by S/τ.*
(d) *A set S together with a tolerance relation τ is called a* **tolerance space**, *denoted (S,τ); and Tol_S denotes the set of all tolerances on S.*

A block of ρ thus is a maximal non-void subset of $B \subseteq S$ such that the restriction of ρ to B is the universal relation U_B. Looking at a tolerance as an undirected graph, its blocks correspond to its maximal cliques and the tolerance class of a node corresponds to the set of its neighbors, together with the node itself. As for equivalences, due to reflexivity, the blocks and the classes of a tolerance on a set S form coverings of S, but they do not have to be mutually disjoint, and so these coverings need not be partitions. If, however, one of these coverings is a partition, then the tolerance is an equivalence and the coverings by blocks and by classes coincide. More on tolerances can be found in [45, Section 4.3.2].

2.8.5.3 *Partial Equivalence Relations*

Note that the difference between a PER and an equivalence relation is that the former need not be reflexive. So every equivalence relation is a PER, but not conversely: $\{\langle 0,0\rangle\}$

is a PER on {0,1} only, since it is not reflexive. But by Prop. 2.15(5) we know that a PER ρ is reflexive on its field $\mathsf{fld}(\rho)$, so ρ is an equivalence on $\mathsf{fld}(\rho)$. We begin by listing some elementary properties of PERs in

Proposition 2.70. *Let ρ and σ be partial equivalence relations on a set S. Then the following holds:*

(1) *$\rho \cap \sigma$ is a partial equivalence relation.*
(2) *$\rho^2 = \rho$.*
(3) *ρ is a partial equivalence iff it is symmetric and idempotent.*
(4) *$\rho\sigma$ is a partial equivalence iff $\rho\sigma = \sigma\rho$.*
(5) *ρ is a partial equivalence iff $\rho\rho^{\smile} = \rho$.*
(6) *$\rho\rho^{\smile}\rho = \rho$.*
(7) *$\rho \subseteq \sigma$ implies $\rho\sigma = \sigma$.*

Proof. (1): By Prop. 2.23(4) and (6).

(2): By Prop. 2.25(1) we have $\rho \subseteq \rho\rho^{\smile}\rho$. Using symmetry and transitivity of ρ we get $\rho\rho^{\smile}\rho = \rho^2\rho \subseteq \rho\rho$, so $\rho \subseteq \rho\rho$, which together with transitivity of ρ gives the desired result.

(3): Obvious from (2) and the definition of transitivity.

(4): If $\rho\sigma$ is a PER, then $\rho\sigma = (\rho\sigma)^{\smile} = \sigma^{\smile}\rho^{\smile} = \sigma\rho$, by the symmetry of each of $\rho\sigma, \rho$ and σ. Conversely, assume $\rho\sigma = \sigma\rho$. By the symmetry of ρ and σ we get $\rho\sigma = \sigma\rho = \sigma^{\smile}\rho^{\smile} = (\rho\sigma)^{\smile}$, whence $\rho\sigma$ is symmetric; and, finally, $\rho\sigma$ is transitive: $\rho\sigma\rho\sigma = \rho\sigma\sigma\rho = \rho\sigma\rho = \rho\rho\sigma = \rho\sigma$ by (2).

(5): Suppose ρ to be a partial equivalence . Then $\rho\rho = \rho$ by (2) and $\rho^{\smile} = \rho$, whence $\rho\rho^{\smile} = \rho$. To show the other direction, let $\rho\rho^{\smile} = \rho$. Then $\rho^{\smile} = (\rho\rho^{\smile})^{\smile} = \rho^{\smile\smile}\rho^{\smile} = \rho\rho^{\smile} = \rho$; whence ρ is symmetric. So $\rho\rho = \rho\rho^{\smile} = \rho$; thus ρ is transitive.

(6): Applying (5) and (2) we get $\rho\rho^{\smile}\rho = \rho^2 = \rho$.

(7): Obviously, $\sigma = I\sigma \subseteq \rho\sigma$; by assumption we also have $\rho\sigma \subseteq \sigma\sigma = \sigma$ by monotonicity of composition and (2). □

Concerning the algebraic structure of the set of all partial equivalences on a set, denoted by PEq_S, we have the following results which, as well as their proofs, are very similar to the ones of Prop. 2.67:

Proposition 2.71. *Let $\rho, \sigma \in \mathsf{PEq}_S$. Then the following holds:*

(1) *$(\mathsf{PEq}_S, \subseteq)$ is a complete lattice.*
(2) *$\rho \vee \sigma = \mathbf{t}(\rho \cup \sigma)$.*
(3) *$\rho \vee \sigma = \rho \cup \rho\sigma \cup \rho\sigma\rho \cup \ldots \cup \sigma \cup \sigma\rho \cup \sigma\rho\sigma \ldots$.*
(4) *If $\rho\sigma = \sigma\rho$, then $\rho \vee \sigma = \rho \cup \sigma \cup \rho\sigma$.*

Proof. (4): $\rho \vee \sigma = \rho \cup \rho\sigma \cup \rho\sigma\rho \cup \ldots \cup \sigma \cup \sigma\rho \cup \sigma\rho\sigma \ldots$ reduces to $\rho \cup \sigma \cup \rho\sigma$ because the products with more than two factors are smaller than $\rho\sigma = \sigma\rho$ by transitivity. □

Partial equivalences also appear under the names *symmetric idempotents*, especially in connection with *splittings* (see [52, Section 2.3.2] or [48, 49]), *weak congruences* (see [26, 46]), and in the context of *partial partitions* (see [35]).

2.9 Some Algebraic Properties of Relations

From the definitions and results given earlier in this book we do already know the following:

- $(\mathcal{B}_X, \cdot, \breve{}, I_X)$ is an involution monoid,
- $(\mathcal{B}_X, \cdot)$ is a semigroup with zero element $\emptyset$,
- $(\mathcal{B}_X, \cdot; \subseteq)$ is a partially ordered semigroup.

We are, of course, interested in how these and other concepts of semigroup theory manifest themselves in semigroups of binary relations. Before we take a closer look at algebraic properties of $\mathcal{B}_X$, we turn to the perhaps not so popular fact that $\mathcal{B}_X$ forms a Baer semigroup. The **left** and **right annihilators** of a subset $A \subseteq S$ of a semigroup S with zero are defined as

$$L(A) = \{x \in S : \forall a \in A[xa = 0]\} \text{ and } R(A) = \{x \in S : \forall a \in A[ax = 0]\}.$$

Definition 2.72. *A semigroup S with zero element 0 is called a* **Baer semigroup** *if for each $x \in S$ there exist idempotents $e, f \in S$ such that $R(x) = \{y \in S : xy = 0\} = eS$ and $L(x) = \{y \in S : yx = 0\} = Sf$.*

A semigroup with zero is thus a Baer semigroup if and only if the right annihilator of every $x \in S$ is an idempotent-generated principal right ideal, and analogous for the left annihilator. In particular we have $S = R(0) = L(0)$, whence there exist idempotents $e, f \in S$ with $S = eS = Sf$. Then e is a left, and f is a right identity; therefore $e = f$ by Prop. 1.28 and so S has an identity.

Proposition 2.73. ([6, Ex. 2.26]) *$\mathcal{B}_X$ is a Baer semigroup with zero element $\emptyset$.*

Proof. Let $\sigma, \tau \in \mathcal{B}_X$ with $\sigma\tau = \emptyset$. $z \in \mathsf{rng}(\sigma)$ implies $z \notin \mathsf{dom}(\tau)$, i.e. $\mathsf{rng}(\sigma) \subseteq \overline{\mathsf{dom}(\tau)}$, or equivalently $\mathsf{rng}(\sigma) \cap \mathsf{dom}(\tau) = \emptyset$. Conversely, if $\mathsf{rng}(\sigma) \cap \mathsf{dom}(\tau) = \emptyset$ then clearly there are no $x, y \in X$ with $x\sigma\tau y$, whence

$$\sigma\tau = \emptyset \quad \text{iff} \quad \mathsf{rng}(\sigma) \cap \mathsf{dom}(\tau) = \emptyset \quad \text{iff} \quad \mathsf{rng}(\sigma) \subseteq \overline{\mathsf{dom}(\tau)}.$$

Let $S \subseteq X$ and $\tau \in \mathcal{B}_X$ such that $\tau = I_S\tau$; then, using Th. 2.9(9), $\mathsf{rng}(\tau) = \mathsf{rng}(I_S\tau) = \tau[\mathsf{rng}(I_S)] = \tau[S] = I_S\tau[S] \subseteq I_S[X] = S$, since $\tau[S] \subseteq X$. Otherwise $\mathsf{rng}(\tau) \subseteq S$ implies $x\tau y \to yI_Sy$, whence $\tau = \tau I_S$. Thus $\tau \in \mathcal{B}_X I_S$ iff $\mathsf{rng}(\tau) \subseteq S$.

Therefore, for every $\sigma \in \mathcal{B}_X$, for the left annihilator we have $L(\sigma) = \mathcal{B}_X I_S$, with $S = \overline{\mathsf{dom}(\sigma)}$. Since $\tau\sigma = \emptyset \leftrightarrow \breve{\sigma}\breve{\tau} = \emptyset$ a dual argument gives $R(\sigma) = I_T\mathcal{B}_X$, with $T = \overline{\mathsf{dom}(\breve{\sigma})} = \overline{\mathsf{rng}(\sigma)}$. Note that for each $S \subseteq X$, $I_S \in \mathcal{B}_X$ is coreflexive and thus idempotent by Prop. 2.24(5). Hence $\mathcal{B}_X$ is a Baer semigroup. $\square$

Definition 2.74. *A binary relation $\alpha \in \mathcal{B}_X$ is said to be*

(a) **row reduced** *if its nonempty rows form an independent set.*
(b) **column reduced** *if its nonempty columns form an independent set.*
(c) *a* **reduced relation** *if it is both row and column reduced.*

So ρ is row reduced if none of its rows $x\rho$ can be written as a union of other rows in ρ, i.e. for all $x \in X$ and $A \subseteq X$, $\emptyset \neq x\alpha = A\alpha = \bigcup_{a\in A} a\rho$ implies $x \in A$. As an example take the relation $\rho := \begin{pmatrix} 1 & 0 & 1 & 0 \\ 0 & 1 & 0 & 1 \\ 1 & 1 & 0 & 0 \\ 1 & 1 & 1 & 0 \end{pmatrix}$ on $X = \{x_1, x_2, x_3, x_4\}$. It is column reduced but not row reduced ($x_4\rho = x_1\rho \cup x_3\rho$), and has column rank 4 and row rank 3.

Remember that a partial order ρ on a set X is characterized by the following properties: $I_X \subseteq \rho, \rho^2 \subseteq \rho$, and $\rho \cap \rho^{\smile} \subseteq I_X$. The set of partial orders on X is non-empty, since $\emptyset$ and I_X are examples which have two properties in common: they are reduced and idempotent. That this is not accidental is shown in

Proposition 2.75. ([31, THEOREM 2.4]) *The class of reduced idempotents in $\mathcal{B}_X$ coincides with the class of partial order relations on subsets of X.*

Proof. Let α be a partial order on $Y \subseteq X$; then α is a preorder that is additionally antisymmetric, and therefore idempotent by Prop. 2.24(3). If $x\alpha = H\alpha$ for some $x \in Y$ and $H \subseteq Y$, then, by reflexivity, $x \in x\alpha = H\alpha$, whence $x \in h\alpha$ for some $h \in H$. But, again by reflexivity, $h \in H\alpha = x\alpha$. Thus $x\alpha h$ and $h\alpha x$, so $x\, \alpha \cap \alpha^{\smile}\, h$ and thus $x = h$ by the antisymmetry of α. Therefore, $x \in H$, whence α is row reduced. Similarly we get that α is also column reduced.

Conversely, let α be a reduced idempotent. Then $x\alpha^2 = (x\alpha)\alpha = x\alpha$, and since α is reduced we get $x \in x\alpha$ for all $x \in \alpha X = \mathsf{dom}(\alpha)$. Hence $x \in \bigcup_{x\in X} x\alpha = X\alpha = \mathsf{rng}(\alpha)$. So $\alpha X \subseteq X\alpha$, and similarly $X\alpha \subseteq \alpha X$ follows from $\alpha(\alpha x) = \alpha x$. We let $Y = X\alpha = \alpha X$. From $x \in x\alpha$ for all $x \in Y$ it follows that $I_Y \subseteq \alpha$. And since idempotent relations clearly are transitive, it remains to show that α is antisymmetric. Assume $x\alpha y$ with $x \neq y$; then $x\alpha \neq y\alpha$ since α is assumed reduced. Thus there is a $z \in X$ such that either $x\alpha z \wedge \neg y\alpha z$ or $\neg x\alpha z \wedge y\alpha z$. If we also had $y\alpha x$, the first case would give $y\alpha x\alpha z$, i.e. $y\alpha z$, a contradiction to $\neg y\alpha z$. So we conclude that the second alternative must hold; but for it we get $x\alpha y\alpha z$, i.e. $x\alpha z$, a contradiction to $\neg x\alpha z$. Altogether we cannot keep the assumption $y\alpha x$, whence α must be antisymmetric. □

The following proposition refers to an early classical result on binary Boolean relations, which we present with its original proof. Later on we will learn about a corresponding generalization for incline matrices.

Let $B \in \mathbb{B}_n$. We sometimes have need to write B^m as $[b_{ij}^{(m)}]$. Note that for any $i, j \in \underline{n}$ one then as

$$b_{ij}^{(m)} = \sum_{i_1, i_2, \ldots, i_{m-1} \in \underline{n}} b_{ii_1} b_{i_1 i_2} \ldots b_{i_{m-2} i_{m-1}} b_{i_{m-1} j}. \tag{2.11}$$

Proposition 2.76. ([20, Theorem 3.2.1]) *Let $B \in \mathbb{B}_n$ be reflexive. Then*

$$I_n \subseteq B \subseteq B^2 \subseteq \ldots \subseteq B^{n-1} = B^n = B^{n+1} = \ldots.$$

Proof. Due to isotonicity of relation product we get $I_n \subseteq B \subseteq B^2 \subseteq \ldots \subseteq B^{n-1} \subseteq B^n \subseteq B^{n+1} \subseteq \ldots$ .Thus it suffices to prove $B^n \subseteq B^{n-1}$.

Note that for all $h \geq 1, 1 \leq j \leq n$ we have $b_{jj}^{(h)} = 1$ and that

$$b_{ij}^{(h+1)} = \sum_k b_{ik}^{(h)} b_{kj} = b_{ij}^{(h)} + \sum_{k \neq j} b_{jk}^{(h)} b_{kj}. \tag{2.12}$$

Consider a term

$$b_{ii_1} b_{i_1 i_2} \ldots b_{i_{n-2} i_{n-1}} b_{i_{n-1} j} \tag{2.13}$$

of the sum (2.11) for $k = n$. Since there are only n distinct values, at least two of $i = i_0, i_1, \ldots, i_{n-1}, i_n = j$ must be equal. Suppose $j = i_s$ for some $s \leq n-1$; then (2.13) takes the form

$$b_{ii_1} b_{i_1 i_2} \ldots b_{i_{s-1} j} b_{j i_{s+1}} \ldots b_{i_{n-1} j}. \tag{2.14}$$

We clearly have (2.14) $\leq b_{ii_1} b_{i_1 i_2} \ldots b_{i_{s-1} j}$, a term of $b_{ij}^{(s)}$, and hence also (2.14) $\leq b_{ij}^{(n-1)}$ by (2.12). A similar argument works in the case $i = i_s$.

If neither i nor j is equal to any i_k, then there are $1 \leq r < s \leq n-1$ such that $i_r = i_s$, and (2.13) becomes

$$b_{ii_1} \ldots b_{i_{r-1} i_s} b_{i_s i_{r+1}} \ldots b_{i_{s-1} i_s} b_{i_s i_{s+1}} \ldots b_{i_{n-1} j} \leq b_{ii_1} \ldots b_{i_{r-1} i_s} b_{i_s i_{s+1}} \ldots b_{i_{n-1} j}$$

with the last term again being smaller than $b_{ij}^{(n-1)}$. Since the conclusions hold for all possible value sequences $i_1, \ldots, i_{n-1}$ and all $1 \leq i, j \leq n$, $B^n \subseteq B^{n-1}$ has been proved. □

Corollary 2.77. *Let $B \in \mathbb{B}_n$ a reflexive square Boolean matrix.*

(1) *For the index k of B we have $k \leq n-1$.*
(2) *k is the index of B iff $B \subset B^2 \subset \ldots \subset A^{k-1} \subset B^k = B^{k+1} = \ldots = B^n = \ldots$.*
(3) *B^m is idempotent for all $m \geq k$.*

2.9.1 *Subsemigroups of $\mathcal{B}_X$*

Remember that due to Def. 2.36 an element $\rho \in \mathcal{B}_X$ is called a **partial mapping** on X if for all $x \in \mathsf{dom}(\rho)$ we have $|x\rho| = 1$, that is, if

$$\forall x, y_1, y_2 \in X[(\langle x, y_1 \rangle \in \rho \wedge \langle x, y_2 \rangle \in \rho) \rightarrow y_1 = y_2]. \tag{2.15}$$

Note that this condition vacuously holds for the empty relation $\emptyset$, which is therefore included into the set of the partial maps on X. The set of all partial maps of X is denoted by $\mathcal{P}_X$. For $\alpha \in \mathcal{P}_X$ it is a convenient and common habit to let $x\alpha$ denote the unique y such that $x\alpha y$. In semigroup theory, this concept runs under the name of a **partial transformation** on a set X, defined as a function $\alpha : A \longrightarrow B$, with $A, B \subseteq X$.

A partial transformation is called **transformation**, if $\mathsf{dom}(\alpha) = X$, and a **permutation** if it is a bijective transformation. The set of all transformations of X is called the **full transformation semigroup** on X, denoted by $\mathcal{T}_X$; and the permutations on a set X form the **symmetric group** $\mathcal{S}_X$. Much more on basic results on transformations than is included below can be found in [16]. The set $\mathcal{I}_X := \{\alpha \in \mathcal{P}_X : \alpha \text{ is injective}\}$ is called the **symmetric inverse semigroup.**

Since the nature of elements of X is not important for the current considerations, instead of X the set $\underline{n}$ can be used, where $n = |X|$. In a similar way as for Boolean matrices $\mathbb{B}_n$ we shall also write $\mathcal{P}_n$, $\mathcal{T}_n$, and $\mathcal{S}_n$. A common notation for partial transformations is in form of a two-line array: As an example, let $\epsilon_n : \underline{n} \longrightarrow \underline{n}$ be defined by $i\epsilon_n = i$, or $\epsilon_n(i) = i$, for all $i \in \underline{n}$. This is often written as an

$$\epsilon_n = \begin{pmatrix} 1\,2 \cdots n \\ 1\,2 \cdots n \end{pmatrix}, \text{ and generally: } \alpha = \begin{pmatrix} 1 & 2 & \cdots & n \\ 1\alpha & 2\alpha & & n\alpha \end{pmatrix}.$$

For denoting the composition of transformations one can apply either the relational or the functional notation, and we will use them at liberty. For composition we get $\beta(\alpha(x)) = x\alpha\beta$, whence right properties of functions will become left properties of relations, and vice versa.

Obviously, ϵ_n is the identity element in both $\mathcal{P}_n$ and $\mathcal{T}_n$ and corresponds to the relation $I_{\underline{n}}$. The partial transformation $\mathbf{0}_n : \emptyset \longrightarrow \underline{n}$ is the zero element of $\mathcal{P}_n$, since $\alpha\mathbf{0}_n = \mathbf{0}_n\alpha = \mathbf{0}_n$.

The importance of $\mathcal{T}_X$ in semigroup theory stems from the following embedding property (see e.g. [21, Theorem 1.1.2] or [23, Chapter I.3.9]):

Proposition 2.78. *If S is a semigroup and $X = S^1$ then there is an isomophism $\phi : S \longrightarrow \mathcal{T}_X$.*

Proof. For each $a \in S$, we define a map $\rho_a : S^1 \longrightarrow S^1$ by $x\rho_a = xa$, $x \in S^1$. Thus $\rho_a \in \mathcal{T}_X$, whence there is a map $\alpha : S \longrightarrow \mathcal{T}_X$ given by $a\alpha = \rho_a$, $a \in S$. This map is bijective since

$$\forall a, b \in S[a\alpha = b\alpha \rightarrow \rho_a = \rho_b \rightarrow \forall x \in S^1[xa = xb]] \rightarrow 1a = 1b \rightarrow a = b.$$

α is a homomorphism, since

$$\forall x \in S^1[x(\rho_a\rho_b) = (x\rho_a)\rho_b = (xa)b = x(ab) = x\rho_{ab},$$

hence $(a\alpha)(b\alpha) = (ab)\alpha$. □

Theorem 2.79. ([16, Theorem 2.6.3]) *The semigroups $\mathcal{P}_n$, $\mathcal{T}_n$, and $\mathcal{I}_n$ are regular.*

(1) *$\mathcal{P}_n$ is a regular submonoid of $\mathcal{B}_X$, $|X| = n$.*
(2) *$\mathcal{T}_n$ is a regular submonoid of $\mathcal{P}_n$.*
(3) *$\mathcal{I}_n$ is a regular submonoid of $\mathcal{P}_n$.*
(4) *$\mathcal{S}_n$ is a subsemigroup of $\mathcal{T}_n$.*

Proof. (1): Let $\alpha, \beta \in \mathcal{P}_n$, and assume $x\alpha\beta y_1$ and $x\alpha\beta y_2$. Then

$$\exists z_1, z_2 \in X[x\alpha z_1 \wedge z_1\beta y_1 \wedge x\alpha z_2 \wedge z_2\beta y_2].$$

Condition (2.15) applied to α implies that $z_1 = z_2$, and then applied to β that $y_1 = y_2$. Thus, $\alpha\beta \in \mathcal{P}_n$. (Compare this with the relational proof in Cor. 2.38(1).) Furthermore, $I_n \in \mathcal{P}_n$, and composition is associative.

To see that $\mathcal{P}_n$ is regular, let $\alpha \in \mathcal{P}_n$. We then construct a $\beta \in \mathcal{P}_n$ as follows: for each $x \in \mathsf{rng}(\alpha)$ we take some $y \in X$ with $\alpha(y) = x$ and set $\beta(x) = y$, and for $x \notin \mathsf{rng}(\alpha)$ we define

$\beta(x) = x_1$. Then, with $x = \alpha(z)$, i.e. $x \in \mathsf{rng}(\alpha)$, we get $\alpha(\beta(\alpha(z))) = \alpha(\beta(x)) = \alpha(z)$. Thus $\alpha\beta\alpha = \alpha$.

(2): By (1).

(3): Obviously, $\emptyset, I_n \in \mathcal{I}_n$. For $\alpha \in \mathcal{I}_n$ we define $\beta \in \mathcal{I}_n$ as follows: $\mathsf{dom}(\beta) := \mathsf{rng}(\alpha)$, and for each $x \in \mathsf{rng}(\alpha)$ we define β as in (1). Since α is a partial bijection, viz. from $\mathsf{dom}(\alpha)$ to $\mathsf{rng}(\alpha)$, β is the inverse bijection from $\mathsf{rng}(\alpha)$ to $\mathsf{dom}(\alpha)$, so $\beta \in \mathcal{I}_n$. As above in (1), we have $\alpha\beta\alpha = \alpha$.

(4): It is well known that the permutations of a set form a group. □

2.10 Invertibility

The notion of an inverse shows up in the context of square boolean matrices as early as 1934 in [47]. Following Luce [25], we proceed with a definition, followed by some early classical results on orthogonal relations and inverses.

Definition 2.80. *Let ρ be a given finite square Boolean relation of order n, i.e. $\rho \in \mathbb{B}_n$. Then ρ is called*

(a) **orthogonal** *(or* **permutation relation***) if $\rho^{-1} = \rho^\smile$, i.e. $I = \rho\rho^\smile = \rho^\smile\rho$.*
(b) **involutory** *if $\rho^{-1} = \rho$.*
(c) **row (column) consistent** *if $\rho U = U$ $(U\rho = U)$.*

Corollary 2.81. ([14, p. 250]) *For a relation ρ the following holds:*

(1) $I \cap U\rho = I \cap \rho^\smile\rho$,
(2) $I \cap \rho U = I \cap \rho\rho^\smile$.

Proof. (1) By applying the Dedekind rule 2.20(3) we get $U\rho \cap I \subseteq (U \cap I\rho^\smile)(\rho \cap U^\smile I) = \rho^\smile(\rho \cap U) = \rho^\smile\rho$; thus $U\rho \cap I \subseteq \rho^\smile\rho \cap I$. The converse inclusion follows by $\rho^\smile\rho \subseteq U\rho$, whence $\rho^\smile\rho \cap I \subseteq U\rho \cap I$. (Compare this to Prop. 2.11(1) and (2).) □

Proposition 2.82. ([25, 38]) *Let ρ be a given finite square Boolean relation of order n, i.e. $\rho \in \mathbb{B}_n$. Then the following holds:*

(1) ρ is row consistent iff $I \subseteq \rho\rho^\smile$, and column consistent iff $I \subseteq \rho^\smile\rho$.
(2) ρ is orthogonal iff it is both row and column consistent, and

$$\forall i,j,k[i \neq j \rightarrow \rho_{ik} \wedge \rho_{jk} = 0 = \rho_{ki} \wedge \rho_{kj}].$$

(3) ρ has an inverse iff it is orthogonal.

Proof. (1): We show the case of row consistency: By Prop. 2.26(3) we have $I \subseteq \rho \rightarrow \rho U = U = U\rho$, and by Cor. 2.81(1) $I \cap \rho U = I \cap \rho\rho^\smile$. So $I \subseteq \rho\rho^\smile$ implies $I = I \cap \rho\rho^\smile = I \cap \rho U$, whence $I \subseteq \rho U$. Thus $U = IU \subseteq \rho UU = \rho U \subseteq U$, and therefore $U = \rho U$.

Conversely, let $U = \rho U$. Then, using the DEDEKIND rule Th. 2.20(3), we get $I = \rho U \cap I \subseteq (\rho \cap IU^\smile)(U \cap \rho^\smile I) = (\rho \cap U)(U \cap \rho^\smile) = \rho\rho^\smile$, i.e. $I \subseteq \rho\rho^\smile$.

(2): If ρ is orthogonal, then by (1) it is both row and column consistent. From $I = \rho\rho^{\smallsmile}$ we get for every $i, j, i \neq j$: $0 = \bigvee_k(\rho_{ik} \wedge \rho^{\smallsmile}_{kj})$, whence for every k: $\rho_{ik} \wedge \rho_{jk} = 0$. Similarly, for every $i, j, k, i \neq j$ we get $\rho_{ki} \wedge \rho_{kj} = 0$ from $I = \rho^{\smallsmile}\rho$.

For the reverse direction assume that $\rho_{ik} \wedge \rho_{jk} = 0$ for every $i, j, k, i \neq j$; then $\bigvee_k(\rho_{ik} \wedge \rho^{\smallsmile}_{kj}) = 0$. Moreover, since ρ is assumed row consistent, we have $I \subseteq \rho\rho^{\smallsmile}$ by (1), i.e. $\bigvee_k(\rho_{ik} \wedge \rho^{\smallsmile}_{ki}) = 1$. Thus $\rho\rho^{\smallsmile} \subseteq I$, hence, $I = \rho\rho^{\smallsmile}$; the second case, $I = \rho^{\smallsmile}\rho$, can be shown in an analogous way.

(3): (As noted in [25], this result appears already in [47].) We only have to show the direction that is not obvious by definition. So assume that ρ has an inverse, (say $\sigma := \rho^{-1}$); then $I = \rho\sigma = \sigma\rho$ and by Prop. 2.26(3) we get $\rho\sigma U = U\rho\sigma = U$ and $\sigma\rho U = U\sigma\rho = U$. Now $\rho U \subseteq U$ and $\sigma U \subseteq U$ always hold. Thus $U = (\rho\sigma)U = \rho(\sigma U) \subseteq \rho U \subseteq U$, i.e. ρ is row consistent. The other cases, $U = U\rho, \sigma U = U$, and $U = U\sigma$ can be shown analogously, whence by (1) we get $I \subseteq \rho\rho^{\smallsmile}$, $I \subseteq \rho^{\smallsmile}\rho$, $I \subseteq \sigma\sigma^{\smallsmile}$, and $I \subseteq \sigma^{\smallsmile}\sigma$. Thus, both ρ and $\sigma := \rho^{-1}$ are row and column consistent. Using (2) it remains to show that $\forall i, j, k[i \neq j \rightarrow \rho_{ik} \wedge \rho_{jk} = 0 = \rho_{ki} \wedge \rho_{kj}]$. Let $I = \rho\rho^{-1}$; then we have, for every $i, j, i \neq j$, $\bigvee_k(\rho_{ik} \wedge \rho^{-1}_{kj}) = 0$, whence, for every k, $\rho_{ik} \wedge \rho^{-1}_{kj} = 0$. Thus, for k fixed, $\bigvee_{i\neq j}(\rho_{ik} \wedge \rho^{-1}_{kj}) = 0$, and therefore

$$\rho_{ik} \wedge \rho^{-1}_{ki} = \bigvee_{i\neq j}(\rho_{ik} \wedge \rho^{-1}_{kj}) \vee (\rho_{ik} \wedge \rho^{-1}_{ki}) = \bigvee_j(\rho_{ik} \wedge \rho^{-1}_{kj}) = \rho_{ik} \wedge \bigvee_j \rho^{-1}_{kj} = \rho_{ik}.$$

(Note that $\bigvee_j \rho^{-1}_{kj} = 1$ because ρ^{-1} is row consistent.) So, for every $i, j, k, i \neq j$ we have

$$\rho_{ik} \wedge \rho_{jk} = (\rho_{ik} \wedge \rho^{-1}_{ki}) \wedge (\rho_{jk} \wedge \rho^{-1}_{kj}) = (\rho_{ik} \wedge \rho^{-1}_{kj}) \wedge (\rho_{jk} \wedge \rho^{-1}_{ki}) = 0,$$

whence ρ is orthogonal. □

In [38] it has been shown that if a square Boolean matrix α possesses a one-sided inverse, that inverse is also a two-sided inverse. Furthermore such an inverse, if it exists, is unique and is $\alpha^{\smallsmile}$. Permutation matrices in $\mathbb{B}_n$ are the only matrices which have inverses. We need the following

Lemma 2.83. *Let α be a square Boolean matrix such that*

$$\bigvee_j a_{ij} = 1, \quad (i = 1, \ldots, n), \tag{2.16}$$

and

$$a_{ij} \wedge a_{kj} = 0, \quad (i \neq k). \tag{2.17}$$

Then equations (2.16) *and* (2.17) *are equivalent to* (2.18) *and* (2.19).

Proof. From (2.16) it follows that $1 = \bigwedge_i(\bigvee_j a_{ij}) = \bigvee_{f\in\Phi}(\bigwedge a_{if(i)})$, the last equation being a well known result from lattice theory, where Φ is the set of functions from $\{1, 2, \ldots, n\}$ into $\{1, 2, \ldots, n\}$; see [44, p. 124]. A typical intersection $a_{1j_1} \wedge a_{2j_2} \wedge \ldots \wedge a_{nj_n}$ evaluates to 0 according to (2.17), unless all the j_k are distinct. This means that only permutations of $\{1, 2, \ldots, n\}$ contribute a 1, whence for the permanent we have $\mathsf{per}(\alpha) = 1$ (see Def. 3.30(a)). Expanding $\mathsf{per}(\alpha)$ according to Prop. 3.31(4) and using monotonicity results in $\bigvee_i a_{ij} \geq \bigvee_i(a_{ij} \wedge \mathsf{per}(\alpha(i|j))) = 1$, and so

$$\bigvee_i a_{ij} = 1, \quad (j = 1, \ldots, n). \tag{2.18}$$

So what we have shown is that there are 1's in each row and each column, by (2.16) and (2.18), and that there is exactly one 1 in each column by (2.17) together with (2.18). Now assume $a_{ij} = a_{ik} = 1, j \neq k$. Then there are $n-1$ rows which must be assigned a 1 such that each column contains exactly one. But, since columns j and k are no longer available, there remain only $n-2$ columns for $n-1$ required occurrences of 1's. Thus there is one line with no 1 in it, a contradiction to (2.16), and therefore

$$a_{ki} \wedge a_{kj} = 0, \quad (i \neq j). \tag{2.19}$$

The other direction follows using the above argumentation for $\alpha^{\smile}$ instead of α. □

Theorem 2.84. ([38, p. 51f]) *If a square relation (square Boolean matrix) β possesses a one-sided inverse, this inverse is also a two-sided inverse. Furthermore, such an inverse, if it exists, is unique and equals $\beta^{\smile}$.*

Proof. Consider the equation $\beta\gamma^{\smile} = I$, with β, γ square Boolean matrices of order n. This equation implies the following relationships:

$$\bigvee_j (b_{ij} \wedge c_{ij}) = 1, \quad (i = 1, \ldots, n), \tag{2.20}$$

$$\bigvee_j (b_{ij} \wedge c_{kj}) = 0, \quad (i \neq k).$$

From the second equation we get

$$b_{ij} \wedge c_{kj} = 0, \quad (i \neq k, j - 1, \ldots, n). \tag{2.21}$$

Setting $a_{ij} := b_{ij} \wedge c_{ij}$ obviously gives

$$a_{ij} \wedge a_{kj} = b_{ij} \wedge c_{ij} \wedge b_{kj} \wedge c_{kj} = 0, \quad (i \neq k). \tag{2.22}$$

From (2.20) and (2.22) we see that $\alpha = [a_{ij}]$ satisfies (2.16) and (2.17), and therefore also (2.18) and (2.19) by Lemma 2.83. Since $a_{ij} \leq b_{ij}$ and $a_{ij} \leq c_{ij}$ we have

$$\bigvee_i b_{ij} = 1, \quad \bigvee_i c_{ij} = 1, \quad (i = 1, \ldots, n),$$

by (2.18). And applying (2.21) we get

$$\begin{aligned} b_{kj} = b_{kj} \wedge \left(\bigvee_i c_{ij}\right) &= \bigvee_i (b_{kj} \wedge c_{ij}) = b_{kj} \wedge c_{kj} \\ &= \bigvee_i (b_{ij} \wedge c_{kj}) = \left(\bigvee_i b_{ij}\right) \wedge c_{kj} = c_{kj}. \end{aligned}$$

So, by the definition of a_{ij} above, we have shown $\alpha = \beta = \gamma$, whence the equation $\beta\gamma^{\smile} = I$ can only be satisfied if $\gamma = \beta$. Thus, if β has a right inverse, this must be $\beta^{\smile}$, and β is orthogonal. Then, (2.20) and (2.21) reduce to (2.16) and (2.17), which, by Lemma 2.83, imply (2.18) and (2.19). These in turn imply $\beta^{\smile}\beta = I$. Moreover, $\beta^{\smile}$ is the only left inverse of β: Let $\chi\beta = I$, then $\beta = \chi^{\smile}$ and so $\chi = \beta^{\smile}$. □

2.11 Generalized Inverses and Regularity

In analogy to the standard classification of matrix theory using the Moore-Penrose equations (see Def. 2.85(f) below), the following kinds of inverses are of interest in the context of binary relations:

Definition 2.85. *Let α be a given relation; a relation γ is called*

(a) a **left** *(***right***)* **inverse** *of α, and α is called* left (right) invertible*, if*
$\gamma\alpha = I$ $(\alpha\gamma = I)$.

(b) an **inverse** *of α, denoted as α^{-1}, and α is called* invertible*, if*
$\gamma\alpha = I = \alpha\gamma$.

(c) a **subinverse** *of α if*
$\alpha\gamma\alpha \subseteq \alpha$.

(d) a **generalized inverse** *(or* **g-inverse***) of α if*
$\alpha\gamma\alpha = \alpha$.

(e) a **pseudoinverse** *of α (sometimes also* **Vagner inverse***) if*
$\alpha\gamma\alpha = \alpha,\ \gamma\alpha\gamma = \gamma$.

(f) a **Moore-Penrose inverse** *of α if*
(1) $\alpha\gamma\alpha = \alpha$, *(2)* $\gamma\alpha\gamma = \gamma$, *(3)* $(\alpha\gamma)^{\smile} = \alpha\gamma$, *(4)* $(\gamma\alpha)^{\smile} = \gamma\alpha$.

In the following, let

$$\tilde{\alpha} := \overline{\alpha^{\smile}\overline{\alpha}\alpha^{\smile}} = \overline{\alpha\overline{\alpha}^{\smile}\alpha}^{\smile}. \tag{2.23}$$

Clearly, subinverses do always exist, since $\emptyset = \rho\emptyset\rho \subseteq \rho$ for every relation ρ; and since the union of two subinverses is again a subinverse, one may ask for the greatest subinverse of a relation. This question is answered in

Proposition 2.86. ([40,41]) *$\tilde{\alpha}$ is the greatest subinverse of α.*

Proof. Let χ be a subinverse of α; then $\alpha\chi\alpha \subseteq \alpha$. Using Prop. 2.8(5) and the SCHRÖDER equivalences Th. 2.20(1) we get $\alpha\chi\alpha \subseteq \alpha \leftrightarrow (\alpha\chi)^{\smile}\overline{\alpha} = \chi^{\smile}(\alpha^{\smile}\overline{\alpha}) \subseteq \overline{\alpha} \leftrightarrow \alpha(\alpha^{\smile}\overline{\alpha})^{\smile} = \alpha\overline{\alpha}^{\smile}\alpha \subseteq \overline{\chi}^{\smile}$. But the last inequality is equivalent to $\chi^{\smile} \subseteq \overline{\alpha\overline{\alpha}^{\smile}\alpha}$, whence $\chi \subseteq \overline{\alpha\overline{\alpha}^{\smile}\alpha}^{\smile}$ by Prop. 2.8(6). □

A relation ρ is called **regular**, if it has a generalized inverse. g-inverses are not uniquely determined: as an example take U_X, for which both I_X and U_X are generalized inverses. Moreover, the union of two g-inverses is again a g-inverse; a regular relation thus has a greatest one.

Proposition 2.87. ([40,41])

(1) α is regular iff $\alpha \subseteq \alpha\tilde{\alpha}\alpha$.

(2) If α is regular, its maximum pseudoinverse is $\tilde{\alpha}\alpha\tilde{\alpha}$.

(3) Moore-Penrose inverses are uniquely determined if they exist.

Proof. (1): If α is regular then there is a relation χ such that $\alpha\chi\alpha = \alpha$, whence χ is a subinverse and $\chi \subseteq \tilde{\alpha}$ by Prop. 2.86. Then $\alpha = \alpha\chi\alpha \subseteq \alpha\tilde{\alpha}\alpha$. Conversely, assume $\alpha \subseteq \alpha\tilde{\alpha}\alpha$; since $\tilde{\alpha}$ is a subinverse of α, we have $\alpha\tilde{\alpha}\alpha \subseteq \alpha$ by definition, whence equality follows, and α is regular.

(2): $\alpha' := \tilde{\alpha}\alpha\tilde{\alpha}$ is a pseudoinverse of α: $\alpha\alpha'\alpha = \alpha$ and $\alpha'\alpha\alpha' = \alpha'$ can easily be shown using (1). Let χ be a pseudoinverse of α; then $\chi = \chi\alpha\chi \subseteq \tilde{\alpha}\alpha\tilde{\alpha} = \alpha'$, since $\chi \subseteq \tilde{\alpha}\alpha\tilde{\alpha}$ by Prop. 2.86.

(3): This is a special case of Prop. 6.59. □

An **ideal** $(A, \leq)$ is a subset $\emptyset \neq D \subseteq A$ such that for $y \in A$ we have $x \in D \wedge y \leq x$ implies $y \in D$. If L is a complete lattice and $a \in L$, let $K(a) := \bigcap\{D : D \text{ ideal in } L \wedge a \leq \bigvee D\}$. Then the following holds:

Proposition 2.88. ([33, Theorem 1]) *Let L be a complete lattice. L is completely distributive iff* $a = \bigvee K(a)$, *for every* $a \in L$.

Regularity in $\mathbb{B}_n$ was first investigated by Zaretskiĭ [50], who showed the remarkable result of Th. 2.90 below, the proof of which uses

Lemma 2.89. ([2], [24, p. 495], [34, proof of Theorem 4]) *A complete lattice L is completely distributive iff* $x = \bigvee\{a \in L : x \not\leq \bigvee_{b \in L,\ a \not\leq b} b\}$, *for all* $x \in L$.

Proof. Let $i(a) := \{b \in L : a \not\leq b\}$; then, for a non-zero $a \in L$, $i(a)$ is an ideal in L with $a \notin i(a)$. Furthermore, we set $a^* := \bigvee i(a)$; so, due to Prop. 2.88, we have to show $x = \bigvee K(x) = \bigvee\{a \in L : x \not\leq a^*\}$.

Following [34] we give an indirect proof that for $x \in L$, $a \in K(x)$ iff $x \not\leq a^*$: Let $x \leq a^*$; then by definition of a^* we have $x \leq \bigvee i(a) = \bigvee\{b \in L : a \not\leq b\}$. Since $i(a)$ is an ideal in L not containing a, a cannot be in $K(x)$. Therefore, $a \in K(x)$ implies $x \not\leq a^*$. Conversely, if $a \notin K(x)$, then, by definition of $K(x)$, there is an ideal i_a such that $x \leq \bigvee i_a$ and $a \notin i_a$. Then $i_a \subseteq \{b \in L : a \not\leq b\} = i(a)$, whence $\bigvee i_a \leq \bigvee i(a) = a^*$. Therefore, $x \leq a^*$, so $x \not\leq a^*$ implies $a \in K(x)$. □

For $\alpha \in \mathcal{B}_X$ we call $V(\alpha) := \{A\alpha : A \subseteq X\}$ the **row space** of α. Then $V(\alpha) \subseteq V_n$ is the collection consisting of the zero vector $\mathbf{0}$ and all possible unions of (one or more) non-zero rows of α. For every $\alpha \in \mathbb{B}_n$, $V(\alpha)$ is a complete lattice with respect to inclusion of Boolean vectors where joins are set-theoretic unions (see [2, p. 123]).

Theorem 2.90. *A relation* $\alpha \in \mathcal{B}_X$ *is regular iff* $V(\alpha)$ *forms a completely distributive lattice under inclusion.*

Proof. (The detailed proof given here follows [2], but note that relation composition $\rho \circ \sigma$ used in this paper is our $\sigma\rho$).

Given three relations ρ, σ, τ we have

$$\begin{aligned} x(\rho\sigma\tau)y &\iff \exists s, t \in X\,[x\rho s\sigma t\tau y] \\ &\iff \exists s, t \in X\,[s \in x\rho \wedge s \in t\sigma^{\smile} \wedge y \in t\tau] \\ &\iff \exists t \in X\,[x\rho \not\subseteq \overline{t\sigma^{\smile}} \wedge y \in t\tau]. \end{aligned}$$

Thus $x(\rho\sigma\tau) = \bigcup_{t \in X}\{t\tau : x\rho \not\subseteq \overline{t\sigma^{\smile}}\}$, $x \in X$, whence

$$A(\rho\sigma\tau) := (\rho\sigma\tau)[A] = \bigcup_{B \subseteq X} \{\tau[B] : \rho[A] \not\subseteq \overline{\sigma^{\smile}}[B]\}, \quad A \subseteq X. \tag{2.24}$$

Applying (2.24) to $\rho\tilde{\rho}\rho$ gives

$$\rho\tilde{\rho}\rho[A] = \bigcup_{B\subseteq X}\{\rho[B] : \rho[A] \not\subseteq \overline{\tilde{\rho}^{\smile}}[B]\}, \quad A \subseteq X. \tag{2.25}$$

But using (2.23) we get $\overline{\tilde{\rho}^{\smile}} = \overline{\tilde{\rho}}^{\smile} = \big(\overline{\overline{\rho^{\smile}\overline{\rho}\rho^{\smile}}}\big)^{\smile} = (\rho^{\smile}\overline{\rho}\rho^{\smile})^{\smile} = \rho\overline{\rho}^{\smile}\rho$, which, by applying (2.24) to it, yields

$$\overline{\tilde{\rho}^{\smile}}[B] = \rho\overline{\rho}^{\smile}\rho[B] = \bigcup_{C\subseteq X}\{\rho[C] : \rho[B] \not\subseteq \overline{\overline{\rho}^{\smile\smile}}[C]\}, \quad B \subseteq X. \tag{2.26}$$

Further note that $\overline{\overline{\rho}^{\smile\smile}} = \rho$ and that $\rho\tilde{\rho}\rho = \rho$ if ρ is regular (see proof of Prop. 2.87(1)); so from (2.25) we get by substitution of (2.26)

$$\rho[A] = \bigcup_{B\subseteq X}\Big\{\rho[B] : \rho[A] \not\subseteq \bigcup_{C\subseteq X}\{\rho[C] : \rho[B] \not\subseteq \rho[C]\}\Big\}, \quad A \subseteq X. \tag{2.27}$$

Now apply Lemma 2.89 with $x := \rho[A]$, $a := \rho[B]$, $b := \rho[C]$, and $L := V(\rho)$ to see that $V(\rho)$ is a completely distributive lattice if ρ is regular (noting that $S \subseteq X$ means $\rho[S] \in V(\rho)$).

Conversely, if $V(\rho)$ is a completely distributive lattice then (2.27) holds by Lemma 2.89. Using (2.25) and (2.26), and since $\overline{\tilde{\rho}^{\smile}} = \rho\overline{\rho}^{\smile}\rho$ and $\overline{\overline{\rho}^{\smile\smile}} = \rho$, we get

$$\rho\tilde{\rho}\rho[A] = \bigcup_{B\subseteq X}\{\rho[B] : \rho[A] \not\subseteq \overline{\tilde{\rho}^{\smile}}[B]\}, \quad A \subseteq X,$$

and

$$\overline{\tilde{\rho}^{\smile}}[B] = \rho\overline{\rho}^{\smile}\rho[B] = \bigcup_{C\subseteq X}\{\rho[C] : \rho[B] \not\subseteq \rho[C]\}, \quad B \subseteq X.$$

Therefore,

$$\rho\tilde{\rho}\rho[A] = \bigcup_{B\subseteq X}\Big\{\rho[B] : \rho[A] \not\subseteq \bigcup_{C\subseteq X}\{\rho[C] : \rho[B] \not\subseteq \rho[C]\}\Big\}, \quad A \subseteq X, \tag{2.28}$$

and by comparison with (2.27) (for which we did not use the regularity of ρ but Lemma 2.89!) we get $\rho\tilde{\rho}\rho[A] = \rho[A]$, $A \subseteq X$. This equality holds for arbitrary subsets of X, thus for all singleton sets $\{x\}_{x\in X}$. So $x(\rho\tilde{\rho}\rho) = x\rho$ for all $x \in X$, whence $\rho\tilde{\rho}\rho = \rho$, i.e. ρ is regular. □

2.12 Commuting and Idempotent Relations

Two relations ρ and σ are called **commuting** in case $\rho\sigma = \sigma\rho$. For sake of completeness, we begin with recalling some results that we have reported earlier in this book:

(1) Let ρ and σ be symmetric, and $\rho\sigma \subseteq \sigma\rho$; then $\rho\sigma = \sigma\rho$. (Prop. 2.24(4)).

(2) *Let ρ and σ be two preorder relations on a set S. Then ρ and σ commute iff both $\rho\sigma$ and $\sigma\rho$ are preorders.* (Prop. 2.52).

We continue with some easy to prove properties of commuting relations.

Proposition 2.91.

(1) If a relation ρ commutes with two relations σ and τ then it also commutes with their union.

(2) If both ρ and ρ' commute with a relation σ then also their union does.

Proof. (1): $\rho(\sigma \cup \tau) = \rho\sigma \cup \rho\tau = \sigma\rho \cup \tau\rho = (\sigma \cup \tau)\rho$.
(2): Just a reformulation of (1). □

Proposition 2.92. *For two commuting relations ρ and σ on a set S the following holds:*

(1) If ρ and σ are symmetric, then also their composition.
(2) If ρ and σ are transitive, then also their composition.

Proof. (1): $(\rho\sigma)^{\smile} = \sigma^{\smile}\rho^{\smile} = \sigma\rho = \rho\sigma$.
(2): $(\rho\sigma)^2 = \rho\sigma\rho\sigma = \rho^2\sigma^2 \subseteq \rho\sigma$. □

For sake of completeness we first repeat two results concerning commuting equivalences which have already been mentioned in Prop. 2.66(3) and Prop. 2.67(4):

Proposition 2.93. *Let* $\rho, \sigma \in \mathsf{Eq}_S$. *Then the following holds:*

(1) ρσ is an equivalence iff ρ and σ commute.
(2) ρ and σ commute iff $\rho \vee \sigma = \rho\sigma$ *iff* $\rho\sigma \in \mathsf{Eq}_S$.

Obviously, idempotent relations are a special case of both commuting and transitive relations. We did already meet an important class of idempotent relations, viz. equivalences. In fact it has been shown in Prop. 2.24(3), that all preorders, i.e. the reflexive and transitive relations, exhibit this property, and also coreflexive relations are idempotent by Prop. 2.24(5). It is known that preorders are a proper subset of idempotent relations which are a proper subset of the set of transitive relations (see [22, p. 83] or [7, p. 169]). Examples are given immediately after Prop. 2.98.

In the following we will use the terms "*relation*" and "*matrix*" interchangeably, depending on where the concepts and terminology under consideration arise from. In Def. 1.34 we introduced the notation E_A for the subset of idempotents of a semigroup A. In the following, we will simply use Idm_X to denote the set of idempotent binary relations on an arbitrary finite, but fixed set X, $|X| \geq 2$, instead of the more cumbersome $E_{\mathcal{B}_X}$. In fact, thinking in terms of Boolean matrices, only their order $|X| = n$ is relevant.

Before we start investigating characteristic properties of idempotent relations themselves, we will have a look at the structure of Idm_X.

Corollary 2.94. *The set of idempotents* Idm_X *forms a complete lattice.*

Proof. $\{\emptyset, I_X, U_X\} \subseteq \mathsf{Idm}_X$, whence, for all idempotents χ, $\emptyset \subseteq \chi \subseteq U_X$; so Idm_X is a poset with a least and a greatest element. For $A = \{\chi_1, \ldots, \chi_K\} \subseteq \mathsf{Idm}_X$ we define $\bigvee A = \mathbf{t}(\chi_1 \cup \ldots \cup \chi_K)$, for which, by Prop. 2.35(3), we have

$$\mathbf{t}(\chi_1 \cup \ldots \cup \chi_K) = \textstyle\bigcup_{1 \leq i_k \leq K,\, n \geq 1} \chi_{i_1}\chi_{i_2}\cdots\chi_{i_n}.$$

Next we have to ensure that $\mathbf{t}(\chi_1 \cup \ldots \cup \chi_K) \in \mathsf{Idm}_X$. Once again, for sake of simplicity we demonstrate the proof idea for $K = 2$:

$$\begin{aligned}
[\mathbf{t}(\rho \cup \sigma)]^2 &= [\textstyle\bigcup_{i \geq 1}(\rho \cup \sigma)^i]^2 \\
&= [(\rho \cup \sigma) \cup (\rho^2 \cup \sigma\rho \cup \rho\sigma \cup \sigma^2) \cup (\rho^3 \cup \sigma\rho^2 \cup \rho\sigma\rho \cup \ldots) \cup \ldots]^2 \\
&= (\rho \cup \sigma)^2 \cup (\rho^2 \cup \sigma\rho \cup \rho\sigma \cup \sigma^2)(\rho \cup \sigma) \cup (\rho^3 \ldots
\end{aligned}$$

$$\begin{aligned}
&= (\rho^2 \cup \sigma\rho \cup \rho\sigma \cup \sigma^2) \cup (\rho^2 \cup \sigma\rho \cup \rho\sigma \cup \sigma^2)(\rho \cup \sigma) \cup (\rho^3 \dots \\
&= (\rho \cup \sigma\rho \cup \rho\sigma \cup \sigma) \cup (\rho \cup \sigma\rho \cup \rho\sigma \cup \sigma)(\rho \cup \sigma) \cup (\rho^3 \dots \\
&= (\rho \cup \sigma\rho \cup \rho\sigma \cup \sigma) \cup (\rho\sigma\rho \cup \sigma\rho\sigma) \cup (\rho^3 \dots \\
&= \textstyle\bigcup_{1\le i_k \le 2,\ n\ge 1} \chi_{i_1}\chi_{i_2}\cdots\chi_{i_n} \qquad \text{(with } \chi_1 := \rho, \chi_2 := \sigma) \\
&= \mathbf{t}(\rho \cup \sigma).
\end{aligned}$$

So we have proved that, for any $\emptyset \neq A \subseteq \mathsf{Idm}_X$ the join $\bigvee A$ exists, whence Idm_X is a complete join semilattice. From lattice theory it is known that if for a poset P $\bigvee A$ exists for all $A \subseteq P$, then P is a complete lattice (see [17, p. 24, Lemma 14] for the dual result): Let $B \subseteq P$ be the set of lower bounds of A; then, by hypothesis, $l := \bigvee B$ exists. If $a \in A$, then $a \geq b$ for all $b \in B$, whence $a \geq l$, and $l \in B$. Thus l is the greatest member of B, that is, $l = \bigwedge A$. □

Idm_X is however not a sublattice of Eq_X since the meet operation differs as we will see soon. In order to establish this fact, we need some preparation: In the following we write $I_\rho := \{a \in X : a\rho a\} \subseteq X$, $a \sim b := a(\rho \cap \sigma)b$, and $\hat{\sim} := a(\rho \wedge \sigma)b$.

Lemma 2.95. *Let $\rho, \sigma \in \mathsf{Idm}_X$, with X finite and $|X| = k$.*

(1) $a\rho b$ iff $\exists m \in I_\rho\, [a\rho m \wedge m\rho b]$; and $\rho \neq \emptyset$ implies $I_\rho \neq \emptyset$ ([43, Lemma 1]).
(2) $\rho \cap \sigma$ is transitive.

Proof. (1): By the idempotency of ρ, for any a, b such that $a\rho b$ and each $n \geq 1$, we have a sequence $a = x_0, x_1, \dots, x_n = b$ with $x_i\rho x_{i+1}$ for $0 \leq i \leq n-1$. And since ρ is transitive by definition, $x_i \rho x_j$ whenever $i < j$. Taking $n > k$, $x_i = x_j$ must hold for some $i < j$. So $a\rho x_i \wedge x_i\rho b$, and $x_i \in I_\rho$.
(2): $(\rho \cap \sigma)^2 = \rho^2 \cap \rho\sigma \cap \sigma\rho \cap \sigma^2 = \rho \cap \sigma \cap \rho\sigma \cap \sigma\rho \subseteq \rho \cap \sigma$. □

Following [43, p. 298,(3)] we define

$$\rho \wedge \sigma := \{\langle a, b\rangle \in \rho \cap \sigma : \text{for some } m \in I_{\rho\cap\sigma} \text{ both } \langle a, m\rangle \text{ and } \langle m, b\rangle \in \rho \cap \sigma\},$$

and translated to our shorthand notation this gives

$$a \hat{\sim} b :\iff a \sim b \ \wedge\ \exists m \in I_{\rho\cap\sigma}\, [a \sim m \wedge m \sim b].$$

Proposition 2.96. *Let $\rho, \sigma \in \mathsf{Idm}_X$.*

(1) $\rho \wedge \sigma$ is idempotent.
(2) $\rho \wedge \sigma$ is the greatest lower bound of ρ and σ in Idm_X.

Proof. (1): Let $a \hat{\sim}^2 b$; then $\exists c \in X\, [a \hat{\sim} c \hat{\sim} b]$, whence by definition of $\hat{\sim}$ we get

$$\exists c \in X \big[\exists m \in I_{\rho\cap\sigma}\, [a \sim m \wedge m \sim c] \ \wedge\ \exists m' \in I_{\rho\cap\sigma}\, [c \sim m' \wedge m' \sim b]\big],$$

that is $\exists c \in X \exists m, m' \in I_{\rho\cap\sigma}$ such that

$$a\rho m \wedge a\sigma m \wedge m\rho c \wedge m\sigma c \ \wedge\ c\rho m' \wedge c\sigma m' \wedge m'\rho b \wedge m'\sigma b.$$

By transitivity, $m\rho c \wedge c\rho m' \ \rightarrow\ m\rho m'$, so that $a\rho m\rho m'\rho b$, i.e. $a\rho m \wedge m\rho^2 b$; and similarly we have $a\sigma m \wedge m\sigma^2 b$. Thus, $a(\rho\cap\sigma)m \wedge m(\rho^2 \cap \sigma^2)b$ which means $\exists m \in I_{\rho\cap\sigma}\, [a \sim m \wedge m \sim b]$. Furthermore, since $\rho \wedge \sigma \subseteq \rho \cap \sigma$ by definition, we get $(\rho \wedge \sigma)^2 \subseteq (\rho \cap \sigma)^2 \subseteq$

$\rho \cap \sigma$ by monotonicity and Lemma 2.95(2); so $a \mathrel{\hat{\sim}}^2 b$ implies $a \sim b$. Altogether we thus have $a \mathrel{\hat{\sim}}^2 b \subseteq a \mathrel{\hat{\sim}} b$.

For the other inclusion, assume $a \mathrel{\hat{\sim}} b$; this gives $a \sim b \wedge \exists m \in I_{\rho \cap \sigma} [a \sim m \wedge m \sim b]$. Since also $m \sim m$ we get $a \mathrel{\hat{\sim}} m \wedge m \mathrel{\hat{\sim}} b$, whence $a \mathrel{\hat{\sim}}^2 b$.

(2): We have to show that, for any $\lambda \in \mathsf{Idm}_X$ with $\lambda \subseteq \rho \cap \sigma$, also $\lambda \subseteq \rho \wedge \sigma$. Let $a\lambda b$; then by Lemma 2.95(1) $\exists m \in I_\lambda [a\lambda m \wedge m\lambda b]$. Thus, since $\lambda \subseteq \rho \cap \sigma$ has been assumed, we get $a \sim b \wedge \exists m \in I_{\rho \cap \sigma} [a \sim m \wedge m \sim b]$. Therefore, $a \mathrel{\hat{\sim}} b$ and together with (1) the assertion follows. □

In the following, we will present various characterizations of idempotent relations of finite order to be found in literature, as well as a few of their properties. We begin with listing obvious and easy to prove facts:

Proposition 2.97.

(1) Idempotent relations are transitive.
(2) Converses of idempotents are idempotent.
(3) Idempotent relations are regular.
(4) An idempotent $\chi \neq I$ cannot posses an inverse.
(5) Let π be a permutation relation; then ρ is idempotent iff $\pi\rho\pi^\smile$ is.

Proof. (2): $\lambda = \lambda^2$ implies $\lambda^\smile = (\lambda\lambda)^\smile = \lambda^\smile\lambda^\smile$.
(4): Assuming $\chi\chi^{-1} = I$ implies $\chi = \chi I = \chi\chi\chi^{-1} = \chi\chi^{-1} = I$.
(5): $\rho^2 = \rho$ implies $\pi\rho\pi^\smile\pi\rho\pi^\smile = \pi\rho^2\pi^\smile = \pi\rho\pi^\smile$. Conversely, from $(\pi\rho\pi^\smile)^2 = \pi\rho\pi^\smile$ we get $\pi\rho^2\pi^\smile = \pi\rho\pi^\smile$ as above. Multiplying the last equation with $\pi^\smile$ from the left and with π from the right we get $\rho^2 = \rho$. □

We have already met examples of idempotent relations in former chapters, and for sake of completeness we repeat these results in

Proposition 2.98.

(1) Preorders are idempotent.
(2) Equivalence relations are idempotent.
(3) Coreflexive relations are idempotent.
(4) Let ρ and σ be partial equivalence relations on a set S, then $\rho\sigma$ is idempotent.
(5) If $\rho^\smile\rho = \rho$ then ρ is idempotent.

Proof. (1): By Prop. 2.24(3).
(2): By (1).
(3) By Prop. 2.24(5).
(4) By Prop. 2.70(5).
(5) From the hypothesis we get $\rho^\smile = (\rho^\smile\rho)^\smile = \rho^\smile\rho = \rho$, whence ρ is symmetric. Therefore, $\rho^2 = \rho^\smile\rho = \rho$. □

The inclusions preorders ⊂ idempotents ⊂ transitive relations are proper: $\begin{pmatrix}0&1\\0&1\end{pmatrix}$ is idempotent but not reflexive, and $\begin{pmatrix}0&1\\0&0\end{pmatrix}$ is transitive but not idempotent.

Various characterizations of the idempotency of relations of finite order can be found in literature. Among the earliest ones are [36], based on graph-theoretic considerations, and [39], based on a viewpoint from order relations. The latter one is presented in Th. 2.99 below, but before we can do so we need some new notions.

Let ρ be a preorder, i.e. a reflexive and transitive relation on a set A. An element $a \in A$ is called *ρ-strict* if $a\rho a' \wedge a'\rho a$ imply $a = a'$. Generalizing the notion from orders to arbitrary relations we say that b **covers** a in ρ, denoted as $a \prec_\rho b$, if $a\rho b$ and $a\rho c\rho b \to c = a \vee c = b$. Two elements $a, b \in A$ are called *ρ-neighbors* if $a \prec_\rho b \vee b \prec_\rho a$. A subset $B \subseteq A$ is called *ρ-permissible*, if all $b \in B$ are ρ-strict and B does not contain ρ-neighbors.

A binary relation $\sigma \in \mathcal{B}_A$ is called a **pseudo-order** if $\sigma = \rho - I_B$, with $\rho \in \mathcal{B}_A$ a preorder and $B \subseteq A$ a ρ-permissible subset of A.

Theorem 2.99. *$\sigma \in \mathcal{B}_A$ is idempotent iff it is a pseudo-order relation.*

Proof. [$\Leftarrow$] Let $\sigma = \rho - I_B$ be a pseudo-order relation, with ρ a preorder and $B \subseteq A$ a ρ-permissible subset of A, and let $a\sigma b \wedge b\sigma c$. Then, since clearly $\sigma \subseteq \rho$, $a\rho b \wedge b\rho c$, whence $a\rho c$ by transitivity of ρ. Now assume aI_Bc; then $a \in B$ and $a = c$ by definition, whence we get $c\sigma b \wedge b\sigma c$. Since $a = c \in B$ is ρ-strict, we have to conclude that $a = b$ and aI_Bb, a contradiction to our hypothesis that $a\sigma b$. So $\neg aI_Bc$ and $a\sigma c$, i.e. $\sigma^2 \subseteq \sigma$.

Conversely, let $a\sigma b$. If $a \notin B \vee b \notin B$ then either $a\sigma a$ or $b\sigma b$. In both cases we get $a\sigma^2 b$. And assuming the contrary, i.e. $\{a, b\} \subseteq B$, we get $a\rho b$, since $\sigma \subseteq \rho$. As a preorder, ρ is idempotent by Prop. 2.24(3), so $a\rho c\rho b$ for some $c \in A$. Since a, b are not ρ-neighbors (due to $B \subseteq A$ being assumed to be ρ-permissible) it follows that $a \neq c$ and $b \neq c$, whence $a\sigma c\sigma b$. Thus $\sigma \subseteq \sigma^2$, and so σ is idempotent.

[$\Rightarrow$] Let σ be idempotent such that $\rho = \sigma \cup I_A$, and $B = A - \mathsf{fld}(\sigma \cap I_A)$. Then ρ is reflexive and $\rho^2 = \sigma^2 \cup \sigma I_A \cup I_A\sigma \cup I_A^2 = \sigma \cup I_A = \rho$, i.e. ρ is also transitive and thus a preorder. Clearly, $\sigma = \rho - I_B$.

We have to show that B is ρ-permissible: Let $a \in B$ and $a\rho b \wedge b\rho a$. $a \neq b$ implies $a\sigma b \wedge b\sigma a$, whence $a\sigma a$ since σ was assumed to be idempotent, a contradiction to $\sigma = \rho - I_B$. Therefore, $a = b$, i.e. a is ρ-strict. Now we consider two distinct elements $a, b \in B$ with $a\rho b$. We have $a\sigma b$, whence $a\sigma c\sigma b$ for some $c \in A$. $c = a \vee c = b$ implies $\neg a\sigma c \vee \neg c\sigma b$ so that we get $a \neq c \wedge c \neq b$, whence a and b are not ρ-neighbors. □

Anticipating more general results, we conclude this section with some further characterizations of the idempotency of Boolean relations and matrices:

Proposition 2.100. *Let $B \in \mathbb{B}_n$ be reflexive with index k. Then B is idempotent iff $B = \mathsf{adj}(B)$.*

Proof. Assume $B = \mathsf{adj}(B)$. By Cor. 2.77(1) we have $k \le n-1$, whence, by Cor. 2.77(3), B^{n-1} is idempotent. And in the proof of Prop. 8.14(1) it is shown that $\mathsf{adj}(B) = B^{n-1}$. Conversely, if B is idempotent, then $B = B^{n-1} = \mathsf{adj}(B)$ by Prop. 8.14. □

Proposition 2.101. ([28, Lemmas 1.7, 1.8])

(1) *Let $\alpha \in \mathcal{B}_X$; then $\alpha^2 = \alpha$ iff $A\alpha = A$ for every $A \in V(\alpha)$ iff $\alpha A = A$ for every $A \in W(\alpha)$.*

(2) *Let $\emptyset \neq A, B \subseteq X$, such that $A \cap B \neq \emptyset$, and let $\alpha = A \times B$. Then $V(\alpha) = \{\emptyset, B\}$ and $W(\alpha) = \{\emptyset, A\}$; hence $\alpha \in \mathsf{Idm}_X$.*

(3) *Partial order relations on subsets of X are reduced idempotents.*

Proof. (See page 81 for the definitions of the row and column space of a relation.) (1): Assume $A \in V(\alpha)$; then there is a $B \subseteq X$ such that $A = B\alpha$. Thus, $A\alpha = B\alpha\alpha = B\alpha = A$. Conversely, $x\alpha \in V(\alpha)$ for every $x \in X$. By assumption, $x\alpha = (x\alpha)\alpha = x(\alpha\alpha)$, whence $\alpha^2 = \alpha$.

(2): By definition of α, $a\alpha = B$ for all $a \in A$ and $a\alpha = \emptyset$ for all $a \in \overline{A}$. Thus, $V(\alpha) = \{\emptyset, B\}$. Since $A \cap B \neq \emptyset$ we get

$$B\alpha = \bigcup_{x \in B} x\alpha = \Big(\bigcup_{x \in A \cap B} x\alpha\Big) \cup \Big(\bigcup_{x \in B - A} x\alpha\Big) = B \cup \emptyset = B.$$

and α is idempotent by (1).

(3): See Prop. 2.75. □

References

[1] Bartol, W., Miró, J., Pióro, K., Rosselló, F.: On the coverings by tolerance classes, Information Sciences **166**, 193-211 (2004).

[2] Bandelt, H.-J.: Regularity and complete distributivity, Semigroup Forum **19**, 123-126 (1980).

[3] Berghammer R.: *Ordnungen, Verbände und Relationen mit Anwendungen*, 2nd ed., Springer Vieweg, Wiesbaden (2012).

[4] Berghammer, R., Winter, M.: Embedding mappings and splittings with applications, Acta Informatica **47**, 77-110 (2010).

[5] Berghammer, R., Winter, M.: Decomposition of Relations and Concept Lattices, Fundamenta Informaticae **126**, 37-82 (2013).

[6] Blyth, T.S.: *Lattices and Ordered Algebraic Structures*, Springer-Verlag, London (2005).

[7] Boyd, J.P.: Structural similarity, semigroups and idempotents, Social Networks **5**, 157-172 (1983).

[8] Carnap, R.: *Der Logische Aufbau der Welt*, Weltkreis-Verlag (1928).

[9] Caspard, N., Leclerc, B., Monjardet, B.: *Finite Ordered Sets - Concepts, Results and Uses*, Cambridge University Press, Cambridge, UK (2012).

[10] Chajda,I.: Partitions, coverings and blocks of binary relations, Glasnik Matematički **14**, 21-26 (1979).

[11] Chajda, I., Duda, J.: Blocks of binary relations, Annales Univ. Sci. Budapest, sectio Math **22**, 3-9 (1979).

[12] Chajda, I., Niederle, J., Zelinka, B.:. On existence conditions for compatible tolerances, Czechoslovak Math. J. **26**, 304-311 (1976).

[13] Clifford, A.H., Preston, G.B.: *The Algebraic Theory of Semigroups, Vol. 1*, 2nd ed., reprinted with corrections, American Mathematical Society, Providence, Rhode Island (1977).

[14] Desharnais, J., Jaoua, A.: A relational division operator: the conjugate kernel, Theoretical Computer Science **114**, 247-272 (1993).

[15] Gallier, J.: *Discrete Mathematics*, Springer Science+Business Media (2011).

[16] Ganyushkin, O., Mazorchuk, V.: *Classical Finite Transformation Semigroups - An Introduction*, Springer-Verlag London Limited (2009).

[17] Grätzer, G.: *General Lattice Theory*, Birkhäuser Verlag, Basel und Stuttgart (1978).

[18] Gries, D., Schneider, F.B.: *A Logical Approach to Discrete Math*, Springer-Verlag, New York (1993).

[19] Harzheim, E.: *Ordered Sets*, Springer Science+Business Media (2005).

[20] Hohn, P.E., Schissler, L.R.: Boolean matrices and combinational circuit design, Bell System Tech. J. **34**, 177-202 (1955).

[21] Howie, J.M: *Fundamentals of Semigroup Theory*, Clarendon Press, Oxford (1995).

[22] Kim, K.H.: *Boolean Matrix Theory and Applications*, Marcel Dekker, Inc., New York and Basel (1982).

[23] Ljapin, E.S.: *Semigroups*, American Mathematical Society, Providence, Rhode Island (1963).

[24] Longstaff: Strongly reflexive lattices, J. London Math. Soc. (2) **11**, 491-498 (1975).

[25] Luce, R. D.: A note on Boolean matrix theory, Proc. of the American Math. Soc. **3**, 382-388 (1952).

[26] Madarász, R.: On the partial algebras of weak congruences, Rev. Res. Fac. Sci. Univ. Novi Sad **25**, 89-97 (1995).

[27] Mirin, A.C.: Relation Algebras, Master of Arts in Mathematics Thesis, Mills College (2013).

[28] Namnak, C., Preechasilp, P.: Natural partial orders on the semigroup of binary relations, Thai J. Mathematics **4**, 39-50 (2006).

[29] Ono, K.: On some properties of binary relations, Nagoya Math. J. **12**, 161-170 (1957).

[30] Peters, J. F., Wasilewski, P.: Tolerance spaces: Origins, theoretical aspects and applications. Information Sciences **195**, 211-225 (2012).

[31] Plemmons, R., West, M.T.: On the semigroup of binary relations, Pacific Journal of Mathematics **35**, 743-753 (1970).

[32] Poincaré, J.H. : *La Valeur de la Science*, Flammarion (1905).

[33] Raney, G.N.: A Subdirect-Union Representation for Completely Distributive Complete Lattices, Proc. of the American Math. Soc. **4**, 518-522 (1953).

[34] Raney, G.N.: Tight Galois connections and complete distributivity, Trans. of the American Math. Soc. **97**, 418-426 (1960).

[35] Ronse, C.: Idempotent Block Splitting on Partial Partitions, I: Isotone Operators, Order **28**, 273-306 (2011).

[36] Rosenblatt, D.: On the graphs of finite idempotent Boolean relation matrices, J. Res. Nat. Bureau of Standards **67B**, 249-256 (1963).

[37] Ross, K.A., Wright, C.R.B.: *Discrete Mathematics*, 4th ed., Prentice-Hall, New Jersey (1999).

[38] Rutherford, D.: Inverses of Boolean matrices, Proc. of the Glasgow Math. Assoc. **6**, 49-53 (1963).

[39] Schein, B.M.: A construction for idempotent binary relations, Proc. Japan Acad. **46**, 246-247 (1970).

[40] Schein, B.M.: Regular elements of the semigroup of all binary relations, Semigroup Forum **13**, 95-102 (1976).

[41] Schmidt, G.: Rectangles, fringes, and inverses, in: Berghammer, R., Möller, B., Struth, G. (Eds.), *Relations and Kleene Algebra in Computer Science, RelMiCS/AKA 2008*. LNCS 4988, Springer-Verlag Berlin Heidelberg, pp. 352-366 (2008).

[42] Schwarz, S.: On the semigroup of binary relations on a finite set, Czech. Math. J. **20**, 632-679 (1970).

[43] Shmuely, Z.: The lattice of idempotent binary relations, Algebra Universalis **9**, 297-303 (1979).

[44] Skornjakov, L.A.: *Elements of Lattice Theory*, Hindustan Publishing Corporation (India) (1977).

[45] Toth, H.: *Relations: Concrete, Abstract and Applied - An Introduction*, World Scientific Publishing, Singapore (2020).

[46] Vojvodić, G. Šešelja, B.: On the lattice of weak congruence relations, Algebra Universalis **25**, 121-130 (1988).

[47] Wedderburn, J.H.M.: Boolean Linear Associative Algebra, Annals of Mathematics **35**, 185-194 (1934).

[48] Winter, M.: Weak Relational Products, in: Schmidt, R.A. (Ed.), *Relations and Kleene Algebra in Computer Science, RelMiCS/AKA 2006*, LNCS 4136, Springer-Verlag Berlin Heidelberg, pp. 417–431 (2006).

[49] Winter, M.: Products in categories of relations, The Journal of Logic and Algebraic Programming **76**, 145-159 (2008).

[50] Zaretskiĭ, K.A.: The semigroup of binary relations, Mat. Sbornik **61**, 291-305 (1963).

[51] Zeeman, E.C.: The topology of the brain and the visual perception, in: Fort, K.M. (Ed.), *Topology of 3-manifolds and Selected Topics*. Prentice Hall, Englewood Cliffs, pp. 240–256 (1965).

[52] Zhang, S.: Generating Finite Integral Relation Algebras, Master Thesis, Department of Computer Science, Brock Univ. (2010).

CHAPTER 3

BOOLEAN MATRICES

The considerations on Boolean matrices have been split in two parts according to the two roles they have in this text: The one as a standard example of matrices which are neither real nor complex, and possibly known fom the context of graph theory as e.g. an adjacency matrix; the other as a special case within the various kinds of semirings that will be considered later on in Section 8.3.2.

Let m and n be positive integers; a rectangular array

$$A = [a_{ij}] = [a_{ij}]_{m,n} \qquad (i = 1,2,\dots,m;\ j = 1,2,\dots,n)$$

is said to be a *matrix* of size m by n, and to have m *rows* and n *columns*. In case $m = n$ A is called a *square* matrix of order n. An m by n matrix contains $m \cdot n$ *entries* or *scalars* out of a certain set, like e.g. the set of rational or real numbers, in the context of engineering problems or in the field of statistics, or the sets of states, or of input or output alphabets in automata theory etc. In case $a_{ij} \in \mathbf{2}$, the binary Boolean algebra, then A is called a Boolean matrix; see Def. 2.12(a).

Remark. It is important to be aware of the fact that a matrix with all its entries from the set $\{0,1\}$, also called "*(0,1)-matrices*", does *not* automatically qualify as a Boolean matrix by this property alone, but only if the Boolean operations $\vee$ and $\wedge$ are chosen for the matrix operations of addition and multiplication.

In other words: If you search for additional information, you may also have a look for the terms "*relation matrices*", "*binary Boolean matrices*", "*zero-one-matrices*", or "*non-negative matrices*" to possibly find properties of these kinds of matrices which do not depend on the usual arithmetical nature of addition and multiplication and which are valid in the Boolean context.

As explicated in [18, p. viii], Boolean matrices "require a completely separate treatment from that of matrices over the real field $\mathbb{R}$ or complex field $\mathbb{C}$ due to the fact that Boolean matrices do not satisfy many of the fundamental properties of matrices over $\mathbb{R}$ or $\mathbb{C}$". For example, in contrast to the real or complex case, Boolean matrices need not always have a g-inverse, and their row and column rank may not be equal.

We first introduce some basic matrix terminology, mainly following [6]. You may also wish to consult one of [1, 2, 22] for more details. Various other sources that have been used to compile the material in this section will be mentioned accordingly.

It is customary to denote zero matrices with an O, matrices with each entry equal to 1 with J, and the identity matrix with $I := [\delta_{ij}]$, where δ is the so-called **Kronecker delta** defined as $\delta_{ij} = 1$, if $i = j$, and $\delta_{ij} = 0$ otherwise. As we did it already for relations, we will sometimes use subscripts to emphasize the size of matrices.

In connection with matrices it is common to speak of rows and columns: Rows and columns of any $M \in \mathbb{B}_{m\times n}$ are elements of a set $V_k := \{0,1\}^k$, i.e. *Boolean vectors* of length k, for some $k \in \mathbb{N}$. Let $M \in \mathbb{B}_{m\times n}$ be a non-zero Boolean matrix; if we let $\boldsymbol{m}_{i*}$ denote the i-th row, and $\boldsymbol{m}_{*j}$ the j-th column of M, then we get $M = [\boldsymbol{m}_{1*}\boldsymbol{m}_{2*}\ldots\boldsymbol{m}_{m*}]^T = [\boldsymbol{m}_{*1}\boldsymbol{m}_{*2}\ldots\boldsymbol{m}_{*n}]$.

Many problems in the theory and application of non-negative matrices only depend on where zero entries are located. So in these cases the relevant property of an entry m_{ij} of a matrix M is if $m_{ij} = 0$. The problem to be investigated can be often simplified by substituting for M the $(0,1)$-matrix with exactly the same zero pattern.

Definition 3.1. *Let A and B be two $m \times n$ non-negative matrices.*

(a) *A and B are said to have the same* **zero pattern** *if $a_{ij} = 0 \leftrightarrow b_{ij} = 0$.*
(b) *Let $A_0^1 \in \mathbb{B}_{m\times n}$ denote the matrix $C = [c_{ij}]_{m,n}$ with $c_{ij} = 1$ if $a_{ij} > 0$ and $c_{ij} = 0$ otherwise. We call A_0^1 the* **support matrix** *of A.*

Thus, a non-negative matrix A and its support matrix have the same zero pattern. Clearly, for the support matrix of a Boolean matrix B one has $B_0^1 = B$.

3.1 Boolean Vector Spaces

It is a well established result that $\mathbb{B}_n$ with Boolean matrix multiplication and $(\mathcal{B}_X, \cdot)$, with $|X| = n$, are isomorphic semigroups. Therefore, we will sometimes denote elements from both $\mathcal{B}_X$ and $\mathbb{B}_n$ with small Greek letters, and usually will use the same letter for a relation and its Boolean matrix representation, that is, we will henceforth also use the simple notation $\rho \in \mathbb{B}_n$ instead of the more complicated $M_\rho \in \mathbb{B}_n$ (see Def. 2.12(b)).

In the context of Boolean matrices the notion of a *Boolean vector space* arises in a natural way: The rows of any $\beta \in \mathbb{B}_n$ are elements of the set $V_n := \{0,1\}^n$, i.e. *Boolean vectors* of dimension n, naturally corresponding to the aftersets $x\beta$. Analogously we can define $V_n^T = \{v^T : v \in V_n\}$ as the set of all Boolean column vectors of length n comprising the columns of any $\beta \in \mathbb{B}_n$, corresponding to the foresets βy. In the sequel we will introduce the basic notions and properties in terms of row vectors.

If we define addition and multiplication for **2** in the usual arithmetical way with the exception that we set $1 + 1 = 1$, then $+$ is called *Boolean addition.* We thus have $1+1 = 1+0 = 0+1 = 1 = 1\cdot 1$ and $0\cdot 0 = 1\cdot 0 = 0\cdot 1 = 0 = 0+0$, whence $+$ and $\cdot$ become equivalent to the $\vee$ and $\wedge$ operations on **2**. As a consequence you will find both kinds of notation in literature (and also in the following). From a computational point of view one can add and multiply as usual, but has to replace all positive entries in the final result by ones. With **0** we will denote the zero row vector $(0,0,\dots,0)$ and also the zero column vector vector $(0,0,\dots,0)^T$, and with **1** the vectors consisting of 1's only; row vectors $(v_1, v_2,\dots, v_n)$ will be denoted by $\boldsymbol{v}$.

Definition 3.2. ([18, Section 1.1])

(a) *The set V_n together with the operation of component-wise addition is called the* **Boolean vector space** *(of dimension n):*

$$\boldsymbol{u}+\boldsymbol{v}=(u_1,u_2,\dots,u_n)+(v_1,v_2,\dots,v_n)=(u_1+v_1,u_2+v_2,\dots,u_n+v_n).$$

(b) *$\emptyset \neq W \subseteq V_n$ is called a* **subspace** *of V_n, if it contains* **0** *and is closed under addition.*

(c) *The* **span** *of $W \subseteq V_n$, denoted by $\langle W\rangle$, is the set of all linear combinations of elements of W, $\langle W\rangle = \{\sum_i b_i w_i : b_i \in \{0,1\}, w_i \in W\}$, with the convention that $\sum_\emptyset = \mathbf{0}$.*

If we define **scalar multiplication** by $a\boldsymbol{u} = (a\cdot u_1, a\cdot u_2,\dots, a\cdot u_n)$, for $a \in \mathbf{2}$ and $\boldsymbol{u} \in V_n$, then it is easy to see that the following properties hold for $(V_n, +,\cdot)$: For all $s,t \in \mathbf{2}$ and all $\boldsymbol{u},\boldsymbol{v},\boldsymbol{w} \in V_n$ we have

(A1) $\boldsymbol{u}+\boldsymbol{v} \in V_n$ (*Closure under addition*).
(A2) $(\boldsymbol{u}+\boldsymbol{v})+\boldsymbol{w} = \boldsymbol{u}+(\boldsymbol{v}+\boldsymbol{w})$ (*Associativity*).
(A3) $(\boldsymbol{u}+\boldsymbol{v}) = (\boldsymbol{v}+\boldsymbol{u})$ (*Commutativity*).
(A4) V_n has a zero vector **0** such that, for every $\boldsymbol{u} \in V_n$, $\boldsymbol{u}+\mathbf{0} = \boldsymbol{u}$ (*Additive identity*).

(M1) $s\boldsymbol{u} \in V_n$ (*Closure under scalar multiplication*).
(M2) $(s\cdot t)\boldsymbol{u} = s(t\boldsymbol{u})$ (*Associativity*).
(M3) $s(\boldsymbol{u}+\boldsymbol{v}) = s\boldsymbol{u}+s\boldsymbol{v}$ (*Distributivity with respect to vector addition*).
(M4) $(s+t)\boldsymbol{u} = s\boldsymbol{u}+t\boldsymbol{u}$ (*Distributivity with respect to scalar addition*).
(M5) $1\boldsymbol{u} = \boldsymbol{u}$ (*Scalar identity*).

If we compare this with the definition of a **vector space** as given in Def. 5.1, then the only missing of the constituting properties is

(A5) For each $\boldsymbol{a} \in V_n$ there is an element $(-\boldsymbol{a}) \in V_n$ such that $\boldsymbol{a}+(-\boldsymbol{a}) = \mathbf{0}$,

i.e. the existence of an *additive inverse.* But this clearly does not make sense for the binary Boolean algebra **2** and V_n, respectively. So the economical definition of a Boolean vector space becomes comprehensible. As in linear algebra we have

Corollary 3.3. *Let $\emptyset \neq W \subseteq V_n$; then W is a subspace of V_n iff the two closure conditions (A1) and (M1) hold for W.*

Proof. Assume W to be a subspace of V_n, then W is a Boolean vector space and therefore it is closed under addition and scalar multiplication. If $\boldsymbol{u},\boldsymbol{v},\boldsymbol{w} \in W$ then $\boldsymbol{u},\boldsymbol{v},\boldsymbol{w} \in V_n$, whence all the above mentioned conditions hold in W except (A4). But, since W is closed under scalar multiplication, $0\boldsymbol{u} = \mathbf{0} \in W$, whence (A4) also holds in W. □

Definition 3.4. ([5, p. 226] or [15]). *Let $B \subseteq V_n$ be a non-empty set of n-dimensional Boolean vectors.*

(a) *$\boldsymbol{v} \in V_n$ is said to be* **linearly dependent** *on B if $\boldsymbol{v} \in \langle B \rangle$, i.e. if $\boldsymbol{v}$ can be written as a sum $\boldsymbol{v} = \sum_i b_i \boldsymbol{w}_i$, $b_i \in \{0,1\}, \boldsymbol{w}_i \in B$.*

(b) *A set $B \subseteq V_n$ is said to be* **linearly dependent** *if there is a $\boldsymbol{v} \in B$ linearly dependent on B, or $\boldsymbol{0} \in B$. Otherwise, B is said to be* **linearly independent**; *i.e. if $\boldsymbol{0} \notin B \neq \emptyset$ and if no vector is the sum of some of the remaining vectors: $\boldsymbol{v} \notin \langle B - \{\boldsymbol{v}\} \rangle$, for each $\boldsymbol{v} \in B$.*

(c) *A* **basis** *of a subspace W of V_n is an linearly independent set B such that $W = \langle B \rangle$. (Note that the basis of $\boldsymbol{0}$ is $\emptyset$.) We also say that B generates W.*

(d) *Moreover, for $\boldsymbol{u}, \boldsymbol{v} \in V_n$ we define $\boldsymbol{u} \leq \boldsymbol{v}$ iff $u_i = 1$ implies $v_i = 1$, for each $1 \leq i \leq n$.*

The set of Boolean vectors $\{(111),(110),(011)\}$ is linearly dependent since $(111) = 1 \cdot (110) + 1 \cdot ((011)$, whereas $\{(100),(011),(110),(001)\}$ are independent. Various concepts of linear (in)dependence over more general structures, so-called *bottleneck algebras*, are considered in [9].

Corollary 3.5. ([18, Theorem 1.1.1], [5, Theorem 5.6.1]) *Let W be a subspace of V_n; then there exists a unique basis $B \subseteq V_n$ of W.*

Proof. Let $B = \{\boldsymbol{b}_j\}$ be the set of all vectors of W that cannot be expressed as sum of some other vectors in W. Then B is linearly independent. Suppose that $\langle B \rangle \subset W$ and that $\boldsymbol{v}$ is a vector with the least number of ones in $W - \langle B \rangle$. Since $\boldsymbol{v} \notin B$, it can be written as a linear combination of other vectors $\boldsymbol{w}_i$. But these vectors must be in $\langle B \rangle$ since $\boldsymbol{v}$ was the vector with minimal number of ones in $W - \langle B \rangle$. Thus $\boldsymbol{v} \in \langle B \rangle$, a contradiction. So $\langle B \rangle = W$ and B is a basis of W.

To prove uniqueness note that by independence $B \subseteq B'$ for every other basis B'. Assume $\boldsymbol{u}$ to be minimal in $B' - B$; then by the above $\boldsymbol{u}$ must be a sum of vectors in B, a contradiction. So $B = B'$. □

A vector space V is said to be of finite **dimension** n, written as $\dim V = n$, if and only if V has a basis $B = \{\boldsymbol{v}_1, \boldsymbol{v}_2, \ldots, \boldsymbol{v}_n\}$ consisting of n vectors. If $V = \{\boldsymbol{0}\}$, we say V has dimension 0.

We have not yet defined the notion of multiplication of two Boolean vectors and related concepts which we add in:

Definition 3.6.
Let $\boldsymbol{u} = (u_1, u_2, \ldots, u_n)$ and $\boldsymbol{v} = (v_1, v_2, \ldots, v_n)$ be two Boolean vectors.

(a) *The (***inner***)* **product** *of $\boldsymbol{u}$ and and $\boldsymbol{v}$ is given by $\boldsymbol{uv} := (u_1 v_1 + \ldots + u_n v_n)$.*

(b) *Two non-zero vectors $\boldsymbol{u}$ and $\boldsymbol{v}$ are called* **orthogonal** *if $\boldsymbol{uv} = 0$.*

(c) *A set of vectors $C = \{\boldsymbol{v}_1, \boldsymbol{v}_2, \ldots, \boldsymbol{v}_n\}$ in the Boolean vector space V_n is said to be* **orthogonal** *if $\boldsymbol{v}_i \boldsymbol{v}_j = 1$ for $i = j$, and $\boldsymbol{v}_i \boldsymbol{v}_j = 0$ for $i \neq j$, i.e. if $\boldsymbol{v}_i \boldsymbol{v}_j = \delta_{ij}$.*

(d) *A basis that is an orthogonal set is said to be an* **orthogonal basis**.

We will now have a short look at some properties that are well known from linear algebra and check if they are retained for Boolean vector spaces (see [25, Section 4.3]):

Proposition 3.7. *Let V be a vector space, $W = \{\boldsymbol{v}_1, \boldsymbol{v}_2, \ldots, \boldsymbol{v}_m\} \subseteq V$, and $c_i \in \{0,1\}$, for all $i \in \underline{m}$.*

(1) *If W is a set of orthogonal vectors and $\sum_{i\in\underline{m}} c_i \boldsymbol{v}_i = \boldsymbol{0}$, then $c_i = 0$, for all $i \in \underline{m}$.*
(2) *An orthogonal set of nonzero vectors in V_n is linearly independent.*
(3) *Any set of vectors $W \subseteq V$ that contains $\boldsymbol{0}$ is linearly dependent.*
(4) *Let $W = \{\boldsymbol{v}_1, \boldsymbol{v}_2, \ldots, \boldsymbol{v}_m\} \subseteq V$ be linearly dependent in the vector space V. Any set of vectors $W' \supset W$ is linearly dependent.*
(5) *If W is a linearly independent set then also any subset $W' \subseteq W$.*

Proof. (1): Multiplying $\sum_{i\in\underline{m}} c_i \boldsymbol{v}_i = \boldsymbol{0}$ with $\boldsymbol{v}_j$ gives

$$c_j = c_j \boldsymbol{v}_j \boldsymbol{v}_j = \sum_{i\in\underline{m}} c_i \boldsymbol{v}_i \boldsymbol{v}_j = (\sum_{i\in\underline{m}} c_i \boldsymbol{v}_i)\boldsymbol{v}_j = \boldsymbol{0}\boldsymbol{v}_j = 0.$$

This conclusion holds for all $j \in \underline{m}$.

(2): Assume that $\boldsymbol{v}_k$ is dependent on $W - \{\boldsymbol{v}_k\}$. Then $\boldsymbol{v}_k = \sum_{i\neq k} c_i \boldsymbol{v}_i$, whence $\boldsymbol{v}_k\boldsymbol{v}_k = 1 = \sum_{i\neq k} c_i \boldsymbol{v}_i \boldsymbol{v}_k = 0$, a contradiction. Since $\boldsymbol{v}_k$ has been chosen arbitrarily, this holds for all $\boldsymbol{v} \in W$.

(3): By Def. 3.4(b).

(4): Let $W = \{\boldsymbol{v}_1, \boldsymbol{v}_2, \ldots, \boldsymbol{v}_m\}$ be a set of linearly dependent vectors. Then, for some $k \in \underline{m}$, $\boldsymbol{v}_k$ can be written as a sum $\boldsymbol{v}_k = \sum_{k\neq i\in\underline{m}} c_i^v \boldsymbol{v}_i$. Assume that $W' = W \cup \{\boldsymbol{v}_{m+1}, \ldots, \boldsymbol{v}_n\}$. Then any $\boldsymbol{v}_k \in W$ we get by using the linear combination $\boldsymbol{v}_k = \sum_{k\neq i\in\underline{m}} c_i^v \boldsymbol{v}_i + \sum_{m+1\leq i\leq n} 0\boldsymbol{v}_i$.

(5): Assuming W' to be a set of linearly dependent vectors, (4) would imply that also W is linearly dependent, a contradiction. □

(Note that a set of orthogonal vectors $W = \{\boldsymbol{v}_1, \boldsymbol{v}_2, \ldots, \boldsymbol{v}_m\}$ is linearly independent in the usual sense of linear algebra: W is said to be *linearly independent* if $\sum_{i\in\underline{m}} c_i \boldsymbol{v}_i = \boldsymbol{0}$ can only be satisfied for $c_i = 0$, for all $i \in \underline{m}$.)

The row and column vectors of the identity matrix have a very special role to play in linear algebra. They are called the **standard unit vectors** and are denoted as follows: We let $I = [\boldsymbol{e}_1 \boldsymbol{e}_2 \ldots \boldsymbol{e}_m]$, where for $e_{ij} = \delta_{ij}$ for all $i, j \in \underline{m}$. Thus the column vector $\boldsymbol{e}_i$ has a single 1 at position i and zeros elsewhere. So the $\boldsymbol{e}_i$'s point in the direction of the axes in a Cartesian coordinate system. Since $I = I^T$ the same holds true for the row vectors. The standard unit vectors clearly form an orthogonal basis for V_n.

3.2 Basics of Boolean Matrix Algebra

Quite naturally, our first steps into matrix algebra are done by defining the most basic, but at the same time most important, operations that can be performed on matrices:

Definition 3.8. *Let $A = [a_{ij}], A' = [a'_{ij}] \in \mathbb{B}_{m\times n}, B = [b_{ij}] \in \mathbb{B}_{m\times p}$ and $C = [c_{ij}] \in \mathbb{B}_{p\times n}$ be Boolean matrices, and $\gamma \in \mathbf{2} := \{0,1\}$, called a* **scalar**.

(a) *The* **transpose** *of A is the matrix $A^T = [a_{ji}] \in \mathbb{B}_{n\times m}$ which is obtained by interchanging the rows and columns of A.*
(b) *The* **scalar multiplication** *of A with γ is the matrix $\gamma A = [\gamma \cdot a_{ij}] \in \mathbb{B}_{m\times n}$.*
(c) *The* **sum** *of A and A' is the matrix $A + A' = [a_{ij} + a'_{ij}] \in \mathbb{B}_{m\times n}$.*
(d) *The* **product** *of B and C, denoted by $B \cdot C$ (or just by juxtaposition BC), is the matrix*
$$B \cdot C = D = [d_{ij}] \in \mathbb{B}_{m\times n}, \quad \text{with} \quad d_{ij} = \sum_{k=1}^{p} b_{ik}c_{kj},$$
for $i \in \underline{m} = \{1, 2, \ldots, m\}$ and $j \in \underline{n}$.
(e) *The* **direct sum** *of $M \in \mathbb{B}_{m\times n}$ and $N \in \mathbb{B}_{p\times q}$ is the $(m+p)\times(n+q)$ matrix given by*
$$M \oplus N = \begin{bmatrix} M & O_{m,q} \\ O_{p,n} & N \end{bmatrix}.$$

Note: If you compare the definition of matrix product with Def. 2.12(c), you can see that $\vee$ has the role of Boolean addition, and $\wedge$ the one of multiplication. In general, matrix product depends on the operations defined in the underlying algebra from which the entries are taken.

Two matrices A and B are called **conformable** for multiplication in the order AB, if A has exactly as many columns as B has rows. Below it will be tacitly assumed that the required size restrictions are fulfilled for the matrices involved in various operations.

Obviously, an $m \times 1$ column vector $\boldsymbol{u}$ and a $1 \times n$ row vector $\boldsymbol{v}^T$ a conformable for multiplication, and their product $[\boldsymbol{uv}] := [u_{i1}]_{m,1} \cdot [v_{1j}]_{1,n}$ is called the **outer product**, yielding the $m \times n$ matrix $[u_i v_j]_{i\in\underline{m}, j\in\underline{n}}$. Obviously, the outer differs from the inner product: the latter gives a scalar value and is defined only for vectors of the same dimension; moreover $[\boldsymbol{uv}]$ is not symmetric (i.e. in general $[\boldsymbol{uv}] \neq [\boldsymbol{vu}]$), whereas $\boldsymbol{uv}$ is.

From the definition above we can see that the (i, j)-entry of AB is determined only by the i-th row of A and the j-th column of B. Sometimes it is convenient to break down matrix multiplication AB into multiplications of the matrix $A_{m\times n}$ and $n \times 1$ column vectors $\boldsymbol{b}$:

$$A\boldsymbol{b} = \begin{bmatrix} \sum_{j\in\underline{n}} a_{1j}b_j \\ \sum_{j\in\underline{n}} a_{2j}b_j \\ \vdots \\ \sum_{j\in\underline{n}} a_{mj}b_j \end{bmatrix} = b_1 \begin{bmatrix} a_{11} \\ a_{21} \\ \vdots \\ a_{m1} \end{bmatrix} + b_2 \begin{bmatrix} a_{12} \\ a_{22} \\ \vdots \\ a_{m2} \end{bmatrix} + \ldots b_n \begin{bmatrix} a_{1n} \\ a_{2n} \\ \vdots \\ a_{mn} \end{bmatrix} = \sum_{j\in\underline{n}} b_j \boldsymbol{a}_j. \tag{3.1}$$

So we get

$$AB = [A\boldsymbol{b}_{*1}\, A\boldsymbol{b}_{*2} \ldots A\boldsymbol{b}_{*n}] \tag{3.2}$$

$$= \begin{bmatrix} \sum_{j\in\underline{n}} a_{1j}b_{j1} & \ldots & \sum_{j\in\underline{n}} a_{1j}b_{jn} \\ \sum_{j\in\underline{n}} a_{2j}b_{j1} & \ldots & \sum_{j\in\underline{n}} a_{2j}b_{jn} \\ \vdots & \vdots & \vdots \\ \sum_{j\in\underline{n}} a_{mj}b_{j1} & \ldots & \sum_{j\in\underline{n}} a_{mj}b_{jn} \end{bmatrix} = \begin{bmatrix} \boldsymbol{a}_{1*}^T \boldsymbol{b}_{*1} & \ldots & \boldsymbol{a}_{1*}^T \boldsymbol{b}_{*n} \\ \boldsymbol{a}_{2*}^T \boldsymbol{b}_{*1} & \ldots & \boldsymbol{a}_{2*}^T \boldsymbol{b}_{*n} \\ \vdots & \vdots & \vdots \\ \boldsymbol{a}_{m*}^T \boldsymbol{b}_{*1} & \ldots & \boldsymbol{a}_{m*}^T \boldsymbol{b}_{*n} \end{bmatrix} \tag{3.3}$$

for two Boolean matrices $A \in \mathbb{B}_{m\times p}$ and $B \in \mathbb{B}_{p\times n}$. And noting that each matrix-vector product $A\boldsymbol{b}_{*j}$ is an $m \times 1$ column vector, we can see that AB has in fact dimension $m \times n$ as required.

The operations defined above satisfy some properties analogous to some we have already seen to hold for relations (see Section 2.1). Though many of the proofs for results presented below are straightforward and can easily be found in standard textbooks, we insert some sample proofs in the terminology of linear algebra and matrices to give a flavor of their techniques.

Corollary 3.9. *Let A, B and C be Boolean matrices conformable for the respective operations, and let $+$ and juxtaposition stand for the Boolean $(\vee, \wedge)$ operations. Furthermore, let $\gamma, \delta \in \mathbf{2}$. Then the following holds:*

(1) $(A^T)^T = A$.
(2) $(A+B)^T = A^T + B^T$.
(3) $(AB)^T = B^T A^T$ *(note that if A and B are conformable then B^T and A^T also are).*
(4) $A^T A$ *and* AA^T *are symmetric matrices (see* Def. 3.14(e)*).*
(5) $A+B = B+A$ *and* $(A+B)+C = A+(B+C)$.
(6) $(\gamma+\delta)A = \gamma A + \delta A$ *and* $\gamma(A+B) = \gamma A + \gamma B$.
(7) $(\gamma A + \delta B)^T = \gamma A^T + \delta B^T$, *in particular* $(\gamma A)^T = \gamma A^T$.
(8) $A(BC) = (AB)C$.
(9) $A(B+C) = AB + AC$, *and* $(B+C)A = BA + CA$.
(10) *Assume* $A \in \mathbb{B}_{m\times n}$*; then* $A \leq AA^T A$.

Proof. (2): A special case of (7) with $\gamma = \delta = 1$.

(3): Using mixed, but convenient and obvious notation we get

$$(B^T A^T)_{ij} = \sum_h (B^T)_{ih}(A^T)_{hj} = \sum_h b_{hi} a_{jh} = \sum_h a_{jh} b_{hi} = (AB)_{ji} = (AB)^T_{ij}.$$

(7): $(\gamma A + \delta B)^T = [\gamma a_{ij} + \delta b_{ij}]^T = [\gamma a_{ji} + \delta b_{ji}] = [\gamma a_{ji}] + [\delta b_{ji}] = \gamma A^T + \delta B^T$. The second assertion fololows easily with $\delta = 0$.

(8): From the considerations above we know that BC can be written as $BC = [B\boldsymbol{c}_{*1}\, B\boldsymbol{c}_{*2} \ldots B\boldsymbol{c}_{*n}]$. Then,

$$\begin{aligned} A(BC) &= A[B\boldsymbol{c}_{*1}\, B\boldsymbol{c}_{*2} \ldots B\boldsymbol{c}_{*n}] = [AB\boldsymbol{c}_{*1}\, AB\boldsymbol{c}_{*2} \ldots AB\boldsymbol{c}_{*n}] \\ &= (AB)[\boldsymbol{c}_{*1}\, \boldsymbol{c}_{*2} \ldots \boldsymbol{c}_{*n}] \quad = (AB)C. \end{aligned}$$

(10): If we set $B := AA^T A$, then we get $b_{ij} = \sum_{k=1}^n \sum_{l=1}^m a_{ik} a_{lk} a_{lk}$. We must show that $a_{ij} \leq b_{ij}$, for all i, j. This clearly holds if $a_{ij} = 0$. So assume that $a_{ij} = 1$. If we set $k = j$ and $l = i$, then $a_{ik} a_{lk} a_{lk} = a_{ij}^3 = a_{ij} = 1$. (Compare this argument with the one for Prop. 2.25(1)). □

It is important to have in mind that in general, even for square matrices, multiplication is not commutative, i.e. $AB \neq BA$, so $(A+B)^2 = A^2 + BA + AB + B^2$. Furthermore we let $A \le B :\leftrightarrow a_{ij} \le b_{ij}$ for all $i, j \in \underline{n}$. Then it is easy to show that $\mathbb{B}_n$ forms a partially ordered monoid (whence Th. 1.59 applies).

A matrix A is said to be **partitioned** if horizontal and vertical lines have been introduced subdividing it into submatrices, called **blocks**. Partitioning allows A to be written as a matrix whose entries are its blocks. As an example consider the permutation matrix

$$A = \begin{bmatrix} O & I_2 \\ I_2 & O \end{bmatrix} = \left[\begin{array}{cc|cc} 0 & 0 & 1 & 0 \\ 0 & 0 & 0 & 1 \\ \hline 1 & 0 & 0 & 0 \\ 0 & 1 & 0 & 0 \end{array}\right].$$

In properly partitioned matrices blocks that appear to the left or right of other blocks have the same number of rows, and blocks that appear above or below other blocks have the same number of columns.

If matrices are very large, as it is often the case in real-life applications of matrix theory, the computational overhead and the risk of accumulated numerical effects due to rounding and truncation may become substantial. Partitioning of matrices not only frequently reveals underlying structures, but it often speeds up computation, especially when matrices are being multiplied, and when they have many blocks of zeros.

To be familiar with some simple facts about operations with partitioned matrices is very useful. It turns out that the multiplication of partitioned matrices is just like ordinary matrix multiplication. To avoid cumbersome notation fin the general case, we restrict our presentation of standard matrix operations for block matrices having a few blocks only:

Corollary 3.10. *For scaling, sum, product, and transposition of block matrices the following holds:*

(1) $\lambda \begin{bmatrix} A_{11} & A_{12} \end{bmatrix} = \begin{bmatrix} \lambda A_{11} & \lambda A_{12} \end{bmatrix}.$

(2) $\begin{bmatrix} A_{11} & A_{12} \\ A_{21} & A_{22} \end{bmatrix} + \begin{bmatrix} B_{11} & B_{12} \\ B_{21} & B_{22} \end{bmatrix} = \begin{bmatrix} A_{11}+B_{11} & A_{12}+B_{12} \\ A_{21}+B_{21} & A_{22}+B_{22} \end{bmatrix}.$

(3) $\begin{bmatrix} A_{11} & A_{12} \\ A_{21} & A_{22} \\ A_{31} & A_{32} \end{bmatrix} \cdot \begin{bmatrix} B_{11} & B_{12} \\ B_{21} & B_{22} \end{bmatrix} = \begin{bmatrix} A_{11}B_{11}+A_{12}B_{21} & A_{11}B_{12}+A_{12}B_{22} \\ A_{21}B_{11}+A_{22}B_{21} & A_{21}B_{12}+A_{22}B_{22} \\ A_{31}B_{11}+A_{32}B_{21} & A_{31}B_{12}+A_{32}B_{22} \end{bmatrix}.$

(4) $\begin{bmatrix} A_{11} & A_{12} \\ A_{21} & A_{22} \\ A_{31} & A_{32} \end{bmatrix}^T = \begin{bmatrix} A_{11}^T & A_{21}^T & A_{31}^T \\ A_{12}^T & A_{22}^T & A_{32}^T \end{bmatrix}.$

This gives us the following rules:

(1) To multiply partitioned matrices, treat its blocks as though they were elements of a matrix, and multiply the corresponding submatrices. Thus if A is an $(m+n) \times (p+q)$ matrix and B a $(p+q) \times (r+s)$ matrix such that A_{11} is of order $m \times p$

and B_{11} is of order $p \times r$, then all submatrices are well-defined and AB is an $(m+n) \times (r+s)$ partitioned matrix,

(2) To transpose a partitioned matrix, treat its submatrices as though they were elements of a matrix, and transpose each submatrix. Thus the transpose of an $m \times n$ partitioned matrix results in an $n \times m$ partitioned form.

Definition 3.11. *Let $A = [a_{ij}] \in \mathbb{B}_n$, and $i, j \in \underline{n} := \{1, 2, \ldots, n\}$. A is called*

(a) **upper triangular** *if $a_{ij} = 0$ for all $i > j$, i.e. if all its elements below the diagonal are zero.*

(b) **lower triangular** *if A^T is upper triangular, i.e. if $a_{ij} = 0$ for all $i < j$.*

(c) **triangular** *if A is upper or lower triangular.*

(d) **strictly triangular** *if A is triangular and $a_{ii} = 0$ for all $i \in \underline{n}$.*

(e) **block diagonal** *if it is partitioned into submatrices $A_{ij} \in \mathbb{B}_{n_i \times n_j}$ such that $A = [A_{ij}]$, $\sum_{i=1}^{k} n_i = n$, and $A_{ij} = O$, for all $i \neq j$.*

(f) **block upper triangular** *f it is partitioned into submatrices $A_{ij} \in \mathbb{B}_{n_i \times n_j}$ such that $A = [A_{ij}]$, $\sum_{i=1}^{k} n_i = n$, and $A_{ij} = O$, for all $i > j, 1 \leq i, j \leq k$.*

(g) **block lower triangular** *if A^T is block upper triangular.*

Thus, a square matrix A is lower triangular if all its entries above the main diagonal equal zero, and block-diagonal if it takes the form

$$\begin{bmatrix} A_{11} & O & \cdots & O \\ O & A_{22} & \cdots & O \\ \vdots & & \ddots & \vdots \\ O & O & \cdots & A_{kk} \end{bmatrix}.$$

From their structures it is obvious that e.g. the sum of two upper(lower)-triangular matrices is again upper(lower)-triangular. An analogous result holds for products of upper(lower)-triangular matrices:

Proposition 3.12. ([4, p. 23, Theorem 1.8]) *If two square matrices A and B of order n are upper(lower)-triangular, then also their product $C = AB$.*

Proof. The (i, j)-th element of C is given by

$$c_{ij} = \sum_{k=1}^{n} a_{ik} b_{kj} = \sum_{k=1}^{i} a_{ik} b_{kj} + \sum_{k=i+1}^{n} a_{ik} b_{kj}.$$

Suppose $i < j$. Then, since B is assumed lower-triangular, $b_{kj} = 0$ for each k in the second sum, because $k \leq i < j$. For the third sum note that $a_{ik} = 0$ for all $k > i$ since also A is lower-triangular, whence both sums give zero. Therefore, $c_{ij} = 0$ for all $i < j$. □

Proposition 3.13. ([4, p. 23, Lemma 1.2]) *A strictly triangular square matrix $A \in \mathbb{B}_n$ is nilpotent with $A^n = O$.*

Proof. We prove the assertion for upper-triangular matrices using induction. The case $n = 1$ is obvious: $[0] = [0]^1$. Now let $n > 1$, and assume that the assertion holds for

$(n-1) \times (n-1)$ strictly triangular matrices, and assume B to be such a matrix. Then consider the an $n \times n$ upper-triangular matrix

$$A = \left[\begin{array}{c|c} 0 & u_{1,n-1} \\ \hline \mathbf{0}_{n-1,1} & B \end{array}\right].$$

A clearly is strictly triangular since B is. Then by straightforward calculation using Cor. 3.10(3) we get (note that 0 is of course a shorthand for $\mathbf{0}_{1,1}$)

$$A^2 = \left[\begin{array}{c|c} 0\cdot 0 + u_{1,n-1}\mathbf{0}_{n-1,1} & 0\cdot u_{1,n-1} + u_{1,n-1}B \\ \hline \mathbf{0}_{n-1,1}\cdot 0 + B\mathbf{0}_{n-1,1} & \mathbf{0}_{n-1,1}u_{1,n-1} + BB \end{array}\right] = \left[\begin{array}{c|c} 0 & u_{1,n-1}B \\ \hline \mathbf{0}_{n-1,1} & B^2 \end{array}\right],$$

and finally

$$A^n = \left[\begin{array}{c|c} 0 & u_{1,n-1}B^{n-1} \\ \hline \mathbf{0}_{n-1,1} & BB^{n-1} \end{array}\right] = \left[\begin{array}{c|c} 0 & u_{1,n-1}\mathbf{0}_{n-1,n-1} \\ \hline \mathbf{0}_{n-1,1} & B\mathbf{0}_{n-1,n-1} \end{array}\right] = \left[\begin{array}{c|c} 0 & \mathbf{0}_{1,n-1} \\ \hline \mathbf{0}_{n-1,1} & \mathbf{0}_{n-1,n-1} \end{array}\right].$$

□

3.3 Some Basic Kinds of Matrices

Beside the properties mentioned above in Def. 3.11, there are the following which sometimes are considered in various aspects of investigations; see e.g. [3, 17].

Definition 3.14. *Let $A = [a_{ij}] \in \mathbb{B}_n$, and $\underline{n} := \{1,2,\ldots,n\}$. A is called*

(a) **primitive** *if $A^k = J$, for some positive integer k.*
(b) **nilpotent** *if $A^k = O$, for some positive integer k.*
(c) **involutory** *if $A^{-1} = A$.*
(d) **orthogonal** *if $A^{-1} = A^T$.*
(e) **symmetric** *if $A = A^T$.*
(f) **diagonal** *if $a_{ij} = 0$ whenever $i \neq j$.*
(g) **scalar** *if it is diagonal and if $a_{ii} = c$ for all $i \in \underline{n}$.*
(h) **cross diagonal** *if all the entries are zero except those on the diagonal going from the lower left corner to the upper right corner, forming the* **cross diagonal.**
(i) the **backward identity** *(or* **exchange matrix***), denoted as $\mathbf{E}_n$ if the only 1's are on the cross diagonal, i.e. $e_{i,n-i+1} = 1$, and $e_{ij} = 0$ otherwise, for all $i, j \in \underline{n}$.*

From the many results which have been obtained for matrices of the various kinds defined above during the last few decades, only some general ones can be presented in the remainder of this section - just enough to give a flavor of the investigation efforts. Most of these results are immediate consequences from the definitions, even if some of their proofs need straightforward calculations. In order to avoid a 'never ending story' we have to refrain from reporting about applications.

Corollary 3.15. *For the exchange matrix $\mathbf{E}_n$ the following holds:*

(1) $\mathbf{E}_n$ is a special form of both a permutation and a cross diagonal matrix (in the Boolean case the only one).
(2) $\mathbf{E}_n = [\delta_{i,n-i+1}]_{ij}$.
(3) The rows of $\mathbf{E}_n A$ are the rows of A in reverse order; the columns of $A\mathbf{E}_n$ are the columns of A in reverse order.
(4) $\mathbf{E}_n = \mathbf{E}_n{}^T$; $\mathbf{E}_n{}^k = I$ for even k, and $\mathbf{E}_n{}^k = \mathbf{E}_n$ for odd k. So $\mathbf{E}_n$ is involutory, i.e. $\mathbf{E}_n{}^{-1} = \mathbf{E}_n$.

Corollary 3.16. *Let A be an involutory matrix. Then the following holds:*

(1) A^T is involutory.
(2) A^m, $m \in \mathbb{N}$, is involutory.
(3) If B is another involutory matrix which commutes with A, i.e. $AB = BA$, then their product is also involutory.
(4) iI A is both symmetric and orthogonal, then A is involutory.
(5) If A is involutory then it is symmetric iff it is orthogonal.
Equivalently: An involutory matrix which is also symmetric is an orthogonal matrix, and conversely every orthogonal involutory matrix is symmetric.

Proof. (1): $A^2 = I$ implies $(AA)^T = A^T A^T = I^T = I$, using Cor. 3.9(3).

(3): $ABAB = ABBA = ABB^{\ 1}A^{\ 1} = I$.

(4): By assumption we get $A^T = A = A^{-1}$; so clearly $A^2 = AA^{-1} = I$.

(5): Assume A to be involutory, so $A = A^{-1}$. If A is also symmetric we get $A^T = A = A^{-1}$, whence A is orthogonal. And if A is involutory and orthogonal then again $A^{-1} = A = A^T$, whence A is symmetric.

So we have shown (with obvious abbreviations) that $(I \wedge S \to O) \wedge (I \wedge O \to S)$. Using Th. 1.17(12),(14) we get

$$(I \to (S \to O)) \wedge (I \to (O \to S)), \text{ so } I \to (S \to O) \wedge (O \to S)$$

which shows the asserted equivalence. □

(Note that items (4): $S \wedge O \to I$, and (5): $I \to (S \leftrightarrow O)$, of the above lemma are different from the corollary given on ([20, p. 386]) which asserts $I \leftrightarrow S \wedge O$.)

We have met orthogonal relations already earlier, and have shown the following results in Prop. 2.82, which we partly repeat below in terms of matrices. Recall that according to Def. 2.80 a Boolean matrix A is **row** (**column**) **consistent** if $AU = U$ ($UA = U$).

Proposition 3.17. ([20,24]) *For a finite square Boolean matrix $A \in \mathbb{B}_n$ the following holds:*

(1) A is orthogonal iff it is both row and column consistent, and

$$\forall i,j,k [i \neq j \to a_{ik} \wedge a_{jk} = 0 = a_{ki} \wedge a_{kj}].$$

(2) A has an inverse iff it is orthogonal.

We continue with some easy to prove properties of square Boolean matrices:

Corollary 3.18. *Let $A = [a_{ij}] \in \mathbb{B}_n$; then the following holds:*

(1) *A is involutory iff $A^2 = I$.*
(2) *A is orthogonal iff $AA^T = A^T A = I$.*
(3) *If A is orthogonal, then its columns and rows are orthogonal unit vectors.*

Proof. (1): Let A be involutory; then $A = A^{-1}$, so $A^2 = I$. Conversely, let $A^2 = I$; then A is both a left and a right inverse of itself, and by Prop. 1.29 it is the unique inverse for A in $\mathbb{B}_n$.

(2): Similar to (1).

(3): By (2) we have $AA^T = A^T A = I$. By (a non-circular) anticipation of results shown later, from Cor. 3.35 we get that A is a permutation matrix, so its columns and rows are unit vectors by definition. Orthogonality follows from similar arguments as used in the proof of the Prop. 3.34. □

3.3.1 *(Bi-/Centro-/Per-) Symmetric Matrices*

Next we turn to the various kinds of symmetry a matrix can exhibit, and report some of their properties and relationships. See e.g. [23] or [17, Section 0.9.10].

Definition 3.19. *Let $A = [a_{ij}] \in \mathbb{B}_n$. A is called*

(a) **persymmetric** *if it is symmetric with respect to its cross diagonal.*
(b) **bisymmetric** *if it is symmetric and persymmetric.*
(c) **centrosymmetric** *if it is symmetric about its center.*

In Cor. 3.15(3) it has been stated that in $\boldsymbol{E}_n A$ the rows and in $A\boldsymbol{E}_n$ the columns of A are in reversed order. So, an element a_{ij} changes its position as follows:

$$(i,j) \overset{A\boldsymbol{E}_n}{\rightsquigarrow} (i, n-j+1) \overset{\boldsymbol{E}_n(A\boldsymbol{E}_n)}{\rightsquigarrow} (n-i+1, n-j+1), \text{ and}$$

$$(i,j) \overset{A^T}{\rightsquigarrow} (j,i) \overset{A^T\boldsymbol{E}_n}{\rightsquigarrow} (j, n-i+1) \overset{\boldsymbol{E}_n(A^T\boldsymbol{E}_n)}{\rightsquigarrow} (n-j+1, n-i+1).$$

It has been shown that there are equivalent characterizations of the kinds of symmetric matrices under consideration; see e.g. [23, p. 241]. As it is the case for relations, we can avoid the cumbersome and error-prone indexed-element descriptions, and instead use point-free algebraic characterizations given in the table below.

A is	iff, for all $i, j \in \underline{n}$	iff	iff
symmetric	$a_{ij} = a_{ji}$	$A = A^T$	
persymmetric	$a_{ij} = a_{n-j+1,n-i+1}$	$A = \boldsymbol{E}_n A^T \boldsymbol{E}_n$	$A\boldsymbol{E}_n = \boldsymbol{E}_n A^T$
bisymmetric	$a_{ij} = a_{ji} = a_{n-j+1,n-i+1}$	$A = \boldsymbol{E}_n A^T \boldsymbol{E}_n = A^T$	
centrosymmetric	$a_{ij} = a_{n-i+1,n-j+1}$	$A = \boldsymbol{E}_n A \boldsymbol{E}_n$	$A\boldsymbol{E}_n = \boldsymbol{E}_n A$

It is well known that the above kinds of symmetries do not occur isolated: Obviously, if a symmetric matrix A is additionally per- / centrosymmetric, then it is also centro- / persymmetric; and if A is per- and centrosymmetric then also symmetric: By assumption we have $\boldsymbol{E}_n A \boldsymbol{E}_n = \boldsymbol{E}_n A^T \boldsymbol{E}_n$. Multiplying this equation from both sides with $\boldsymbol{E}_n$ we get $A = A^T$ since $\boldsymbol{E}_n$ is involutory. In other words, if a matrix has two out of the three properties symmetric, persymmetric, centrosymmetric, then it has also the third one. So bisymmetric matrices are centrosymmetric, and a symmetric per- / centrosymmetric matrix is bisymmetric.

Corollary 3.20. *For two centrosymmetric matrices $A, B \in \mathbb{B}_n$ the following holds:*

(1) Both product and sum of two centrosymmetric matrices give a centrosymmetric matrix, as does scalar multiplication.
(2) If A is centrosymmetric and invertible, then A^{-1} is centrosymmetric.
(3) The product of two bisymmetric matrices gives a centrosymmetric matrix.

Proof. (1): $A+B = \boldsymbol{E}_n A \boldsymbol{E}_n + \boldsymbol{E}_n B \boldsymbol{E}_n = \boldsymbol{E}_n (A+B) \boldsymbol{E}_n$, using Cor. 3.9(2),(6). Clearly, $0 \cdot A = [0 \cdot a_{ij}] = O$ and $1 \cdot A = [1 \cdot a_{ij}] = A$, and O is centrosymmetric.
(2): $\boldsymbol{E}_n A^{-1} = (A\boldsymbol{E}_n)^{-1} = (\boldsymbol{E}_n A)^{-1} = A^{-1}\boldsymbol{E}_n$. Using Cor. 3.15(4), we get the first equality from $(A\boldsymbol{E}_n)(\boldsymbol{E}_n A^{-1}) = AIA^{-1} = I$ and $(\boldsymbol{E}_n A^{-1})(A\boldsymbol{E}_n) = \boldsymbol{E}_n I \boldsymbol{E}_n = I$. Similar for the third equality.
(3): $AB = \boldsymbol{E}_n A^T \boldsymbol{E}_n \boldsymbol{E}_n B^T \boldsymbol{E}_n = \boldsymbol{E}_n A^T B^T \boldsymbol{E}_n = \boldsymbol{E}_n AB \boldsymbol{E}_n$. □

Since the identity matrix is also centrosymmetric, it follows that the centrosymmetric matrices form a semilinear space which is a subspace of the semilinear space of all $A \in \mathbb{B}_n$ (see Def. 5.18). Centrosymmetric matrices have a special structure:

Proposition 3.21. ([10, Lemma 1]) *A centrosymmetric matrix $M \in \mathbb{B}_n$ can be partitioned as follows:*

(1) $n = 2m+1$ (odd order):

$$M = \left[\begin{array}{c|c|c} A & \boldsymbol{s} & B \\ \hline \boldsymbol{t}^T & b & \boldsymbol{t}^T \boldsymbol{E}_m \\ \hline \boldsymbol{E}_m B \boldsymbol{E}_m & \boldsymbol{E}_m \boldsymbol{s} & \boldsymbol{E}_m A \boldsymbol{E}_m \end{array}\right] \tag{3.4}$$

(2) $n = 2m$ (even order):

$$M = \left[\begin{array}{c|c} A & B \\ \hline \boldsymbol{E}_m B \boldsymbol{E}_m & \boldsymbol{E}_m A \boldsymbol{E}_m \end{array}\right] \tag{3.5}$$

where $b \in \mathbf{2} = \{0,1\}$ is a scalar, A and B are square matrices of order m, and $\boldsymbol{s}$ and $\boldsymbol{t}$ are vectors in $\mathbf{2}^m$.

Proof. (1): Let M be partitioned as follows:

$$M = \left[\begin{array}{c|c|c} A & \boldsymbol{s} & B \\ \hline \boldsymbol{t}^T & b & \boldsymbol{v}^T \\ \hline C & \boldsymbol{u} & D \end{array}\right]. \tag{3.6}$$

Noting that

$$\boldsymbol{E}_n = \left[\begin{array}{c|c|c} \boldsymbol{0}_{m,m} & \boldsymbol{0}_{m,1} & \boldsymbol{E}_m \\ \hline \boldsymbol{0}_{1,m} & 1 & \boldsymbol{0}_{1,m} \\ \hline \boldsymbol{E}_m & \boldsymbol{0}_{m,1} & \boldsymbol{0}_{m,m} \end{array}\right]$$

and that $M \in \mathbb{B}_n$ is centrosymmetric, using Cor. 3.10(3) we get

$$M = \boldsymbol{E}_n B \boldsymbol{E}_n = \left[\begin{array}{c|c|c} \boldsymbol{E}_m D \boldsymbol{E}_m & \boldsymbol{E}_m \boldsymbol{u} & \boldsymbol{E}_m C \boldsymbol{E}_m \\ \hline \boldsymbol{v}^T \boldsymbol{E}_m & b & \boldsymbol{t}^T \boldsymbol{E}_m \\ \hline \boldsymbol{E}_m B \boldsymbol{E}_m & \boldsymbol{E}_m \boldsymbol{s} & \boldsymbol{E}_m A \boldsymbol{E}_m \end{array}\right]. \tag{3.7}$$

Comparing the two matrices (3.6) and (3.7) gives the following correspondences: $D = \boldsymbol{E}_m A \boldsymbol{E}_m$, $C = \boldsymbol{E}_m B \boldsymbol{E}_m$, $\boldsymbol{u} = \boldsymbol{E}_m \boldsymbol{s}$, and $\boldsymbol{t}^T = \boldsymbol{v}^T \boldsymbol{E}_m$. But $\boldsymbol{E}_m$ is involutory, so we have $\boldsymbol{E}_m D \boldsymbol{E}_m = \boldsymbol{E}_m \boldsymbol{E}_m A \boldsymbol{E}_m \boldsymbol{E}_m = A$ etc., and finally obtain the partitioning of (3.4).

(2): Analogous to (1); an explicit proof is given in [8, Lemma 2]. □

Corollary 3.22. *For $A \in \mathbb{B}_n$ the following holds:*

(1) The sum of two persymmetric matrices A, B , and also the scalar product give persymmetric matrices.

(2) The product of two persymmetric matrices A, B is persymmetric if A and B commute.

(3) If A is persymmetric and invertible, then A^{-1} is persymmetric.

Proof. (2): Since $\boldsymbol{E}_n$ is involutory by Cor. 3.15(4) we can conclude as follows:
$AB = \boldsymbol{E}_n A^T \boldsymbol{E}_n \boldsymbol{E}_n B^T \boldsymbol{E}_n = \boldsymbol{E}_n A^T B^T \boldsymbol{E}_n = \boldsymbol{E}_n (BA)^T \boldsymbol{E}_n = \boldsymbol{E}_n (AB)^T \boldsymbol{E}_n$.

(3): $AA^{-1} = I$ implies $(AA^{-1})^T = (A^{-1})^T A^T = I$. Since $\boldsymbol{E}_n$ is involutory we get

$$I = \boldsymbol{E}_n \boldsymbol{E}_n = \boldsymbol{E}_n[(A^{-1})^T A^T]\boldsymbol{E}_n = \boldsymbol{E}_n[(A^{-1})^T \boldsymbol{E}_n \boldsymbol{E}_n A^T]\boldsymbol{E}_n = (\boldsymbol{E}_n (A^{-1})^T \boldsymbol{E}_n)(\boldsymbol{E}_n A^T \boldsymbol{E}_n).$$

But by assumption $\boldsymbol{E}_n A^T \boldsymbol{E}_n = A$; and since inverses are unique, it follows that $\boldsymbol{E}_n (A^{-1})^T \boldsymbol{E}_n = A^{-1}$. □

Since the zero matrix is persymmetric, it follows that the persymmetric matrices form a semilinear subspace of the semilinear space of all $A \in \mathbb{B}_n$. Note that bisymmetric matrices are persymmetric by definition, so Cor. 3.22 applies to them.

3.3.2 *Permutation Matrices*

Square matrices $P = [p_{ij}]_m$ with (at most) exactly one 1 in each row and column are called **(partial) permutation matrices.** We have already met two special examples of permutation matrices, viz. the identity and exchange matrix, and also have had a look at some of their properties. Alternatively we can define a permutation matrix as a matrix obtained by permuting the columns or rows of the $m \times m$ identity matrix. This section adds a few more results.

Let $\mathcal{S}_m$ denote the group of permutations on $\underline{m}$. The permutation matrix corresponding to $\pi = (l_1, l_2, \ldots, l_m) \in \mathcal{S}_m$ is given by (recall from page 95 that standard unit vectors $\boldsymbol{e}_i$ have been defined as column vectors)

$$P_\pi = \begin{bmatrix} \boldsymbol{e}_{l_1}^T \\ \boldsymbol{e}_{l_2}^T \\ \vdots \\ \boldsymbol{e}_{l_m}^T \end{bmatrix} = [\delta_{\pi(i)j}]_{i,j \in \underline{m}} = [\delta_{i\pi^{-1}(j)}]_{i,j \in \underline{m}}$$

with δ_{ij} the Kronecker delta. So we get $P_\pi = [p_{ij}]_m$ with $p_{ij} = 1$, if $j = l_i$, and $p_{ij} = 0$ otherwise. For an $m \times n$ matrix A, $P_\pi A$ is obtained by permuting the rows of A so that in $P_\pi A$ they appear in the order $l_1, l_2, \ldots, l_m$:

$$\text{If } A = \begin{bmatrix} \boldsymbol{r}_1 \\ \boldsymbol{r}_2 \\ \vdots \\ \boldsymbol{r}_m \end{bmatrix} \quad \text{then} \quad P_\pi A = \begin{bmatrix} \boldsymbol{r}_{l_1} \\ \boldsymbol{r}_{l_2} \\ \vdots \\ \boldsymbol{r}_{l_m} \end{bmatrix}.$$

In other words, $P_\pi A = [a_{\pi(i)j}]_{i \in \underline{m}, j \in \underline{n}}$. Similarly, for a $n \times m$ matrix B, we get $BP_\pi = [b_{i\pi^{-1}(j)}]_{i \in \underline{n}, j \in \underline{m}}$.

The definition of a permutation matrix treats rows and columns in the same way. So, to permute the columns of the matrix $A_{m,n}$ such that they appear in the order $l_1, l_2, \ldots, l_n$, one has to multiply it from the right by the permutation matrix Q^T, where Q corresponds to the permutation $(l_1, l_2, \ldots, l_n)$ of $(1, 2, \ldots, n)$. If A is a square matrix of order n, then the matrix QAQ^T is obtained from A by putting its rows in the order $l_1, l_2, \ldots, l_n$, and then by putting the columns in the order $l_1, l_2, \ldots, l_n$. We say that the matrix QAQ^T is obtained from A by *simultaneous permutations* of its rows and columns.

Note that the set $\boldsymbol{P}_n \subseteq \mathbb{B}_n$ of all permutation matrices forms a subgroup of the semigroup $\mathbb{B}_n$.

Definition 3.23. *Two matrices A and B are called* **permutation equivalent** *if there exist $P, Q \in \boldsymbol{P}_n$ such that $A = PBQ$; and* **permutation similar** *if for some $P \in \boldsymbol{P}_n$ $A = P^{-1}BP$.*

In Cor. 3.15(4) we have mentioned some properties of the exchange matrix: $\boldsymbol{E}_n = \boldsymbol{E}_n{}^T = \boldsymbol{E}_n{}^{-1}$; $\boldsymbol{E}_n{}^k = I$ for even k, and $\boldsymbol{E}_n{}^k = \boldsymbol{E}_n$ for odd k. We will now turn to analog properties of general permutation matrices in

Corollary 3.24. *Let $P = [p_{ij}], Q = [q_{ij}] \in \boldsymbol{P}_n$ be permutation matrices; then the following holds:*

(1) P^T is a permutation matrix.

(2) $PP^T = I = P^T P$, i.e. $P^{-1} = P^T$.

(3) *P is orthogonal.*
(4) *PQ is a permutation matrix.*
(5) *P^k is a permutation matrix for each $k \in \mathbb{N}$.*
(6) *$P^k = I$ for some $k \in \mathbb{N}$.*

Proof. (2): Using (3.3) we get

$$PP^T = \begin{bmatrix} e_{l_1}^T \\ e_{l_2}^T \\ \vdots \\ e_{l_m}^T \end{bmatrix} \begin{bmatrix} e_{l_1} \ e_{l_2} \ \dots \ e_{l_n} \end{bmatrix} = \begin{bmatrix} e_{l_1}^T e_{l_1} & e_{l_1}^T e_{l_2} & \dots & e_{l_1}^T e_{l_n} \\ e_{l_2}^T e_{l_1} & e_{l_2}^T e_{l_2} & \dots & e_{l_2}^T e_{l_n} \\ \vdots & \vdots & \ddots & \vdots \\ e_{l_n}^T e_{l_1} & e_{l_n}^T e_{l_2} & \dots & e_{l_n}^T e_{l_n} \end{bmatrix}.$$

Since $e_i^T e_j = \delta_{ij}$, the first identity follows.

(3): The column vectors are orthogonal because the only 1 appears in a different place in each column.

(4): Postmultiplication by Q permutes the columns of P, and since the columns of P are a permutation of the identity matrix, this also holds for PQ.

(5): Immediate from (4).

(6): Recall from Section 1.4.2 that $\langle P\rangle = \{P, P^2, P^3, \dots\}$ is the cyclic subsemigroup of $\mathbb{B}_n$ generated by P. And since there are only finitely many permutation matrices in $\mathbb{B}_n$, $\langle P\rangle$ is finite, so Th. 1.59(2)(d) applies: There exist the index k and the period p of P such that $K_P := \{P^k, \dots, P^{k+p-1}\}$ is a cyclic subgroup of $\langle P\rangle$ of order p, i.e. $I \in K_P$. □

3.3.3 *Irreducible and Fully Indecomposable Matrices*

Definition 3.25. ([13]) *Let $A \in \mathbb{B}_n$, $n \geq 2$, and $1 \leq r < n$). A is called*

(a) **irreducible** *(or* **indecomposable***) if no permutation matrix P exists such that*

$$PAP^T = \begin{bmatrix} A_{11} & B \\ O_{n-r,r} & A_{22} \end{bmatrix}, \tag{3.8}$$

where A_{11} and A_{22} are nonempty square matrices (with size $r \times r$ and $(n-r) \times (n-r)$, respectively).

(b) **fully indecomposable** *if no permutation matrices P and Q exist such that*

$$PAQ = \begin{bmatrix} A_{11} & B \\ O_{n-r,r} & A_{22} \end{bmatrix}, \tag{3.9}$$

where A_{11} and A_{22} are nonempty square matrices.

By convention, both of the 1×1 binary matrices are said to be irreducible. If A is not irreducible it is called **reducible** (or **decomposable**), and if A is not fully indecomposable it is called **partly decomposable.** Obviously, every fully indecomposable matrix is indecomposable, and every decomposable matrix is partly decomposable.

The definitions given above can be restated in terms of indices as follows: A square $n \times n$ matrix $A = [a_{ij}]$ is *fully indecomposable* if and only if there are no proper subsets

$K, L \subseteq \underline{n}$ with $|K| + |L| = n$ such that $a_{kl} = 0$ for $k \in K$ and $l \in L$. If additionally $K \cap L = \emptyset$ then we get an alternative definition for A being *indecomposable.*

The diagonal entries of an irreducible permutation matrix are all equal to 0. The converse is not true , as can be seen from the following example from [26, p. 155]:

$$\begin{bmatrix} 0 & 1 & 0 & 0 \\ 1 & 0 & 0 & 0 \\ 0 & 0 & 0 & 1 \\ 0 & 0 & 1 & 0 \end{bmatrix}.$$

If A is reducible and in the form (3.8), and if a diagonal block is also reducible, then this block can be reduced further; if this process is continued then finally there exists a suitable permutation matrix P such that PAP^T is block upper triangular

$$PAP^T = \begin{bmatrix} A_{11} & A_{12} & \cdots & A_{1k} \\ O & A_{22} & \cdots & A_{2k} \\ \vdots & & \ddots & \vdots \\ O & O & \cdots & A_{kk} \end{bmatrix}, \tag{3.10}$$

and each block A_{ii} is a square and irreducible submatrix. (3.10) is called the **Frobenius normal form** of A. For $A \in \mathbb{B}_n$, there is always a permutation matrix P such that PAP^T is block upper-triangular with irreducible diagonal blocks. See e.g. [14, p. 106] and [7, Theorem 3.2.4].

Before you proceed you may possibly need a short look at Def. 13.13 and its context.

Corollary 3.26. ([6, Theorem 1.2.3(i)]) *Let $G = (V, E)$ be a digraph. Then G is strongly connected iff there is no partition $\{U, W\}$ of V such that all edges between U and W have their initial vertex in U and their target vertex in W.*

Proposition 3.27. ([7, Theorem 3.2.1]) *$A \in \mathbb{B}_n$ is irreducible if and only if its digraph $D := G_A$ is strongly connected.*

Proof. If A is reducible, then, by Cor. 3.26, the vertex set V of D can be partitioned into two nonempty sets U and W such that there is no arc from a $u \in U$ to a $w \in W$. So if $a \in U$ and $b \in W$ there is no path from a to b, whence D is not strongly connected.

Conversely assume that D is not strongly connected. Then there are $a, b \in V, a \neq b$, for which there is no path from a to b. Let $W_1 := \{b\} \cup \{$all vertices from which there is a directed walk to $b\}$ and let $W_2 := \{a\} \cup \{$all vertices to which there is a directed walk from $a\}$. Then $W_1 \cap W_2 = \emptyset$ and $W_1, W_2 \neq \emptyset$. Furthermore we set $W_3 := V - (W_1 \cup W_2)$. If we now simultaneously permute the lines of A so that the lines are ordered according to the vertices in W_2 followed by those in W_3 and W_1 we get the following matrix:

$$\begin{array}{c} \\ W_2 \\ W_3 \\ W_1 \end{array} \begin{array}{c} \begin{array}{ccc} W_2 & W_3 & W_1 \end{array} \\ \left[\begin{array}{cc|c} X_{11} & X_{12} & O \\ X_{21} & X_{22} & O \\ \hline X_{31} & X_{32} & X_{33} \end{array}\right] \end{array}.$$

Since there is no directed path from a to b there is no arc from a vertex in W_2 to a vertex in W_1, whence $X_{13} = O$. There is no arc from a vertex c in W_3 to a vertex in W_1, because such an arc implies that c belongs to W_1; therefore, we also have $X_{23} = O$. And since the blocks on the diagonal are square submatrices, A is reducible. □

If A is irreducible then for each pair of distinct nodes a and b in D there is a path from a to b of length at most equal to $n-1$. Thus it follows from Prop. 3.27 that A is irreducible if and only if $(I+A)^{n-1} = J$.

3.3.4 *Prime Boolean Matrices*

Just to let them be mentioned we insert this very short section on prime Boolean matrices. More details and sufficient conditions and characterization theorems can be found in [11].

Definition 3.28. ([11,21]) *A matrix $A \in \mathbb{B}_n$ is called*

(a) **prime**, *if it is not a permutation matrix, and whenever $A = BC$, with $B, C \in \mathbb{B}_n$, then either B or C is a permutation matrix.*
(b) **factorizable**, *if it is neither prime nor invertible.*

The permutation matrices are the units in $\mathbb{B}_n$, i.e. the matrices $A \in \mathbb{B}_n$ such that for some $B \in \mathbb{B}_n$, $AB = BA = I_n$. So we get: A is prime if it is not a unit and $A = BC$ implies that B or C is a unit, which is is in accordance with how prime elements in monoids have been defined in Def. 1.57(c).

Lemma 3.29. *If $A \in \mathbb{B}_n$ is a prime matrix then there are no two columns $\boldsymbol{a}_{*j}$ and $\boldsymbol{a}_{*k}$, $j \neq k$, such that $\boldsymbol{a}_{*j} \leq \boldsymbol{a}_{*k}$.*

Proof. By reordering the columns of A, we may assume without loss of generality that $\boldsymbol{a}_{*1} \geq \boldsymbol{a}_{*2}$. Then $A \notin \boldsymbol{P}_n$, so it possibly is prime. But we have

$$A = A \cdot \left(\begin{bmatrix} 1 & 0 \\ 1 & 1 \end{bmatrix} \oplus I_{n-2} \right)$$

with none of the two factors being a permutation matrix. So, A cannot be prime. □

Similarly, if $A \in \mathbb{B}_n$ is prime no row can contain another. In particular, A does not have zero rows or columns, nor a row or column with all entries equal to 1.

As shown in [21], there are no primes in $\mathbb{B}_1$ and $\mathbb{B}_2$: $[1]_{1,1}$ clearly is a permutation matrix contrary to $[0]_{1,1}$. But $[0]_{1,1}$ is idempotent, thus not prime. If $A \in \mathbb{B}_2$ is not a permutation matrix, then one column of A contains the other, whence there are no primes in $\mathbb{B}_2$ either. Consequently, when discussing primes in $\mathbb{B}_n$, it is always assumed that $n \geq 3$. And the only candidate for being prime in $\mathbb{B}_3$, i.e. which is no permutation matrix and without row or column containment, is (up to equivalence)

$$B = \begin{bmatrix} 0 & 1 & 1 \\ 1 & 0 & 1 \\ 1 & 1 & 0 \end{bmatrix}.$$

3.4 Permanents

We now turn to the notions of a permanent and an adjoint of a Boolean matrix, and mention some of their properties.

Definition 3.30. *Let $A \in \mathbb{B}_n$ be a square Boolean matrix.*

(a) *The* **permanent** *of A is defined by*

$$\mathsf{per}(A) = \sum_{\sigma \in S_n} a_{1\sigma(1)}\, a_{2\sigma(2)} \ldots a_{n\sigma(n)},$$

where $\sigma \in S_n$, the set of all permutations of $\underline{n}$.

(b) *The (i,j)-***minor matrix** *of A, denoted with $A(i|j)$, is the $(n-1)\times(n-1)$ submatrix of A obtained by deleting the i-th row and the j-th column of A.* $\mathsf{per}(A(i|j))$ *is called the* **minor** *of the entry a_{ij}.*

(c) *The* **adjoint** *of A is the matrix* $\mathsf{adj}(A) = [b_{ij}] \in \mathbb{B}_n$, *with* $b_{ij} = \mathsf{per}(A(j|i))$ *for all* $i, j \in \underline{n}$.

Proposition 3.31. ([12, Prop. 2], [16, Lemmas 4.1 and 4.3(3)]) *Permanents of Boolean matrices $A, B \in \mathbb{B}_n$ have the following properties:*

(1) $\mathsf{per}(A) = \mathsf{per}(A^T)$.

(2) *If P and Q are permutation matrices then* $\mathsf{per}(PAQ) = \mathsf{per}(A)$, *i.e the value of* $\mathsf{per}(A)$ *invariant under arbitrary permutations of its rows or columns.*

(3) $\mathsf{per}(A) = \sum_j a_{ij}\, \mathsf{per}(A(i|j)) = \sum_i a_{ij}\, \mathsf{per}(A(i|j))$.

(4) *If $A \le B$, then* $\mathsf{per}(A) \le \mathsf{per}(B)$.

Proof. Special cases of Prop. 6.7(2),(3), Prop. 6.8(4), and Prop. 7.16(1), respectively. □

3.5 Invertible Matrices

We did already have a short look on the topic of invertibility in Section 2.10 in terms of relations. Now we will consider it from the viewpoint of linear algebra and matrices.

Definition 3.32. *A square matrix $A \in \mathbb{B}_n$ is called* **invertible** *or* **nonsingular** *iff there exists a matrix $B \in \mathbb{B}_n$ with $AB = I_n$, where I is the identity matrix. B is the unique matrix with $AB = BA = I_n$, and is called the* **inverse** *of A denoted by A^{-1}. A matrix that does not have an inverse is called a* **singular**.

Note that an invertible matrix must be a square matrix for the above definition to make sense and A^{-1} must have the same order as A. A square matrix with a row or a column of only zeroes does not have an inverse, so the null matrix O is a rather extreme example of a singular matrix. Moreover, from Prop. 3.34 and Cor. 3.35 below we know that permutation matrices are the only matrices in $\mathbb{B}_n$ which have inverses.

In Def. 3.32 we made two assertions that have still to be shown valid: $BA = I_n$ and B is the unique matrix with $AB = BA = I_n$.

Corollary 3.33. *Let A, B be square Boolean matrices. Then the following holds:*

(1) $I^{-1} = I$.
(2) $(A^{-1})^{-1} = A$.
(3) $(A^{-1})^T = (A^T)^{-1}$.
(4) $(AB)^{-1} = B^{-1}A^{-1}$.
(5) $BA = I$ *iff* $AB = I$.
(6) *The inverse of A is unique.*

Proof. (1): $IA = A$ for any matrix A, so $II = I$.
(2): We have $AA^{-1} = A^{-1}A = I$, so by definition A is the inverse of A^{-1}.
(3): $A^T(A^{-1})^T = (A^{-1}A)^T = I^T = I$ and $(A^{-1})^T A^T = (AA^{-1})^T = I$, using Cor. 3.9(3).
(4): $ABB^{-1}A^{-1} = AA^{-1} = I$ and $B^{-1}A^{-1}AB = B^{-1}B = I$, so $B^{-1}A^{-1}$ is the unique inverse of AB.
(5): Assume that $BX = I$ and $AB = I$. Then $X = IX = ABX = AI = A$, whence $AB = I$ implies $BA = I$. To show the converse direction one simply has to interchange A and B.
(6): Let C, C' be two inverses of A. Then $C = CI = C(AC') = (CA)C' = IC' = C'$. □

Up to here, and Section 2.10 included, we have dealt with the invertibility of *square* Boolean matrices. We will now take a look at the general case, following the presentation in [19], and start with examples which demonstrate that not every matrix has an inverse, and that inverses need not be unique in the general case: Let $A \in \mathbb{B}_{2\times 2}$ and $B \in \mathbb{B}_{3\times 2}$ be given as

$$A = \begin{bmatrix} 0 & 0 \\ 1 & 1 \end{bmatrix} \quad \text{and} \quad B = \begin{bmatrix} 0 & 0 \\ 0 & 1 \\ 1 & 0 \end{bmatrix}.$$

Then A has neither a left nor right inverse; on the other hand, each of

$$\begin{bmatrix} 0 & 0 & 1 \\ 0 & 1 & 0 \end{bmatrix}, \quad \begin{bmatrix} 1 & 0 & 1 \\ 0 & 1 & 0 \end{bmatrix}, \quad \begin{bmatrix} 0 & 0 & 1 \\ 1 & 1 & 0 \end{bmatrix}, \quad \text{and} \quad \begin{bmatrix} 1 & 0 & 1 \\ 1 & 1 & 0 \end{bmatrix}$$

is a left inverse of B.

Proposition 3.34. ([19, Theorem 2-1]) *If a square Boolean matrix A has a left inverse A', then A' is a permutation matrix.*

Proof. Let $A \in \mathbb{B}_n$, and $A'A = I_n = [I_{ij}]$. Then A' must be of order $n \times n$. Each 1 in I_n, say 1_{ii}, arises from the ith row of A' and the ith column of A, in particular from, say, $a'_{ik} = 1$ and $a_{ki} = 1$. For any other row $j \neq i$ of A' we must have $a'_{jk} = 0$, since otherwise we would get $i_{ji} = 1$, contradicting the definition of I. Thus, the kth column of A' must have a single 1, and a similar argument holds for the rows of A'. Therefore, A' is a permutation. □

Corollary 3.35. *If a square Boolean matrix A has a left inverse A', then A is a permutation matrix.*

Proof. Let $A'A = I$, then $(A'A)^T = A^T A'^T = I^T = I$. So by Prop. 3.34 A^T is a permutation, whence A is a permutation too. □

Corollary 3.36. ([19, Corollary 2-2]) *If a left inverse A' exists for a matrix $A \in \mathbb{B}_{m\times n}$, then $n \le m$.*

Proof. First note that $A' \in \mathbb{B}_{n\times m}$, and $I \in \mathbb{B}_{n\times n}$. Assuming $n > m$ we can partition as follows: $A = \begin{bmatrix} S & E \end{bmatrix}$ and $A' = \begin{bmatrix} S' \\ E' \end{bmatrix}$, where $S, S' \in \mathbb{B}_{m\times m}$, $E \in \mathbb{B}_{m\times(n-m)}$, and $E' \in \mathbb{B}_{(n-m)\times m}$. Then $A'A = \begin{bmatrix} S' \\ E' \end{bmatrix} \begin{bmatrix} S & E \end{bmatrix} = \begin{bmatrix} S'S & S'E \\ E'S & E'E \end{bmatrix} = I_n$, where $S'S$ and $E'E$ are square, whence they contain all the 1's of I_n; so $S'S = I_m$ and $E'E = I_{(n-m)\times(n-m)}$. Thus, S' is a left inverse and by Prop. 3.34, and Cor. 3.35 both S and S' are permutations. Moreover we have $E'S = O_{(n-m)\times m}$ and $S'E = O_{m\times(n-m)}$. Since S, S' are permutations it follows that E' and E must have all elements equal 0, so $E'E = O_{(n-m)\times(n-m)}$, i.e. we got a contradiction, since we have just shown that $E'E$ is a unit matrix. Thus $n \le m$. □

Corollary 3.37. ([19, Corollary 2-3]) *If a Boolean matrix $A \in \mathbb{B}_{m\times n}$ has a left inverse $A' \in \mathbb{B}_{n\times m}$, then A can be partitioned into a permutation matrix $S \in \mathbb{B}_{n\times n}$ and a remaining matrix $E \in \mathbb{B}_{(m-n)\times n}$.*

Proof. By Cor. 3.36 we have $n \le m$, so A' can be partitioned as follows: $A' = \begin{bmatrix} S' & E' \end{bmatrix}$ and $A = \begin{bmatrix} S \\ E \end{bmatrix}$, where $S, S' \in \mathbb{B}_{n\times n}$, $E \in \mathbb{B}_{(m-n)\times n}$, and $E' \in \mathbb{B}_{n\times(m-n)}$. Then $A' \cdot A = \begin{bmatrix} S' & E' \end{bmatrix} \cdot \begin{bmatrix} S \\ E \end{bmatrix} = \begin{bmatrix} S' \cdot S + E' \cdot E \end{bmatrix} = I_n$.

We now turn to the construction of the columns of S': Each row i of A' induces a 1 in I_n, so we must have $a'_{ik} = 1$ for at least one k. We choose a $k \in \underline{n}$ such that the corresponding column is in S'. Due to reasons analogous to those used in the proof of Prop. 3.34 no two rows can have their contributing unit in the same column, whence S' is a permutation. Thus, we get $S'S = I_n$, i.e. S' is a left inverse of S, and S therefore a permutation by Cor. 3.35. □

Theorem 3.38. ([19, Theorem 2-2]) *A necessary and sufficient condition that a matrix $A \in \mathbb{B}_{m\times n}$ has a left inverse is that A can be partitioned as $A = \begin{bmatrix} S \\ E \end{bmatrix}$ where $S \in \mathbb{B}_{n\times n}$ is a permutation matrix and $E \in \mathbb{B}_{(m-n)\times n}$ is a remaining matrix. Then the left inverse is given by $A' = \begin{bmatrix} S^T & E' \end{bmatrix}$, with E' any matrix such that $E'E \subseteq I_n$.*

Proof. Necessity follows from Cor. 3.37. Conversely, assume that $A = \begin{bmatrix} S \\ E \end{bmatrix}$ is a partition, with $S \in \mathbb{B}_{n\times n}$ is a permutation matrix. Then

$$A' \cdot A = \begin{bmatrix} S^T & E' \end{bmatrix} \cdot \begin{bmatrix} S \\ E \end{bmatrix} = \begin{bmatrix} S^T \cdot S + E' \cdot E \end{bmatrix}.$$

But S is a permutation matrix, so $S^T S = I_n$, whence $A'A = I_n + E'E = I_n$ since by assumption $E'E \subseteq I_n$. □

Analogous results hold for a right inverse of a matrix, with the final result as follows:

Theorem 3.39. ([19, Theorem 2-3]) *A necessary and sufficient condition that a matrix $A \in \mathbb{B}_{m\times n}$ has a right inverse is that A can be partitioned as $A = \begin{bmatrix} S & E \end{bmatrix}$ where $S \in \mathbb{B}_{m\times m}$ is a permutation matrix and $E \in \mathbb{B}_{m\times(n-m)}$ is a remaining matrix. Then the right inverse is given by $A' = \begin{bmatrix} S^T \\ E' \end{bmatrix}$, with E' any matrix such that $EE' \subseteq I_n$.*

References

[1] Abadir, K.M., Magnus, J.R.: *Matrix Algebra,* Cambridge University Press (2005).

[2] Agarwal, R.P., Flaut, C.: *An Introduction to Linear Algebra,* CRC Press, Boca Raton (2017).

[3] Antony, R., Alemayehu, H.: A note on special matrices, Italian Journal of Pure and Applied Mathematics **35**, 587-604 (2015).

[4] Banerjee, S., Roy, A.: *Linear Algebra and Matrix Analysis for Statistics,* CRC Press, Taylor & Francis Group, Boca Raton (2014).

[5] Bapat, R.B., Ragvahan, T.E.S.: *Nonnegative Matrices and Applications,* Cambridge Univ. Press, Cambridge, UK (1997).

[6] Brualdi, R.A., Cvetković, D.: *A Combinatorial Approach to Matrix Theory and Its Applications,* Taylor & Francis Group, Boca Raton (2009).

[7] Brualdi, R.A., Ryser, H.J.: *Combinatorial Matrix Theory,* Cambridge University Press (1991).

[8] Cantoni, A., Butler, P.: Eigenvalues and Eigenvectors of Symmetric Centrosymmetric Matrices, Linear Algebra and its Applications **13**, 275-288 (1976).

[9] Cechlárová, K., Plávka, J.: Linear independence in bottleneck algebras, Fuzzy Sets and Systems, **77**, 337-348 (1996).

[10] Datta, L., Morgera, S.D.: On the reducibility of centrosymmetric matrices - Applications in engineering problems, Circuits, Systems, and Signal Processing **8**, 71-96 (1989).

[11] de Caen, D., Gregory, D.A.: Prlmes in the Semigroup of Boolean Matrlces, Linear Algebra and its Applications **37**, 119-134 (1981).

[12] Duan, J.-S.: The transitive closure, convergence of powers and adjoint of generalized fuzzy matrices, Fuzzy Sets and Systems **145**, 301-311 (2004).

[13] Fenner, T.I., Loizou, G.: On Fully Indecomposable Matrices, J. of Computer and System Sciences **5**, 607-622 (1971).

[14] Gregory, D.A., Kirkland, S., Pullman, N.J.: Power Convergent Boolean Matrices, Linear Algebra and its Applications, **179**, 105-117 (1993).

[15] Hall, F.J., Li, Z., Rao, B.: From Boolean to sign pattern matrices, Linear Algebra and its Applications **393**, 233-251 (2004).

[16] Han, S.C., Li, H.X.: Invertible incline matrices and Cramer's rule over inclines, Linear Algebra and its Applications **389**, 121-138 (2004).

[17] Horn, R.A., Johnson, C.R.: *Matrix Analysis,* 2nd ed., Cambridge University Press (2013).

[18] Kim, K.H.: *Boolean Matrix Theory and Applications,* Marcel Dekker Inc., New York and Basel (1982).

[19] Ledley, R.S.: The Inverse of a Boolean Matrix, National Biomedical Research Foundation, Maryland, 10 pp. (1965).

[20] Luce, R. D.: A note on Boolean matrix theory, Proc. of the American Math. Soc. **3**, 382-388 (1952).

[21] Marenich, E.E.: Factorization properties of $(n \times n)$ Boolean matrices, Journal of Mathematical Sciences **163**, 732-738 (2009).

[22] Meyer, C.D.: *Matrix Analysis and Applied Linear Algebra*, Soc. for Industrial and Applied Mathematics, Philadelphia (2001).

[23] Pressman, I.S.: Matrices with multiple symmetry properties - applications of centrohermitian and perhermitian matrices, Linear Algebra and its Applications **284**, 239-258 (1998).

[24] Rutherford, D.: Inverses of Boolean matrices, Proc. of the Glasgow Math. Assoc. **6**, 49-53 (1963).

[25] Williams, G.: *Linear Algebra with Applications*, 9th ed., Jones & Bartlett Learning, Burlington (2019).

[26] Zhang, F.: *Matrix Theory - Basic Results and Techniques*, 2nd ed., Springer Science+Business Media (2011).

PART 2

(Generalized) Fuzzy Matrices

CHAPTER 4

VARIOUS STRUCTURES FOR MANY-VALUED ALGEBRA

Despite of their modeling power, binary relations are only a special two-valued (or crisp) case of a more general class of relations where the two-element Boolean algebra $\mathbf{2} := \{0 < 1\}$ is replaced by a more general algebraic structure as valuation space. In order to be well prepared for many-valued extensions of classical two-valued concepts, an extension of the basic prerequisites from Chapter 1 becomes necessary.

Dealing with many-valued concepts often leads to situations where definitions and properties result in formulas identical to their two-valued counterparts. For sake of completeness the most basic facts will be repeated in their many-valued version. Moreover, it is interesting to see which of the classical results carry over to the many-valued case; or, on which level of generality can well-known classical results already be shown.

Our overview on the various kinds of structures used for generalizing the two-element Boolean algebra **2** will proceed from the general to the more special cases.

As stated in [17], besides the "algebra stars" groups, rings, and fields, semirings and dioids sometimes are considered as more exotic and less known notions. But in fact they are rather common, and they pop up in various applications (see e.g. [18]). To mention two of them: For some decades already, semirings play an important role in both the theory of formal languages and automata (see [24]), and in the investigation of various kinds of *path problems* ([3, Chapter 2] or [26] may serve as good starting points). The well known prototype example for semirings is the set $\mathbb{N}$ of natural numbers which, endowed with ordinary addition and multiplication, neither has the structure of a ring or a field, but that of a dioid.

As an early motivation we mention a few relationships that hold between the various kinds of algebraic structures to be introduced later in this chapter:

(1) Boolean algebras $\subset$ bounded distributive lattices $\subset$ 0-1-inclines $\subset$ path algebras $\subset$ antirings, dioids $\subset$ semirings.
(2) So the two element Boolean algebra $(\{0,1\}, \vee, \wedge)$, any power set $(\mathcal{P}(A), \cup, \cap)$, and any fuzzy algebra $(\mathbb{F}, \max, \min)$, where $\{0,1\} \subseteq \mathbb{F} \subseteq [0,1] \subset \mathbb{R}$, bounded distributive lattices, and 0-1-inclines are all commutative path algebras.

Thus, literature using path algebras as underlying structure for matrix entries may also contain results pertinent to the topics under consideration.

We will present results and properties at the most general level, and, for sake of convenience, will usually *not* repeat them in the context of the more special cases. So to get all the properties and results valid for a certain underlying structure, readers will have to proceed by topic instead of the usual reading order. An overview on the various kinds of algebras is given in Section 4.5 at the end of this chapter.

4.1 Semirings

Definition 4.1. *A* **semiring** *is an algebra* $(R,+,\cdot,0,1)$ *of type* $\langle 2,2,0,0\rangle$ *such that*

(SR1) $(R,+,0)$ *is an Abelian monoid with neutral element* 0

(SR2) $(R,\cdot,1)$ *is a monoid with neutral element* 1.

(SR3) $x\cdot(y+z)=(x\cdot y)+(x\cdot z)$,
$(x+y)\cdot z=(x\cdot z)+(y\cdot z)$, *for all* $x,y,z\in R$;.

(SR4) $x0=0x=0$, *for all* $x\in R$.

(SR5) $1\neq 0$.

Note that in a semiring as defined above, (SR4) means that 0 acts as a zero in the monoid $(R,\cdot,1)$. This condition follows from the other requirements: we have $0x+x=(0+1)x=1x=x$, whence $0x=0$ since, by Prop. 1.28, 0 is the unique identity in $(R,+,0)$. So the difference between a ring and a semiring is that in the former addition obeys the laws of a group, whereas in the latter only those of a semigroup. Do also note that 0 is the only element of R satisfying (SR4): if for $z\in R$ we have $zr=z=rz$, for all $r\in R$, then $0=0z=z$. So 0 is an absorbing element. Requirement (SR5) serves to avoid the trivial case $R=\{0\}$: if $1=0$ then $r=r1=r0=0$ for each element $r\in R$.

One of the most important applications of semiring theory occurs in formal languages and algebraic automata theory. Let $A=\{a_1,\ldots,a_n\}$ be a finite alphabet; the a_i are called *symbols* or *letters*. *Words* over A are finite sequences $w=x_1x_2\ldots x_m$ such that each $x_i\in A$. The length of a word $w=x_1x_2\ldots x_m$ is $|w|=m$. The special word with length 0 is called the *empty word* and is denoted by $\Box$. The set of all finite words over A (including $\Box$) is usually denoted by A^* and called the *free monoid on* A[1] . On A^* we can define the binary operation of **concatenation** as follows: If $u=u_1u_2\ldots u_n$ and $w=w_1w_2\ldots w_m$ are two words in A^*, then their concatenation is $u\cdot w=u_1u_2\ldots u_nw_1w_2\ldots w_m$. We set $w\cdot\Box=\Box\cdot w=w$. Obviously, $(A^*,\cdot,\Box)$ is a monoid from which we can construct a semiring that is at the heart of algebraic automata theory in the following way: Let

[1] **Some short remarks**: The alphabet $A=\{|\}$ gives rise to a free monoid that probably everyones knows: Let us call | a stick, then A^* helps childen to visualize natural numbers and their addition. For more examples and algebraic characterizations of free monoids see [25, Chapter 5] or [7, 21]. Only two more facts shall be mentioned: (1) The definition of A^* implies that $a_1a_2\cdots a_n=b_1b_2\cdots b_m\rightarrow n=m$ and $a_i=b_i$, for all $i\in\underline{n}$, i.e. each word has a unique written form as a product of letters. (2) The free monoid $(A^*,\cdot,\Box)$ is characterized by the following (in category-theoretical terms) *universal property*: Let $(S,\cdot)$ be an arbitrary semigroup, and $f:A\longrightarrow S$ be an arbitrary map. Then there is a unique homomorphism $f':A^*\longrightarrow S$ extending f.

$A, B \in \mathcal{P}(A^*)$; taking $A * B = \{a \cdot b \mid a \in A, b \in B\}$ makes $(\mathcal{P}(A^*), \cup, *, \emptyset, \{\square\})$ an additively idempotent and entire semiring (if $A * B = \emptyset$, then at least one of A, B must also be equal $\emptyset$), which is also zerosumfree ($A \cup B = \emptyset$ clearly implies $A = B = \emptyset$).

There are a number of special kinds of semirings which can be found in literature, among them the following ones:

Definition 4.2. *A semiring R is called*

(a) **commutative** *if $(R, \cdot, 1)$ is an Abelian monoid;*
(b) **additively idempotent** *(or a* **path algebra***) if $a + a = a$;*
(c) **idempotent** *if it is a multiplicatively idempotent path algebra: $a + a = a = a \cdot a$;*
(d) **zerosumfree** *(or* **antinegative***, or an* **antiring***) if $a + b = 0$ implies $a = b = 0$;*
(e) **cancellative** *if $a \neq 0$ and $ab = ac$ imply $b = c$;*
(f) **entire** *(or* **zero-divisor-free***) if $ab = 0$ implies $a = 0$ or $b = 0$;*
(g) **division semiring** *if all elements in $R - \{0\}$ are units;*
(h) **semifield** *if it is a commutative division semiring;*
(i) **join semiring** *if R is a partially ordered set such that $0 \le a \le 1$ and $a + b = a \vee b$ (i.e. $a + b$ is the least upper bound of a and b in R);*
(j) **partially ordered** *if there is a partial order $\le$ on R compatible with both $+$ and $\cdot$;*
(k) **positive** *if it is partially ordered by $\le$ and $0 \le a$,*

for all $a, b \in R$.

Note that in a zerosumfree semiring the zero element is the only element with an additive inverse. Two elements $a \neq 0$ and $b \neq 0$ in R are called **zero divisors** if $ab = 0$. Thus, an entire semiring is a semiring which does not have zero divisors. If multiplication is cancellative, the semiring is entire: $ab = 0 \rightarrow ab = a0 \rightarrow b = 0$, by cancellation of $a \neq 0$.

Semirings are abundant, as can be seen from the following list of examples:

- Any ring with unit (in particular, any field) is a semiring.
- The most popular, and the simplest example of a semiring which is not a ring, is the **Boolean semiring** $\mathbb{B} = (\{0,1\}, +, \cdot, 0, 1)$ with addition defined as $1 + 0 = 0 + 1 = 1 + 1 = 1$ and $0 + 0 = 0$; and with multiplication given by $1 \cdot 0 = 0 \cdot 1 = 0 \cdot 0 = 0$ and $1 \cdot 1 = 1$. It is easy to see that $a + b = a \vee b$ and $a \cdot b = a \wedge b$. It is additively idempotent, commutative and zerosumfree.
- $(\mathcal{B}_X, \cup, \circ, \emptyset, I)$, where $\mathcal{B}_X$ denotes the set of relations over X, is an additively idempotent semiring with set inclusion as its natural ordering.
- $(\mathbb{N}, +, \cdot)$, $(\mathbb{Q}^+, +, \cdot)$ and $(\mathbb{R}^+, +, \cdot)$ are commutative and zerosumfree path algebras, with zero element 0 and multiplicative unit 1.
- Any **fuzzy algebra** $(\mathbb{F}, \vee, \wedge)$, where $\{0,1\} \subseteq \mathbb{F} \subseteq [0,1] \subset \mathbb{R}$ is a commutative path algebra.
- Boolean algebras, the fuzzy algebra $([0,1], +, \cdot)$ under the operations $a + b = \sup\{a, b\}$ and $ab = \inf\{a, b\}$, and every bounded distributive lattice are commutative, additively idempotent, and zerosumfree semirings (in fact, all of them

are also join-semirings).

- The set of nonnegative integers under the operations $a+b = g.c.d.\{a,b\}$ and $a \cdot b = l.c.m.\{a,b\}$ is a commutative, and with usual addition and multiplication a commutative entire antiring.
- $(\mathbb{R} \cup \{\infty\}, \min, +, \infty, 0)$, called the **tropical semiring** (or **min-plus algebra**), is the algebraic structure underlying many classical shortest-paths algorithms, where ∞ is an extra element that represents infinity.
- Similar to the above is $(\mathbb{R} \cup \{-\infty\}, \max, +, -\infty, 0)$, called the **arctic semiring** (also **max-plus algebra** or **schedule algebra**).

Proposition 4.3.

(1) ([15, (4.28)]) $\mathbb{B} = (\{0,1\}, +, \cdot)$ *is an additively idempotent division semiring, in fact the only finite one.*

(2) *It is also the only multiplicatively idempotent semifield.*

Proof. (1): Assume that R is a finite additively idempotent division semiring, with the sum of all elements being d. Then, by construction, $r + d = d$ for all $r \in R$, in particular $d^2 + d = d$. But we also have $1 + d = d$, so multiplying this equation with d we get $(1+d)d = d + d^2 = d^2$, and so $d = d^2$ follows. That R is a division semiring implies that d is a unit, i.e. $cd = 1$ for some $c \in R$, so $cd^2 = d$. But, as just shown, $d = d^2$, whence $d = cd^2 = cd = 1$. If $0 \neq r \in R$ then $rr' = 1$ for some $r' \in R$; and since $r' + 1 = r' + d = d = 1$ we get $1 = d = d + r = 1 + r = rr' + r = r(r'+1) = r$. Therefore, $R = (\{0,1\}, +, \cdot) = \mathbb{B}$.

(2): We have $\mathbb{B}_0 = \mathbb{B} - \{0\} = \{1\}$, and 1 certainly is a unit which makes $\mathbb{B}$ a semifield. Indeed, if R is a multiplicatively idempotent semifield and if $0 \neq a \in R$ then $a = a1_R = aaa^{-1} = aa^{-1} = 1_R$ and so $R = \mathbb{B}$. □

Proposition 4.4. ([15, pp. 3,4]) *For a semiring S the following holds:*

(1) *S is additively idempotent iff* $1 + 1 = 1$.

(2) *If S is a path algebra then it is zerosumfree.*

(3) *If there is an element* $c \in S$ *with* $c = c + 1$*, then S is zerosumfree.*

(4) *If S is multiplicatively idempotent then* $xy = 1$ *implies* $x = y = 1$.

Proof. (1): For the nontrivial direction note that $a = a1 = a(1+1) = a1 + a1 = a + a$.

(2): Assume that S is a path algebra and that $a + b = 0$; then $a = a + 0 = a + (a+b) = (a+a) + b = a + b = 0$, and similarly $b = 0$. So S is zerosumfree.

(3): If $a + b = 0$ then

$$\begin{aligned} 0 &= (a+b)c = ac + bc = a(c+1) + b(c+1) \\ &= a(c+1) + b(c+1+1) = (a+b) + b + (a+b)c = b, \end{aligned}$$

whence $a = a + b = 0$ as well.

(4): Assume $ab = 1$; then $a = a1 = a(ab) = a^2 b = ab = 1$, and similarly $b = 1$. □

Definition 4.5. ([16, p. 14]) *Let S be a semiring.*

(a) An element $e \in S$ is called **infinite** *if $a + e = e$ for all $a \in S$.*
(b) S is called **simple** *if 1 is infinite.*

Proposition 4.6. ([16, p. 14]) *Let S be a semiring.*

(1) If S has an infinite element e, then it is unique.
(2) If S is simple and $a \neq 1$, then $ab \neq 1$ for all $b \in S$.
(3) Simple semirings are additively idempotent and hence zerosumfree.

Proof. (1): Assume that both e and e' are infinite; then $e = e' + e = e + e' = e'$.
(2): If there is a $b \in S$ with $ab = 1$, then $a = a1 = a(b+1) = ab + a = 1 + a = 1$, a contradiction to $a \neq 1$.
(3): By definition we have $1 + 1 = 1$, whence by Prop. 4.4(1),(2) S is additively idempotent and zerosumfree. □

Definition 4.7. *Let R be a semiring. An element $a \in R$ is called*

(a) **invertible** *in R if $ab = ba = 1$ for some $b \in R$.*
(b) **additively invertible** *in R if $a + b = b + a = 0$ for some $b \in R$.*
(c) **additively irreducible** *in R if $a = b + c$ implies $a = b$ or $a = c$.*

$U(R)$ denotes the set of all invertible, and $V(R)$ the set of all additively invertible elements in R.

In both cases, if an invertible element exists, it is unique and is denoted by a^{-1} and $-a$, respectively. Note that $\{1\} \subseteq U(R)$ is a subgroup of the semigroup $(R, \cdot)$. And $V(R) = \{0\}$ if and only if R is a zerosumfree semiring.

Lemma 4.8. ([41, Lemma 2.1]) *Let R be a semiring. Then*

(1) For any $a, b \in R$, $a + b \in V(R)$ iff $a, b \in V(R)$.
(2) For any $a \in V(R)$ and any $r \in R$, $ra, ar \in V(R)$.
(3) If R is commutative, then $ab \in U(R)$ iff $a, b \in U(R)$.

Proof. (1): Assume that $a + b \in V(R)$; then $a + b + x = b + a + x = 0$ for some $x \in R$, whence $a, b \in V(R)$. Conversely, if $a, b \in V(R)$ then $a + x = 0$ and $b + y = 0$ for some $x, y \in R$; thus $(a + b) + (x + y) = 0$.

(2): If $a \in V(R)$, then $a + x = 0$ for some $x \in R$. So, for any $r \in R$, $ra + rx = r(a + x) = 0$ and $ar + xr = (a + x)r = 0$, i.e. $ra, ar \in V(R)$.

(3): Assume that $ab \in U(R)$; then $abx = bax = 1$ for some $x \in R$, whence $a, b \in U(R)$. Conversely, if $a, b \in U(R)$ then $ax = 1$ and $by = 1$ for some $x, y \in R$; thus $(ax)(by) = (ab)(xy) = 1$. □

Lemma 4.9. ([42, Lemma 2.1]) *Let R be a commutative semiring. Then for any $a, b \in V(R)$ and $r \in R$, we have $(-a)r = -ar$ and $(-a)(-b) = ab$.*

Proof. For any $a \in V(R)$ we clearly have $-a \in V(R)$ and $-(-a) = a$. Then $(-a)r + ar = ((-a) + a)r = 0r = 0$, so $(-a)r = -(ar)$ and $(-a)(-b) = -a(-b) = -(-ab) = ab$. □

Two examples for how to construct semirings out of given algebraic structures are as follows:

(a) Let $(R, +, \cdot, 0, 1)$ be a semiring; then $(\mathcal{P}(R), \oplus, *, \emptyset, \{1\})$ becomes a semiring by setting $A \oplus B = \{a + b \mid a \in A, b \in B\}$ and $A * B = \{ab \mid a \in A, b \in B\}$.

(b) Let $(M, \cdot, 1_M)$ be a monoid; then $(\mathcal{P}(M), \oplus, *, \emptyset, \{1_M\})$ becomes a semiring with addition and multiplication defined as $A \oplus B = A \cup B$ and $A * B = \{ab | a \in A, b \in B\}$.

Definition 4.10. *Let $(R, +, 0)$ be a commutative monoid; let the relation $\le$ on R be defined as $a \le b :\leftrightarrow \exists c \in R[b = a + c]$, called the* **canonical preorder** *relation on $(R, +, 0)$. $\le$ is called* **canonical order** *if it is a partial order.*

Proposition 4.11. *The canonical preorder $\le$ on a commutative monoid R has the properties its name suggests. Moreover, if R is additively idempotent, then $\le$ is also antisymmetric, i.e. a partial order.*

Proof. It is easy to show that $\le$ is indeed a preorder: Reflexivity follows by $a + 0 = a$; and assuming $a \le b$ and $b \le d$ we get $b = a + c$ and $d = b + c'$, for some $c, c' \in R$. Thus, $d = a + c + c'$, so $\le$ is transitive.

Now assume $a \le b$ and $b \le a$; then there are $c, c' \in R$ such that $a = b + c$ and $b = a + c'$. So, $a = a + c' + c$. Using the assumption that R is additively idempotent we get $b = a + c' = a + c' + c + c' = a + c' + c = a$, and antisymmetry has been shown. □

Proposition 4.12. *Let S be a semiring, and $a, b, c \in S$. Then the following holds for the canonical preorder $\le$ on S:*

(1) $\le$ is compatible with $+$.
(2) $\le$ is compatible with $\cdot$.
(3) If S is positive, then it is zerosumfree.

Proof. (1): Assume $a \le b$ and $c \le d$; then, since $+$ is commutative, $b + d = a + a' + c + c' = a + c + (a' + c')$, for some $a', c' \in S$. Therefore, $a + c \le b + d$.

(3): If $a, b \in S$ satisfy $a + b = 0$, then $0 \le a \le a + b = 0$ and so $a = 0$ and hence $b = 0$. □

4.2 Path Algebras and Dioids

Most of the material in this section is taken from [1, 18, 19]. But be careful: What is called a dioid in [1, Definition 4.1] and [19] is called a *path algebra*, i.e. an additively idempotent semiring, here and elsewhere (see e.g. [9, Chapter 3] or [22]). In Cor. 4.15(1) below we will learn that path algebras are dioids. Since they play an important role in path evaluation problems, we will focus on the former, but will introduce the notion of a dioid for sake of completeness. Before we can do so we need the following result:

Theorem 4.13. ([18, Theorem 1, p. 13)]) *A monoid* $(R,+,0)$ *cannot both be a group and canonically ordered.*

Proof. Assume that $(R,+,0)$ is a group, with a^{-1} denoting the inverse of a, and also canonically ordered. Let $a, b \in R$ with $a \neq b$. Then, by a well known result in group theory[2], there exist $c, d \in R$ with $a = b + c$ (take $c = b^{-1} + a$), and $b = a + d$ (take $d = a^{-1} + b$). So $b \leq a$ and $a \leq b$, and by the antisymmetry of $\leq$ we get $a = b$, a contradiction. □

Definition 4.14. ([17, Definition 2], [18]) *A semiring* $\mathcal{S} = (S,+,\cdot,0,1)$ *such that* $(S,+)$ *is canonically ordered is called a* **dioid**.

Corollary 4.15.

1. *Path algebras are dioids.*
2. *Dioids and rings are two disjoint subclasses of semirings.*

Proof. (1): Immediate from Prop. 4.11 since path algebras are additively idempotent.
(2): By Def. 1.35 and Th. 4.13. □

The position of dioids within the scope of algebraic structures is described in [1, p. 159] as follows:

> Dioids are structures that lie somewhere between conventional linear algebra and semilattices endowed with an internal operation generally called multiplication. With the former, it shares combinatorial properties such as associativity and commutativity of addition, associativity of multiplication, distributivity of multiplication with respect to addition, and of course the existence of zero and identity elements. With the latter, it shares the features of an ordered structure (adding is then simply taking the upper bound) endowed with another 'compatible' operation. Therefore, one may expect that the results of linear algebra which depend only on combinatorial properties will generalize to dioids.

Some examples:

1. The two-element Boolean algebra $(\{0,1\}, \vee, \wedge)$ and the power set $(\mathcal{P}(A), \cup, \cap)$ are both commutative path algebras.
2. Fuzzy algebras $(\mathbb{F}, \vee, \wedge)$, where $\{0,1\} \subseteq \mathbb{F} \subseteq [0,1] \subset \mathbb{R}$, every bounded distributive lattice, and any 0-1-incline are commutative path algebras.
3. The max-plus dioid $\mathbb{R}_{\max} := (\mathbb{R} \cup -\infty, \max, +, \mathbb{0}, \mathbb{1})$, with $\mathbb{0} := -\infty$ and $\mathbb{1} := 0$ is commutative path algebra.
4. Let D be a dioid, and $M_n(D)$ denote the set of $n \times n$ matrices with entries in D, and with matrix addition and multiplication as the operations. Then $(M_n(D), +, \cdot)$ is a dioid.

[2]Let G be a group and $a, b \in G$; then $\exists x \in G[ax = b]$: If $ax = b$, then $x = 1x = a^{-1}ax = a^{-1}b$. Conversely, if $x = a^{-1}b$, then $ax = aa^{-1}b = 1b = b$.

(5) This is Lemma 4.24 of [1]: Let D be a dioid, The quotient of D by a congruence is a path algebra with addition and multiplication induced by those of D, i.e. $[a]+[b]:=[a+b]$ and $[a]\cdot[b]:=[a\cdot b]$. (A **congruence** in a dioid D is an equivalence relation $\equiv$ on D which is compatible with $+$ and $\cdot$, that is, $\forall a,b,c \in D[a\equiv b \rightarrow a+c\equiv b+c \wedge a\cdot c\equiv b\cdot c)$.

Idempotency of addition is a distinguishing feature of semirings. It is at the basis of the introduction of an order relation which has been used in Th. 4.13 above to show that the class of semirings contains two disjoint sub-classes depending on whether addition $+$ satisfies one of the following two properties: (i) $+$ imposes a group structure on $(D,+,0)$, or (ii) $+$ makes $(D,+,0)$ a canonically ordered monoid.

Theorem 4.16. ([1, Theorem 4.28]) *Let D be a path algebra, and $a,b,c \in D$. Then the following holds:*

(1) $a+b=b \leftrightarrow \exists c[b=a+c]$.
(2) $a\leq b \leftrightarrow a+b=b$ is a partial order on D.
(3) $\leq$ is compatible with $+$.
(4) $\leq$ is compatible with $\cdot$.
(5) two elements $a,b \in D$ always have the least upper bound $a+b$.
(6) D is positive.

Proof. (1): If $a+b=b$ then trivially there is a $c\in D$ with $b=a+c$, namely $c=b$. Conversely, if $b=a+c$ for some $c\in D$, then adding a on both sides of this equality yields $a+b=a+(a+c)=a+c=b$.

(2): By Prop. 4.11.

(3),(4): By Prop. 4.12.

(5): Obviously, by definition $a\leq a+b$ and $b\leq b+a$. $a+b$ is also the least upper bound because for every other upper bound c of a and b, from $a\leq c$ and $b\leq c$ we get $a+b\leq c+c=c$ by (3).

(6): For all $a\in D$, $0+a=a$, and therefore $0\leq a$ by definition. Moreover, $\leq$ is a partial order on D by (2). □

This theorem essentially states that the idempotent addition in a path algebra D induces a structure of a join semilattice over D.

4.3 Inclines

Definition 4.17. ([8]) *An* **incline** *is an algebra $(L,+,\cdot)$ of type $\langle 2,2\rangle$ such that*

(In1) *$(L,+)$ is a semilattice;*
(In2) *$(L,\cdot)$ is a commutative semigroup;*
(In3) *$a\cdot(b+c)=a\cdot b+a\cdot c$, for all $a,b,c\in L$;*
(In4) *$a+a\cdot b=a$, for all $a,b\in L$.*

If L has an additive identity 0, it is called the *zero* of L, and If L has an multiplicative identity 1, it is called the *identity* of L. In this section, L will always denote an incline with zero and identity, also called a 0-1-**incline**.

From the respective definitions we can see that

- the notion of an incline generalizes that of a distributive lattice in that multiplication need not be idempotent, i.e. we have $a^2 \le a$ for all $a \in L$ due to (In4);
- the notion of an incline specializes that of a dioid by requesting multiplication to be commutative;
- an incline is a commutative, additively idempotent and zerosumfree semiring due to Cor. 4.18(8),(9), in which products are less than or equal to factors: $ab \le a$ due to (In4).

Every Boolean algebra, fuzzy algebras $(\mathbb{F}, \vee, \wedge)$, where $\{0,1\} \subseteq \mathbb{F} \subseteq [0,1] \subset \mathbb{R}$, and any bounded distributive lattice are examples of 0-1-inclines. Also $([0,1], \vee, \cdot)$ is an incline, but not a distributive lattice, if $a \cdot b$ denotes ordinary multiplication.

Corollary 4.18. ([13, Proposition 1]) *Let L be an 0-1-incline and $a, b, c \in L$; then*

(1) $0 \le a \le 1$.
(2) if $b \le c$ *then* $a + b \le a + c$, $ab \le ac$, *and* $ba \le ca$.
(3) $a \le a + b$ *and* $a + b = \sup\{a, b\}$.
(4) $ab \le a$ *and* $ab = \inf\{a, b\}$.
(5) $acb \le ab$.
(6) $a + 0 = 0 + a = a$ *and* $a + 1 = 1 + a = 1$.
(7) $a0 = 0a = 0$ *and* $a1 = 1a = a$.
(8) $a + b = 0$ *iff* $a = b = 0$, *and* $ab = 1$ *iff* $a = b = 1$.
(9) $a = a + a$.
(10) $(a + b) \cdot c = a \cdot c + b \cdot c$.

Proof. The proofs are immediate consequences of the definition and straightforward:
(1): By definition, $0 + a = a$, and by (In4) $a + 1 = 1a + 1 = 1$.
(2): By Prop. 4.12 and commutativity of multiplication.
(4): From (1) we get $b \le 1$, whence by (2), $ab \le a1 = a$, and similar for $ab \le b$. Moreover, let c be another lower bound, i.e. $ac = c$ and $bc = c$; then $abc = a(bc) = ac = c$, whence $c \le ab$.
(5): $acb = abc = (ab)c \le ab$, using commutativity of $\cdot$ and (4).
(7): $0a = 0 + 0a = 0$ by (In4).
(8): By Prop. 4.4(2); moreover, $a = b = 1$ implies $ab = a = b = 1$ by the identity property of a, b. Conversely, $1 = ab \le a \le 1$ by (4) and (1).
(9): $a = a1 = a(1+1) = a1 + a1 = a + a$, using (6) for the second equality.
(10): $(a + b) \cdot c = c \cdot (a + b) = c \cdot a + c \cdot b = a \cdot c + b \cdot c$, by (In3) and commutativity. □

Corollary 4.19.

(1) A commutative path algebra R is a 0-1-incline if $a + 1 = 1$ for all $a \in R$.

(2) Any incline L is a path algebra.

Proof. (1): By Th. 4.16(2)(5) $a+b$ is the supremum of a and b, so $(R,+)$ is a semilattice. By hypothesis multiplication is commutative, so $(R,\cdot,1)$ is an Abelian monoid, and (In3) holds by Def. 4.1(SR3). Finally, $1+b=1$ implies $a=a\cdot 1=a\cdot(1+b)=a+a\cdot b$, for all $a,b\in R$, yielding (In4).

(2): By Cor. 4.18(9). □

4.4 Relatively Pseudocomplemented and Residuated Lattices

We have already learned about the most basic concepts of lattice theory in Def. 1.39, and also have presented related properties subsequently. Here we will add some concepts needed in our analysis of fuzzy relations later on.

Definition 4.20. *Let L be a lattice, and* $a,b\in L$.

(a) *If the set* $T_{ab}:=\{x\in L: a\wedge x\le b\}$ *has a greatest element, it is called the* **relative pseudocomplement** *of a in b, denoted by* $a \overset{\wedge}{\rightharpoonup} b$, *and a is called relatively pseudocomplemented. L is called a* **relatively pseudocomplemented lattice** *if* $a \overset{\wedge}{\rightharpoonup} b$ *exists for all* $a,b\in L$.

(b) *If L is a lattice with bottom element 0 then an element* $x\in L$ *is* **pseudocomplemented** *if* $T_{a0}:=\{x\in L: a\wedge x\le 0\}$ *has a greatest element* $a \overset{\wedge}{\rightharpoonup} 0$, *denoted by* $\neg a$. *L is called a* **pseudocomplemented lattice** *if* $\neg a$ *exists for all* $a\in L$.

(c) *A* **Heyting algebra** *or* **Heyting lattice** *is relatively pseudocomplemented lattice with a smallest element 0.*

(d) *A* **Brouwerian lattice** *is the dual of a Heyting algebra (we have* $\vee$ *instead of* $\wedge$ *and* $\ge$ *instead of* $\le$*), i.e. a lattice L such that, for every pair* $a,b\in L$, *the set* $\{x\in L: a\vee x\ge b\}$ *has a smallest element* $a \overset{\wedge}{\leftharpoonup} b$.

CAVEAT 2: Unfortunately you cannot necessarily rely on the names of the concepts involved. Whereas *relatively pseudocomplemented* is unambiguous, the name *Brouwerian* is not: What is usually called a *relatively pseudocomplemented* lattice got the name *Brouwerian* lattice in Birkhoff [5, p. 45], a very influential book by an authority in the field. This term has become very popular in the fuzzy community starting with [36], and is still in use (see [39]). The definition of *Heyting algebra* as given above appears in e.g. in [2, p. 174], or [6, p. 112].

Another synonym for *relatively pseudocomplemented* is **implicative** (in e.g. [27, 37] or [11]), and Heyting algebras are sometimes called **pseudo-Boolean algebras** (in e.g. [20, 34, 35]).

Heyting algebras serve as the algebraic models of propositional intuitionistic logic in the same way Boolean algebras model propositional classical logic. They generalize

Boolean algebras in the sense that a Heyting algebra satisfying the excluded middle law ($a \vee \overline{a} = 1$) or, equivalently, the double negation law ($\overline{\overline{a}} = a$), is a Boolean algebra.

Examples for relatively pseudocomplemented and Heyting lattices are:

- Every chain becomes a relatively pseudocomplemented lattice by defining $a \mathrel{\dot{\rightarrow}} b = 1$ whenever $a \leq b$, and $a \mathrel{\dot{\rightarrow}} b = b$ otherwise.
- By the above, every totally ordered set that has a least element 0 and a greatest element 1 is a Heyting algebra.
- Every Boolean algebra is a Heyting lattice when putting $a \mathrel{\dot{\rightarrow}} b = \overline{a} \vee b$.
- (For logicians only.) The Lindenbaum algebra of propositional intuitionistic logic is a Heyting algebra.

We will now have a look at some properties of relatively pseudocomplemented lattices:

Proposition 4.21.

(1) Any relatively pseudocomplemented lattice L is distributive.
(2) Any finite distributive lattice L is relatively pseudocomplemented.

Proof. (1): For given $a, b, c \in L$ let $d = (a \wedge b) \vee (a \wedge c)$. Since $a \wedge b \leq d$ and $a \wedge c \leq d$ we have $b \leq a \mathrel{\dot{\rightarrow}} d$ and $c \leq a \mathrel{\dot{\rightarrow}} d$, whence $b \vee c \leq a \mathrel{\dot{\rightarrow}} d$. Thus, $a \wedge (b \vee c) \leq a \wedge a \mathrel{\dot{\rightarrow}} d \leq d = (a \wedge b) \vee (a \wedge c)$. Together with Prop. 1.45(L5) this implies $a \wedge (b \vee c) = (a \wedge b) \vee (a \wedge c)$ which is Prop. 1.46(L6), and this is equivalent to the second distributive law (L6') of the same proposition, whence we have shown the distributivity of L.

(2): In a finite distributive lattice the join $u = \bigvee\{x_u : a \wedge x_u \leq b\}$ satisfies $a \wedge u = a \wedge \bigvee x_u = \bigvee(a \wedge x_u) \leq b$. □

Proposition 4.22. *Let L be a relatively pseudocomplemented lattice, and $a, b \in L$. Then*

(1) L has a greatest element 1.
(2) $a \leq b$ iff $a \mathrel{\dot{\rightarrow}} b = 1$.
(3) $a \mathrel{\dot{\rightarrow}} a = 1, \quad a \mathrel{\dot{\rightarrow}} 1 = 1, \quad 0 \mathrel{\dot{\rightarrow}} a = 1$.
(4) $b \leq a \mathrel{\dot{\rightarrow}} b$;
(5) $1 \mathrel{\dot{\rightarrow}} a = a$.
(6) A Heyting algebra is a bounded relatively pseudocomplemented lattice.

Proof. (1): For any b we have $a \wedge b \leq a$, and thus $b \leq a \mathrel{\dot{\rightarrow}} a$ by definition of $\mathrel{\dot{\rightarrow}}$; so $a \mathrel{\dot{\rightarrow}} a$ is the greatest element in L, for any a.
(2): $a \mathrel{\dot{\rightarrow}} b = 1 \;\leftrightarrow\; 1 \leq a \mathrel{\dot{\rightarrow}} b \;\leftrightarrow\; a \wedge 1 \leq b$.
(3): Direct consequences of (2).
(4): $a \wedge b \leq b \;\rightarrow\; b \leq a \mathrel{\dot{\rightarrow}} b$.
(5): By (1) and Def. 4.20(c).
(6): By Def. 4.20(c), a Heyting algebra is a relatively pseudocomplemented lattice with a smallest element 0, and by (1) it has a greatest element 1. □

Proposition 4.23. ([40, Lemmas 2.2 and 2.7], [45, Lemmas 6 and 7]) *Let L be a relatively pseudocomplemented lattice, and $a, b, c \in L$ and $\{a_i | i \in I\} \subseteq L$. Then*

(1) $a \dot{\rightarrow} b = a \dot{\rightarrow} (a \wedge b)$.

(2) $a \wedge (a \dot{\rightarrow} (b \wedge c)) = a \wedge b \wedge c$.

(3) $\bigwedge_{i\in I}(a_i \dot{\rightarrow} b) = \left(\bigvee_{i\in I} a_i\right) \dot{\rightarrow} b$.

(4) $\bigwedge_{i\in I}(b \dot{\rightarrow} a_i) = b \dot{\rightarrow} \left(\bigwedge_{i\in I} a_i\right)$.

(5) $\left(\left(\bigvee_{j\in I} a_j\right) \dot{\rightarrow} b\right) \wedge a_i = a_i \wedge b$, *for all* $i \in I$.

Proof. (1): By definition of $\dot{\rightarrow}$, $a \wedge a \dot{\rightarrow} b \le b$, whence $a \wedge a \dot{\rightarrow} b \le a \wedge b$ by forming the meet with a on both sides. Thus, $a \dot{\rightarrow} b \le a \dot{\rightarrow} (a \wedge b)$. Conversely, $a \dot{\rightarrow} (a \wedge b)$ is the greatest element such that $a \wedge (a \dot{\rightarrow} (a \wedge b)) \le a \wedge b \le b$; so we get $a \dot{\rightarrow} (a \wedge b) \le a \dot{\rightarrow} b$.

(2): By Prop. 4.22(4) we have $b \wedge c \le a \dot{\rightarrow} (b \wedge c)$, so $a \wedge b \wedge c \le a \wedge (a \dot{\rightarrow} (b \wedge c))$. Conversely, $a \wedge (a \dot{\rightarrow} (b \wedge c)) \le b \wedge c$, implies $a \wedge (a \dot{\rightarrow} (b \wedge c)) = a \wedge (a \wedge (a \dot{\rightarrow} (b \wedge c))) \le a \wedge b \wedge c$.

(3): For all $i \in I$, we have $a_i \wedge (a_i \dot{\rightarrow} b) \le b$. With $r := \bigwedge_{i\in I}(a_i \dot{\rightarrow} b)$ we get $r \le a_i \dot{\rightarrow} b$, whence $a_i \wedge r \le a_i \wedge (a_i \dot{\rightarrow} b) \le b$, for all $i \in I$. Therefore, $\left(\bigvee_{i\in I} a_i\right) \wedge r \le b$, so $r \le \left(\bigvee_{i\in I} a_i\right) \dot{\rightarrow} b$. On the other hand, let $s := \left(\bigvee_{i\in I} a_i\right) \dot{\rightarrow} b$. Then $\left(\bigvee_{i\in I} a_i\right) \wedge s \le b$, whence $a_i \wedge s \le b$ and thus $y \le a_i \dot{\rightarrow} b$, for all $i \in I$. So we have $s \le \bigwedge_{i\in I}(a_i \dot{\rightarrow} b)$.

(4): $b \wedge \bigwedge_{i\in I}(b \dot{\rightarrow} a_i) = \bigwedge_{i\in I}(b \wedge (b \dot{\rightarrow} a_i)) \le \bigwedge_{i\in I} a_i$ implies $\bigwedge_{i\in I}(b \dot{\rightarrow} a_i) \le b \dot{\rightarrow} \bigwedge_{i\in I} a_i$. Conversely, for all $i \in I$ we have $b \wedge b \dot{\rightarrow} \bigwedge_{i\in I} a_i \le \bigwedge_{i\in I} a_i \le a_i$. Therefore, for all $i \in I$, $b \dot{\rightarrow} \bigwedge_{i\in I} a_i \le a \dot{\rightarrow} a_i$, whence $b \dot{\rightarrow} \bigwedge_{i\in I} a_i \le \bigwedge_{i\in I}(a \dot{\rightarrow} a_i)$.

(5): Since $\left(\bigvee_{j\in I} a_j\right) \dot{\rightarrow} b \ge b$ by Prop. 4.22(4), we get, for all $j \in I$, $\left(\left(\bigvee_{j\in I} a_j\right) \dot{\rightarrow} b\right) \wedge a_i \ge a_i \wedge b$. On the other hand, by the definition of pseudocomplement, $\left(\bigvee_{j\in I} a_j\right) \wedge \left(\left(\bigvee_{j\in I} a_j\right) \dot{\rightarrow} b\right) \le b$ holds which implies $a_i \wedge \left(\bigvee_{j\in I} a_j\right) \wedge \left(\left(\bigvee_{j\in I} a_j\right) \dot{\rightarrow} b\right) \le a_i \wedge b$ whence $a_i \wedge \left(\left(\bigvee_{j\in I} a_j\right) \dot{\rightarrow} b\right) \le a_i \wedge b$ for all $j \in I$. □

Corollary 4.24. *Let L be a relatively pseudocomplemented lattice, and $a, b, c \in L$. Then*

(1) $a \dot{\rightarrow} (b \vee c) \ge a \dot{\rightarrow} b$.

(2) $a \dot{\rightarrow} (a \wedge b) \ge b$.

Proof. (1): Obvious from the definition of $\dot{\rightarrow}$.

(2): Using Prop. 4.23(1) and Prop. 4.22(4) we get $a \dot{\rightarrow} (a \wedge b) = a \dot{\rightarrow} b \ge b$. □

Definition 4.25. *Let Q be a complete lattice, $S \subseteq Q$ any subset of Q, and $a \in Q$. Then Q is called a*

(a) **quantale** *if it has an associative binary operation $*$ that distributes over arbitrary joins, i.e.* $a * \left(\bigvee_{s\in S} s\right) = \bigvee_{s\in S}(a * s)$ $\left(\bigvee_{s\in S} s\right) * a = \bigvee_{s\in S}(s * a)$.

(b) **frame** *(or* **locale***) if Q is a quantale with $* = \wedge$.*

(c) *A quantale Q is called* **unital** *if there exists $e \in Q$ such that $(Q, *, e)$ is a monoid,* **commutative** *if $*$ is commutative, and* **idempotent** *if $*$ is idempotent.*

(d) *A unital quantale Q is called* **integral** *if e is the greatest element in Q.*

Frames are exactly complete Heyting algebras with the Heyting implication given by $a \rightharpoonup b := \bigvee\{s \in L : a \wedge s \leq b\}$, and also a typical example of an integral commutative quantale. Quantales are sometimes referred to as *complete residuated semigroups.*

Definition 4.26. *A* **residuated lattice** *is an algebra* $\mathscr{L} = (L, \wedge, \vee, \otimes, \rightharpoonup, 0, 1)$ *of type* $\langle 2,2,2,2,0,0\rangle$ *such that*

(1) $(L, \wedge, \vee, 0, 1)$ *is a bounded lattice;*
(2) $(L, \otimes, 1)$ *is a commutative monoid;*
(3) *the operation* $\rightharpoonup$ *is a* **residuation** *operation with respect to* $\otimes$*, i.e*
$$a \otimes b \leq c \;\leftrightarrow\; a \leq b \rightharpoonup c \qquad \textbf{(adjointess property)}.$$

In comparison with an ordinary lattice, this structure is enriched by the two operations $\otimes, \rightharpoonup : L \times L \longrightarrow L$ called an **adjoint pair.** The operations $\otimes$ (called *multiplication*) and $\rightharpoonup$ (called *residuum*) are intended for modeling conjunction and implication of the corresponding logical calculus, and supremum ($\bigvee$) and infimum ($\bigwedge$) are intended for modeling the existential and universal quantifier, respectively.

If $(L, \wedge, \vee, 0, 1)$ is a complete lattice then $\mathscr{L}$ is called a complete residuated lattice. Another operation in residuated lattices is **negation** $\neg : L \longrightarrow L$ defined by $\neg a := a \rightharpoonup 0$.

The examples given below show that residuated lattices offer a structure that is both general and rich enough to include Boolean and Heyting algebras as well as the most often applied structures used in many-valued approaches (see [28] and [31]):

- Important examples of residuated lattices are induced by t-norms: A **triangular norm** (short: **t-norm**) is a mapping $\otimes : [0,1] \times [0,1] \longrightarrow [0,1]$ which makes $([0,1], \otimes, 1)$ a commutative ordered monoid. If $\otimes$ is a left-continuous t-norm then $x \rightharpoonup y := \bigvee_z \{z \otimes x \leq y\}$ turns $([0,1], \min, \max, \otimes, \rightharpoonup, 0, 1)$ into a residuated lattice. For more details on t-norms and their properties see Section 9.2.
- Heyting algebras can be interpreted as residuated lattices if we take $a \otimes b = a \wedge b$ and $a \rightharpoonup b = \bigvee\{c : a \wedge c \leq b\}$.
- Every Boolean algebra is a residuated lattice when putting $a \rightharpoonup b = \overline{a} \vee b$. Note that in this case $a \otimes b = a \wedge b$.
- The **Łukasiewicz algebra** $([0,1], \wedge, \vee, \otimes_Ł, \rightharpoonup_L, 0, 1)$ with $a \otimes_Ł b = 0 \vee (a + b - 1)$ and $a \rightharpoonup_Ł b = 1 \wedge (1 - a + b)$.
- The **Gödel algebra** $([0,1], \wedge, \vee, \otimes_G, \rightharpoonup_G, 0, 1)$ with $\otimes_G = \wedge$ and $a \rightharpoonup_G b = 1$ if $a \leq b$ and $a \rightharpoonup_G b = b$ otherwise.
- A **BL-algebra** is a residuated lattice which satisfies *divisibility*: $a \wedge b = a \otimes (a \rightharpoonup b)$, and *prelinearity*: $(a \rightharpoonup b) \vee (b \rightharpoonup a) = 1$.
- An **MV-algebra** is a BL-algebra satisfying the law of double negation: $a = \neg\neg a$.
- A Boolean algebra is a residuated lattice which is both a Heyting algebra and an MV-algebra.
- Boolean, Łukasiewicz (with negation operation defined as $\neg_L\, a = 1 - a$), and Gödel algebras (with $\neg_G\, a = 1$, if $a = 0$ and $\neg_G\, a = 0$, if $a > 0$, respectively) are all BL-algebras.

4.4.1 *Various Properties of Residuated Lattices*

Next we give some important properties of residuated lattices, mainly following the presentation in [4, Section 2.3.2], where also the proofs omitted below can be found.

Proposition 4.27. ([4, Theorem 2.17]) *Let* $(L, \wedge, \vee, \otimes, \curvearrowright, 0, 1)$ *be a residuated lattice.*

(1) *For a given multiplication* $\otimes$*, there is at most one operation* $\curvearrowright$ *satisfying the adjointness property (see* Def. 4.26(3)*).*
(2) $x \otimes (x \curvearrowright y) \leq y, \quad y \leq x \curvearrowright (x \otimes y), \quad x \leq (x \curvearrowright y) \curvearrowright y.$
(3) $x \leq y$ *iff* $x \curvearrowright y = 1$.
(4) $x \curvearrowright x = 1, \quad x \curvearrowright 1 = 1, \quad 0 \curvearrowright x = 1.$
(5) $1 \curvearrowright x = x$.
(6) $x \otimes y \leq x, \quad x \leq y \curvearrowright x, \quad x \otimes y \leq x \wedge y.$
(7) $x \otimes 0 = 0 = 0 \otimes x$.
(8) $(x \otimes y) \curvearrowright z = x \curvearrowright (y \curvearrowright z)$.
(9) $x \curvearrowright y$ *is the greatest element in the set* $\{z : x \otimes z \leq y\}$.
(10) $x \otimes y$ *is the least element in the set* $\{z : x \leq y \curvearrowright z\}$.

Proof. (1): Assume $\rightsquigarrow$ is another operation satisfying the adjointness property; then, for any x, $x \leq y \curvearrowright z$ iff $x \otimes y \leq z$ iff $x \leq y \rightsquigarrow z$, whence obviously $y \curvearrowright z = y \rightsquigarrow z$. Similarly, $\curvearrowright$ uniquely determines $\otimes$.

(2): Using commutativity of $\otimes$ and adjointness we have $x \curvearrowright y \leq x \curvearrowright y$ iff $(x \curvearrowright y) \otimes x = x \otimes (x \curvearrowright y) \leq y$, from which it follows by adjointess that $x \leq (x \curvearrowright y) \curvearrowright y$.

(3): $x \leq y$ iff $1 \otimes x \leq y$ iff $1 \leq x \curvearrowright y$ iff $1 = x \curvearrowright y$.

(4): Direct consequences of (3).

(5): We have $x \leq 1 \curvearrowright x$ iff $x = 1 \otimes x \leq x$. The other inequality we get as follows: $1 \curvearrowright x \leq 1 \curvearrowright x$ gives $1 \curvearrowright x = 1 \otimes (1 \curvearrowright x) \leq x$, using (2).

(6): By (4) we have $x \curvearrowright x = 1$, whence $y \leq x \curvearrowright x$, and $x \otimes y \leq x$ by adjointness. Thus, we also have $y \otimes x \leq y$. Moreover, $x \otimes y = y \otimes x$, whence $x \otimes y \leq x \wedge y$, and again by adjointness $x \leq y \curvearrowright x$.

(7): $x \otimes 0 = 0$ iff $x \otimes 0 \leq 0$ iff $0 \leq x \curvearrowright 0$.

(8): We prove both inequalities and begin with

$$(x \otimes y) \curvearrowright z \leq x \curvearrowright (y \curvearrowright z)$$
$$\text{iff} \quad x \otimes [(x \otimes y) \curvearrowright z] \leq y \curvearrowright z$$
$$\text{iff} \quad (x \otimes y) \otimes [(x \otimes y) \curvearrowright z] \leq z,$$

which holds by the first inequality of (2). Assuming that the other inequality holds, we get

$$x \curvearrowright (y \curvearrowright z) \leq (x \otimes y) \curvearrowright z$$
$$\text{iff} \quad (x \otimes y) \otimes [x \curvearrowright (y \curvearrowright z)] \leq z \qquad \text{(commutativity, adjointness)}$$
$$\text{iff} \quad y \otimes x \otimes [x \curvearrowright (y \curvearrowright z)] \leq z \qquad \text{(commutativity, associativity, adjointness)}$$
$$\text{iff} \quad x \otimes [x \curvearrowright (y \curvearrowright z)] \leq (y \curvearrowright z). \qquad \text{(adjointness)}$$

But this last inequality again holds by (2).

(9): By (2) $x \otimes (x \curvearrowright y) \leq y$, whence $x \curvearrowright y \in \{z : x \otimes z \leq y\}$, and, by adjointness, $x \otimes z \leq y$ implies $z \leq x \curvearrowright y$.
(10): By (2) $x \leq y \curvearrowright (x \otimes y)$, whence $x \otimes y \in \{z : x \leq y \curvearrowright z\}$, and, by adjointness, $x \leq y \curvearrowright z$ implies $x \otimes y \leq z$. □

A class of algebras of the same type is called a **variety** if it is the class of all algebras satisfying a given set of identities. Equivalently, a variety is a class of algebras which is closed under taking homomorphic images, subalgebras, and direct products of algebras of this class. See e.g. [43, Section 10.2.4].

Proposition 4.28. ([4, Theorem 2.18]) *The class of all residuated lattices is a variety of algebras.*

Proposition 4.29. (Monotonicity properties of $\otimes$ and $\curvearrowright$; [4, Theorem 2.23]) *In a residuated lattice $(L, \wedge, \vee, \otimes, \curvearrowright, 0, 1)$ the following holds:*

(1) $y \leq y' \to x \otimes y \leq x \otimes y'$.
(2) $y \leq y' \to x \curvearrowright y \leq x \curvearrowright y'$.
(3) $x \leq x' \to x' \curvearrowright z \leq x \curvearrowright z$.

Proof. (1): From Prop. 4.27(2) we get $y' \leq x \curvearrowright (x \otimes y')$, whence, by the assumption $y \leq y'$, also $y \leq x \curvearrowright (x \otimes y')$, and this is equivalent to $x \otimes y \leq x \otimes y'$.
(3) By adjointness we have $x' \curvearrowright z \leq x \curvearrowright z$ iff $x \otimes (x' \curvearrowright y) \leq y$. But $x \leq x'$ implies $x \otimes (x' \curvearrowright y) \leq x' \otimes (x' \curvearrowright y) \leq y$, by (1) and Prop. 4.27(2). □

So, by commutativity of $\otimes$, it is isotone in both arguments, and therefore has all the constituting properties of a t-norm on L (see Def. 9.17). $\curvearrowright$ is antitone in the first and isotone in the second argument.

Proposition 4.30. (Distributivity of $\otimes, \curvearrowright$ over $\wedge, \vee$; [4, Theorem 2.25]) *Let $(L, \wedge, \vee, \otimes, \curvearrowright, 0, 1)$ be a residuated lattice, and I an index set. Then*

(1) $x \otimes \bigvee_{i \in I} y_i = \bigvee_{i \in I}(x \otimes y_i)$.
(2) $x \otimes \bigwedge_{i \in I} y_i \leq \bigwedge_{i \in I}(x \otimes y_i)$.
(3) $x \curvearrowright \bigwedge_{i \in I} y_i = \bigwedge_{i \in I}(x \curvearrowright y_i)$.
(4) $\bigvee_{i \in I} x_i \curvearrowright y = \bigwedge_{i \in I}(x_i \curvearrowright y)$.

Proof. (1): We have $x \otimes y_i \leq x \otimes \bigvee_{i \in I} y_i$, for each $i \in I$, whence $\bigvee_{i \in I}(x \otimes y_i) \leq x \otimes \bigvee_{i \in I} y_i$. On the other hand, $x \otimes y_i \leq \bigvee_{i \in I}(x \otimes y_i)$; so $y_i \leq x \curvearrowright \bigvee_{i \in I}(x \otimes y_i)$ for each $i \in I$. Therefore, $\bigvee_{i \in I} y_i \leq x \curvearrowright \bigvee_{i \in I}(x \otimes y_i)$, whence $x \otimes \bigvee_{i \in I} y_i \leq \bigvee_{i \in I}(x \otimes y_i)$.
(3): Since $\curvearrowright$ is monotone in the second argument, we get $x \curvearrowright \bigwedge_{i \in I} y_i \leq x \curvearrowright y_i$, for all $i \in I$; so $x \curvearrowright \bigwedge_{i \in I} y_i \leq \bigwedge_{i \in I}(x \curvearrowright y_i)$. Let $z := \bigwedge_{i \in I}(x \curvearrowright y_i)$; then, $z \leq x \curvearrowright y_i$, hence $z \otimes x \leq y_i$, for all $i \in I$. Thus, $z \otimes x \leq \bigwedge_{i \in I} y_i$. Therefore, $z \leq x \curvearrowright \bigwedge_{i \in I} y_i$, which gives the converse inequality. □

Corollary 4.31. *For a residuated lattice* $\mathscr{L} = (L, \wedge, \vee, \otimes, \curvearrowright, 0, 1)$ *its reduct* $(L, \vee, \otimes, 0, 1)$ *is an incline.*

Proof. By Def. 4.26 $(L, \vee)$ is a semilattice, and $(L, \otimes)$ is a commutative semigroup. By Prop. 4.30(3) we get $a \otimes (b \vee c) = (a \otimes b) \vee (a \otimes c)$. Finally, from Prop. 4.27(6) we get $(a \otimes b) \leq a$, so $a \vee (a \otimes b) = a$. □

Another important operation in the context of residuated lattices is **biresiduation** defined as $x \leftrightharpoons y = x \curvearrowright y \wedge y \curvearrowright x$ with the following properties:

Corollary 4.32.

(1) $x \leftrightharpoons y = y \leftrightharpoons x.$
(2) $x \leftrightharpoons x = 1.$
(3) $x \leftrightharpoons 1 = x.$
(4) $x = y$ *iff* $x \leftrightharpoons y = 1.$
(5) $(x \leftrightharpoons y) \otimes (y \leftrightharpoons z) \leq x \leftrightharpoons z.$
(6) $x \otimes y \leq x \wedge y \leq x \leftrightharpoons y.$

Proof. (2): By Prop. 4.27(4).

(3): By Prop. 4.27(4) and (5).

(4): Assuming $x = y$, then $x \leftrightharpoons y = 1$ by (2). For the other direction assume $x \leftrightharpoons y = x \curvearrowright y \wedge y \curvearrowright x = 1$. Then $x \curvearrowright y = 1$ and $y \curvearrowright x = 1$, and by Prop. 4.27(3) $x \leq y \wedge y \leq x$.

(5): By adjointness, $(x \leftrightharpoons y) \otimes (y \leftrightharpoons z) \leq x \leftrightharpoons z$ is equivalent to $x \otimes (x \leftrightharpoons y) \otimes (y \leftrightharpoons z) \leq z$. Using Prop. 4.29(1) and Prop. 4.27(2) twice, the second inequality follows from $x \otimes (x \leftrightharpoons y) \otimes (y \leftrightharpoons z) \leq x \otimes (x \curvearrowright y) \otimes (y \curvearrowright z) \leq y \otimes (y \curvearrowright z) \leq z$.

(6): Prop. 4.27(6) implies $x \otimes y \leq x \wedge y$ and $x \leq y \curvearrowright x, y \leq x \curvearrowright y$, whence $x \wedge y \leq x \leftrightharpoons y$. □

Proposition 4.33. ([4, Theorem 2.30]) *Let* $(L, \wedge, \vee, \otimes, \curvearrowright, 0, 1)$ *be a complete residuated lattice; for the negation operation* $\neg a := a \curvearrowright 0$ *the following holds:*

(1) $a \leq b \rightarrow \neg b \leq \neg a \rightarrow \neg\neg a \leq \neg\neg b.$
(2) $a \leq \neg\neg a, \quad a \leq b \rightarrow a \leq \neg\neg b.$
(3) $\neg 0 = 1, \quad \neg 1 = 0.$
(4) $a \otimes \neg a = 0;$
(5) $\neg(a \otimes b) = a \curvearrowright \neg b = b \curvearrowright \neg a.$
(6) $\neg\neg 0 = 0, \quad \neg\neg 1 = 1.$
(7) $a \otimes b = 0 \leftrightarrow a \leq \neg b.$
(8) $a \leq b \rightarrow a \otimes \neg b = 0.$
(9) $\neg a = \neg\neg\neg a.$

Proof. (1): By Prop. 4.29(3), $a \leq b$ implies $b \curvearrowright 0 \leq a \curvearrowright 0$.

(2): By Prop. 4.27(2), $a \leq (a \curvearrowright 0) \curvearrowright 0 = \neg a \curvearrowright 0 = \neg\neg a$; the implication follows by application of this result to b.

(3): By Prop. 4.27(4) and (5), $\neg 0 = 0 \curvearrowright 0 = 1$, and $\neg 1 = 1 \curvearrowright 0 = 0$.

(4): By Prop. 4.27(2), $a \otimes (a \curvearrowright 0) \leq 0$. By definition of $\neg$ we have $a = 0$, or $\neg a = 0$ otherwise.

(5): By Prop. 4.27(7), $\neg(a \otimes b) = (a \otimes b) \curvearrowright 0 = a \curvearrowright (b \curvearrowright 0)$.

(6): By (3), $0 = \neg 1 = \neg\neg 0$.

(7): $a \leq \neg b$ follows by the definition of the latter as the greatest element such that $a \otimes b = 0$. Conversely, $a \leq \neg b$ implies $a \otimes b \leq \neg b \otimes b = 0$ by (4).

(8): $a \leq b$ implies $\neg b \leq \neg a$ by (1); thus $\neg b \otimes a = 0$ by (7).

(9): By (2) $\neg a \leq \neg\neg\neg a$. Conversely, by monotonicity of $\wedge$, (2), and (4), $\neg\neg\neg a \wedge a \leq \neg\neg\neg a \wedge \neg\neg a = 0$. So by (7) we get $\neg\neg\neg a \leq \neg a$. □

(As can be seen from (2) the law of double negation does not hold in general for residuated lattices, i.e. $\neg$ is not involutive. However, by (6) we have $\neg\neg 0 = 0$ and $\neg\neg 1 = 1$, i.e. the restriction to the two-valued case would work. In contrast, MV-algebras satisfy the double negation property, whence they are used for investigations of fuzzy relation algebras in [32, 33]; see also [29].)

Proposition 4.34. ([4, Theorem 2.28]) *Let $(L, \wedge, \vee, \otimes, 0, 1)$ be a complete bounded lattice, such that $(L, \otimes, 1)$ is a commutative monoid. Then the following conditions are equivalent:*

(1) *There exists an operation $\curvearrowright : L \times L \longrightarrow L$ satisfying adjointness: $x \otimes y \leq z$ iff $x \leq y \curvearrowright z$.*
(2) *For any $x, y \in L$ the set $\{z : x \otimes z \leq y\}$ has a greatest element.*
(3) *$x \otimes \bigvee_{i\in I} y_i = \bigvee_{i\in I}(x \otimes y_i)$ holds in L.*
(4) *With $x \curvearrowright y := \bigvee\{z : x \otimes z \leq y\}$, $\otimes$ and $\curvearrowright$ satisfy adjointness.*

Proof. (4) $\rightarrow$ (1): Obvious, i.e. L is a residuated lattice.

(1) $\rightarrow$ (2): By Prop. 4.27(9).

(1) $\rightarrow$ (3): By Prop. 4.30(1).

(2) $\rightarrow$ (4): Assume (2) and define $\curvearrowright$ as done in (4). Then $x \curvearrowright y$ is the greatest element in $\{z : x \otimes z \leq y\}$, i.e. $x \otimes z \leq y$ implies $z \leq x \curvearrowright y$. And if $z \leq x \curvearrowright y$ then $x \otimes z \leq x \otimes (x \curvearrowright y) \leq y$. Hence, $\otimes$, $\curvearrowright$ satisfy adjointness.

(3) $\rightarrow$ (4): Assume (3) and define $\curvearrowright$ as done in (4). If $x \otimes z \leq y$ then $z \leq x \curvearrowright y$ by definition of $x \curvearrowright y$. And if $z \leq x \curvearrowright y$ then $x \otimes z \leq x \otimes (x \curvearrowright y) = x \otimes \bigvee\{z' : a \otimes z' \leq y\} = \bigvee\{a \otimes z' : a \otimes z' \leq y\} \leq y$, whence (4) holds. □

In order to understand why residuated lattices get their prominent role in the area of fuzzy relations, note that Sanchez (following Birkhoff [5] in his seminal paper [36]) used the term *Brouwerian* for what usually is called a *relatively pseudocomplemented* lattice (see Def. 4.20). This terminology has been prevalent since then in literature on fuzzy relation equations and inequalities; see e.g. [12, 14, 30] for usage in books. The term is still used in many papers, e.g. in [38]; but in rare cases you can find the other name, viz. *Heyting* lattice as e.g. in [23]. Now recall that, in the terminology of [36], a **Heyting algebra** is nothing else than a Brouwerian, i.e. relatively pseudocomplemented, lattice with a zero element (standing for falsehood, or "*is not*"), and thus a residuated lattice with $a \otimes b := a \wedge b$.

4.4.2 *Residuated Lattices and Semirings*

Though residuated lattice are not required to be distributive, and semirings need not be residuated, the two concepts are very similar. It is therefore interesting to have a short look at how they interfere. Our exposition follows the one in [10]. But before we will focus on technical details, we insert some informal characterizations that provide us with first orientation marks:

> Loosely speaking, residuation is the ability to distinguish best, i.e., closest, approximate solutions to linear equations, and as such it gives us a way to carry out division (to a greater or lesser degree of accuracy).
>
> ...
>
> A residuated lattice is essentially an idempotent semiring which is simultaneously a lattice, and in which each linear equation has a maximal approximate solution. In particular, this means that if a and b are elements of a residuated lattice then there is a maximal approximate solution to the linear equation $ax = b$, that is, there is a maximal element x satisfying $ax \leq b$.
> [44, p. 16]

Theorem 4.35. ([10, Theorem 4]) *Let* $(L, \vee, \wedge, \odot, \curvearrowright, 0, 1)$ *be a residuated lattice. Then* $(L, \vee, \odot, 0, 1)$ *is a commutative, additively idempotent, and simple semiring.*

Proof. Let $a, b, c \in L$. Since $(L, \vee, \wedge, 0, 1)$ is a bounded lattice, its reduct $(L, \vee, 0)$ is a commutative monoid, with $\vee$ idempotent. Further, $(L, \odot, 1)$ is a commutative monoid by definition. Since 0 is the least element of L, we have $0 \leq a \curvearrowright 0$ and according to the adjointness property $0 \odot a \leq 0$ and thus $0 \odot a = 0$. By Prop. 4.30(1) we have $a \odot (b \vee c) = (a \odot b) \vee (a \odot c)$. And finally, $a \vee 1 = 1$. □

In order to obtain a residuated lattice from a given semiring we need the following concept.

Definition 4.36. ([10, Definition 7]) *A commutative additively idempotent semiring* $(S, +, \cdot, 0, 1)$ *is called* **completely distributive** *if the following conditions hold (note that both conditions hold in case S is finite):*

(1) *In the join-semilattice* $(S, +)$*, the sum* $\sum_{i \in I} y_i$ *exists for all sets* I*,* $y_i \in S$*, and* $i \in I$*;*
(2) $x \cdot (\sum_{i \in I} y_i) = \sum_{i \in I} (x \cdot y_i)$ *for all* $x, y_i \in S$ *and* $i \in I$*.*

Theorem 4.37. ([10, Theorem 8]) *Let* $(S, +, \cdot, 0, 1)$ *be a commutative idempotent simple completely distributive semiring. Then there exist binary operations* $\wedge$ *and* $\curvearrowright$ *on S such that* $(S, +, \wedge, \cdot, \curvearrowright, 0, 1)$ *is a residuated lattice.*

Proof. Let $a, b, c \in S$. Clearly, $(S, +, 0)$ is a join-semilattice with least element 0. Since S is simple, $a + 1 = 1$, i.e. $a \leq 1$, where $\leq$ denotes the induced semilattice order. In other words, $(S, +, 0, 1)$ is a bounded join-semilattice. According to Def. 4.36(i), in the poset $(S, \leq)$ there exist arbitrary suprema and hence also arbitrary infima. Therefore, we obtain a bounded complete lattice $(S, +, \wedge, 0, 1)$. Putting $a \curvearrowright b := \sum_{a \cdot x \leq b} x$, it follows that $a \cdot b \leq c$ implies $a \leq b \curvearrowright c$.

Conversely, according to complete distributivity of $(S, +, \cdot, 0, 1)$ and isotonicity of $\cdot$, the assertion $a \leq b \curvearrowright c$ implies $a \cdot b \leq (b \curvearrowright c) \cdot b = \left(\sum_{b \cdot x \leq c} x\right) \cdot b = \sum_{b \cdot x \leq c} (x \cdot b) \leq c$. □

Corollary 4.38. ([10, Corollary 9]) *Let $\mathcal{S} = (S, +, \cdot, 0, 1)$ be a commutative idempotent simple finite semiring. Then there exist binary operations $\wedge$ and $\curvearrowright$ on S such that $(S, +, \wedge, \cdot, \curvearrowright, 0, 1)$ is a residuated lattice.*

4.5 Summary

TABLE 4.1: Comparison of structures (• means '*by definition*' and ⊢ '*derived*').

	+ assoc-iative	+ commu-tative	+ 0	+ idem-potent	· assoc-iative	· commu-tative	· 1	· idem-potent	· zero	distri-butive
semiring	•	•	•		•		•		0	•
path algebra	•	•	•	•	•		•		0	•
incline	•	•	•	•	•	•	•		⊢	•, ⊢
resid. lattice	•	•	•	•	•	•	•	•	⊢	•

For inclines, see the derivations in Cor. 4.18, and for residuated lattices Prop. 4.27(7). Further note that according to Def. 4.17 inclines obey the additional condition (In4) $a + a \cdot b = a$, for all $a, b \in L$ (not included in Table 4.1 above).

References

[1] Baccelli, F., Cohen, G., Olsder, G.J., Quadrat, J.-P.: *Synchronization and Linearity*, Web-Edition (2001).

[2] Balbes, R., Dwinger, P.: *Distributive Lattices*, University of Missouri Press (1974).

[3] Baras, J.S., Theodorakopoulos, G.: *Path Problems in Networks*, Synthesis Lectures on Communication Networks 3, Morgan & Claypool (2010).

[4] Bělohlàvek, R.: *Fuzzy Relational Systems - Foundations and Principles*, Springer Science + Business Media, New York (2002).

[5] Birkhoff, G.: *Lattice Theory*, 3rd ed., American Mathematical Society, Providence, Rhode Island (1973).

[6] Blyth, T.S.: *Lattices and Ordered Algebraic Structures*, Springer-Verlag, London (2005).

[7] Cain, A.J.: Nine Chapters on the Semigroup Art, version 0.66.62, (2020).

[8] Cao, Z.Q., Kim, K.H., Roush, F.W.: *Incline Algebra and Applications*, Ellis Horwood Limited, Chichester (1984).

[9] Carre, B.: *Graphs and Networks*, Oxford University Press (1979).

[10] Chajda, I., Länger, H.: When does a semiring become a residuated lattice?, Asian-European J. of Mathematics **9**, 1650088-1 - 1650088-10 (2016).

[11] Curry, H.B.: *Foundations of Mathematical Logic*, Dover Publications Inc. New York (1977).

[12] Di Nola, A., Sessa, S., Pedrycz, W., Sanchez, E.: *Fuzzy Relation Equations and Their Applications to Knowledge Engineering*, Kluwer Academic Publishers, Dordrecht (1989).

[13] Duan, J.-S.: The transitive closure, convergence of powers and adjoint of generalized fuzzy matrices, Fuzzy Sets and Systems **145**, 301-311 (2004).

[14] Dubois, D., Prade, H.: *Fuzzy Sets and Systems: Theory and Applications*, Academic Press, New York (1980).

[15] Golan, J.S.: *Semirings and their Applications*, Springer Science+Business Media Dordrecht (1999).

[16] Golan, J.S.: *Semirings and Affine Equations over Them: Theory and Applications*, Springer Science+Business Media Dordrecht (2003).

[17] Gondran, M., Minoux, M.: Dioids and semirings - Links to fuzzy sets and other applications, Fuzzy Sets and Systems **158**, 1273-1294 (2007).

[18] Gondran, M., Minoux, M.: *Graphs, Dioids and Semirings - New Models and Algorithms*, Springer Science & Business Media (2008).

[19] Gunawardena, J.: An introduction to idempotency, HPL-BRIMS-96-24, HP Laboratories Bristol (1996).

[20] Hazewinkel, M. (Ed.): *Encyclopaedia of Mathematics* Vol 7, Kluwer Academic Publishers, Dordrecht (1991).

[21] Howie, J.M: *Fundamentals of Semigroup Theory*, Clarendon Press, Oxford (1995).

[22] Jiang, J., Tian, X.: The composition, convergence and transitivity of powers and adjoint of generalized fuzzy matrices, Italian J. of Pure and Applied Mathematics **36**, 377-398 (2016).

[23] Jiménez, J., Montes, S., Šešelja, B., Tepavčević, A.: Fuzzy correspondence inequations and equations. Fuzzy Sets and Systems **239**, 81-90 (2014).

[24] Kuich, W., Salomaa, A.: *Semirings, Automata, Languages*, Springer-Verlag Berlin Heidelberg (1986).

[25] Lallement, G.: *Semigroups and Combinatorial Applications*, John Wiley & Sons (1979).

[26] Mohri, M.: Semiring frameworks and algorithms for shortest-distance problems, Journal of Automata, Languages and Combinatorics **7**, 321-350 (2002).

[27] Nemitz, W.C.: Implicative semi-lattices, Trans. of the American Math. Soc. **117**, 128-142 (1965).

[28] Novák, V., Perfilieva, I., Močkoř, J.: *Mathematical Principles of Fuzzy Logic*, Kluwer Academic Publishers, Boston/Dordrecht (1999).

[29] Pan, X., Xu, Y.: On the algebraic structure of binary lattice-valued fuzzy relations. Soft Computing, **17**, 411-420 (2013).

[30] Peeva, K., Kyosev, Y.: *Fuzzy Relational Calculus. Theory, Applications and Software*, World Scientific, Singapore (2004).

[31] Perfilieva, I., Nosková, L.: System of fuzzy relation equations with inf- $\rightarrow$ composition: Complete set of solutions. Fuzzy Sets and Systems **159**, 2256-2271 (2008).

[32] Popescu, A.: Many-valued relation algebras, Algebra universalis **53**, 73-108 (2005).

[33] Popescu, A.: Some algebraic theory for many-valued relation algebras, Algebra universalis **56**, 211-235 (2007).

[34] Rasiowa, H.: *Algebraic Approach to Non-classical Logics*, North Holland, Amsterdam (1974).

[35] Rybakow, V.V.: *Admissibility of Logical Inference Rules*, Elsevier Science B.V., Amsterdam (1997).

[36] Sanchez, E.: Resolution of Composite Fuzzy Relation Equations, Information and Control **30**, 38-48 (1976).

[37] Smith, D.P.: Meet-Irreducible elements in implicative lattices, Proc. of the American Math. Soc. **34**, 57-62 (1972).

[38] Sun, F., Wang, X.P., Qu, X.B.: Minimal join decompositions and their applications to fuzzy relation equations over complete Brouwerian lattices, Information Sciences **224**, 143-151 (2013).

[39] Winter, M.: *Goguen Categories: a Categorical Approach to L-fuzzy Relations*, Springer Science & Business Media (2007).

[40] Tan, Y.-J.: Eigenvalues and eigenvectors for matrices over distributive lattices, Linear Algebra and its Applications **283**, 257-272 (1998).

[41] Tan, Y.-J.: On invertible matrices over commutative semirings, Linear and Multilinear Algebra **61**, 710-724 (2013).

[42] Tan, Y.-J.: Bases in semimodules over commutative semirings, Linear Algebra and its Applications **443**, 139-152 (2014).

[43] Toth, H.: *Relations: Concrete, Abstract and Applied - An Introduction*, World Scientific Publishing, Singapore (2020).

[44] Wilding, D.: Linear Algebra Over Semirings, PhD Thesis, School of Mathematics, University of Manchester (2014).

[45] Zhao, C.-K.: On matrix equations in a class of complete and completely distributive lattices, Fuzzy Sets and Systems **22**, 303-320 (1987).

CHAPTER 5

ALGEBRAIC PRELIMINARIES

Investigation on matrices over lattices and semirings started earlier than fuzzy set theory did emerge and has, at least for a longer period, evolved independently of it; see e.g. [11, 20, 25]. The development of fuzzy relation and matrix theory began in 1971, when Zadeh [26] introduced fuzzy equivalence relations, and emerged as a mean of representing fuzzy equivalences and fuzzy relations between finite sets. Fuzzy matrix theory in its proper sense has been initiated in 1980 by Kim and Roush [17] who introduced a general framework as a generalization of Boolean matrix theory and the theories of matrices over nonnegative real numbers, nonnegative integers, and over other types of semirings. They established basic concepts of linear algebra over fuzzy matrices.

The first part of this chapter is a short recapitulation of basic concepts from linear algebra. In the second section we will have a look on semilinear spaces.

5.1 Linear Algebra: A Very Short Summary

Before we start the presentation of the proper topics of this chapter, it seems appropriate to provide something like a dictionary of "classical" concepts and results in order to enable their comparison with what is suitable for and holds in the context of the different types of semiring we are going to consider. So this section contains a list of standard definitions and facts relevant for the considerations below, but with only few comments, and without proofs. (For more details you may look into one of [1, 4, 6, 8].) This also supports one of our main goals: to keep the book self-contained to a fair degree in order to enhance convenience of its usage.

We start with the definition of *vector space*. In linear algebra vector spaces are usually considered over sets that form a division ring or a field (see Def. 1.35), with $\mathbb{R}$ as the standard example. So $\mathbb{R}^n$ becomes the paradigmatic vector space, with the usual componentwise addition and scalar multiplication of n-tuples. If nothing else is specified, matrices and vectors are assumed to range over some field F. Combined with basic properties of real numbers, vector addition and scalar multiplication in $\mathbb{R}^n$ have the properties listed in

Definition 5.1. *Let R be a division ring, $V_n \subseteq R^n$, and $\mathbf{0}$ denote the n-tuple $(0,0,\ldots 0)$. If for all $s,t \in R$ and all $\mathbf{u},\mathbf{v},\mathbf{w} \in V_n$ we have*

(A1) $\mathbf{u}+\mathbf{v} \in V_n$ *(Closure under addition),*
(A2) $(\mathbf{u}+\mathbf{v})+\mathbf{w} = \mathbf{u}+(\mathbf{v}+\mathbf{w})$ *(Associativity),*
(A3) $(\mathbf{u}+\mathbf{v}) = (\mathbf{v}+\mathbf{u})$ *(Commutativity),*
(A4) $\mathbf{u}+\mathbf{0} = \mathbf{u}$ *(Additive identity),*
(A5) *there is* $(-\mathbf{u}) \in V_n$ *such that* $\mathbf{u}+(-\mathbf{u}) = \mathbf{0}$,

(M1) $s\mathbf{u} \in V_n$ *(Closure under scalar multiplication),*
(M2) $(st)\mathbf{u} = s(t\mathbf{u})$ *(Associativity),*
(M3) $s(\mathbf{u}+\mathbf{v}) = s\mathbf{u}+s\mathbf{v}$ *(Distributivity),*
(M4) $(s+t)\mathbf{u} = s\mathbf{u}+t\mathbf{u}$ *(Distributivity),*
(M5) $1\mathbf{u} = \mathbf{u}$ *(Scalar identity),*

then V_n is called a **vector space** *(with addition and scalar multiplication defined componentwise in the usual way).*

Remark: Vector spaces over division rings are also called **linear spaces**; this explains the name *semilinear space* which have semirings as underlying structure.

Vector spaces, well known to most of us, are a special case of a more general concept, called *modules* (see e.g. [7, Section 1.4] or [19, Chapter 9]), defined in

Definition 5.2. *Let R be a ring with 1_R its multiplicative identity. A (***left***) R-***module** *M consists of an Abelian group $(M,+)$ and an operation $\cdot : R \times M \longrightarrow M$ such that for all $r,s \in R$ and $x,y \in M$ the following holds:*

(a) $r \cdot (x+y) = r \cdot x + r \cdot y$,
(b) $(r+s) \cdot x = r \cdot x + s \cdot x$,
(c) $(rs) \cdot x = r \cdot (s \cdot x)$,
(d) $1_R \cdot x = x$.

The operation of the ring on M is called **scalar multiplication**, and is usually also written by juxtaposition, i.e. as rx for $r \in R$ and $x \in M$, since it is clear which kind of multiplication is to be used. If R denotes a field, then vector spaces over R and (left) R-modules are identical.

Some examples of modules are: (i) Every ring R is itself an R-module with usual multiplication in R as scalar multiplication. (ii) Every ideal I of a ring R is an R-module with $\cdot : R \times I \longrightarrow I$, $(r,x) \mapsto rx$, for $r \in R$ and $x \in I$. (iii) Every Abelian group is a module over the ring of integers $\mathbb{Z}$: For $n > 0$ we let $nx = x+x+\cdots+x$ (n summands), $0x = 0$, and, for $n < 0$, $nx = -(nx)$. (iv) Let M denote the set of all $m \times n$ matrices with entries in R. Then M is an R-module, where addition is ordinary matrix addition, and multiplication of $A \in M$ with the scalar $r \in R$ is multiplication of each entry with r.

In using V_n to denote n-tuples as $1 \times n$ matrices, called *row vectors,* we follow [17]. The isomorphic set of $n \times 1$-matrices is called the set of (the n-dimensional) *column vectors*, denoted as V^n. For any result about V_n there is a corresponding one that holds in V^n, so we need not always be overly strict in using rows or columns in our wording.

As demonstrated by the examples given above, the definition of "vectors" need not be restricted to n-tuples of real numbers, but may e.g. also comprise $m \times n$ matrices with real entries and matrix addition and scalar multiplication as vector space operations, $m, n \geq 1$; or even real-valued functions. See [4, Section 4.1].

Often we are interested in those subsets of $W \subseteq R^n$ whose members can be added or multiplied by a scalar without yielding new vectors $\boldsymbol{v} \in R^n - W$:

Definition 5.3. *A* **linear subspace**, *or simply* **subspace**, *of a vector space V is a subset $\emptyset \neq W \subseteq V$ that is closed under addition and scalar multiplication, i.e. for $\boldsymbol{v}, \boldsymbol{w} \in W$ we have $s\boldsymbol{v} \in W$ and $\boldsymbol{v} + \boldsymbol{w} \in W$, for all $s \in R$. Equivalently: $\emptyset \neq W \subseteq V$ is a subspace of V if $\boldsymbol{v} + s\boldsymbol{w} \in W$ for every $\boldsymbol{v}, \boldsymbol{w} \in W$ and every $s \in R$.*

5.1.1 *Linear (In)Dependence, Dimension, Basis*

Definition 5.4. ([4, p.196, Definition 3], [22, Definition 3.2]) *Let V be a vector space over a division ring R.*

(a) A finite sum $\sum_{i \in \underline{k}} s_i \boldsymbol{v}_i$ is called a **linear combination** *of the vectors $\boldsymbol{v}_i \in V$, where $s_i \in R$.*
(b) The **span** *of $W \subseteq V$, denoted by $\langle W \rangle$, is the set of all such linear combinations of elements of W, with the convention that $\sum_\emptyset = \boldsymbol{0}$. If $\langle W \rangle = T$ then W is called a* **spanning set** *or* **set of generators** *for T.*
(c) A subspace $W \subseteq V$ is said to be **finitely generated** *if there exists a finite set of generators for W.*
(d) Let $W = \{\boldsymbol{w}_1, \boldsymbol{w}_2, \ldots, \boldsymbol{w}_m\} \subseteq V$ and $U = \{\boldsymbol{u}_1, \boldsymbol{u}_2, \ldots, \boldsymbol{u}_n\} \subseteq V$ be two sets of vectors. If every $\boldsymbol{w}_i, i \in \underline{m}$, can be represented as a linear combination of vectors in U, and every $\boldsymbol{u}_j, j \in \underline{n}$ can be represented as a linear combination of vectors in W, then U and W are said to be **equivalent**.

Definition 5.5. ([4, p.203, Definition 1]) *Let V be a vector space.*

(a) A set $W = \{\boldsymbol{w}_1, \boldsymbol{w}_2, \ldots, \boldsymbol{w}_n\}, n \geq 2$, of vectors in V is said to be a **linearly independent** *if no vector in W can be expressed as a linear combination of the others, i.e. if $\boldsymbol{w}_i \notin \langle W - \{\boldsymbol{w}_i\} \rangle$, for any $i \in \underline{n}$. A set that is not linearly independent is said to be* **linearly dependent**.
(b) A **basis** *of a subspace $W \subseteq V$ is a linearly independent set B of generators for W, i.e. $W = \langle B \rangle$.*
(c) If V contains a finite set of n linearly independent vectors and any set of $n+1$ vectors is linearly dependent, then V is said to be of finite **dimension** *n, written as $\dim V = n$.*

In classical linear algebra (i.e. over fields) we have the following result:

Theorem 5.6. ([8, Theorem 3.8]) *The vectors $\boldsymbol{v}_1, \boldsymbol{v}_2, \ldots, \boldsymbol{v}_m$, $m \geq 2$, are linearly dependent if and only if one of them can be expressed as a linear combination of the others.*

Proof. Suppose that $\boldsymbol{v}_1, \boldsymbol{v}_2, \ldots, \boldsymbol{v}_m$, $m \geq 2$, are linearly dependent; then there exist scalars $c_i, i \in \underline{m}$ such that $c_1 \boldsymbol{v}_1 + \cdots + c_m \boldsymbol{v}_m = 0$, with at least one of the c_i not zero , say $c_k \neq 0$. Then this equation can be rewritten as $\boldsymbol{v}_k = -\frac{c_1}{c_k} \boldsymbol{v}_1 - \cdots - \frac{c_m}{c_k} \boldsymbol{v}_m$, i.e. $\boldsymbol{v}_k$ is a linear combination of the other $m-1$ vectors.

Conversely, assume that $\boldsymbol{v}_k = b_1 \boldsymbol{v}_1 + \cdots + b_{k-1} \boldsymbol{v}_{k-1} + b_{k+1} \boldsymbol{v}_{k+1} + \cdots + b_m \boldsymbol{v}_m$. Then we can write this as $b_1 \boldsymbol{v}_1 + \cdots + b_{k-1} \boldsymbol{v}_{k-1} + (-1) \boldsymbol{v}_k + b_{k+1} \boldsymbol{v}_{k+1} + \cdots + b_m \boldsymbol{v}_m = 0$, where the left hand side of this equation is a non-trivial linear combination of $\boldsymbol{v}_1, \boldsymbol{v}_2, \ldots, \boldsymbol{v}_m$, whence these vectors are linearly dependent. □

Proposition 5.7. ([4, Section 4.3]) *In a vector space V the following holds:*

(1) *A set of vectors $\{\boldsymbol{v}_1, \boldsymbol{v}_2, \ldots, \boldsymbol{v}_m\} \subseteq V$ is linearly independent if and only if the only coefficients satisfying the equation $\sum_{i \in \underline{m}} c_i \boldsymbol{v}_i = \boldsymbol{0}$ are $c_1 = c_2 = \ldots = c_m = 0$.*
(2) *Any set of vectors containing the null vector $\boldsymbol{0}$ is linearly dependent.*
(3) *A singleton set $\{\boldsymbol{v}\}$ is linearly independent if and only if $\boldsymbol{v} \neq \boldsymbol{0}$.*

Proof. (1): Follows from Th. 5.6.

(2): Suppose that $\boldsymbol{v}_1 = \boldsymbol{0}$, then $1\boldsymbol{v}_1 + 0\boldsymbol{v}_2 + \cdots + 0\boldsymbol{v}_n = \boldsymbol{0}$, for all $n \geq 1$.

(3): Immediate from (2). □

Proposition 5.8. ([4, Section 4.5; p. 216, Theorem 4.4.1]) *Let V be a vector space, and $\emptyset \neq S \subseteq V$. Then the following holds:*

(1) *All bases for a finite-dimensional vector space have the same number of vectors.*
(2) *In a finite dimensional vector space V with $\dim V = n$ (a) no set of more than n vectors can be linearly independent, and (b) no set of fewer than n vectors can generate V.*
(3) *If S is a linearly independent set and $\boldsymbol{v} \in V$ a vector such that $\boldsymbol{v} \notin \langle S \rangle$, then $S \cup \{\boldsymbol{v}\}$ is still linearly independent.*
(4) *If $\boldsymbol{v} \in \langle S \rangle$ then S and $S - \{\boldsymbol{v}\}$ span the same space: $\langle S \rangle = \langle S - \{\boldsymbol{v}\} \rangle$.*
(5) *If S spans V but is not a basis for it, then S can be reduced to a basis for V by discarding vectors if necessary.*
(6) *If S is a linearly independent set that is not already a basis for V, then S can be extended to a basis for V by adding more vectors if necessary.*
(7) *Every nonzero vector in a finite-dimensional vector space V has a unique representation in terms of a given basis.*

It is well known that every vector space V (i.e. a R-module over a field R) has a basis, and that all bases of V have equal cardinality. None of these properties is true in general for modules: Unlike vector spaces, modules need not have a basis. A module which has a basis is called a **free module.** For more details on free modules see e.g. [7, 19].

5.1.2 *Ranks of Matrices*

An $m \times n$ matrix $A = [\boldsymbol{a}_{1*}\boldsymbol{a}_{2*}\ldots\boldsymbol{a}_{m*}]^T = [\boldsymbol{a}_{*1}\boldsymbol{a}_{*2}\ldots\boldsymbol{a}_{*n}]$ can clearly be viewed as a table of m rows of length n, but also as a table of n columns of length m. We let A_{col} and A_{row} denote the sets of columns and rows of A, i.e. $A_{\mathsf{col}} = \{\boldsymbol{a}_{*1}, \boldsymbol{a}_{*2}, \ldots, \boldsymbol{a}_{*k}\}, k \leq n$, and $A_{\mathsf{row}} = \{\boldsymbol{a}_{1*}, \boldsymbol{a}_{2*}, \ldots, \boldsymbol{a}_{l*}\}, l \leq m$. Accordingly, the following two vector spaces are associated with A (see [6, Section 4.6]:

Definition 5.9. *Let A be an $m \times n$ matrix over a field F.*

(a) The **column space** *of A is* $\mathsf{col}(A) = \{\boldsymbol{y} \in F^m : \boldsymbol{y} = A\boldsymbol{x}$ *for some* $\boldsymbol{x} \in F^n\}$.
(b) The **row space** *of A is* $\mathsf{row}(A) = \mathsf{col}(A^T) = \{\boldsymbol{y} \in F^n : \boldsymbol{y} = A^T\boldsymbol{x}$ *for some* $\boldsymbol{x} \in F^m\}$.

So $\mathsf{col}(A)$ is the set obtained by taking all possible linear combinations of its columns, i.e. $\mathsf{col}(A) = \langle A_{\mathsf{col}} \rangle$, and likewise $\mathsf{row}(A) = \langle A_{\mathsf{row}} \rangle$ is the span of all rows of A. In particular, every column vector of A is in $\mathsf{col}(A)$ since $\boldsymbol{a}_{*j} = A\boldsymbol{e}_j$, with $\boldsymbol{e}_j$ the j-th standard unit vector. Also, $\boldsymbol{0}_{m\times 1} = A\boldsymbol{0}_{n\times 1}$, whence $\boldsymbol{0}_{m\times 1} \in \mathsf{col}(A)$.

Definition 5.10. *Let $A \neq \boldsymbol{0}_{m\times n}$ be an $m \times n$ matrix.*

(a) The **column rank** *of A, denoted by* $\mathsf{cr}(A)$, *is the maximum number of linearly independent columns in A, i.e.* $\mathsf{cr}(A) = \dim \mathsf{col}(A)$.
(b) The **row rank** *of A, denoted by* $\mathsf{rr}(A)$, *is the maximum number of linearly independent rows in A, i.e.* $\mathsf{rr}(A) = \dim \mathsf{row}(A)$.

Proposition 5.11. ([2, Theorem 11.1]) *Let A be an $m \times n$ matrix over a division ring; then* $\mathsf{cr}(A) = \mathsf{rr}(A)$. *This common number is called the* **rank** *of A, denoted as* $\mathsf{rk}(A)$.

In the following corollary we list some prominent properties of matrix rank; for more details and proofs see e.g. [1, Section 4.1], [2, Chapter 11], or [6, Chapters 4 and 5].

Corollary 5.12. *Let A be an $m \times n$ matrix, and A, B compatible matrices over a division ring; then*

(1) $\mathsf{rk}(A) = \mathsf{rk}(A^T)$.
(2) $\mathsf{rk}(AB) \leq \min\{\mathsf{rk}(A), \mathsf{rk}(B)\}$.
(3) $\mathsf{rk}(A+B) \leq \mathsf{rk}(A) + \mathsf{rk}(B)$.
(4) $\mathsf{rk}(A+B) \geq |\mathsf{rk}(A) - \mathsf{rk}(B)|$.
(5) $\mathsf{rk}(A) = 1$ *if and only if* $A = [\boldsymbol{a}\boldsymbol{b}^T]$ *for some vectors* $\boldsymbol{a}, \boldsymbol{b} \neq \boldsymbol{0}$.
(6) If $\mathsf{rk}(A) = p$ *then A can be written as $A = BC^T$, with B an $m \times p$ and C an $n \times p$ matrix, and* $\mathsf{rk}(B) = \mathsf{rk}(C) = p$.
(7) $\mathsf{rk}(A') \leq \mathsf{rk}(A)$ *for submatrices A' of A.*

We conclude this section with some remarks clarifying the synonymity of the terms "*invertible*" and "*nonsingular*" which we did already meet in Def. 3.32, and will also find later on. For this purpose we present some well known results from linear algebra and matrix theory over fields, omitting their proofs. (These, and much more details, can be found in e.g. [3, Chapters 2 and 3], [6, Chapters 4 and 5], or [15, Section 4.4].)

Definition 5.13. *An $m \times n$ matrix A is said to*

(a) *have* **full row rank** *if* $\mathsf{rk}(A) = m$*, that is, if its rank equals the number of rows.*
(b) *have* **full column rank** *if* $\mathsf{rk}(A) = n$*, that is, if its rank equals the number of columns.*
(c) *be* **nonsingular** *if it has both full row rank and full column rank.*

Thus, for nonsingular matrices we have $\mathsf{rk}(A) = m = n$, whence they are square by definition. Since invertible matrices must also be square, nonsingular matrices possibly have an inverse. In fact, there are a number of characterizations for nonsingular matrices as shown in

Proposition 5.14. ([3, Table 3.1]) *The following are equivalent for an $n \times n$ matrix A over a field F:*

(1) *A is nonsingular.*
(2) *The inverse A^{-1} exists.*
(3) $\mathsf{rk}(A) = n$.
(4) $|A| \neq 0$.
(5) *$A\boldsymbol{x} = \boldsymbol{b}$ has a unique solution $\boldsymbol{x} = A^{-1}\boldsymbol{b}$, for each $\boldsymbol{b} \in F^n$.*
(6) *The only solution to $A\boldsymbol{x} = \boldsymbol{0}$ is $x = \boldsymbol{0}$.*
(7) *The rows of A are linearly independent.*
(8) *The columns of A are linearly independent.*

5.1.3 *Eigenvalues and Eigenvectors*

For sake of simplicity and clarity we will introduce the concepts mentioned in the header of this section in the familiar context of matrices over $\mathbb{C}$ or $\mathbb{R}$, the fields of complex or real numbers. Then a well known result in linear algebra is

Theorem 5.15. ([4, p. 277, Theorem 4.10.2]) *Let A be a square matrix of order n, and $\boldsymbol{x}$ be a vector of size n. Then the following statements are equivalent:*

(1) *A is invertible.*
(2) *$A\boldsymbol{x} = \boldsymbol{0}$ has only the trivial solution.*
(3) $\mathsf{det}(A) \neq 0$.

Of particular interest in many settings is the following question: What are the nonzero vectors $\boldsymbol{x}$ satisfying the equation $A\boldsymbol{x} = \lambda\boldsymbol{x}$, with $\lambda \in \mathbb{C}$ a scalar. When we multiply a vector $\boldsymbol{x}$ with a matrix A, the resulting vector $A\boldsymbol{x}$ usually has a direction different from that of $\boldsymbol{x}$. So what we are looking for are very special members in the set of vectors $A\boldsymbol{x}$ (those that remain parallel to $\boldsymbol{x}$) which therefore got a special name:

Definition 5.16. *Let A be an $n \times n$-matrix.*

(a) *If $\boldsymbol{x} \neq \boldsymbol{0}$ is a vector such that the equation $A\boldsymbol{x} = \lambda\boldsymbol{x}$ holds, then the scalar λ is called an* **eigenvalue** *of A.*
(b) *Any $\boldsymbol{x} \neq \boldsymbol{0}$ for which $A\boldsymbol{x} = \lambda\boldsymbol{x}$ is called an* **eigenvector** *associated with the eigenvalue λ. $(\boldsymbol{x}, \lambda)$ is also referred to as an* **eigenpair** *of A.*
(c) *The set $E_\lambda^A := \{\boldsymbol{x} | A\boldsymbol{x} = \lambda\boldsymbol{x}\}$ is called the* **eigenspace** *of A associated with the eigenvalue λ.*

Now, if $\boldsymbol{x}$ is an eigenvector for A corresponding to the eigenvalue λ, then we have

$$A\boldsymbol{x} = \lambda\boldsymbol{x} = \lambda I_n \boldsymbol{x} \quad \text{or} \quad (\lambda I_n - A)\boldsymbol{x} = \boldsymbol{0}. \tag{5.1}$$

Since $\lambda I_n - A$ is a square matrix, Th. 5.15 implies that if its determinant has a value different from zero, then the last equation in (5.1) will have $\boldsymbol{x} = \boldsymbol{0}$ as its only solution. So the equation $A\boldsymbol{x} = \lambda\boldsymbol{x}$ has solutions $\boldsymbol{x} \neq \boldsymbol{0}$ if and only if

$$p_A(\lambda) := \det(\lambda I_n - A) = \begin{vmatrix} \lambda - a_{11} & -a_{12} & \cdots & -a_{1n} \\ -a_{21} & \lambda - a_{22} & \cdots & -a_{2n} \\ \vdots & \vdots & \cdots & \vdots \\ -a_{n1} & -a_{n2} & \cdots & \lambda - a_{nn} \end{vmatrix} = 0. \tag{5.2}$$

When $\det(\lambda I_n - A)$ is expanded, the result is known as the **characteristic polynomial** of A, with λ as indeterminate. This is a **monic** polynomial of degree n (i.e. the coefficient of λ^n is 1) and can be written as

$$p_A(\lambda) = \lambda^n + a_1\lambda^{n-1} + \cdots + a_{n-1}\lambda + a_n. \tag{5.3}$$

The *fundamental theorem of algebra* ensures that any monic non-zero, single variable polynomial of degree $n \geq 1$ with complex coefficients can be represented as a product of exactly n complex or real linear factors:

$$p_A(\lambda) = (\lambda - \lambda_1)(\lambda - \lambda_2)\cdots(\lambda - \lambda_n). \tag{5.4}$$

So each of the λ_i, $i \in \underline{n}$, is a root of $p_A(\lambda)$, hence an eigenvalue of A, but not all of them must be different. Counted with their multiplicities, (5.2) can be rewritten as

$$p_A(\lambda) = (\lambda - \lambda_1)^{n_1}(\lambda - \lambda_2)^{n_2}\cdots(\lambda - \lambda_m)^{n_m} = 0. \tag{5.5}$$

where $\lambda_i \neq \lambda_j$, $n_i \geq 1$, and $n_1 + \cdots + n_m = n$.

Then, if we know the distinct eigenvalues λ_i, $i = 1, \ldots, m \leq n$ from (5.5), we can obtain the corresponding eigenvectors by computing nontrivial solutions of the homogenous systems $(\lambda_i I_n - A)\boldsymbol{x} = \boldsymbol{0}$.

(Just to mention two examples for the usage of eigenvalues and eigenvectors: Basic solutions of ordinary differential and difference equations with constant coefficients are expressed in terms of eigenvalues and eigenvectors of the coefficient matrices. Another important aspect is that almost all matrix results associated with square matrices can be derived from representations of square matrices in terms of their eigenvalues and eigenvectors.)

5.2 Semilinear Spaces

Our considerations on Boolean matrices started out and are based on the notion of a Boolean vector space; recall Def. 3.2. Moving forward to matrices over more general structures we also have to generalize the concept of a vector space.

Definition 5.17. ([12, p. 149], [22, Def. 2.2]) *Let* $\mathcal{S} = (S, +, \cdot, 0, 1)$ *be a semiring,* $(M, +_M, 0_M)$ *a commutative monoid, and* $\cdot : S \times M \longrightarrow M$*,* $(s, m) \mapsto sm$*, a left action on* M*, called* **scalar multiplication**. *If, for all* $s, s' \in S$ *and* $m, m' \in M$*, we have*

(1) $(ss')m = s(s'm)$,
(2) $s(m +_M m') = sm +_M sm'$,
(3) $(s + s')m = sm +_M s'm$,
(4) $1_S m = m$,
(5) $s0_M = 0_M = 0m$,

then $(\mathcal{S}; M, +_M, 0_M)$ *is called a (left)* $\mathcal{S}$**-semimodule**.

Right $\mathcal{S}$-semimodules are defined in an analogous way. From now on $\mathcal{S}$-*semimodules* will always mean left $\mathcal{S}$-semimodules. In fact, semimodules over semirings are the analogs of vector spaces over fields and modules over rings. Each semiring $\mathcal{S}$ is also an $\mathcal{S}$-semimodule, if we take $M := S$ and the operation from $S \times S$ into S as the product in S.

The elements of an $\mathcal{S}$-semimodule are called *vectors* and will be denoted by bold letters if they are vectors in the usual sense. But one also has group and endomorphism algebras, and, as the prominent example we are interested in, the ring of square matrices over a field with the usual matrix multiplication (see [18, Chapter 1]).

$\mathcal{S}$-semimodules are also called ($\mathcal{S}$-)**semilinear spaces**, e.g. in [21, 23], and originally in a more narrow sense in e.g. [10]. Much of research runs under this notion, whence we follow this naming.

Definition 5.18. ([22, Def. 3.3]) *A* **semilinear subspace** *of a semilinear space* $\mathcal{L} = ((S, +, \cdot, 0, 1); M, +_M, 0_M)$ *is a subset* $W \subseteq M$ *that is closed under addition and scalar multiplication, i.e. for* $\boldsymbol{v}, \boldsymbol{w} \in M$ *we have* $s\boldsymbol{v} \in M$ *and* $\boldsymbol{v} +_M \boldsymbol{w} \in M$*, for all* $s \in S$.

We did already have a short look on Boolean vector spaces in Def. 3.2. The definitions and results presented there will serve us as a guideline, since the Boolean notions must be included in the more general setting as a special case.

According to our notation $\mathbb{B}_n$ or $\mathbb{B}_{m\times n}$ for sets of square and rectangular Boolean matrices we will e.g. use $\mathbb{S}_n$, $\mathbb{L}_{m\times n}$ or $\mathbb{I}_{m\times n}$ for sets of matrices over semirings, lattices or inclines, respectively.

Definition 5.19. *Let $\mathcal{S} = (S, +, \cdot, 0, 1)$ be a semiring.*

(a) For each $n \geq 1$, we denote by $V_n(S)$ the set $\{\mathbf{s} = (s_1, s_2, \ldots, s_n) | s_i \in S, 1 \leq i \leq n\}$ of row vectors of length n over S.
(b) The operation $+$ on $V_n(S)$ is defined by $\mathbf{r} + \mathbf{s} = (r_1 + s_1, r_2 + s_2, \ldots, r_n + s_n)$.
(c) $\cdot : S \times V_n(S) \longrightarrow V_n(S)$ is defined as $r \cdot \mathbf{s} = (r \cdot s_1, r \cdot s_2, \ldots, r \cdot s_n)$.

In an analogous way we can define $V^n(S) = \{v^T : v \in V_n(S)\}$, the set of column vectors and the corresponding operations. We will simply write $\mathbf{0}$ and $\mathbf{1}$ to denote row or column vectors of appropriate length. Then both $(\mathcal{S}; V_n(S), +, \mathbf{0})$ and $(\mathcal{S}; V^n(S), +, \mathbf{0})$ are semilinear spaces (over S), usually abbreviated to $V_n(\mathcal{S})$ and $V^n(\mathcal{S})$; operation symbols will be overloaded, since their nature is determined by the operands. Note that for any result about $V_n(\mathcal{S})$, there exists a corresponding one about $V^n(\mathcal{S})$.

Note: To make our discussion more convenient, below we will confine our terminology to $V_n(S)$. As is customary, we will call $\boldsymbol{v} \in V_n(S)$ vectors, and elements $s \in S$ scalars, and will use the shorthand V_n (or even just V) for $V_n(S)$ when the rest is clear from the context.

With the definitions given above, it is easy to see that in semilinear spaces $(\mathcal{S}; V_n(S), +, \mathbf{0})$ all of the constituting properties of a classical vector space in linear algebra (see Def. 5.1) are valid with the exception of (A5) which requires the existence of an *additive inverse*; but this clearly is no surprise since only semigroups are involved.

As stated in [10], the notions of linear dependence and independence can be extended to semilinear spaces. But the well known condition, according to which linear dependent and independent vectors are separated on the basis of the representability of the zero vector as a linear combination with some non-zero coefficients, does not work here because the difference between rings and semirings is the lack of additive inverses in the latter. Instead we can use the definition as given in Def. 5.20 just below, which is equivalent to the just mentioned representability condition in the context of fields (see Prop. 5.7(1)). This definition of linear (in)dependence in (generalized) fuzzy algebras is used in e.g. [10, 13, 16, 24].

Definition 5.20. *Let $\mathcal{S} = (S, +, \cdot, 0, 1)$ be a semiring and $W \subseteq V_n(S)$.*

(a) $\boldsymbol{v} \in V_n$ is called **dependent** *on W if $\boldsymbol{v} \in \langle W \rangle$.*
(b) A set $W \neq \emptyset$ is called a **linearly independent set** *if $\boldsymbol{v} \notin \langle W - \{\boldsymbol{v}\} \rangle$, for any $\boldsymbol{v} \in W$, i.e. if none of the vectors can be represented by a linear combination of the others.*
(c) $\emptyset$ is defined to be linearly independent. If W is not linearly independent then it is called **linearly dependent**.
(d) A **basis** *of a subspace $W \subseteq V_n$ is a linearly independent set B such that $W = \langle B \rangle$.*

Note that $\langle W \rangle = \left\{\sum_{i \in \underline{m}} s_i \boldsymbol{v}_i \,|\, \boldsymbol{v}_i \in V_n, s_i \in S, m \in \underline{k}\right\}$ if $W = \{\boldsymbol{v}_1, \boldsymbol{v}_2, \ldots, \boldsymbol{v}_k\}$, and that we have $\langle \emptyset \rangle = \{\mathbf{0}\}$ by the convention of Def. 5.4(b), i.e. the basis of $\mathbf{0}$ is $\emptyset$. Thus, any singleton $\{v \neq \mathbf{0}\}$ always forms a linearly independent set since $\boldsymbol{v} \notin \langle \emptyset \rangle$. The set $\{\mathbf{0}\}$ is not linearly independent, since always $\mathbf{0} \in \langle \emptyset - \{\mathbf{0}\} \rangle = \langle \emptyset \rangle$.

Recall that in Cor. 3.5 it has been shown that a subspace W of the Boolean vector space V_n has a unique basis $B \subseteq V_n$. It is apparent that every spanning set contains at least one basis. But without further restrictions, and contrary to the Boolean case, in the context of semirings neither is a basis unique in general, nor is it true that all bases have the same cardinality. Noticing that Boolean algebras are the most special kind of structure in our hierarchy this must not come as a surprise, and one must be prepared for weaker results in the more general cases. This is illustrated by some examples taken from the above mentioned papers:

- For a semiring $(S,+,\cdot,0,1)$, both $\{(1,1)\}$ and $\{(1,0),(0,1)\}$ are bases for V_2 since every $\boldsymbol{v} \in V_2$ is a multiple of $(1,1)$, and also a linear combination of $(1,0)$ and $(0,1)$.
- For the fuzzy algebra, $\{(0.5,1),(1,0.6),(0.7,0.7)\}$ is linearly dependent since $(0.7,0.7) = 0.7\cdot(0.5,1)+0.7\cdot(1,0.6) = (0.5,0.7)+(0.7,0.6)$. (Note: $\cdot$ is min and $+$ is max.)

Before we will turn to first results for semiring based structures, we have to remark that almost all research on matrices on semirings is done by considering semirings with at least one additional property, in most cases the ones of being commutative and/or zerosumfree.

For sake of completeness we now recapitulate some definitions we have already met for the Boolean case in Def. 3.2, possibly in a generalized form.

Definition 5.21. *Let $\boldsymbol{u}, \boldsymbol{v}, \boldsymbol{u}_i \in V_n$.*

(a) The (**inner**) **product** *of $\boldsymbol{u}$ and and $\boldsymbol{v}$ is given by $\boldsymbol{uv} := (u_1v_1 + \ldots + u_nv_n)$.*
(b) Two non-zero vectors $\boldsymbol{u}$ and $\boldsymbol{v}$ are called **orthogonal** *if $\boldsymbol{uv} = 0$.*
(c) Let $\boldsymbol{u}_1, \boldsymbol{u}_2, \ldots, \boldsymbol{u}_n \in V_n$. If $\boldsymbol{u}_i\boldsymbol{u}_j = 0$, for all $i \neq j$, and $\boldsymbol{u}_i\boldsymbol{u}_i \in U(R)$, with $i, j \in \underline{n}$, then the set $\{\boldsymbol{u}_1, \boldsymbol{u}_2, \ldots, \boldsymbol{u}_n\}$ is said to be **standard orthogonal**.
(d) A basis that is a standard orthogonal set is said to be an **orthogonal basis**.

Note that by the definition of the product of two vectors, the following holds for all $\boldsymbol{u}, \boldsymbol{v} \in V_n$ and $r \in S$ in commutative semirings S:

$$\boldsymbol{uv} = \boldsymbol{vu}, \quad \text{and} \quad r(\boldsymbol{uv}) = (r\boldsymbol{u})\boldsymbol{v} = \boldsymbol{u}(r\boldsymbol{v}). \tag{5.6}$$

Proposition 5.22. ([21, Theorem 4.1]) *In zerosumfree semirings standard orthogonal sets are linearly independent.*

Proof. Let $\{\boldsymbol{u}_1, \boldsymbol{u}_2, \ldots, \boldsymbol{u}_m\}$ be a standard orthogonal set in V_n, and assume that, for some $i \in \underline{m}$, $\boldsymbol{u}_i = \sum_{j\in\underline{m}, j\neq i} r_j\boldsymbol{u}_j$, with $r_j \in R$, then $\boldsymbol{u}_i\boldsymbol{u}_i = \sum_{j\in\underline{m}, j\neq i} r_j\boldsymbol{u}_j\boldsymbol{u}_i = 0$. This is a contradiction since $\boldsymbol{u}_i\boldsymbol{u}_i \in U(R)$ by definition, whence it must be different from 0. Thus the vectors $\{\boldsymbol{u}_1, \boldsymbol{u}_2, \ldots, \boldsymbol{u}_n\}$ form a linearly independent set. □

References

[1] Abadir, K.M., Magnus, J.R.: *Matrix Algebra*, Cambridge University Press, New York (2005).

[2] Agarwal, R.P., Flaut, C.: *An Introduction to Linear Algebra*, CRC Press, Boca Raton (2017).

[3] Andrilli, S., Hecker, D.: *Elementary Linear Algebra*, 5th ed., Academic Press/Elsevier Inc., London (2016).

[4] Anton, H., Rorres, C.: *Elementary Linear Algebra*, 11th ed., Wiley (2014).

[5] Ash, R.B.: *Basic Abstract Algebra*, Robert B. Ash (2000).

[6] Banerjee, S., Roy, A.: *Linear Algebra and Matrix Analysis for Statistics*, CRC Press, Boca Raton (2014).

[7] Bland, P.E.: *Rings and Their Modules*, Walter de Gruyter GmbH & Co. KG, Berlin/New York (2011).

[8] Blyth, T.S., Robertson, E.F.: *Basic Linear Algebra*, 2nd ed., Springer-Verlag London (2002).

[9] Butkovič, P.: Max-algebra: the linear algebra of combinatorics?, Linear Algebra and its Applications **367**, 313-335 (2003).

[10] Di Nola, A., Lettieri, A., Perfilieva, I., Novák, V.: Algebraic analysis of fuzzy systems, Fuzzy Sets and Systems **158**, 1-22 (2007).

[11] Give'on, Y.: Lattice matrices, Information and Control **7**, 477-484 (1964).

[12] Golan, J.S.: *Semirings and their Applications*, Springer Science+Business Media Dordrecht (1999).

[13] Guo, S.-Z., Wang, P.-Z., Di Nola, A., Sessa, S.: Further contributions to the study of finite fuzzy relation equation, Fuzzy Sets and Systems **26**, 93-104 (1988).

[14] Guterman, A.E.: Rank and determinant functions for matrices over semirings, in: Young, N., Choi, Y. (Eds.), *Surveys in Contemporary Mathematics* (London Mathematical Society Lecture Note Series, Vol. 347), Cambridge University Press, Cambridge, pp. 1-33 (2008).

[15] Harville, D.A.: *Matrix Algebra From a Statistician's Perspective*, Springer-Verlag New York (1997).

[16] Kim, K.H.: *Boolean Matrix Theory and Applications*, Marcel Dekker, Inc., New York and Basel (1982).

[17] Kim, K.H., Roush, F.W.: Gerneralized fuzzy matrices, Fuzzy Sets and Systems **4**, 293-315 (1980).

[18] Pierce, R.S.: *Associative Algebras*, Springer-Verlag New York Inc. (1982).

[19] Robinson, D.J.S.: *An Introduction to Abstract Algebra*, 2nd ed., Walter de Gruyter GmbH, Berlin/Boston (2015).

[20] Rutherford, D.E.: The Cayley-Hamilton Theorem for Semi-rings, Proc. Roy. Soc. Edinburgh Sec A **66**, 211-215 (1964).

[21] Shu, Q-Y., Wang, X-P.: Bases in semilinear spaces over zoresumfree semirings, Linear Algebra and its Applications **435**, 2681-692 (2011).

[22] Shu, Q-Y., Wang, X-P.: Standard orthogonal vectors in semilinear spaces and their applications, Linear Algebra and its Applications **437**, 2733-2754 (2012).

[23] Shu Q.-Y., Wang X-P.: Dimensional formulas of semilinear subspaces over semirings. Linear and Multilinear Algebra **62**, 1475-1490 (2014).

[24] Tan, Y.-J.: Bases in semimodules over commutative semirings, Linear Algebra and its Applications **443**, 139-152 (2014).

[25] Yoeli, M.: A Note on a Generalization of Boolean Matrix Theory, The American Mathematical Monthly **68**, 552-557 (1961).

[26] Zadeh, L.A.: Similarity relations and fuzzy orderings, Information Science **3**, 177-200 (1971).

CHAPTER 6

MATRICES OVER ARBITRARY AND COMMUTATIVE SEMIRINGS

The theory of matrices over semirings has a long history and has become an object of intensive study over the last few decades in both theoretical and applied mathematics. It has broad applications in optimization theory, models of discrete event networks, graph theory, and automata theory; see e.g. [2, 48] for some early classical, or [4, 9, 17, 32, 39] for some recent papers.

An illustrative example has been given in [9, p. 314]: Suppose two trains leaving two different stations have to arrive at a third station where things have to be ready for further actions at time t. Let us denote the departure times of the trains as x_1, x_2, and the duration of their journeys (including the necessary times for other preparations) by a_1 and a_2, respectively. For proper organization we have to solve the non-linear problem $\max\{x_1 + a_1, x_2 + a_2\} \leq t$. However, using the semiring $(\mathbb{R}, \oplus, \odot)$ with $\oplus = \max$ and $\odot = +$, we can reformulate this as a linear equation $x_1 \odot a_1 \oplus x_2 \odot a_2 \leq t$.

Another example for the application of semirings is given in [18, p. 1] and describes a situation that arises constantly in parallel computations, disguised in a similar real life analogue as follows (it is in fact a multidimensional version of the above) Consider k seaports, and ships arriving at the i-th port at the moment $x_i, i = 1, \ldots, k$. Assume that there are also l airports with an airplane in each airport which has to depart at the time $b_j, j = 1, \ldots, l$ and that certain goods must be delivered from the sea-ports to the airports, needing the travel time t_{ij} from the i-th port to the j-th airport. The problem is to find $x_i, i = 1, \ldots, k$, (or certain relations between them) such that all goods coming on ships will be delivered in due time. This problem can be formulated in a linear form using the semiring $(\mathbb{R}, \oplus, \odot)$ as above: Given an $l \times k$ matrix $T = [t_{ij}]$ and a vector $\boldsymbol{b} = (b_1, \ldots, b_l)^T$, we need to find a vector $\boldsymbol{x} = (x_1, \ldots, x_k)^T$ such that $T\boldsymbol{x} \leq \boldsymbol{b}$.

Recall from the begin of Chapter 4 that the following inclusion relationships hold among the various structures introduced in that chapter:

Boolean algebras $\subset$ bounded distributive lattices $\subset$ 0-1-inclines $\subset$ path algebras $\subset$ antirings, dioids $\subset$ semirings.

Accordingly, we will organize our presentation of results starting at the most general level of semirings and proceed down to more special structures. Usually, we will place results on the most general level for which they are valid. But sometimes, in order to enable the reader to pick up a structure at will, we will make results available also in the more special context, providing references to their proofs.

For the results reported below in this and the following chapters we will mention the source where they are taken from and where more details can be found, and we will explicitly state for which kind of underlying structure they hold. Though we will not do it for each assertion, we hope to offer enough proofs to give an impression of the techniques and the kind of argumentation applied.

In analogy to the Boolean case let $\mathbb{S}_{m\times n}(R)$ denote the set of all $m \times n$ matrices with entries in a semiring R. If $m = n$, we use the notation $\mathbb{S}_n(R)$. And we simply write $\mathbb{S}_{m\times n}$ or $\mathbb{S}_n$ if we let R be any semiring or if R is obvious from the context.

For matrices $A = [a_{ij}], A' = [a'_{ij}] \in \mathbb{S}_{m\times p}$ and $B = [b_{ij}] \in \mathbb{S}_{p\times n}$, their product $A \cdot B \in \mathbb{S}_{m\times n}$ is defined as

$$A \cdot B = [c_{ij}], \quad \text{with} \quad c_{ij} := \sum_{k\in \underline{p}} a_{ik} b_{kj},\ i \in \underline{m}, j \in \underline{n};$$

and their sum $A + A' \in \mathbb{S}_{m\times p}$ as

$$A + A' = [a_{ij} + a'_{ij}],\ i \in \underline{m}, j \in \underline{p}.$$

O, I and U are the same as in the Boolean case. With $E_{ij} \in \mathbb{S}_{m\times n}$ we denote the matrix with $e_{ij} \neq 0$, and all other entries equal to zero. E_{ij} is called a **cell**, and $E_{ii} \in \mathbb{S}_n$ is called a **diagonal cell**. Furthermore, we let $A \le B :\leftrightarrow\ a_{ij} \le b_{ij}$ for all $i \in \underline{m}, j \in \underline{n}$ for matrices over partially ordered semirings.

$A \in \mathbb{S}_n(S)$ is called a **diagonal matrix** if $a_{ij} = 0$ for all $i, j \in \underline{n}$ such that $i \neq j$. Let $a_1, a_2, \ldots, a_n \in S$; then $\text{diag}(a_1, a_2, \ldots, a_n)$ denotes the diagonal matrix with $a_{ii} = a_i$. O is a special diagonal matrix.

6.1 Various Matrix Properties

Proposition 6.1. *$\mathbb{S}_n(R)$, the set of square matrices over an arbitrary semiring R, is a semiring with O_n as zero and I_n as identity.*

Proof. For $A, B, C \in \mathbb{S}_n(R)$ it follows from the respective definitions that $A + B, AB \in \mathbb{S}_n$. Since matrix addition can easily be seen to be both commutative and associative and $O_n + A = A$, $(\mathbb{S}_n, +, O_n)$ is a commutative monoid. Moreover, we have associativity by $\sum_k \left(\sum_j a_{ij} b_{jk}\right) c_{kl} = \sum_j a_{ij} \left(\sum_k b_{jk} c_{kl}\right) = \sum_{j,k} a_{ij} b_{jk} c_{kl}$ and $AI_n = I_n A = A$; so $(\mathbb{S}_n, \cdot, I_n)$ is also a monoid. The distributivity laws can also be shown by straightforward calculation. □

Proposition 6.2. *Let $A \in \mathbb{S}_{m\times p}$ and $B \in \mathbb{S}_{p\times n}$ be matrices over a commutative semiring R. Then $(AB)^T = B^T A^T$.*

Proof. Let $C = AB$ and $c'_{ij} = c_{ji}$ etc; then $c'_{ij} = c_{ji} = \sum_k a_{jk}b_{ki} = \sum_k b_{ki}a_{jk} = \sum_k b'_{ik}a'_{kj}$, whence $(AB)^T = B^T A^T$. □

Note that by the last two propositions, $(\mathbb{S}_n, \cdot, I_n)$ is an involution monoid. Of course, $(\mathbb{S}_n(S), \cdot, I_n)$ over a commutative semiring S is not commutative in general.

Lemma 6.3. ([41, Lemma 2.2]) *For matrices $A, B \in \mathbb{S}_{m\times n}$ over an arbitrary semiring the following holds:*

(1) $\mathsf{row}(B) \subseteq \mathsf{row}(A)$ *iff there exists a matrix C such that $B = CA$.*
(2) $\mathsf{col}(B) \subseteq \mathsf{col}(A)$ *iff there exists a matrix D such that $B = AD$.*
(3) $\mathsf{row}(A) = \mathsf{row}(B)$ *iff there exist C, D such that $B = CA$ and $A = DB$.*
(4) $\mathsf{col}(A) = \mathsf{col}(B)$ *iff there exist C, D such that $B = AC$ and $A = BD$.*

Proof. (1): The i-th row of CA is given by $\sum_{j\in \underline{n}} x_{ij}\boldsymbol{a}_{j*} \in \mathsf{row}(A)$, whence $\mathsf{row}(CA) \subseteq \mathsf{row}(A)$. Conversely assume $\mathsf{row}(B) \subseteq \mathsf{row}(A)$ and let x_{ij} be such that $\boldsymbol{b}_{i*} = \sum_{j\in\underline{n}} x_{ij}\boldsymbol{a}_{j*}$. Then $CA = B$.

(3): Follows directly from (1). □

(**Green's equivalence relations** $\mathcal{L}$ and $\mathcal{R}$ on a monoid S are defined as (i) $s\mathcal{L}t$ iff $\exists x, y \in S[xs = t \wedge yt = s]$, and (ii) $s\mathcal{R}t$ iff $\exists x, y \in S[sx = t \wedge ty = s]$. Then (3) can be written as $A\mathcal{L}B$ iff $\mathsf{row}(A) = \mathsf{row}(B)$, and (4) as $A\mathcal{R}B$ iff $\mathsf{col}(A) = \mathsf{col}(B)$; see [24, Theorem 2.5].)

6.2 Determinants, Permanents and Adjoints

Whereas determinants are well known from what we usually learn about linear algebra, permanents are not. From a formal point of view, they a very similar as we will shortly see, but they differ considerably in other aspects. The concept of a permanent made its first appearance in 1812 already, but it gained greater interest not earlier than in the late fifties of the last century. See the first paragraph of the preface and the bibliography in [30].

First we have to recall some basic properties of permutations (on finite sets): Assume X to be a set with n elements; then any bijective mapping $\pi : X \longrightarrow X$ is called a **permutation** of X. Since the nature of elements of X is not important for the current considerations, instead of X the set $\underline{n}$ can be used, with $n = |X|$.

A common notation for permutations is in form of a two-line array: As an example, let $\epsilon_n : \underline{n} \longrightarrow \underline{n}$ be defined $\epsilon_n(i) = i$, for all $i \in \underline{n}$. This is often written as

$$\epsilon_n = \begin{pmatrix} 1\,2\cdots n \\ 1\,2\cdots n \end{pmatrix}, \text{ or generally: } \pi = \begin{pmatrix} 1 & 2 & \cdots & n \\ \pi(1) & \pi(2) & \cdots & \pi(n) \end{pmatrix}.$$

It is well known that the permutations on a set X form the so-called **symmetric group** $\mathcal{S}_X$, with ϵ_n as identity.

A pair $\langle i, j\rangle$ is called an **inversion** if $i \leq j$ but $\pi(i) > \pi(j)$. π is called **even**(**odd**) if the number of inversions in π, denoted as $\mathrm{inv}(\pi)$, is even (odd). If we set $\mathrm{sgn}(\pi) = (-1)^{\mathrm{inv}(\pi)}$, then clearly $\mathrm{sgn}(\pi) = 1$ for even, and $\mathrm{sgn}(\pi) = -1$ for odd permutations.

Now we are ready to recall some basic folklore facts on determinants. Let A be an $n \times n$ matrix over a field $\mathbb{F}$; a *determinant* is a function $\det() : A \longrightarrow \mathbb{F}$ having a few remarkable properties as e.g. (i) $\det(A) \neq 0$ if and only if A is invertible, (ii) if also $B \in \mathbb{F}_{n\times n}$ then $\det(AB) = \det(A)\det(B)$, and (iii) $\det(A^T) = \det(A)$.

But there is more behind this which is interesting and worthwhile to take a short look at (see e.g. [7, Chapter 8] or [26, Chapter 6] for details and proofs):

Definition 6.4. *Let $A = [\boldsymbol{a}_{*1}\boldsymbol{a}_{*2}\ldots\boldsymbol{a}_{*n}]$ and $B = [\boldsymbol{b}_{*1}\boldsymbol{b}_{*2}\ldots\boldsymbol{b}_{*n}]$ be square matrices over $\mathbb{F}$, $\boldsymbol{c}$ be a column vector, and $\gamma \in \mathbb{F}$. A mapping $d : \mathbb{F}_{n\times n} \longrightarrow \mathbb{F}$ is called* **determinantal** *if*

(1) *it is multilinear (i.e. a linear function of each column):*

(D1) $d([\ldots,\boldsymbol{a}_{*i}+\boldsymbol{c},\ldots]) = d([\ldots,\boldsymbol{a}_{*i},\ldots]) + d([\ldots,\boldsymbol{c},\ldots])$;

(D2) $d([\ldots,\gamma\boldsymbol{a}_{*i},\ldots]) = \gamma d([\ldots,\boldsymbol{a}_{*i},\ldots])$;

(2) *it is alternating:*

(D3) $d([\ldots,\boldsymbol{a}_{*i},\ldots,\boldsymbol{a}_{*j},\ldots]) = -d([\ldots,\boldsymbol{a}_{*j},\ldots,\boldsymbol{a}_{*i},\ldots])$;

(3) *it fulfills*

(D4) $d(I_n) = 1$.

Theorem 6.5. ([7, p. 137, Theorem 8.5]) *There is a unique determinantal mapping $d : \mathbb{F}_{n\times n} \longrightarrow \mathbb{F}$, such that for each $A \in \mathbb{F}_{n\times n}$*

$$d(A) = \textstyle\sum_{\sigma\in S_n} \mathrm{sgn}(\sigma)\ a_{1\sigma(1)}\, a_{2\sigma(2)} \ldots a_{n\sigma(n)}, \tag{6.1}$$

where $\sigma \in S_n$, the set of all permutations of $\{1,2,\ldots,n\}$. This unique mapping is called the **determinant** *of A, denoted by* $\det(A)$.

If you compare (6.1) with (6.2) below, then, as announced above, the definitions of determinants and permanents are very similar: they differ only in the factor attached to the products forming the $n!$ summands. There are however major differences of which we will mention only a few: E.g., the multiplicativity property $\mathrm{per}(AB) = \mathrm{per}(A)\,\mathrm{per}(B)$ is not valid in general for permanents, and they also fail to obey the invariance under certain elementary operations on matrices. This deficiency is a reason for the fact that proving theorems on permanents, and evaluating them, is substantially more difficult in general than solving analogous problems for determinants ([30, p. 15]).

So much about determinants and permanents over rings and fields, which is the context that we usually get presented in the course of our linear algebra education. If we change over to matrices and related concepts over semirings then we have to face the fact that we can no longer use the sign $\mathrm{sgn}(\sigma)$ of permutations in the definition of determinants and that we will have to look for alternatives, because for both the addition and multiplication operation of semirings only semigroup constraints are required.

The next definition is a straightforward generalization of concepts and notation which we have already introduced for the Boolean case; it is in fact a copy of Def. 3.30 in which $\mathbb{B}_n$ has been replaced by $\mathbb{S}_n$:

Definition 6.6. *Let $A \in \mathbb{S}_n$ be a square semiring matrix.*

(a) *The* **permanent** *of A is defined by*

$$\mathrm{per}(A) = \sum_{\sigma \in S_n} a_{1\sigma(1)}\, a_{2\sigma(2)} \dots a_{n\sigma(n)}, \tag{6.2}$$

where $\sigma \in S_n$, the set of all permutations of $\{1,2,\dots,n\}$.

(b) *The (i,j)-***minor matrix** *of A, denoted with $A(i|j)$, is the $(n-1)\times(n-1)$ submatrix of A obtained by deleting the ith row and the jth column of A. $\mathrm{per}(A(i|j))$ is called the* **minor** *of the entry a_{ij}.*

(c) *The* **adjoint** *of A is the matrix $\left(\mathrm{adj}(A)\right)_{ij} = [\mathrm{per}(A(j|i))] = [\mathrm{per}(A(i|j))]^T \in \mathbb{S}_n$, for all $i,j \in \underline{n}$.*

(Once again a warning hint: What we and many others call a *permanent* is often named *determinant* in the context of structures weaker than rings or fields, when there are no additive inverses, i.e. negative elements are missing. As an example take [23], a paper that defines $\mathrm{det}(A)$ as our $\mathrm{per}(A)$, but also states that "*we may call* $\mathrm{det}(A)$ *the permanent of A*", or [13, Definition 3], [36, Definition 3.1]. So you have to be careful about the underlying definitions.)

Proposition 6.7. ([16, (12.3) Proposition], [20, Lemmas 4.1, 4.2]) *For a commutative semiring S, permanents of matrices $A \in \mathbb{S}_n(S)$ have the following properties:*

(1) $\mathrm{per}(rA) = r^n \mathrm{per}(A)$, $r \in R$.
(2) $\mathrm{per}(A) = \mathrm{per}(A^T)$.
(3) *If P and Q are permutation matrices then* $\mathrm{per}(PAQ) = \mathrm{per}(A)$.

Proof. (2): Consider any term $t = a_{1i_1} a_{2i_2} \dots a_{ni_n}$ of $\mathrm{per}(A)$. Since the subscripts $i_1, i_2, \dots, i_n$ form a permutation of $\underline{n}$, all factors $a_{1i_1}, a_{2i_2}, \dots, a_{ni_n}$ are in different rows and different columns of A^T, whence t occurs also as a term of $\mathrm{per}(A^T)$. Since the converse also holds, $\mathrm{per}(A)$ and $\mathrm{per}(A^T)$ consist of the same terms and are therefore equal.

(3): Let $A = [a_{kj}]$ and $P = [p_{ik}]$, and assume $p_{ik_i} = 1$ for any $i \in \underline{n}$, where $k_1, k_2, \dots, k_n$ are all different. Then the (i,j)-entry of PA is $a_{k_i j}$, whence

$$\mathrm{per}(PA) = \sum_{\sigma \in S_n} \prod_{i \in \underline{n}} a_{k_i\sigma(k_i)} = \sum_{\sigma \in S_n} \prod_{k_i \in \underline{n}} a_{k_i\sigma(k_i)} = \mathrm{per}(A).$$

Similarly, we have $\mathrm{per}(AQ) = \mathrm{per}(A)$. Thus, $\mathrm{per}(PAQ) = \mathrm{per}(AQ) = \mathrm{per}(A)$. □

Proposition 6.8. ([20, Lemma 4.3(3)], [36, Proposition 3.3]) *For a commutative semiring S, permanents of matrices $A \in \mathbb{S}_n(S)$ have the following properties:*

(1) *If A contains a zero row or column, then* $\mathrm{per}(A) = 0$.
(2) *If A is triangular, then* $\mathrm{per}(A) = \prod_{i \in \underline{n}} a_{ii}$*; so* $\mathrm{per}(I_n) = 1$*, for all $n \in \mathbb{N}$.*
(3) $\mathrm{adj}(I_n) = I_{n-1}$.
(4) $\mathrm{per}(A) = \sum_j a_{ij}\, \mathrm{per}(A(i|j)) = \sum_i a_{ij}\, \mathrm{per}(A(i|j))$.

Proof. (1): Each term $t = a_{1\sigma(1)}\, a_{2\sigma(2)} \ldots a_{n\sigma(n)}$ in $\mathrm{per}(A)$ contains a factor of each row and column, and hence contains a factor 0; so $t = 0$.

(2): Suppose A is triangular from below, i.e. $a_{ij} = 0$ for $i < j$, and consider a term $t = a_{1\sigma(1)}\, a_{2\sigma(2)} \ldots a_{n\sigma(n)}$. Now, let $\sigma(1) \neq 1$, so $1 < \sigma(1)$ and $a_{1\sigma(1)} = 0$; thus $t = 0$. If $\sigma(1) = 1$ but $\sigma(2) \neq 2$, then $2 < \sigma(2)$ and again we get $a_{2\sigma(2)} = 0$ and $t = 0$. In a similar manner we see that for all $k \leq n-1$ we have $t = 0$ for each term t with $\sigma(k) \neq k$ and $\sigma(j) = j$ for $1 \leq j \leq k-1$. And if $\sigma(j) = j$, for all $1 \leq j \leq n-1$ then clearly $\sigma(n) = n$ since $\sigma[\underline{n}] = \underline{n}$.

(3): It is easy to see that $I_n(i|i) = I_{n-1}$ and that $I_n(i|j)$ is an $(n-1) \times (n-1)$ matrix which contains a zero row or column. So, from (1) and (2) we get $(\mathrm{adj}(A))_{ij} = [\mathrm{per}(A(j|i))] = 1$ for $i = j$ and $= 0$ otherwise.

(4): Let $S_{i|j}$ denotes the set of bijections from $\underline{n} - \{i\}$ to $\underline{n} - \{j\}$. We have

$$\begin{aligned}\mathrm{per}(A) &= \sum_{j \in \underline{n}} \sum_{\sigma \in S_n, \sigma(i)=j} a_{1\sigma(1)}\, a_{2\sigma(2)} \ldots a_{n\sigma(n)} \\ &= \sum_{j \in \underline{n}} a_{ij} \sum_{\pi \in S_{i|j}} a_{1\pi(1)} \ldots a_{i-1\pi(i-1)}\, a_{i+1\pi(i+1)} \ldots a_{n\pi(n)}.\end{aligned}$$

But from the definition of $\mathrm{per}(A)$ we get that the inner sum is just $\mathrm{per}(A_{i|j})$. For the second equality note that by Prop. 6.7(2) $\mathrm{per}(A) = \mathrm{per}(A^T)$ and that the columns of A are the rows in A^T. So the result follows by application of what we have just shown to $\mathrm{per}(A^T)$. □

Now we turn to the topic of determinants over semirings. To develop linear algebra over semirings certainly requires an analog of the determinant function. However, the determinant cannot be defined in the classical way as in the case of fields and commutative rings, since in semirings which are not rings, not all elements possess an additive inverse. And due to the lack of a multiplicative inverse we dot no longer have various relationships between determinants and matrix properties.

We first recall (6.1), the definition of the determinant over a field $\mathbb{F}$, in a slightly rewritten form:

$$\det(A) = \sum_{\sigma \in S_n} \mathrm{sgn}(\sigma) \prod_{i \in \underline{n}} a_{i\sigma(i)} = \sum_{\sigma \text{ even}} \prod_{i \in \underline{n}} a_{i\sigma(i)} - \sum_{\sigma \text{ odd}} \prod_{i \in \underline{n}} a_{i\sigma(i)}, \tag{6.3}$$

where $\sigma \in S_n$, the set of all permutations of $\{1, 2, \ldots, n\}$.

As one can easily recognize, a plausible replacement of the determinant function for matrices over commutative semirings is the *bideterminant*; see e.g. [15, Chapter 19] or [19, 45, 46]:

Definition 6.9. *Let R be a commutative semiring, and $A \in \mathbb{S}_n(R)$. The* **bideterminant** *of A is an ordered pair $\langle |A|^+, |A|^- \rangle \in R^2$, denoted as $|A|^\pm$, where*

$$|A|^+ = \sum_{\sigma \text{ even}} \prod_{i \in \underline{n}} a_{i\sigma(i)} \quad \textit{and} \quad |A|^- = \sum_{\sigma \text{ odd}} \prod_{i \in \underline{n}} a_{i\sigma(i)}.$$

$|A|^+$ is called the **positive** *and $|A|^-$ the* **negative determinant** *of A.*

With this definition it is clear that we have $\mathsf{per}(A) = |A|^+ + |A|^-$, and $\det(A) = |A|^+ - |A|^-$ if R is a commutative semiring.

We call a square matrix A **nonsingular** if it has an inverse B, i.e. if we have $AB = BA = I$. This B is unique, and is denoted as A^{-1}. In usual linear algebra, i.e. with a field as underlying value set, it is known that

(1) if A is nonsingular then $\det(A^{-1}) = \det(A)^{-1}$,
(2) A is nonsingular iff $\det(A) \neq 0$, or equivalently: A is singular iff $\det(A) = 0$.
(3) $\det(AB) = \det(A)\det(B)$.

Clearly, we cannot generally check a condition like (1) in the semiring context due to the lack of multiplicative inverses. For invertibility, however, we could use the condition in (2) in the form $|A|^+ = |A|^-$. Following [34], we will say that $|A|^\pm$ is zero, written $|A|^\pm \equiv 0$, if $|A|^+ = |A|^-$. We list some simple properties of the bideterminant in

Proposition 6.10. ([14, 19, 34]) *Let R be a commutative semiring, $A, B, C \in \mathbb{S}_n(R)$ and $r \in R$; then the following holds:*

(1) $|I_n|^\pm = \langle 1, 0\rangle$.
(2) Let $A = [\boldsymbol{a}_{*1} \ldots r \cdot \boldsymbol{a}_{*k} \ldots \boldsymbol{a}_{*n}]$, *and* $B = [\boldsymbol{a}_{*1} \ldots \boldsymbol{a}_{*k} \ldots \boldsymbol{a}_{*n}]$. *Then* $|A|^\pm = r|B|^\pm$. *An analogous result holds for rows.*
(3) $|rA|^\pm = \langle r^n|A|^+, r^n|A|^-\rangle = r^n|A|^\pm$.
(4) Let $A = [\boldsymbol{a}_{*1} \ldots \boldsymbol{0} \ldots \boldsymbol{a}_{*n}]$. *Then* $|A|^\pm = \langle 0, 0\rangle$.
(5) $|A^T|^\pm = \langle |A^T|^+, |A^T|^-\rangle = \langle |A|^+, |A|^-\rangle = |A|^\pm$.
(6) Let $A = [\boldsymbol{a}_{*1} \ldots (\boldsymbol{b}_{*k} + \boldsymbol{c}_{*k}) \ldots \boldsymbol{a}_{*n}]$, $B = [\boldsymbol{a}_{*1} \ldots \boldsymbol{b}_{*k} \ldots \boldsymbol{a}_{*n}]$, *and* $C = [\boldsymbol{a}_{*1} \ldots \boldsymbol{c}_{*k} \ldots \boldsymbol{a}_{*n}]$. *Then* $|A|^\pm = |B|^\pm + |C|^\pm$.

Proof. (1): From Prop. 6.8(2) we know that $\mathsf{per}(I_n) = \prod_{i \in \underline{n}} a_{ii} = 1$, and as mentioned above, $\mathsf{per}(A) = |A|^+ + |A|^-$. Since $\epsilon_n(i) = i$ is an even permutation we thus get $|I_n|^+ = 1$.

(2): Each term $t = a_{1\sigma(1)}\, a_{2\sigma(2)} \ldots a_{n\sigma(n)}$ in $\mathsf{per}(A)$ contains a factor of each row (column) and hence contains a factor $r \cdot a_{kl}$, for some $k, l \in \underline{n}$.

(4): Follows from (2) with $r = 0$.

(5): Similar to Prop. 6.7(2). □

The bideterminant is not multiplicative in general; only the following weaker version holds:

Proposition 6.11. ([35, Theorem 4.5]) *Let R be a commutative semiring. For $A, B \in \mathbb{S}_n(R)$ we have*

$$|AB|^+ + |A|^+|B|^- + |A|^-|B|^+ = |AB|^- + |A|^+|B|^+ + |A|^-|B|^-.$$

The next task is to define analogies to minors, for which we need some new notation (following [35, Definition 2.2]): Let $n \in \mathbb{N}$, and σ be a permutation of $\underline{n}$. Then we set

$$\pi(n) := \begin{cases} + & \text{if } n \text{ is even} \\ - & \text{if } n \text{ is odd} \end{cases} \quad \text{and} \quad \pi(\sigma) := \begin{cases} + & \text{if } \sigma \text{ is even} \\ - & \text{if } \sigma \text{ is odd} \end{cases}.$$

If $Q = [\mathbf{e}_{\sigma(1)}\mathbf{e}_{\sigma(2)}\dots\mathbf{e}_{\sigma(m)}]$ is the permutation matrix associated with σ, let $\pi(Q) := \pi(\sigma)$. Then we have $\pi(Q) := \pi(\sigma) = \pi(\sigma^{-1}) = \pi(Q^T)$, and $\pi(m+n) = \pi(m)\cdot\pi(n)$.

In the proof of the following lemma we need the concepts of the **positive** and **negative** (i,j)**-minor** of A, defined as $|A_{i|j}|^+$ and $|A_{i|j}|^-$, respectively, which for convenience we abbreviate as $A^+_{i|j}$ and $A^-_{i|j}$. Below we list some properties of positive and negative minors:

Lemma 6.12. (Cofactor expansion) ([35, Lemma 3.2]) *Let $A \in \mathbb{S}_n(R)$, with R a commutative semiring; then*

(1) $|A|^+ = a_{p1}A^{\pi(p+1)}_{p|1} + a_{p2}A^{\pi(p+2)}_{p|2} + \cdots + a_{pn}A^{\pi(p+n)}_{p|n}$ *(by row p)*

(2) $|A|^+ = a_{1q}A^{\pi(q+1)}_{1|q} + a_{2q}A^{\pi(q+2)}_{2|q} + \cdots + a_{nq}A^{\pi(q+n)}_{n|q}$ *(by column q)*

(3) $|A|^- = a_{p1}A^{\pi(p)}_{p|1} + a_{p2}A^{\pi(p+1)}_{p|2} + \cdots + a_{pn}A^{\pi(p+n-1)}_{p|n}$ *(by row p)*

(4) $|A|^- = a_{1q}A^{\pi(q)}_{1|q} + a_{2q}A^{\pi(q+1)}_{2|q} + \cdots + a_{nq}A^{\pi(q+n-1)}_{n|q}$ *(by column q)*

Proof. (1): Note that (a) the number of terms in the expansion $A^+_{i|k}$, $k \in \underline{n}$, is $(n-1)!/2$, and each of these terms also appears in $|A|^+$, and (b) there are no common terms in $a_{ik}A^+_{i|k}$ and $a_{il}A^+_{i|l}$ for $k \neq l$. The same also holds for negative minors. So there are $n(n-1)!/2 = n!/2$ distinct terms in each sum $a_{p1}A^{\pi(p+1)}_{p|1} + a_{p2}A^{\pi(p+2)}_{p|2} + \cdots + a_{pn}A^{\pi(p+n)}_{p|n}$ all of which are also in $|A|^+$, whence we must have equality. Analogous arguments hold for the other cases. □

Lemma 6.13. (Zero column/row; Transpose) ([35, Lemmas 3.6, 3.7]) *Let $A \in \mathbb{S}_n(R)$, with R a commutative semiring.*

(1) If A has a zero row or column, then $|A|^+ = |A|^- = 0$.
(2) $|A|^+ = |A^T|^+$ and $|A|^- = |A^T|^-$.

Lemma 6.14. (Interchanging columns/rows) ([35, Lemma 3.9]) *Let $A \in \mathbb{S}_n(R)$, with R a commutative semiring, and suppose that B is obtained from A by interchanging two rows (or columns). Then $|A|^+ = |B|^-$ and $|A|^- = |B|^+$.*

Proof. The proof uses induction on n and proceeds by expansion of each positive (negative) determinant by another row or column. As the basis step we prove the assertion for $n = 2$: If $A = \begin{bmatrix} a_{11} & a_{12} \\ a_{21} & a_{22} \end{bmatrix}$ then $B = \begin{bmatrix} a_{12} & a_{11} \\ a_{22} & a_{21} \end{bmatrix}$. So $|A|^+ = a_{11}a_{22} = |B|^-$ and $|B|^+ = a_{12}a_{21} = |A|^-$.

We suppose that column 3 will not be one of the columns to be interchanged. Using this column for expansion, we know from Lemma 6.12(2) that $|A|^+ = \sum_{i\in\underline{n}} a_{i3}A^{\pi(3+i)}_{i|3}$. If we now switch two columns in A, then we must do so also in each $(i,3)$-minor matrix $A_{i|3}$. By induction hypothesis, the assertion holds for matrices in $\mathbb{S}_{n-1}$; so $A^{\pi(3+i)}_{i|3} = B^{\pi(3+i-1)}_{i|3}$. Therefore, $|A|^+ = \sum_{i\in\underline{n}} a_{i3}A^{\pi(3+i)}_{i|3} = \sum_{i\in\underline{n}} a_{i3}B^{\pi(3+i-1)}_{i|3} = |B|^-$, with the last equality following from Lemma 6.12(4). Analogous arguments show that $|A|^- = |B|^+$

by switching the roles of A and B, and that the assertion also holds if rows are interchanged. □

Lemma 6.15. (Equal columns/rows) ([35, Lemma 3.11]) *Let $A \in \mathbb{S}_n(R)$, with R a commutative semiring, such that two columns (rows) in A are equal, say $\boldsymbol{a}_{*k} = \boldsymbol{a}_{*l}$, $k \neq l$. Then $|A|^{\pm} \equiv 0$.*

Proof. Interchanging $\boldsymbol{a}_{*k}$ and $\boldsymbol{a}_{*l}$ results in a matrix B with $|A|^+ = |B|^-$ by Lemma 6.14. But since $A = B$ we also have $|B|^- = |A|^-$, which gives $|A|^+ = |A|^-$. □

Lemma 6.16. (False cofactor expansion) ([14, Lemma 1.2], [35, Lemma 3.14]) *Let $A \in \mathbb{S}_n(R)$, with R a commutative semiring, and $p \neq q$. Then*

$$\sum_{k \in \underline{n}} a_{pk} A_{q|k}^{\pi(q+k)} = \sum_{k \in \underline{n}} a_{pk} A_{q|k}^{\pi(q+k-1)}.$$

Proof. Let C be the matrix obtained from A by replacing row q by another copy of row p. Then C has two equal rows, and by Lemma 6.15 we have $|C|^+ = |C|^-$. From Lemma 6.12 we get

$$|C|^+ = \sum_{k \in \underline{n}} c_{qk} C_{q|k}^{\pi(p+k)} \qquad \text{and} \qquad |C|^- = \sum_{k \in \underline{n}} c_{qk} C_{q|k}^{\pi(p+k-1)}.$$

And since $A_{q|k} = C_{q|k}$ and $a_{pk} = c_{qk}$, for all $k \in \underline{n}$, the assertion follows. □

Corollary 6.17. ([15, p. 214]) *Let $A \in \mathbb{S}_n(R)$, with R a commutative semiring. If one column (row) of A is a linear combination of the other columns (rows) of A then $|A|^{\pm} \equiv 0$.*

Proof. Let $A = [\boldsymbol{a}_{*1}\boldsymbol{a}_{*2} \ldots \boldsymbol{a}_{*n}]$ with say $\boldsymbol{a}_{*1} = \sum_{i \neq 1} \lambda_i \boldsymbol{a}_{*i}$. Then, using Prop. 6.10(2) and (6) we get (mentioning only the first column of summands) $|A[\sum_{i \neq 1} \lambda_i \boldsymbol{a}_{*i}]|^+ = \sum_{i \neq 1} \lambda_i |A[\boldsymbol{a}_{*i}]|^+$. Now each of the matrices $A[\boldsymbol{a}_{*i}]$ has two identical columns, viz, $\boldsymbol{a}_{*i}$. So by Lemma 6.15 $|A[\boldsymbol{a}_{*i}]|^+ = |A[\boldsymbol{a}_{*i}]|^-$, whence $\sum_{i \neq 1} \lambda_i |A[\boldsymbol{a}_{*i}]|^+ = \sum_{i \neq 1} \lambda_i |A[\boldsymbol{a}_{*i}]|^- = |A[\sum_{i \neq 1} \lambda_i \boldsymbol{a}_{*i}]|^-$. Therefore, $|A|^+ = |A|^-$. □

That the converse of this statement does not generally hold is shown in the following example, borrowed from the source just mentioned: Consider the semiring $(\mathbb{R}^+ \cup \{\infty\}, \max, \min)$. Then for $A = \begin{bmatrix} 12 & 14 & 3 \\ 11 & 15 & 7 \\ 6 & 8 & 10 \end{bmatrix}$ we have $|A|^+ = |A|^- = 10$ but no column (row) is a linear combination of the other columns (rows).

6.2.1 *The Concept of ϵ-Determinants*

A still more general concept of determinant has been proposed in [43, 45], called the *epsilon determinant*, the name being inspired by the notation used by its inventor. The papers mentioned use two variants for the definition of an ϵ-function, from which we use the version of the first one including clause (4):

Definition 6.18. *Let R be a semiring. A bijection $\epsilon : R \longrightarrow R$ is called an* **ϵ-function of** *R if*

(1) $\epsilon(\epsilon(a)) = a$,
(2) $\epsilon(a+b) = \epsilon(a) + \epsilon(b)$,
(3) $\epsilon(ab) = a \cdot \epsilon(b) = \epsilon(a) \cdot b$,
(4) $\epsilon(a) = -a$ *for all* $a \in V(R)$ (in [43] only).

An identical concept has been introduced a few years earlier in [1, Definition 4.1] under the name **symmetry on** R; see also [31, Section 2]. It is easy to check that, in case we require only (1) - (3), every semiring has at least one symmetry, viz. the identity map $\epsilon(a) = a$; and if R is a ring then $\epsilon(a) = -a$ is a symmetry.

Definition 6.19. ([43, Definition 2.1]) *Let R be a commutative semiring, and ϵ be a symmetry on R. The* **ϵ-determinant** *of $A \in \mathbb{S}_n(R)$ is given by*

$$\det_\epsilon(A) := \sum_{\sigma \in S_n} \epsilon^{\mathrm{inv}(\sigma)}\big(a_{1\sigma(1)} a_{2\sigma(2)} \cdots a_{n\sigma(n)}\big), \tag{6.4}$$

where ϵ^k is defined by $\epsilon^0(a) = a$ and $\epsilon^k(a) = \epsilon^{k-1}(\epsilon(a))$, for all $k \geq 1$.

Proposition 6.20. *Let R be a semiring, and ϵ a symmetry on R. Then*

(1) $\epsilon(0) = 0$.
(2) $\epsilon^1(a) = \epsilon(a)$ *and* $\epsilon^2(a) = \epsilon(\epsilon(a)) = a$.
(3) $\epsilon(a)\epsilon(b) = ab$.

Proof. (1): $\epsilon(0) = \epsilon(0a) = 0 \cdot \epsilon(a) = 0$.

(2): We have $\epsilon^0(a) = a$ by definition; so $\epsilon^1(a) = \epsilon^0(\epsilon(a)) = \epsilon(a)$, and therefore $\epsilon^2(a) = \epsilon^1(\epsilon(a)) = \epsilon(\epsilon(a))$.

(3): Using Def. 6.18(3) and (2) we get $\epsilon(\epsilon(a)\epsilon(b)) = \epsilon^2(a)\epsilon(b) = a\epsilon(b) = \epsilon(ab)$; together with (2) this gives $\epsilon(a)\epsilon(b) = \epsilon^2(\epsilon(a)\epsilon(b)) = \epsilon^2(ab) = ab$. □

Lemma 6.21. ([45, Lemma 2.1]) *Let R be a commutative semiring and ϵ be a symmetry on R and $a, b \in R$. Then*

(1) $\epsilon^{2k}(a) = a$ *and* $\epsilon^{2k+1}(a) = \epsilon(a)$, *for* $k \geq 0$.
(2) $\epsilon^{k+2l} = \epsilon^k$, *for* $k \geq 0$ *and* $l \geq 1$.
(3) $\epsilon^{k+l}(a) = \epsilon^k(\epsilon^l(a))$, *for any* $k, l \geq 0$.
(4) ϵ^k *is a symmetry on R for any* $k \geq 0$.
(5) $\epsilon^{k+l}(ab) = \epsilon^k(a)\epsilon^l(b)$, *for any* $k, l \geq 0$.
(6) $\epsilon^k(a+b) = \epsilon^k(a) + \epsilon^k(b)$ *and* $\epsilon^k(ab) = \epsilon^k(a)b = a\epsilon^k(b)$, *for any* $k, l \geq 0$.

Proof. (1): For the base cases we have $\epsilon^0(a) = a$ by definition and $\epsilon^1(a) = \epsilon(a)$ by Prop. 6.20(2). The induction step, given that $\epsilon^{2k}(a) = a$, follows by simple calculation: $\epsilon^{2(k+1)}(a) = \epsilon^{2k+1}(\epsilon(a)) = \epsilon^{2k}(\epsilon(\epsilon(a))) = \epsilon(\epsilon(a)) = a$. Similar for odd exponents of ϵ.

(2): $\epsilon^{k+2l}(a) = \epsilon^{k+2l-1}(\epsilon(a)) = \epsilon^{k+2(l-1)}(\epsilon(\epsilon(a))) = \epsilon^{k+2(l-1)}(a)$, for all $l \geq 1$.

(3): For $l = 0$, we get $\epsilon^k(a) = \epsilon^k(\epsilon^0(a))$ by definition. Assuming $\epsilon^{k+m}(a) = \epsilon^k(\epsilon^m(a))$ yields $\epsilon^{k+m+1}(a) = \epsilon^{k+m}(\epsilon(a)) = \epsilon^k(\epsilon^m(\epsilon(a))) = \epsilon^k(\epsilon^{m+1}(a))$.

(4): By (1), we have $\epsilon^k = id_R$ for even k, which is a symmetry, and $\epsilon^k = \epsilon$ for odd k.

(5): By (4), all of ϵ^k, ϵ^l and ϵ^{k+l} are ϵ-functions of R. Then using (3) $\epsilon^{k+l}(ab) = \epsilon^k(\epsilon^l(ab)) = \epsilon^k(a\epsilon^l(b)) = \epsilon^k(a)\epsilon^l(b)$.

(6): By (4), ϵ^k is an ϵ-functions of R, so the first assertion follows from Def. 6.18(2). And from (5) we get $\epsilon^{k+l}(ab) = \epsilon^k(a)\epsilon^l(b)$, so $\epsilon^k(ab) = \epsilon^0(a)\epsilon^l kb) = a\epsilon^k(b)$. □

From Lemma 6.21(1) and Def. 6.18(2) we can see that $\det_\epsilon(A)$ can be rewritten as follows:

$$\begin{aligned}\det_\epsilon(A) &= \sum_{\sigma\in S_n} \epsilon^{\mathrm{inv}(\sigma)}(a_{1\sigma(1)}a_{2\sigma(2)}\cdots a_{n\sigma(n)}) \\ &= \sum_{\sigma \text{ even}} a_{1\sigma(1)}a_{2\sigma(2)}\cdots a_{n\sigma(n)} + \sum_{\sigma \text{ odd}} \epsilon(a_{1\sigma(1)}a_{2\sigma(2)}\cdots a_{n\sigma(n)}) \\ &= \sum_{\sigma \text{ even}} a_{1\sigma(1)}a_{2\sigma(2)}\cdots a_{n\sigma(n)} + \epsilon\Big(\sum_{\sigma \text{ odd}} a_{1\sigma(1)}a_{2\sigma(2)}\cdots a_{n\sigma(n)}\Big) \\ &= |A|^+ + \epsilon(|A|^-). \end{aligned} \tag{6.5}$$

So, choosing $\epsilon = id_R$, for any commutative semiring R we get

$$\det_\epsilon(A) = \sum_{\sigma\in S_n} a_{1\sigma(1)}a_{2\sigma(2)}\cdots a_{n\sigma(n)} = \mathrm{per}(A),$$

and for R a commutative ring with $\epsilon(a) = -a = (-1)a$ we have

$$\det_\epsilon(A) = \sum_{\sigma\in S_n} (-1)^{\mathrm{inv}(\sigma)} a_{1\sigma(1)}a_{2\sigma(2)}\cdots a_{n\sigma(n)} = \det(A).$$

We continue with some ϵ-determinantal identities over a commutative semiring R which definitely are hints that this concept is a reasonable generalization (compare with Prop. 6.10).

Theorem 6.22. ([45, Theorem 3.1]) *Let R be a commutative semiring, ϵ be a symmetry on R, and $r \in R$. Then, for any $A, B, C \in \mathbb{S}_n(R)$, we have:*

(1) $\det_\epsilon(I_n) = 1$.
(2) Let $A = [\boldsymbol{a}_{*1} \ldots r\cdot\boldsymbol{a}_{*k} \ldots \boldsymbol{a}_{*n}]$, *and* $B = [\boldsymbol{a}_{*1} \ldots \boldsymbol{a}_{*k} \ldots \boldsymbol{a}_{*n}]$, *then* $\det_\epsilon(A) = r\cdot\det_\epsilon(B)$.
(3) Let $A = [\boldsymbol{a}_{*1} \ldots \boldsymbol{0} \ldots \boldsymbol{a}_{*n}]$. *Then* $\det_\epsilon(A) = 0$.
(4) $\det_\epsilon(A^T) = \det_\epsilon(A)$.
(5) Let $A = [\boldsymbol{a}_{*1} \ldots (\boldsymbol{b}_{*k} + \boldsymbol{c}_{*k}) \ldots \boldsymbol{a}_{*n}]$, $B = [\boldsymbol{a}_{*1} \ldots \boldsymbol{b}_{*k} \ldots \boldsymbol{a}_{*n}]$, *and* $C = [\boldsymbol{a}_{*1} \ldots \boldsymbol{c}_{*k} \ldots \boldsymbol{a}_{*n}]$. *Then* $\det_\epsilon(A) = \det_\epsilon(B) + \det_\epsilon(C)$.

The above results also hold for rows instead of columns.

Proof. (1): By (6.5) we have $\det_\epsilon(I_n) = |I_n|^+ + \epsilon(|I_n|^-)$, so $\det_\epsilon(I_n) = 1 + \epsilon(0) = 1$, using Prop. 6.10(1) and Prop. 6.20(1).

(2): From Prop. 6.10(2) we have $|A|^\pm = r|B|^\pm$. So, using (6.5), a straightforward calculation gives $\det_\epsilon(A) = |A|^+ + \epsilon(|A|^-) = r|B|^+ + \epsilon(r|B|^-) = r|B|^+ + r\epsilon(|B|^-) = r\big(|B|^+ + \epsilon(|B|^-)\big)$.

(3): Follows from (2) with $r = 0$. □

We use $A^{i\Downarrow j}$ to denote the matrix obtained from A by replacing row j by row i, and $A^{i\Rightarrow j}$ correspondingly for columns.

Theorem 6.23. ([45, Theorem 3.1]) *Let R be a commutative semiring and ϵ be a symmetry on R. Then, for any $A \in \mathbb{S}_n(R)$, the following holds:*

(1) *Suppose that B is obtained from A by interchanging two rows (or columns); then* $\mathsf{det}_\epsilon(B) = \epsilon(\mathsf{det}_\epsilon(A))$.
(2) *If A has two identical rows (or columns), then* $\mathsf{det}_\epsilon(A) = |A|^+ + \epsilon(|A|^+)$.
(3) *If one column (row) of A is a linear combination of the other columns (rows) of A then* $\mathsf{det}_\epsilon(A) = |A|^+ + \epsilon(|A|^+)$.
(4) *Let $B = [\boldsymbol{a}_{*1} \ldots (\boldsymbol{a}_{*i} + r\cdot \boldsymbol{a}_{*j})\,\boldsymbol{a}_{*j+1}\ldots \boldsymbol{a}_{*n}]$, then* $\mathsf{det}_\epsilon(B) = \mathsf{det}_\epsilon(A) + r\cdot|A^{i\Rightarrow j}|^+ + \mathsf{det}_\epsilon(|A^{i\Rightarrow j}|^+)$.

Proof. (1): By Lemma 6.14 we have $|A|^+ = |B|^-$ and $|A|^- = |B|^+$. Thus, $\mathsf{det}_\epsilon(B) = |B|^+ + \epsilon(|B|^-) = |A|^- + \epsilon(|A|^+) = \epsilon(|A|^+ + \epsilon(|A|^-))$, the last equation following from the definition of symmetries and the commutativity of addition. Applying (6.5) shows the assertion.

(2): By Lemma 6.15 $|A|^+ = |A|^-$.

(3): Let $A = [\boldsymbol{a}_{*1}\boldsymbol{a}_{*2}\ldots\boldsymbol{a}_{*n}]$ with say $\boldsymbol{a}_{*1} = \sum_{i\neq 1}\lambda_i \boldsymbol{a}_{*i}$. Then, as argued in the proof of Cor. 6.17, we get $|A[\sum_{i\neq 1}\lambda_i\boldsymbol{a}_{*i}]|^+ = |A[\sum_{i\neq 1}\lambda_i\boldsymbol{a}_{*i}]|^-$. Therefore, $|A|^+ = |A|^-$.

(4): Follows from Th. 6.22(2), (5) and (2) above. □

There are some more, and more advanced versions of determinantal identities for commutative semirings with symmetry, e.g. the Laplace and the Cauchy-Binet theorem, the Cramer identity, or the Cayley-Hamilton formula. For these we refer to [1, Section 3.2] and [45, Theorems 3.3 and 3.5].

6.3 Invertibility

So far we have dealt with addition and multiplication of matrices. The topic of this section is to look for matrix analogues of the reciprocal of a number. Once again for sake of completeness and of convenience, we start with repeating some simple facts on invertibility of semiring matrices. Contrary to e.g. the multiplication of real numbers, matrix multiplication is not commutative in general. This fact is mirrored in

Definition 6.24. *A matrix $A \in \mathbb{S}_{m\times n}$ is called*

(a) **left invertible** *if $BA = I_n$ for some $B \in \mathbb{S}_{n\times m}$, and B a* **left inverse** *of A;*
(b) **right invertible** *if $AB = I_n$ for some $B \in \mathbb{S}_{m\times n}$, and B a* **right inverse** *of A;*
(c) **invertible** *(or* **nonsingular***) if it is both right and left invertible in $\mathbb{S}_n$;*
(d) **singular** *if it does not have an inverse;*
(e) **monomial** *if there is exactly one nonzero entry in each row and each column.*

Proposition 6.25. *Let $A \in \mathbb{S}_n$ be square matrices over an arbitrary semiring. Then the following holds:*

(1) *If A has an inverse then it is unique.*
(2) $I^{-1} = I$.
(3) $(A^{-1})^{-1} = A$.
(4) *If A is invertible, then $AB = AC$ implies $B = C$.*

Proof. (1): Let C, C' be two inverses of A. Then $C = CI = C(AC') = (CA)C' = IC' = C'$. Consequently, the inverse of A can be denoted by A^{-1}.

(2): $IA = A$ for each matrix A, so $II = I$. And since the inverse is unique by (1), we get $I^{-1} = I$.

(3): We have $AA^{-1} = A^{-1}A = I$, so by definition A is the inverse of A^{-1}. □

(**Note:** In what follows below we will present some results without including their lengthy proofs which sometimes even comprise a few pages and, therefore, are beyond the scope of this text. But we will always give the references where the interested reader can find the arguments.)

Since, as mentioned above, matrix multiplication is not commutative, a matrix may have a one sided inverse only. The situation is different for square matrices as shown in the next proposition.

Proposition 6.26. *Let $A, B \in \mathbb{S}_n$ be square matrices over a commutative semiring. Then the following holds:*

(1) $AB = I$ *implies* $BA = I$.
(2) *Let A be invertible; then* $(A^{-1})^T = (A^T)^{-1}$.
(3) *Let A, B be invertible; then AB is also invertible, and* $(AB)^{-1} = B^{-1}A^{-1}$. *Conversely, if AB is invertible, then also A and B.*

Proof. (1): See [38] for two different proofs, each of them needing several pages. (The assertion holds for square matrices over a field, and also for matrices over a commutative ring, but it is not generally true for matrices over non-commutative rings.)

(2): $A^T(A^{-1})^T = (A^{-1}A)^T = I^T = I$ and $(A^{-1})^T A^T = (AA^{-1})^T = I$, using Prop. 6.2.

(3): Assume that A and B be invertible; so both A^{-1} and B^{-1} exist. Thus $(AB)(B^{-1}A^{-1}) = A(BB^{-1})A^{-1} = AIA^{-1} = AA^{-1} = I$. By (1) and Prop. 6.25(1) we get $(AB)^{-1} = B^{-1}A^{-1}$. Conversely, let AB be invertible. Then $I = (AB)(AB)^{-1} = A(B(AB)^{-1})$ and $I = (AB)^{-1}(AB) = ((AB)^{-1}A)B$. By (1) it follows that $I = (B(AB)^{-1})A$, so A is invertible. But $(B(AB)^{-1})A = B((AB)^{-1}A)$ whence also B is invertible. □

Corollary 6.27. *Let $A, A', B \in \mathbb{S}_n$ be square matrices over a commutative semiring.*

(1) *A is right invertible iff A is left invertible.*
(2) *A is invertible if and only if A^T is.*

Proof. (1): Assume that A is right invertible, i.e. there is a B such that $AB = I$. Then $BA = I$ by Prop. 6.26(1), i.e. A is left invertible. By an analogous argument the converse direction can be shown.

(2): By Prop. 6.26(2) if A is invertible, then $(A^{-1})^T$ is the inverse of A^T. Conversely, if A^T is invertible, then again by Prop. 6.26(2) $(A^T)^{-1}$ is the transpose of the inverse of A, i.e. $((A^T)^{-1})^T = A^{-1}$. □

Corollary 6.28. ([28, Theorem 3.1]) *Let S be a commutative semiring and $A_1, A_2 \dots, A_m \in \mathbb{S}_n(S)$ such that $A_1 A_2 \cdots A_m = I_n$. Then each A_i is invertible.*

Proof. From the assumption we get $A_1(A_2 \cdots A_m) = I_n$, whence using Prop. 6.26(1) $(A_2 \cdots A_m)A_1 = I_n$. So A_1 is invertible. But $(A_2 \cdots A_m)A_1 = I_n = A_2(A_3 \cdots A_m A_1) = (A_3 \cdots A_m A_1)A_2$, so A_2 is invertible too. This argument can be repeated for all other matrices A_j, $3 \le j \le m$. □

Lemma 6.29. ([44, Lemma 2.3]) *Let S be a commutative semiring such that $1 \notin V(S)$ and $1 = u + v$ implies $u \in U(S)$ or $v \in U(S)$, for all $u, v \in S$. If $a_{ii} \in U(S)$ for all $i \in \underline{n}$, and $a_{ij} \in V(S)$ for all $i, j \in \underline{n}$, with $i \neq j$, then $A \in \mathbb{S}_n(S)$ is invertible.*

Lemma 6.30. ([42, Lemma 2.2]) *Let S be a commutative semiring, and $a_{ii} \in S$, $i \in \underline{n}$.*

(1) *The matrix diag$(a_{11}, a_{22}, \dots, a_{nn})$ is invertible in $\mathbb{S}_n(S)$ if and only if all a_{ii} are invertible in S.*
(2) *The inverse of any invertible diagonal matrix in $\mathbb{S}_n(S)$ is a diagonal matrix.*

Proof. (1): Let $\mathrm{diag}(a_{11}, a_{22}, \dots, a_{nn})$ be invertible in $\mathbb{S}_n(S)$. Then

$$\begin{bmatrix} a_{11} & 0 & \cdots & 0 \\ 0 & a_{22} & \cdots & 0 \\ \vdots & & \ddots & \vdots \\ 0 & 0 & \cdots & a_{nn} \end{bmatrix} \cdot \begin{bmatrix} b_{11} & b_{12} & \cdots & b_{1n} \\ b_{21} & b_{22} & \cdots & b_{2n} \\ \vdots & & \ddots & \vdots \\ b_{n1} & b_{n2} & \cdots & b_{nn} \end{bmatrix} = I_n \tag{6.6}$$

So, for each $i \in \underline{n}$, we have

$$c_{ii} = a_{i1}b_{1i} + a_{i2}b_{2i} + \cdots + a_{in}b_{ni} = a_{ii}b_{ii} = 1. \tag{6.7}$$

Since S is commutative, $a_{ii}b_{ii} = b_{ii}a_{ii} = 1$, i.e. all a_{ii} are invertible in S. Note that this implies $a_{ii} \neq 0$ and $b_{ii} \neq 0$, for all $i \in \underline{n}$. And for all $i, j \in \underline{n}, i \neq j$, we have

$$c_{ij} = a_{i1}b_{1j} + a_{i2}b_{2j} + \cdots + a_{in}b_{nj} = a_{ii}b_{ij} = 0, \tag{6.8}$$

whence $b_{ij} = 0$.

Conversely, let all a_{ii} be invertible in S, say $a_{ii}b_{ii} = b_{ii}a_{ii} = 1$. So $a_{ii} \neq 0$ and $b_{ii} \neq 0$, for all $i \in \underline{n}$. Then from (6.7) we see that $C := \mathrm{diag}(a_{11}, a_{22}, \dots, a_{nn}) \cdot B \geq I_n$, for any matrix B with diagonal elements b_{ii}. If we choose $B := \mathrm{diag}(b_{11}, b_{22}, \dots, b_{nn})$ then $AB = I_n = BA$.
(2): Follows from (6.7) and (6.8). □

Lemma 6.31. ([15, (19.6) Proposition]) *Let S be a commutative semiring and $A, B \in \mathbb{S}_n(S)$; then there exists an $s \in S$ with*

$$|AB|^+ = |A|^+|B|^+ + |A|^-|B|^- + s \quad \text{and} \quad |AB|^- = |A|^+|B|^- + |A|^-|B|^+ + s$$

Proposition 6.32. ([14, Proposition 2.1, Example 2.8]) *Let S be a commutative semiring and $A \in \mathbb{S}_n(S)$ be invertible; then $|A|^\pm \neq 0$, but the converse is not always true.*

Proof. By hypothesis we have $AB = I_n$ for some $B \in \mathbb{S}_n$. Assuming $|A|^+ = |A|^-$, using Lemma 6.31 we get $|AB|^+ = |AB|^- = |A|^+(|B|^+ + |B|^-) + r$. But $|AB|^+ = |I_n|^+ = 1$ and $|AB|^- = |I_n|^- = 0$, a contradiction. Thus $|A|^+ \neq |A|^-$.

Let $S = (\mathbb{Q}_0^+, \max, \cdot)$, i.e. the set of non-negative rationals with standard multiplication and addition defined as maximum. For $A = \begin{bmatrix} 1 & 2 \\ 3 & 4 \end{bmatrix} \in \mathbb{S}_2(S)$ we have $|A|^+ = 4 \neq 6 = |A|^-$. But there is no $B \in \mathbb{S}_2(S)$ such that $AB = I_2$. □

Corollary 6.33. ([49, Lemma 3.1]) *Let S be a commutative semiring and $A \in \mathbb{S}_n(S)$ be invertible; then the column (row) vectors of A are linearly independent.*

Proof. For A invertible we have $|A|^+ \neq |A|^-$ by Prop. 6.32. Therefore, none of the columns (rows) can be a linear combination of the other columns (rows) of A due to Cor. 6.17. □

We have already mentioned the *standard unit vectors* at the end of Def. 3.2: Writing the identity matrix as $I_n = [\boldsymbol{e}_1 \boldsymbol{e}_2 \ldots \boldsymbol{e}_n]$, then each column vector $\boldsymbol{e}_i$ has a single 1 at position i and zeroes elsewhere: i.e. for $e_{ij} = \delta_{ij}$ for all $i, j \in \underline{n}$.

Proposition 6.34. ([49, Theorem 3.1]) *Let R be a commutative semiring. $A \in \mathbb{S}_n(R)$ is invertible if and only if the set of the column vectors of A is a basis of V_n.*

Proof. If A is invertible, then there is a matrix $X \in \mathbb{S}_n(R)$ such that $AX = I_n$. Let $A = [\boldsymbol{a}_{*1} \boldsymbol{a}_{*2} \ldots \boldsymbol{a}_{*n}]$; then $A\boldsymbol{x}_{*1} = \boldsymbol{e}_1$, $A\boldsymbol{x}_{*2} = \boldsymbol{e}_2, \ldots, A\boldsymbol{x}_{*n} = \boldsymbol{e}_n$. Therefore

$$A\boldsymbol{x}_{*k} = [\boldsymbol{a}_{*1} \boldsymbol{a}_{*2} \ldots \boldsymbol{a}_{*n}] \begin{bmatrix} x_{1k} \\ x_{2k} \\ \vdots \\ x_{nk} \end{bmatrix} = \begin{bmatrix} a_{11}x_{1k} + a_{12}x_{2k} + \ldots + a_{1n}x_{nk} \\ a_{21}x_{1k} + a_{22}x_{2k} + \ldots + a_{2n}x_{nk} \\ \vdots \\ a_{n1}x_{1k} + a_{n2}x_{2k} + \ldots + a_{nn}x_{nk} \end{bmatrix} = \boldsymbol{e}_k,$$

and keeping in mind commutativity of the underlying semiring this gives

$$x_{1k}\boldsymbol{a}_{*1} + x_{2k}\boldsymbol{a}_{*2} + \cdots + x_{nk}\boldsymbol{a}_{*n} = \boldsymbol{e}_k$$

for all $k \in \underline{n}$. And since $\{\boldsymbol{e}_1, \boldsymbol{e}_2, \ldots, \boldsymbol{e}_n\}$ is a standard basis of V_n, every $\boldsymbol{b} \in V_n$ can be written as a sum $\boldsymbol{b} = c_1\boldsymbol{e}_1 + c_2\boldsymbol{e}_2 + \cdots + c_n\boldsymbol{e}_n$, with coefficients $c_i \in R$. Using the last two equalities we get

$$\begin{aligned} \boldsymbol{b} &= c_1(x_{11}\boldsymbol{a}_{*1} + \cdots + x_{n1}\boldsymbol{a}_{*n}) + \cdots + c_n(x_{1n}\boldsymbol{a}_{*1} + \cdots + x_{nn}\boldsymbol{a}_{*n}) \\ &= (c_1x_{11} + \cdots + c_nx_{1n})\boldsymbol{a}_{*1} + \cdots + (c_1x_{n1} + \cdots + c_nx_{nn})\boldsymbol{a}_{*n} \\ &= \left(\textstyle\sum_{k\in\underline{n}} c_kx_{1k}\right)\boldsymbol{a}_{*1} + \cdots + \left(\textstyle\sum_{k\in\underline{n}} c_kx_{nk}\right)\boldsymbol{a}_{*n}, \end{aligned}$$

i.e. the vector $\boldsymbol{b}$ can be represented as a linear combination of the vectors $\boldsymbol{a}_{*1}, \boldsymbol{a}_{*2}, \ldots, \boldsymbol{a}_{*n}$ since $\sum_{k\in n} c_kx_{jk} \in R$ for every $j \in \underline{n}$. Thus, $\{\boldsymbol{a}_{*1}, \boldsymbol{a}_{*2}, \ldots, \boldsymbol{a}_{*n}\}$ is a set of generators of V_n which is linear independent by Cor. 6.33, and hence a basis by Def. 5.5.

Conversely, assume that $\boldsymbol{a}_{*1}, \boldsymbol{a}_{*2}, \ldots, \boldsymbol{a}_{*n}$ are the column vectors of matrix A and form a basis of V_n. Then each unit vector $\boldsymbol{e}_j$ can be represented as a linear combination of the vectors $\boldsymbol{a}_{*1}, \boldsymbol{a}_{*2}, \ldots, \boldsymbol{a}_{*n}$ (again using commutativity) as

$$\boldsymbol{e}_k = c_{1k}\boldsymbol{a}_{*1} + c_{2k}\boldsymbol{a}_{*2} + \cdots + c_{nk}\boldsymbol{a}_{*n} = [\boldsymbol{a}_{*1}\boldsymbol{a}_{*2}\ldots\boldsymbol{a}_{*n}]\begin{bmatrix} c_{1k} \\ c_{2k} \\ \vdots \\ c_{nk} \end{bmatrix}$$

with coefficients $c_{1k}, c_{2k}, \ldots, c_{nk} \in R$. So

$$I_n = [\boldsymbol{e}_1\boldsymbol{e}_2\ldots\boldsymbol{e}_n] = [\boldsymbol{a}_{*1}\boldsymbol{a}_{*2}\ldots\boldsymbol{a}_{*n}]\begin{bmatrix} c_{11} & c_{12} & \ldots & c_{1n} \\ c_{21} & c_{22} & \ldots & c_{2n} \\ & & \vdots & \\ c_{n1} & c_{n2} & \ldots & c_{nn} \end{bmatrix} = AC,$$

i.e. A is right invertible, and by Prop. 6.26(1) invertible. □

Theorem 6.35. ([49, Theorem 3.2]) *Let R be a commutative semiring and $A \in \mathbb{S}_n(R)$ be invertible. Then the following statements are equivalent:*

(1) A is invertible.
(2) A is right invertible.
(3) A is left invertible.
(4) The column vectors of A form a basis of V_n.
(5) All unit vectors of V_n can be written as linear combination of the column vectors of A.
(6) Each vector of an arbitrary basis of V_n can be written as linear combination of the column vectors of A.

Proof. The equivalence of the statements (1) - (4) follows from Prop. 6.26(1) and Cor. 6.27. (5) $\leftrightarrow$ (6) and the implication (4) $\rightarrow$ (5) are obvious. Finally we have to show (5) $\rightarrow$ (1): Every unit vector of can be represented by a linear combination of the column vectors of A, so A is invertible by the proof of Cor. 6.27. □

6.3.1 *Invertibility and ϵ-Determinants*

This section is included to show how much concepts have been generalized during the last 15 years, and to give you an impression of the argumentation needed to proceed thus far. To keep the presentation self-contained, and as an exception to the rule, we also reproduce a rather lengthy proof for illustration purposes. All results in this section are taken from [43].

For what follows we need the following concepts (recall that $A(i|j)$ is the (i,j)-*minor matrix* of A obtained by deleting the ith row and the jth column of A):

Definition 6.36. *Let $A \in \mathbb{S}_n(R)$ be a square semiring matrix, and $i, j \in \underline{n}$.*

(a) The **ϵ-minor** *of the element a_{ij} is defined as $M_{ij}^A := \mathsf{det}_\epsilon(A(i|j)) \in R$.*
(b) The **ϵ-cofactor** *of the element a_{ij}, is defined as $A_{ij}^\epsilon := \epsilon^{i+j}(M_{ij}^A) \in R$.*
(c) The **ϵ-adjoint** *of A is the matrix $\mathsf{adj}_\epsilon(A) = [A_{ji}^\epsilon] = [A_{ij}^\epsilon]^T \in \mathbb{S}_n$.*

Lemma 6.37. ([43, Lemma 2.5]) *Let R be a commutative semiring with ϵ a symmetry on it, and $A \in \mathbb{S}_n(R)$. Then, for any $i, j \in \underline{n}$, the following holds:*

$$\sum_{k\in\underline{n}} a_{ik}A^{\epsilon}_{jk} = \begin{cases} \det_{\epsilon}(A) & \text{for } i = j \\ |A^{i\Downarrow j}|^{+} + \epsilon(|A^{i\Downarrow j}|^{+}) & \text{for } i \neq j, \end{cases} \tag{6.9}$$

and, similarly,

$$\sum_{k\in\underline{n}} a_{ki}A^{\epsilon}_{kj} = \begin{cases} \det_{\epsilon}(A) & \text{for } i = j \\ |A^{i\Rightarrow j}|^{+} + \epsilon(|A^{i\Rightarrow j}|^{+}) & \text{for } i \neq j, \end{cases} \tag{6.10}$$

Proof. Every term of $\det_\epsilon(A)$ contains an element of the i-th row. If the term $t^n_\sigma :=$ $\epsilon^{\mathrm{inv}(\sigma)}(a_{1\sigma(1)}a_{2\sigma(2)}\cdots a_{n\sigma(n)})$ contains a_{ik}, i.e. $\sigma(i) = k$, then, by Lemma 6.21(6),

$$t^n_\sigma = a_{ik}\epsilon^{\mathrm{inv}(\sigma)}(a_{1\sigma(1)}\cdots a_{i-1,\sigma(i-1)}a_{i+1,\sigma(i+1)}\cdots a_{n\sigma(n)}) =: a_{ik}t^n_{ik/\sigma}. \tag{6.11}$$

Therefore, taking the sum over all terms containing a_{ik}, $k \in \underline{n}$, we get

$$\begin{aligned} \sum_{k\in\underline{n}} a_{ik}t^n_{ik/\sigma} &= \sum_{\substack{k\in\underline{n} \\ \sigma\in S_n:\sigma(i)=k}} a_{ik}t^n_{ik/\sigma} \\ &= \sum_{k\in\underline{n}} a_{ik}\Big(\sum_{\substack{\sigma\in S_n: \\ \sigma(i)=k}} t^n_{ik/\sigma}\Big) \qquad (6.12) \\ &= \sum_{\sigma\in S_n} a_{ik}\Big(\sum_{\sigma(i)=k} t^n_{ik/\sigma}\Big) \\ &= \sum_{\sigma\in S_n} t^n_\sigma \\ &= \det_\epsilon(A). \qquad (6.13) \end{aligned}$$

So, in order to establish $\sum_{k\in\underline{n}} a_{ik}A^{\epsilon}_{ik} = \det_\epsilon(A)$ we have to show that (see (6.12))

$$A^{\epsilon}_{ik} = \hat{A}^{\epsilon}_{ik} := \Big(\sum_{\sigma\in S_n:\sigma(i)=k} t^n_{ik/\sigma}\Big), \tag{6.14}$$

for all $i, k \in \underline{n}$.

We first consider the case $i = k = 1$: Using (6.11) we get

$$\hat{A}^{\epsilon}_{11} = \sum_{\sigma\in S_n:\sigma(1)=1} \epsilon^{\mathrm{inv}(\sigma)}(a_{2\sigma(2)}\cdots a_{n\sigma(n)}), \tag{6.15}$$

where $\mathrm{inv}(\sigma)$ is the number of the inversions in the permutation $(1, \sigma(2), \ldots, \sigma(n))$ or, equivalently, of the permutation $(2\sigma(2), 3\sigma(3), \ldots, n\sigma(n))$ of $(2, 3, \ldots, n)$. Thus, with Def. 6.36 and Prop. 6.20(2),

$$\hat{A}^{\epsilon}_{11} = M^A_{11} = \epsilon(\epsilon(M^A_{11})) = \epsilon^{1+1}(M^A_{11}) = A^{\epsilon}_{11}. \tag{6.16}$$

In order to prove (6.14) for the general case (i.e. $2 \le i, j \le n$), we shift the element a_{ik} to the position $(1,1)$ by means of $i-1$ successive interchanges of adjacent rows followed by $k-1$ interchanges of adjacent columns, and call he resulting matrix B. Then Th. 6.23(1) and Lemma 6.21(2) imply

$$\det_\epsilon(B) = \epsilon^{k-1}(\epsilon^{i-1}(\det_\epsilon(A))) = \epsilon^{i+k-2}(\det_\epsilon(A)) = \epsilon^{i+k}(\det_\epsilon(A)), \tag{6.17}$$

whence, by Lemma 6.21(1),

$$\epsilon^{i+k}(\det_\epsilon(B)) = \epsilon^{2(i+k)}(\det_\epsilon(A)) = \det_\epsilon(A). \tag{6.18}$$

Since (6.16) holds for whatever is on position $(1,1)$, where we now have the (i,j)-row/column pair, we get

$$\hat{A}^\epsilon_{ik} = M^A_{ik} = \epsilon(\epsilon(M^A_{ik})) = \epsilon^{1+1}(M^A_{ik}) = A^\epsilon_{ik}.$$

But $M^B_{ik} = \det_\epsilon(B(i|k)) = \det_\epsilon(A(i|k)) = M^A_{ik}$ because the relative positions of the other rows and columns in both submatrices remain unchanged, and the number of inversions thus is the same. So with $\overline{t_{ik}}$ denoting the sum of all terms not involving a_{ik}, and using (6.12) and (6.14) we get $\det_\epsilon(B) = a_{ik}M^A_{ik} + \overline{t_{ik}} = a_{ik}\hat{A}^\epsilon_{ik} + \overline{t_{ik}}$. Applying this in (6.18) and using Lemma 6.21(6) we obtain

$$\det_\epsilon(A) = \epsilon^{i+k}\big(a_{ik}M^A_{ik} + \overline{t_{ik}}\big) = a_{ik}\epsilon^{i+k}\big(M^A_{ik}\big) + \epsilon^{i+k}\big(\overline{t_{ik}}\big).$$

Thus, $\hat{A}^\epsilon_{ik} = \epsilon^{i+k}\big(M^A_{ik}\big) = A^\epsilon_{ik}$, whence finally $\det_\epsilon(A) = \sum_{k\in\underline{n}} a_{ik}A^\epsilon_{jk}$.

To finish the proof of (6.9) we let $i \neq j$; then, by Th. 6.23(2), we have $\sum_{k\in\underline{n}} a_{ik}A^\epsilon_{jk} = \det_\epsilon(A^{i\Downarrow j}) = |A^{i\Downarrow j}|^+ + \epsilon(|A^{i\Downarrow j}|^+)$. Equation (6.10) can be shown using equation (6.9) and Th. 6.22(4). □

Proposition 6.38. ([43, Proposition 3.1]) *Let R be a commutative semiring and $A \in \mathbb{S}_n(R)$ be invertible. Then the following holds:*

(1) *$\sum_{i\in\underline{n}} a_{ij}a_{ik} \in V(R)$, for all $j,k \in \underline{n}$ with $j \neq k$.*
(2) *$\sum_{j\in\underline{n}} a_{ij}a_{kj} \in V(R)$, for all $i,k \in \underline{n}$ with $i \neq k$.*

Proof. (1): By assumption $AB = BA = I_n$ for some $B \in \mathbb{S}_n(R)$, whence $\sum_{k\in\underline{n}} a_{ik}b_{kj} = 0$ for all $i,j \in \underline{n}$ with $i \neq j$, and $\sum_{k\in\underline{n}} a_{ik}b_{ki} = 1$ for all $i \in \underline{n}$. So $a_{ik}b_{kj} \in V(R)$ for all $i,j \in \underline{n}$ with $i \neq j$ by Lemma 4.8(1), and $\prod_{i\in\underline{n}}\big(\sum_{k\in\underline{n}} a_{ik}b_{ki}\big) = 1$. Using the last equation, for all $i,j,k \in \underline{n}$ with $j \neq k$, we get

$$\begin{aligned}
a_{ij}a_{ik} &= a_{ij}a_{ik}\prod_{p\in\underline{n}}\big(\sum_{q\in\underline{n}} a_{pq}b_{qp}\big)\\
&= a_{ij}a_{ik}\big(\sum_{s_1,s_2,\ldots,s_n\in\underline{n}} a_{1s_1}b_{s_1 1}\cdot a_{2s_2}b_{s_2 2}\cdots a_{ns_n}b_{s_n n}\big)\\
&= a_{ij}a_{ik}\big(\sum_{s_1,s_2,\ldots,s_n\in\underline{n}} t^n_{ab}\big)\\
&= \big(\sum_{s_1,s_2,\ldots,s_n\in\underline{n}} a_{ij}a_{ik}t^n_{ab}\big),
\end{aligned}$$

with t^n_{ab} as an obvious abbreviation. For any $s_1,s_2,\ldots,s_n \in \underline{n}$ we have: (i) if $s_u = s_v$ for some $u \neq v$, then $a_{us_u}b_{s_v v} = a_{us_u}b_{s_u v} \in V(R)$, whence $a_{ij}a_{ik}t^n_{ab} \in V(R)$ by Lemma 4.8(2); and (ii) if $s_1,s_2,\ldots,s_n$ are mutually different, then there must be $u,v \in \underline{n}$ with $u \neq v$ such that $j = s_u$ and $k = s_v$. So, $a_{ij}b_{s_u u}a_{ik}b_{s_v v} = a_{ij}b_{ju}a_{ik}b_{kv} \in V(R)$, and again by Lemma 4.8(2) $a_{ij}a_{ik}t^n_{ab} \in V(R)$. Therefore, applying Lemma 4.8(1) twice, $a_{ij}a_{ik} = \big(\sum_{s_1,s_2,\ldots,s_n\in\underline{n}} a_{ij}a_{ik}t^n_{ab}\big) \in V(R)$, and hence also $\sum_{i\in\underline{n}} a_{ij}a_{ik} \in V(R)$ for all $j,k \in \underline{n}$ with $j \neq k$. □

We omit the proof of the next proposition, which needs tedious but straightforward calculations of two pages length:

Proposition 6.39. ([43, Proposition 3.2]) *Let R be a commutative semiring with ϵ a symmetry on it, and $A \in \mathbb{S}_n(R)$ be invertible. Then* $\mathrm{det}_\epsilon(AB) = \mathrm{det}_\epsilon(BA) = \mathrm{det}_\epsilon(A)\mathrm{det}_\epsilon(B)$ *for any $B \in \mathbb{S}_n(R)$.*

Proposition 6.40. ([43, Theorem 3.1]) *Let R be a commutative semiring with ϵ a symmetry on it. Then, for any $A \in \mathbb{S}_n(R)$ the following statements are equivalent:*

(1) A is invertible in $\mathbb{S}_n(R)$.
(2) $\mathrm{det}_\epsilon(A) \in U(R)$ and $\sum_{i\in\underline{n}} a_{ij}a_{ik} \in V(R)$, for all $j,k \in \underline{n}$ with $j \neq k$.
(3) $\mathrm{det}_\epsilon(A) \in U(R)$ and $\sum_{j\in\underline{n}} a_{ij}a_{kj} \in V(R)$, for all $i,k \in \underline{n}$ with $i \neq k$.

And in this case

$$A^{-1} = \mathrm{adj}_\epsilon(A)(\mathrm{det}_\epsilon(A))^{-1} = (\mathrm{det}_\epsilon(A))^{-1}\mathrm{adj}_\epsilon(A), \tag{6.19}$$

the second equality holding by the commutativity of R.

Proof. We show the equivalence of (1) and (2): First assume that A is invertible in $\mathbb{S}_n(R)$; then $AB = BA = I_n$ for some $B\mathbb{S}_n(R)$, and $1 = \mathrm{det}_\epsilon(I_n) = \mathrm{det}_\epsilon(AB) = \mathrm{det}_\epsilon(A)\mathrm{det}_\epsilon(B)$ by Prop. 6.39. Thus, $\mathrm{det}_\epsilon(A) \in U(R)$. The second part of (2) follows by Prop. 6.38(1).

Conversely, assume that $\mathrm{det}_\epsilon(A) \in U(R)$ and $\sum_{i\in\underline{n}} a_{ij}a_{ik} \in V(R)$, for all $j,k \in \underline{n}$ with $j \neq k$. From Lemma 4.8(1) we get $a_{ij}a_{ik} \in V(R)$ for all $i,j,k \in \underline{n}$ with $j \neq k$, so

$$a_{ij}a_{ik} + \epsilon(a_{ij})a_{ik} = a_{ij}a_{ik} + \epsilon(a_{ij}a_{ik}) = a_{ij}a_{ik} + (-(a_{ij}a_{ik})) = 0. \tag{†}$$

(Note it is here where we need clause (4) from Def. 6.18.)

With $C := A\,\mathrm{adj}_\epsilon(A)$ and using Lemma 6.37 we have for any $i,j \in \underline{n}$: (i) if $i = j$ then $c_{ii} = \sum_{k\in\underline{n}} a_{ik}A^\epsilon_{ik} = \mathrm{det}_\epsilon(A)$; (ii) if $i \neq j$ then

$$\begin{aligned}
c_{ij} &= \sum_{k\in\underline{n}} a_{ik}A^\epsilon_{jk} \\
&= |A^{i\Downarrow j}|^+ + \epsilon(|A^{i\Downarrow j}|^+) \\
&= \sum_{\sigma \text{ even}} a_{1\sigma(1)}\cdots a_{i\sigma(i)}\cdots a_{i\sigma(j)}\cdots a_{n\sigma(n)} + \epsilon\Big(\sum_{\sigma \text{ even}} a_{1\sigma(1)}\cdots a_{i\sigma(i)}\cdots a_{i\sigma(j)}\cdots a_{n\sigma(n)}\Big) \\
&= \sum_{\sigma \text{ even}} a_{1\sigma(1)}\cdots a_{i\sigma(i)}\cdots a_{i\sigma(j)}\cdots a_{n\sigma(n)} + \sum_{\sigma \text{ even}} \epsilon\big(a_{1\sigma(1)}\cdots a_{i\sigma(i)}\cdots a_{i\sigma(j)}\cdots a_{n\sigma(n)}\big) \\
&= \sum_{\sigma \text{ even}} a_{1\sigma(1)}\cdots a_{i\sigma(i)}\cdots a_{i\sigma(j)}\cdots a_{n\sigma(n)} + \sum_{\sigma \text{ even}} a_{1\sigma(1)}\cdots \epsilon(a_{i\sigma(i)})\cdots a_{i\sigma(j)}\cdots a_{n\sigma(n)} \\
&= \sum_{\sigma \text{ even}} \big(a_{i\sigma(i)}a_{i\sigma(j)} + \epsilon(a_{i\sigma(i)})a_{i\sigma(j)}\big) \prod_{k\neq i,j} a_{k\sigma(k)} \\
&= 0 \qquad \text{(because } \sigma(i) \neq \sigma(j)\text{, and using (†)).}
\end{aligned}$$

Therefore, $C = A\,\mathrm{adj}_\epsilon(A) = I_n\,\mathrm{det}_\epsilon(A)$. By assumption $\mathrm{det}_\epsilon(A) \in U(R)$, so we can take $B := \mathrm{adj}_\epsilon(A)(\mathrm{det}_\epsilon(A))^{-1}$ to get $AB = A\,\mathrm{adj}_\epsilon(A)(\mathrm{det}_\epsilon(A))^{-1} = I_n\,\mathrm{det}_\epsilon(A)(\mathrm{det}_\epsilon(A))^{-1} = I_n$. This implies $BA = I_n$ by Prop. 6.26(1), which means that A is invertible.

In a similar way one proves the equivalence of (1) and (3). And from the proof of (2) ⇒ (1) we get $A^{-1} = \mathrm{adj}_\epsilon(A)(\mathrm{det}_\epsilon(A))^{-1}$. □

The last result we present in this section establishes Cramer's rule for a matrix equation over a commutative semiring R.

Proposition 6.41. ([43, Theorem 4.1]) *Let R be a commutative semiring with ϵ a symmetry on it, $\boldsymbol{b} = (b_1, b_2, \ldots, b_n)^T \in V(R)$, and $A \in \mathbb{S}_n(R)$. If A is invertible in $\mathbb{S}_n(R)$ then the matrix equation $A\boldsymbol{x} = \boldsymbol{b}$ has the unique solution $\boldsymbol{x} = (d^{-1}d_1, d^{-1}d_2, \ldots, d^{-1}d_n)^T$, where $d = \mathsf{det}_\epsilon(A)$, $d_j = \mathsf{det}_\epsilon(A^{\boldsymbol{b}\Rightarrow j})$, $j \in \underline{n}$, and $A^{\boldsymbol{b}\Rightarrow j}$ is the matrix formed by replacing the j-th column of A by the column vector $\boldsymbol{b}$.*

Proof. Obviously, $\boldsymbol{x} = A^{-1}\boldsymbol{b}$ is a solution for the matrix equation $A\boldsymbol{x} = \boldsymbol{b}$. Let $\boldsymbol{y}$ be any other solution; then $A\boldsymbol{y} = \boldsymbol{b}$ implies $\boldsymbol{y} = A^{-1}\boldsymbol{b}$, which means that $A\boldsymbol{x} = \boldsymbol{b}$ has a unique solution.

Let $(d_1, d_2, \ldots, d_n)^T := \mathsf{adj}(A)\boldsymbol{b}$; then, for any $j \in \underline{n}$, $d_j = \sum_{i\in\underline{n}} b_i A^\epsilon_{ij}$. On the other hand,

$$\mathsf{det}_\epsilon(A^{\boldsymbol{b}\Rightarrow j}) = \mathsf{det}_\epsilon([\boldsymbol{a}_{*1} \ldots \boldsymbol{a}_{*(j-1)}\ \boldsymbol{b}\ \boldsymbol{a}_{*(j+1)} \ldots \boldsymbol{a}_{*n}]) = \textstyle\sum_{i\in\underline{n}} b_i A^\epsilon_{ij},$$

by Lemma 6.37. Then, using (6.19), we get

$$\boldsymbol{x} = A^{-1}\boldsymbol{b} = (\mathsf{det}_\epsilon(A))^{-1}\mathsf{adj}_\epsilon(A)\boldsymbol{b} = d^{-1}\mathsf{adj}_\epsilon(A)\boldsymbol{b} = (d^{-1}d_1, d^{-1}d_2, \ldots, d^{-1}d_n)^T.$$

□

6.4 Bases

In Cor. 3.5 it has been shown that Boolean vector spaces V have a unique basis, and V was said to be of finite dimension n, written as $\mathsf{dim}\, V = n$, if and only if V has a basis $B = \{\boldsymbol{v}_1, \boldsymbol{v}_2, \ldots, \boldsymbol{v}_n\}$ consisting of n vectors. A straightforward generalization of this very special case results in the following

Definition 6.42. ([40, Definition 3.1]) *Let S be a commutative semiring and $\mathcal{L}$ be a semilinear space over S. If all bases of $\mathcal{L}$ have the same number n of elements, then we call this number the* **dimension** *of $\mathcal{L}$, written as $\mathsf{dim}\, \mathcal{L} = n$.*

Proposition 6.43. ([44, Theorem 3.1]) *Let S be a commutative semiring and $\mathcal{L}$ be a semilinear space over S. If $\mathcal{L}$ has an infinite basis then every basis for $\mathcal{L}$ is infinite.*

Proof. Let A be an infinite basis for $\mathcal{L}$. If $\mathcal{L}$ has a finite basis B then each $\boldsymbol{b} \in B$ can be expressed as a linear combination of elements in A: $\boldsymbol{b} = \sum_{i\in\underline{n}} s_i \boldsymbol{a}_i$, where $\boldsymbol{a}_i \in A$ and $s_i \in S$. Put $\boldsymbol{b}(S) = \{\boldsymbol{a}_1, \boldsymbol{a}_2, \ldots, \boldsymbol{a}_n\}$ and $B' = \bigcup_{\boldsymbol{b}\in B} \boldsymbol{b}(S)$. Then $B' \subseteq A$ and $A - B' \neq \emptyset$ since B' is finite. Each $\boldsymbol{b} \in B$ can be written as a linear combination of elements in B', and any $\boldsymbol{a} \in A$ as a linear combination of elements in B. So, any $\boldsymbol{a} \in A$ can be written as a linear combination of elements in B'. Then there exists $\boldsymbol{a} \in A - B' \subseteq A$ such that $\boldsymbol{a} \in SB' \subseteq S(A - \{\boldsymbol{a}\})$, but this contradicts the definition of A, whence every basis for $\mathcal{L}$ is infinite. □

Corollary 6.44. ([44, Remark 3.1]) Prop. 6.43 *shows that if a semilinear space $\mathcal{L}$ over a commutative semiring has a finite basis then any basis for $\mathcal{L}$ is finite.*

At the end of Def. 3.2 we did already learn about the very special role of the row and column vectors of the identity matrix in linear algebra: We have $I_n = [\boldsymbol{e}_1 \boldsymbol{e}_2 \ldots \boldsymbol{e}_n]$, where for $e_{ij} = \delta_{ij}$ for all $i, j \in \underline{n}$. The standard unit vectors $\boldsymbol{e}_i$ clearly form an orthogonal basis for V^n.

Theorem 6.45. ([40, Theorem 3.1]) *Let S be a commutative semiring, and $\mathcal{L}$ be a semi-linear space over S. Then* $\dim \mathcal{L} = n$ *iff every vector can be uniquely represented as a linear combination of a basis.*

Proof. Assume $\dim \mathcal{L} = n$ and let $\{\boldsymbol{x}_1, \boldsymbol{x}_2, \ldots, \boldsymbol{x}_n\}$ a basis of $\mathcal{L}$. For any $\boldsymbol{a} \in \mathcal{L}$ suppose that

$$\boldsymbol{a} = r_1\boldsymbol{x}_1 + r_2\boldsymbol{x}_2 + \cdots + r_n\boldsymbol{x}_n = r_1'\boldsymbol{x}_1 + r_2'\boldsymbol{x}_2 + \cdots + r_n'\boldsymbol{x}_n$$

with $r_i, r_i' \in$, for all $i \in \underline{n}$. Noting that $I_n = [\boldsymbol{e}_1 \boldsymbol{e}_2 \ldots \boldsymbol{e}_n]$ and $X = [\boldsymbol{x}_1 \boldsymbol{x}_2 \ldots \boldsymbol{x}_n]$ we then get, for some $C, D \in \mathbb{S}_n(S)$, $I_n = XC$ since $\{\boldsymbol{x}_1, \boldsymbol{x}_2, \ldots, \boldsymbol{x}_n\}$ is a basis of $\mathcal{L}$, and $X = I_n D$, since $\{\boldsymbol{e}_1, \boldsymbol{e}_2, \ldots, \boldsymbol{e}_n\}$ also forms a basis of $\mathcal{L}$. So $I_n = I_n DC = DC$, whence $CD = I_n$ by Prop. 6.26(1), i.e. D is invertible. By assumption we have $\boldsymbol{a} = X\boldsymbol{r} = X\boldsymbol{r}' = I_n D\boldsymbol{r} = I_n D\boldsymbol{r}'$; therefore $\boldsymbol{r} = \boldsymbol{r}'$ since D has just been shown to be invertible.

Conversely, by Cor. 6.44 each basis of $\mathcal{L}$ is finite, since the standard unit vectors $\boldsymbol{e}_i$ clearly form a finite orthogonal basis for $\mathcal{L}$. Let $\{\boldsymbol{x}_1, \boldsymbol{x}_2, \ldots, \boldsymbol{x}_n\}$ and $\{\boldsymbol{y}_1, \boldsymbol{y}_2, \ldots, \boldsymbol{y}_m\}$ be two bases of $\mathcal{L}$. If $m \neq n$ then either $m < n$ or $n < m$. Assume $m < n$, then each of the vectors $\boldsymbol{y}_i, i \in \underline{m}$, can be written as a linear combination of $\boldsymbol{x}_1, \boldsymbol{x}_2, \ldots, \boldsymbol{x}_n$, i.e. for each $i \in \underline{m}$ there are $a_{ij} \in S$ such that $\boldsymbol{y}_i - \sum_{j \in \underline{n}} a_{ij}\boldsymbol{x}_j$. With $A = [a_{ij}] \in \mathbb{S}_{m\times n}$, we then get that $\boldsymbol{y}_k = A\boldsymbol{x}_k$. Thus, with $Y = [\boldsymbol{y}_1 \boldsymbol{y}_2 \cdots \boldsymbol{y}_m]^T \in \mathbb{S}_{m\times n}$, and $X = [\boldsymbol{x}_1 \boldsymbol{x}_2 \ldots \boldsymbol{x}_n]^T \in \mathbb{S}_n$, $Y = AX$. Likewise there exist elements $b_{jk} \in S$ such that $\boldsymbol{x}_j = \sum_{k \in \underline{m}} b_{jk}\boldsymbol{y}_k$ for any $j \in \underline{n}$. Thus, as before, we get $X = BY$ with $B = [b_{ij}] \in \mathbb{S}_{n\times m}$, whence $X = BAX$. Since, by assumption, every vector can be uniquely represented as a linear combination of a basis, this implies $BA = I_n$.

Adding $n - m$ rows $\boldsymbol{0}$ to A and $n - m$ columns $\boldsymbol{0}$ to B gives two matrices $A_0 = \begin{bmatrix} A \\ \boldsymbol{0} \end{bmatrix}$ and $B_0 = \begin{bmatrix} B\boldsymbol{0} \end{bmatrix}$ such that $B_0 A_0 = I_n$. Then by Prop. 6.26(1) both B_0 and A_0 are invertible. And from Prop. 6.10(1) and Prop. 6.11 we get $1 + |B_0|^+|A_0|^- + |B_0|^-|A_0|^+ = |B_0|^+|A_0|^+ + |B_0|^-|A_0|^-$. But since all the determinants evaluate to 0 (see Prop. 6.26(4)), we get the contradiction $1 = 0$. An analogous result can be obtained in case $n < m$, whence we must have $n = m$. □

6.5 Ranks of Matrices

We have already mentioned various rank functions and their properties in Section 5.1.2 and will now consider ranks of semiring matrices, repeating notions given already in Def. 5.10 for sake of easier look up. We also need some new notions: A *line* in a matrix is a row or a column of that matrix. A matrix $A \in \mathbb{S}_{m\times n}(S)$ is a rectangular scheme with mn positions $(i, j), i \in \underline{m}, j \in \underline{n}$. A set of $\min\{m, n\}$ positions in A is called *independent* or a **generalized diagonal** if it contains no two positions that occur on the same line; see

[18, 19], two papers which give very good overviews on the topic, including many hints to original sources.

If $C = AB$, the matrices A and B will be called *factors* of C, and (A, B) will be called a **decomposition** or **factorization** of C. In case $A \in \mathbb{S}_{m\times k}$ and $B \in \mathbb{S}_{k\times n}$, (A, B) is called a *k-decomposition* of matrix C, and k is called the **factorization size** of C into A and B.

Definition 6.46. ([5, Section 2]) *Let $O_{m\times n} \neq A \in \mathbb{S}_{m\times n}$ be given.*

(a) The **column rank** *of A, $\mathsf{cr}(A)$, is the smallest possible size of a spanning set for the column space of A, i.e. $\mathsf{cr}(A) = \dim \mathsf{col}(A)$.*
(b) The **row rank** *of A, $\mathsf{rr}(A)$, is the smallest possible size of a spanning set for the row space of A, i.e. $\mathsf{rr}(A) = \dim \mathsf{row}(A)$.*
(c) The **rank** *(also* **factor, semiring** *or* **Schein rank***), of A is the smallest integer r such that $A = LR$, with $L \in \mathbb{S}_{m\times r}$ and $R \in \mathbb{S}_{r\times n}$, denoted by $\mathsf{fr}(A)$.*
(d) The **term rank** *is the minimal number of lines required to contain all the nonzero entries of A, denoted by $\mathsf{tr}(A)$.*
(e) The **spanning row (column) rank** *is the minimal number of rows (columns) of A that span its row (column) space, denoted by $\mathsf{sr}(A)$ $(\mathsf{sc}(A))$.*

The zero matrices O are assigned the factor, column (row), and spanning column (row) rank 0.

As you can see, there are some more possibilities to define rank functions for matrices over semirings than we usually learn about. They are mentioned here to demonstrate the diversity of concepts which collapse in the case when the underlying semiring is a field in which case we have $\mathsf{fr}(A) = \mathsf{cr}(A) = \mathsf{rr}(A) = \mathsf{sc}(A) = \mathsf{sr}(A)$. (To see that e.g. $\mathsf{cr}(A) = \mathsf{sc}(A)$ for matrices A over some field, let $\mathsf{cr}(A) = k$, i.e. $\dim \mathsf{col}(A) = k$. So any r columns of A are linearly dependent for $r > k$ by Prop. 5.8(2). Hence $\mathsf{sc}(A) \le k = \mathsf{cr}(A)$, and the assertion follows using Prop. 6.49(1).)

As to be expected and shown below, the situation is quite different for matrices over general semirings. We begin with some basic facts:

Lemma 6.47. ([5, Proposition 2.6], [49, Lemmas 4.1 and 4.2]) *Let $O \neq A \in \mathbb{S}_{m\times n}$ be a matrix over an arbitrary semiring. Then*

(1) $\mathsf{fr}(A)$ is equal to the minimal number of rank-1 matrices whose sum is A.
(2) If $A = BC$ then $\mathsf{fr}(A) \le \min\{\mathsf{fr}(B), \mathsf{fr}(C)\}$.
(3) $\mathsf{fr}(I_m) = m$.

Proof. (1): Let $\mathsf{fr}(A) = k$ and $A = BC$ be a factorization of A, with $B \in \mathbb{S}_{m\times k}$, $C \in \mathbb{S}_{k\times n}$. Recall that $\boldsymbol{b}_{*j}$ denotes the j-th column of B and $\boldsymbol{c}_{j*}$ the j-th row of C. Then $A = [\boldsymbol{b}_{*1}\boldsymbol{c}_{1*}] + [\boldsymbol{b}_{*2}\boldsymbol{c}_{2*}] + \cdots + [\boldsymbol{b}_{*k}\boldsymbol{c}_{k*}]$. Each of the outer products $[\boldsymbol{b}_{*i}\boldsymbol{c}_{i*}]$ is a matrix of rank 1, so $\mathsf{fr}(A)$ is at least the minimum number of rank 1 matrices whose sum is A. On the other hand, if $A = \sum_{j\in \underline{l}} \boldsymbol{b}_{*j}\boldsymbol{c}_{j*}$ with $\boldsymbol{b}_{*j} \in \mathbb{S}_{m\times 1}$ and $\boldsymbol{c}_{j*} \in \mathbb{S}_{1\times n}$, then $A = BC$ where $B = [\boldsymbol{b}_{*1}\, \boldsymbol{b}_{*2} \dots \boldsymbol{b}_{*l}]$ and $C = [\boldsymbol{c}_{1*}\, \boldsymbol{c}_{2*} \dots \boldsymbol{c}_{l*}]^T$. Thus, $\mathsf{fr}(A)$ is at most the minimal number of rank 1 matrices whose sum is A.

(2): Let $\mathsf{fr}(A) = k > l = \mathsf{fr}(B)$. Then there exist matrices $B_1 \in \mathbb{S}_{m\times l}$ and $B_2 \in \mathbb{S}_{l\times n}$ such that $B = B_1B_2$. Therefore, $A = B_1B_2C = B_1(B_2C)$, whence $\mathsf{fr}(A) \le l < k$, a contradiction. $\mathsf{fr}(A) \le \mathsf{fr}(C)$ can be shown in an analogous way.

(3): $I_m = \sum_{k\in \underline{m}} E_{ii}$, and obviously m is the least number of cells whose sum gives I_m (see page 152). The assertion follows by (1) since $E_{ii} = [\boldsymbol{e}_i\, \boldsymbol{e}_i]$, $i \in \underline{m}$. □

Proposition 6.48. ([49, Proposition 4.1]) *Let $A \in \mathbb{S}_n$ be an invertible matrix over a commutative semiring. Then $\mathsf{fr}(A) = n$.*

Proof. By assumption there exists a matrix $B \in \mathbb{S}_n$ such that $AB = I_n$. We get $\mathsf{fr}(A) \ge \mathsf{fr}(I_n) = n$ from Lemma 6.47(2),(3). Therefore, by the definition of the factor rank, $\mathsf{fr}(A) = n$, since $A = I_nA$. □

(That the converse of this proposition is not true in general, is demonstrated by [49, Example 4.1].)

Proposition 6.49. ([6, Section 2]) *Let S be an arbitrary semiring and $A \in \mathbb{S}_{m\times n}(S)$. Then the following holds:*

(1) $0 \le \mathsf{fr}(A) \le \mathsf{cr}(A) \le \mathsf{sc}(A) \le n$*, and analogous for rows.*
(2) $\mathsf{cr}(B) \le \mathsf{cr}(A)$*, if B is obtained by deleting some rows from A.*
(3) $0 \le \mathsf{fr}(A) \le \min\{m, n\}$.
(4) $\mathsf{fr}(B) \le \mathsf{fr}(A)$ *for all submatrices B of A.*
(5) $\mathsf{fr}(A) \le \min\{\mathsf{cr}(A), \mathsf{rr}(A)\}$.
(6) $\mathsf{fr}(A) = \mathsf{fr}(A^T)$*, if S is commutative.*

Proof. (1): If not a smaller set of vectors, then certainly all the rows (columns) of A span $\mathsf{row}(A)$ ($\mathsf{col}(A)$). $\mathsf{cr}(A) \le \mathsf{sc}(A)$ follows directly from their definitions, since maybe fewer rows (columns) are needed if one can choose from not only the rows (columns) of A. So it remains to show $\mathsf{fr}(A) \le \mathsf{cr}(A)$: From the proof of Lemma 6.47(1) we see that, if not less, then the smallest possible size of a spanning set for the column space of A, i.e. $\mathsf{cr}(A) = \dim \mathsf{col}(A)$, would give us a suitable matrix for a factorization of A.

(2): [6, Lemmas 2.2, 2.4, and 2.6].

(3): Follows from (1), or from $A = AI_n$ and $A = I_mA$.

(4): Let $B \in \mathbb{S}_{k\times l}(S)$ be a submatrix of A. Then $k \le m$ and $l \le n$ and $\mathsf{fr}(B) \le \min\{k, l\}$.

(5): By (1) we have $\mathsf{fr}(A) \le \mathsf{cr}(A)$ and $\mathsf{fr}(A) \le \mathsf{rr}(A)$.

(6): Let $\mathsf{fr}(A) = k$; then there exists a k-decomposition $A = BC$ for some $B \in \mathbb{S}_{m\times k}$ and $C \in \mathbb{S}_{k\times n}$. Since $A^T = C^TB^T$ by Prop. 6.2, with $C^T \in \mathbb{S}_{n\times k}$ and $B^T \in \mathbb{S}_{k\times m}$, $\mathsf{fr}(A^T) = k$. □

Lemma 6.50. ([41, Lemma 3.1]) *For matrices $A, B \in \mathbb{S}_{m\times n}$ over an arbitrary semiring the following holds:*

(1) If A is a factor of B, and B is a factor of A, then $\mathsf{fr}(A) = \mathsf{fr}(B)$.
(2) If $\mathsf{row}(A) = \mathsf{row}(B)$ *then* $\mathsf{rr}(A) = \mathsf{rr}(B)$ *and* $\mathsf{fr}(A) = \mathsf{fr}(B)$.

(3) *If* $\mathsf{col}(A) = \mathsf{col}(B)$ *then* $\mathsf{cr}(A) = \mathsf{cr}(B)$ *and* $\mathsf{fr}(A) = \mathsf{fr}(B)$.

Proof. (1): By hypothesis there there exist matrices C and D such that $A = BC$ and $B = AD$. If $\mathsf{fr}(A) = k$ then there is a k–decomposition (A', A'') of A, so $B = AD = (A'A'')D = A'(A''D)$. Since $(A', A''D)$ is a k–decomposition of B, we get $\mathsf{fr}(B) \le k = \mathsf{fr}(A)$. Similarly one can prove $\mathsf{fr}(A) \le l = \mathsf{fr}(B)$ for some l-decomposition (B', B'') of B. Since k, l have been chosen arbitrarily, we have shown the assertion.

(2): If $\mathsf{row}(A) = \mathsf{row}(B)$, then both spaces are spanned by the same set, whence $\mathsf{rr}(A) = \mathsf{rr}(B)$. By Lemma 6.3(3) we get that A and B are factors of each other, so $\mathsf{fr}(A) = \mathsf{fr}(B)$ by (1). □

Proposition 6.51. ([41, Theorem 3.2]) *For a matrix* $A \in \mathbb{S}_{m\times n}$ *over an arbitrary semiring the following characterizations of the row, column, and factor rank hold:*

(1) $\mathsf{rr}(A) = k$ *iff* k *is the least integer such that there exist a* k*-decomposition* $A = CB$ *and a matrix* $D \in \mathbb{S}_{k\times m}$ *for which* $B = DA$.
(2) $\mathsf{cr}(A) = k$ *iff* k *is the least integer such that there exists a* k*-decomposition* $A = BC$ *and a matrix* $D \in \mathbb{S}_{n\times k}$ *for which* $B = AD$.
(3) *If* $\mathsf{fr}(A) = k$ *and* (B, C) *is a* k*-decomposition of* A*, then*
$$\mathsf{fr}(B) = \mathsf{fr}(C) = k = \mathsf{cr}(B) = \mathsf{rr}(C).$$

Proof. (1): Let R be an arbitrary basis for $\mathsf{row}(A)$ with k elements, and $B \in \mathbb{S}_{k\times n}$ a matrix whose rows are the row vectors from R. Then $\mathsf{row}(B) = \mathsf{row}(A)$, and by Lemma 6.3(3) we have a k-decomposition $A = CB$ and $B = DA$ for some matrices $C \in \mathbb{S}_{m\times k}$ and $D \in \mathbb{S}_{k\times m}$. Then, if $\mathsf{rr}(A) = k$, a spanning set R for $\mathsf{row}(A)$ with k elements has the smallest size if and only if k is the smallest integer such that for some $B \in \mathbb{S}_{k\times n}$, $C \in \mathbb{S}_{m\times k}$, and $D \in \mathbb{S}_{k\times m}$ we have $A = CB$ and $B = DA$.

(2): Similar to (1).

(3): Let $\mathsf{fr}(A) = k$, and let (B, C) a k-decomposition of A. Then $\mathsf{fr}(B) \le k$ by Prop. 6.49(3). Now suppose $\mathsf{fr}(B) = k' < k$; then there is a k'-decomposition (B', B'') of B, and we get $A = B'(B''C)$, which results in the contradiction $\mathsf{fr}(A) \le k' < k$. Thus $\mathsf{fr}(B) = k$. But k is the number of columns of B; so, by Prop. 6.49(1), we have $k = \mathsf{fr}(B) \le \mathsf{cr}(B) \le k$, whence $\mathsf{cr}(B) = k$. Similar for C. □

Corollary 6.52. ([41, Theorem 3.3]) *Let* $A \in \mathbb{S}_{m\times n}$ *be a matrix over an arbitrary semiring,* $m_r := |A_{\mathsf{row}}|$*, and* $n_c := |A_{\mathsf{col}}|$*. For* $k \in \{\mathsf{fr}(A), \mathsf{rr}(A), \mathsf{cr}(A), m_r, n_c\}$ *there exists a* k*-decomposition of* A*.*

Proof. From its definition it is immediate that there is always an $\mathsf{fr}(A)$-decomposition of A. If $k \in \{\mathsf{rr}(A), \mathsf{cr}(A)\}$, then a k-decomposition of A exists by Prop. 6.51. Let $A_{\mathsf{col}} = \{\boldsymbol{c}_1, \boldsymbol{c}_2, \dots, \boldsymbol{c}_{n_c}\}$, and let $A^C \in \mathbb{S}_{m\times n_c}$ denote a matrix whose columns are just $\boldsymbol{c}_j$, $1 \le j \le n_c$. Since every column of A is a linear combination of columns of A^C, we get $A^C B = A$ for some matrix $B \in \mathbb{S}_{n_c\times n}$. So (A^C, B) is an n_c-decomposition of A. Similar arguments show that there also exists a m_r-decomposition of A. □

Proposition 6.53. ([41, Theorem 3.4]) *Let $A \in \mathbb{S}_m$ be an idempotent matrix over an arbitrary semiring, Then* $\mathsf{fr}(A) = \mathsf{rr}(A) = \mathsf{cr}(A)$.

Proof. Assume $\mathsf{fr}(A) = k$, and let $A = BC$ for some $B \in \mathbb{S}_{m\times k}$ and $C \in \mathbb{S}_{k\times m}$. Then, with $B' = AB$ and $C' = CA$, one gets $B'C' = A(BC)A = A^3 = A$. So we have $A = BC = B'C'$ and $B' = AB$ whence from Prop. 6.51(3) we get $\mathsf{fr}(B') = \mathsf{fr}(A) = \mathsf{cr}(B')$. Since $B' = AB'$ and $A = B'C'$, using Lemma 6.3(4) and Lemma 6.50(3) we have $\mathsf{col}(A) = \mathsf{col}(B')$ and thus $\mathsf{cr}(A) = \mathsf{cr}(B') = \mathsf{fr}(A)$. Similarly for $\mathsf{cr}(A) = \mathsf{fr}(A)$. □

Proposition 6.54. ([41, Theorem 3.5]) *Let $A \in \mathbb{S}_n$ be a preorder matrix over a path algebra, i.e. $I_n \le A$ and $A^2 \le A$, Then* $\mathsf{fr}(A) = \mathsf{rr}(A) = \mathsf{cr}(A)$.

Proof. By monotonicity of matrix multiplication we get $A = AI_n \le AA$ which together with transitivity implies idempotence, whence the result follows by Prop. 6.53. □

We continue with a (famous) characterization of the term rank which dates back to 1936 and which is of pure combinatorial nature, because we are concerned only with whether an entry is zero or not. That is, an arbitrary property may be substituted for "nonzero", which means we have to do with a partition of the matrix entries into two classes. So, for sake of convenience, we adhere to $(0,1)$-matrices in our terminology.

Theorem 6.55. ([8, Theorem 1.2.1], [25, p. 240]) *Let S be an arbitrary semiring and $A \in \mathbb{S}_{m\times n}(S)$. Then the term rank of A is the maximal number of nonzero entries in some generalized diagonal of A.*

Proof. For sake of brevity, let $t := \mathsf{tr}(A)$ and d denote the maximal number of nonzero entries in a generalized diagonal of A. A covering of (the 1's of) A consisting of t lines is called *minimal*, and a minimal covering of A is called *proper* if it does not consist of all m rows of A or of all n columns of A.

Then clearly $t \ge d$ since, otherwise, a contradiction would arise: t lines do not suffice to cover $d > t$ 1's in a generalized diagonal. So it remains to show the converse inequality $t \le d$, which is done by a proof using induction.

As base consider the case that $m = 1$ or $n = 1$. Then A is either a zero matrix, and $t = d = 0$, or it has a single 1 in a generalized diagonal, i.e. $d = 1$, and this 1 is covered by a single line, i.e. $t = 1$. In both cases we have $t = d$. So we suppose $m \ge 2$ and $n \ge 2$ and assume that the theorem holds for matrices with the number of lines $\le m+n$. We consider two cases:

Case 1: A does not have a proper minimal covering. Then $t = \min\{m,n\}$. Take an entry $a_{kl} \ne 0$ in A, and delete the k-th row and l-th column from A, resulting in a matrix $A' \in \mathbb{S}_{m-1\times n-1}$. Then $t' \le \min\{m-1, n-1\} = t-1$. But A' cannot have a covering composed of fewer than $t-1$ lines because such a covering of A' together with the two deleted lines would yield a proper covering for A, contradicting the assumption. So $t' = t-1$. Applying the induction hypothesis we have $t' = d' = t-1$. a_{kl} clearly does not belong to the t' lines covering the 1's in A', whence a_{kl} in addition to the d' entries of A', no two of which are on a line, gives $d \ge d'+1 = t$.

Case 2: A has a proper minimal covering. Assume that the 1's in A are covered by $p < m$ rows and $q < n$ columns, with $t = p+q$. If $p = 0$ or $q = 0$, then the induction hypothesis gives the result. So let $p, q \geq 1$. Since permutation of rows and columns does not affect the quantities of interest, we permute the lines of A so that the p rows and q columns occupy the initial positions. Then A takes the form $\begin{bmatrix} X & B \\ C & O \end{bmatrix}$, where O has dimension $(m-p)\times(n-q)$, $B \in \mathbb{S}_{p\times(n-q)}$, and $C \in \mathbb{S}_{(m-p)\times q}$. Then $t_B = p$ and $t_C = q$: otherwise we would have $t < p+q$, contradicting our assumption. Applying the induction hypothesis to B and C gives $d_B = p$ and $d_C = q$. Thus, $d \geq d_B + d_C = p+q = t$. □

(Note that there is also a widespread statement of König's theorem in terms of graphs: In a bipartite graph the number of edges in a maximal matching is equal to the number of elements of a minimal vertex covering.)

Proposition 6.56. ([5, Proposition 3.1(3)]) *Let S be an arbitrary semiring, and* $O \neq A \in \mathbb{S}_{m\times n}(S)$. *Then* $\mathsf{fr}(A) \leq \mathsf{tr}(A)$.

Proof. Suppose $\mathsf{tr}(A) = k$. By Th. 6.55 (proof of *Case 2*) there exist permutation matrices $P \in \mathbb{S}_m$ and $Q \in \mathbb{S}_n$ such that

$$PAQ = \begin{bmatrix} A_1 & A_2 \\ A_3 & O_{(m-p)\times(n-q)} \end{bmatrix},$$

where $A_1 \in \mathbb{S}_{p\times q}$, $A_2 \in \mathbb{S}_{p\times(n-q)}$, $A_3 \in \mathbb{S}_{(m-p)\times q}$, and $p+q = k$. Next we define

$$X = \begin{bmatrix} I_p & O_{p\times q} \\ O_{(m-p)\times p} & A_3 \end{bmatrix} \quad \text{and} \quad Y = \begin{bmatrix} A_1 & A_2 \\ I_q & O_{q\times(n-q)} \end{bmatrix}.$$

Then we get $PAQ = XY$, so for $B = P^T X \in \mathbb{S}_{m\times k}$ and $C = YQ^T \in \mathbb{S}_{k\times n}$ we have $BC = P^T XYQ^T = P^T PAQQ^T = A$. Thus, $\mathsf{fr}(A) \leq k = \mathsf{tr}(A)$. □

6.6 Generalized Inverses and Regularity

Remember our short look on g-inverses and regularity in Section 2.11 where some results have been obtained from the viewpoint and in the terminology of Boolean relations. Here we will investigate these two concepts by considering matrices over more general algebraic structures, and by defining various kinds of generalized inverses more systematically.

Be aware that the classical context for the consideration of generalized inverses is the one of matrices over the complex numbers, or over fields in general. Two recent monographs, [11] and [47], offer a lot of information. You may also want to have a look into [37], where generalized inverses are analyzed for matrices over commutative rings. As it is common practice, the familiar (or standard) case is the benchmark for the generalized concepts.

We quote "A Rationale for Generalized Inverses" from [10, p. 4]: "Loosely speaking, the theory of generalized inverses of matrices is concerned with extending the concept of an inverse of a square nonsingular matrix to singular matrices and, more generally, to rectangular matrices by considering various sets of equations which A and X may be required to satisfy."

Recall from Def. 2.85(f) that the **Moore-Penrose inverse** of a relation has been defined using the four equations listed below. These equations have their origin in matrix theory in Theorem 1 of [33], stating that they have a unique solution for any A over the complex numbers, called the **Moore-Penrose inverse:**

(1) $AXA = A$
(2) $XAX = X$
(3) $AX = (AX)^T$
(4) $XA = (XA)^T$,

where $A \in \mathbb{S}_{m\times n}$ is a given, and $X \in \mathbb{S}_{n\times m}$ is an unknown matrix over S. Moreover, we can require that A and X commute, i.e.

(5) $AX = XA$, and
(1^k) $A^k = A^{k+1}X$, $k \geq 1$.

where $A, X \in \mathbb{S}_n$ are square matrices. Note that in view of (5) we have $A^k = A^{k+1}X = A^k XA$, whence this condition is usually denoted by (1^k).

Based on different subsets of the above mentioned equations, various kinds of inverses are defined in matrix theory, so a convenient notation for the various types of generalized inverses is needed. The one commonly used is as follows: For any A and $\gamma \subseteq \{1,2,3,4,5,1^k\}$ we let $A\gamma$ denote the set of all matrices X which satisfy the equations (i), (j), ..., (k), $i, j, \ldots, k \in \gamma$, and if $X \in A\gamma$ we say it is an γ-inverse of A, denoted by $A^{(i,j,\ldots,k)}$. We will, however, adhere to the 'non-numerical' names if such are commonly used.

Definition 6.57. *Let $A \in \mathbb{S}_{m\times n}$. A γ-inverse $X \in \mathbb{S}_{n\times m}$ of A is called*

(a) **generalized inverse** *(or* **g-inverse***), and also* **inner inverse** *for $\gamma = \{1\}$.*
(b) **outer inverse** *for $\gamma = \{2\}$.*
(c) **semi-inverse** *of A, also* **Thierrin-Vagner inverse** *for $\gamma = \{1,2\}$.*
(d) **least-squares g-inverse** *for $\gamma = \{1,3\}$.*
(e) **minimum-norm g-inverse** *for $\gamma = \{1,4\}$;*
(f) **Moore-Penrose inverse** *for $\gamma = \{1,2,3,4\}$.*
(g) **group inverse** *for $\gamma = \{1,2,5\}$, with $A, X \in \mathbb{S}_n$.*
(h) **Drazin inverse** *for $\gamma = \{1^k,2,5\}$, with $k \geq 1$ and $A, X \in \mathbb{S}_n$.*

As it is done for relations, matrices which have a g-inverse are called **regular.** Note that any idempotent matrix is regular. Obviously, also $J, O \in \mathbb{S}_{m\times n}$ are regular because $JEJ = J$ and $OEO = O$ for any cell $E \in \mathbb{S}_{n\times m}$. (In case you are interested in the reason for the naming of least-squares and minimum-norm g-inverses see [3, Section 4.2].)

If A has a Drazin inverse, say A^D, the **index of** A, denoted by $\text{ind}(A)$, is defined by $\text{ind}(A) = \min\{k \in \mathbb{N} \cup \{0\} : A^k = A^{k+1}A^D\}$.

In the following we list some algebraic properties of these kinds of matrices, which are easy to prove by straightforward calculations in most cases. These properties are usually formulated for matrices over the complex numbers, or over fields or rings in general, but as long as only matrix multiplication (and transposition) is used in the definitions and arguments involved, they do already hold on the general level of arbitrary (commutative) semirings. (Note that commutativity is essential, since it entitles us to use Prop. 6.2, viz. $(AB)^T = B^T A^T$. And remember: If nothing else is specified, commutativity of the underlying semiring is assumed.)

6.6.1 *Generalized and Moore-Penrose Inverses*

Clearly, every idempotent matrix A is regular, since A itself is a g-inverse, and also every invertible matrix is: $AA^{-1}A = A$. But a regular matrix need not be invertible as the following example from [29, p. 79] shows: For $A = \begin{bmatrix} 1 & 1 \\ 1 & 0 \end{bmatrix}$ the matrix $X = \begin{bmatrix} 0 & 1 \\ 1 & 1 \end{bmatrix}$ is a g-inverse, but A is not invertible since it is not a permutation matrix; recall Cor. 3.24(2) and Cor. 3.33(6).

Proposition 6.58. *Let A be a regular matrix over an arbitrary semiring and X be a g-inverse of A. Then*

(1) $A^+ := XAX$ is a semi-inverse of A, and $AA^+ = AX$.
(2) AX and XA are idempotent and regular.

Proof. (1): We have $AA^+A = AXAXA = AXA = A$, and also $A^+AA^+ = (XAX)A(XAX) = XA(XAX) = XAX = A^+$. Moreover, $AA^+ = AXAX = AX$.

(2): $(AX)(AX) = (AXA)X = AX$, therefore $(AX)(AX)(AX) = AX$. □

Proposition 6.59. *If a Moore-Penrose inverse of a matrix A over a commutative semiring exists, it is uniquely determined and denoted by $A^\dagger$.*

Proof. Let B and C be two Moore-Penrose inverses of A. Then, using Prop. 6.2, we have $C = CAC = C(AC)^T = CC^TA^T = CC^T(ABA)^T = C(C^TA^T)(B^TA^T) = C(AC)^T(AB)^T = CACAB = CAB = CABAB = A^TC^TA^TB^TB = (ACA)^TB^TB = A^TB^TB = BAB = B$. □

Proposition 6.60. *Let A be a regular matrix over a commutative semiring and X be a g-inverse of A. Then*

(1) X^T is a g-inverse of A^T.
(2) $A^\dagger(A^\dagger)^TA^T = A^\dagger = A^T(A^\dagger)^TA^\dagger$.

Proof. (1): From $AXA = A$ we get $A^TX^TA^T = A^T$.

(2): $A^\dagger(A^\dagger)^TA^T = A^\dagger(AA^\dagger)^T = A^\dagger AA^\dagger = A^\dagger$; similar for the other equality. □

6.6.2 Group Inverses

Proposition 6.61. ([21, Lemma 3.1]) *If a group inverse of a matrix A over an arbitrary semiring exists, it is uniquely determined and denoted by $A^\#$.*

Proof. (The proof in this paper is given for dioids, but it holds in fact for arbitrary semirings since only matrix multiplication is used in both definition and the proof below.) Assume that A has two group inverses B and C; then by definition of the group inverse we have

$$ABA = A,\ BAB = B,\ AB = BA, \quad \text{and} \quad ACA = A,\ CAC = C, \text{ and } AC = CA.$$

So, $AB = (ACA)B = (CA)(BA) = C(ABA) = CA = AC$, and therefore $B = B(AB) = B(AC) = (AB)C = (CA)C = C$. □

Obviously, for $k = 1$, the Drazin gives the group inverse: We have $A^k = A^{k+1}D = A^kAD = A^kDA$, $k \geq 1$. So taking $k = 1$ gives equation (1) as required.

The proposition below is given in its original form, i.e. formulated for an arbitrary semiring S, as an example of results that can directly be applied to the matrix semiring $\mathbb{S}_n$.

Proposition 6.62. ([22, Lemma 3.1]) *An element a of a semiring S has the group inverse if and only if $a^2x = a$ and $ya^2 = a$ both have solutions. Then $a^\# = yax$.*

Proof. Given that $a^\#$ exists, we get $a^2a^\# = aa^\#a = a$ and $a^\#a^2 = aa^\#a = a$, whence the equations $a^2x = a$ and $ya^2 = a$ both have solutions. Conversely, let $x, y \in S$ be such that $a^2x = a$ and $ya^2 = a$, which implies $ax = ya^2x = ya$. Denoting $g := yax$ we then get

$$aga = ayaxa = a^2xxa = axa = ya^2 = a,$$

$$gag = (yax)a(yax) = y(ax)ayax = y(yaa)yax = ya(ya)x = y(aax)x = yax = g, \text{ and}$$

$$ag = ayax = aaxx = ax = ya = yyaa = yaxa = ga.$$

□

Corollary 6.63. *Let R be an arbitrary semiring and $A \in \mathbb{S}_n(R)$. Then $A^\#$ exists if and only if $A^2X = A$ and $YA^2 = A$ both have solutions. Then $A^\# = YAX$.*

A great number of results on the various kinds of inverses can be found in [29, Chapter 3]. Although they are given in the framework of the classical fuzzy algebra $([0,1], \max, \min)$, many of them hold for more general, and some of them even for arbitrary semirings, as we will see below.

Lemma 6.64. ([29, p. 135, Lemma 3.7.2]) *Let R be an arbitrary semiring and $A \in \mathbb{S}_n(R)$. If $A^\#$ exists, then also $(A^n)^\#$ for any $n \geq 1$, and $(A^n)^\# = (A^\#)^n$.*

Proof. By straightforward calculations, using $AA^\# = A^\#A$ and induction on n, we argue as follows for e.g. condition (1): The assertion holds for $n = 1$ by the assumption that $A^\#$ exists. Assuming the validity for $n - 1$ we get, repeatedly using $A^\#A = AA^\#$:

$$\begin{aligned} A^n(A^\#)^n A^n &= A^{n-1}A(A^\#)^{n-1}A^\# A^n \\ &= A^{n-1}(A^\#)^{n-1}A^\# AA^n \\ &= (A^{n-1}(A^\#)^{n-1}A^{n-1})A^\# A^2 \\ &= A^{n-1}(AA^\# A) \\ &= A^{n-1}A. \end{aligned}$$

□

6.6.3 *Drazin Inverses*

Proposition 6.65. ([12, Theorem 1], [29, p. 140, Theorem 3.7.7]) *A matrix $A \in \mathbb{S}_n(S)$, with S an arbitrary semiring, has a unique Drazin inverse, denoted by A^D.*

Proof. For the proof we use Th. 1.59 which states that for any element a of a semigroup S the set $\langle a \rangle = \{a, a^2, a^3, \ldots\}$ contains only different powers of a and is thus isomorphic to $(\mathbb{N}, +)$, i.e. infinite; or that $\langle a \rangle$ is finite and that $a^i = a^{i+mp}$, for each $m \geq 1$, with i the index and p the period of a. Now, all the entries of $A \in \mathbb{S}_n(S)$ are taken from a finite set with at most n^2 different elements. So there is also a finite number of possible patterns only for the powers A^k, $k \geq 1$, i.e. $\langle A \rangle$ is finite. Therefore, we have $A^{k+d} = A^k$ for some $k, d \geq 1$. We choose $d' \geq k+1$ such that $A^{k+d'} = A^k$ (such a d' exists: we can choose any m' such that $d' = m'p \geq k+1$). First we show that $X = A^{d'-1}$ is a Drazin inverse of A (recall from Def. 6.57 that Drazin inverses are characterized by $\gamma = \{1^k, 2, 5\}$):

- $A^{k+1}X = A^{k+1}A^{d'-1} = A^{k+d'} = A^k$, so (1^k) holds.
- $AX = AA^{d'-1} = A^{d'} = A^{d'-1}A = XA$, so (5) holds.
- $X^2A = (A^{d'-1})^2 A = A^{2d'-2}A = A^{2d'-1} = A^{d'}A^{d'-1} = A^{d'}X = A^{d'-1} = X$, by (1^k); and with (5) we get $X^2A = XAX = X$, so (2) is valid.

Let X and Y be two Drazin inverses of A. By several times applying (1^k) to X we get

$$A^{2k+1}X^{k+1} = A^k(A^{k+1}X)X^k = A^kA^kX^k = A^{k-1}(A^{k+1}X)X^{k-1} = \ldots = A^k.$$

And from this result and the equivalent $Y^2A = Y$ of (2) we get

$$Y^{k+1}A^{2k+1}X^{k+1} = Y^{k+1}A^k = Y^{k-1}(Y^2A)A^{k-1} = Y^{k-1}YA^{k-1} = Y^{k-2}(Y^2A)A^{k-2} = \ldots = Y.$$

Repeatedly using (5) gives $Y^{k+1}A^{2k+1} = A^{2k+1}Y^{k+1}$, so similarly as above for Y

$$Y^{k+1}A^{2k+1}X^{k+1} = A^kX^{k+1} = A^{k-1}(AX^2)X^{k-1} = \ldots = X,$$

whence $X = Y$. □

Proposition 6.66. ([12, Theorems 1 and 4]

(1) *If a Drazin inverse of a matrix A over an arbitrary semiring exists, it commutes with every matrix B that commutes with A.*
(2) *A^D, $A \neq O$, exists if and only if there are $p, q \in \mathbb{N}_0$ and $B, C \in \mathbb{S}_n$ such that*

$$A^p = A^{p+1}B \quad \textit{and} \quad A^q = CA^{q+1}. \tag{6.20}$$

(3) *If there exist $p, q \in \mathbb{N}_0$ and $B, C \in \mathbb{S}_n$ such that* (6.20) *holds then* $\operatorname{ind}(A) \leq \min\{p, q\}$. (See bottom of page 177.)

Proof. (1): Assume $X = A^D$ and $AB = BA$; then

$$XA^m B = XBA^m = XBA^{m+1}X = XA^{m+1}BX = A^{m+1}XBX = A^m BX.$$

So $X^{m+1}A^m B = A^m BX^{m+1}$, since the above exchanges can be performed for each occurrence of X. From $XAX = X$ and $AX = XA$ we get $X^2A = AX^2 = X$, whence $X = X^{m+1}A^m$ by trivial induction on m. Using the commuting property of A with both B and X it follows that

$$XB = X^{m+1}A^m B = A^m BX^{m+1} = BX^{m+1}A^m = BX.$$

(2): It is easy to see that the existence of A^D implies (6.20): The first equation follows immediately from (1^k) by $B = A^D$, and the second one from the first by applying the commuting property (5).

Conversely, let $m = \max\{p, q\}$, and assume $p < q$. Then from (6.20) we have $A^m = CA^{m+1}$, and $A^m = A^q = A^{q-p}A^p$, whence $A^{m+1}B = A^{q-p}(A^{p+1}B) = A^{q-p}A^p = A^q = A^m$. Therefore, (6.20) implies $A^{m+1}B = A^m = CA^{m+1}$, so $A^m B = CA^{m+1}B = CA^m$, and, by induction over k, $A^m B^k = C^k A^m$, $k \geq 1$. Taking $X := A^m B^{m+1} = C^{m+1}A^m$ one gets

(5): $AX = AA^m B^{m+1} = A^{m+1}BB^m = A^m B^m = C^m A^m = C^m CA^{m+1} = XA$;

(1^k): Again by induction on k with base case $A^{m+1}B = A^m$ we get $A^m = A^{m+k}B^k$, whence $A^{m+1}X = A^{m+1}A^m B^{m+1} = A^{m+(m+1)}B^{m+1} = A^m$;

(2): Using the two results just obtained we have $X^2A = XAX = XAA^m B^{m+1} = A^{m+1}XB^{m+1} = A^m B^{m+1} = X$.

(3): Assume that there are $p, q \in \mathbb{N}_0$ and $B, C \in \mathbb{S}_n$ such that (6.20) holds; then A^D exists by (2). In its proof for (1^k) we have shown that $A^m = A^{m+k}B^k$, $k \geq 1$. Now assume $p < q$; then $A^p = A^{p+(q-p)}B^{q-p} = A^q B^{q-p}$, whence, using (6.20), we get

$$CA^{p+1} = CA(A^q B^{q-p}) = (CA^{q+1})B^{q-p} = A^q B^{q-p} = A^p.$$

□

6.6.4 *Least-Squares and Minimum Norm Generalized Inverses*

We close with a result on least-squares and minimum norm g-inverses:

Proposition 6.67. ([21, Theorems 2.1 and 2.2]) *Let A and X be matrices over a commutative semiring.*

(1) X is a least-squares g-inverse of A if and only if $A^T AX = A^T$.
(2) X is a minimum norm g-inverse of A if and only if $XAA^T = A^T$.

Proof. (1): If X is a least-squares g-inverse of A, then, using Prop. 6.2, we have

$$A^T AX = A^T(AX)^T = A^T X^T A^T = (AXA)^T = A^T.$$

Conversely, assume $A^T AX = A^T$. Then $X^T A^T A = A$, and

$$(AX)^T = X^T A^T = X^T(A^T AX) = (X^T A^T A)X = AX. \tag{6.21}$$

Moreover, $A = X^T A^T A = (AX)^T A = AXA$, using (6.21) for the last equality.

(2): Similar to (1). □

References

[1] Akian, M., Gaubert, S., Guterman, A.: Linear independence over tropical semirings and beyond, in: Litvinov, G.L., Sergeev, S.N. (Eds), *Tropical and Idempotent Mathematics and Applications,* American Mathematical Society, pp. 1-38 (2009).

[2] Backhouse, R.C., Carré, B.: Regular Algebra Applied to Path-Finding Problems, IMA Journal of Applied Mathematics **15**, 161-186 (1975).

[3] Bapat, R.B.: *Linear Algebra and Linear Models,* 3rd ed., Springer-Verlag London Limited (2012).

[4] Baras, J.S., Theodorakopoulos, G.: *Path problems in networks.* Synthesis Lectures on Communication Networks 3, Morgan & Claypool (2010).

[5] Beasley, L.B., Guterman, A.E.: Rank inequalities over semirings, J. Korean Math. Soc. **42**, 223-241 (2005).

[6] Beasley, L.B., Pullman, N.J.: Semiring rank versus column rank, Linear Algebra and its Applications **101**, 33-48 (1988).

[7] Blyth, T.S., Robertson, E.F.: *Basic Linear Algebra,* 2nd ed., Springer-Verlag London (2002).

[8] Brualdi, R.A., Ryser, H.J.: *Combinatorial Matrix Theory,* Cambridge University Press (1991).

[9] Cerinšek, M., Batagelj, V.: Semirings and Matrix Analysis of Networks, in: Alhajj, R., Rokne, J. (Eds.), *Encyclopedia of Social Network Analysis and Mining,* Springer Science+Business Media LLC, pp. 2379–2386 (2017).

[10] Cline, R.R.: *Elements of the Theory of Generalized Inverses of Matrices,* Education Development Center, Inc. (1979).

[11] Cvetković Ilić, D.S., Wei, Y.: *Algebraic Properties of Generalized Inverses,* Springer Nature Singapore Pte Ltd., (2017).

[12] Drazin, M.P.: Pseudo-inverses in associative rings and semigroups, The American Math. Monthly **65**, 506–514 (1958).

[13] Duan, J.-S.: The transitive closure, convergence of powers and adjoint of generalized fuzzy matrices, Fuzzy Sets and Systems **145**, 301-311 (2004).

[14] Ghosh, S.: Matrices over Semirings, Information Sciences **90**, 221-230 (1996).

[15] Golan, J.S.: *Semirings and their Applications,* Springer Science+Business Media Dordrecht (1999).

[16] Golan, J.S.: *Semirings and Affine Equations over Them: Theory and Applications,* Springer Science+Business Media Dordrecht (2003).

[17] Griffin, T.G., Gurney, A.J.: Increasing bisemigroups and algebraic routing, in: Berghammer, R., Möller, B., Struth, G. (Eds.), *Relations and Kleene Algebra in Computer Science, RelMiCS/AKA 2008,* LNCS 4988, pp. 123-137 (2008).

[18] Guterman, A.E.: Rank and determinant functions for matrices over semirings, in: Young, N., Choi, Y. (Eds.), *Surveys in Contemporary Mathematics* (London Mathematical Society Lecture Note Series, Vol. 347), Cambridge University Press, Cambridge, pp. 1-33 (2008).

[19] Guterman, A.E.: Matrix invariants over semirings, in: Hazewinkel, M. (Ed.), *Handbook of Algebra, Vol. 6,* Elsevier, North-Holland, pp. 3-33 (2009).

[20] Han, S.C., Li, H.X.: Invertible incline matrices and Cramer's rule over inclines, Linear Algebra and its Applications **389**, 121-138 (2004).

[21] Hashemi, B., Tavakolipour, H., Nasrollahi Shirazi, M. : Explicit formulae for the greatest least-squares and minimum norm g-inverses and the unique group inverse of matrices over commutative residuated dioids, Semigroup Forum **92**, 541-550 (2016).

[22] Khan, I.A., Wang, Q.W.: The Drazin inverse in an arbitrary semiring, Linear and Multilinear Algebra **59**, 1019-1029 (2011).

[23] Kim, J.B., Baartmans, A., Sahadin, N.S.: Determinant theory for fuzzy matrices, Fuzzy Sets and Systems **29**, 349-356 (1989).

[24] Kim, K.H., Roush, F.W.: Gerneralized Fuzzy Matrices, Fuzzy Sets and Systems **4**, 293-315 (1980).

[25] König, D.: *Theorie der endlichen und unendlichen Graphen*, Akademische Verlagsgesellschaft m.b.H, Leipzig (1936), reprinted by BSB B. G. Teubner Verlagsgesellschaft, Leipzig (1986).

[26] Lang, S.: *Linear algebra*, 3rd ed., Springer-Verlag New York (1987).

[27] Liao, Y.l., Wang, X.P.: Note on invertible matrices over commutative semirings, Linear and Multilinear Algebra **64**, 477-483 (2016).

[28] Luo, W., Weng, C.: The matrix in semilinear spaces over commutative semirings, Mathematica Aeterna **7**, 211-221 (2017).

[29] Meenakshi, A.R.: *Fuzzy Matrix Theory and Applications*, MJP Publishers, Chennai, India (2008).

[30] Minc, H.: *Permanents*, Addison-Wesley Publishing Company (1978).

[31] Mohindru, P., Pereira, R.: Tan's Epsilon-Determinant and Ranks of Matrices over Semirings, International Scholarly Research Notices, Volume 2015, Article ID 242515 (2015).

[32] Mohri, M.: Semiring frameworks and algorithms for shortest-distance problems, J. of Automata, Languages and Combinatorics **7**, 321-350 (2002).

[33] Penrose, R.: A generalized inverse for matrices, Math. Proc. of the Cambridge Philosophical Soc. **51**, 406-413 (1955).

[34] Perfilieva, I., Kupka, J.: Kronecker-Capelli theorem in semilinear spaces, in: Ruan, D., Li, T.R. et.al. (Eds.), *Computational Intelligence: Foundations and Applications*, World Scientific, pp. 43-51 (2010).

[35] Poplin, P.L., Hartwig, R.E.: Determinantal identities over commutative semirings, Linear Algebra and its Applications **387**, 99-132 (2004).

[36] Ragab, M.Z., Emam, E.G.: The determinant and adjoint of a square fuzzy matrix, Fuzzy Sets and Systems **61**, 297-307 (1994).

[37] Rao, K.P.S.B.: *Theory of Generalized Inverses Over Commutative Rings*, Taylor & Francis, London (2002).

[38] Reutenauer, C., Straubing, H.: Inversion of Matrices over a Commutative Semiring, J. of Algebra **88**, 350-360 (1984).

[39] Schönfelder, R., Leucker, M.: Abstract Routing Models and Abstractions in the Context of Vehicle Routing, in: *Proceedings of the 24th International Joint Conference on Artificial Intelligence* (IJCAI'15), pp. 2639-2645, AAAI Press, (2015).

[40] Shu, Q.Y., Wang, X.P.: The cardinality of bases in semilinear spaces over commutative semirings, Linear Algebra and its Applications **459**, 83-100 (2014).

[41] Stamenković, A., Ćirić, M., Basić, M.: Ranks of fuzzy matrices. Applications in state reduction of fuzzy automata, Fuzzy Sets and Systems **333**, 124-139 (2018).

[42] Tan, Y.J.: On invertible matrices over antirings, Linear Algebra and its Applications **423**, 428-444 (2007).

[43] Tan, Y.J.: On invertible matrices over commutative semirings, Linear and Multilinear Algebra **61**, 710-724 (2013).

[44] Tan, Y.J.: Bases in semimodules over commutative semirings, Linear Algebra and its Applications **443**, 139-152 (2014).

[45] Tan, Y.J.: Determinants of matrices over semirings, Linear and Multilinear Algebra **62**, 498-517 (2014).

[46] Wang, X.P., Shu, Q.Y.: The bideterminants of matrices over semirings, Soft Computing **18**, 729-742 (2014).

[47] Wang, G., Wei, Y., Qiao, S.: *Generalized Inverses: Theory and Computation*, 2nd ed., Springer Nature Singapore Pte Ltd. and Science Press (2018).

[48] Wongseelashote, A.: Semirings and path spaces, Discrete Mathematics **26**, 55-78 (1979).

[49] Zhao, S., Wang, X.P.: Invertible matrices and semilinear spaces over commutative semirings, Information Sciences **180**, 5115-5124 (2010).

CHAPTER 7

MATRICES OVER ANTIRINGS AND PATH ALGEBRAS

Recall from Section 4.2 that path algebras generalize the concepts of Boolean algebra, distributive lattice, incline, max-plus algebra and min-plus algebra. In fact, every Boolean algebra, any fuzzy algebra $(\mathbb{F}, \vee, \wedge)$, where $\{0,1\} \subseteq \mathbb{F} \subseteq [0,1] \subset \mathbb{R}$, every bounded distributive lattice, and any 0-1-incline are commutative path algebras. And path algebras are zerosumfree by Prop. 4.4(2). So matrices over antirings and commutative path algebras comprise a group of matrices of interest within the scope of this text which seems large enough to look at their properties more closely.

7.1 Various Matrix Properties

In view of Prop. 6.1 it is easy to see that for a path algebra S $(\mathbb{S}_n(S), +, \cdot)$, the set of $n \times n$ square matrices over S, is a path algebra, and that $A + C \leq B + D$ and $AC \leq BD$ for all $A, B, C, D \in \mathbb{S}_n(S)$ with $A \leq B$ and $C \leq D$. (Note that

$$\textstyle\sum_{k \in \underline{n}} a_{ik} c_{kj} \leq \sum_{k \in \underline{n}} (a_{ik} + a'_{ik})(c_{kj} + c'_{kj}) = \sum_{k \in \underline{n}} b_{ik} d_{kj}$$

using Prop. 4.12(1) and (2)).

The following properties are important for matrices over a path algebra, defined in full analogy to the case of Boolean relations:

Definition 7.1. *Let D be a path algebra and $A \in \mathbb{S}_n(D)$. A is called*

(a) **reflexive** *if $I_n \leq A$;*
(b) **irreflexive** *if $a_{ii} = 0$ for all $i \in \underline{n}$;*
(c) **transitive** *if $A^2 \leq A$;*
(d) **increasing** *if $A \leq A^2$;*
(e) **idempotent** *if $A = A^2$;*
(f) **Hall matrix** *if there exists an invertible matrix C with $C \leq A$.*

Analogous to Def. 2.29 a transitive matrix B is called the **transitive closure** of A if $A \leq B$ and if for any transitive matrix $C \in \mathbb{S}_n(S)$ satisfying $A \leq C$ we have $B \leq C$. The transitive closure of A is denoted by $\mathbf{t}(A)$. It is clear that if A has a transitive closure then it is unique.

Let $A \in \mathbb{S}_n$. We sometimes need to write A^m as $[a_{ij}^{(m)}]$. Note that for any $i, j \in \underline{n}$ it holds that

$$a_{ij}^{(m)} = \sum_{i_1, i_2, \ldots, i_{m-1} \in \underline{n}} a_{ii_1} a_{i_1 i_2} \ldots a_{i_{m-2} i_{m-1}} a_{i_{m-1} j}. \tag{7.1}$$

Proposition 7.2. ([17, Proposition 3.1]) *Let L be a commutative path algebra and $A \in \mathbb{S}_n(L)$. Then*

(1) A is transitive if and ony if $a_{ik}a_{kj} \leq a_{ij}$ for all $i, j, k \in \underline{n}$.
(2) A is transitive if and ony if PAP^T is transitive for any permutation matrix P.

Proof. (1): Assuming $A^2 \leq A$ we have for any $i, j \in \underline{n}$ that $a_{ij}^{(2)} = \sum_{k \in \underline{n}} a_{ik}a_{kj} \leq a_{ij}$. Since the sum is the supremum of all the product summands (see Th. 4.16), the assertion $a_{ik}a_{kj} \leq a_{ij}$ for all $i, j, k \in \underline{n}$ follows. The converse direction is also straightforward.
(2): Let A be transitive. We have $(PAP^T)^2 = (PA)(P^TP)(AP^T) = (PA)I_n(AP^T) = PA^2P^T \leq PAP^T$. Conversely, assume $(PAP^T)^2 \leq PAP^T$; then, similar to the above, we get $PA^2P^T \leq PAP^T$. This implies $P^TPA^2P^TP \leq P^TPAP^TP$, i.e. $I_nA^2I_n \leq I_nAI_n$, so $A^2 \leq A$. □

Proposition 7.3. ([17, Theorem 5.1]) *Let L be a commutative path algebra and $A \in \mathbb{S}_n(L)$. If $A^{m+1} \leq \sum_{k \in \underline{m}} A^k$ for some positive integer m, then A has a transitive closure $\mathbf{t}(A) = \sum_{k \in \underline{m}} A^k$.*

Proof. We first prove $A^{m+k} \leq \sum_{l \in \underline{m}} A^l$ for any $k \geq 1$: For $k = 1$ this is obvious by our hypothesis. Assuming $A^{m+k} \leq \sum_{l \in \underline{m}} A^l$ by induction we get $A^{m+k+1} = AA^{m+k} \leq A(\sum_{l \in \underline{m}} A^l) = A^{m+1} + \sum_{l=2}^{m} A^l \leq \sum_{l \in \underline{m}} A^l + \sum_{l=2}^{m} A^l = \sum_{l \in \underline{m}} A^l$, observing that $(\mathbb{S}_n(L), +, \cdot)$ is a path algebra.

We now set $B := \sum_{l \in \underline{m}} A^l$. Then, obviously, $A \leq B$. and

$$B^2 = \sum_{l=2}^{2m} A^l = \sum_{l=2}^{m} A^l + \sum_{l=m+1}^{2m} A^l \leq \sum_{l=2}^{m} A^l + \sum_{l=1}^{m} A^l = \sum_{l=1}^{m} A^l = B,$$

where the inequality follows from the result just shown above. Thus B is transitive. If there exists a transitive matrix $C \in \mathbb{S}_n(L)$ such that $A \leq C$, then from $C^2 \leq C$ we get $C^k \leq C$ for all $k \geq 1$. Therefore, $A^k \leq C^k \leq C$, whence $B \leq C$. Finally we have thus shown $\mathbf{t}(A) = B$ as desired. □

Corollary 7.4. ([17, Corollary 5.1]) *Let L be a commutative path algebra and let $A \in \mathbb{S}_n(L)$ have index k and period d; then A has a transitive closure $\mathbf{t}(A) = \sum_{l=1}^{k+d-1} A^l$.*

Proof. We have $A^{k+d} = A^k \leq \sum_{l=1}^{k+d-1} A^l$. By Prop. 7.3 we then get $\mathbf{t}(A) = \sum_{l=1}^{k+d-1} A^l$. □

Definition 7.5. ([8, Definition 3.1]) *A matrix $A = [a_{ij}] \in \mathbb{S}_n(L)$ over a commutative path algebra L is said to be*

(a) **row diagonally dominant** *if $a_{ij} = a_{ii}a_{ij}$, for all $i, j \in \underline{n}$.*
(b) **column diagonally dominant** *if $a_{ij} = a_{ij}a_{jj}$, for all $i, j \in \underline{n}$.*

(c) **weakly diagonally dominant** *if for any* $i \in \underline{n}$,
either $a_{ij} = a_{ii}a_{ij}$, *for all* $j \in \underline{n}$ *or* $a_{ji} = a_{ii}a_{ji}$, *for all* $j \in \underline{n}$.

(d) **diagonally dominant** *if it is both row and column diagonally dominant, i.e.*
$a_{ij} = a_{ii}a_{ij} = a_{ij}a_{jj}$, *for all* $i, j \in \underline{n}$.

Proposition 7.6. ([8, Theorem 3.1]) *Let L be a commutative path algebra and $A \in \mathbb{S}_n(L)$. If A is transitive and row diagonally dominant, then $C := [c_{ij}]$ is idempotent, where $c_{ij} = a_{ij}a_{ji}$.*

Proof. For any $i, j \in \underline{n}$ we have

$$c_{ij}^{(2)} = \sum_{k\in\underline{n}} c_{ik}c_{kj} = \sum_{k\in\underline{n}} a_{ik}a_{ki}a_{kj}a_{jk} = \sum_{k\in\underline{n}} (a_{ik}a_{kj})(a_{jk}a_{ki}) \le \sum_{k\in\underline{n}} (a_{ij}a_{ji}) = c_{ij}$$

because A is transitive. So $C^2 \le C$. Furthermore we have $c_{ij}^{(2)} = \sum_{k\in\underline{n}} c_{ik}c_{kj} \ge c_{ii}c_{ij} = a_{ii}a_{ii}a_{ij}a_{ji} = a_{ij}a_{ji} = c_{ij}$, since A is row diagonally dominant. Thus $C^2 \ge C$, and equality is proved. □

An element a of a path algebra D is called **transitive** in D if $a^2 \le a$. Without its lengthy proof we mention

Lemma 7.7. ([17, Theorem 4.3(1)]) *Let D be a commutative path algebra and $A \in \mathbb{S}_n(D)$. If $a_{i_0i_1}a_{i_1i_1}\cdots a_{i_{m-1}i_0}$ is transitive for any $m, i_0, i_1, \ldots, i_{m-1} \in \underline{n}$, then A^s is transitive for some positive integer s.*

Proposition 7.8. ([17, Theorem 4.3(2)]) *Let D be a commutative path algebra and $A \in \mathbb{S}_n(D)$. If every element of A is transitive in D, then A^s is transitive for some positive integer s.*

Proof. If every element of A is transitive in D, then clearly $a_{i_0i_1}a_{i_1i_1}\cdots a_{i_{m-1}i_0}$ is transitive in D for any $m, i_0, i_1, \ldots, i_{m-1} \in \underline{n}$. By Lemma 7.7 there exists a positive integer s such that A^s is transitive. □

7.2 Powers, Periods, and Convergence

(Before you proceed you may want to recall some things related to the current topic which have already been mentioned before in Section 1.4.2 and Section 2.7.)

When a matrix A (or a relation ρ) is constructed with its entries taken from an infinite set S, then all its powers $A^i, i \in \mathbb{N}$, may be different. If, however, the underlying set S is finite, then only a finite number of different matrices (or relations) on S is available for the A^i; so the question for possible patterns in the sequence $A, A^2, A^3, \ldots$ arises in a natural way. In the following assume that A is any matrix on a finite set $S = \{s_1, s_2, \ldots, s_n\}$.

Let L be a path algebra. An element a in L is said to be **idempotent** if $a^2 = a$; we set $I(L) = \{a \in L : a^2 = a\}$. An element $a \ne 0$ in L is said to be **nilpotent** if $a^k = 0$ for some positive integer k. The least positive integer k satisfying $a^k = 0$ is called the **nilpotency**

index of a and denoted by $h(a)$. And $a \in L$ is said to be **almost periodic** if there exist positive integers k and d such that $a^k = a^{k+d}$. The least such positive integers k and d are called the **index** and the **period** of a, and denoted by $k(a)$ and $d(a)$, respectively. As mentioned above, $(\mathbb{S}_n(L), +, \cdot)$ is a path algebra; so the above concepts apply to matrices $A \in \mathbb{S}_n(L)$ as well. In particular, if $d(A) = 1$ then A is said to **converge in a finite number of steps.**

Proposition 7.9. ([16, Proposition 3.4(1)]) *Let D be a commutative path algebra and $A \in \mathbb{S}_n(D)$ be almost periodic. Then A has transitive closure and* $\mathbf{t}(A) = \sum_{j \in \underline{m}} A^j$ *for all* $m \geq k(A) + d(A) - 1$.

Proof. Let $B = \sum_{l=1}^{k(A)+d(A)-1} A^l$. Obviously, $A \leq B$. Since $(\mathbb{S}_n(D), +, \cdot)$ is a path algebra, and $2(k(A) + d(A) - 1) \geq k(A) + d(A) - 1$,

$$B^2 = \sum_{l=2}^{2k(A)+2d(A)-2} A^l = \sum_{l=2}^{k(A)+d(A)-1} A^l \leq B,$$

in which the second equality holds because $A^{l+d(A)} = A^l$ for $l \geq k(A)$, whence B is transitive.

Now let $C \in \mathbb{S}_n(D)$ be any transitive matrix such that $A \leq C$. Then $C^2 \leq C$, therefore $C^k \leq C$, for all $k \geq 1$, and $A^k \leq C^k \leq C$. So $B = \sum_{l=1}^{k(A)+d(A)-1} A^l \leq C$, and $\mathbf{t}(A) = B$ by the definition of transitive closure. Since $\sum_{l=1}^{m} A^l = \sum_{l=1}^{k(A)+d(A)-1} A^l = B$ for any $m \geq k(A) + d(A) - 1$, we get $\mathbf{t}(A) = \sum_{l=1}^{m} A^l$. □

Lemma 7.10. ([16, Proposition 3.4(2),(3)]) *Let D be a commutative path algebra and $A \in \mathbb{S}_n(D)$ be almost periodic. Then*

(1) $\mathbf{t}(A)^s = A^{s-1}\mathbf{t}(A)$ *for any* $s \geq 1$.
(2) $\mathbf{t}(A)$ *converges to* $A^{k(A)-1}\mathbf{t}(A)$ *in a finite number of steps and* $k(\mathbf{t}(A)) \leq k(A)$.

Proof. (1): For any positive integer s we have

$$\left(\mathbf{t}(A)\right)^s = \left(\sum_{l=1}^{k(A)+d(A)-1} A^l\right)^s = \sum_{l=s}^{(k(A)+d(A)-1)s} A^l = A^{s-1} \sum_{l=1}^{(k(A)+d(A)-1)s-(s-1)} A^l,$$

where the second equality holds because $(\mathbb{S}_n(D), +, \cdot)$ is a path algebra. Now, $(k(A) + d(A) - 1)s - (s-1) \geq k(A) + d(A) - 1$, so by Prop. 7.9

$$\left(\mathbf{t}(A)\right)^s = A^{s-1} \sum_{l=1}^{(k(A)+d(A)-1)s-(s-1)} A^l = A^{s-1}\mathbf{t}(A).$$

□

Proposition 7.11. ([17, Theorem 4.1]) *Let D be a commutative path algebra and $A \in \mathbb{S}_n(D)$. Then there exists a positive integer s such that A^s is transitive if and only if $A^{k+d} \leq A^k$ for some positive integers k and d.*

Proof. Suppose $A^{k+d} \leq A^k$ for some positive integers k and d. Then

$$A^{(k+l)d} = A^{k+d}A^{k(d-1)+(l-1)d} \leq A^k A^{k(d-1)+(l-1)d} = A^{kd}A^{(l-1)d} = A^{(k+l-1)d},$$

for all $l \in \underline{k}$. Therefore, $(A^{kd})^2 = A^{(k+k)d} \leq A^{(2k-1)d} \leq A^{(2k-2)d} \leq \cdots \leq A^{(k+1)d} \leq A^{kd}$. Thus, A^s with $s = kd$ is transitive. Conversely, if A^s is transitive for some positive integer s, then $A^{s+s} \leq A^s$. □

Proposition 7.12. ([8, Lemma 4.2]) *Let D be a commutative path algebra and $A \in \mathbb{S}_n(D)$ be almost periodic. Then there exists a positive integer s such that A^s is idempotent.*

Proof. Assume that $A^k = A^{k+d}$ for some positive integers k and d; then similar as above in Prop. 7.11 we have $A^{(k+l)d} = A^{(k+l-1)d}$ for all $l \in \underline{k}$. Therefore, $(A^{kd})^2 = A^{2kd} = A^{(2k-1)d} = A^{(2k-2)d} = \cdots = A^{kd}$, so A^{kd} is idempotent. □

Proposition 7.13. ([8, Proposition 4.1]) *Let D be a commutative path algebra and $A \in \mathbb{S}_n(D)$ be almost periodic. If A is row (or column) diagonally dominant, then $\mathbf{t}(A) = A^s$, where A^s is idempotent.*

Proof. We consider the case that A is row diagonally dominant; the other proof is similar. By Prop. 7.12 A^s with $s = kd$ is idempotent, so $A^{ms} = A^s$ for all $m \geq 1$, i.e. $a_{ij}^{(ms)} = a_{ij}^{(s)}$, for all $i, j \in \underline{n}$. Since A is assumed row diagonally dominant (i.e. $a_{ij} = a_{ii}a_{ij}$ for all i, j), we get $a_{ii} = a_{ii}a_{ii} = a_{ii}^n$, for all $n \geq 1$. So (7.1) yields

$$a_{ij}^{(m)} = \textstyle\sum_{i_1, i_2, \ldots, i_{m-1} \in \underline{n}} a_{ii_1} a_{i_1 i_2} \cdots a_{i_{m-2} i_{m-1}} a_{i_{m-1} j} = a_{ii}\, a_{ij}^{(m)} = a_{ii}^n\, a_{ij}^{(m)},$$

for all $n \geq 1$. Therefore, $a_{ij}^{(m)} = a_{ii}^{m(s-1)}\, a_{ij}^{(m)}$. But $a_{ii}^{m(s-1)}\, a_{ij}^{(m)}$ is one of the summands in

$$a_{ij}^{(ms)} = \textstyle\sum_{i_1, i_2, \ldots, i_{ms-1} \in \underline{n}} a_{ii_1} a_{i_1 i_2} \cdots a_{i_{ms-2} i_{ms-1}} a_{i_{ms-1} j},$$

whence $a_{ij}^{(m)} \leq a_{ij}^{(ms)} = a_{ij}^{(s)}$. So, using Prop. 4.12(1), $\sum_{p \in \underline{n}} a_{ij}^{(p)} \leq \sum_{p \in \underline{n}} a_{ij}^{(s)} = a_{ij}^{(s)}$, for all $n \geq k + d - 1$, whence $\mathbf{t}(A) = \sum_{j \in \underline{n}} A^j \leq A^s$ by Prop. 7.9. Conversely, since, for $k, d \geq 1$, $s = kd = (k-1)(d-1) + k + d - 1 \geq k + d - 1$ and, again using Prop. 7.9, $\mathbf{t}(A) = \sum_{j \in \underline{m}} A^j$ for all $m \geq k + d - 1$, we get $\mathbf{t}(A) = \sum_{j \in \underline{s}} A^j \geq A^s$. □

Recall that $h(A)$ denotes least positive integer k satisfying $A^k = O$. The next lemma shows that any nilpotent matrix over a path algebra has the transitive closure.

Lemma 7.14. ([15, Lemma 2.4]) *Let D be a path algebra and $A \in \mathbb{S}_n(D)$ be nilpotent. Then $\mathbf{t}(A) = \sum_{k=1}^{h(A)-1} A^k$.*

Proof. Let $B = \sum_{k=1}^{h(A)-1} A^k$; then $A \leq B$. Since $(\mathbb{S}_n(D), +, \cdot)$ is a path algebra, it follows that $B^2 = \sum_{k=2}^{2h(A)-2} A^k = \sum_{k=2}^{h(A)-1} A^k \leq B$, because $A^l = O$ for all $l \geq h(A)$. If $C \in \mathbb{S}_n(D)$ is a transitive matrix with $A \leq C$, then $C^2 \leq C$, hence $C^k \leq C$ for all $k \geq 1$. Therefore, $A^k \leq C^k \leq C$, whence $B \leq C$. So we conclude that $\mathbf{t}(A) = B$. □

7.3 Determinants, Permanents and Adjoints

Proposition 7.15. ([7, Proposition 2.3]) *Let D be a commutative path algebra, and $A, B \in \mathbb{S}_n(D)$; then* $\mathrm{per}(AB) \geq \mathrm{per}(A)\mathrm{per}(B)$.

Proof. We set $C = AB \in \mathbb{S}_n(D)$, and let T_n denote the set of all mappings from the set $\underline{n}$ to itself. By the definition of a permanent we have

$$\begin{aligned}
\mathsf{per}(AB) &= \sum_{\sigma\in S_n} c_{1\sigma(1)}c_{2\sigma(2)}\cdots c_{n\sigma(n)} \\
&= \sum_{\sigma\in S_n}\Big(\sum_{k\in\underline{n}} a_{1k}b_{k\sigma(1)}\Big)\Big(\sum_{k\in\underline{n}} a_{2k}b_{k\sigma(2)}\Big)\cdots\Big(\sum_{k\in\underline{n}} a_{nk}b_{k\sigma(n)}\Big) \\
&= \sum_{\sigma\in S_n}\ \sum_{k_1,k_2,\dots,k_n\in\underline{n}} a_{1k_1}a_{2k_2}\cdots a_{nk_n}b_{k_1\sigma(1)}b_{k_2\sigma(2)}\dots b_{k_n\sigma(n)} \\
&= \sum_{\tau\in T_n}\Big(a_{1\tau(1)}a_{2\tau(2)}\cdots a_{n\tau(n)}\sum_{\sigma\in S_n} b_{\tau(1)\sigma(1)}b_{\tau(2)\sigma(2)}\cdots b_{\tau(n)\sigma(n)}\Big) \\
&= \sum_{\tau\in S_n}\Big(a_{1\tau(1)}a_{2\tau(2)}\cdots a_{n\tau(n)}\sum_{\sigma\in S_n} b_{\tau(1)\sigma(1)}b_{\tau(2)\sigma(2)}\cdots b_{\tau(n)\sigma(n)}\Big) \\
&\quad + \sum_{\tau\in T_n-S_n}\Big(a_{1\tau(1)}a_{2\tau(2)}\cdots a_{n\tau(n)}\sum_{\sigma\in S_n} b_{\tau(1)\sigma(1)}b_{\tau(2)\sigma(2)}\cdots b_{\tau(n)\sigma(n)}\Big) \\
&\ge \sum_{\tau\in S_n}\Big(a_{1\tau(1)}a_{2\tau(2)}\cdots a_{n\tau(n)}\sum_{\sigma\in S_n} b_{\tau(1)\sigma(1)}b_{\tau(2)\sigma(2)}\cdots b_{\tau(n)\sigma(n)}\Big) \qquad (\dagger) \\
&= \sum_{\tau\in S_n}\Big(a_{1\tau(1)}a_{2\tau(2)}\cdots a_{n\tau(n)}\,\mathsf{per}(B)\Big) \qquad (*) \\
&= \Big(\sum_{\tau\in S_n} a_{1\tau(1)}a_{2\tau(2)}\cdots a_{n\tau(n)}\Big)\mathsf{per}(B) \\
&= \mathsf{per}(A)\mathsf{per}(B).
\end{aligned}$$

Note that we need Th. 4.16(5) to get the line marked with (†), and that we need commutativity of multiplication for concluding the third from the second line, and to show that $\sum_{\sigma\in S_n} b_{\tau(1)\sigma(1)}b_{\tau(2)\sigma(2)}\cdots b_{\tau(n)\sigma(n)}$ equals $\mathsf{per}(B)$ for each $\tau\in S_n$ for (∗). So, D must be a commutative path algebra. □

Proposition 7.16. ([7, Proposition 3.1]) *Adjoints of square matrices over a commutative path algebra D have the following properties:*

(1) *If* $A \le B$, *then* $\mathsf{per}(A) \le \mathsf{per}(B)$.
(2) *If* $A \le B$, *then* $\mathsf{adj}(A) \le \mathsf{adj}(B)$.
(3) $\mathsf{adj}(A) + \mathsf{adj}(B) \le \mathsf{adj}(A+B)$.
(4) $(\mathsf{adj}(A))^T = \mathsf{adj}(A^T)$.

Proof. (1): By assumption, $a_{ij} \le b_{ij}$, for all $i, j \in \underline{n}$, i.e. there exist elements c_{ij} such that $a_{ij} + c_{ij} = b_{ij}$; recall Def. 4.10. So

$$\mathsf{per}(B) = \sum_{\sigma\in S_n}\prod_{i\in\underline{n}} b_{i\sigma(i)} = \sum_{\sigma\in S_n}\prod_{i\in\underline{n}}\big(a_{i\sigma(i)} + c_{i\sigma(i)}\big).$$

The assertion follows by Prop. 4.12.

(2): By definition of adjoints and (1) (see Def. 6.6).

(3): Note that for a path algebra D, also $\mathbb{S}_n(D)$ is a path algebra, and $\mathsf{adj}(A) \in \mathbb{S}_n(D)$ for $A \in \mathbb{S}_n(D)$. So, we can apply Th. 4.16(3) and argue as follows: Since both $A \le A+B$ and $B \le A+B$, using (2) one gets $\mathsf{adj}(A) \le \mathsf{adj}(A+B)$ and $\mathsf{adj}(B) \le \mathsf{adj}(A+B)$, whence

$\mathrm{adj}(A)+\mathrm{adj}(B) \le \mathrm{adj}(A+B)+\mathrm{adj}(A+B) = \mathrm{adj}(A+B)$.

(4): Immediate consequence of the definition. □

For the following we need some further notation and concepts: Let S be a semiring and $A \in \mathbb{S}_n(S)$. We set $\Omega_{r,n} := \{\alpha = (i_1, i_2, \dots, i_r) \mid 1 \le i_1 < i_2 < \dots < i_r \le n\}$, $r \in \underline{n-1}$. For $\alpha = (i_1, i_2, \dots, i_r)$, $\beta = (j_1, j_2, \dots, j_r) \in \Omega_{r,n}$ we denote by $A[\alpha|\beta] \in \mathbb{S}_r$ the submatrix of A whose (u, v)-entry is equal to $a_{i_u j_v}$, $u, v \in \underline{r}$, and by $A(\alpha|\beta) \in \mathbb{S}_{n-r}$ the submatrix of A obtained by deleting rows $i_1, i_2, \dots, i_r$ and columns $j_1, j_2, \dots, j_r$ from A. In case $|\alpha| = |\beta| \le 2$ we simply write $A(i|j)$, $A_{i|j}$ or $A(ij|kl)$, dependent on typographical considerations. A matrix $A[\alpha|\alpha]$ is called a **principal submatrix** of order r of A, and $\mathrm{per}(A[\alpha|\alpha])$ is called a **principal minor** of order r of A. Then, similar to the Laplace expansion theorem for matrices over a field, one has the following:

Proposition 7.17. ([7, Proposition 2.2]) *Let D be a commutative path algebra, $A \in \mathbb{S}_n(D)$, and $\alpha = (i_1, i_2, \dots, i_k) \in \Omega_{k,n}$. Then*

$$\mathrm{per}(A) = \sum_{\beta \in \Omega_{k,n}, k \in \underline{n}} \mathrm{per}(A[\alpha|\beta])\mathrm{per}(A(\alpha|\beta)). \tag{7.2}$$

In particular, for $i \in \underline{n}$, we have $\mathrm{per}(A) = \sum_{j \in \underline{n}} a_{ij} \mathrm{per}(A(i|j))$.

Proof. Recall from Def. 6.6 that $\mathrm{per}(A) = \sum_{\sigma \in S_n} a_{1\sigma(1)} a_{2\sigma(2)} \dots a_{n\sigma(n)}$, where $\sigma \in S_n$, the set of all permutations of $\underline{n}$. Let $S_{\alpha,\beta}$ denote the set of all bijective mappings from α to β; then, for any $\beta \in \Omega_{k,n}$, we have $\mathrm{per}(A[\alpha|\beta]) = \sum_{\sigma \in S_{\alpha,\beta}} \prod_{i \in \underline{k}} a_{\alpha_i, \sigma(\alpha_i)}$. So there are $k!$ terms in the expansion of $\mathrm{per}(A[\alpha|\beta])$, and similarly, there are $(n-k)!$ terms in the expansion of $\mathrm{per}(A(\alpha|\beta))$. The product of any term of $\mathrm{per}(A[\alpha|\beta])$ and any term of $\mathrm{per}(A(\alpha|\beta))$ is a term of $\mathrm{per}(A)$, and there are $k!(n-k)!$ distinct terms in $\mathrm{per}(A[\alpha|\beta])\mathrm{per}(A(\alpha|\beta))$. Thus, in total there are $\binom{n}{k} k!(n-k)! = n!$ distinct terms in $\sum_{\beta \in \Omega_{k,n}, k \in \underline{n}} \mathrm{per}(A[\alpha|\beta])\mathrm{per}(A(\alpha|\beta))$. And since any term of this sum is a term of $\mathrm{per}(A)$ and $\mathrm{per}(A)$ has $n!$ terms, equality (7.2) is proved. Taking $\alpha = (i)$ in (7.2) gives the second equality. □

Lemma 7.18. ([17, Lemma 2.3]) *Let D be a commutative path algebra, and $A, B \in \mathbb{S}_n(D)$; then for any $\alpha, \beta \in \Omega_{k,n}$, $k \in \underline{n}$, we have*

$$\mathrm{per}(AB[\alpha|\beta]) \ge \textstyle\sum_{\lambda \in \Omega_{k,n}, k \in \underline{n}} \mathrm{per}(A[\alpha|\lambda])\mathrm{per}(B[\lambda|\beta]).$$

Proposition 7.19. ([17, Proposition 3.2]) *Let D be a commutative path algebra, and $A \in \mathbb{S}_n(D)$ be transitive; then*

(1) $(\mathrm{per}(A))^2 \le \mathrm{per}(A^2) \le \mathrm{per}(A)$, *i.e.* $\mathrm{per}(A)$ *is a transitive element in D.*

(2) $\mathrm{adj}(A)$ *is a transitive matrix.*

Proof. (1): For the first inequality use Prop. 7.15, and for the other $A^2 \le A$ together with Prop. 7.16(1).

(2): Setting $C := \mathsf{adj}(A^2)$ we get, for any $i, j \in \underline{n}$

$$\begin{aligned} c_{ij} &= \mathsf{per}(A^2(j|i)) \\ &\geq \sum_{\delta\in\Omega_{n-1,n}} \mathsf{per}(A[\underline{n}-\{j\}|\delta])\,\mathsf{per}(A[\delta|\underline{n}-\{i\}]) \qquad \text{by Lemma 7.18} \\ &= \sum_{k\in\underline{n}} \mathsf{per}(A(j|k))\,\mathsf{per}(A(k|i)) \\ &= \sum_{k\in\underline{n}} [\mathsf{adj}(A)]_{ik}\,[\mathsf{adj}(A)]_{kj} \\ &= [\mathsf{adj}(A)\mathsf{adj}(A)]_{ij}, \end{aligned}$$

i.e. $\mathsf{adj}(A)^2 \leq \mathsf{adj}(A^2)$, and from the transitivity of A we get $\mathsf{adj}(A^2) \leq \mathsf{adj}(A)$ by Prop. 7.16(2). □

Proposition 7.20. ([17, Proposition 3.3]) *Let D be a commutative path algebra, and $A \in \mathbb{S}_n(D)$ be transitive; then*

(1) all principal submatrices of A are transitive.
(2) all principal minors of A are transitive elements in D.
(3) all main diagonal entries of A^k are transitive elements in D for any integer $k \geq 1$.

Proof. (1): Let B be a principal submatrix of order r of A, say $B = A[\alpha|\alpha]$ with $\alpha = (i_1, i_2, \ldots, i_r)$. Then there exists a permutation matrix P such that $PAP^T = \begin{bmatrix} B & C \\ E & F \end{bmatrix}$, where $C \in \mathbb{S}_{r\times(n-r)}(D)$, $E \in \mathbb{S}_{(n-r)\times r}(D)$, and $F \in \mathbb{S}_{n-r}(D)$: Take any permutation $\pi \in \boldsymbol{P}_n$ such that $\pi = \begin{pmatrix} 1 & 2 & \cdots & r & r+1 & \cdots & n \\ i_1 & i_2 & \cdots & i_r & \pi(r+1) & \cdots & \pi(n) \end{pmatrix}$. By Prop. 7.2(2) PAP^T is transitive, whence

$$\begin{bmatrix} B & C \\ E & F \end{bmatrix}^2 = \begin{bmatrix} B^2+CE & BC+CF \\ EB+FE & EC+F^2 \end{bmatrix} \leq \begin{bmatrix} B & C \\ E & F \end{bmatrix}.$$

This implies $B^2 \leq B$.

(2): By (1) and Prop. 7.19(1).

(3): Since A is transitive, $A^2 \leq A$ implies $(A^2)^k = (A^k)^2 \leq A^k$, so A^k is transitive for any positive integer k, and by (2) all main diagonal entries of A^k are transitive. □

Following [7, p. 5], a matrix $A \in \mathbb{S}_n(R)$, with $n \geq 2$ and R a path algebra, is said to satisfy condition (*) if

$$a_{ii} \geq a_{jk}, \text{ for all } i, j, k \in \underline{n}. \tag{*}$$

Obviously, we then have

$$a_{11} = a_{22} = \cdots = a_{nn} \geq a_{ij} \text{ for all } i, j \in \underline{n}. \tag{7.3}$$

We will call A a matrix with **maximal diagonal** in this case.

Lemma 7.21. ([7, Lemma 3.2]) *Let D be a commutative path algebra and $A \in \mathbb{S}_n(D)$ be a matrix with maximal diagonal. Then, for any $l \geq 1$, $\mathsf{per}(A^l) = \mathsf{per}(A)^l$.*

Proof. From (7.3) we get $\mathsf{per}(A) = \sum_{\sigma\in S_n} a_{1\sigma(1)}a_{2\sigma(2)}\cdots a_{n\sigma(n)} = \sum_i a_{ii}^n = a_{11}^n$. Furthermore, for $l \geq 2$, we have

$$a_{ij}^{(l)} = \sum_{j_1,j_2,\cdots,j_{l-1}\in \underline{n}} a_{ij_1}a_{j_1j_2}\cdots a_{j_{l-1}j} \leq a_{11}^l$$

and

$$a_{ii}^{(l)} = \sum_{j_1,j_2,\cdots,j_{l-1}\in \underline{n}} a_{ij_1}a_{j_1j_2}\cdots a_{j_{l-1}i} = a_{11}^l,$$

for $i,j \in \underline{n}$. Thus, $\mathsf{per}(A^l) = \sum_{\sigma\in S_n} a_{1\sigma(1)}^{(l)}a_{2\sigma(2)}^{(l)}\cdots a_{n\sigma(n)}^{(l)} = a_{11}^{ln} = (\mathsf{per}(A))^l$. □

Theorem 7.22. ([7, Theorem 3.3(1)]) *Let D be a commutative path algebra and $A \in \mathbb{S}_n(D)$ be a matrix with maximal diagonal. Then $A^{n-1} = \mathsf{adj}(A)$.*

Proof. The proof proceeds in two steps: We show that (a): $\mathsf{per}(A(j|i)) \leq a_{ij}^{(n-1)}$, and that (b): $\mathsf{per}(A(j|i)) \geq a_{ij}^{(n-1)}$. From this we get $A^{n-1} = \mathsf{adj}(A)$ by Def. 6.6(c).

(a): Assume $i = j$; then

$$\begin{aligned}
\mathsf{per}(A(i|i)) &= \sum_{\sigma\in S_n} a_{1\sigma(1)}\ldots a_{i-1\sigma(i-1)}\,a_{i+1\sigma(i+1)}\ldots a_{n\sigma(n)} \\
&\leq \sum_{\sigma\in S_n} a_{11}\ldots a_{i-1,i-1}\,a_{i+1,i+1}\ldots a_{nn} \\
&= a_{11}\ldots a_{i-1,i-1}\,a_{i+1,i+1}\ldots a_{nn} && \text{by idempotency of } + \\
&= a_{ii}^{n-1} && \text{since } a_{ii} = a_{jj} \\
&\leq \sum_{j_1,j_2,\ldots,j_{n-2}\in \underline{n}} a_{ij_1}\,a_{j_1j_2}\cdots a_{j_{n-2}i} \\
&= a_{ii}^{(n-1)}. && a_{ii}^{n-1} \text{ is a term in} = a_{ii}^{(n-1)}
\end{aligned}$$

For $i \neq j$ we get

$$\mathsf{per}(A(j|i)) = \sum_{\sigma\in S_n,\sigma(j)=i} a_{1\sigma(1)}\cdots a_{i-1\sigma(i-1)}\,a_{i+1\sigma(i+1)}\cdots a_{n\sigma(n)}.$$

Let $t_\sigma = a_{1\sigma(1)}\cdots a_{i-1\sigma(i-1)}\,a_{i+1\sigma(i+1)}\cdots a_{n\sigma(n)}$ denote any term of $\mathsf{per}(A(j|i))$. Since $\sigma(j) = i$, there is a $k \geq 1$ such that $\sigma^k(i) = j$, $0 \leq k \leq n-1$. Let l be the least such k; then $i, \sigma(i), \cdots, \sigma^{l-1}(i), j$ are all different. Thus

$$t_\sigma \leq a_{i\sigma(i)}\,a_{\sigma(i)\sigma^2(i)}\cdots a_{\sigma^{l-1}(i)j}\,a_{jj}^{n-1-l} \leq a_{ij}^{(n-1)}.$$

Since this holds for any term t_σ we get

$$\mathsf{per}(A(j|i)) = \sum_{\sigma\in S_n,\sigma(j)=i} t_\sigma \leq \sum_{\sigma\in S_n,\sigma(j)=i} a_{ij}^{(n-1)} = a_{ij}^{(n-1)}.$$

(b): Assume $i = j$; then

$$\begin{aligned}
a_{ii}^{(n-1)} &= \sum_{j_1,j_2,\ldots,j_{n-2}\in \underline{n}} a_{ij_1}\,a_{j_1j_2}\cdots a_{j_{n-2}i} \\
&\leq \sum_{j_1,j_2,\ldots,j_{n-2}\in \underline{n}} a_{11}\cdots a_{i-1,i-1}\,a_{i+1,i+1}\cdots a_{nn} \\
&= a_{11}\cdots a_{i-1,i-1}\,a_{i+1,i+1}\cdots a_{nn}
\end{aligned}$$

$$\leq \mathrm{per}(A(i|i)).$$

Now we let $i \neq j$, and denote by $t = a_{ij_1} a_{j_1 j_2} \cdots a_{j_{n-2} j}$ any summand of $a_{ij}^{(n-1)}$. If the subscripts $i = j_0, j_1, \ldots, j_{n-2}, j_{n-1} = j$ are all different, then t is a term of $\mathrm{per}(A(j|i))$; thus $t \leq \mathrm{per}(A(j|i))$. If $j_r = j_s$ for some $r, s \in \{0, 1, \ldots, n-1\}$, $r < s$, we delete the subproduct $a_{j_r j_{r+1}} \cdots a_{j_{s-1} j_s}$ from t and get $a_{ij_1} \cdots a_{j_{r-1} j_r} a_{j_s j_{s+1}} \cdots a_{j_{n-2} j}$. This process is repeated until we get $a_{il_1} a_{l_1 l_2} \cdots a_{l_m j}$, where $0 \leq m < n-2$ and $i, l_1, \ldots, l_m, j$ are all different. We let $\{p_1, p_2, \ldots, p_{n-m-2}\} = \underline{n} - \{i, l_1, \ldots, l_m, j\}$; then

$$t \leq a_{il_1} a_{l_1 l_2} \cdots a_{l_m j} a_{p_1 p_1} a_{p_2 p_2} \cdots a_{p_{n-m-2} p_{n-m-2}} \leq \mathrm{per}(A(j|i)).$$

Since t has been arbitrarily chosen, $a_{ij}^{(n-1)} \leq \mathrm{per}(A(j|i))$. □

The two-valued counterpart of the above theorem for Boolean matrices has been shown already in the early fifties in Lunts [9] as reported in Hammer and Rudeanu [5, p. 15]. Some more properties of adjoints over inclines are given in Section 8.1.3 below.

Corollary 7.23. ([7, Theorem 3.3(2)]) *Let D be a commutative path algebra and $A \in \mathbb{S}_n(D)$ be a matrix with maximal diagonal. Then* $\mathrm{per}(\mathrm{adj}(A)) = \mathrm{per}(A)^{n-1}$.

Proof. By Lemma 7.21 we have $\mathrm{per}(A^{n-1}) = \mathrm{per}(A)^{n-1}$. Thus, $\mathrm{per}(\mathrm{adj}(A)) = \mathrm{per}(A)^{n-1}$ follows from Th. 7.22. □

As a final result for commutative path algebras (given without its proof of nearly three pages length) we mention

Corollary 7.24. ([7, Theorem 3.5]) *Let D be a commutative path algebra and $A \in \mathbb{S}_n(D)$ be a matrix with maximal diagonal. Then* $\mathrm{per}(A\mathrm{adj}(A)) = \mathrm{per}(\mathrm{adj}(A)A) = \mathrm{per}(A)^n$.

7.4 Invertibility

We first present some results needed later on, and then characterize invertibility for idempotent semirings, followed by some basic results for commutative antirings.

Proposition 7.25. ([2, (19.4) PROPOSITION], [10, Theorem 2.3(i)]) *Let S be a commutative antiring, $n \geq 1$, and $A, B \in \mathbb{S}_n(S)$ such that $AB = I_n$. Then*

(1) *$a_{ik} b_{kj} = a_{ki} b_{jk} = 0$ for any $i, j, k \in \underline{n}$, with $i \neq j$.*
(2) *$a_{ik} a_{jk} = 0$ for any $i, j, k \in \underline{n}$, with $i \neq j$, i.e. the product of any two elements in the same column is 0.*
(3) *$a_{ik} a_{jk} = a_{ki} a_{kj} = b_{ki} b_{kj} = b_{ik} b_{jk} = 0$ for any $i, j, k \in \underline{n}$, with $i \neq j$.*
(4) *$\left(\sum_{k \in \underline{n}} a_{ik}\right)\left(\sum_{l \in \underline{n}} b_{li}\right) = \left(\sum_{k \in \underline{n}} a_{kj}\right)\left(\sum_{l \in \underline{n}} b_{jl}\right) = 1$, for any $i \in \underline{n}$.*

Proof. (1): Let $i \neq j$; then $(AB)_{ij} = \sum_{k \in \underline{n}} a_{ik} b_{kj} = 0$. Since S is zerosumfree $a_{ik} b_{kj} = 0$ for all $i, j, k \in \underline{n}$ with $i \neq j$. By Prop. 6.26(1) we also have $BA = I_n$, so as before $b_{jk} a_{ki} = a_{ki} b_{jk} = 0$ for all $i, j, k \in \underline{n}$ with $i \neq j$.

(2): Let $B \in \mathbb{S}_n(S)$ such that $AB = BA = I_n$. Then by (1) $a_{ik}b_{kj} = 0$ for any $i, j, k \in \underline{n}$, with $i \neq j$. Moreover, for any $k \in \underline{n}$ we have

$$\begin{aligned} a_{ik}a_{jk} &= a_{ik}a_{jk}\cdot 1 = a_{ik}a_{jk}(BA)_{kk} = \\ &= a_{ik}a_{jk}\big(\textstyle\sum_{l\in\underline{n}} b_{kl}a_{lk}\big) \\ &= \textstyle\sum_{l\in\underline{n}} a_{ik}a_{jk}b_{kl}a_{lk} \\ &= a_{ik}a_{jk}b_{kj}a_{jk} + \textstyle\sum_{l\in\underline{n}, l\neq j} a_{ik}a_{jk}b_{kl}a_{lk} \\ &= (a_{ik}b_{kj})a_{jk}^2 + \textstyle\sum_{l\in\underline{n}, l\neq j} a_{ik}(a_{jk}b_{kl})a_{lk} \\ &= 0 + 0. \end{aligned}$$

(3): The first equality, $a_{ik}a_{jk} = 0$ has been shown in (2). We now turn to the third one, $b_{ki}b_{kj} = 0$. By (1) $a_{ik}b_{kj} = 0$ for all $i, j, k \in \underline{n}$ with $i \neq j$; moreover $1 = \sum_{k\in\underline{n}} a_{ik}b_{ki}$. So,

$$\begin{aligned} 1 = \prod_{i\in\underline{n}}\big(\sum_{k\in\underline{n}} a_{ik}b_{ki}\big) &= \big(\sum_{k\in\underline{n}} a_{1k}b_{k1}\big)\big(\sum_{k\in\underline{n}} a_{2k}b_{k2}\big)\cdots\big(\sum_{k\in\underline{n}} a_{nk}b_{kn}\big) \\ &= \sum_{k_1,\dots,k_n\in\underline{n}} (a_{1k_1}b_{k_1 1})(a_{2k_2}b_{k_2 2})\cdots(a_{nk_n}b_{k_n n}) \\ &= \sum_{\sigma\in S_n} (a_{1\sigma(1)}b_{\sigma(1)1})(a_{2\sigma(2)}b_{\sigma(2)2})\cdots(a_{n\sigma(n)}b_{\sigma(n)n}). \end{aligned}$$

For the correctness of the last equality note that $a_{ik}b_{ki}a_{jk}b_{kj} = a_{ik}a_{jk}b_{ki}b_{kj} = 0$ holds for all such terms with $i \neq j$, using commutativity of S and applying (2). So in order that terms of the form $(a_{1k_1}b_{k_1 1})(a_{2k_2}b_{k_2 2})$ do not take the value 0, k_1 and k_2 must be different, whence the permutations σ of $\underline{n}$ come into play.

Let us abbreviate the last sum with Σ_σ^{ab}. By (1) $a_{ik}b_{kj} = 0$ for all $i, j, k \in \underline{n}$ with $i \neq j$; so, using commutativity, for all terms with $i \neq j$ we get $b_{ki}b_{kj}a_{ik} = b_{ki}a_{ik}b_{kj} = 0$. Therefore, for $i \neq j$, $b_{ki}b_{kj} = b_{ki}b_{kj}\Sigma_\sigma^{ab} = 0$, since each summand contains a factor $b_{ki}b_{kj}a_{i\sigma(i)}$ such that $\sigma(i) = k$. By similar reasoning the remaining cases can be shown.

(4): By (1) we have $a_{ki}b_{jk} = 0$ for any $i, j, k \in \underline{n}$, with $i \neq j$. Thus $\big(\sum_{k\in\underline{n}} a_{ik}\big)\big(\sum_{l\in\underline{n}} b_{li}\big) = \sum_{k\in\underline{n}} a_{ik}b_{ki} = 1$. □

Recall that a semiring S is called idempotent in case it is a path algebra which is also multiplicatively idempotent, i.e. $a + a = a = a^2$ for $a \in S$.

Lemma 7.26. ([10, Lemma 2.2]) *Let S be a commutative idempotent semiring and $A \in \mathbb{S}_n(S)$, $n \geq 2$, be invertible. Then $|A|^+ + |A|^- = 1$.*

Proof. Let $B \in \mathbb{S}_n(S)$ such that $AB = BA = I_n$. By Prop. 6.10(1) we know that $|I_n|^\pm = \langle 1, 0\rangle$. So, using Lemma 6.31 we get

$$|AB|^+ = |I_n|^+ = 1 = |A|^+|B|^+ + |A|^-|B|^- + r \text{ for some } r \in S,$$

$$|AB|^- = |I_n|^- = 0 = |A|^+|B|^- + |A|^-|B|^+ + r \text{ for some } r \in S.$$

The last equality and Prop. 4.4(2) imply

$$|A|^+|B|^- = |A|^-|B|^+ = r = 0.$$

Then

$$\begin{aligned} 1 &= |A|^+|B|^+ + |A|^-|B|^- \\ &= |A|^+|B|^+ + |A|^-|B|^- + |A|^+|B|^- + |A|^-|B|^+ \\ &= (|A|^+ + |A|^-)(|B|^+ + |B|^-), \end{aligned}$$

and from Prop. 4.4(4) the assertion follows. □

Theorem 7.27. ([10, Theorem 2.3]) *Let S be a commutative idempotent semiring and $A \in \mathbb{S}_n(S)$, $n \geq 2$. Then A is invertible if and only if*

(a) the product of any two elements in the same column is 0, and
(b) the sum of all elements in each row is 1.

Proof. Assume A is invertible and let $B \in \mathbb{S}_n(S)$ such that $AB = BA = I_n$. Then (a) follows by Prop. 7.25(2). By Lemma 7.26 $|A|^+ + |A|^- = 1$, and from (a) we get that $a_{1k_1}a_{2k_2}\cdots a_{nk_n} = 0$, if $k_1,\ldots,k_n \in \underline{n}$ are not all distinct. Thus

$$\begin{aligned} \Big(\sum_{k\in\underline{n}} a_{1k}\Big)\Big(\sum_{k\in\underline{n}} a_{2k}\Big)\cdots\Big(\sum_{k\in\underline{n}} a_{nk}\Big) &= \sum_{k_1,\ldots,k_n\in\underline{n}} a_{1k_1}a_{2k_2}\cdots a_{nk_n} \\ &= \sum_{\sigma\in S_n} a_{1\sigma(1)}a_{2\sigma(2)}\cdots a_{n\sigma(n)} \\ &= |A|^+ + |A|^- = 1. \end{aligned}$$

Hence, $\sum_{k\in\underline{n}} a_{1k} = \sum_{k\in\underline{n}} a_{2k} = \cdots = \sum_{k\in\underline{n}} a_{nk} = 1$ by Prop. 4.4(4), and (b) is proved.

Conversely, assume that both (a) and (b) hold. Then, from (b) we get

$$(AA^T)_{ii} = \sum_{k\in\underline{n}} a_{ik}a^T_{ki} = \sum_{k\in\underline{n}} a_{ik}a_{ik} = \sum_{k\in\underline{n}} a_{ik} = 1,$$

and from (a) it follows that, for $i \neq j$,

$$(AA^T)_{ij} = \sum_{k\in\underline{n}} a_{ik}a^T_{kj} = \sum_{k\in\underline{n}} a_{ik}a_{jk} = 0.$$

Hence, A is invertible over S. □

Corollary 7.28. ([10, Corollary 2.4]) *Let S be a commutative idempotent semiring and $A \in \mathbb{S}_n(S)$, $n \geq 2$. Then A is invertible if and only if*

(a) the product of any two elements in the same row is 0, and
(b) the sum of all elements in each column is 1.

Proof. By Cor. 6.27 A is invertible over S if and only if A^T is. □

The results given in the remainder of this section all hold for commutative path algebras too, since path algebras are antirings by Prop. 4.4(2).

Proposition 7.29. ([13, Proposition 3.2]) *Let R be a commutative antiring, $n \geq 1$, and $A \in \mathbb{S}_n(R)$. If A is invertible, then AA^T and A^TA are invertible diagonal matrices.*

Proof. Suppose A is invertible in $\mathbb{S}_n(R)$; then there is a $B \in \mathbb{S}_n(R)$ with $AB = I_n$. By Prop. 7.25(4) we have $\left(\sum_{k\in\underline{n}} a_{ik}\right)\left(\sum_{l\in\underline{n}} b_{li}\right) = 1$, hence $u_i := \sum_{k\in\underline{n}} a_{ik} \in U(R)$ for all $i \in \underline{n}$. Then, by Prop. 7.25(3), $u_i^2 = \left(\sum_{k\in\underline{n}} a_{ik}\right)^2 = \sum_{k\in\underline{n}} a_{ik}^2 + \sum_{k,l\in\underline{n}} 2a_{ik}a_{il} = \sum_{k\in\underline{n}} a_{ik}^2$, and by Lemma 4.8 $u_i^2 \in U(R)$. We have

$$(AA^T)_{ij} = \sum_{k\in\underline{n}} a_{ik}a_{jk} = \begin{cases} u_i^2 & i = j \\ 0 & i \neq j \end{cases} \tag{7.4}$$

by Prop. 7.25(3). And, by Lemma 6.30, $AA^T = \mathrm{diag}(u_1^2, u_2^2, \ldots, u_n^2)$ is invertible in $\mathbb{S}_n(R)$. Similar for $A^T A$. □

Corollary 7.30. ([13, Corollary 3.1]) *Let R be a commutative antiring with $U(R) = \{1\}$, $n \geq 1$, and $A \in \mathbb{S}_n(R)$. If A is invertible, then $AA^T = A^T A = I_n$.*

Proof. Obvious from (6.7). □

Proposition 7.31. ([13, Proposition 3.3]) *Let R be a commutative antiring and $A \in \mathbb{S}_n(R)$. If A is right invertible in $\mathbb{S}_n(R)$, then $A^{[n]}$ is an invertible diagonal matrix in $\mathbb{S}_n(R)$, where $[n]$ denotes the least common multiple of the integers $1,2,\ldots,n$.*

Proof. Let A is right invertible in $\mathbb{S}_n(R)$; then it is invertible by Prop. 6.26(1). So $A^{[n]}$ is invertible. We will now prove that $A^{[n]}$ is a diagonal matrix. For $n = 2$ we get $[n] = 2$, and thus

$$A^2 = \begin{bmatrix} a_{11} & a_{12} \\ a_{21} & a_{22} \end{bmatrix}^2 = \begin{bmatrix} a_{11}^2 + a_{12}a_{21} & a_{11}a_{12} + a_{12}a_{22} \\ a_{11}a_{21} + a_{21}a_{22} & a_{21}a_{12} + a_{22}^2 \end{bmatrix}$$

And according to Prop. 7.25(3), $a_{11}a_{12} + a_{12}a_{22} = a_{11}a_{21} + a_{21}a_{22} = 0$.

We now assume $n \geq 3$ and let $w := [n] > n$. For any $i, j \in \underline{n}$, if $a_{ij}^{(w)} \neq 0$ then there exist $i_1, i_2, \ldots, i_{w-1} \in \underline{n}$ such that $a_{ii_1}a_{i_1i_2}\cdots a_{i_{w-1}j} \neq 0$. It will be proved in four steps that $i = j$:

(1) Taking $i = i_0$ and $j = i_w$, let $i_s = i_t$ for some s, t with $0 \leq s < t < w$; then $i_{s+1} = i_{t+1}$: For assuming the contrary, we would have $a_{i_si_{s+1}}a_{i_ti_{t+1}} = a_{i_si_{s+1}}a_{i_si_{t+1}} = 0$ by Prop. 7.25(3), whence $a_{ii_1}a_{i_1i_2}\cdots a_{i_{w-1}j} = a_{i_0i_1}\cdots a_{i_si_{s+1}}\cdots a_{i_ti_{t+1}}\cdots a_{i_{w-1}i_w} = 0$, a contradiction.

(2) If $i_s = i_t$ for some s, t with $0 < s < t \leq w$, then $i_{s-1} = i_{t-1}$: Assuming $i_{s-1} \neq i_{t-1}$ gives $a_{i_{s-1}i_s}a_{i_{t-1}i_t} = a_{i_{s-1}i_s}a_{i_{t-1}i_s} = 0$ by Prop. 7.25(3), whence $a_{ii_1}a_{i_1i_2}\cdots a_{i_{w-1}j} = a_{i_0i_1}\cdots a_{i_{s-1}i_s}\cdots a_{i_{t-1}i_t}\cdots a_{i_{w-1}i_w} = 0$, again a contradiction.

(3) $i_0 = i_d$ for some $d \in \underline{n}$: Remember $w > n$; and since $i_1, i_2, \ldots, i_n \in \underline{n}$ there are u, v in $\{0,1,\ldots,n\}$ such that $i_u = i_v$ with $u < v$. If $u = 0$, then $i_0 = i_v$, and $i_0 = i_d$ is proved. If $0 < u$, then by (2) we have $i_{u-1} = i_{v-1}$; and if $0 < u-1$, then, again using (2), we get $i_{u-2} = i_{v-2}$. Repeating this argument, we get $i_0 = i_{v-u}$. So, taking $d = v - u$ gives $d \in \underline{n}$ and $i_0 = i_d$.

(4) From $i_0 = i_d$ it follows that $d|w$. Let $w = gd$, $g \geq 1$. Then, by(1) and (3),we get $i = i_0 = i_d = i_{2d} = \cdots = i_{gd} = i_w = j$.

Therefore, $a_{ij}^{(w)} = 0$ for all $i, j \in \underline{n}$ with $i \neq j$, whence $A^{[n]}$ is a diagonal matrix. □

Compare the following proposition to Th. 6.35, which does not presuppose the underlying semiring to be zerosumfree.

Proposition 7.32. ([13, Theorem 3.1]) *Let R be a commutative antiring, $n \geq 1$, and $A \in \mathbb{S}_n(R)$. Then the following statements are equivalent:*

(1) A is right invertible.
(2) A is left invertible.
(3) A is invertible.
(4) AA^T is an invertible diagonal matrix.
(5) A^TA is an invertible diagonal matrix.
(6) AA^T and A^TA are invertible diagonal matrices.

Proof. By Prop. 6.26(1), Cor. 7.30 and Prop. 7.31. □

Let S be an antiring, and $T := \{a_1, a_2, \dots, a_m\} \subseteq S$. T is called a **decomposition** of 1 in S if $\sum_{i \in \underline{m}} a_i = 1$. and called **orthogonal** if $a_i a_j = 0$ for any $i, j \in \underline{m}$ with $i \neq j$. $T \subseteq S$ is called an **orthogonal decomposition** of 1 in S if it is both orthogonal and a decomposition of 1 in S. Let $\{A_1, A_2, \dots, A_m\} \subseteq \mathbb{S}_n(S)$. A matrix $A \in \mathbb{S}_n(S)$ is called an **orthogonal combination** of the matrices $A_1, A_2, \dots, A_m$ if there exists an orthogonal decomposition $\{a_1, a_2, \dots, a_m\}$ of 1 in S such that $A = \sum_{i \in \underline{m}} a_i A_i$. The next result is a generalization of [1, Theorem 8].

Proposition 7.33. ([13, Proposition 3.7]) *Let S be a commutative antiring with $U(S) = \{1\}$, and $A \in \mathbb{S}_n(S)$. Then the following statements are equivalent:*

(1) A is invertible.
(2) A is an orthogonal combination of some permutation matrices of order n.

Proof. Let A be invertible in $\mathbb{S}_n(S)$; then, from the assumption $U(S) = \{1\}$, we have $AA^T = A^TA = I_n$ by Cor. 7.30. Therefore, $a_{ij}a_{ik} = a_{ji}a_{ki} = 0$ for all $i, j, k \in \underline{n}$ with $j \neq k$ and $\sum_{s \in \underline{n}} a_{is}^2 = \sum_{t \in \underline{n}} a_{tj}^2 = 1$ for all $i, j \in \underline{n}$. So

$$\left(\sum_{s \in \underline{n}} a_{is}\right)^2 = \sum_{s \in \underline{n}} a_{is}^2 + \sum_{s_1 \neq s_2} a_{is_1} a_{is_2} = \sum_{s \in \underline{n}} a_{is}^2 = 1$$

and

$$\left(\sum_{t \in \underline{n}} a_{tj}\right)^2 = \sum_{t \in \underline{n}} a_{tj}^2 + \sum_{t_1 \neq t_2} a_{t_1 j} a_{t_2 j} = \sum_{t \in \underline{n}} a_{tj}^2 = 1.$$

Since, by assumption, 1 is the only invertible element of S (see Def. 4.7(a)), we get $\sum_{s \in \underline{n}} a_{is} = \sum_{t \in \underline{n}} a_{tj} = 1$, i.e. each row and column of A is an orthogonal decomposition of 1 in S. And for any $i, j \in \underline{n}$, $a_{ij} = a_{ij}\left(\sum_{s \in \underline{n}} a_{is}\right) = a_{ij}^2 + \sum_{s \neq j} a_{is} a_{ij} = a_{ij}^2$ by Prop. 7.25(3), so each entry in A is idempotent.

We let $a_\sigma := a_{1\sigma(1)} a_{2\sigma(2)} \cdots a_{n\sigma(n)}$, for a permutation $\sigma \in S_n$, and $G := \{\sigma \in S_n : a_\sigma \neq 0\}$. As just shown, $\sum_{s \in \underline{n}} a_{is} = 1$ for all $i \in \underline{n}$; so $1 = \prod_{i \in \underline{n}} \left(\sum_{s \in \underline{n}} a_{is}\right) = \sum_{1 \leq s_1, s_2, \dots, s_n \leq n} a_{1s_1} a_{2s_2} \cdots a_{ns_n}$. If $s_i = s_j$ for some $i, j \in \underline{n}$ with $i \neq j$, then $a_{is_i} a_{js_j} = 0$ by Prop. 7.25(3), whence $a_{1s_1} a_{2s_2} \cdots a_{ns_n} = 0$. So $\sum_{\sigma \in G} a_\sigma = 1$, and $a_\sigma a_\tau = 0$ for any $\sigma, \tau \in G$ with $\sigma \neq \tau$ (since there exists an index i such that $a_{ij}a_{ik}$, with $j \neq k$, is part of $a_\sigma a_\tau$);

hence $\{a_\sigma : \sigma \in G\}$ is an orthogonal decomposition of 1 in S. And for any $\sigma \in G$ and $i, j \in \underline{n}$ we have $a_\sigma a_{ij} = \begin{cases} a_\sigma & \text{if } j = \sigma(i) \\ 0 & j \neq \sigma(i), \end{cases}$ because $a_{ij} = a_{ij}^2$ as shown above. Hence, for any $\sigma \in G$, the matrix $a_\sigma A$ has exactly one nonzero element in each row and column. Thus, there exists a unique $(0,1)$-matrix $P_\sigma \in \mathbb{S}_n(S)$ such that $a_\sigma A = a_\sigma P_\sigma$, with $(P_\sigma)_{ij} = 1$ iff $\sigma(i) = j$. It follows that $a_\sigma^2 P_\sigma P_\sigma^T = (a_\sigma P_\sigma)(a_\sigma P_\sigma)^T = (a_\sigma A)(a_\sigma A)^T = a_\sigma^2 AA^T = a_\sigma^2 I_n$. Hence $P_\sigma P_\sigma^T = I_n$, and P_σ thus is a permutation matrix for each $\sigma \in G$ (see the arguments in the proof of Prop. 3.34). Furthermore, $\sum_{\sigma \in G} a_\sigma P_\sigma = \sum_{\sigma \in G} a_\sigma A = \left(\sum_{\sigma \in G} a_\sigma\right) A = A$, i.e. A is an orthogonal combination of permutation matrices.

Conversely, if A is an orthogonal combination of some permutation matrices, say $A = \sum_{s \in \underline{m}} a_s P_s$, then

$$\begin{aligned}
AA^T &= \Big(\sum_{s \in \underline{m}} a_s P_s\Big)\Big(\sum_{t \in \underline{m}} a_t P_t\Big)^T \\
&= \sum_{s \in \underline{m}} \sum_{t \in \underline{m}} a_s a_t P_s P_t^T \\
&= \sum_{s \in \underline{m}} a_s^2 P_s P_s^T \qquad\qquad \text{(since } a_s a_t = 0 \text{ for } s \neq t) \\
&= \sum_{s \in \underline{m}} a_s^2 I_n \\
&= \sum_{s \in \underline{m}} \sum_{t \in \underline{m}} a_s a_t I_n \qquad\qquad \text{(since } a_s a_t = 0 \text{ for } s \neq t) \\
&= \Big(\sum_{s \in \underline{m}} a_s\Big)^2 I_n \\
&= I_n,
\end{aligned}$$

since $\sum_{s \in \underline{m}} a_s$ has been assumed an orthogonal decomposition of 1 in S. □

Lemma 7.34. ([13, Lemma 4.1, Corollary 4.1]) *Let S be a commutative antiring, $A \in \mathbb{S}_m(S)$ and $C \in \mathbb{S}_n(S)$ be invertible, and $B \in \mathbb{S}_{m\times n}(S)$, $m \leq n$. Then*

(1) $\mathsf{per}(ABC) = \mathsf{per}(A)\mathsf{per}(B)\mathsf{per}(C)$.
(2) $\mathsf{per}(A) \in U(S)$.

Proof. (1): By straightforward, but somewhat lengthy calculations.

(2): Since A is invertible there is an $A' \in \mathbb{S}_m(S)$ with $AA' = I_m$. Then by (1) we have $1 = \mathsf{per}(I_m) = \mathsf{per}(AA') = \mathsf{per}(AI_mA') = \mathsf{per}(A)\mathsf{per}(A)' \in U(S)$, whence $\mathsf{per}(A) \in U(S)$ by Lemma 4.8(3). □

Lemma 7.35. ([13, Lemma 4.2]) *Let R be a commutative antiring, $n \geq 1$, and $A \in \mathbb{S}_n(R)$ be invertible. Then $A^{-1} = (\mathsf{per}(A))^{-1} \cdot \mathsf{adj}(A)$.*

Proof. By Prop. 7.25(3) $a_{ik}a_{il} = 0$ for any $i, k, l \in \underline{n}$ with $k \neq l$. Let $B := A \cdot \mathsf{adj}(A)$; then $b_{ij} = \sum_{k \in \underline{n}} a_{ik}\mathsf{per}(A(j|k))$ for any $i, j \in \underline{n}$. So $b_{ii} = \sum_{k \in \underline{n}} a_{ik}\mathsf{per}(A(i|k)) = \mathsf{per}(A)$

by Prop. 6.8(4). For $i \neq j$ we have

$$\begin{aligned} b_{ij} &= \sum_{k\in \underline{n}} a_{ik}\Big(\sum_{\substack{l\in\underline{n}\\ l\neq k}} a_{il}\mathsf{per}(A(ij|kl))\Big) && \text{by Prop. 6.8(4)} \\ &= \sum_{k\in\underline{n}} \sum_{\substack{l\in\underline{n}\\ l\neq k}} a_{ik}a_{il}\mathsf{per}(A(ij|kl)) \\ &= 0 && \text{by Prop. 7.25(3)} \end{aligned}$$

Therefore, $B = A\cdot \mathsf{adj}(A) = \mathsf{per}(A)\cdot I_n$. By Lemma 7.34(2) we have $\mathsf{per}(A) \in U(R)$, so $(\mathsf{per}(A))^{-1}A\cdot\mathsf{adj}(A) = (\mathsf{per}(A))^{-1}\mathsf{per}(A)\cdot I_n = I_n$; but $(\mathsf{per}(A))^{-1}A\cdot\mathsf{adj}(A) = A\cdot(\mathsf{per}(A))^{-1}\mathsf{adj}(A)$, whence $A^{-1} = (\mathsf{per}(A))^{-1}\mathsf{adj}(A)$. □

The following theorem gives Cramer's rule for a matrix equation over commutative antirings. Compare this to Prop. 6.41 and Th. 8.21, the analogous results for commutative semirings and inclines.

Theorem 7.36. ([13, Theorem 4.1]) *Let R be a commutative antiring, $A \in \mathbb{S}_n(R)$, and $\boldsymbol{b} = (b_1, b_2, \ldots, b_n)^T \in V_n(R)$. If A is invertible in $\mathbb{S}_n(R)$, then the matrix equation $A\boldsymbol{x} = \boldsymbol{b}$ has a unique solution $\boldsymbol{x} = (d^{-1}d_1, d^{-1}d_2, \ldots, d^{-1}d_n)^T \in V_n(R)$, where $d = \mathsf{per}(A)$ and $d_j = \mathsf{per}(A^{\boldsymbol{b}\Rightarrow j})$, $j \in \underline{n}$.*

Proof. Obviously, $\boldsymbol{x} = A^{-1}\boldsymbol{b}$ is a solution for the matrix equation $A\boldsymbol{x} = \boldsymbol{b}$. Let $\boldsymbol{y}$ be any other solution; then $\boldsymbol{y} = I_n\boldsymbol{y} = (A^{-1}A)\boldsymbol{y} = A^{-1}(A\boldsymbol{y}) = A^{-1}\boldsymbol{b}$, which means that $A\boldsymbol{x} = \boldsymbol{b}$ has a unique solution.

Let $(d_1, d_2, \ldots, d_n)^T := \mathsf{adj}(A)\boldsymbol{b}$; then, for any $j \in \underline{n}$, $d_j = \sum_{i\in\underline{n}} \mathsf{per}(A(i|j))b_i$. On the other hand, by Prop. 6.8(4), we have $\mathsf{per}(A^{\boldsymbol{b}\Rightarrow j}) = \sum_{i\in\underline{n}} \mathsf{per}(A(i|j))b_i$. Therefore, using Lemma 7.35, $\boldsymbol{x} = A^{-1}\boldsymbol{b} = (\mathsf{per}(A))^{-1}\mathsf{adj}(A)\boldsymbol{b} = d^{-1}\mathsf{adj}(A)\boldsymbol{b} = (d^{-1}d_1, d^{-1}d_2, \ldots, d^{-1}d_n)^T$. □

The next corollary is a generalization of [6, Theorem 4.3]:

Corollary 7.37. ([13, Corollary 4.4]) *Let R be a commutative antiring such that $U(R) = \{1\}$, $A \in \mathbb{S}_n(R)$, and $\boldsymbol{b} = (b_1, b_2, \ldots, b_n)^T \in V_n(R)$. If A is invertible in $\mathbb{S}_n(R)$, then the matrix equation $A\boldsymbol{x} = \boldsymbol{b}$ has a unique solution $\boldsymbol{x} = (d_1, d_2, \ldots, d_n)^T \in V_n(R)$, where $d_j = \mathsf{per}(A^{\boldsymbol{b}\Rightarrow j})$, $j \in \underline{n}$.*

7.5 Bases

Lemma 7.38. [19, Lemma 3.2]) *Let S be a commutative semiring. If every unit vector of $\mathcal{V}_n$ can be represented by a linear combination of $\boldsymbol{a}_1, \boldsymbol{a}_2, \ldots, \boldsymbol{a}_m \in \mathcal{V}_n$, then $\{\boldsymbol{a}_1, \boldsymbol{a}_2, \ldots, \boldsymbol{a}_m\}$ contains a set of generators for $\mathcal{V}_n$, and $m \geq n$.*

Proof. The first part can be shown in a way analogous to the first half of the proof of Prop. 6.34. Assume that $m < n$. Since every unit vector of $\mathcal{V}_n$ can be represented by a linear combination of $\boldsymbol{a}_1, \boldsymbol{a}_2, \ldots, \boldsymbol{a}_m$, there is a matrix $C = [c_{ij}] \in \mathbb{S}_{m\times n}$ such that $I_n = [\boldsymbol{a}_1 \boldsymbol{a}_2 \ldots \boldsymbol{a}_m]C$. Let $A' = [\boldsymbol{a}_1 \boldsymbol{a}_2 \ldots \boldsymbol{a}_m \boldsymbol{e}_1 \boldsymbol{e}_2 \ldots \boldsymbol{e}_{n-m}]$. Then

$$A'\begin{bmatrix} C \\ O_{n-m,n}\end{bmatrix} = \begin{bmatrix} a_{11} & a_{12} & \ldots & a_{1m} & 1\,0\ldots 0 \\ & & \vdots & & \\ a_{(n-m)1} & a_{(n-m)2} & \ldots & a_{(n-m)m} & 0\,0\ldots 1 \\ & & \vdots & & \\ a_{n1} & a_{n2} & \ldots & a_{nm} & 0\,0\ldots 0 \end{bmatrix}\begin{bmatrix} c_{11} & c_{12} & \ldots & c_{1n} \\ & & \vdots & \\ c_{m1} & c_{m2} & \ldots & c_{mn} \\ & & \vdots & \\ 0_{n1} & 0_{n2} & \ldots & 0_{nn}\end{bmatrix} = I_n.$$

Therefore, A' is right invertible, and invertible by Prop. 6.26(1). Thus, the set $\{\boldsymbol{a}_1, \boldsymbol{a}_2, \ldots, \boldsymbol{a}_m, \boldsymbol{e}_1, \boldsymbol{e}_2, \ldots, \boldsymbol{e}_{n-m}\}$ is a basis of $\mathcal{V}_n$ by Prop. 6.34. But by assumption, each $\boldsymbol{e}_i, i \in \underline{n-m}$, can be represented by a linear combination of $\boldsymbol{a}_1, \boldsymbol{a}_2, \ldots, \boldsymbol{a}_m$, which is a contradiction to the definition of a basis (see Def. 5.20(d)). □

Corollary 7.39. ([19, Corollary 3.1]) *Let S be a commutative semiring. For the cardinality c of each basis of $\mathcal{V}_n$ $c \geq n$ holds.*

Proof. Follows directly from Lemma 7.38 since each basis of $\mathcal{V}_n$ is a set of generators. □

Proposition 7.40. (([11, Theorem 3.1], [19, Lemma 3.3, Theorem 3.3]) *Let $(L, +, \cdot, 0, 1)$ be a commutative zerosumfree semiring with $U(L) = \{1\}$ and 1 additively irreducible. Then the cardinality of each basis of $\mathcal{V}_n$ is n.*

Proof. (The prove below follows the sufficiency part of Theorem 3.1 in [11].) Let $\boldsymbol{a}_1, \boldsymbol{a}_2, \ldots, \boldsymbol{a}_m$ form a basis of $\mathcal{V}_n$, where $\boldsymbol{a}_i = (a_{1i}, a_{2i}, \ldots, a_{ni})^T$. Then every unit vector $\boldsymbol{e}_j$, $j \in \underline{n}$, can be represented by a linear combination of $\boldsymbol{a}_1, \boldsymbol{a}_2, \ldots, \boldsymbol{a}_m$, so by Lemma 7.38 $m \geq n$.

Suppose $\{\boldsymbol{a}_1, \boldsymbol{a}_2, \ldots, \boldsymbol{a}_m\} \neq \{\boldsymbol{e}_1, \boldsymbol{e}_2, \ldots, \boldsymbol{e}_n\}$. Then, since $\boldsymbol{a}_1, \boldsymbol{a}_2, \ldots, \boldsymbol{a}_m$ are linearly independent, there exists at least one unit vector $\boldsymbol{e}_j \notin \{\boldsymbol{a}_1, \boldsymbol{a}_2, \ldots, \boldsymbol{a}_m\}, j \in \underline{n}$ (because otherwise $\boldsymbol{a}_1, \boldsymbol{a}_2, \ldots, \boldsymbol{a}_m$ would be lincarly dependent). Let $\boldsymbol{e}_j = \sum_{k\in\underline{m}} \lambda_k \boldsymbol{a}_k$. Then we get

$$1 = \lambda_1 a_{j1} + \lambda_2 a_{j2} + \cdots + \lambda_m a_{jm} \quad \text{and} \tag{7.5}$$

$$0 = \lambda_1 a_{i1} + \lambda_2 a_{i2} + \cdots + \lambda_m a_{im} \text{ for any } i \in \underline{n}, i \neq j. \tag{7.6}$$

Since 1 is additively irreducible, (7.5) implies that $\lambda_s a_{js} = 1$ for some $s \in \underline{m}$, whence $\lambda_s, a_{js} \in U(L) = \{1\}$, i.e. $\lambda_s = a_{js} = 1$. From (7.6) we get $\lambda_s a_{is} = 0$ since L is zerosumfree, i.e. $\lambda_s a_{is} = 1 \cdot a_{is} = a_{is} = 0$ for $i \neq j$. Therefore, $a_{is} = \delta_{ij}$, which means $\boldsymbol{a}_s = \boldsymbol{e}_j \in \{\boldsymbol{a}_1, \boldsymbol{a}_2, \ldots, \boldsymbol{a}_m\}$, a contradiction.

This argument holds for any $\boldsymbol{e}_j \notin \{\boldsymbol{a}_1, \boldsymbol{a}_2, \ldots, \boldsymbol{a}_m\}, j \in \underline{n}$; so $\{\boldsymbol{e}_1, \boldsymbol{e}_2, \ldots, \boldsymbol{e}_n\} \subseteq \{\boldsymbol{a}_1, \boldsymbol{a}_2, \ldots, \boldsymbol{a}_m\}$. But then we must have $m = n$, i.e. $\{\boldsymbol{e}_1, \boldsymbol{e}_2, \ldots, \boldsymbol{e}_n\} = \{\boldsymbol{a}_1, \boldsymbol{a}_2, \ldots, \boldsymbol{a}_m\}$, because otherwise $\boldsymbol{a}_1, \boldsymbol{a}_2, \ldots, \boldsymbol{a}_m$ would be linearly dependent. □

Corollary 7.41. ([11, Corollary 3.2]) *Let S be a commutative zerosumfree semiring with $U(L) = \{1\}$ and 1 additively irreducible. Then $\{\boldsymbol{e}_1, \boldsymbol{e}_2, \ldots, \boldsymbol{e}_n\}$ is the unique basis of $\mathcal{V}_n$.*

Proof. Follows immediately from the proof of Prop. 7.40. □

Theorem 7.42. ([11, Theorem 3.2]) *Let S be a commutative zerosumfree semiring. Then each basis of $\mathcal{V}_n$ has the same number of elements if and only if every vector can uniquely be represented by a linear combination of each basis.*

Proof. Let $\{\boldsymbol{x}_1,\boldsymbol{x}_2,\dots,\boldsymbol{x}_n\}$ and $\{\boldsymbol{y}_1,\boldsymbol{y}_2,\dots,\boldsymbol{y}_m\}$ be two bases of $\mathcal{V}_n$, and $m \neq n$. Assume that each vector in $\mathcal{V}_n$ can uniquely be represented by a linear combination of each basis, and let $n > m$. Since $\{\boldsymbol{x}_1,\boldsymbol{x}_2,\dots,\boldsymbol{x}_n\}$ is a basis of $\mathcal{V}_n$, every $\boldsymbol{y}_i$, $i \in \underline{m}$, can be represented as $\boldsymbol{y}_i = \sum_{j\in\underline{n}} a_{ij}\boldsymbol{x}_j$. So $(\boldsymbol{y}_1,\boldsymbol{y}_2,\dots,\boldsymbol{y}_m) = (\boldsymbol{x}_1,\boldsymbol{x}_2,\dots,\boldsymbol{x}_n)A$, where $A = [a_{ij}]^T \in \mathbb{S}_{n\times m}$. In a similar way we get $(\boldsymbol{x}_1,\boldsymbol{x}_2,\dots,\boldsymbol{x}_n) = (\boldsymbol{y}_1,\boldsymbol{y}_2,\dots,\boldsymbol{y}_m)B$, where $B = [b_{ij}]^T \in \mathbb{S}_{m\times n}$.

Therefore, $(\boldsymbol{x}_1,\boldsymbol{x}_2,\dots,\boldsymbol{x}_n) = (\boldsymbol{x}_1,\boldsymbol{x}_2,\dots,\boldsymbol{x}_n)AB$, whence $AB = I_n$ since every vector can uniquely be represented by a linear combination of each finite basis. Now add $n-m$ columns $\mathbf{0}$ to A and $n-m$ rows $\mathbf{0}$ to B; then $\begin{bmatrix} A & \mathbf{0}\end{bmatrix}\begin{bmatrix} B \\ \mathbf{0}\end{bmatrix} = I_n$. From Th. 6.35 we know that both $\begin{bmatrix} A & \mathbf{0}\end{bmatrix}$ and $\begin{bmatrix} B \\ \mathbf{0}\end{bmatrix}$ are invertible. But from Prop. 6.8(1) and Lemma 7.34(1) it follows that

$$1 = \mathsf{per}(I_n) = \mathsf{per}\left(\begin{bmatrix} A & \mathbf{0}\end{bmatrix}\begin{bmatrix} B \\ \mathbf{0}\end{bmatrix}\right) = \mathsf{per}\left(\begin{bmatrix} A & \mathbf{0}\end{bmatrix} I_n \begin{bmatrix} B \\ \mathbf{0}\end{bmatrix}\right) = \mathsf{per}(\begin{bmatrix} A & \mathbf{0}\end{bmatrix})\mathsf{per}\left(\begin{bmatrix} B \\ \mathbf{0}\end{bmatrix}\right) = 0,$$

a contradiction. Similarly, a contradiction follows from the assumption $n < m$. Thus we have to conclude that $m = n$.

Conversely, assume that each basis of $\mathcal{V}_n$ has n elements, and that $\{\boldsymbol{x}_1,\boldsymbol{x}_2,\dots,\boldsymbol{x}_n\}$ is one of them. For $\boldsymbol{a} \in \mathcal{V}_n$ suppose that

$$\boldsymbol{a} = r_1\boldsymbol{x}_1 + r_2\boldsymbol{x}_2 + \cdots + r_n\boldsymbol{x}_n = r_1'\boldsymbol{x}_1 + r_2'\boldsymbol{x}_2 + \cdots + r_n'\boldsymbol{x}_n$$

with $r_i, r_i' \in S$, and let $(\boldsymbol{e}_1,\boldsymbol{e}_2,\dots,\boldsymbol{e}_n) = (\boldsymbol{x}_1,\boldsymbol{x}_2,\dots,\boldsymbol{x}_n)C$ and $(\boldsymbol{x}_1,\boldsymbol{x}_2,\dots,\boldsymbol{x}_n) = (\boldsymbol{e}_1,\boldsymbol{e}_2,\dots,\boldsymbol{e}_n)D$ for some matrices $C, D \in \mathbb{S}_n$. Then, $(\boldsymbol{e}_1,\boldsymbol{e}_2,\dots,\boldsymbol{e}_n) = (\boldsymbol{e}_1,\boldsymbol{e}_2,\dots,\boldsymbol{e}_n)DC$, whence $DC = I_n$ and thus invertible by Th. 6.35. Hence,

$$\begin{aligned}\boldsymbol{a} &= (\boldsymbol{x}_1,\boldsymbol{x}_2,\dots,\boldsymbol{x}_n)(r_1,r_2,\dots,r_n)^T = (\boldsymbol{x}_1,\boldsymbol{x}_2,\dots,\boldsymbol{x}_n)(r_1',r_2',\dots,r_n')^T \\ &= (\boldsymbol{e}_1,\boldsymbol{e}_2,\dots,\boldsymbol{e}_n)D(r_1,r_2,\dots,r_n)^T = (\boldsymbol{e}_1,\boldsymbol{e}_2,\dots,\boldsymbol{e}_n)D(r_1',r_2',\dots,r_n')^T.\end{aligned}$$

Thus, $D(r_1,r_2,\dots,r_n)^T = D(r_1',r_2',\dots,r_n')^T$ and, since D is invertible as shown above, we get $(r_1,r_2,\dots,r_n)^T = (r_1',r_2',\dots,r_n')^T$. This proves that every vector $\boldsymbol{a}$ can uniquely be represented by a linear combination of each basis. □

Proposition 7.43. ([19, Theorem 3.4]) *Let S be a commutative zerosumfree semiring. Then any two different vectors of a basis of $\mathcal{V}_n$ are orthogonal.*

Proof. Let $\{\boldsymbol{a}_1,\boldsymbol{a}_2,\dots,\boldsymbol{a}_n\}$ be a basis of $\mathcal{V}_n$ with cardinality n. Then, by Prop. 6.34, the $\boldsymbol{a}_i$ can be regarded as the column vectors of an invertible matrix $A \in \mathbb{S}_n(S)$. Since A^TA is an invertible diagonal matrix due to Prop. 7.29, we have

$$\begin{bmatrix} \boldsymbol{a}_1^T \\ \vdots \\ \boldsymbol{a}_n^T \end{bmatrix}\begin{bmatrix} \boldsymbol{a}_1 & \cdots & \boldsymbol{a}_n\end{bmatrix} = \begin{bmatrix} u_1 & 0 & \cdots & 0 \\ 0 & u_2 & \cdots & 0 \\ \vdots & \vdots & \vdots & \vdots \\ 0 & 0 & \cdots & u_n \end{bmatrix}$$

with $u_i \in S$. So $\boldsymbol{a}_i^T\boldsymbol{a}_i = u_i$, and $\boldsymbol{a}_i^T\boldsymbol{a}_j = 0$, for all $i, j \in \underline{n}$ and $i \neq j$. □

Proposition 7.44. ([12, Theorem 3.1]) *Let L be a commutative zerosumfree semiring and $\{\boldsymbol{u}_1,\boldsymbol{u}_2,\ldots,\boldsymbol{u}_m\}$ be a standard orthogonal set in $V^n(L)$. If $\{\boldsymbol{u}_1,\boldsymbol{u}_2,\ldots,\boldsymbol{u}_m,\boldsymbol{a}\}\subseteq V^n(L)$, are linearly dependent, then $\boldsymbol{a}$ can be uniquely represented by a linear combination of the vectors $\boldsymbol{u}_1,\boldsymbol{u}_2,\ldots,\boldsymbol{u}_m$.*

Proof. If $\boldsymbol{u}_1,\boldsymbol{u}_2,\ldots,\boldsymbol{u}_m,\boldsymbol{a}$, are linearly dependent, then either one of the $\boldsymbol{u}_i$, $i\in\underline{m}$, or $\boldsymbol{a}$ can be represented by a linear combination of the other vectors; see Def. 5.20. Suppose that

$$\boldsymbol{u}_i = r\boldsymbol{a}+\sum_{j\in\underline{m},j\neq i} s_j\boldsymbol{u}_j, \tag{7.7}$$

for $r,s_j\in L$. Since $\{\boldsymbol{u}_1,\boldsymbol{u}_2,\ldots,\boldsymbol{u}_m\}$ is standard orthogonal, this implies

$$\boldsymbol{u}_i\boldsymbol{u}_i = \boldsymbol{u}_i(r\boldsymbol{a})+\textstyle\sum_{j\in\underline{m},j\neq i}\boldsymbol{u}_i(s_j\boldsymbol{u}_j) = r(\boldsymbol{u}_i\boldsymbol{a})+\sum_{j\in\underline{m},j\neq i}s_j(\boldsymbol{u}_i\boldsymbol{u}_j) = r(\boldsymbol{u}_i\boldsymbol{a})\in U(L).$$

Thus, $1=(\boldsymbol{u}_i\boldsymbol{u}_i)(\boldsymbol{u}_i\boldsymbol{u}_i)^{-1}=r(\boldsymbol{u}_i\boldsymbol{a})(\boldsymbol{u}_i\boldsymbol{u}_i)^{-1}$, i.e. $r\in U(L)$. But from (7.7) we also get

$$0=\boldsymbol{u}_j\boldsymbol{u}_i = r(\boldsymbol{u}_j\boldsymbol{a})+s_j(\boldsymbol{u}_j\boldsymbol{u}_j),\ i\neq j, j\in\underline{m}.$$

Thus, $s_j(\boldsymbol{u}_j\boldsymbol{u}_j)=0$ since L is commutative and zerosumfree, whence $s_j=0$ for $i\neq j, j\in\underline{m}$ since $\boldsymbol{u}_j\boldsymbol{u}_j\in U(L)$. Therefore, (7.7) means that $\boldsymbol{a}=r^{-1}\boldsymbol{u}_i$, i.e. $\boldsymbol{a}$ can be represented as a linear combination of $\boldsymbol{u}_1,\boldsymbol{u}_2,\ldots,\boldsymbol{u}_m$.

To show uniqueness assume that $\boldsymbol{a}=\sum_{i\in\underline{m}}r_i\boldsymbol{u}_i=\sum_{i\in\underline{m}}s_i\boldsymbol{u}_i$, $r_i,s_i\in L$. Then

$$\left(\textstyle\sum_{i\in\underline{m}}r_i\boldsymbol{u}_i\right)\boldsymbol{u}_j=\left(\sum_{i\in\underline{m}}s_i\boldsymbol{u}_i\right)\boldsymbol{u}_j,\ \text{so}\ r_i(\boldsymbol{u}_i\boldsymbol{u}_i)=s_i(\boldsymbol{u}_i\boldsymbol{u}_i), i\in\underline{m},$$

and $r_i=s_i$ since $\boldsymbol{u}_i\boldsymbol{u}_i\in U(L)$, i.e. invertible, by the definition of standard orthogonality in Def. 5.21(c). □

A generalization of the above proposition to commutative semirings has already been given in Th. 6.45.

7.6 Eigenvalues

There are only few results on eigenvalues for general variants of semirings which date back to [3] from the late seventies. The results thereof are reported in [4, Chapter 6], and are also shortly mentioned in [18, Section 2.3] and [20, p. 171f]. Among them is

Proposition 7.45. *Let S be a commutative semiring, and $A\in\mathbb{S}_n$.*

(1) The eigenspace E^A_λ, i.e. the set of eigenvectors of A associated with the eigenvalue λ, is a (left) sub-semimodule (or semilinear subspace).

(2) If S is multiplicatively idempotent, and $(\boldsymbol{x},1)$ is an eigenpair for A, then also $(\lambda\boldsymbol{x},\lambda)$.

Proof. (1): Let $r,s\in S$ and $\boldsymbol{v},\boldsymbol{w}\in E^A_\lambda$. Then we have to show that $r\boldsymbol{v}+s\boldsymbol{w}\in E^A_\lambda$, which follows by straightforward calculation.

(2): By assumption, we have $A\boldsymbol{x}=\boldsymbol{x}$; so $A(\lambda\boldsymbol{x})=\lambda(A\boldsymbol{x})=\lambda\boldsymbol{x}=\lambda^2\boldsymbol{x}=\lambda(\lambda\boldsymbol{x})$. □

Let $A \in \mathbb{S}_n(S)$ be a square matrix with values in the semiring S. Vectors in E_1^A are called **standard eigenvectors** of A; so $E_1^A = \{\boldsymbol{x} \mid A\boldsymbol{x} = \boldsymbol{x}\}$. For convenience we sometimes write $E_1^A = E(A)$, and let $C(A)$ denote the span of the column vectors of A, i.e. $C(A) = \langle\{A_{*j} \mid j \in \underline{n}\}\rangle = \left\{\sum_{j\in\underline{n}} s_i A_{*j} \mid s_i \in S\right\}$. Following [16] we will now present some results of the standard eigenvectors of matrices with period over a commutative path algebra starting with

Lemma 7.46. ([16, Lemma 5.1]) *Let D be a commutative path algebra and $A \in \mathbb{S}_n(D)$ be almost periodic. Then*

(1) For any $\boldsymbol{x}, \boldsymbol{y} \in E_1^A$ and $\lambda \in D$, we have $\boldsymbol{x}+\boldsymbol{y} \in E_1^A$ and $\lambda\boldsymbol{x} \in E_1^A$.
(2) $E(A) \subseteq C(A)$.
(3) For any positive integers $s_1, s_2, \ldots, s_r$, $E(A) \subseteq C(A^{s_1} + A^{s_2} + \cdots + A^{s_r})$.

Proof. (1): $\boldsymbol{x}, \boldsymbol{y} \in E_1^A$ and $\lambda \in D$ implies $A(\boldsymbol{x}+\boldsymbol{y}) = A\boldsymbol{x} + A\boldsymbol{y} = \boldsymbol{x}+\boldsymbol{y}$ and $A(\lambda\boldsymbol{x}) = \lambda(A\boldsymbol{x}) = \lambda\boldsymbol{x}$.
(2): For $\boldsymbol{x} \in E(A)$ we have

$$\boldsymbol{x} = A\boldsymbol{x} = \begin{bmatrix} \sum_{j\in\underline{n}} a_{1j}x_j \\ \sum_{j\in\underline{n}} a_{2j}x_j \\ \vdots \\ \sum_{j\in\underline{n}} a_{mj}x_j \end{bmatrix} = x_1 \begin{bmatrix} a_{11} \\ a_{21} \\ \vdots \\ a_{m1} \end{bmatrix} + x_2 \begin{bmatrix} a_{12} \\ a_{22} \\ \vdots \\ a_{m2} \end{bmatrix} + \ldots x_n \begin{bmatrix} a_{1n} \\ a_{2n} \\ \vdots \\ a_{mn} \end{bmatrix} = \sum_{j\in\underline{n}} x_j \boldsymbol{a}_j.$$

(3): Clearly, we have $E(A) \subseteq E(A^k)$, for any $k \geq 1$, so $(\sum_{k\in\underline{r}} A^{s_k})\boldsymbol{x} = \sum_{k\in\underline{r}} A^{s_k}\boldsymbol{x} = \sum_{k\in\underline{r}} \boldsymbol{x} = \boldsymbol{x}$, for any $\boldsymbol{x} \in E(A)$, i.e. $E(A) \subseteq E(A^{s_1} + A^{s_2} + \cdots + A^{s_r})$. Using (2) proves the assertion. □

Proposition 7.47. ([16, Theorem 5.1]) *Let D be a commutative path algebra and $A \in \mathbb{S}_n(D)$ be almost periodic with index k and period d. Then $E(A) = C(A')$, where $A' = \sum_{j=k}^{k+d-1} A^j$.*

Proof. From

$$AA' = \sum_{j=k+1}^{k+d} A^j = \Big(\sum_{j=k+1}^{k+d-1} A^j\Big) + A^{k+d} = \Big(\sum_{j=k+1}^{k+d-1} A^j\Big) + A^k = A'$$

we get that all column vectors of A' belong to $E(A)$. Thus, by Lemma 7.46(1), $C(A') \subseteq E(A)$. And by Lemma 7.46(2) $E(A) \subseteq C(A')$, whence equality is proved. □

Corollary 7.48. ([16, Corollary 5.1]) *Let D be a commutative path algebra and $A \in \mathbb{S}_n(D)$ be almost periodic with index k and period d. If A converges in a finite number of steps, then $E(A) = C(A^k)$.*

Proposition 7.49. ([16, Theorem 5.2]) *Let D be a commutative path algebra and $A \in \mathbb{S}_n(D)$ be almost periodic with index k and period d. Then $E(\mathbf{t}(A)) = E(A)$, i.e. the set of standard eigenvectors of $\mathbf{t}(A)$ is the same as that of A.*

Proof. Since A is almost periodic, by Lemma 7.10(2) $\mathbf{t}(A)$ converges in a finite number of steps and $k(\mathbf{t}(A)) \leq k$. Then we have

$$
\begin{aligned}
E(\mathbf{t}(A)) &= C\big(\mathbf{t}(A)^{k(\mathbf{t}(A))}\big) && \text{by Cor. 7.48} \\
&= C\big(\mathbf{t}(A)^{k}\big) && \text{because } k(\mathbf{t}(A)) \le k \text{ by Lemma 7.10(2)} \\
&= C\Big(\textstyle\sum_{j=k}^{k(k+d-1)} A^j\Big) && \text{by Prop. 7.9} \\
&= C\Big(\textstyle\sum_{j=k}^{k+d-1} A^j\Big) && \text{because } k(k+d-1)-(k-1) \ge d \\
& E(A) && \text{by Prop. 7.47.}
\end{aligned}
$$

□

7.7 Nilpotent Matrices

In this section we list some basic facts around the concept of nilpotency, both for semiring elements as well as for semiring matrices. Recall that $h(a)$ is the least positive integer k such that $a^k = 0$.

Lemma 7.50. ([14, Lemmas 2.1, 2.2]) *Let S be a commutative semiring, $A \in \mathbb{S}_n(S)$, and $a \in S$. Then*

(1) if a is nilpotent then ar and ra are nilpotent for any $r \in S$.
(2) A is nilpotent if and only if PAP^T is nilpotent for any permutation matrix $P \in \mathbb{S}_n$, and $h(A) = h(PAP^T)$.

Proof. (2): Assuming $A^k = O$ for some $k \ge 1$ we get $(PAP^T)^k = PA^kP^T = O$, since P is orthogonal, i.e. $P^T = P^{-1}$. Conversely, from $(PAP^T)^k = PA^kP^T = O$ we get $PA^kP^T = P^TPA^kP^TP = A^k = P^TOP = O$. □

Lemma 7.51. ([14, Lemmas 2.1, 2.2]) *Let S be a commutative antiring, $A \in \mathbb{S}_n(S)$, and $a_1, a_2, \ldots, a_m \in S$. Then*

(1) $\sum_{i\in \underline{m}} a_i$ is nilpotent if and only if $a_1, a_2, \ldots, a_m$ are all nilpotent.
(2) if $A+B$ is nilpotent, then A and B are nilpotent.

Proof. (1): Recall that $x^0 = 1 \in S$ for all $x \in S$ and that by the binomial theorem $(x+y)^k = \sum_{l=0}^{k}\binom{k}{l}x^{k-l}y^l$, for all $k \ge 1$. We first assume that $\sum_{i\in\underline{m}} a_i$ is nilpotent; then, for some k, $0 = (a_1 + a_2 + \cdots + a_m)^k = a_1^k + \ldots$. By the antiring property we get $a_1^k = 0$, so a_1 is nilpotent. Using the commutativity of + each of the a_i can take the first position and thus be shown to be nilpotent.

We now prove the converse direction by induction over m, The base case $m = 1$ is obvious: $\sum_{i=1}^{1} a_i = a_1$, with a_1 being nilpotent by assumption. So we assume that the assertion holds for $m \ge 2$. Let $a_1, a_2, \ldots, a_m$ be nilpotent, then $S_m := \sum_{i=1}^{m-1} a_i$ is nilpotent by induction hypothesis. Let $h_1 = h(S_m)$ and $h_2 = h(a_m)$ and $h = h_1 + h_2$. Then $(\sum_{i=1}^{m} a_i)^h = (S_m + a_m)^h = \sum_{l=0}^{h}\binom{h}{l}(S_m)^{h-l}a_m^l$. If $l \ge h_2$ then $a_m^l = 0$, and if $l < h_2$ then $h - l \ge h_1$, whence $(S_m)^{h-l} = 0$. Thus, $\binom{h}{l}(S_m)^{h-l}a_m^l = 0$ for all $l, 0 \le l \le h$, and therefore $(\sum_{i=1}^{m} a_i)^h = 0$.

(2): By assumption $(A+B)^k = O_n$ for some $k \ge 1$. A^k and B^k are two terms in the expansion of $(A+B)^k$, so $A^k = B^k = O_n$ since $(\mathbb{S}_n(S), +, \cdot)$ is an antiring. □

Lemma 7.52. ([15, Lemma 2.1]) *Let D be a commutative path algebra and $a, b, a_1, a_2, \ldots, a_m \in D$. Then*

(1) *if a is nilpotent and $b \le a$ then b is nilpotent.*
(2) *$\sum_{i\in \underline{m}} a_i$ is nilpotent if and only if $a_1, a_2, \ldots, a_m$ are all nilpotent.*

Proof. (2): Follows from Lemma 7.51(1), since path algebras are antirings. □

Proposition 7.53. ([15, Lemma 2.4]) *Let D be a commutative path algebra and $A \in \mathbb{S}_n(D)$ be nilpotent. Then $\mathbf{t}(A) = \sum_{k=1}^{h(A)-1} A^k$.*

Proof. Obviously, $A \le B := \sum_{k=1}^{h(A)-1} A^k$. Since $(\mathbb{S}_n(D), +, \cdot)$ is a path algebra and $A^k = O$ for all $k \ge h(A)$, we have

$$B^2 = \sum_{k=2}^{2h(A)-2} A^k = \sum_{k=2}^{h(A)-1} A^k \le B,$$

i.e. B is transitive. If there exists a transitive matrix C in $\mathbb{S}_n(D)$ such that $A \le C$, then $C^2 \le C$, thus $C^k \le C$ for all $k \in \mathbb{N}$, whence $A^k \le C^k \le C$. Thus $B = \sum_{k=1}^{h(A)-1} A^k \le C$, i.e. $\mathbf{t}(A) = B$. □

The following two results are proved in [14] for commutative antirings. But checking the proofs it turns out that the first assertion actually holds in arbitrary commutative semirings. Since by Prop. 4.4(2) path algebras are antirings, the results are also valid in commutative path algebras; see [15, Propositions 3.1 and 3.2].

Lemma 7.54. ([14, Proposition 3.1]) *Let D be a commutative semiring and $A \in \mathbb{S}_n(D)$ such that every element of A is nilpotent. Then A is nilpotent.*

Proof. Suppose that a_{ij} is nilpotent for all $i, j \in \underline{n}$. We let $k = \max\{h(a_{ij}) | i, j \in \underline{n}\}$, then $a_{ij}^k = 0$ for any $i, j \in \underline{n}$. We put $l = n^2 k$. Taking $i_0 = i$ and $i_l = j$, from (7.1) we know that $a_{ij}^{(l)} = \sum_{i_1, i_2, \ldots, i_{l-1} \in \underline{n}} a_{ii_1} a_{i_1 i_2} \cdots a_{i_{l-2} i_{l-1}} a_{i_{l-1} j}$. Let $p = a_{ii_1} a_{i_1 i_2} \cdots a_{i_{l-2} i_{l-1}} a_{i_{l-1} j}$ be any product in this sum. Since $a_{ii_1}, a_{i_1 i_2}, \ldots, a_{i_{l-2} i_{l-1}}, a_{i_{l-1} j} \in \{a_{rs} | r, s \in \underline{n}\}$ and this set contains at most n^2 different elements, there exist $t_0, t_1, t_2, \ldots, t_{k-1} \in \{0, 1, \ldots, l-1\}$ such that $t_0 < t_1 < \cdots < t_{k-1}$ and $a_{i_{t_0} i_{t_0}+1} = a_{i_{t_1} i_{t_1}+1} = \cdots = a_{i_{t_{k-1}} i_{t_{k-1}}+1}$. So p contains the factor $(a_{i_{t_0} i_{t_0}+1})^k$ (it is here where commutativity is used). But $(a_{i_{t_0} i_{t_0}+1})^k = 0$, so $p = 0$ and since $a_{ij}^{(l)}$ and p have been chosen arbitrarily, $a_{ij}^{(l)} = 0$ for all $i, j \in \underline{n}$, i.e. $A^l = O_n$. □

Proposition 7.55. ([14, Proposition 3.2]) *Let D be a commutative antiring and $A \in \mathbb{S}_n(D)$. If all main diagonal entries of A^k are nilpotent for any $k \in \underline{n}$, then*

(1) *all elements of A^n are nilpotent.*
(2) *A is nilpotent.*

Proof. (1): $a_{ij}^{(n)} = \sum_{i_1, i_2, \ldots, i_{n-1} \in \underline{n}} a_{ii_1} a_{i_1 i_2} \cdots a_{i_{n-2} i_{n-1}} a_{i_{n-1} j}$, for any $i, j \in \underline{n}$. Let $p = a_{ii_1} a_{i_1 i_2} \cdots a_{i_{n-2} i_{n-1}} a_{i_{n-1} j}$ be any product in this sum. Since $i, i_1, i_2, \ldots, i_{n-1}, j \in \underline{n}$, there are $s, t \in \{0, 1, \ldots, n\}$ such that $i_s = i_t$ with $s < t$, taking $i_0 = i$ and $i_l = j$; so

$a_{i_s i_{s+1}} \dots a_{i_{t-1} i_t} = a_{i_s i_{s+1}} \dots a_{i_{t-1} i_s}$. But $a_{i_s i_{s+1}} \dots a_{i_{t-1} i_s}$ is a term in $a_{i_s i_s}^{(t-s)}$ which is nilpotent by hypothesis. Thus $a_{i_s i_{s+1}} \dots a_{i_{t-1} i_t}$ is nilpotent by Lemma 7.51(1) which uses the antiring property. And since $a_{i_s i_{s+1}} \dots a_{i_{t-1} i_t}$ is a factor of p, p is nilpotent by Lemma 7.50(1). Since p has been chosen arbitrarily, $a_{ij}^{(n)} = 0$ for all $i, j \in \underline{n}$, i.e. $A^l = O_n$.

(2): By (1) and Lemma 7.54 A^n is nilpotent, say $(A^n)^k = O$; then $A^{nk} = O$. □

A matrix $A \in \mathbb{S}_n(D)$ is called **weakly irreflexive** if a_{ii} is nilpotent for any $i \in \underline{n}$. Clearly, every irreflexive matrix is weakly irreflexive.

Corollary 7.56. ([15, Corollary 3.1]) *Let D be a commutative antiring and $A \in \mathbb{S}_n(D)$ be weakly irreflexive and transitive; then A is nilpotent.*

Proof. Since A is transitive, we get $A^k \le A$ for all $k \in \underline{n}$, so $a_{ij}^{(k)} \le a_{ij}$ for any $i, j \in \underline{n}$. By weak irreflexivity we have that a_{ii} is nilpotent for any $i \in \underline{n}$. Therefore, by Lemma 7.52(1), $a_{ii}^{(k)}$ is nilpotent for any $i, k \in \underline{n}$, and by Prop. 7.55(2) A is nilpotent. □

Proposition 7.57. ([15, Proposition 3.3]) *Let D be a commutative path algebra and $A \in \mathbb{S}_n(D)$. If all principal minors of A are nilpotent then*

(1) *all elements of A^n are nilpotent.*
(2) *A is nilpotent.*

Proof. (1): Consider any term $T = a_{ii_1} a_{i_1 i_2} \cdots a_{i_{n-1} j}$ of $a_{ij}^{(n)}$ for some $i, j \in \underline{n}$. Let $i_0 = i$ and $i_n = j$; then there exist $s, t \in \{0, 1, \dots, n\}$ such that $s < t$ and $i_s = i_t$. So $K := \{(p, q) | p < q \,\&\, i_p = i_q\} \neq \emptyset$, and there exists a pair $(k, l) \in K$ with $l - k = \min\{q - p | (p, q) \in K\}$. Let $\alpha = (i_k, i_{k+1}, \dots, i_{l-1})$. The product $a_{i_k i_{k+1}} \cdots a_{i_{l-1} i_l} = a_{i_k i_{k+1}} \cdots a_{i_{l-1} i_k}$ is a term of the principal minor $\mathsf{per}(A[\alpha|\alpha])$. By assumption, $\mathsf{per}(A[\alpha|\alpha])$ is nilpotent, whence $a_{i_k i_{k+1}} \cdots a_{i_{l-1} i_l}$ is also nilpotent by Lemma 7.52(2). Since $a_{i_k i_{k+1}} \cdots a_{i_{l-1} i_l}$ is a factor of T, T is nilpotent by Lemma 7.50(1). As T is an arbitrary term of $a_{ij}^{(n)}$, we get that $a_{ij}^{(n)}$ is nilpotent for any $i, j \in \underline{n}$ by Lemma 7.52(2).

(2): By (1) and Lemma 7.54 A^n is nilpotent, so A is nilpotent too. □

Proposition 7.58. ([15, Proposition 3.4]) *Let D be a commutative path algebra and $A \in \mathbb{S}_n(D)$ be nilpotent. Then*

(1) *all main diagonal entries of A^m are nilpotent for any $m \ge 1$.*
(2) *$a_{i_1 i_2} a_{i_2 i_3} \dots a_{i_m i_1}$ is nilpotent for any $m \ge 1$ and any $i_1, i_2, \dots, i_m \in \underline{n}$.*
(3) *$\mathsf{per}(A)$ is nilpotent.*

Proof. (1): Since $A^k = O_n$ for some $k \ge 1$, $(A^m)^k = A^{km} = O_n$ for any positive m. Thus $a_{ii}^{(mk)} = 0$ for all $i \in \underline{n}$. But $(a_{ii}^{(m)})^k$ is some term in $a_{ii}^{(mk)}$, whence $(a_{ii}^{(m)})^k = 0$ by Lemma 7.52(2), i.e. $a_{ii}^{(m)}$ is nilpotent for all $i \in \underline{n}$ and any positive integer m.

(2): $a_{i_1 i_2} a_{i_2 i_3} \dots a_{i_m i_1}$ is a term in $a_{i_1 i_1}^{(m)}$, and by (1) and Lemma 7.52(2) this product is nilpotent.

(3): Let $t_\sigma = a_{1\sigma(1)}a_{2\sigma(2)}\cdots a_{n\sigma(n)}$ any term of per(A), with $\sigma \in S_n$. Since $\sigma^m(1) \in \underline{n}$ for all $m \geq 1$, there exist $s,t \in \{0,1,\ldots,n\}$ such that $s < t$ and $\sigma^s(1) = \sigma^t(1)$, and $\sigma^s(1),\sigma^{s+1}(1),\ldots,\sigma^{t-1}(1)$ are mutually different (note that $\sigma^0(1) = 1$). Then $a_{\sigma^s(1)\sigma^{s+1}(1)}\cdots a_{\sigma^{t-1}(1)\sigma^s(1)} = a_{\sigma^s(1)\sigma^{s+1}(1)}\cdots a_{\sigma^{t-1}(1)\sigma^t(1)}$ is a factor in t_σ, where the second product is nilpotent by Lemma 7.50(1). So t_σ is nilpotent for any $\sigma \in S_n$ by (2), and as a consequence per(A) is nilpotent by Lemma 7.51(1). □

Proposition 7.59. ([15, Theorem 3.1]) *Let D be a commutative path algebra and $A \in \mathbb{S}_n(D)$. Then the following statements are equivalent:*

(1) A is nilpotent.
(2) All principal submatrices of A are nilpotent.
(3) All principal minors of A are nilpotent.
(4) All elements of A^n are nilpotent.
(5) All main diagonal entries of A^k are nilpotent, for any $k \in \underline{n}$.

Proof. (1) → (2): Let $B \in \mathbb{S}_r(D)$ be any principal submatrix of A. Then there exists a permutation matrix $P \in \mathbb{S}_n(D)$ such that

$$PAP^T = \begin{bmatrix} B & C \\ E & F \end{bmatrix} = \begin{bmatrix} B & O \\ O & O \end{bmatrix} + \begin{bmatrix} O & C \\ E & F \end{bmatrix},$$

where $C \in \mathbb{S}_{r\times(n-r)}(D)$, $E \in \mathbb{S}_{(n-r)\times r}(D)$, and $F \in \mathbb{S}_{n-r}(D)$. Since PAP^T is nilpotent by Lemma 7.50(2), $\begin{bmatrix} B & O \\ O & O \end{bmatrix}$ is nilpotent by Lemma 7.51(2). Therefore, B is nilpotent.

(2) → (3): By Prop. 7.58(3).

(3) → (4): By Prop. 7.57(1).

(4) → (5): Obvious.

(5) → (1): By Prop. 7.55(2), since path algebras are antirings. □

Lemma 7.60. ([16, Theorem 5.4]) *Let D be a commutative path algebra and $A \in \mathbb{S}_n(D)$ be almost periodic with index k and period d. Then A is nilpotent if and only if $E(A) = \{0\}$.*

Proof. If A is nilpotent; then $A^k = O$. Therefore, $E(A) = C(A^k) = C(O) = \{0\}$ by Cor. 7.48. Conversely, let $E(A) = \{0\}$. Assuming that A is not nilpotent implies that $A^l \neq O$ for all positive integers l. Thus, $A' = \sum_{j=k}^{k+d-1} A^j \neq O$. But by Prop. 7.47 we have $E(A) = C(A') \neq \{0\}$. which is a contradiction. (You might want to recall Th. 1.17(8).) □

References

[1] Give'on, Y.: Lattice matrices, Information and Control **7**, 477-484 (1964).

[2] Golan, J.S.: *Semirings and their Applications*, Springer Science+Business Media Dordrecht (1999).

[3] Gondran, M., Minoux, M.: Valeurs propres et vecteurs propres dans les dioïdes et leur interprétation en théorie des graphes, EDF, Bull. Dir. Etudes Rech., Ser. C, Math. Inform. **2**, 25-41 (1977),

[4] Gondran, M., Minoux, M.: *Graphs, Dioids and Semirings - New Models and Algorithms*, Springer Science & Business Media (2008).

[5] Hammer P.L., Rudeanu, S.: *Boolean Methods in Operations Research and Related Areas*, Springer-Verlag Berlin Heidelberg (1968).

[6] Han,S.C., Li, H.X.: Invertible incline matrices and Cramer's rule over inclines, Linear Algebra and its Applications **389**, 121-138 (2004).

[7] Huang, Y., Lian, H.: On permanents of matrices over a commutative additively idempotent semiring, `https://arxiv.org/abs/1205.6891` (2012).

[8] Jiang, J., Tian, X.: The composition, convergence and transitivity of powers and adjoint of generalized fuzzy matrices, Italian J. of Pure and Applied Mathematics **36**, 377-398 (2016).

[9] Lunts, A.G.: Algebraic methods of analysis and synthesis of relay contact networks, Izv. Akad. Nauk SSSR Ser. Mat. **16**, 405-426 (1952) (in Russian).

[10] Mora, W., Wasanawichit, A., Kemprasit, Y.: Invertible Matrices over Idempotent Semirings, Chamchuri J. of Mathematics **1**, 55-61 (2009).

[11] Shu, Q.Y., Wang, X.P.: Bases in semilinear spaces over zoresumfree semirings, Linear Algebra and its Applications **435**, 2681-2692 (2011).

[12] Shu, Q.Y., Wang, X.P.: Standard orthogonal vectors in semilinear spaces and their applications, Linear Algebra and its Applications **437**, 2733-2754 (2012).

[13] Tan, Y.J.: On invertible matrices over antirings, Linear Algebra and its Applications **423**, 428-444 (2007).

[14] Tan, Y.J.: On nilpotency of matrices over antirings, Linear Algebra and its Applications **433**, 1541-1554 (2010).

[15] Tan, Y.J.: On nilpotency of generalized fuzzy matrices, Fuzzy Sets and Systems **161**, 2213-2226 (2010).

[16] Tan, Y.J.: On generalized fuzzy matrices with periods, Fuzzy Sets and Systems **172**, 87-103 (2011).

[17] Tan, Y.J.: On transitivity of generalized fuzzy matrices, Fuzzy Sets and Systems **210**, 69-88 (2013).

[18] Valverde-Albacete, F.J., Peláez-Moreno, C.: The spectra of irreducible matrices over completed idempotent semifields, Fuzzy Sets and Systems **271**, 46-69 (2015).

[19] Zhao, S., Wang, X..P.: Invertible matrices and semilinear spaces over commutative semirings, Information Sciences **180**, 5115-5124 (2010).

[20] Zimmermann, U.: *Linear and Combinatorial Optimization in Ordered Algebraic Structures*, North-Holland Publishing Company (1981).

CHAPTER 8

MATRICES OVER INCLINES AND VARIOUS KINDS OF LATTICES

8.1 Matrices over Inclines

First of all, recall that, due to Cor. 4.18(8) and (9), inclines are commutative path algebras in which products are less than or equal to factors, i.e. $ab \le a$ due to Def. 4.17(In4). Moreover, path algebras are antirings by Prop. 4.4(2). Thus, besides the results given below, those given in Chapter 7 apply to matrices over inclines. Our presentation of some basic facts on incline matrices is mainly based on [21–23].

In analogy to our notation used for less special semirings let $\mathbb{I}_{m\times n}(L)$ denote the set of all $m \times n$ matrices with entries in an incline L, shortened to $\mathbb{I}_n(L)$ for $m = n$. We simply write $\mathbb{I}_{m\times n}$ or $\mathbb{I}_n$ if we let L be any incline or if L is obvious from the context. The definitions for sum and product of matrices remain the same as for semirings, as do the meanings of O, I and U. Furthermore, we let $A \le B :\Longleftrightarrow a_{ij} \le b_{ij}$ for all $1 \le i, j \le n$. It is then easy to check that $\mathbb{I}_n$ is a partially ordered, but not commutative, semiring with O_n as zero and I_n as identity.

8.1.1 *Various Matrix Properties*

Let $A \in \mathbb{I}_n$. In the following we write $A^m = \left[a_{ij}^{(m)}\right]$, and let $k = k(A)$ denote the index and $p = p(A)$ the period of A. Furthermore, recall that we use $\underline{n}$ as abbreviation for $\{1, 2, \ldots, n\}$, and that we write $[n]$ for the least common multiple of the numbers in $\underline{n}$.

Lemma 8.1. ([21, Lemma 3.1]) *If $A \in \mathbb{I}_n$ then $A^{n^2[n]} \le A^{n^2[n]-[n]}$.*

Proof. Let $m = n^2[n]$. Now consider any summand of $a_{ij}^{(m)}$ and divide it into $n[n]$ products of length n, yielding

$$a_{ii_1} \cdots a_{i_{n-1}i_n} \mid a_{i_n i_{n+1}} \cdots a_{i_{2n-1}i_{2n}} \mid \cdots \mid a_{i_{m-n}i_{m-n+1}} \cdots a_{i_{m-1}j}$$

In every product of n elements there are $n+1$ subscripts taken out of $\underline{n}$, whence two of them must be equal. Therefore, each of the n-element products contains a sequence

$S_p^r := a_{pp_1} \cdots a_{p_r p}$. From each of these $n[n]$ n-element products select exactly one such sequence S_p^r. Then there exists $r \in \underline{n}$ such that at least $[n]$ subproducts S_p^r of r elements are chosen. (Otherwise, for every $r \in \underline{n}$, the number of selected sequences S_p^r is smaller than $[n]$, and hence the total number of selected sequences S_p^r smaller than $n[n]$.)

Now select exactly $[n]/r$ sequences S_p^r and delete them from $a_{ii_1} a_{i_1 i_2} \dots a_{i_{m-1} j}$, yielding a term of $a_{ij}^{(m-[n])}$ (since clearly $r \cdot [n]/r = [n]$ elements are removed). Thus $a_{ii_1} a_{i_1 i_2} \dots a_{i_{m-1} j} \le a_{ij}^{(m-[n])}$. Since the last inequality holds for an arbitrary summand of $a_{ij}^{(m)}$, we have shown that $a_{ij}^{(m)} \le a_{ij}^{(m-[n])}$. □

Proposition 8.2. ([22, Theorem 3.1]) *If $A \in \mathbb{L}_n$ is right invertible then $A^{[n]} = I$.*

Proof. Since A is right invertible we have $AB = I$ for some $B \in \mathbb{L}_n$. Therefore, for any $m \ge 1$,

$$A^m B^m = \underbrace{A \cdots (A(A}_{m \text{ times}} \underbrace{B)B) \cdots B}_{m \text{ times}} = AB = I.$$

Let $m = n^2[n]$ and $d = [n]$; then $A^{m+d} \le A^m$ by Lemma 8.1. Hence

$$A^d = A^d \cdot I = A^d \cdot AB = A^d \cdot A^m B^m = A^{m+d} B^m \le A^m B^m = I.$$

For the other inequality note that $A^d B^d = I$, whence $\sum_{s \in \underline{n}} a_{is}^{(d)} b_{si}^{(d)} = 1$. Since $A^d \le I$, $a_{ij}^{(d)} = 0$, for $i \ne j$, and, for every $i \in \underline{n}$, $a_{ii}^{(d)} = \sum_{s \in \underline{n}} a_{is}^{(d)} \ge \sum_{s \in \underline{n}} a_{is}^{(d)} b_{si}^{(d)} = 1$, i.e. $a_{ii}^{(d)} = 1$. Thus, we get $A^d = I$. □

Proposition 8.3. *Let L be an incline and $A \in \mathbb{L}_n$ be diagonally dominant. Then*

(1) $a_{ii} = a_{jj}$ for all $i, j \in \underline{n}$.
(2) A is weakly diagonally dominant.

Proof. (1): By assumption, $a_{ii} a_{jk} = a_{jk}$ for all $i, j, k \in \underline{n}$. Thus $a_{jj} = a_{ii} a_{jj} \le a_{ii}$; similarly, $a_{ii} = a_{ii} a_{jj} \le a_{jj}$. So $a_{ii} = a_{jj}$ for all $i, j \in \underline{n}$.
(2): Recall Def. 7.5(c). Since $a_{ii} a_{jk} = a_{jk}$ for all $i, j, k \in \underline{n}$, we have, for any $i \in \underline{n}$, $a_{ii} a_{ij} = a_{ij}$ for all $j \in \underline{n}$. And since $A \to A \vee B$ by Th. 1.17(2), the assertion is proved. □

If we denote by $P_n(L), G_n(L)$ and $H_n(L)$ the sets of all permutation, invertible, and Hall matrices, respectively, then the following holds:

- $P_n(L)$ is a subgroup of $G_n(L)$ and is isomorphic to the symmetric group of degree n.
- $G_n(L)$ is a subgroup of $\mathbb{L}_n$.
- $H_n(L)$ is a subsemigroup of $\mathbb{L}_n$.
- $P_n(L) \subseteq G_n(L) \subseteq H_n(L) \subseteq \mathbb{L}_n$. These inclusions are proper in general (see [23, Example 2.1]).

Recall that $A \in \mathbb{L}_n$ is called a *Hall matrix* if $A \ge C \in G_n(L)$. Thus incline Hall matrices are a generalization of incline reflexive matrices.

Proposition 8.4. ([22, Theorem 5.1]) *A subsemigroup $H \subseteq G_n(L)$ is a subgroup of $G_n(L)$.*

Proof. Let $A, B \in H$. Then by Prop. 8.2 there is an $l \geq 2$ such that $A^l = I$ So $A^{-1} = A^{l-1} \in H$, whence $A^{-1}B \in H$, and this means that H is a subgroup of $G_n(L)$. □

Lemma 8.5. ([23, Lemma 2.3]) *Let $A \in \mathbb{I}_n$ and $B, C \in G_n(L)$. If $BAC \leq A$ or $BAC \geq A$, then $BAC = A$.*

Proof. By Prop. 8.2 there are a positive integers k, l such that $B^k = C^l = I$, whence we get $B^{kl} = C^{kl} = I$. So $BAC \leq A$ implies $A \geq BAC \geq \ldots \geq B^{kl}AC^{kl} = A$. Similar for $BAC \geq A$, □

We will now turn to some results on Hall matrices over inclines, following [23], because they are of some interest on their own, and because some of the proofs are based on pure algebraic results from (semi)group theory, without need to resort to the matrix entries involved. For this purpose we let $S_A := \{C \in G_n(L) \mid C \leq A\}$ for a matrix $A \in H_n(L)$. Let $A, B \in \mathbb{I}_n$. A is said to be **equivalent** to B if $B = PAQ$ for some $P, Q \in G_n(L)$. In particular, if $B = PAP^T$ for some $P \in G_n(L)$, then A is said to be **similar** to B.

Lemma 8.6. ([23, Lemma 2.10]) *If $A \in H_n(L)$, then A is equivalent to a reflexive matrix in $H_n(L)$.*

Proof. Assume that $A \geq P$ for some $P \in G_n(L)$. Then by Prop. 8.2 there is a positive integer $l \geq 2$ such that $P^l = I$. Thus $AP^{l-1} \geq P^l = I$, $P^{l-1} \in G_n(L)$ and clearly $I \in G_n(L)$. And since $AP^{l-1} = IAP^{l-1}$, A is equivalent to the reflexive matrix $I \leq AP^{l-1} \in H_n(L)$. □

Proposition 8.7. ([23, Lemma 3.1]) *If $A, G \in H_n(L)$ and $AGA = A$, then $APA = A$ and $P^T \in S_A$ for all $P \in S_G$.*

Proof. For $Q \in S_A$ and $P \in S_G$ we have $APQ \leq APA \leq AGA = A$. Thus, since $PQ \in G_n(L)$ by Prop. 6.26(3), $IAPQ \leq A$ implies $APQ = A$ by Lemma 8.5. Therefore, $APA = A$. If $Q_1, Q_2 \in S_A$, then $Q_1PQ_2 \leq APA = A$ and, again by Prop. 6.26, $Q_1PQ_2 \in G_n(L)$, so $Q_1PQ_2 \in S_A$. Thus, $(PQ_1)(PQ_2) \in P \cdot S_A = \{PQ \mid Q \in S_A\}$. Hence, $P \cdot S_A$ is a subsemigroup of $G_n(L)$, and a subgroup of $G_n(L)$ by Prop. 8.4. Therefore, $I \in P \cdot S_A$ and $PQ = I$ for some $Q \in S_A$. As shown later on in Prop. 8.18, for any invertible matrix $A \in \mathbb{I}_n$ we have $A^TA = AA^T = I$. So $P^{-1} = P^T = Q \in S_A$ since the inverse is unique for matrices in arbitrary semirings by Prop. 6.25(1). □

Proposition 8.8. ([23, Theorem 3.1]) *If $A \in H_n(L)$ then A is regular in $H_n(L)$ iff A is equivalent to an idempotent matrix in $\mathbb{I}_n$.*

Proof. Let $G \in H_n(L)$ be a generalized inverse of A; then $AGA = A$. By Prop. 8.7 we have $APA = A$ for all $P \in S_G = \{C \in G_n(L) \mid C \leq G\}$. Thus, $PAPA = PA$, so PA is idempotent, and $PAI = PA$ with $P, I \in G_n(L)$, whence A is equivalent to an idempotent matrix.

Conversely, assume that $P, Q \in G_n(L)$ such that $(PAQ)^2 = PAQ$. Since P and Q are invertible, $PAQPAQ = PAQ$ implies $AQPA = A$, and $QP \in G_n(L) \subseteq H_n(L)$, since the product of two invertible matrices is invertible by Prop. 6.26. Thus, A is regular in $H_n(L)$. □

8.1.2 *Powers, Periods, and Convergence*

Proposition 8.9. ([13, Lemma 1]) *Let $A \in \mathbb{I}_n$; then, for all $m \geq n$, $A^m \leq B := \sum_{k=0}^{n-1} A^k$, where $A^0 = I$.*

Proof. By Cor. 4.18(1) we have $a_{ii}^{(m)} \leq 1 = b_{ii}$. Note that from (7.1) we get

$$a_{ij}^{(m)} = \sum_{i_1,\ldots,i_{m-1}\in \underline{n}} a_{ii_1}\, a_{i_1 i_2} \cdots a_{i_{m-2}i_{m-1}}\, a_{i_{m-1}j}.$$

For the case $i \neq j$ we consider an arbitrary product in this sum. Since $m+1 > n$ and $\{i, i_1, i_2, \ldots, i_{m-1}, j\} \subseteq \underline{n}$, there are r, s such that $i_r = i_s$ and $0 \leq r < s \leq m$ (where $i_0 = i, i_m = j$). Deleting $a_{i_r i_{r+1}}\, a_{i_{r+1} i_{r+2}} \cdots a_{i_{s-1} i_s}$ from the summand, by Cor. 4.18(4) and (5) one obtains

$$a_{ii_1}\, a_{i_1 i_2} \cdots a_{i_{m-2}i_{m-1}}\, a_{i_{m-1}j} \leq a_{ii_1} \cdots a_{i_{r-1} i_r}\, a_{i_s i_{s+1}} \cdots a_{i_{m-1}j}.$$

As long as the number of subscripts on the right hand side of the resulting inequality is still $> n$, the same deletion method can be applied. Finally, we then find a $t \leq n-1$ such that $a_{ii_1}\, a_{i_1 i_2} \cdots a_{i_{m-2}i_{m-1}}\, a_{i_{m-1}j} \leq a_{il_1}\, a_{l_1 l_2} \cdots a_{l_{t-1}j}$. By Cor. 4.18(3) we get

$$a_{ii_1}\, a_{i_1 i_2} \cdots a_{i_{m-2}i_{m-1}}\, a_{i_{m-1}j} \leq \sum_{l_1,\ldots,l_{t-1}\in \underline{n}} a_{il_1}\, a_{l_1 l_2} \cdots a_{l_{t-1}j} = a_{ij}^{(t)} \leq \sum_{k\in \underline{n-1}} a_{ij}^{(k)}.$$

Thus, for the case $i \neq j$, for any summand s of $a_{ij}^{(m)}$ we have $s \leq \sum_{k=1}^{n-1} a_{ij}^{(k)}$. So, $a_{ij}^{(m)} \leq \sum_{k=1}^{n-1} a_{ij}^{(k)} = b_{ij}$, (since $A^0 = I$ does not contribute to the off-diagonal entries, and using Prop. 4.12(1) and additive idempotency). □

Corollary 8.10. ([13, Theorem 1]) *For $A \in \mathbb{I}_n$, $\mathbf{t}(A) = \sum_{k\in \underline{n}} A^k$.*

Proof. Follows from Prop. 7.3 and Prop. 8.9. □

We now turn to the generalization of Prop. 2.76 to incline matrices:

Proposition 8.11. ([21, Theorem 5.3], [29, Proposition 3.8]) *Let L be an incline and $A \in \mathbb{I}_n$ be weakly diagonally dominant. Then the following holds:*

(1) $A^2 \leq A^3$;
(2) a_{ii} is idempotent and $a_{ii} = a_{ii}^{(2)} = \cdots = a_{ii}^{(n)}$, for all $i \in \underline{n}$;
(3) A converges to $A^{k(A)}$ with $k(A) \leq n-1$.

Proof. (1): We have $a_{ij}^{(2)} = \sum_{k\in \underline{n}} a_{ik}a_{kj}$, for all $i, j, k \in \underline{n}$. Consider any term $a_{ik}a_{kj}$ of this sum: Assuming $a_{kl} = a_{kk}a_{kl}$, for all $l \in \underline{n}$, we get $a_{ik}a_{kj} = a_{ik}(a_{kk}a_{kj})$ with the latter one being a term in $a_{ij}^{(3)}$. So

$$a_{ij}^{(2)} = \sum_{k\in \underline{n}} a_{ik}(a_{kk}a_{kj}) \leq a_{ij}^{(3)}.$$

And if $a_{lk} = a_{lk}a_{kk}$, for all $l \in \underline{n}$, then again

$$a_{ij}^{(2)} = \sum_{k\in\underline{n}} (a_{ik}a_{kk})a_{kj} \leq a_{ij}^{(3)}.$$

So $a_{ij}^{(2)} \leq a_{ij}^{(3)}$ for arbitrary $i, j \in \underline{n}$, whence $A^2 \leq A^3$.

(2): Assume as first case that, for all $j \in \underline{n}$, $a_{ij} = a_{ii}a_{ij}$; then, taking $i = j$, $a_{ii} = a_{ii}a_{ii}$ follows, i.e. a_{ii} is an idempotent element of L. Thus, $a_{ii}^k = a_{ii}, k \geq 1$, and using Cor. 4.18 we get

$$a_{ii} = a_{ii}^s \leq a_{ii}^{(s)} = \sum_{i_1,\dots,i_{s-1}\in\underline{n}} a_{ii_1}a_{i_1i_2}\cdots a_{i_{s-1}i} \leq \sum_{k\in\underline{n}} a_{ik} = \sum_{k\in\underline{n}} a_{ii}a_{ik} \leq \sum_{k\in\underline{n}} a_{ii} = a_{ii}.$$

As an analogous result holds for the second case, for all $j \in \underline{n}$, $a_{ji} = a_{ji}a_{ii}$, $a_{ii} = a_{ii}^{(s)}$ for all $s \in \underline{n}$.

(3): From (1) we get $A^2 \leq A^3 \leq \cdots \leq A^{n-1} \leq A^n$. To prove equality we have to show $A^{n-1} \geq A^n$. Considering (2), it suffices to show $a_{ij}^{(n-1)} \geq a_{ij}^{(n)}$ for $i \neq j$.

We let $i = k_0$ and $j = k_n$ and begin with considering any term $T := a_{ik_1}a_{k_1k_2}\cdots a_{k_{n-1}j}$ of $a_{ij}^{(n)}$. T has $n+1$ subscripts, all of them in $\underline{n}$, so two of them must be equal, say $k_s = k_t$, $0 \leq s < t \leq n$. Then $T \leq a_{ik_1}a_{k_1k_2}\cdots a_{k_{s-1}k_s}a_{k_tk_{t+1}}\cdots a_{k_{n-1}j} =: T'$, the latter being a term in $a_{ij}^{(n-(t-s))}$. Since T was chosen arbitrarily, this gives $T \leq T' \leq a_{ij}^{(n-(t-s))}$, whence $a_{ij}^{(n)} \leq a_{ij}^{(n-(t-s))}$. Thus, for $m = n-(t-s) < n$, we have $A^n \leq A^m \leq A^{n-1}$, since from (1) $A^{k-1} \leq A^k$ for $k \geq 3$. So, finally, $A^{n-1} = A^n$ and $k(A) \leq n-1$. □

Corollary 8.12. *Let L be an incline and $A \in \mathbb{L}_n$.*

(1) If A is row (or column) diagonally dominant, then $k(A) \leq n-1$.
(2) If A is reflexive, then $k(A) \leq n-1$.
(3) If A is reflexive, then $I_n \subseteq A \subseteq A^2 \subseteq \ldots \subset A^{n-1} = A^n = A^{n+1} = \ldots$.

Proof. (1): By Prop. 8.11, since a row (or column) diagonally dominant matrix is weakly diagonally dominant.

(2): By (1), since a reflexive matrix is row (and column) diagonally dominant.

(3): Reformulation of (2). □

Note that Cor. 8.12(2),(3) are the general counterparts to Cor. 2.77(1) and Prop. 2.76 announced above.

8.1.3 *Determinants, Permanents and Adjoints*

Lemma 8.13. ([13, Theorem 5]) *Let L be a commutative 0-1-incline and $A \in \mathbb{L}_n$ be a reflexive matrix, i.e. $I_n \leq A$. Then $A^{n-1} = \mathsf{adj}(A)$.*

Proof. This is a special case of Th. 7.22. □

Proposition 8.14. ([23, Lemma 2.7]) *Let L be a 0-1-incline and $A \in \mathbb{L}_n$ be a reflexive matrix, i.e. $I_n \leq A$. Then the following holds:*

(1) $\mathsf{adj}(A)$ is idempotent.

(2) $A \cdot \mathsf{adj}(A) = \mathsf{adj}(A) \cdot A = \mathsf{adj}(A)$.
(3) $A^2 = A$ *implies* $\mathsf{adj}(A) = A$.
(4) $\mathsf{adj}(A)$ *is reflexive.*

Proof. (1): $[\mathsf{adj}(A)]^2 = [A^{n-1}]^2 = A^{2(n-1)} = A^{n-1} = \mathsf{adj}(A)$ by Lemma 8.13 and Cor. 8.12(3), since $2(n-1) \geq n$.

(2): Again using Lemma 8.13 and Cor. 8.12(3) we get $A \cdot \mathsf{adj}(A) = A \cdot A^{n-1} = A^n = A^{n-1} = \mathsf{adj}(A)$. Similarly, $\mathsf{adj}(A) \cdot A = \mathsf{adj}(A)$.

(3): By Lemma 8.13 $\mathsf{adj}(A) = A^{n-1}$, and since A is assumed idempotent, we get $A^{n-1} = A$.

(4): Take $B := \mathsf{adj}(A)$; then by Def. 6.6

$$b_{ii} = \mathsf{per}(A(i|i)) = \textstyle\sum_{\sigma \in S_{n-1}} a_{1\sigma(1)} \dots a_{(i-1)\sigma(i-1)}\, a_{(i+1)\sigma(i+1)} \dots a_{n\sigma(n)}.$$

So, taking the identity permutation $\sigma(i) = i$, we get

$$b_{ii} \geq a_{11} \dots a_{(i-1)(i-1)}\, a_{(i+1)(i+1)} \dots a_{nn} = 1.$$

□

Lemma 8.15. ([23, Lemma 2.8]) *For $A \in \mathbb{I}_n$ and $P, Q \in G_n(L)$ we have* $\mathsf{adj}(PAQ) = Q^T \mathsf{adj}(A) P^T$.

Lemma 8.16. ([23, Lemma 3.2]) *If $A \in H_n(L)$ then the following holds:*

(1) *If A is transitive then it is reflexive and* $\mathsf{adj}(A) = A$.
(2) $\mathsf{adj}(A) \in H_n(L)$ *and it is regular in $H_n(L)$.*

Proof. (1): If A is transitive then for any $P, Q \in S_A$ we get $PQ \leq AA \leq A$ and $PQ \in G_n(L)$. Hence $PQ \in S_A$. Thus, S_A is a subsemigroup of $G_n(L)$, and by Prop. 8.4 even a subgroup of $G_n(L)$, whence $I \in S_A$. Therefore, A is reflexive, hence $A \leq A^2$ by Cor. 8.12(3) and so $A = A^2$. Finally, $\mathsf{adj}(A) = A$ by Prop. 8.14(3).

(2): Let $P \in S_A$. Then $I = P^T P \leq P^T A$ using Prop. 8.18. So $P^T A$ is reflexive and $\mathsf{adj}(P^T A)$ is idempotent by Prop. 8.14(1). By Lemma 8.15 we get $\mathsf{adj}(P^T A) = \mathsf{adj}(P^T A I) = I\mathsf{adj}(A)P$, whence $\mathsf{adj}(A)$ is equivalent to the idempotent matrix $\mathsf{adj}(P^T A)$. Moreover, by Prop. 7.16(2) and Prop. 6.8(4) $\mathsf{adj}(P^T A) \geq \mathsf{adj}(I) = I$, whence $P^T \leq \mathsf{adj}(P^T A)P^T$. But we have just shown that $\mathsf{adj}(P^T A) = I\mathsf{adj}(A)P$, whence $\mathsf{adj}(P^T A)P^T = I\mathsf{adj}(A) = \mathsf{adj}(A)$, and so $\mathsf{adj}(A) \in H_n(L)$. Therefore, $\mathsf{adj}(A)$ is regular in $H_n(L)$ by Prop. 8.8. □

As the final result in this section we mention the following one, providing another characterization of regularity in addition to Prop. 8.8:

Proposition 8.17. ([23, Theorem 3.3]) *If $A \in H_n(L)$ then A is regular in $H_n(L)$ iff* $\mathsf{adj}(\mathsf{adj}(A)) = A$.

Proof. Let A is regular in $H_n(L)$, then there are $P, Q \in G_n(L)$ such that $D = PAQ$ is idempotent by Prop. 8.8. Thus, $D \in H_n(L)$ and $\mathsf{adj}(D) = D$ by Lemma 8.16(1), and $A = P^T D Q^T$. From Lemma 8.15 we get

$$\mathsf{adj}(\mathsf{adj}(A)) = \mathsf{adj}(Q\mathsf{adj}(D)P) = \mathsf{adj}(QDP) = P^T \mathsf{adj}(D) Q^T = P^T D Q^T = A.$$

Conversely, applying Lemma 8.16(2) first to A we get $\mathsf{adj}(A) \in H_n(L)$, and then to $\mathsf{adj}(A)$ gives $\mathsf{adj}(\mathsf{adj}(A)) \in H_n(L)$ and $\mathsf{adj}(\mathsf{adj}(A))$ is regular in $H_n(L)$. Since $\mathsf{adj}(\mathsf{adj}(A)) = A$, A is also regular in $H_n(L)$. □

8.1.4 *Invertibility*

Proposition 8.18. ([22, Theorem 3.2]) *If $A \in \mathbb{I}_n$ is an invertible matrix, then $A^T A = AA^T = I$.*

Proof. By assumption, $AB = BA = I$ for some $B \in \mathbb{I}_n$. From $AB = I$ we have $\sum_{s\in \underline{n}} a_{is}b_{sj} = 0$ for $i \neq j$; so, for all $s \in \underline{n}$, $a_{is}b_{sj} = 0$. And from $BA = I$ one gets

$$\textstyle\sum_{s\in \underline{n}} a_{si} \geq \sum_{s\in \underline{n}} b_{is}a_{si} = 1 \quad \text{and} \quad \sum_{s\in \underline{n}} b_{is} \geq \sum_{s\in \underline{n}} b_{is}a_{si} = 1,$$

for all $i \in \underline{n}$. Thus

$$\begin{aligned} a_{ij} &= a_{ij} \cdot 1 = a_{ij} \cdot \textstyle\sum_{s\in \underline{n}} b_{js} = \sum_{s\in \underline{n}} a_{ij}\, b_{js} = a_{ij}\, b_{ji} + \sum_{s\neq i} a_{ij}\, b_{js} \\ &= a_{ij}\, b_{ji} + 0 = a_{ij}\, b_{ji} + \textstyle\sum_{s\neq i} a_{sj}\, b_{ji} = \sum_{s\in \underline{n}} a_{sj}\, b_{ji} \\ &= \left(\textstyle\sum_{s\in \underline{n}} a_{sj}\right) \cdot b_{ji} = 1 \cdot b_{ji} \\ &= b_{ji}, \end{aligned}$$

for all $i, j \in \underline{n}$. Hence, $B = A^T$. □

To relate this proposition to some early classical results: It generalizes [20, Theorem 6] and [38, Theorem 4.2]; moreover, it is a special case of Cor. 7.30.

Theorem 8.19. ([22, Theorem 3.4]) *For $A \in \mathbb{I}_n$ the following statements are equivalent:*

(1) A is right invertible.
(2) A is left invertible.
(3) A is invertible.
(4) $AA^T = I$.
(5) $A^T A = I$.
(6) $AA^T = A^T A = I$.
(7) $A^l = I$ for some $l \geq 1$.

Proof. We have (1) → (7) by Prop. 8.2, and can show (2) → (7) in an analogous way. We can prove the implications (1) → (7) → (3) → (6) → (4) → (1) by application of Prop. 8.2 and Prop. 8.18; the second implication is clear from $A^l = AA^{l-1} = A^{l-1}A = I$. Similarly we get (2) → (7) → (3) → (6) → (5) → (2), so the asserted equivalences have been proved. □

There are also the following interrelationships between invertibility and the permanent and adjoint of an incline matrix $A \in \mathbb{I}_n$.

Proposition 8.20. ([22, Theorems 4.1, 4.2], [57, Lemma 2.10]) *Let $A \in \mathbb{I}_n$ be invertible. Then*

(1) $\mathsf{per}(A) = 1$.
(2) $\mathsf{adj}(A) = A^T$.

Proof. (1): By Th. 8.19(4), $AA^T = I$. Then for any $i \in \underline{n}$, $1 = \sum_{s\in\underline{n}} a_{is}a_{is} \leq \sum_{s\in\underline{n}} a_{is}$, i.e. $\sum_{s\in\underline{n}} a_{is} = 1$. And for $i \neq j$ and $t \in \underline{n}$ we have $a_{it}a_{jt} \leq \sum_{k\in\underline{n}} a_{ik}a_{jk} = 0$, i.e. $a_{it}a_{jt} = 0$. Thus

$$1 = \prod_{i\in\underline{n}} 1 = \prod_{i\in\underline{n}} \Big(\sum_{s\in\underline{n}} a_{is}\Big) = \sum_{\sigma\in S_n} a_{1\sigma(1)}\, a_{2\sigma(2)} \dots a_{n\sigma(n)} = \mathsf{per}(A),$$

where S_n is the set of all permutations of $\underline{n}$.

(2): By Th. 8.19(6), $AA^T = A^TA = I$. Then, as in (1) above, for any $i \in \underline{n}$ we have $\sum_{s\in\underline{n}} a_{is} = 1$. Since $\sum_{k\in\underline{n}} a_{ik}a_{jk} = 0$ for $i \neq j$ we have $a_{ij}a_{ik} = a_{ji}a_{ki} = 0$ for all $i, j, k \in \underline{n}$, $j \neq k$. From the definition of $\mathsf{per}(A)$ we get

$$\mathsf{per}(A_{i|j}) = \textstyle\sum_{\pi\in S_{i|j}} a_{1\pi(1)} \dots a_{i-1\pi(i-1)}\, a_{i+1\pi(i+1)} \dots a_{n\pi(n)},$$

where $S_{i|j}$ denotes the set of bijections from $\underline{n} - \{i\}$ to $\underline{n} - \{j\}$. For $t \neq j$, every term of $\mathsf{per}(A_{i|t})$ has a factor which is an entry of the jth column of A and is different from a_{ij}, say $a_{ij'}$, $j \neq j'$. Hence, every term in $a_{ij}\mathsf{per}(A_{i|t})$ contains a product $a_{ij}a_{ij'} = 0$, whence $a_{ij}\mathsf{per}(A_{i|t}) = 0$.

By (1), Prop. 6.8(4) and Cor. 4.18(2) we obtain

$$1 = \mathsf{per}(A) = \textstyle\sum_{j\in\underline{n}} a_{ij}\, \mathsf{per}(A_{i|j}) \leq \sum_{j\in\underline{n}} \mathsf{per}(A_{i|j}),$$

i.e. $\sum_{j\in\underline{n}} \mathsf{per}(A_{i|j}) = 1$. Thus,

$$\begin{aligned}
a_{ij} &= a_{ij}\cdot 1 = a_{ij}\cdot\big(\textstyle\sum_{t\in\underline{n}} \mathsf{per}(A_{i|t})\big)\\
&= a_{ij}\,\mathsf{per}(A_{i|j}) + \textstyle\sum_{t\in\underline{n}, t\neq j} a_{ij}\,\mathsf{per}(A_{i|t})\\
&= a_{ij}\,\mathsf{per}(A_{i|j})\\
&= a_{ij}\,\mathsf{per}(A_{i|j}) + \textstyle\sum_{t\in\underline{n}, t\neq j} a_{it}\,\mathsf{per}(A_{i|j})\\
&= \textstyle\sum_{t\in\underline{n}} a_{it}\,\mathsf{per}(A_{i|j})\\
&= \big(\textstyle\sum_{t\in\underline{n}} a_{it}\big)\cdot\mathsf{per}(A_{i|j})\\
&= 1\cdot\mathsf{per}(A_{i|j}).
\end{aligned}$$

Thus, $a_{ij} = \mathsf{per}(A_{i|j})$ for all $i, j \in \underline{n}$, whence $\mathsf{adj}(A) = \big[\mathsf{per}(A_{j|i})\big] = A^T$. □

The following theorem gives Cramer's rule for a matrix equation over inclines. Compare this to Prop. 6.41 and Th. 7.36, the analogous results for commutative semirings and commutative antirings.

Theorem 8.21. ([22, Theorem 4.3]) *Let L be an incline, $A \in \mathbb{M}_n(L)$, and $\boldsymbol{b} = (b_1, b_2, \dots, b_n)^T \in V_n(L)$. If A is invertible in $\mathbb{M}_n(L)$, then the matrix equation $A\boldsymbol{x} = \boldsymbol{b}$ has a unique solution $\boldsymbol{x} = (d_1, d_2, \dots, d_n)^T \in V_n(L)$, where $d_j = \mathsf{per}(A^{\boldsymbol{b}\Rightarrow j})$, $j \in \underline{n}$.*

Proof. Using Th. 8.19(6) we get $A(A^T\boldsymbol{b}) = I_n\boldsymbol{b} = \boldsymbol{b}$, so the equation $A\boldsymbol{x} = \boldsymbol{b}$ has the solution $A^T\boldsymbol{b}$. Let $\boldsymbol{y}$ be any other solution; then $A\boldsymbol{y} = \boldsymbol{b}$ implies $\boldsymbol{y} = A^T\boldsymbol{b}$, which means that $A\boldsymbol{x} = \boldsymbol{b}$ has a unique solution.

Let $(d_1, d_2, \dots, d_n)^T := A^T\boldsymbol{b}$; then by Prop. 8.20(2) $d_j = \sum_{i\in\underline{n}} a_{ij}b_i = \sum_{i\in\underline{n}} \mathsf{per}(A_{i|j})b_i$, for any $j \in \underline{n}$. On the other hand, by Prop. 6.8(4), we have $\mathsf{per}(A^{\boldsymbol{b}\Rightarrow j}) = \sum_{i\in\underline{n}} b_i\mathsf{per}(A_{i|j})$. □

8.1.5 *Eigenvalues*

In addition to the results on standard eigenvalues of matrices over a commutative path algebra reported above in Section 7.6 some more facts are known about the greatest standard eigenvalue in case inclines are used as a basis. Recall that by $C(A)$ we denote the set of all possible linear combinations of the columns $\boldsymbol{a}_{*1}, \boldsymbol{a}_{*2}, \ldots, \boldsymbol{a}_{*n}$ of A, and by $E(A)$ the set of its standard eigenvectors. Then clearly $\sum_{j\in \underline{n}} \boldsymbol{a}_{*j}$ is the greatest element of $C(A)$, denoted by $\tau(A)$.

Lemma 8.22. ([25, Lemma 2.2]) *Let L be an incline and $A, B \in \mathbb{M}_n(L)$; then the following holds:*

(1) $\tau(AB) = A\cdot\tau(B)$.
(2) $\tau(AB) \le \tau(A)$.
(3) $\tau(A+B) = \tau(A)+\tau(B)$.

Proof. (Maybe you want to have a look at Section 3.2.)
(1), (2): $\tau(AB) = \sum_{j\in\underline{n}}(AB)_{*j} = \sum_{j\in\underline{n}} A\cdot \boldsymbol{b}_{*j} = A\cdot\sum_{j\in\underline{n}} \boldsymbol{b}_{*j} = A\cdot\tau(B) \le \sum_{j\in\underline{n}} \boldsymbol{a}_{*j} = \tau(A)$.
(3): $\tau(A+B) = \sum_{j\in\underline{n}}(A+B)_{*j} = \sum_{j\in\underline{n}}(A_{*j}+B_{*j}) = \sum_{j\in\underline{n}} A_{*j} + \sum_{j\in\underline{n}} B_{*j} = \tau(A)+\tau(B)$. □

Lemma 8.23. ([25, Lemma 3.1]) *Let L be an incline and $A \in \mathbb{M}_n(L)$ be almost periodic with index i and period p. Then $\tau(A^i) = \tau(A^{i+m})$ for all positive integers m.*

Proof. From Th. 1.59(2)(a) we know that $A^i = A^{i+mp}$, for each $m \ge 1$, and from Lemma 8.22(2) we get $\tau(A^i) \ge \tau(A^{i+m}) \ge \tau(A^{i+mp}) = \tau(A^i)$. □

Proposition 8.24. ([25, Theorem 3.1]) *Let L be an incline and $A \in \mathbb{M}_n(L)$ be almost periodic with index index i and period p. Then, with $A^{s,t} := A^s + A^{s+1} + \cdots + A^{s+t-1}$, the following holds:*

(1) $E(A) = C(A^{i,p})$.
(2) $\tau(A^i)$ *is the greatest standard eigenvector of A.*

Proof. (1): Since $A^i = A^{i+p}$, we get

$$\begin{aligned} A\cdot A^{i,p} &= A\cdot(A^i + A^{i+1} + \cdots + A^{i+p-1}) \\ &= A^{i+1} + A^{i+2} + \cdots + A^{i+p-1} + A^{i+p} \\ &= A^{i+1} + A^{i+2} + \cdots + A^{i+p-1} + A^{i} \\ &= A^{i,p}. \end{aligned}$$

Thus, all the column vectors of $A^{i,p}$ belong to $E(A)$. By Lemma 7.46(1) and (3) we have $C(A^{i,p}) \subseteq E(A)$, and $E(A) \subseteq C(A^{i,p})$, whence equality holds.

(2): Using Lemma 8.22(3) and Lemma 8.23 we have

$$\begin{aligned} \tau(A^{i,p}) &= \tau(A^i + A^{i+1} + \cdots + A^{i+p-1}) \\ &= \tau(A^i) + \tau(A^{i+1}) + \cdots + \tau(A^{i+p-1}) \\ &= \tau(A^i) + \tau(A^i) + \cdots + \tau(A^i) \end{aligned}$$

$$= \tau(A^i),$$

i.e. $\tau(A^{i,p}) = \tau(A^i)$. Since $\tau(A^{i,p})$ is the greatest element of $C(A^{i,p})$, $\tau(A^i)$ is the greatest element in $C(A^{i,p})$. Finally, by (1), $\tau(A^i)$ is the greatest element in $E(A)$. □

8.1.6 *Nilpotent Incline Matrices*

In addition to the ones presented in Section 7.7, some more results on nilpotent matrices have been reported in [26], which forms the basis of this section. Let L be an incline and $A \in \mathbb{I}_n(L)$ be almost periodic with index i and period p. If A is nilpotent, then it converges in finite steps and its index is called the **nilpotency index** of A: $h(A) = \min\{k \in \mathbb{N} \mid A^k = O\}$.

Proposition 8.25. ([26, Theorem 3.1]) *$A \in \mathbb{I}_n(L)$ is nilpotent iff A is almost periodic and the zero vector $\mathbf{0}$ is its only standard eigenvector.*

Proof. If A is nilpotent then $A^{h(A)} = O$. Let $\boldsymbol{x} \in L^n$ be a standard eigenvector of A. Then $x = A\boldsymbol{x} = A^2\boldsymbol{x} = \cdots = A^{h(A)}\boldsymbol{x} = O\boldsymbol{x} = \mathbf{0}$. Conversely, assume $\mathbf{0}$ to be the only standard eigenvector of A. Then, by Prop. 8.24(2) we get $\tau(A^{h(A)}) = \mathbf{0}$, and since $\tau(A^{h(A)})$ is the sum of all column vectors of $A^{h(A)}$, each of its columns equals $\mathbf{0}$. □

Proposition 8.26. ([26, Theorem 3.2]) *Let $A \in \mathbb{I}_n(L)$; if all main diagonal entries of A^k are* 0 *for any $k \in \underline{n}$, then $A^n = O$.*

Proof. Assume $A^n \neq O$. Then there exist $i_0, i_n \in \underline{n}$ such that $a^{(n)}_{i_0 i_n} > 0$, whence $a_{i_0 i_1} a_{i_1 i_2} \cdots a_{i_{n-1} i_n} > 0$ for some $i_1, i_2, \ldots, i_{n-1} \in \underline{n}$. But since $\{i_0, i_1, i_2, \ldots, i_n\} \subseteq \underline{n}$, there exist r, s such that $0 \le r < s \le n$ and $i_r = i_s$. So $s - r \in \underline{n}$ and $a_{i_s i_{r+1}} \cdots a_{i_{s-1} i_s} \ge a_{i_0 i_1} a_{i_1 i_2} \cdots a_{i_{n-1} i_n} > 0$. Thus, $a^{(s-r)}_{i_s i_s} > 0$, a contradiction to the assumption about diagonal elements. □

Proposition 8.27. ([26, Corollary 3.1]) *If $A \in \mathbb{I}_n(L)$ is irreflexive and transitive, then A is nilpotent.*

Proof. $A^2 \le A$ implies $A^k \le A$ for each $k \in \underline{n}$. Thus, from irreflexivity we get $a^{(k)}_{ii} \le a_{ii} = 0$ for any $i, k \in \underline{n}$. So A is nilpotent by Prop. 8.26. □

Proposition 8.28. ([13, Theorem 3(i)], [26, Theorem 3.3]) *If all principal minors of $A \in \mathbb{I}_n(L)$ are* 0, *then $A^n = O$.*

Proof. We consider any term $T = a_{i i_1} a_{i_1 i_2} \cdots a_{i_{n-1} j}$ of $a^{(n)}_{ij}$ for some $i, j \in \underline{n}$. Let $i_0 = i$ and $i_n = j$; since $\{i_0, i_1, i_2, \ldots, i_n\} \subseteq \underline{n}$, there exist r, s such that $0 \le r < s \le n$ and $i_r = i_s$. So $K := \{(p,q) | p < q \;\&\; i_p = i_q\} \neq \emptyset$, and there exists a pair $(k, l) \in K$ with $l - k = \min\{q - p | (p,q) \in K\}$. By Cor. 4.18(4),(5) we have $T \le a_{i_k i_{k+1}} \cdots a_{i_{l-1} i_l}$. The right hand side of this inequalilty is a summand of the principal minor $\mathrm{per}(A[\alpha|\alpha])$, where $\alpha = (i_k, i_{k+1}, \ldots, i_{l-1})$. By assumption, $\mathrm{per}(A[\alpha|\alpha]) = 0$, whence also $a_{i_k i_{k+1}} \cdots a_{i_{l-1} i_l} = 0$ by Cor. 4.18(8). Thus, $T \le a_{i_k i_{k+1}} \cdots a_{i_{l-1} i_l} = 0$ and therefore $a^{(n)}_{ij} = 0$ □

Lemma 8.29. ([26, Lemma 3.2]) *If L has no nilpotent elements and $A \in \mathbb{I}_n(L)$ is nilpotent, then $a_{i_1 i_2} a_{i_2 i_3} \cdots a_{i_{m-1} i_m} a_{i_m i_1} = 0$ for any positive integer m and any $i_1, i_2, \ldots, i_m$.*

Proof. From $A^k = O$ for some $k \in \mathbb{N}$ we get $A^{mk} = (A^k)^m = O$ for any $m \in \mathbb{N}$. Therefore,

$(a_{i_1 i_2} a_{i_2 i_3} \cdots a_{i_{m-1} i_m} a_{i_m i_1})^k \leq a^{(mk)}_{i_1 i_1} = \sum_{i_1, i_2, \ldots, i_{mk-1} \in \underline{n}} a_{i i_1} a_{i_1 i_2} \ldots a_{i_{mk-2} i_{mk-1}} a_{i_{mk-1} i_1} = 0,$

for any $i_1, i_2, \ldots, i_m \in \underline{n}$. Since L is assumed to have no nilpotent elements, we get $a_{i_1 i_2} a_{i_2 i_3} \cdots a_{i_{m-1} i_m} a_{i_m i_1} = 0$. □

Proposition 8.30. ([26, Theorem 3.4]) *If L has no nilpotent elements and $A \in \mathbb{I}_n(L)$, then the following statements are equivalent:*

(1) *A is nilpotent.*
(2) *All principal minors of A are nilpotent.*
(3) *$A^n = O$.*
(4) *All main diagonal entries of A^k are 0, for any $k \in \underline{n}$.*

Proof. (1) → (2): Due to Cor. 4.18(8),(9), inclines are commutative path algebras. Apply Prop. 7.59(1),(3).

(2) → (3): By Prop. 8.28.

(3) → (4): By Lemma 8.29.

(4) → (1): By Prop. 8.26. □

8.2 Lattice Matrices

(Non-numerical) Matrix theory over lattice structures more general than the two-valued Boolean algebra $\{0 < 1\}$ has been widely investigated during the last few decades because of its purely algebraic interest and its importance in numerous applications. Research on this kind of matrices started in the early sixties with [20]: In this paper the author investigates "the particular properties of the algebra of square matrices over an arbitrary distributive lattice with 0 and 1". So, from a formal point of view, "*generalized fuzzy matrices*" have been tackled a few years earlier than the notion of fuzzy sets and fuzzy relations have been proposed as a new semantical paradigm in Zadeh's seminal papers [62, 63].

The two kinds of lattices over which matrix properties have been intensively studied are, as just indicated, bounded distributive lattices on the one hand, and residuated lattices on the other. From Chapter 4 we know about the following hierarchical relationships: bounded distributive lattices ⊂ 0-1-inclines ⊂ path algebras ⊂ antirings, dioids ⊂ semirings. So many facts valid for matrices over distributive lattices are covered by those for more general types of matrices.

Some of them will be mentioned in a kind of retrospect by taking a look at both early and later investigations on matrices beyond classical Boolean matrices. We will provide the original references, but the proofs for the results presented below are given in a derived form as consequences of more general facts that have been found.

Contrary to the case of distributive lattices, residuated lattices have only recently been considered as an underlying structure for matrices. First of all we have to stress the change of context for matrix multiplication: Let $(L, \vee, \wedge, \odot, \curvearrowright, 0, 1)$ be a residuated lattice and $A \in \mathbb{L}_{m\times p}$ and $B \in \mathbb{L}_{p\times n}$ two matrices over L; their composition is defined as

$$C := A \odot B \in \mathbb{L}_{m\times n} \quad \text{with} \quad c_{ij} = \bigvee_{k\in \underline{p}} a_{ik} \odot b_{kj},\ i \in \underline{m}, j \in \underline{n}. \tag{8.1}$$

Recall that by Th. 4.35, if $(L, \vee, \wedge, \odot, \rightarrow, 0, 1)$ is a residuated lattice, then $(L, \vee, \odot, 0, 1)$ is a commutative, additively idempotent, and simple semiring (so it is a path algebra and thus also an antiring). That means, multiplication of matrices over a residuated lattice does in fact take place in a special kind of semiring, whence again a lot of results already obtained can be applied.

8.2.1 *Invertibility*

Theorem 8.31. ([20, THEOREM 6]) *If $A \in \mathbb{L}_n$ is an invertible matrix over a bounded distributive lattice, then it is orthogonal.*

Proof. Since a bounded distributive lattice L is a commutative antiring $(L, \vee, \wedge, 0, 1)$ with $U(L) = \{1\}$, by Cor. 7.30 we get $AA^T = A^T A = I_n$. And since inverses in semigroups are unique by Prop. 1.29, $A^T = A^{-1}$, so A is orthogonal due to Def. 3.14(d). □

Theorem 8.32. ([20, THEOREM 7]) *If $A, B \in \mathbb{L}_n$ are such that $AB = I$ then A and B are roots of I (this means that there are k, l with $A^k = B^l = I$).*

Proof. By assumption, A is right invertible, so by Prop. 8.2 it is a root of I. From Prop. 6.26(1) we get $BA = I$, i.e. B is right invertible too, and thus also a root of I. □

The following notions have been introduced in [53, p. 290] (we follow their formulation in [40, p. 3969]): Let L be a bounded distributive lattice and $A \in \mathbb{L}_n$. A is said to be a **row orthogonal matrix** over L if

$$a_{ik}a_{jk} = 0 \text{ for any } i, j, k \in \underline{n}, \text{ with } i \neq j \quad \text{and} \quad \textstyle\bigvee_{j\in\underline{n}} a_{ij} = 1, i \in \underline{n},$$

i.e. the product of any two elements in the same column is 0, and the supremum of each row is 1. A is called a **column orthogonal matrix**

$$a_{ki}a_{kj} = 0 \text{ for any } i, j, k \in \underline{n}, \text{ with } i \neq j \quad \text{and} \quad \textstyle\bigvee_{i\in\underline{n}} a_{ij} = 1, j \in \underline{n}.$$

Theorem 8.33. ([53, Theorem 1]) *Let L be a bounded distributive lattice and $A \in \mathbb{L}_n$. Then the following properties of A are equivalent:*

(1) *A is right invertible.*
(2) *A is left invertible.*
(3) *A is invertible.*
(4) *A is row orthogonal.*

(5) *A is column orthogonal.*
(6) *A is row and column orthogonal.*
(7) $A^k = I$ *for some* $k \geq 1$.

Proof. (1)-(3): Because a bounded distributive lattice is a commutative semiring, the equivalence of the first three assertions follows from Cor. 6.27(1) and Def. 6.24.

Now assume that (3) holds. Since L is a commutative and idempotent semiring the equivalence of (3)-(6) follows by Th. 7.27 and Cor. 7.28.

Concerning (7), from Th. 8.32 we get that a right invertible matrix is a root of I. And from $A^k = I$ we obviously get hat $A^{-1} = A^{k-1}$, i.e. A is invertible. □

Contrary to the case of distributive lattices, invertibility of matrices over residuated lattices has been investigated only recently, at least in an explicit way. To the author's knowledge, [1] is the only paper on this topic which, therefore, serves as our source of information, and to which we refer the reader interested in more than the most basic properties. Remember Th. 4.35 and (8.1).

Proposition 8.34. ([1]) *Let L be a residuated lattice and $A, B \in \mathbb{L}_n$. Then*

(1) $(\mathbb{L}_n, \odot, I_n)$ *is a monoid.*
(2) *If A has an inverse then it is unique.*
(3) *If A has an inverse then* $(A^{-1})^T = (A^T)^{-1}$.
(4) *If A, B are invertible, then also $A \odot B$, and* $(A \odot B)^{-1} = B^{-1} \odot A^{-1}$.

Proof. By Prop. 6.25 and Prop. 6.26. □

8.2.2 *Generalized Inverses*

Note that residuated lattices need not be distributive, and distributive lattices need not be residuated. Accordingly, results on generalized inverses for both kinds have been reported. Some of them we shall present below and relate them to the ones given earlier in Section 6.6.

8.2.2.1 *Distributive Lattices*

Anticipating some results from Section 12.2, and adapting them to the matrix context, we are able to show the following characterization of regular matrices in complete BROUWERian lattices (in the terminology of Sanchez [50]), or HEYTING lattice, in our terminology (see Def. 4.20), i.e. complete residuated lattices $(L, \wedge, \vee, \otimes, \curvearrowright, 0, 1)$ with $\otimes = \wedge$. First of all, take a look at Def. 12.1 and Th. 12.3. Furthermore, recall that any relatively pseudocomplemented lattice is distributive by Prop. 4.21(1), so we are in the right place.

Theorem 8.35. ([64, Theorem 2]) *Let L be a complete bounded Brouwerian lattice and $A \in \mathbb{L}_{m\times n}$, and let $\hat{Y} := A^T @ A$, $X' := (A @ \hat{Y}^T)^T$. Then A is regular iff X' is a g-inverse of A. In case this holds, X' is the greatest g-inverse of A.*

Proof. To show the non-trivial direction assume that A is regular and X is a g-inverse of A. Then clearly there is a Y such that $AY = A$ and $XA = Y$. By Th. 12.3(1), $\hat{Y}$ is the greatest solution of the matrix equation $AY = A$, whence $Y \le \hat{Y}$ and thus $Y^T \le \hat{Y}^T$. Similarly, by (2) of this theorem, $\hat{X} := (A @ Y^T)^T$ is the greatest solution of $XA = Y$; so using Lemma 12.4 we get

$$X \le (A @ Y^T)^T \le (A @ \hat{Y}^T)^T = X'. \tag{8.2}$$

It remains to show that X' is a g-inverse of A: Let $\hat{Y} = [\hat{y}_{ij}]$, and $X'A = [(X'A)_{ij}]$; then we get for $i, j \in \underline{n}$

$$(X'A)_{ij} = \bigvee_{k \in \underline{m}} \big((\bigwedge_{l \in \underline{n}} (a_{kl} \curvearrowright \hat{y}_{il})) \wedge a_{kj} \big) \le \bigvee_{k \in \underline{m}} \big(a_{kj} \wedge (a_{kj} \curvearrowright \hat{y}_{ij}) \big) \le \bigvee_{k \in \underline{m}} \hat{y}_{ij} = \hat{y}_{ij},$$

since the definition of the relative pseudocomplement implies $a \wedge a \curvearrowright b \le b$, which proves the second inequality. Thus $X'A \le \hat{Y}$. X is assumed to be a g-inverse of A, so by (8.2) $A = AXA \le AX'A \le A\hat{Y} = A$. Hence X' is a g-inverse of A, and again by (8.2) it is the greatest one. □

8.2.2.2 *Residuated Lattices*

The contents of this section are based on the results given in [12] (but see also [28]). Before plunging into medias res we have to give some preliminary remarks:

(1) It has already been mentioned above in the context of (8.1) that matrix multiplication over residuated lattices $(L, \vee, \wedge, \odot, \curvearrowright, 0, 1)$ in fact takes place in a special kind of semirings $(L, \vee, \odot, 0, 1)$ which are commutative, idempotent and simple, whence commutative path algebras, and thus also antirings (see Section 4.1 and Th. 4.35). For the considerations in this section we need the notion of a residual of two matrices to be introduced below.
(2) Take a look at the beginning of Section 2.8.3 where the notions of *right residual* σ / ρ and *left residual* $\sigma \backslash \rho$ of a relation σ by ρ and some of their properties have been introduced. Among the latter are Prop. 2.46(3) and (4):

$$x(\sigma/\rho)y \leftrightarrow \rho x \subseteq \sigma y, \text{ i.e. } \forall z[z\rho x \rightarrow z\sigma y],$$

$$x(\sigma\backslash\rho)y \leftrightarrow x\rho \subseteq y\sigma, \text{ i.e. } \forall z[x\rho z \rightarrow y\sigma z].$$

(3) In connection to Def. 4.26 it has been stated that the operations $\odot$ and $\curvearrowright$ are intended for modeling conjunction and implication, respectively.

In view of the above it can easily be seen that the following definition is a straightforward generalization of the Boolean case. Note that, as it is often the case, notation of residuals differs in literature (for some examples see [59, Table 3.1]): The authors of [12, 28] write $A\backslash C$ for the right residual of C by A and C/B for the left residual of C by B. So going to the original papers will, unfortunately, cause some translation tasks. But recall the rationale for our choice of notation in the second paragraph of Section 2.8.3.

Definition 8.36. *Let L be a complete residuated lattice and $A \in \mathbb{L}_{m\times n}$, $B \in \mathbb{L}_{n\times p}$, and $C \in \mathbb{L}_{m\times p}$.*

(a) *The* **right residual** *of* C *by* A *is* $C/A \in \mathbb{L}_{n\times p}$ *defined as* $(C/A)_{ij} = \bigwedge_{s\in \underline{m}} a_{si} \curvearrowright c_{sj}$.

(b) *The* **left residual** *of* C *by* B *is* $C\backslash B \in \mathbb{L}_{m\times n}$ *defined as* $(C\backslash B)_{ij} = \bigwedge_{s\in \underline{p}} b_{js} \curvearrowright c_{is}$.

Note that the following proposition is the counterpart to the residuation properties given in Prop. 2.46(7) for binary Boolean relations.

Proposition 8.37. *Let* $(L, \vee, \wedge, \odot, \curvearrowright, 0, 1)$ *be a complete residuated lattice and* $A \in \mathbb{L}_{m\times p}$ *and* $B \in \mathbb{L}_{p\times n}$ *two matrices over* L*. Then* $AB \le C \;\leftrightarrow\; A \le C\backslash B \;\leftrightarrow\; B \le C/A$.

Proof. We have $AB \le C \leftrightarrow \forall i, j\big[\bigvee_{k\in \underline{p}}(a_{ik} \odot b_{kj}) \le c_{ij}\big] \leftrightarrow \forall i, j, k\big[a_{ik} \odot b_{kj} \le c_{ij}\big] \leftrightarrow \forall i, j, k\big[a_{ik} \le b_{kj} \curvearrowright c_{ij}\big] \leftrightarrow \forall i, k\big[a_{ik} \le \bigwedge_j (b_{kj} \curvearrowright c_{ij})\big] \leftrightarrow A \le C\backslash B$. Using the commutativity of $\odot$, a similar argument gives the second equivalence. □

Proposition 8.38. *Let* $(L, \vee, \wedge, \odot, \curvearrowright, 0, 1)$ *be a complete residuated lattice and* $A \in \mathbb{L}_{m\times p}$ *and* $B \in \mathbb{L}_{p\times n}$ *two matrices over* L*. Then* $(C/A)\backslash B = (C\backslash B)/A$.

Proof.

$$
\begin{aligned}
\big((C\backslash B)/A\big)_{ij} &= \textstyle\bigwedge_s a_{si} \curvearrowright (C\backslash B)_{sj} \\
&= \textstyle\bigwedge_s (a_{si} \curvearrowright \bigwedge_k (b_{jk} \curvearrowright c_{sk})) \\
&= \textstyle\bigwedge_s \bigwedge_k (a_{si} \curvearrowright (b_{jk} \curvearrowright c_{sk})) && \text{by Prop. 4.30(3)} \\
&= \textstyle\bigwedge_s \bigwedge_k ((a_{si} \odot b_{jk}) \curvearrowright c_{sk}) && \text{by Prop. 4.27(8)} \\
&= \textstyle\bigwedge_s \bigwedge_k ((b_{jk} \odot a_{si}) \curvearrowright c_{sk}) && \text{since } \odot \text{ commutative} \\
&= \textstyle\bigwedge_s \bigwedge_k (b_{jk} \curvearrowright (a_{si} \curvearrowright c_{sk})) && \text{by Prop. 4.27(8)} \\
&= \textstyle\bigwedge_k (b_{jk} \curvearrowright \bigwedge_s (a_{si} \curvearrowright c_{sk})) && \text{by Prop. 4.30(3)} \\
&= \textstyle\bigwedge_k b_{jk} \curvearrowright (C/A)_{ik} \\
&= \big((C/A)\backslash B\big)_{ij}.
\end{aligned}
$$

□

(That this result shows a kind of associativity becomes apparent if one uses the definition of the right residual given at the bottom of [12, p. 21] as $(A/C)_{ij} = \bigwedge_s a_{si} \curvearrowright c_{sj}$. Then the equality in the above proposition becomes $(A/C)\backslash B = A/(C\backslash B)$.)

Before we carry on with further results, for sake of a self-contained presentation, it seems advisable to insert a short remark on fixed points in partial orders (see e.g. [48, Chapter 12]): Let P be a poset, $a \in P$, and $f : P \longrightarrow P$ be isotone. Then a is called a **pre-**, a **post-** or a **fixed point** if $f(a) \le a$, $f(a) \ge a$, or $f(a) = a$. The corresponding sets of fixpoints are denoted as $\mathrm{Pre}(f)$, $\mathrm{Post}(f)$, and $\mathrm{Fix}(f)$ respectively.

Theorem 8.39. (Knaster-Tarski) *Let* L *be a complete lattice and* f *be an isotone function on* L*; then the following holds:*

(1) *The least fixed and least pre-fixed point of* f *exist and are identical to* $\bigwedge \mathrm{Pre}(f)$.

(2) *The greatest fixed and greatest post-fixed point of* f *exist and are identical to* $\bigvee \mathrm{Post}(f)$.

(3) *The fixed points of* f *form a complete lattice.*

Moreover, the following holds for matrices over a complete residuated lattice: Two matrices $A, B \in \mathbb{L}_{m\times n}$, with $m, n \in \mathbb{N}$, can be ordered coordinatewise, i.e. $A \le B$ iff $a_{ij} \le b_{ij}$ for all $i \in \underline{m}$ and $j \in \underline{n}$. Endowed with this ordering, $\mathbb{L}_{m\times n}$ forms a complete lattice with meet and join defined as

$$\left[\bigwedge_k A_k\right]_{ij} = \bigwedge_k([A_k]_{ij}) \quad \text{and} \quad \left[\bigvee_k A_k\right]_{ij} = \bigvee_k([A_k]_{ij}).$$

For what follows below you may want to recall Def. 6.57.

Proposition 8.40. ([12, Theorem 1]) *Let L be a complete residuated lattice and A be an arbitrary matrix over L. Then the following holds:*

(1) A has the greatest $\{2\}$-inverse.
(2) A has the greatest γ-inverse, for each $\gamma \subseteq \{3,4,5\}$.

Proof. (1): Let $A \in \mathbb{L}_{m\times n}$ for some $m,n \in \mathbb{N}$, and let $\phi : \mathbb{L}_{n\times m} \longrightarrow \mathbb{L}_{n\times m}$ be defined as $\phi(X) = XAX$, for $X \in \mathbb{L}_{n\times m}$. Then ϕ is isotone and the set of $\{2\}$-inverses of A is the set of fixpoints of ϕ. $\mathbb{L}_{n\times m}$ is a complete lattice, so by Th. 8.39 there is the greatest fixpoint of ϕ.

(2): As an example the assertion for $\gamma = \{3,4,5\}$ will be proved. First note that due to equation (5): $AX = XA$ the matrices must be square, say $A, X \in \mathbb{L}_n$.

Since $(AX)^T \geq AX$ implies $((AX)^T)^T = AX \geq (AX)^T$ it follows that $(AX)^T = AX$ iff $(AX)^T \geq AX$. But by Prop. 8.37 the latter inequality is equivalent to $X \leq (AX)^T/A$. Similar arguments show that $(XA)^T = XA$ is equivalent to $X \leq (XA)^T \backslash A$. Moreover, equation (5) holds iff $AX \leq XA$ and $XA \leq AX$, which is equivalent to $X \leq (XA)/A$ and $X \leq (AX)\backslash A$.

So equations (3)-(5) are equivalent to the system of the four inequalities just given:

$$X \leq (AX)^T/A, \quad X \leq (XA)^T\backslash A, \quad X \leq (XA)/A, \quad X \leq (AX)\backslash A.$$

which in turn are equivalent to the single inequality

$$X \leq (AX)^T/A \ \wedge\ (XA)^T\backslash A \ \wedge\ (XA)/A \ \wedge\ X \leq (AX)\backslash A.$$

Defining $\phi : \mathbb{L}_n \longrightarrow \mathbb{L}_n$ by $\phi(X) = (AX)^T/A \ \wedge\ (XA)^T\backslash A \ \wedge\ (XA)/A \ \wedge\ (AX)\backslash A$ gives an isotone mapping, and the set of all $\{3,4,5\}$-inverses of A is the set of all post-fixed points of ϕ, and by the Knaster-Tarski theorem Th. 8.39 we obtain that there exists the greatest post-fixed point of ϕ. □

Proposition 8.41. ([12, Theorem 2]) *A fuzzy matrix $A \in \mathbb{L}_{m\times n}$ has a g-inverse iff $A^\diamond := (A/A)\backslash A \in A\{1\}$. If this is the case, then $A^\diamond$ is the greatest g-inverse and $A^\diamond AA^\diamond$ is the greatest Thierrin-Vagner inverse of A.*

Proof. To show the non-trivial direction, assume that A has a g-inverse, say B; so B is a solution to the inequality $AXA \leq A$. Using Prop. 8.37 we then get $ABA \leq A \leftrightarrow BA \leq A/A \leftrightarrow B \leq (A/A)\backslash A = A^\diamond$. Since B was arbitrary, $A^\diamond$ is the greatest solution for $AXA \leq A$. Moreover, $A = ABA \leq AA^\diamond A \leq A$, whence $A^\diamond$ is the greatest g-inverse of A.

Let $A' = A^\diamond AA^\diamond$; then $A'AA' = A'$ and $AA'A = A$ can also be shown by easy computation: we have $(A^\diamond AA^\diamond)A(A^\diamond AA^\diamond) = A^\diamond(AA^\diamond A)(A^\diamond AA^\diamond) = A^\diamond A(A^\diamond AA^\diamond) = A'$ since $A^\diamond$ is a g-inverse of A. Let B be an arbitrary $\{1,2\}$-inverse of A, then it is a g-inverse of A and thus $B \leq A^\diamond$. Therefore, $B = BAB \leq A^\diamond AA^\diamond$. □

The corresponding result for Boolean relations has been shown in Section 2.11: Recall from Def. 2.45 that $\rho/\rho := \overline{\rho^{\smile}\,\overline{\rho}}$ and $\rho\backslash\rho := \overline{\overline{\rho}\,\rho^{\smile}}$. Then $\widetilde{\rho} := (\rho/\rho)\backslash\rho = \overline{\overline{(\rho/\rho)}\rho^{\smile}} = \overline{\rho^{\smile}\,\overline{\rho}\rho^{\smile}}$. Prop. 2.86 and Prop. 2.87(b) state the analog properties for $\widetilde{\rho}$ that have been shown above for $A^{\diamond}$.

8.2.3 *Eigenvalues and Eigenvectors*

The results in this section are taken from the work of one of the leading investigators in the area of generalized fuzzy matrices during the last 25 years, and we present them together with their proofs. It is assumed that L is a complete and completely distributive lattice with the greatest element 1 and the least element 0. Thus, by a well known result from lattice theory (see [7, p.128, THEOREM 24]), L is relatively pseudocomplemented. We let $\boldsymbol{e} = (1, 1, \ldots, 1)^T$ (so it is the maximal vector), and first consider the case that an eigenvector $\boldsymbol{\xi}$ of a matrix $A \in \mathbb{L}_n$ is given and determine the range of its eigenvalues.

Theorem 8.42. ([56, Theorem 3.1]) *Let $\boldsymbol{\xi} = (x_1, x_2, \ldots, x_n)^T$ be an eigenvector of $A \in \mathbb{L}_n$. Then the eigenvalues of $\boldsymbol{\xi}$ form a sublattice of L consisting of the interval $[\lambda^0, \lambda^*]$, where $\lambda^0 = \boldsymbol{e}^T A\boldsymbol{\xi}$ and $\lambda^* = (\boldsymbol{e}^T\boldsymbol{\xi}) \mathrel{\dot{\rightharpoonup}} \lambda^0$.*

Proof. Assume $A\boldsymbol{\xi} = \lambda\boldsymbol{\xi} = \mu\boldsymbol{\xi}$; then

$$A\boldsymbol{\xi} = \lambda\boldsymbol{\xi} \vee \mu\boldsymbol{\xi} = (\lambda \vee \mu)\boldsymbol{\xi} \qquad \text{and} \qquad A\boldsymbol{\xi} = \lambda\boldsymbol{\xi} \wedge \mu\boldsymbol{\xi} = (\lambda \wedge \mu)\boldsymbol{\xi},$$

so the eigenvalues of $\boldsymbol{\xi}$ form a sublattice of L. Since

$$\boldsymbol{e}^T A\boldsymbol{\xi} = \boldsymbol{e}^T(\lambda\boldsymbol{\xi}) = \lambda(\boldsymbol{e}^T\boldsymbol{\xi}) = \lambda(x_1 \vee \ldots \vee x_n) \le \lambda,$$

$\boldsymbol{e}^T A\boldsymbol{\xi}$ is a lower bound for λ^0 which is attained due to

$$(\boldsymbol{e}^T A\boldsymbol{\xi})\boldsymbol{\xi} = \lambda(\boldsymbol{e}^T\boldsymbol{\xi})\boldsymbol{\xi} = \lambda(x_1 \vee \ldots \vee x_n)\boldsymbol{\xi} = \lambda\boldsymbol{\xi} = A\boldsymbol{\xi}.$$

Therefore, $\lambda^0 = \boldsymbol{e}^T A\boldsymbol{\xi}$.

Assume $A\boldsymbol{\xi} = \lambda^0\boldsymbol{\xi} = \lambda^*\boldsymbol{\xi}$; then, for all $j \in \underline{n}$, $\lambda^0 x_j = \lambda^* x_j$, so we must have $\lambda^* \le x_j \mathrel{\dot{\rightharpoonup}} (\lambda^0 x_j)$, i.e. λ^* is less than or equal to the greatest $l \in L$ such that $l x_j \le \lambda^0 x_j$. Since by Prop. 4.23(1) $x_j \mathrel{\dot{\rightharpoonup}} (\lambda^0 x_j) = x_j \mathrel{\dot{\rightharpoonup}} \lambda^0$, we get $\lambda^* \le x_j \mathrel{\dot{\rightharpoonup}} \lambda^0$, for all $j \in \underline{n}$, and therefore by Prop. 4.23(3)

$$\lambda^* \le \bigwedge_{j\in\underline{n}}(x_j \mathrel{\dot{\rightharpoonup}} \lambda^0) = \Big(\bigvee_{j\in\underline{n}} x_j\Big) \mathrel{\dot{\rightharpoonup}} \lambda^0 = (\boldsymbol{e}^T\boldsymbol{\xi}) \mathrel{\dot{\rightharpoonup}} \lambda^0.$$

So $(\boldsymbol{e}^T\boldsymbol{\xi}) \mathrel{\dot{\rightharpoonup}} \lambda^0$ is an upper bound for λ^* which is attained since, applying Prop. 4.23(5) to $\Big(\big(\bigvee_{j\in\underline{n}} x_j\big) \mathrel{\dot{\rightharpoonup}} \lambda^0\Big) \wedge x_i = \lambda^0 \wedge x_i$ to get the second equality below, we have

$$((\boldsymbol{e}^T\boldsymbol{\xi}) \mathrel{\dot{\rightharpoonup}} \lambda^0)\boldsymbol{\xi} = \Big(\big(\bigvee_{j\in\underline{n}} x_j\big) \mathrel{\dot{\rightharpoonup}} \lambda^0\Big)\boldsymbol{\xi} = \lambda^0\boldsymbol{\xi} = A\boldsymbol{\xi}.$$

Therefore, $\lambda^* = (\boldsymbol{e}^T\boldsymbol{\xi}) \mathrel{\dot{\rightharpoonup}} \lambda^0$. □

Next, we turn to the case that an eigenvalue λ of a matrix $A \in \mathbb{L}_n$ is given and determine its eigenvectors. Note that the set $V^n(L)$ of all column vectors over L forms a complete and completely distributive lattice if we define $\boldsymbol{\xi} \vee \boldsymbol{\eta} = (x_1 \vee y_1, \ldots, x_n \vee y_n)^T$ and $\boldsymbol{\xi} \wedge \boldsymbol{\eta} = (x_1 \wedge y_1, \ldots, x_n \wedge y_n)^T$.

Theorem 8.43. ([56, Theorem 3.3]) *Let λ be an eigenvalue of $A \in \mathbb{L}_n$. Then the eigenvectors of λ form a subspace of $V^n(L)$, consisting of the interval $[\boldsymbol{\xi}^0, \boldsymbol{\xi}^*]$, where $\boldsymbol{\xi}^0 = \mathbf{0}$ and $\boldsymbol{\xi}^* = \bigvee\{\boldsymbol{\xi} | \boldsymbol{\xi} \in E_\lambda^A\}$.*

Proof. Assume that $A\boldsymbol{\xi} = \lambda\boldsymbol{\xi}$ and $A\boldsymbol{\eta} = \lambda\boldsymbol{\eta}$; then $A(\boldsymbol{\xi} \vee \boldsymbol{\eta}) = A\boldsymbol{\xi} \vee A\boldsymbol{\eta} = \lambda\boldsymbol{\xi} \vee \lambda\boldsymbol{\eta} = \lambda(\boldsymbol{\xi} \vee \boldsymbol{\eta})$. Moreover, for all $a \in L$, $A(a\boldsymbol{\xi}) = a(A\boldsymbol{\xi}) = a(\lambda\boldsymbol{\xi}) = \lambda(a\boldsymbol{\xi})$. It can be easily seen that the remaining subspace properties hold. □

Theorem 8.44. ([56, Theorem 3.5]) *For any given $\lambda \in L$ and $\boldsymbol{\xi} = (x_1, x_2, \ldots, x_n)^T$ we define $T(\lambda, \boldsymbol{\xi}) = \{A \in \mathbb{L}_n | A\boldsymbol{\xi} = \lambda\boldsymbol{\xi}\}$. Then $T(\lambda, \boldsymbol{\xi})$ is both a subsemigroup with respect to matrix multiplication and a $\vee$-subsemilattice of $\mathbb{L}_n$. The maximal element in $T(\lambda, \boldsymbol{\xi})$ is the matrix $M = M(\lambda, \boldsymbol{\xi})$ with $m_{ij} = x_j \stackrel{\wedge}{\to} (\lambda x_i)$.*

Proof. Let $A = [a_{ij}], B = [b_{ij}] \in T(\lambda, \boldsymbol{\xi})$; then $(AB)\boldsymbol{\xi} = A(B\boldsymbol{\xi}) = A(\lambda\boldsymbol{\xi}) = \lambda(A\boldsymbol{\xi}) = \lambda(\lambda\boldsymbol{\xi}) = \lambda\boldsymbol{\xi}$. Hence $AB \in T(\lambda, \boldsymbol{\xi})$ and $T(\lambda, \boldsymbol{\xi})$ is a subsemigroup of $\mathbb{L}_n$. Moreover,

$$\begin{aligned} ((A \vee B)\boldsymbol{\xi})_i &= \bigvee_{j \in \underline{n}}\big((a_{ij} \vee b_{ij}) \wedge x_j\big) = \big(\bigvee_{j \in \underline{n}}(a_{ij} \wedge x_j)\big) \vee \big(\bigvee_{j \in \underline{n}}(b_{ij} \wedge x_j)\big) \\ &= (A\boldsymbol{\xi})_i \vee (B\boldsymbol{\xi})_i = (\lambda\boldsymbol{\xi})_i \vee (\lambda\boldsymbol{\xi})_i = (\lambda\boldsymbol{\xi})_i, \end{aligned}$$

whence $(A \vee B)\boldsymbol{\xi} = \lambda\boldsymbol{\xi}$, so $A \vee B \in T(\lambda, \boldsymbol{\xi})$.

For the matrix M with $m_{ij} := x_j \stackrel{\wedge}{\to} (\lambda x_i)$ we get $m_{ij}x_j = (x_j \stackrel{\wedge}{\to} (\lambda x_i))x_j = \lambda x_i x_j$ by Prop. 4.23(2). Thus, for each $i \in \underline{n}$, $\bigvee_{j \in \underline{n}}(m_{ij}x_j) = (\lambda x_i) \wedge \big(\bigvee_{j \in \underline{n}} x_j\big) = \lambda x_i$, i.e. $M\boldsymbol{\xi} = \lambda\boldsymbol{\xi}$, whence $M \in T(\lambda, \boldsymbol{\xi})$. Finally we show that M is the greatest element in $T(\lambda, \boldsymbol{\xi})$:

$$\begin{aligned} A \in T(\lambda, \boldsymbol{\xi}) &\to \bigvee_{j \in \underline{n}}(a_{ij}x_j) = \lambda x_i, \text{ for all } i \in \underline{n} \\ &\to a_{ij}x_j \le \lambda x_i, \text{ for all } i, j \in \underline{n} \\ &\to a_{ij} \le x_j \stackrel{\wedge}{\to} (\lambda x_i), \text{ for all } i, j \in \underline{n} \\ &\to A \le M. \end{aligned}$$

□

More results can be found in both [56] and [58], e.g. on the solutions of (an adapted form of) the characteristic equation in Section 4 of the first paper; and in the second one, among others, the following is shown

Theorem 8.45. ([58, Theorems 3.1, 3.6 and 3.9])

(1) *Let λ be an eigenvalue of $A \in \mathbb{L}_n$; then the maximal eigenvector $\boldsymbol{\xi}^*(\lambda)$ associated with λ is $\boldsymbol{\xi}^*(\lambda) = (\lambda \boldsymbol{e} \vee A^T e) \stackrel{\wedge}{\to} (\lambda A^n \boldsymbol{e})$.*
(2) *For any given $\boldsymbol{\xi} \in V_n(L)$, the set $M(\boldsymbol{\xi}) = \{M(\lambda, \boldsymbol{\xi}) | \lambda \in L\}$ is a subsemigroup of $\mathbb{L}_n$ in which multiplication coincides with intersection.*
(3) *Given any interval $[\lambda^0, \lambda^*] \subseteq L$ such that $\lambda^0 \le \lambda^*$, $M(\lambda, \boldsymbol{\xi})$ is invariant over $[\lambda^0, \lambda^*]$ (i.e. it is the same for any $\lambda \in [\lambda^0, \lambda^*]$) iff $\bigvee_{i \in \underline{n}} x_i \le \lambda^* \stackrel{\wedge}{\to} \lambda^0$.*

Some results on standard eigenvectors of matrices over a distributive lattice can be found in [25, Section 4].

8.3 Matrices over Chain Semirings

Just to note it explicitly: Here is the place in the structure of this book where fuzzy matrices/relations in their original sense, i.e. over sets $L \subseteq [0,1] \subseteq \mathbb{R}$, have their logical position. From the results valid for this kind of matrices/relations many are dealt with elsewhere in this book since they do already hold in underlying structures which are more general. So only a few findings will be mentioned here, or only hinted at, that (to the author's knowledge) do not have generalized counterparts, do not fit well into another place, or are "classical" results and thus interesting from a historical perspective.

As already mentioned at various occasions, also Boolean matrices have a chain semiring as underlying valuation structure: $\{0 < 1\}$. Hence some related aspects will be reported below which have not been mentioned in Chapter 3, are of interest of their own, and, from a historical point of view, are the starting point for later investigations.

8.3.1 *Fuzzy Matrices*

By a **fuzzy algebra** we mean a triple $(F,+,\cdot)$, where $(F,\leq,0,1)$ is a bounded linearly ordered set with the binary operations maximum and minimum, denoted as $+$ and $\cdot$, and a **fuzzy matrix** is one which has all entries from F. So both classical fuzzy relations over $[0,1] \subseteq \mathbb{R}$ and Boolean relations over $\{0,1\}$ are special cases. With $\mathbb{F}$ we denote a fuzzy algebra where $F \subseteq [0,1] \subseteq \mathbb{R}$.

8.3.1.1 *Inverses of Fuzzy Matrices*

By Prop. 2.82(3) a crisp relation R is invertible if and only if it is a permutation. A corresponding result for bounded Brouwerian lattices in which every element has a finite join decomposition has been given in [64, Theorem 1]. We present a somewhat simpler version in

Proposition 8.46. ([42, Proposition 2.1.1]) *A square matrix $A \in \mathbb{F}_n$ is invertible iff A is a permutation matrix. In this case, A^T is the unique inverse of A.*

Proof. Since for any permutation matrix A we have $AA^T = A^TA = I_n$, it follows that A is invertible. Conversely, Let $A = [a_{ij}]$ be invertible; then $AB = BA = I_n$ for some $B \in \mathbb{F}_n$. So $\sum_{k\in\underline{n}} a_{ik}b_{kj} = \sum_{k\in\underline{n}} b_{ik}a_{kj} = 0$ for $i \neq j$ and $\sum_{k\in\underline{n}} a_{ik}b_{ki} = \sum_{k\in\underline{n}} b_{ik}a_{ki} = 1$. Thus,

$$a_{ik}b_{kj} = b_{ik}a_{kj} = 0 \text{ for each } k \in \underline{n} \text{ and } i \neq j, \quad \text{and} \tag{8.3}$$

$$a_{ik}b_{ki} = b_{ik}a_{ki} = 1 \text{ for at least one } k \in \underline{n} \text{ and all } i \in \underline{n}. \tag{8.4}$$

(8.3) means that

$$a_{ik} = 0 \text{ or } b_{kj} = 0 \text{ or } a_{ik} = b_{kj} = 0 \text{ for each } k \in \underline{n} \text{ and all } i \neq j, \tag{8.5}$$

and (8.4) implies

$$a_{ik} = b_{ki} = 1 \text{ and } a_{ki} = b_{ik} = 1 \text{ for at least one } k \in \underline{n} \text{ and all } i \in \underline{n}. \tag{8.6}$$

Now suppose that (8.6) holds for $k = r$, i.e. $a_{ir} = b_{ri} = 1$; then from (8.5) we get $b_{rj} = 0 = a_{jr}$ for all $i \neq j$. Hence, there is exactly one 1 in the rth row of B and the rth column of A, all other entries being zero. Thus, A is a permutation matrix, and $B = A^T$ is its unique inverse. □

Some additional material can be found in [42, Chapter 3], or in the papers [10,11,14].

8.3.1.2 *Eigenvectors of Fuzzy Matrices*

Mainly, but not only, for historical reasons we insert this short section: Among a few other classics, Sanchez's paper [50] on the "Resolution of eigen fuzzy sets equations" of the form $R \circ A = A$ appeared in the very first issue of the *Fuzzy Sets and Systems* journal, describing three "algorithms for the determination of the greatest eigen fuzzy set associated with a given fuzzy relation". So, standard eigenvectors are a very early topic in the context of fuzzy relations. There is the follow up [51], in which it is shown that e.g. the greatest eigen fuzzy set associated with the transitive closure $\mathbf{t}(R)$ is exactly the one associated with R. Some more special results on eigenvectors over the real unit interval can be found in e.g. [18,60].

Whereas the above mentioned papers use $[0,1] \subseteq \mathbb{R}$ as their valuation set, the ones we refer to below use fuzzy algebras in the more general sense as defined above. A group of researchers around M. Gavalec and J. Plavka investigates various eigenvalue problems and the structure of eigenspaces in a number of papers considering interval matrices and interval eigenvectors as e.g. in [19,43,45]. In [17] a complete description of the structure of the whole eigenspace of a square fuzzy matrix is given, for which it is sufficient to describe the structure of the set of all increasing eigenvectors. And in [44] the author characterizes fuzzy algebraic sub-(super-)eigenvectors associated with λ, that is, vectors such that $A \otimes \boldsymbol{x} \leq (\geq)\, \lambda \boldsymbol{x}$.

8.3.2 *Binary Boolean Matrices*

Here is also the place for binary Boolean matrices from the viewpoint of the underlying structure of a chain semiring. Due to its role as a classical reference for the investigation over more general algebraic structures, however, we have already inserted Chapter 3 in Part 1. In this section, we present a few concepts and results that are less familiar in connection to Boolean matrices than the ones included in Chapter 3, so it is a supplement to it.

Research on binary Boolean matrices has the longest history among investigations on matrices over general (non-numerical) structures. Recall Section 2.11 where we have already learned some facts about various kinds of inverses in the context of Boolean relations. But, of course, many more results have been found, some of which we present in the terminology and framework of Boolean matrices.

8.3.2.1 *Ranks of Boolean Matrices*

For the convenience of the reader we shortly repeat two rank notions from Def. 6.46:

- The **column / row rank** of A is the smallest possible size of a spanning set for the column/row space of A, denoted as $\mathsf{cr}(A)$ / $\mathsf{rr}(A)$, i.e. $\mathsf{cr}(A) = \dim \mathsf{col}(A)$ and $\mathsf{rr}(A) = \dim \mathsf{row}(A)$.
- The **rank** (also **factor, semiring** or **Schein rank**), of A is the smallest integer r such that $A = LR$, with $L \in \mathbb{S}_{m\times r}$ and $R \in \mathbb{S}_{r\times n}$, denoted as $\mathsf{fr}(A)$.

In the case of matrices in $\mathbb{B}_{m,n}$ the factor rank is usually referred to as the **Boolean rank.** The zero matrices O are assigned the factor, column (row) rank 0. And from Cor. 3.5 we know that each subspace $W \subseteq V_n$ has a unique basis $B \subseteq V_n$. The **column basis** of M, denoted by $cb(M)$, is the unique basis of $\mathsf{col}(M)$, and the **row basis** of M, denoted by $rb(M)$, is the unique basis of $\mathsf{row}(M)$.

Note that in contrast to the real and complex case, the row and column rank may be different for Boolean matrices of dimension $n \geq 4$ as the following example from [46, p. 427] shows:

$$A = \begin{bmatrix} 1 & 0 & 1 & 0 \\ 0 & 1 & 0 & 1 \\ 1 & 1 & 0 & 0 \\ 1 & 1 & 1 & 0 \end{bmatrix}.$$

A is not row independent (the fourth row is a sum of the first and third), but it is column independent. Thus, $\mathsf{rr}(A) = 3$ while $\mathsf{cr}(A) = 4$. See also [9, Theorem 5].

Another example given in [6, p. 34] demonstrates that the column rank may exceed the Boolean rank. For the matrix

$$B = \begin{bmatrix} 1 & 0 & 1 & 1 \\ 1 & 1 & 0 & 0 \\ 0 & 1 & 1 & 0 \end{bmatrix}$$

we have $\mathsf{cr}(B) = 4$, whereas, by definition, its Boolean rank cannot exceed 3, the number of rows. But for regular matrices M it is known that $\mathsf{fr}(M) = \mathsf{rr}(M) = \mathsf{cr}(M)$, a special case of [11, Theorem 2.3] where these equalities are shown for fuzzy matrices. A direct proof for the Boolean case is given in [32, p. 72].

The following properties hold for the Boolean rank of matrices (see e.g. [3, 6]; you may also want to have a look into [41], [42, Section 1.4] and [52] for further results).

Proposition 8.47. *For all matrices $A \in \mathbb{B}_{m,n}$ we have:*

(1) $0 \leq \mathsf{fr}(A) \leq \min\{m, n\}$.
(2) $0 \leq \mathsf{fr}(A) \leq \min\{\mathsf{rr}(A), \mathsf{cr}(A)\}$.
(3) $\mathsf{fr}(A) = \mathsf{fr}(A^T)$.
(4) $\mathsf{fr}(A)$ *is the minimum number of rank 1 matrices whose sum is A.*
(5) $\mathsf{fr}(A)$ *is the minimum size of a set S of Boolean column vectors with* $\mathsf{col}(A) \subseteq \langle S \rangle$.

Proof. (1): Let $A = LR$; then, setting $L = A$ and taking $R = I_n$, we get $A = AI_n$, so $\mathsf{fr}(A) \leq n$; an analogous argument shows that $\mathsf{fr}(A) \leq m$. A is said to be a **full-rank matrix** if $\mathsf{fr}(A) = \min\{m, n\}$.

(2): This is a special case of Prop. 6.49(5).

(3): Let $A \in \mathbb{B}_{m\times n}$ have rank k; then $A = BC$ for some $B \in \mathbb{B}_{m\times k}$ and $C \in \mathbb{B}_{k\times n}$. From Cor. 3.9(3) we know that $A^T = C^T B^T$, where $C^T \in \mathbb{B}_{n\times k}$ and $B^T \in \mathbb{B}_{k\times m}$. Thus, $\mathsf{fr}(A^T) = k$.

(4): This is a special case of Lemma 6.47(1).

(5): From Def. 5.9 we know that the column space of a matrix $A \in \mathbb{B}_{m\times n}$ is $\mathsf{col}(A) = \{\boldsymbol{y} : \boldsymbol{y} = A\boldsymbol{x}$ for some $\boldsymbol{x}\}$ with $\boldsymbol{x} \in \mathbb{B}_{n\times 1}$, and $\boldsymbol{y} \in \mathbb{B}_{m\times 1}$, i.e. $\mathsf{col}(A)$ is the set obtained by taking all possible linear combinations of the columns of A: $\mathsf{col}(A) = \langle A_{\mathsf{col}}\rangle$. And by Cor. 3.5 there exists a unique basis of $\mathsf{col}(A)$, say $\{\boldsymbol{v}_1, \boldsymbol{v}_2, \ldots, \boldsymbol{v}_l\}$; so there exist scalars $c_{1j}, \ldots, c_{lj}$ such that every $\boldsymbol{a}_j \in A_{\mathsf{col}}$ can be written as linear combination $\boldsymbol{a}_j = c_{1j}\boldsymbol{v}_1 + c_{2j}\boldsymbol{v}_2 + \cdots + c_{lj}\boldsymbol{v}_l$, with $c_{ij} \in \{0,1\}$. So we get (from commutativity of the underlying Boolean algebra) e.g.

$$\begin{bmatrix} a_{11} \\ a_{21} \\ \vdots \\ a_{m1} \end{bmatrix} = \begin{bmatrix} c_{11}v_{11} + c_{21}v_{12} + \cdots + c_{l1}v_{1l} \\ c_{12}v_{21} + c_{22}v_{22} + \cdots + c_{l2}v_{2l} \\ \vdots \\ c_{1l}v_{l1} + c_{2l}v_{l2} + \cdots + c_{mn}v_{nl} \end{bmatrix} = \begin{bmatrix} v_{11}c_{11} + v_{12}c_{21} + \cdots + v_{1l}c_{l1} \\ v_{21}c_{12} + v_{22}c_{22} + \cdots + v_{2l}c_{l2} \\ \vdots \\ v_{l1}c_{1l} + v_{l2}c_{2l} + \cdots + v_{nl}c_{mn} \end{bmatrix}$$

Setting $B := [\boldsymbol{v}_1\boldsymbol{v}_2\ldots\boldsymbol{v}_l] \in \mathbb{B}_{m\times l}$ and $C \in \mathbb{B}_{l\times n}$ by $C = [c_{ij}]_{l,n}$ yields $A = BC$. Clearly, $\dim \mathsf{col}(A) = l = \mathsf{fr}(A)$, □

Corollary 8.48. *For all matrices $A \in \mathbb{B}_{m,n}$ we have:*

(1) $\mathsf{fr}(B) \leq \mathsf{fr}(A)$ *for all submatrices B of A.*
(2) $\mathsf{fr}(AB) \leq \min\{\mathsf{fr}(A), \mathsf{fr}(B)\}$.

Proof. (1): Obvious from the definition of $\mathsf{fr}(\cdot)$.
(2): Let $A = C_kD_k$ and $B = C'_lD'_l$ two factorizations with minimal size, i.e. $\mathsf{fr}(A) = k$ and $\mathsf{fr}(B) = l$. Then $AB = C_kD_kC'_lD'_l = C_k(D_kB) = (AC'_l)D'_l$, where the last two equalities give factorizations of size $\mathsf{fr}(A)$ and $\mathsf{fr}(B)$, respectively. So for the minimal factorization size of AB we get $\mathsf{fr}(AB) \leq \min\{\mathsf{fr}(A), \mathsf{fr}(B)\}$. □

The following lemma shows that permuting the rows, the columns or both simultaneously does not change the factor rank of a square Boolean matrix. (See Section 3.3.2 for permutation matrices.)

Lemma 8.49. ([16, Lemma 2.5(1)]) *For all Boolean matrices $A \in \mathbb{B}_n$, and all permutations $P \in \boldsymbol{P}_n \subseteq \mathbb{B}_n$ we have* $\mathsf{fr}(A) = \mathsf{fr}(PA) = \mathsf{fr}(AP) = \mathsf{fr}(P^{-1}AP)$.

Proof. Let $\mathsf{fr}(A) = k$, $A' = PA$, and $A = BC$ a factorization with $B \in \mathbb{B}_{n,k}$ and $C \in \mathbb{B}_{k,n}$; then $A' = (PB)C$ is a decomposition of factorization size k. So $\mathsf{fr}(A') \leq \mathsf{fr}(A)$. And since $A = P^{-1}A'$, we also get $\mathsf{fr}(A) \leq \mathsf{fr}(A')$. The assertion for $A' = AP$ can be shown in an analogous way. And $\mathsf{fr}(A) = \mathsf{fr}(P^{-1}AP)$ follows immediately, since for $P \in \boldsymbol{P}_n$ also $P^{-1} \in \boldsymbol{P}_n$. □

8.3.2.2 *Idempotency*

Though from the many results existing on idempotent matrices over the binary or a general Boolean algebra only a small part will be mentioned below, this glance nonetheless offers an interesting view on how many distinct approaches have been taken in the course of time. It may be useful to have a look back to Section 2.12 where we have already considered idempotent binary relations.

Lemma 8.50. ([47, Lemma 2.1]) *Let $A \in \mathbb{B}_n$ be idempotent. If $a_{ii} = 0$ then the i-th column of A depends on the other columns and the i-th row of A depends on the other rows. Furthermore, the matrix obtained by deleting the i-th column and the i-th row of A is idempotent.*

Proof. Without loss of generality we set $a_{nn} = 0$; then from $A^2 = A$ one gets

$$A = \begin{bmatrix} B & \boldsymbol{x} \\ \boldsymbol{y}^T & 0 \end{bmatrix} = \begin{bmatrix} B^2 + \boldsymbol{x}\boldsymbol{y}^T & B\boldsymbol{x} \\ \boldsymbol{y}^T B & \boldsymbol{y}^T \boldsymbol{x} \end{bmatrix} = A^2.$$

So $B^2 + \boldsymbol{x}\boldsymbol{y}^T = B$, $B\boldsymbol{x} = \boldsymbol{x}$, $\boldsymbol{y}^T B = \boldsymbol{y}^T$ and $\boldsymbol{y}^T \boldsymbol{x} = 0$, whence $\begin{pmatrix} \boldsymbol{x} \\ 0 \end{pmatrix} = \begin{pmatrix} B \\ \boldsymbol{y}^T \end{pmatrix} \boldsymbol{x}$ and $(\boldsymbol{y}^T \; 0) = \boldsymbol{y}^T (B \; \boldsymbol{x})$. The last two equalities show that the nth column (row) depends on the other columns (rows). Moreover, $B^2 + \boldsymbol{x}\boldsymbol{y}^T = B$ implies $\boldsymbol{x}\boldsymbol{y}^T \le B$, whence $B\boldsymbol{x}\boldsymbol{y}^T \le B^2$. But from $B\boldsymbol{x} = \boldsymbol{x}$ we get $B\boldsymbol{x}\boldsymbol{y}^T = \boldsymbol{x}\boldsymbol{y}^T$, so $\boldsymbol{x}\boldsymbol{y}^T \le B^2$. Thus $B = B^2 + \boldsymbol{x}\boldsymbol{y}^T = B^2$. □

Corollary 8.51. ([47, Corollary 2.1]) *If A is idempotent and nonsingular then $a_{ii} = 1$ for all i.*

Lemma 8.52. ([47, Lemma 2.2]) *Let $A \in \mathbb{B}_n$ be idempotent. If $a_{ii} = 1$ for all $i \in \underline{n}$, then the i-th row of A depends on the other rows if and only if the i-th column of A depends on the other columns. Furthermore, the matrix obtained by deleting the i-th column and the i-th row of A is idempotent.*

Proof. Let $A = \begin{bmatrix} B & \boldsymbol{x} \\ \boldsymbol{y}^T & 1 \end{bmatrix}$. From $A^2 = A$ similar as above we get

(i) $B^2 + \boldsymbol{x}\boldsymbol{y}^T = B$ implies $B^2 \le B$. But by assumption $I \le B$, whence $B \le B^2$. Thus $B^2 = B$.

(ii) $B\boldsymbol{x} + \boldsymbol{x} = \boldsymbol{x}$ implies $B\boldsymbol{x} \le \boldsymbol{x}$; and since $I \le B$ implies $B\boldsymbol{x} \ge I\boldsymbol{x} = \boldsymbol{x}$, equality $B\boldsymbol{x} = \boldsymbol{x}$ follows.

(iii) $\boldsymbol{y}^T B + \boldsymbol{y}^T = \boldsymbol{y}^T$ implies $\boldsymbol{y}^T B \le \boldsymbol{y}^T$, so $\boldsymbol{y}^T B = \boldsymbol{y}^T$ follows as above.

Without loss of generality assume that the n-th row $(\boldsymbol{y}^T \; 1)$ depends on the other rows; then there is a vector $\boldsymbol{z}$ such that $(\boldsymbol{y}^T \; 1) = \boldsymbol{z}^T (B \; \boldsymbol{x})$. Thus, $\boldsymbol{y}^T \boldsymbol{x} = \boldsymbol{z}^T B\boldsymbol{x}$ and $\boldsymbol{z}^T \boldsymbol{x} = 1$. But using (ii) above we have $\boldsymbol{z}^T B\boldsymbol{x} = \boldsymbol{z}^T \boldsymbol{x}$, whence $\boldsymbol{y}^T \boldsymbol{x} = 1$. Therefore, $\begin{pmatrix} \boldsymbol{x} \\ 1 \end{pmatrix} = \begin{pmatrix} B \\ \boldsymbol{y}^T \end{pmatrix} \boldsymbol{x}$, i.e. we have shown that the n-th column of A depends on the other columns. The converse direction follows by similar arguments. □

We now proceed by characterizing idempotent Boolean matrices by what is called the '*Rao normal form*' in [61, Section 3]. Following [47] a matrix $A \in \mathbb{B}_{m,n}$ is called **non-singular** if the row rank of A is m and the column rank of A is n (recall Def. 5.13).

Lemma 8.53. ([61, Lemma 3.7]) *Let $A, B \in \mathbb{B}_n$. If A is reflexive and $ABA \leq A$, then $B \leq A$.*

Proof. Let $C := ABA$. If all main diagonal entries of A are 1 and $ABA \leq A$, then for any $i, j \in \underline{n}$ we have $b_{ij} = 1 \cdot b_{ij} \cdot 1 = a_{ii} b_{ij} a_{jj} \leq \sum_{k \in \underline{n}} \sum_{l \in \underline{n}} (a_{ik} b_{kl} a_{lj}) = c_{ij} \leq a_{ij}$. □

Theorem 8.54. ([47, Theorem 2.2]) *A is an idempotent matrix of rank r if and only if there exists a permutation matrix P such that*

$$A = P \begin{bmatrix} B & BC \\ DB & DBC \end{bmatrix} P^T,$$

where B is an $r \times r$ nonsingular idempotent matrix, and $CD \leq B$.

Proof. Suppose that $A \in \mathbb{B}_n$ is an idempotent matrix of rank $r < n$, then by Lemma 8.50 and Lemma 8.52 we have that the i-th row of A is a linear combination of the remaining rows if and only if the i-th column of A is a linear combination of the remaining columns. Thus by carrying out the same permutations on both rows and columns of A, we can find a matrix A' such that the first r columns and the first r rows of A' are linearly independent. In other words, there exists a permutation matrix $P \in \mathbb{B}_n$ such that $A' = P \begin{bmatrix} B & X \\ Y & Z \end{bmatrix} P^T$, where $B \in \mathbb{B}_r$ is idempotent, with X, Y, and Z of appropriate size. From Prop. 6.53 it follows that an idempotent matrix $B \in \mathbb{B}_m$ has equal column and row rank; moreover we have $\mathsf{rr}(B) = \mathsf{cr}(B) = r$, whence B is nonsingular.

Since $\begin{bmatrix} X \\ Z \end{bmatrix}$ depends on $\begin{bmatrix} B \\ Y \end{bmatrix}$, there is a matrix $C \in \mathbb{B}_{r,n-r}$ such that $\begin{bmatrix} X \\ Z \end{bmatrix} = \begin{bmatrix} B \\ Y \end{bmatrix} C$. So $A' = P \begin{bmatrix} B & BC \\ Y & YC \end{bmatrix} P^T$. Since $\begin{bmatrix} Y & YC \end{bmatrix}$ depends on $\begin{bmatrix} B & BC \end{bmatrix}$, we have $\begin{bmatrix} Y & YC \end{bmatrix} = D \begin{bmatrix} B & BC \end{bmatrix}$ for some $D \in \mathbb{B}_{n-r,r}$. Therefore, $A' = P \begin{bmatrix} B & BC \\ DB & DBC \end{bmatrix} P^T$ and, since A is idempotent, we get $B = B^2 + BCDB$, whence $BCDB \leq B$. Finally, since $I \leq B$ by Cor. 8.51, we conclude $CD \leq B$ using Lemma 8.53.

The converse direction follows by straightforward calculation, using the inequality $CD \leq B$ and the idempotency of B for comparing the various matrix products in the sums occurring in A^2 (e.g. $DBB = DB$ and $DBCDB \leq DB^3 = DB$):

$$\left(P \begin{bmatrix} B & BC \\ DB & DBC \end{bmatrix} P^T \right)^2 = P \begin{bmatrix} B^2 + BCDB & BBC + BCDBC \\ DBB + DBCDB & DBBC + DBCDBC \end{bmatrix} P^T.$$

□

Kang and his colleagues, in their investigations of idempotent Boolean matrices, apply a completely different technique in [4,30,31], using cells and their configurations. We report their main results (which can all be found in [4]), including some typical proofs and always a reference where to find them and also those which we have omitted.

Recall that a Boolean matrix $A \in \mathbb{B}_n$ is called a *cell* if it has exactly one positive entry, denoted by E_{ij}, in case $a_{ij} = 1$. If for $A \in \mathbb{B}_n$ we have $a_{ij} = 1$, then we say that cell E_{ij} *is in* matrix A. So E_{ij} is in A iff $E_{ij} \le A$. E_{ii} is called a **diagonal** cell, and E_{ij} is called **off-diagonal** if $i \neq j$. An immediate consequence of the rules of matrix multiplication is that for all indices i, j and u, v we have

$$E_{ij}E_{uv} = \delta_{ju}E_{iv}. \tag{8.7}$$

Moreover, A can be written uniquely in the canonical form

$$\sum_{i\in\underline{n}}\sum_{j\in\underline{n}} \alpha_{ij}E_{ij}, \tag{8.8}$$

with $\alpha_{ij} \in \{0,1\}$ which shows that A is a sum of cells.

Lemma 8.55. ([31, Lemma 2.2], [4, Corollary 3.22]) *Let $E_1, E_2, \ldots, E_m$ and F be cells in an idempotent matrix $A \in \mathbb{B}_n$, with F being off-diagonal.*

(1) If $E_1 \cdots E_m$ is not zero, then it is a cell in A.
(2) There exist distinct cells G and H in A such that $F = GH$. Moreover, if both cells G and H are off-diagonal, then F, G and H are mutually distinct cells.

Proof. (1): $E_1 \cdots E_m$ is a cell by (8.7). Then, by monotonicity of product, $E_i \le A$ for all $i \in \underline{m}$ implies $E_1 \cdots E_m \le A^m$. And since A is idempotent, we have $A^m = A$.

(2): Let $\Gamma = \{F_1, F_2, \ldots, F_k\}$ denote the set of cells in A. So $A = \sum_{i\in\underline{k}} F_i$. Our assumption $A^2 = A$ implies

$$\sum_{i\in\underline{k}} F_i^2 + \sum_{i,j\in\underline{k}, i\neq j} F_iF_j = A^2 = A = \sum_{i\in\underline{k}} F_i.$$

Since $F \le A$, either $F \le F_i^2$ or $F \le F_iF_j$ for some $i \neq j$, with $i, j \in \underline{k}$. Since F is off-diagonal, the latter applies due to (8.7). So we let $G = F_i$ and $H = F_j$ and get $F = GH$. Furthermore, if G and H are off-diagonal, then F, G and H are mutually distinct, again by (8.7). □

Lemma 8.56. ([4, Lemma 4.1]) *For a nonzero matrix $A \in \mathbb{B}_n$ the following holds:*

(1) If all cells in A are diagonal, then A is idempotent.
(2) If all cells in A are off-diagonal, then A is not idempotent.

Proof. (1): Obviously, by (8.7) we have $E_{ii}E_{ii} = \delta_{ii}E_{ii} = E_{ii}$.

(2): Suppose that all cells in A are off-diagonal, and let $\Gamma = \{F_1, F_2, \ldots, F_m\}$ denote the set of these cells. So we have $A = \sum_{i\in\underline{m}} F_i$. We will show that the assumption $A^2 = A$ leads to the contradiction that Γ becomes infinite.

By Lemma 8.55, there exist three distinct cells F_p, F_q and F_r in Γ such that $F_pF_q = F_r$, and using (8.7) we can write $F_p = E_{ii_1}, F_q = E_{i_1j}$ and $F_r = E_{ij}$ for certain mutually distinct indices i, j and i_1. Applying Lemma 8.55 to $F_p = E_{ii_1} \in \Gamma$ we get that there are $E_{ii_2}, E_{i_2i_1} \in \Gamma$ with $E_{ii_1} = E_{ii_2}E_{i_2i_1}$ for some i_2 different from both i and i_1.

Now assume that for some $k \geq 2$ the set of distinct cells $\{E_{ii_1}, \ldots, E_{ii_k}\} \subseteq \Gamma$ has already been constructed. Then a new element can be added to this set as follows: Since A is idempotent, Lemma 8.55(2) and the assumption that *all* cells are off-diagonal yield $E_{ii_k} = E_{ii_{k+1}} E_{i_{k+1}i_k}$ for some index i_{k+1} different from i and i_k (so $s \neq k$). Given that there is $s \in \{1, \ldots, k-1\}$ such that $i_s = i_{k+1}$, would result in

$$E_{i_s i_s} = E_{i_{k+1} i_s} = E_{i_{k+1} i_j} E_{i_j i_s} = E_{i_{k+1} i_j} \cdot E_{i_j i_{j+1}} E_{i_{j+1} i_s} = \cdots = E_{i_{k+1} i_j} E_{i_j i_{j+1}} \cdots E_{i_{j+m} i_s} \leq A,$$

the inequality following from Lemma 8.55(1). But this contradicts the assumption that all cells in A are off-diagonal, whence $i_s \neq i_{k+1}$ for all $s \in \{1, \ldots, k\}$. Therefore, $E_{ii_s} \in \Gamma$ are distinct for $s \in \{1, \ldots, k+1\}$, so Γ contains an infinite set of distinct cells. □

Let $A \in \mathbb{B}_n$. The ith **row matrix** of A is the matrix whose ith row is the same as the ith row of A and whose other rows are zero, denoted by $\mathbf{R}_i$ (A will always be clear from the context). Similarly, the jth **column matrix** of A is defined and denoted by $\mathbf{C}_j$. A **line matrix** of A is a row or a column matrix of A. The cells $E_1, E_2, \ldots, E_m$ are called **collinear** if $\sum_{i \in \underline{m}} E_i$ is a line matrix. $\mathbf{R}_i$ and $\mathbf{C}_j$ are said to be (i,j)-**disjoint** if $F_{ik} F_{lj} = O$ for all off-diagonal cells $F_{ik} \leq \mathbf{R}_i$ and $F_{lj} \leq \mathbf{C}_j$.

Let E_1, E_2, E_3, and E_4 be four distinct cells in $\mathbb{B}_n$. Then their sum is called a **frame** if the four 1's constitute a rectangle with at least one entry on the diagonal. The **weight** of $A \in \mathbb{B}_n$ is the number of nonzero entries of A, denoted by $|A|$.

Proposition 8.57. ([4, Proposition 4.2]) *Let $A \in \mathbb{B}_n$ be idempotent and $F \leq A$ be a off-diagonal cell such that for any diagonal cell $E \leq A$, E and F are not collinear. Then F is in a frame with one diagonal cell and two additional off-diagonal cells in A.*

Corollary 8.58. ([4, Corollary 4.3]) *Let $A = \sum_{i \in \underline{m}} E_i + F$ be a matrix in $\mathbb{B}_n$ with F the only off-diagonal cell and E_i diagonal cells. Then A is idempotent if and only if F is collinear to some E_i, $i \in \underline{m}$.*

Proof. By Lemma 8.56 we have $m \geq 1$, and from Prop. 8.57 it follows that in an idempotent matrix A, F is in a frame with some other off-diagonal cells if it is not collinear to some E_i, $i \in \underline{m}$. As there are no other off-diagonal cells in A, we get a contradiction, so F is collinear to some E_i.

Assume that E_{ii} and F_{ij} are collinear. By Lemma 8.56(1) all diagonal cells are reproduced in A^2. Moreover, $a_{ij}^{(2)} = \sum_k a_{ik} a_{kj} = E_{ii} F_{ij} = F_{ij}$. So A is idempotent. □

Lemma 8.59. ([4, Lemma 4.5], [31, Lemma 2.6]) *Let $A \in \mathbb{B}_n$ be idempotent with $a_{ii} = 1$ for some i. If $|\mathbf{R}_i| = s+1$ and $|\mathbf{C}_i| = t+1$, then there exist exactly $s \cdot t$ frames in A dominating E_{ii}.*

Proof. If $s = 0$ or $t = 0$ the assertion obviously holds, since no frame can be formed; so we assume $s, t \geq 1$. From $A = A^2$ and Lemma 8.55 we get that for the product of any two cells E_{ki} and E_{il} in A we have $E_{kl} \leq A$. Therefore, the four cells E_{ii}, E_{ki}, E_{il}, and E_{kl} are in a frame in A, for each $k, l \in \underline{n}$ such that $E_{ki} \leq A$ and $E_{il} \leq A$. A thus has at least $s \cdot t$

frames including E_{ii}. And from the definition of a frame it follows that A has at most $s \cdot t$ such frames. □

Definition 8.60. ([4, Definition 3.10], [31, p. 726]) *We say that a matrix $A \in \mathbb{B}_n$ has an* **ith rectangular part** *if*

(1) *there is a frame $F \in \mathbb{B}_n$ such that $E_{ii} \le F \le A$, and*
(2) *for any $k, l \in \underline{n}$, if $E_{li}, E_{ik} \le A$ then $E_{lk} \le A$.*

If $A \in \mathbb{B}_n$ has an ith rectangular part, then the matrix in $\mathbb{B}_n$ with the smallest number of nonzero elements that is dominated by A and dominates all frames of A dominating E_{ii} is called the **ith rectangular part of** *A, denoted by $\mathbf{RP}_i(A)$, or simply by $\mathbf{RP}_i$ if A is clear from the context.*

Let $\mathbf{RP}_i$ be the ith rectangular part of A, and let $\{E_{ii_1}, \ldots, E_{ii_t}\}$ and $\{E_{j_1 i}, \ldots, E_{j_s i}\}$ be the sets of all off-diagonal cells in $\mathbf{R}_i$ and $\mathbf{C}_i$, respectively. Then

$$\mathbf{RP}_i = \sum_{k \in \underline{s}} \sum_{l \in \underline{t}} \left(E_{ii} + E_{ii_l} + E_{j_k i} + E_{j_k i_l}\right).$$

As examples, consider the following matrices (taken from [4] and [31]): Let

$$A = \begin{bmatrix} 1 & 0 & 1 & 1 \\ 0 & 0 & 0 & 0 \\ 1 & 0 & 1 & 1 \\ 0 & 1 & 0 & 1 \end{bmatrix}, \quad B = \begin{bmatrix} 1 & 0 & 1 & 1 \\ 0 & 0 & 0 & 0 \\ 1 & 0 & 1 & 1 \\ 0 & 0 & 0 & 0 \end{bmatrix}, \quad C = \begin{bmatrix} 1 & 1 & 1 \\ 1 & 1 & 0 \\ 0 & 0 & 0 \end{bmatrix}.$$

Then $\mathbf{RP}_1(A) = \mathbf{RP}_3(A) = B = \mathbf{RP}_1(B) = \mathbf{RP}_3(B)$, but there is neither $\mathbf{RP}_2(A)$ nor $\mathbf{RP}_4(A)$. And $\mathbf{RP}_2(C) = E_{11} + E_{12} + E_{21} + E_{22}$. $\mathbf{RP}_1(C)$, however, does not exist.

Definition 8.61. *We say that a matrix $A \in \mathbb{B}_n$ with $a_{ii} = 1$, i.e. $E_{ii} \le A$, has an* **ith line part** *if at least one of $|\mathbf{R}_i|$ and $|\mathbf{C}_i|$ has value 1. In these cases $\mathbf{R}_i + \mathbf{C}_i \in \mathbb{B}_n$ is a line matrix dominating E_{ii}, which is called* **the ith line part of** *A, denoted by $\mathbf{LP}_i(A)$, or simply by $\mathbf{LP}_i$ if A is clear from the context.*

Note that $A \in \mathbb{B}_n$ has the ith line part $\mathbf{LP}_i$ if and only if $\mathbf{R}_i = E_{ii}$ or $\mathbf{C}_i = E_{ii}$. Consider the following matrices (taken from [4]):

$$A = \begin{bmatrix} 1 & 1 \\ 0 & 1 \end{bmatrix}, \quad B = \begin{bmatrix} 1 & 1 \\ 1 & 0 \end{bmatrix}, \quad C = \begin{bmatrix} 0 & 1 & 1 \\ 0 & 0 & 0 \\ 0 & 0 & 0 \end{bmatrix}.$$

Then $\mathbf{LP}_1(A) = E_{11} + E_{12}$, $\mathbf{LP}_2(A) = E_{12} + E_{22}$, and the matrices B and C do not have line parts.

Corollary 8.62. ([4, Corollary 4.7]) *Let $A \in \mathbb{B}_n$ be idempotent; then every cell $E_{ik} \in A$ is either in a rectangular part or in a line part of A.*

Proof. By Prop. 8.57 and Lemma 8.59. □

Proposition 8.63. ([4, Proposition 4.10]) *Let $A \in \mathbb{B}_n$ be idempotent. If $\mathbf{R}_i$ and $\mathbf{C}_j$ are not (i, j)-disjoint, then $E_{ij} \le A$.*

Proof. Suppose that $\mathbf{R}_i$ and $\mathbf{C}_j$ are not (i,j)-disjoint, then there exist off-diagonal cells $X \le \mathbf{R}_i$ and $Y \le \mathbf{C}_j$ with $XY \ne O$. So we may write $X = E_{ix}$ and $Y = E_{yj}$, and from $XY \ne O$, $x = y$ and $XY = E_{ij}$ follows due to (8.7). Since A is idempotent we get $XY \le AA = A$. □

Suppose that $A \in \mathbb{B}_n$ has ith and jth rectangular parts $\mathbf{RP}_i$ and $\mathbf{RP}_j$ for some $i \ne j$. Then $\mathbf{RP}_i$ and $\mathbf{RP}_j$ are called **disjoint** if either $\mathbf{R}_i$ and $\mathbf{C}_j$ are (i,j)-disjoint or $\mathbf{R}_j$ and $\mathbf{C}_i$ are (j,i)-disjoint.

Proposition 8.64. ([4, Proposition 4.14], [31, Proposition 2.10]) *Let $A \in \mathbb{B}_n$ be idempotent. Then any two rectangular parts of A are either disjoint or identical.*

Proof. Suppose that the ith and jth rectangular parts of A are not disjoint. Then, by definition, $\mathbf{R}_i$ and $\mathbf{C}_j$ are not (i,j)-disjoint and $\mathbf{R}_j$ and $\mathbf{C}_i$ are not (j,i)-disjoint. So, $E_{ij} \le A$ and $E_{ji} \le A$ by Prop. 8.63.

Next we show that $E \le \mathbf{RP}_i$ iff $E \le \mathbf{RP}_j$, for any cell $E \le A$. Clearly, the four cells $E_{ii}, E_{jj}, E_{ij}, E_{ji}$ are in both $\mathbf{RP}_i$ and $\mathbf{RP}_j$. Assume $E \le \mathbf{RP}_i$. We first consider the case $E \le \mathbf{R}_i$, say $E = E_{ia}$. Then $E_{ja} = E_{ji}E_{ia} \le A$ by Lemma 8.55(1), and the four cells E_{ia}, E_{ij}, E_{ja} and E_{jj} form a frame. Therefore, $E = E_{ia} \le \mathbf{RP}_j$. Similarly we get $E \le \mathbf{RP}_j$ for the case $E \le \mathbf{C}_i$. Next we consider the case $E \not\le \mathbf{R}_i$ and $E \not\le \mathbf{C}_i$, with say $E = E_{cd}$. By assumption $E \le \mathbf{RP}_i$, so there are two off-diagonal cells $E_{is} \le \mathbf{R}_i$ and $E_{ti} \le \mathbf{C}_i$ such that $E_{cd} = E_{ti}E_{is}$. Therefore, $c = t$ and $d = s$ by (8.7). By Lemma 8.55(1) we get

$$E_{cj} = E_{yj} = E_{yi}E_{ij} \le A \quad \text{and} \quad E_{jd} = E_{jx} = E_{ji}E_{ix} \le A,$$

whence the four cells E_{cd}, E_{cj}, E_{jd} and E_{jj} form a frame, and we finally get $E = E_{cd} \le \mathbf{RP}_j$.

Similar arguments show that for a cell $E \le \mathbf{RP}_j$ we have $E \le \mathbf{RP}_i$. Therefore, the two rectangular parts $\mathbf{RP}_j$ and $\mathbf{RP}_i$ are identical. □

Theorem 8.65. ([4, Theorem 4.16], [31, Theorem 2.11]) *The matrix $A \in \mathbb{B}_n$ is idempotent if and only if the following two conditions hold:*

(1) *There exist integers $r, l \ge 0$ such that A is a sum of r disjoint rectangular parts and l line parts.*
(2) *If, for some $i \ne j$, $\mathbf{R}_i$ and $\mathbf{C}_j$ are not (i,j)-disjoint, then $E_{ij} \le A$.*

Proof. We first show that a matrix satisfying the two conditions is idempotent, beginning with a check that $A \le A^2$ holds: One clearly has that $E_{ii} \le A$ implies $E_{ii}^2 \le A^2$; but $E_{ii}^2 = E_{ii}$, so $E_{ii} \le A^2$. Now consider a cell $E_{ij} \le A$ with $i \ne j$. Then, due to condition (1), E_{ij} lies in a certain rectangular or line part, and from the definitions of rectangular and line parts it follows that either $E_{ii} \le A$, or $E_{jj} \le A$, or there is a k such that $E_{kk} + E_{ik} + E_{kj} \le A$. In all cases we have $E_{ij} \le A^2$ by the rules of matrix multiplication.

To check the reverse inequality $A^2 \le A$, assume $E_{ij} \le A^2$. If $i \ne j$ we get $E_{ii}E_{jj} = O$ whence $\mathbf{R}_i$ and $\mathbf{C}_j$ are not (i,j)-disjoint since there must be two off-diagonal cells whose product is $E_{ij} \ne O$. Thus, $E_{ij} \le A$ by condition (2). And for $E_{ii} \le A^2$ we have that either $E_{ii} \le A$, or $\mathbf{R}_i$ and $\mathbf{C}_i$ are not (i,i)-disjoint. In the latter case $E_{ii} \le A$ as well.

In order to show the opposite implication we assume that A has r rectangular parts and l line parts, with $r, l \geq 0$. An off-diagonal cell F in A is, by Cor. 8.62, in some rectangular part or some line part of A. Therefore, A is a sum of r rectangular and l line parts. By Prop. 8.64 the rectangular parts of A are disjoint, and by Prop. 8.63 $E_{ij} \leq A$ follows if $\mathbf{R}_i$ and $\mathbf{C}_j$ are not (i, j)-disjoint, for $i \neq j$. □

8.3.2.3 *Generalized Inverses and Regularity*

It is known that in contrast to the real or complex case, Boolean matrices do not always have a g-inverse. On the other hand, g-inverses need not be unique: As an example take J_n, for which both I_n and J_n are generalized inverses. Another example is from [37, Section 5.3], with $A \in \mathbb{B}_{4,3}$ given as

$$A = \begin{bmatrix} 1 & 0 & 0 \\ 0 & 1 & 1 \\ 1 & 1 & 1 \\ 0 & 1 & 1 \end{bmatrix}.$$

Then all of the following matrices are g-inverses of A:

$$G_1 = \begin{bmatrix} 1\,0\,0\,0 \\ 0\,0\,0\,1 \\ 0\,1\,0\,1 \end{bmatrix}, \quad G_2 = \begin{bmatrix} 1\,0\,0\,0 \\ 0\,0\,0\,0 \\ 0\,1\,0\,0 \end{bmatrix}, \quad G_3 = \begin{bmatrix} 1\,0\,0\,0 \\ 0\,0\,0\,0 \\ 0\,1\,0\,1 \end{bmatrix}, \quad G_4 = \begin{bmatrix} 1\,0\,0\,0 \\ 0\,0\,0\,1 \\ 0\,0\,0\,0 \end{bmatrix}.$$

Let A^g denote the set of all g-inverses of A. $B \in A^g$ is said to be a *maximal g-inverse* if it has the maximum number of 1's , and to be a *minimal g-inverse*, if B has the minimum number of nonzero entries among all the matrices in A^g. In the example above G_1 is maximal, and both G_2 and G_4 are minimal in A^g; we have $G_2 \leq G_3 \leq G_1$, and $G_4 \leq G_1$, so $G_1 = G_2 \cup G_3 \cup G_4$.

According to [46, p. 427] it is known that each $A \in \mathbb{B}_2$ has a g-inverse. However, for $n \geq 3$ there exist matrices in $\mathbb{B}_n$ for which $AXA = A$ has no solution. Some simple properties of generalized inverses are listed in

Corollary 8.66.

(1) *For a nonsingular matrix A, the inverse A^{-1} is also a g-inverse, and A has no g-inverse other than A^{-1}.*
(2) *If A has a g-inverse, then it also has a pseudoinverse.*
(3) *The union of two g-inverses is again a g-inverse; a regular matrix thus has a greatest one, denoted as A_+.*
(4) *If $A_- \leq G \leq A_+$, for some minimal g-inverse A_- of A, then G is also a g-inverse of A.*

Proof. (1): Clearly, $AA^{-1}A = IA = A$. If G is a g-inverse of A, then $G = A^{-1}AGAA^{-1} = A^{-1}AA^{-1} = A^{-1}$.

(2): Let G be a g-inverse of A. Then, since $AGA = A$, we get $A(GAG)A = AGA = A$, and $(GAG)A(GAG) = GA(GAG) = GAG$.

(3): $A(G_1 \cup G_2)A = AG_1A \cup AG_2A = A$.

(4): Obvious from monotonicity of matrix product. □

Theorem 8.67. ([47, Theorem 2.4]) *A matrix A has a general inverse if and only if there exist a permutation matrices P and Q such that* $A = P\begin{bmatrix} B & BC \\ DB & DBC \end{bmatrix}Q$*, where B is nonsingular and idempotent.*

Proof. Assume $A = P\begin{bmatrix} B & BC \\ DB & DBC \end{bmatrix}Q =: P[A']Q$. Then the partial permutation matrix $G = Q^T\begin{bmatrix} I & 0 \\ 0 & 0 \end{bmatrix}P^T =: Q^T[G']P^T$ is a g-inverse of A, with the blocks of appropriate dimensions. Indeed, we get

$$AGA = P[A']Q \cdot Q^T[G']P^T \cdot P[A']Q = P[A']I[G']I[A']Q.$$

So what we finally are interested in is the product $A' \cdot G' \cdot A'$ which can be calculated as follows:

$$\begin{bmatrix} B & BC \\ DB & DBC \end{bmatrix} \cdot \begin{bmatrix} I & 0 \\ 0 & 0 \end{bmatrix} \cdot \begin{bmatrix} B & BC \\ DB & DBC \end{bmatrix} = \begin{bmatrix} BI & 0 \\ DBI & 0 \end{bmatrix} \cdot \begin{bmatrix} B & BC \\ DB & DBC \end{bmatrix} = \begin{bmatrix} B^2 & B^2C \\ DB^2 & DB^2C \end{bmatrix}.$$

The assertion follows since B is assumed to be idempotent.

Conversely, let G be a g-inverse of A. Then by Prop. 6.58(2) AG is idempotent, and by Th. 8.54 there exists a permutation matrix P such that $AG = P\begin{bmatrix} B & BE \\ DB & DBE \end{bmatrix}P^T$, where B is an $r \times r$ nonsingular idempotent matrix and r the rank of AG. It is known that $\mathsf{col}(A) = \mathsf{col}(AG)$ (see Def. 5.9), and this implies $A = P\begin{bmatrix} B & BC \\ DB & DBC \end{bmatrix}Q$ for some matrix C and permutation matrix Q. □

The following theorem presents several characterizations of matrices admitting the Moore-Penrose inverse. For its proof which needs ten pages we refer to the paper mentioned. Additional material can be found in [46], and on idempotents and also other kinds of inverses in [47].

Theorem 8.68. ([2, Theorem 1.1]) *For $A \in \mathbb{B}_{m,n}$ the following assertions are equivalent:*

(a) The Moore-Penrose inverse of A exists.

(b) The Moore-Penrose inverse of A exists and equals A^T.

(c) $AA^TA = A$.

(d) $AA^TA \le A$.

(e) Any two rows of A are either identical or disjoint (i.e., there is no column with a 1 in both the rows).

(f) Any two columns of A are either identical or disjoint.

(g) The number of 1's in any 2×2 submatrix of A is not 3.

(h) For any 2×2 submatrix of A the Moore-Penrose inverse exists.

(i) *There exist permutation matrices P, Q such that*

$$PAQ = \begin{bmatrix} J_1 & 0 & \cdots & 0 & 0 \\ 0 & J_2 & \cdots & 0 & 0 \\ \vdots & \vdots & \ddots & \vdots & \vdots \\ 0 & 0 & \cdots & J_t & 0 \\ 0 & 0 & \cdots & 0 & 0 \end{bmatrix},$$

where $J_1, J_2, \ldots, J_t$ are matrices (not necessarily square) of all ones.

(j) *There exist permutation matrices P, Q such that*

$$PAQ = \begin{bmatrix} I & C \\ D & DC \end{bmatrix},$$

where $CC^T \le I$ and $D^T D \le I$.

(k) *There exists a matrix G such that $GAA^T = A^T$ and $A^T AG = A^T$.*

8.3.2.4 Eigenvalues

Up to now we have dealt with crisp (or binary) Boolean matrices over the two-element Boolean algebra. But matrices with values in arbitrary Boolean algebras have been a subject of interest already long ago; see [38, 49]. We will present some examples of early results concerning eigenvalues and eigenvectors, with the proofs not in their original but in a form derived from more general results. As before we let $\boldsymbol{e} = (1, 1, \ldots, 1)^T$, so for all vectors $\boldsymbol{\xi}$ we have $\boldsymbol{\xi} \le \boldsymbol{e}$. In this section we let $\mathbb{B}_n$ denote matrices over an arbitrary Boolean algebra.

Proposition 8.69. ([49, Theorem 1]) *Let $\boldsymbol{\xi} = (x_1, x_2, \ldots, x_n)^T$ be an eigenvector of $A \in \mathbb{B}_n$. Then the eigenvalues of $\boldsymbol{\xi}$ form a sublattice of L consisting of the interval $[\lambda^0, \lambda^*]$, where $\lambda^0 = \boldsymbol{e}^T A\boldsymbol{\xi}$ and $\lambda^* = \lambda^0 \vee \overline{(\boldsymbol{e}^T \boldsymbol{\xi})}$.*

Proof. A special case of Th. 8.42: Recall from Section 4.4 that for Boolean algebras we have $a \stackrel{\wedge}{\rightarrow} b = \overline{a} \vee b$. □

Proposition 8.70. ([49, Theorem 2]) *An eigenvector $\boldsymbol{\xi} = (x_1, x_2, \ldots, x_n)^T$ of $A \in \mathbb{B}_n$ has a unique eigenvalue iff $\bigvee_{i \in \underline{n}} x_i = 1$.*

Proof. Assume that $\boldsymbol{\xi}$ has a unique eigenvalue. Then obviously $\lambda^0 = \lambda^*$, so by Prop. 8.69 $\overline{(\boldsymbol{e}^T \boldsymbol{\xi})} \le \boldsymbol{e}^T A\boldsymbol{\xi}$. But since $\boldsymbol{e}^T A \le \boldsymbol{e}^T$, we get $\overline{(\boldsymbol{e}^T \boldsymbol{\xi})} \le \boldsymbol{e}^T \boldsymbol{\xi}$, which is possible only if $\boldsymbol{e}^T \boldsymbol{\xi} = \bigvee_{i \in \underline{n}} x_i = 1$. Conversely, if $\boldsymbol{e}^T \boldsymbol{\xi} = 1$, then, again using Prop. 8.69, $\lambda^* = \lambda^0 \vee \overline{1} = \lambda^0$. □

Proposition 8.71. ([49, Theorem 3]) *Let λ be an eigenvalue of $A \in \mathbb{B}_n$; then the maximal eigenvector $\boldsymbol{\xi}^*(\lambda)$ associated with λ is $\lambda A^n \boldsymbol{e} \vee \overline{\lambda} \overline{(A^T \boldsymbol{e})}$.*

Proof. A special case of Th. 8.45(1) where it has been shown that

$$\boldsymbol{\xi}^*(\lambda) = (\lambda \boldsymbol{e} \vee A^T \boldsymbol{e}) \stackrel{\wedge}{\rightarrow} (\lambda A^n \boldsymbol{e}).$$

For Boolean algebras we have $a \xrightarrow{\cdot} b = \overline{a} \vee b$ (see Section 4.4), so we get

$$\xi^*(\lambda) = \overline{(\lambda e \vee A^T e)} \vee (\lambda A^n e).$$

But $\overline{(\lambda e \vee A^T e)} = \overline{\lambda e} \wedge \overline{(A^T e)}$, and $\overline{\lambda e} = \overline{(\lambda, \lambda, \ldots, \lambda)^T} = (\overline{\lambda}, \overline{\lambda}, \ldots, \overline{\lambda})^T$. Thus, $\overline{\lambda e} \wedge \overline{(A^T e)} = \overline{\lambda}\,\overline{(A^T e)}$. □

The assertion for matrices over Boolean algebras corresponding to Th. 8.44 has been given in [8, THEOREM 1]. Some more results can be found in [33].

8.3.2.5 *Some Early Results on Matrix Equations and Inequalities*

Lemma 8.72. ([38, LEMMA 5.1]) *Let $A, B, C \in \mathbb{B}_n$. Then $ABC \subseteq \overline{I}$ if and only if $\overline{A} \supseteq (BC)^T$.*

Proof. $\overline{I}$ has 1-entries everywhere except on its diagonal. So, by definition of matrix product, $ABC \subseteq \overline{I}$ if and only if $a_{ij} \wedge b_{jk} \wedge c_{ki} = 0$ for all $i, j, k \in \underline{n}$. Rewriting this, we get $\overline{a_{ij}} \geq b_{jk} \wedge c_{ki} = c^T_{ik} \wedge b^T_{kj}$ for all i, j, k, i.e. $\overline{a_{ij}} \geq \bigvee_k \left(c^T_{ik} \wedge b^T_{kj}\right)$. Therefore, $\overline{A} \geq C^T B^T = (BC)^T$. □

Theorem 8.73. ([38, THEOREM 5.2]) *Let $A, B \in \mathbb{B}_n$. Then $XA \subseteq B$ if and only if $X \subseteq \overline{(\overline{B} A^T)}$, and $AX \subseteq B$ if and only if $X \subseteq \overline{(A^T \overline{B})}$.*

Proof. $XA \subseteq B$ implies $\bigvee_j \left(x_{ij} \wedge a_{jk}\right) \leq b_{ik}$ for all $i, k \in \underline{n}$, which is equivalent to $x_{ij} \wedge a_{jk} \leq b_{ik}$, and this is the same as $x_{ij} \wedge a_{jk} \wedge \overline{b_{ik}} = x_{ij} \wedge a_{jk} \wedge \overline{b^T_{ki}} = 0$, for all $i, j, k \in \underline{n}$. Therefore, $\bigvee_j \bigvee_k \left(x_{ij} \wedge a_{jk} \wedge \overline{b^T_{ki}}\right) = 0$, whence $XA\overline{B}^T \subseteq \overline{I}$. But according to Lemma 8.72 this holds if and only if $\overline{X} \supseteq (A\overline{B}^T)^T = \overline{B} A^T$, that is, if and only if $X \subseteq \overline{(\overline{B} A^T)}$. □

Note that a matrix equation $AX = B$ is said to be **consistent** if it has one or more solutions.

Corollary 8.74. ([38, COROLLARY 5.2]) *Let $A, B \in \mathbb{B}_n$. Then $XA = B$ is consistent if and only if $B \subseteq \overline{(\overline{B} A^T)} A$.*

Proof. Assume X^* is a solution to $XA = B$. Then by Th. 8.73 $X^* \subseteq \overline{(\overline{B} A^T)}$. But $X^* A = B$ implies $B \subseteq X^* A$, so $B \subseteq X^* A \subseteq \overline{(\overline{B} A^T)} A$ by monotonicity of matrix multiplication.

Conversely, if $B \subseteq \overline{(\overline{B} A^T)} A$, then $X^* = \overline{(\overline{B} A^T)}$ is a solution since $B \subseteq X^* A$ by the hypothesis, and from $X^* \subseteq \overline{(\overline{B} A^T)}$ we get $X^* A \subseteq B$ by Th. 8.73. □

Proposition 8.75. ([46, LEMMA 1.1]) *Let $A, B \in \mathbb{B}_n$. Then the equation $AX = B$ is consistent for $X \in \mathbb{B}_n$ iff $X = C$ is a solution, with*

$$c_{ij} = \begin{cases} 1 & \text{if for all } 1 \leq k \leq n,\ b_{kj} = 1 \text{ or } a_{ki} = 0, \\ 0 & \text{otherwise.} \end{cases} \tag{8.9}$$

Proof. To prove the non-trivial direction assume that $AX = B$ is solvable for some $X \in \mathbb{B}_n$. We let $D = AC$ and show that $D = B$. $d_{ij} = 1$ implies $a_{ip} = c_{pj} = 1$ for some $p \in \underline{n}$. From $c_{pj} = 1$ we get $b_{qj} = 1$ or $a_{qp} = 0$ for all $q \in \underline{n}$ by (8.9). Since $a_{ip} = 1$, we must have $b_{ij} = 1$, whence $d_{ij} = 1$ implies $b_{ij} = 1$.

Conversely, if $b_{ij} = 1$, then, for some $r \in \underline{n}$, $a_{ir} = x_{rj} = 1$. Assuming $c_{rj} = 0$ gives $b_{qj} = 0$ and $a_{qr} = 1$ for some $q \in \underline{n}$ by (8.9). But $a_{qr} = 1$ and $x_{rj} = 1$, so $b_{qj} = 1$, contradicting $b_{qj} = 0$, whence we must have $c_{rj} = 1$ and thus also $d_{ij} = 1$. Finally, therefore, we have $AC = D = B$. □

Proposition 8.76. (according to [27, p. 108, Theorem 9.1.2]) *Let $A \in \mathbb{B}_{m\times n}$, $G \in \mathbb{B}_{n\times m}$, and $B \in \mathbb{B}_{m\times p}$. Then GB is a solution for $AX = B$ (in X), for every B for which there exists a solution X^*, if and only if G is a generalized inverse of A.*

Proof. Suppose that G is a generalized inverse of A, and take X^* to be any a solution to $AX = B$. Then $A(GB) = (AG)B = AGAX^* = AX^* = B$.

For the converse direction suppose that GB is a solution to $AX = B$ for every B such that $AX = B$ is consistent. Further recall from Section 5.1.2 that every column vector of A is in $\mathrm{col}(A)$ since $\boldsymbol{a}_{*i} = A\boldsymbol{e}_i$, with $\boldsymbol{e}_i$ the i-th standard unit vector. Let $B = [\boldsymbol{a}_{*i}, \boldsymbol{0}, \ldots, \boldsymbol{0}]$, then $X^* = [\boldsymbol{e}_i, \boldsymbol{0}, \ldots, \boldsymbol{0}]$ is a solution to $AX = B$. Then from the assumption it follows that $AGB = AG[\boldsymbol{a}_{*i}, \boldsymbol{0}, \ldots, \boldsymbol{0}] = [\boldsymbol{a}_{*i}, \boldsymbol{0}, \ldots, \boldsymbol{0}]$, whence $AG\boldsymbol{a}_{*i} = \boldsymbol{a}_{*i}$, $i \in \underline{n}$. Thus $AGA = A$. □

8.4 Linear Preservers of Matrix Invariants

In the last section of this chapter we give a short overview on another active research topic in matrix theory, the so-called *linear preserver problems,* just to let it not go unnoticed. By a linear preserver, in general, is meant a linear map of some set of algebraic structures $\mathfrak{A}$ into itself that leaves certain properties of elements $A \in \mathfrak{A}$ invariant. Linear preserver problems concern the characterization of such maps and are regarded to be solved once a complete description of all linear operators T that preserve a certain property, as e.g. ranks or determinants is given.

There are a lot of papers and results, both older and recent ones, from which we will only provide some introductory information by outlining some of the basic principles and we will also present some examples which hopefully suffice to give first impressions and also links for those who want to know more. [35, 36] may serve as good starting points.

The study of linear maps which preserve invariants of matrices goes back to 1897 when Frobenius [15] characterized the linear operators on the $n \times n$ complex matrices that preserve the determinant:

- Let $\mathbb{C}_n$ denote the set of square matrices with complex entries, and $A \in \mathbb{C}_n$. A linear map $\mathbf{T}: \mathbb{C}_n \longrightarrow \mathbb{C}_n$ such that $\det(\mathbf{T}(A)) = \det(A)$ is called a *determinant preserver.* Every determinant preserver $\mathbf{T}(A)$ is either PAQ or PA^TQ, for some $P, Q \in \mathbb{C}_n$ such that $\det(PQ) = 1$.

Sixty years later the following result has been shown (quoted from [34, p. 296]):

- ([39, Theorem 1]) If $\mathbf{T}$ is a linear operator on the $n \times n$ matrices over a field, and $\mathbf{T}$ preserves the rank of any matrix, then $\mathbf{T}(A)$ is either PAQ or PA^TQ for some fixed invertible matrices P and Q.

During the last few decades the study of linear operators on matrix spaces over sets which are not fields has become established. In particular, the study of linear operators on spaces of (0,1)-matrices and matrices over zerosumfree semirings.

Definition 8.77. *Let R be a semiring, $r, s \in R$, and $A, B \in \mathbb{S}_{m,n}(R)$ matrices over R. An operator $\mathbf{T}: \mathbb{S}_{m,n} \longrightarrow \mathbb{S}_{m,n}$ is called*

(a) **linear** *if $\mathbf{T}(rA + sB) = r\mathbf{T}(A) + s\mathbf{T}(B)$.*
(b) a **preserver** *of a certain property of A, if $\mathbf{T}(A)$ also has this property.*
(c) a **strong preserver** *of a certain property of A, if and only if $\mathbf{T}(A)$ also has this property.*
(d) a **(U,V)-operator** *if there exist invertible matrices $U \in \mathbb{S}_m$ and $V \in \mathbb{S}_n$ such that either $\mathbf{T}(A) = UAV$ or $m = n$ and $\mathbf{T}(A) = UA^TV$.*

Let R be a semiring, and let $\mathsf{M} := \mathbb{S}_{m,n}(R)$ denote some matrix space over R. As described in [35, p. 592] or [36, p. 219] the following three general types of linear preserver problems can be identified:

(1) *Linear operators preserving certain functions:* If f is a function defined on M, then $\mathbf{T}$ preserves f if $f(\mathbf{T}(A)) = f(A)$ for all $A \in \mathsf{M}$.
(2) *Linear operators preserving certain subsets:* If $\mathsf{L} \subseteq \mathsf{M}$ and $\mathbf{T}(A) \in \mathsf{L}$ for each $A \in \mathsf{L}$, i.e. $\mathbf{T}(\mathsf{L}) \subseteq \mathsf{L}$, then $\mathbf{T}$ preserves L.
(3) *Linear operators preserving certain relations:* Let $\sim$ be a relation on M. Then $\mathbf{T}$ preserves $\sim$, if $A \sim B$ implies $\mathbf{T}(A) \sim \mathbf{T}(B)$, and strongly preserves $\sim$ if $A \sim B$ if and only if $\mathbf{T}(A) \sim \mathbf{T}(B)$.

The authors also remark that problems of type (1) and (2) are related in some way (see [36, p. 221]): Given a function $f: \mathsf{M} \longrightarrow \mathsf{M}$, one could consider $\mathsf{A} := f^{-1}(\mathsf{B}) \subseteq \mathsf{M}$ for some subset $\mathsf{B} \subseteq f(\mathsf{M})$ and study the linear operators preserving A. If, for example, $f(A) = \det(A)$ then $\mathsf{A} := f^{-1}(0)$ is the set of all singular matrices, and $\mathsf{A} := f^{-1}(F - \{0\})$ is the set of all nonsingular matrices over some field F. On the other hand, if $\mathsf{A} \subseteq \mathsf{M}$ is given, one may look for a suitable function f such that $\mathsf{A} = f^{-1}(\mathsf{B})$ for some $\mathsf{B} \subseteq \mathsf{M}$. In this way linear preservers of f can be related to linear preservers of A. And given a partition P of M, one could define an equivalence relation based on the disjoint subsets in P, relating problems of type (2) and (3).

In the following we will present some typical results on preservers in addition to the two historical ones mentioned above, also using fields as underlying sets to refer to examples from both the classical and the generalized context. Proofs will be omitted, but information where to find them will be included for the interested reader.

- ([35, Theorem 2.1]) Let F be an algebraically closed field. A linear operator on the $n \times n$ matrices over F maps the set of invertible matrices into itself if and only if it is a (U, V)-operator for some invertible matrices U, V over F.

A characterization of the linear operators that strongly preserve Boolean regular matrices is as follows:

- ([55, Theorem 3.2]) Let $\mathbf{T}$ be a linear operator on $\mathbb{B}_{m\times n}$ with $\min\{m, n\} \leq 2$. Then $\mathbf{T}$ strongly preserves all regular matrices.
- ([55, Theorem 3.10]) Let $\mathbf{T}$ be a linear operator on $\mathbb{B}_{m\times n}$ with $\min\{m, n\} \geq 3$. Then $\mathbf{T}$ strongly preserves regularity if and only if it is a (U, V)-operator.

We continue with an analogous result for commutative, zerosumfree and entire semirings, for which we need some additional concepts: The *pattern* of a matrix $A \in \mathbb{S}_{m\times n}$ is the matrix $A^* \in \mathbb{B}_{m\times n}$ whose (i, j)-th entry is 0 if and only if $a_{ij} = 0$. An operator $\mathbf{T}$ on $A \in \mathbb{S}_{m\times n}$ is called a (P, Q, B)-operator it there are permutation matrices P and Q, and a matrix B with $B^* = J$ such $\mathbf{T}(X) = P(XB)Q$ for all $X \in \mathbb{S}_{m\times n}$, or $m = n$ and $\mathbf{T}(X) = P(X^T B)Q$ for all $X \in \mathbb{S}_n$. Then

- ([54, Theorem 3.13]) Let $\mathbf{T}$ be a linear operator on $\mathbb{S}_{m\times n}$ with $\min\{m, n\} \geq 3$. If $\mathbf{T}$ strongly preserves regularity then $\mathbf{T}$ is a (P, Q, B)-operator.

Our final example concerns ranks of matrices:

- ([5, Theorem 4.2]) Let $\mathbf{T}$ be a linear operator on $\mathbb{S}_{m\times n}(S)$ with S a chain semiring and $\min\{m, n\} \geq 2$. Then the following are equivalent:

  (i) $\mathbf{T}$ preserves all ranks.
  (ii) $\mathbf{T}$ preserves ranks 1 and 2.
  (iii) $\mathbf{T}$ is a (U, V)-operator.
  (iv) $\mathbf{T}$ is bijecive and preserves rank 1.

References

[1] Balaei, S., Eslami, E., Saeid, A.B.: Invertible square matrices over residuated lattices, TWMS J. of Pure and Applied Mathematics **11**, 173-188 (2020).

[2] Bapat, R.B., Jain, S.K., Pati, S.: Weighted Moore-Penrose inverse of a boolean matrix, Linear Algebra and its Applications **255**, 267-279 (1997).

[3] Beasley, L.B., Guterman, A.E.: Rank inequalities over semirings, J. Korean Math. Soc. **42**, 223-241 (2005).

[4] Beasley, L.B., Guterman, A.E., Kang, K.T., Song, S.Z.: Idempotent Boolean matrices and majorization, J. of Mathematical Sciences **152**, 456-468 (2008).

[5] Beasley, L.B., Pullman, N.J.: Fuzzy rank-preserving operators, Linear Algebra and its Applications **73**, 197-211 (1986).

[6] Beasley, L.B., Pullman, N.J.: Semiring rank versus column rank, Linear Algebra and its Applications **101**, 33-48 (1988).

[7] Birkhoff, G.: *Lattice Theory*, 3rd ed., American Mathematical Society, Providence, Rhode Island (1973).

[8] Blyth, T.S.: XIV. - On Eigenvectors of Boolean Matrices, Proc. of the Royal Soc. of Edinburgh Section A: Mathematics **67**, 196-204 (1967).

[9] Butler, K. K-H.: Binary Relations, in: Capobianco, M., Frechen, J.B., Krolik, M. (Eds), *Recent Trends in Graph Theory* (Lecture Notes in Math. 186), Springer-Verlag, Berlin-Heidelberg, pp. 25-47 (1971).

[10] Cen, J.: Fuzzy matrix partial orderings and generalized inverses, Fuzzy Sets and Systems **105**, 453-458 (1999).

[11] Cho, H.H.: Regular fuzzy matrices and fuzzy equations, Fuzzy Sets and Systems **105**, 445-451 (1999).

[12] Ćirić, M., Ignjatović, J.: The Existence of Generalized Inverses of Fuzzy Matrices, in: Kóczy, L.T. et al. (Eds.), *Interactions Between Computational Intelligence and Mathematics, Part 2*, pp. 19-38, Springer Nature Switzerland AG (2019).

[13] Duan, J.S.: The transitive closure, convergence of powers and adjoint of generalized fuzzy matrices, Fuzzy Sets and Systems **145**, 301-311 (2004).

[14] Dutta Majumder, D., Rao, P.S.S.N.V.P.: On Generalized Inverses of Fuzzy Matrices,. IETE J. of Research **44**, 149-159 (1998).

[15] Frobenius, G.: Über die Darstellung der endlichen Gruppen durch lineare Substitutionen, 1. Sitzungsbericht Preuss. Akad. Wiss. zu Berlin, 994-1015 (1897).

[16] Froidure, V.: Ranks of binary relations, Semigroup Forum **54**, 381-401 (1997).

[17] Gavalec, M.: Monotone eigenspace structure in max-min algebra, Linear Algebra and its Applications **345**, 149-167 (2002).

[18] Gavalec, M., Němcová, Z.: Steady states of max-Łukasiewicz fuzzy systems, Fuzzy Sets and Systems **325**, 58-68 (2017).

[19] Gavalec, M., Plavka, J., Ponce, D.: Tolerance and weak tolerance of interval eigenvectors in fuzzy algebra, Fuzzy Sets and Systems **410**, 60-74 (2021).

[20] Give'on, Y.: Lattice matrices, Information and Control **7**, 477-484 (1964).

[21] Han,S.C., Li, H.X.: Indices and periods of incline matrices, Linear Algebra and its Applications **387**, 143-165 (2004).

[22] Han,S.C., Li, H.X.: Invertible incline matrices and Cramer's rule over inclines, Linear Algebra and its Applications **389**, 121-138 (2004).

[23] Han,S.C., Li, H.X.: The semigroup of incline Hall matrices, Linear Algebra and its Applications **390**, 183-196 (2004).

[24] Han,S.C., Li, H.X.: Some Conditions for Matrices over an Incline To Be Invertible and General Linear Group on an Incline, Acta Mathematica Sinica **21**, 1093-1098 (2005).

[25] Han,S.C., Li, H.X., Gu, Y.D.: Standard eigenvectors of incline matrices, Linear Algebra and its Applications **389**, 235-248 (2004).

[26] Han,S.C., Li, H.X., Wang, J.Y.: On nilpotent incline matrices, Linear Algebra and its Applications **406**, 201-217 (2005).

[27] Harville, D.A.: *Matrix Algebra From a Statistician's Perspective*, Springer-Verlag New York (1997).

[28] Ignjatović, J., Ćirić, M.: Moore-Penrose Equations in Involutive Residuated Semigroups and Involutive Quantales, Filomat **31**, 183-196 (2017).

[29] Jiang, J., Shu, L., Tian, X.: On Generalized Transitive Matrices, J. Appl. Math. 2011 , Article ID 164371, 16 pages, doi:10.1155/2011/164371 (2011).

[30] Kang, K.T.: General types of idempotent (0,1)-matrices, The J. of Basic Sciences Cheju National Univ. **17**, 89-102 (2004).

[31] Kang, K.T., Song, S.Z., Yang, Y.O.: Structure of idempotent matrices over chain semirings, Bull. Korean Math. Soc. **44**, 721-729 (2007).

[32] Kim, K.H.: *Boolean Matrix Theory and Applications*, Marcel Dekker, Inc., New York and Basel (1982).

[33] Kirkland, S., Pullman, N.J.: Boolean spectral theory, Linear algebra and its applications **175**, 177-190 (1992).

[34] Kirkland, S., Pullman, N.J.: Linear operators preserving invariants of nonbinary Boolean matrices, Linear and Multilinear Algebra, **33**, 295-300 (1993).

[35] Li, C.K., Pierce, S.: Linear Preserver Problems, The American Math. Monthly **108**, 591-605 (2001).

[36] Li, C.K., Tsing, N.K.: Linear preserver problems: A brief introduction and some special techniques, Linear Algebra and its Applications **162-164**, 217-235 (1992).

[37] Liu, B., Lai, H.J.: *Matrices in Combinatorics and Graph Theory*, Kluwer Academic Publishers, Dordrecht (2000).

[38] Luce, R. D.: A note on Boolean matrix theory, Proc. of the American Math, Soc. **3**, 382-388, (1952).

[39] Marcus, M., Moyls, B.N.: Linear transformations on algebras of matrices, Canad. J. Math. **11**, 61-66 (1959).

[40] Marenich, E.E., Kumarov, V.G.: Inversion of matrices over a pseudocomplemented lattice, J. of Mathematical Sciences **144**, 3968-3979 (2007).

[41] Marenich, E.E.: Determining the Schein rank of Boolean matrices, in: Olshevsky, V., Tyrtyshnikov, E. (Eds), *Matrix Methods: Theory, Algorithms and Applications*, World Scientific, pp. 85-103 (2010).

[42] Meenakshi, A.R.: *Fuzzy Matrix: Theory and Applications*, MJP Publishers, Chennai, India (2008).

[43] Myšková, H., Plavka, J.: Reachability of eigenspaces for interval matrices in max-min algebra, Linear Algebra and its Applications **578**, 314-333 (2019).

[44] Plavka, J.: Subeigenvectors and supereigenvectors of fuzzy matrices, Fuzzy Sets and Systems **339**, 62-72 (2018).

[45] Plavka, J., Gazda, M.: Generalized eigenproblem of interval max-min (fuzzy) matrices, Fuzzy Sets and Systems **410**, 27-44 (2021).

[46] Plemmons, R.T.: Generalized Inverses of Boolean Relation Matrices , SIAM J. Appl. Math. **20**, 426-433 (1971).

[47] Rao, P. S.S.N.V.P., Rao, K.P. S.B.: On generalized inverses of Boolean Matrices, Linear Algebra and Applications **11**, 135-153 (1975).

[48] Roman, S.: *Lattices and Ordered Sets*, Springer Science+Business Media, New York (2008).

[49] Rutherford, D.E.: III. -The Eigenvalue Problem for Boolean Matrices, Proc. Roy. Soc. Edinburgh Section A: Mathematics **67**, 25-38 (1965).

[50] Sanchez, E.: Resolution of eigen fuzzy sets equations, Fuzzy Sets and Systems **1**, 69-74 (1978).

[51] Sanchez, E.: Eigen fuzzy sets and fuzzy relations. J. of Mathematical Analysis and Applications **81**, 399-421 (1981).

[52] Shitov, Y.N.: On Boolean matrices with full factor rank, Sbornik: Mathematics **204**, 1691-1699 (2013).

[53] Skornyakov, L.A.: Invertible matrices over distributive lattices, Siberian Mathematical Journal **27**, 289-292 (1986).

[54] Song, S.Z., Kang, K.T., Beasley, L.B., Sze, N.S.: Regular matrices and their strong preservers over semirings. Linear Algebra and its Applications **429**, 209-223 (2008).

[55] Song, S.Z., Kang, K.T., Kang, M.H.: Boolean regular matrices and their strongly preservers. Bull. Korean Math. Soc. **46**, 373-385 (2009).

[56] Tan, Y.J.: Eigenvalues and eigenvectors for matrices over distributive lattices, Linear Algebra and its Applications **283**, 257-272 (1998).

[57] Tan, Y.J.: The Semigroup of Hall Matrices over Distributive Lattices, Semigroup Forum **61**, 303-314 (2000).

[58] Tan, Y.J.: On the eigenproblem of matrices over distributive lattices, Linear Algebra and its Applications **374**, 87-106 (2003).

[59] Toth, H.: *Relations: Concrete, Abstract and Applied - An Introduction*, World Scientific Publishing, Singapore (2020).

[60] Wang, Q., Qin, N., et al.: On Monotone Eigenvectors of a Max-T Fuzzy Matrix, J. of Applied Mathematics and Physics **6**, 1076-1085 (2018).

[61] Yang, L., Yang, S.L.: On Generalized Inverses of $m \times n$ Matrices Over a Pseudoring, Vietnam Journal of Mathematics **50**, 261-274 (2022).

[62] Zadeh, L.A.: Fuzzy Sets, Inform. and Control **8**, 338-353 (1965).

[63] Zadeh, L.A.: Similarity Relations and Fuzzy Orderings, Information Science **3**, 177-200 (1971).

[64] Zhao, C.K.: Inverses of L-fuzzy matrices, Fuzzy Sets and Systems **34**, 103-116 (1990).

PART 3

Fuzzy Relations and Fuzzy Graphs

CHAPTER 9

BASIC PROPERTIES OF FUZZY RELATIONS

For the topic of *many-valued relations* (more convenient and well known as *fuzzy relations*) the following aspects are of relevance in the context of this book: On the syntactic level the concept of (binary) fuzzy relations is a generalization of the classical concept of (binary) crisp relations; so the general formalism must reduce to the two-valued one. On the semantic level fuzzy relations admit a much more fine grained modeling of relationships between two objects under consideration. Instead of only asserting the existence or absence of a relationship, information about its strength can be given. And from the viewpoint of applications, the role of fuzzy relations in approximate reasoning and fuzzy control is the most interesting aspect; it has led to a vivid research in fuzzy relation equations and inequalities.

The current chapter presents many-valued versions of some important crisp concepts around binary relations, and thus becomes something like a generalizing mirror of things mentioned in Chapter 2.

For some readers, it may be advisable at this point to be provided with a few general (informal and formal) remarks on the rationales for and properties of generalized set theoretical and logical concepts and operations. In order to not interrupt the considerations on our proper topic, these remarks have been included in the *first two sections of this chapter, which can possibly be skipped by those with enough background knowledge.* But, of course, I encourage readers to skim through the text, to get a first impression of how carefully one must handle supposedly simple generalizations. Afterwards we will proceed with fuzzy relations, the real topic of this chapter.

9.1 Fuzzy Sets: Basic Definitions and Properties

Technically spoken, the transition from a two-valued to a many-valued conception of the notions of set and relation can best be approached via a generalization of their characteristic function: Let $A = \{a_1, a_2, \ldots, a_m\}$ and $B = \{b_1, b_2, \ldots, b_n\}$ two sets. In Def. 2.12 a Boolean matrix $M_\rho = (m_{ij}) \in \mathbb{B}_{m\times n}$ has been defined to represent a relation $\rho \subseteq A \times B$, in case $m_{ij} = 1$ iff $a_i \rho b_j$. This means that $m_{ij} = 1$ iff $\langle a_i, b_j \rangle \in \rho$ and $m_{ij} = 0$ otherwise. $M_\rho : A \times B \to \mathbf{2}$ can thus be viewed as the (classical) characteristic function of ρ. The crucial steps towards many-valued concepts are (1): the replacement of **2** by another more general poset, usually a complete residuated lattice L, and (2): to choose suitable generalized set-theoretical and logical operations on L for intersection and union, and for conjunction, disjunction, negation and implication, respectively.

On the syntactical level we start from scratch. Let A be a non empty set. It is well-known that there is a bijective correspondence between the family of all subsets of A and the family $\mathbf{2}^A$ of all functions from A to $\mathbf{2} = \{0,1\}$, which assigns to each $B \subseteq A$ its **characteristic function** $\chi_B : A \longrightarrow \mathbf{2}$ defined as

$$\chi_B(a) = \begin{cases} 1, & \text{if } a \in B \\ 0, & \text{otherwise.} \end{cases} \tag{9.1}$$

As just mentioned, the enhancement to many-valued sets can be best achieved by using a lattice L with $|L| > 2$ as valuation space: Using functions $\mu : A \to L$ yields generalized characteristic functions. According to [18], μ is called an ***L*-fuzzy set**, and more specific also an L-fuzzy set over A denoted μ_A. The term *fuzzy* has first been used in a mathematical context for the concept of a **fuzzy set** introduced in Zadeh's 1965 paper [47], where L is taken to be **unit interval** $[0,1] \subseteq \mathbb{R}$.

$\mu_A(x) = l$ is usually read as "*x is element of A with degree l*" or "*the proposition '$x \in A$' has truth degree l*". We will not discuss here the possible interpretations of these statements; instead we are interested in what are desirable and reasonable properties for L as a set of generalized truth values. This is the topic of this section.

We start with introducing the most basic terminology and formal aspects of fuzzy set theory (henceforth often abbreviated to FST)[1]. This is necessary for appreciating the idea and motivation that lies behind the step from the two-valued yes-no or black-and-white description of phenomena to a more fine-grained many-valued modeling mechanism. To first of all give you an impression of what is revolutionary in the fuzzy or graded approach, and which has therefore been felt to be a big challenge to the traditional view, we quote from [49]:

> Given the deeply entrenched tradition of scientific thinking which equates the understanding of a phenomenon with the ability to analyze it in quantitative terms, one is certain to strike a dissonant note by questioning the growing tendency to analyze the behavior of humanistic systems as if they were mechanistic systems governed by difference, differential, or integral equations.
>
> ...
>
> An alternative approach ...is based on the premise that the key elements in human thinking are not numbers, but labels of fuzzy sets, that is, classes of objects in which the transition from membership to non-membership is gradual rather than abrupt.

[1]For more details start with a look into the early classic [13] from 1980 which still carries a lot of Zadeh's original motivation and intents but also does already hint to the formal development directions to emerge later on. You may continue with [26], another renowned book.

Thus, we have a paradigm shift from a quantitative to a qualitative, but only seemingly imprecise, approach for the analysis and description of phenomena, because this approach is still being cast into a classical precise and quantitative language. Nevertheless, FST had to overcome a lot of opposition, especially during its early history, e.g. a heavy dispute if fuzzy sets have a right to exist on their own in addition to probability theory. Let us continue by fixing the new concept in

Definition 9.1. *Let X be a crisp set, and L be a complete residuated lattice with $|L| \geq 2$. An L-***fuzzy set** *(also* **fuzzy set***) A over (in, of) X is (given by) a map $\mu_A : X \to L$. X is called the* **universe of discourse***, comprising all objects under consideration. μ_A is also called the* **membership function** *of A, and $\mu_A(x) \in L$ the* **membership degree** *of x in A, often also denoted by the shorthand $A(x)$.*

Fuzzy sets are defined in a way such that both the classical two-valued and also a number of many-valued variants are subsumed. If we want to emphasize that a set X is to be understood in the classical sense, we will call it a **crisp set**: The universe of discourse of a fuzzy set is an example. The set of all L-fuzzy sets over a universe of discourse X is denoted by $\mathcal{F}^L(X)$, and for $L = [0,1]$ we simply write $\mathcal{F}(X)$.

Note from the definition above that the terminology for this basic notion is a little bit sloppy in literature (or pragmatic, if you want): Sometimes a fuzzy set is identified with, and sometimes is said to be described by, its membership function. But this is not really confusing: To remedy this terminological inconsistency, one may consider the notion of a fuzzy set an extra-mathematical notion that is not defined itself; only the notion of a membership function of a fuzzy set is defined ([4], p. 79).

Henceforth, we will not always distinguish between Zadeh's original version of a fuzzy set (with $L = [0,1] \subseteq \mathbb{R}$) and the more general concept of L-fuzzy sets, and we will often simply use the term '*fuzzy*' to indicate that we are dealing with a many-valued variant. Please note that Def. 9.1 had to be formulated in a way ($|L| \geq 2$) such that it comprises classical sets as a special case with the usual characteristic function acting as membership function.

The use of lattices as valuation spaces imparts to the theory a special flavor which emphasizes order theoretic aspects: It may be helpful to think of fuzzy sets as sets that are ordered according to the membership degrees of their elements. The question now is: What is described with such a generalized characteristic function? A look into the early days of fuzzy set theory shows that many of the first examples of fuzzy sets are in connection with human perceptual equipment like e.g. TALL, COLD, LOUD, HIGH etc. A degree of membership in a fuzzy set HEAVY may be thought of as a degree of compatibility or typicality of a quantitative "objective" measurement like 3KG, 90KG or 3 TONS with a linguistic label '*heavy*' in a given context under consideration. Membership degrees can also be interpreted as degree of truth, e.g. of the statement "*She earns a* HIGH *salary.*". For some more epistemological considerations on fuzziness see [41, 42]. Furthermore, think of e.g. a set of LARGE numbers, a CHEAP car, or of experiencing a GREAT LOSS during a stock exchange crash. All these examples clearly indicate that the choice of a 'fuzzy label' is highly subjective and context dependent. But it is just this flexibility of human language that allows for efficient communication: RED stands for a lot of different wavelengths of the spectrum, among which we cannot discriminate between many pairs of different values when using our inborn perceptual equipment only; and, most of the time in our every-day life, we also have no need to do so.

Just as classical relations are sets, fuzzy relations are fuzzy sets. It makes therefore sense to continue with some basic notions for fuzzy sets. Since (classical) sets are a special case of fuzzy sets, their operations must appropriately be subsumed under the formalism of fuzzy sets. If you translate the definitions for the basic set operations into operations on their characteristic functions, you will more or less automatically arrive at the corresponding definitions for fuzzy sets as can bee seen from

Definition 9.2. *Let S, T be two L-fuzzy sets over a universe X with membership function μ_S and μ_T, respectively, and $(L, \wedge, \vee, \otimes, \curvearrowright, 0, 1)$ a complete residuated lattice.*

(a) *S is called a* **fuzzy subset** *of T, denoted $S \subseteq T$, if $\forall x \in X[\mu_S(x) \leq \mu_T(x)]$.*
(b) *S and T are said to be equal, $S = T$, if $S \subseteq T$ and $T \subseteq S$.*
(c) *We say that S is a* **proper fuzzy subset** *of T, $S \subset T$, if $S \subseteq T$ but $S \neq T$.*
(d) *We say that S is a* **strong fuzzy subset** *of T, $S \dot{\subseteq} T$, if $S \subseteq T$ and $S(x) > 0 \rightarrow S(x) = T(x)$.*
(e) *The* **intersection** *of two fuzzy sets is $S \cap T$ with $\mu_{S \cap T}(x) = \mu_S(x) \wedge \mu_T(x)$.*
(f) *The* **union** *of two fuzzy sets is $S \cup T$ with $\mu_{S \cup T}(x) = \mu_S(x) \vee \mu_T(x)$.*
(g) *The* **product** *of two fuzzy sets is $S \otimes T$ with $\mu_{S \otimes T}(x) = \mu_S(x) \otimes \mu_T(x)$.*
(h) *The* **complement** *of a fuzzy set S is $\overline{S}(x) = 1 - S(x)$, for $L = [0, 1] \subseteq \mathbb{R}$.*

Apart from complement, the above definitions clearly are suitable for all lattices L usually involved (see Def. 1.39(c)), so you should read them in a generic way just as meet and join in a lattice which, however, must obey the involution and De Morgan laws (see Prop. 1.52) to become compatible when restricted to the two-element Boolean algebra. $\overline{S}$ is also called the *standard complement* of S (using the classical standard version of fuzzy sets). A few more concepts will be helpful for later considerations:

Definition 9.3. *Let A be an L-fuzzy set over a universe X, and 0 be the minimal and 1 be the maximal element of L.*

(a) *The* **support** *of A is the set defined as $\mathsf{supp}(A) := \{x \in X : A(x) > 0\} \subseteq X$.*
(b) *The* **level set of** *A is the set defined as $\Lambda(A) := \{A(x) \in L : x \in \mathsf{supp}(A)\} \subseteq L$.*
(c) *The α-***cut** *of A is the crisp set $A_\alpha := \{x \in X : A(x) \geq \alpha\} \subseteq X$.*
(d) *The* **strong** *α-***cut** *of A is the crisp set $A_{\dot{\alpha}} := \{x \in X : A(x) > \alpha\}$, and $A_{=\alpha} := \{x \in X : A(x) = \alpha\} \subseteq A_\alpha$ the α-***level** *of A.*
(e) *The α-***restriction** *of A is the fuzzy set $A_{|\alpha}$ with $A_{|\alpha}(x) = A(x)$ if $A(x) \geq \alpha$ and $A_{|\alpha}(x) = 0$ otherwise. Clearly, $A_{|\alpha}(x) \dot{\subseteq} A$.*
(f) *Provided that $\Lambda(A)$ is a chain: the* **height of** *A is $\mathsf{hgt}(A) := \bigvee_{x \in X} A(x) \in \Lambda(A)$.*
(g) *A fuzzy set A is called* **normal** *if $\mathsf{hgt}(A) = 1$.*
(h) *The* **core** *of A is the set $\mathsf{core}(A) := \{x \in X : A(x) = 1\} \subseteq X$.*

(Concerning (f) note that for L a partial order we might get $\mathsf{hgt}(A) \notin \Lambda(A)$ in which case we do not really learn about A. The requirement of L being a bounded chain would of course do, but is more restrictive than the one on the level set $\Lambda(A)$. In fact, $\bigvee_{x\in X} A(x) \in \Lambda(A)$ suffices for $\mathsf{hgt}(A)$ being a meaningful information about A.)

This definition is a first illustration of the increase in possibilities that comes along with the many-valued approach: Having left behind the severe restriction of two-valued classification, one can now decide to take only those (pairs of) elements into consideration whose degree of compatibility (strength of relatedness) exceeds a certain threshold. And one can do this in two ways, both of which limit the number of candidates for further processing: Applying the concept of α-cut of a fuzzy set you will continue your considerations (or computations) using smaller crisp sets but you will loose information contained in the membership degrees; and applying α-restrictions you get smaller fuzzy sets.

(Strong) α-cuts have some interesting properties, among them a relationship between fuzzy sets and families of crisp subsets of the underlying universe; see e.g. [44, Section 2.3] which is based on the real unit interval (which is a totally ordered set and a complete lattice).

Proposition 9.4. ([44, Proposition 2.8]) *Let $L = [0,1] \subseteq \mathbb{R}$, $A, B \in \mathcal{F}^L(X)$, $K \subseteq L$, $\alpha, \beta \in L$, and $A_L := \{A_\alpha : \alpha \in \Lambda(A)\} \subseteq \mathbf{2}^X$.*

(1) $\mathsf{supp}(A) = A_{\dot{0}}$.
(2) $A_{\dot\alpha} \subseteq A_\alpha$, *for all* $\alpha \in L$.
(3) $\alpha < \beta$ *implies* $A_\beta \subseteq A_\alpha$, $A_{\dot\beta} \subseteq A_{\dot\alpha}$, *and* $A_\beta \subset A_{\dot\alpha}$.
(4) $\bigcap_{\alpha\in K} A_\alpha = A_{\bigvee_{\alpha\in K}\alpha}$.
(5) for each $x \in X$, $\bigcap\{A_\alpha : x \in A_\alpha\} \in A_L$.
(6) $(A\cap B)_\alpha = A_\alpha \cap B_\alpha$ *and* $(A\cup B)_\alpha = A_\alpha \cup B_\alpha$.
$(A\cap B)_{\dot\alpha} = A_{\dot\alpha} \cap B_{\dot\alpha}$ *and* $(A\cup B)_{\dot\alpha} = A_{\dot\alpha} \cup B_{\dot\alpha}$.
(7) $\bigcap_{i\in I}(A_i)_\alpha = \left(\bigcap_{i\in I} A_i\right)_\alpha$ *and* $\bigcup_{i\in I}(A_i)_\alpha \subseteq \left(\bigcup_{i\in I} A_i\right)_\alpha$.
$\bigcap_{i\in I}(A_i)_{\dot\alpha} \subseteq \left(\bigcap_{i\in I} A_i\right)_{\dot\alpha}$ *and* $\bigcup_{i\in I}(A_i)_{\dot\alpha} = \left(\bigcup_{i\in I} A_i\right)_{\dot\alpha}$.
(8) $\overline{A}_\alpha = \overline{A_{1\dot-\alpha}}$ *and* $\overline{A}_{\dot\alpha} = \overline{A_{1-\alpha}}$, *for* $L = [0,1] \subseteq \mathbb{R}$.

Proof. (3): If $a \le \beta$, then $A_\beta(x) = 1$ implies $A_\alpha(x) = 1$; thus $A_\beta \subseteq A_\alpha$.
(4): $x \in A_{\bigvee_{\alpha\in K}\alpha} \leftrightarrow A(x) \ge \bigvee_{\alpha\in K}\alpha \leftrightarrow \forall\alpha \in K[A(x) \ge \alpha] \leftrightarrow x \in \bigcap_{\alpha\in K} A_\alpha$.
(5): Let $x \in X$; then $x \in A_\alpha \leftrightarrow A(x) \ge \alpha \leftrightarrow A_\alpha(x) = 1$. By Prop. 9.5 $A(x) = \bigvee_{\alpha\in\Lambda(A)}\{\alpha : A_\alpha(x) = 1\}$, and by (4) $\bigcap\{A_\alpha : x \in A_\alpha\} = A_{\bigvee\{\alpha\in\Lambda(A):A_\alpha(x)=1\}} \in A_L$.
(6): $x \in (A\cup B)_\alpha \leftrightarrow (A\cup B)_\alpha(x) \ge \alpha \leftrightarrow A_\alpha(x) \vee B_\alpha(x) \ge \alpha \leftrightarrow A_\alpha(x) \ge \alpha \parallel B_\alpha(x) \ge \alpha \leftrightarrow x \in A_\alpha \parallel x \in B_\alpha \leftrightarrow x \in A_\alpha \cup B_\alpha$.
(7): For all $x \in X$ and $\alpha \in L$ we have: $x \in \bigcup_{i\in I}(A_i)_\alpha \to \exists i \in I[x \in (A_i)_\alpha] \to \exists i \in I[A_i(x) \ge \alpha] \to \sup_{i\in I} A_i(x) \ge \alpha \to x \in \left(\bigcup_{i\in I} A_i\right)_\alpha$.
(8): $x \in \overline{A_{1\dot-\alpha}} \leftrightarrow x \notin A_{1\dot-\alpha} \leftrightarrow A(x) \le 1-\alpha \leftrightarrow \overline{A}(x) \ge \alpha \leftrightarrow x \in \overline{A}_\alpha$. □

Next we give a well known classical result on fuzzy sets which needs the following definition: For $\alpha \in [0,1]$ and $A \in \mathcal{F}(x)$ we set $\alpha A \in \mathcal{F}(X)$ by $(\alpha A)(x) = \alpha \wedge A(x)$; so

$(\alpha A)(x) = \alpha$ for $x \in A_\alpha$ and $(\alpha A)(x) = 0$ otherwise.

Proposition 9.5. (First Decomposition Theorem; [44, Theorem 2.7]), [26, Theorem 2.5]) *Any $A \in \mathcal{F}(X)$ can be decomposed using its α-cuts as follows: $A = \bigcup_{\alpha \in [0,1]} \alpha A_\alpha$.*

Proof. $\left(\bigcup_{\alpha\in[0,1]} \alpha A_\alpha\right)(x) = \bigvee_{\alpha\in[0,1]}(\alpha A_\alpha)(x) = \bigvee_{\alpha\in[0,1]}(\alpha \wedge A_\alpha(x))$
$= \max\{\bigvee_{x\in A_\alpha}(\alpha \wedge A_\alpha(x)), \bigvee_{x\notin A_\alpha}(\alpha \wedge A_\alpha(x))\} = \bigvee_{\alpha\leq A(x)} \alpha = A(x)$. □

9.2 Many-valued Connectives on Complete lattices

A lot has been written and there have been many debates about the possible meanings of membership degrees. But these discussions will not be our concern here, where we will give a short overview on syntactical aspects of membership degrees by many-valued connectives.

Subsequently to Def. 1.50 we have already mentioned $\mathbf{2} = (\{0,1\}, \vee, \wedge, \overline{}, 0, 1)$, with $\vee, \wedge$ denoting max, min, and $\overline{x} = 1 - x$, as being the simplest Boolean algebra having more than one element. Replacing $\{0,1\}$ with $[0,1]$ one gets the basic operations of classical fuzzy set theory, but $([0,1], \vee, \wedge, \overline{}, 0, 1)$ is no longer a Boolean but only a De Morgan algebra since 0 and 1 are the only complemented elements.

If one views $\mathbf{2} = (\{0,1\}, \vee, \wedge, \overline{}, 0, 1)$ as an imploded very special case of more general structures, the question is about what are consistent "explosions" of it. E.g. interpreting $\vee = \max$ and $\wedge$ as normal multiplication, we also get operations compatible with the two-valued case. Indeed, there exist infinitely many such binary operations on the unit interval $[0,1]$, both for intersection (called *t-norms*) and union (called *t-conorms*) of fuzzy sets.

In the remainder of this section we will first look on t-norms on the unit interval $[0,1]$, followed by some remarks on their generalization to bounded lattices.

9.2.1 *t-Norms on* $[0,1] \subseteq \mathbb{R}$

As just mentioned above, there are infinitely many t-norms and t-conorms on $[0,1]$ by which min and max can be appropriately replaced. For introductions see e.g. [16, chapter 1], [26, Chapter 3] or [25], and for still more details the monograph [24]. So, though our focus is on binary relations, we have to be aware that whole families of suitably generalized set-theoretical and logical operations for intersection and union, for conjunction and disjunction, and for negation and implication, are available.

Definition 9.6.

(a) *A* **t-norm** *is a mapping* $\otimes : [0,1] \times [0,1] \longrightarrow [0,1]$ *which makes* $([0,1], \otimes, 1)$ *a commutative ordered monoid, i.e.* $x \otimes 1 = x$.

(b) *A* **t-conorm** *is a mapping* $\oplus : [0,1] \times [0,1] \longrightarrow [0,1]$ *which makes* $([0,1], \oplus, 0)$ *a commutative ordered monoid, i.e.* $0 \oplus x = x$.

Examples for important t-norms on the unit interval $[0,1]$ are:

(1) the *Gödel t-norm:* $a \otimes_G b := \min\{a,b\}$,
(2) the *algebraic product* or *product t-norm:* $a \otimes_P b := ab$,
(3) the *bounded difference* or *Łukasiewicz t-norm:* $a \otimes_L b := \max\{a+b-1,0\}$.

Examples for important t-conorms on the unit interval $[0,1]$ are:

(1) the *Gödel t-conorm:* $a \oplus_G b := \max\{a,b\}$,
(2) the *algebraic sum* or *product t-conorm:* $a \oplus_P b := a+b-ab$,
(3) the *bounded sum* or *Łukasiewicz t-conorm:* $a \oplus_L b := \min\{a+b,1\}$.

Definition 9.7.

(a) A t-norm $\otimes$ is **continuous** *if for all convergent sequences $(x_n)_{n\in\mathbb{N}}, (y_n)_{n\in\mathbb{N}}$ one has $\lim_{n\to\infty}\{x_n \otimes y_n\} = \lim_{n\to\infty}\{x_n\} \otimes \lim_{n\to\infty}\{y_n\}$.*
(b) A t-norm $\otimes$ is **left-continuous** *if for each $y \in [0,1]$ and for each non-decreasing sequence $(x_n)_{n\in\mathbb{N}}$ one has $\lim_{n\to\infty}\{x_n \otimes y\} = \lim_{n\to\infty}\{x_n\} \otimes y$.*
(c) Let $\otimes$ be a left-continuous t-norm; $\frown_\otimes$ defined by $x \frown_\otimes y := \sup\{t \in [0,1] : x \otimes t \le y\}$ is called the **residual implication**, *also R-***implication**, *generated by $\otimes$.*

(From Th. 1.17(7) we know that $A \to B \equiv \neg A \vee B$. R-implication is related to the following classical set-theoretic identity: $\overline{A} \cup B = \overline{(A\backslash B)} = \bigcup\{C : A \cap C \subseteq B\}$ where $\backslash$ denotes set difference (see [14]). Another common viewpoint is to interpret a fuzzy logic implication $\to$ in a way that conjunction and implication form an adjoint pair. This is reasonable: for two propositions P and Q, $P \to Q$ should be understood as the weakest proposition such that P together with $P \to Q$ implies Q; see [1, p. 26] or Section 9.2.3.)

All of the three t-norms mentioned above are continuous. Their residua and negations are as follows:

(1) $a \frown_G b = \begin{cases} 1 & \text{if } a \le b, \\ b & \text{otherwise} \end{cases}$ and $\neg_G a = \begin{cases} 1 & \text{if } a = 0, \\ 0 & \text{otherwise.} \end{cases}$

(2) $a \frown_P b = \begin{cases} 1 & \text{if } a \le b, \\ b/a & \text{otherwise} \end{cases}$ and $\neg_P a = \begin{cases} 1 & \text{if } a = 0, \\ 0 & \text{otherwise.} \end{cases}$

(3) $a \frown_L b = \begin{cases} 1 & \text{if } a \le b, \\ 1-a+b & \text{otherwise} \end{cases}$ and $\neg_L a = 1-a$.

A well known result which explains the significance of left continuity is given in

Theorem 9.8. ([20, Proposition 5.4.2], [24, 2.47 Proposition]) *A t-norm $\otimes$ and its residual implication $\frown_\otimes$ form an adjoint pair $(\otimes, \frown_\otimes)$, i.e. $a \otimes b \le c \leftrightarrow b \le a \frown_\otimes c$, iff $\otimes$ is a left-continuous t-norm.*

Negation is another function defined on the real unit interval, generalizing classical two-valued negation, that is of interest in connection with t-norms and t-conorms. In the following we will also need the notion of an automorphism.

Definition 9.9.

(a) *A function $\varphi\colon [0,1] \longrightarrow [0,1]$ is called an* **order automorphism** *if it is continuous, strictly increasing and obeys the border conditions $\varphi(0) = 0$ and $\varphi(1) = 1$.*
(b) *An order automorphism φ is* **reciprocal** *if $\varphi(1-x) = 1-\varphi(x)$ for all $x \in [0,1]$.*
(c) *Given a t-norm T and an order automorphism φ, the* **φ-transform of** *T is the function $T_\varphi(x,y) := \varphi^{-1}(T(\varphi(x),\varphi(y)))$.*

Note that if φ is an order automorphism, then also φ^{-1}; and that φ is reciprocal if and only if φ^{-1} is. By [24, Proposition 2.28(iv)] the φ-transform T_φ of a t-norm T is also a t-norm. For an order automorphism φ, the t-norms T and T_φ are then isomorphic in the sense that for all $x, y \in [0,1]$ $\varphi(T_\varphi(x,y)) = T(\varphi(x),\varphi(y))$.

Definition 9.10.

(a) *Any nonincreasing function $N\colon [0,1] \longrightarrow [0,1]$ satisfying the border conditions $N(0) = 1$ and $N(1) = 0$ is called a* **negation**.
(b) *A negation is called* **strict** *if it is ontinuous and strictly decreasing.*
(c) *A strict negation N is called* **strong** *if N is also involutive, i.e. $N(N(x)) = x$ for all $x \in [0,1]$.*

Since a strict negation N is a strictly decreasing and continuous function, its inverse N^{-1} is also a strict negation, generally different from N. Obviously, we have $N^{-1} = N$ if and only if N is involutive.

The most important negation is the so-called *standard negation* $N_s\colon [0,1] \longrightarrow [0,1]$, given by $N_s(x) = 1-x$. Note that N_s is strong, whereas $N(x) = 1-x^2$ is strict, but not strong. And an example of a negation which is not strict, hence not strong, is the Gödel negation $\neg_G$.

Theorem 9.11. ([43]) *A function $N\colon [0,1] \longrightarrow [0,1]$ is a strong negation if and only if there exists an order automorphism φ such that $N(x) = N_\varphi(x) = \varphi^{-1}(1-\varphi(x)), x \in [0,1]$.*

For a t-norm T, a t-conorm S, and arbitrary negations N_1, N_2 two kinds of De Morgan laws can be formulated (see [16, Section 1.6]):

$$N_1(S(x,y)) = T(N_1(x), N_1(y)), \tag{9.2}$$

$$N_2(T(x,y)) = S(N_2(x), N_2(y)). \tag{9.3}$$

For strong negations these two laws are equivalent; and if T and S are given then (9.2) is true with a strict negation $N_1 = N$ if and only if (9.3) holds with $N_2 = N^{-1}$.

Proposition 9.12. ([16, Proposition 1.9]) *Let N be any strict negation. Then*

(1) *For any t-norm T there exists a t-conorm S with $S(x,y) = N^{-1}(T(N(x),N(y))$ which fulfills (9.2) with $N_1 = N$. If T is continuous, then also S. S is called the* **N-dual** *of T.*

(2) *For any t-conorm S there exists a t-norm T with $T(x,y) = N^{-1}(S(N(x),N(y))$ which fulfills (9.3) with $N_2 = N$. If S is continuous, then also T. T is called the* N-**dual** *of S.*

Definition 9.13. ([24, p. 232]) *Let T be a t-norm, S be a t-conorm and N be a negation.*

(a) *(T,S,N) is called a* **De Morgan triple** *if*

$$N(S(x,y) = T(N(x),N(y)) \quad \text{and} \quad N(T(x,y)) = S(N(x),N(y)),$$

i.e. (9.2) *and* (9.3) *hold with $N_1 = N_2 = N$.*

(b) *A De Morgan triple (T,S,N) is called* **strong** *or a* **Łukasiewicz triple** *if there exists an automorphism φ of the unit interval such that*

$$\begin{aligned} T(x,y) &= W_\varphi(x,y) := \varphi^{-1}(W(\varphi(x),\varphi(y))), \\ S(x,y) &= W'_\varphi(x,y) := N_\varphi(W_\varphi(N_\varphi(x),N_\varphi(y))), \\ N(x) &= N_\varphi(x) := \varphi^{-1}(1-\varphi(x)). \end{aligned}$$

The Łukasiewicz triple $(W_\varphi, W'_\varphi, N_\varphi)$ will be denoted as $\mathbf{Ł}_\varphi$.

Def. 9.13(a) means that for a given t-norm T (T,S,N) is a De Morgan triple if and only if N is a strong negation and S is the N-dual of T.

(A remark on notation: In order to reduce transcription labor we adhere to the widespread habit within the area of preference modeling and denote the Łukasiewicz t-(co)norm by W, thus $W(x,y) = \max(x+y-1,0)$ and $W'(x,y) = \min(x+y,1)$, respectively).

9.2.2 *The General Case: t-Norms on Bounded Lattices*

In the foregoing section we have introduced t-(co)norms, residua and negations on the real unit interval $[0,1]$ as generalizations of the operations min, max and $1-.$, originally used in [47]. This has been done on a pure algebraic basis without any examination if these concepts are suitable candidates for the generalization of the connectives of classical two-valued logic. There are a number of papers and books providing detailed discussions of both intuitive and formal aspects of generalized connectives which will serve as source of information; see e.g. [3,21], [20, Chapter 5], or [23, Section 2.1]. In case you want to get some historical background too, have a look into [5, Chapter 4].

Now we continue with a few general (both formal and informal) remarks on the rationales for and properties of generalized logical connectives, and, as you might perhaps suspect, we will do this on the level of complete residuated lattices $(L, \vee, \wedge, \otimes, \rightsquigarrow, 0, 1)$.

For a complete residuated lattice L, the classical logical operations of conjunction, disjunction, and implication can be extended by a minimal set of requirements in the following way:

Definition 9.14.

(a) *A* **conjunctor** $\mathcal{C} : L^2 \longrightarrow L$ *is a binary operation on L that is isotone in both arguments such that* $\mathcal{C}(0,1) = \mathcal{C}(1,0) = 0$ *and* $\mathcal{C}(1,1) = 1$. *A* **border conjunctor** *additionally satisfies the neutrality principle, i.e.* $\mathcal{C}(1,b) = b$ *for all* $b \in L$.

(b) *A* **disjunctor** $\mathcal{D} : L^2 \longrightarrow L$ *is a binary operation on L that is isotone in both arguments such that* $\mathcal{D}(0,1) = \mathcal{D}(1,0) = 1$ *and* $\mathcal{D}(0,0) = 0$. *A* **border disjunctor** *additionally satisfies the neutrality principle, i.e.* $\mathcal{D}(0,b) = b$ *for all* $b \in L$.

(c) *An* **implicator** $\mathcal{I} : L^2 \longrightarrow L$ *is a binary operation on L that is antitone in the first, and isotone in the second argument such that* $\mathcal{I}(0,0) = \mathcal{I}(1,1) = 1$ *and* $\mathcal{I}(1,0) = 0$. *A* **border implicator** *additionally satisfies the neutrality principle, i.e.* $\mathcal{I}(1,b) = b$ *for all* $b \in L$.

Note that, moreover, for all $a, b \in L$, the following *absorption principles* hold due to the respective monotonicity conditions (e.g. $\mathcal{C}(0,b) \leq \mathcal{C}(0,1) = 0$): For any conjunctor we have $\mathcal{C}(a,0) = \mathcal{C}(0,b) = 0$, for any disjunctor $\mathcal{D}(a,1) = \mathcal{D}(1,b) = 1$, and for any implicator $\mathcal{I}(a,1) = \mathcal{I}(0,b) = 1$. It is evident that $a \wedge b$ is a conjunctor, $a \vee b$ is a disjunctor, and $a \curvearrowright b$ is an implicator on a complete residuated lattice (for the latter see Prop. 4.27(4),(5)).

A commutative, associative border conjunctor $\mathcal{T}$ on the real unit interval is a t-norm as introduced above; or the other way round: any t-norm is a conjunctor. Similarly, a commutative, associative border disjunctor $\mathcal{S}$ is a t-conorm. This shows that t-(co)norms are serious candidates for generalized conjunction (disjunction).

The real unit interval clearly is a chain, i.e. a totally ordered set (additionally bearing suitable arithmetical operations); this surely is a restriction too strong to be met when applying fuzzy sets to human classification abilities and behavior, e.g. in multidimensional decision analysis. Therefore, the idea to use valuation sets more general than $[0,1] \subseteq \mathbb{R}$ emerged as early as in the 1967 paper [18] entitled "*L-fuzzy sets*" by J. Goguen, who uses the concept of a *complete lattice ordered semigroup*, abbreviated as *closg*.

Definition 9.15. *Let* $(L, \wedge, \vee, 0, 1)$ *be a complete (and hence bounded) lattice, and* $*$ *a binary operation on L such that* $(L, *, 1)$ *is an Abelian pomonoid and*

$$x * \Big(\bigvee_{i \in I} y_i\Big) = \bigvee_{i \in I} (x * y_i) \qquad \text{and} \qquad \Big(\bigvee_{i \in I} y_i\Big) * x = \bigvee_{i \in I} (y_i * x) \tag{9.4}$$

holds. Then L is called a **complete lattice-ordered semigroup**.

A binary operation $* : L \times L \longrightarrow L$ fulfilling (9.4) is called **continuous**. A continuous semigroup is usually called a *quantale* (see Def. 4.25). A complete lattice-ordered semigroup as defined above then is a commutative integral quantale with additional monotonicity properties for $*$:

Corollary 9.16. ([18, Proposition 3]) *Let* $(L, *, 1)$ *be a complete lattice-ordered semigroup, and* $a, b, c \in L$. *Then*

(1) $*$ *is isotone in both arguments.*

(2) $(a \wedge b) * (a \vee b) \leq (b * a) \wedge (a * b)$.

(3) $x * 0 = 0 * x = 0$.

Proof. (1): $a \leq b \rightarrow b * c = (a \vee b) * c = (a * c) \vee (b * c) \rightarrow a * c \leq b * c \rightarrow c * a \leq c * b$.
(2): $(a \wedge b) * (a \vee b) = [(a \wedge b) * a] \vee [(a \wedge b) * b] \leq (b * a) \wedge (a * b)$, using (1) for the last inequality.
(3): We have $x * 0 \leq 1 * 0 = 0$ using (1), so also $0 * x = 0$ by commutativity. □

The following considerations borrow from [2, 9, 10]. For mnemonic ($\sqcap$ corresponds to conjunction) and optical reasons (to avoid mixed notation like e.g. $\bigvee_{i \in I} T(a, b_i)$) I will use $\sqcap$ as a generic symbol for t-norms on lattices, and similar $\sqcup$ for t-conorms.

Definition 9.17. *Let $(L, \wedge, \vee, 0, 1)$ be a bounded complete lattice. A* **triangular norm** *(short* **t-norm***) on L is a mapping $\sqcap : L \times L \longrightarrow L$ which is commutative and associative such that for all $a, b, c \in L$*

(1) $1 \sqcap a = a$ *(boundary condition).*
(2) $b \leq c$ *implies* $a \sqcap b \leq a \sqcap c$.

So, algebraically speaking, $\sqcap$ is a t-norm iff $(L, \sqcap, 1, \leq)$ is a commutative partially ordered integral monoid. (A partially ordered monoid is called **integral** if 1 is its neutral element.) If you take the real unit interval for L it is evident that the above coincides with Def. 9.6.

There exist at least two t-norms on L: the meet $a \wedge b$, and the drastic intersection

$$a \sqcap_D b = \begin{cases} a \wedge b & \text{if } 1 \in \{a, b\} \\ 0 & \text{otherwise.} \end{cases}$$

Definition 9.18. *Let $(L, \wedge, \vee, 0, 1)$ be a bounded complete lattice.*

(a) *A t-norm $\sqcap$ is* **left-continuous** *, or* **infinitely $\vee$-distributive**, *if for any non-empty family $\{x_i \in L : i \in I\}$ and $a \in L$ $a \sqcap \bigvee_{i \in I} x_i = \bigvee_{i \in I}(a \sqcap x_i)$.*
(b) *Let $\sqcap$ be a left-continuous t-norm on L; $\curvearrowright_\sqcap$ defined by $x \curvearrowright_\sqcap y := \sup\{z \in L : x \sqcap z \leq y\}$ is called the* **residual implication**, *also R-***implication**, *generated by $\sqcap$.*

If L is a residuated lattice then $\otimes$ is a t-norm on L since it is commutative, associative, isotone in both arguments by Prop. 4.29, and $x \otimes 1 = x$ by Def. 4.26. For complete residuated lattices it has been shown in Prop. 4.27 - Prop. 4.30 that $x \otimes 0 = 0 = 0 \otimes x = x \wedge 0$, $y \leq y' \rightarrow x \otimes y \leq x \otimes y'$, and $x \otimes \left(\bigvee_{i \in I} y_i\right) = \bigvee_{i \in I}(x \otimes y_i)$, i.e. the Abelian monoid $(L, \otimes, 1)$ has a zero, $\otimes$ is a monotone operation and left continuous. Because of commutativity, $\otimes$ is monotone in both arguments and also right continuous.

Definition 9.19. *Let $(L, \wedge, \vee, 0, 1)$ be a bounded complete lattice. A* **triangular conorm** *(short* **t-conorm***) on L is a mapping $\sqcup : L \times L \longrightarrow L$ which is commutative and associative such that for all $a, b, c \in L$*

(1) $0 \sqcup a = a$ *(boundary condition).*

(2) *$b \le c$ implies $a \sqcup b \le a \sqcup c$.*

There exist at least two t-conorms on L: the join $a \vee b$, and the drastic union

$$a \sqcup_D b = \begin{cases} a \vee b & \text{if } 0 \in \{a,b\} \\ 1 & \text{otherwise.} \end{cases}$$

Proposition 9.20. *Let $(L, \wedge, \vee, 0, 1)$ be a bounded complete lattice.*

(1) *For t-norms $\sqcap$ we have $a \sqcap_D b \le a \sqcap b \le a \wedge b$.*
(2) *For t-conorms $\sqcup$ we have $a \vee b \le a \sqcup b \le a \sqcup_D b$.*

Proof. (1): We have $a \sqcap b \le 1 \sqcap b = b$ and $a \sqcap b = b \sqcap a \le 1 \sqcap a = a$, whence $a \sqcap b \le a \wedge b$. Furthermore, $a \sqcap_D 1 = a = a \sqcap 1$, $1 \sqcap_D b = b = 1 \sqcap b$, and $a \sqcap_D b = 0 \le a \sqcap b$ otherwise. Hence, for all $a, b \in L$, $a \sqcap_D b \le a \sqcap b$. □

Proposition 9.21. *Let $(L, \wedge, \vee, 0, 1)$ be a bounded complete lattice, and let $\sqcap$ and $\sqcup$ be a t-norm and a t-conorm on L, respectively. Then the following holds:*

(1) *$\sqcap$ is idempotent, iff $\sqcap = \wedge$.*
(2) *$\sqcup$ is idempotent, iff $\sqcup = \vee$.*
(3) *$\sqcap$ and $\sqcup$ satisfy the absorption law $(a \sqcup b) \sqcap a = a$ iff $\sqcap = \wedge$.*
(4) *$\sqcap$ and $\sqcup$ satisfy the absorption law $(a \sqcap b) \sqcup a = a$ iff $\sqcup = \vee$.*

Proof. (1): If $\sqcap = \wedge$, then obviuosly, $x \wedge x = x$. Conversely, by Prop. 9.20(1), monotonicity and assuming idempotency we get $x \wedge y = (x \wedge y) \sqcap (x \wedge y) \le x \sqcap y \le x \wedge y$.
(3): Taking $b = 0$ yields $a = (a \sqcup 0) \sqcap a = a \sqcap a$, for all $a \in L$, whence $\sqcap = \wedge$ by (1). Now assume $\sqcap = \wedge$; then $(a \sqcup b) \wedge a = a$, since $a \le a \vee b \le a \sqcup b$ by Prop. 9.20(2). □

What is still missing are results about the distributivity properties of t-norms and t-conorms. Let $\sqcap$ and $\sqcup$ be a t-norm and a t-conorm on a lattice L. Following [24, Chapter 2.1], we say that $\sqcup$ *is distributive over* $\sqcap$, if

$$a \sqcup (b \sqcap c) = (a \sqcup b) \sqcap (a \sqcup c),$$

and that $\sqcap$ *is distributive over* $\sqcup$, if

$$a \sqcap (b \sqcup c) = (a \sqcap b) \sqcup (a \sqcap c).$$

Proposition 9.22. *Let $(L, \wedge, \vee, 0, 1)$ be a bounded complete lattice, and let $\sqcap$ and $\sqcup$ be a t-norm and a t-conorm on L, respectively. Then the following holds:*

(1) *$\sqcup$ is distributive over $\sqcap$ iff $\sqcap = \wedge$.*
(2) *$\sqcap$ is distributive over $\sqcup$ iff $\sqcup = \vee$.*
(3) *$\sqcap$ and $\sqcup$ are distributive over each other iff $\sqcap = \wedge$ and $\sqcup = \vee$.*

Proof. (1): Suppose that $\sqcup$ is distributive over $\sqcap$; then we have $x = x \sqcup (0 \sqcap 0) = (x \sqcup 0) \sqcap (x \sqcup 0) = x \sqcap x$, whence $\sqcap = \wedge$ by Prop. 9.21(1). Conversely, let $\sqcap = \wedge$; then $a \sqcup (b \wedge c) \le a \sqcup b$ and $a \sqcup (b \wedge c) \le a \sqcup c$, thus $a \sqcup (b \wedge c) \le (a \sqcup b) \wedge (a \sqcup c)$. Now assume that $a \sqcup (b \wedge c) < (a \sqcup b) \wedge (a \sqcup c)$; then, if $b \le c$, we get $a \sqcup b < (a \sqcup b) \wedge (a \sqcup c)$, and, if $b > c$,

$a \sqcup c < (a \sqcup b) \wedge (a \sqcup c)$, a contradiction. Therefore, any t-conorm on L distributes over the meet operation $\wedge$. □

The notion of a t-norm plays the role of intersection, or in a logical context, that of conjunction; its dual is that of union, or disjunction. In the case of crisp sets (two-valued logic), union and intersection (disjunction and conjunction) are related via complement (negation) by the well-known the De Morgan formulas. So the notion of complement, or negation, is our next candidate for generalization.

Definition 9.23. *Let $(L, \wedge, \vee, 0, 1)$ be a bounded complete lattice.*

(a) A **negator** *on L is an mapping $\eta : L \longrightarrow L$ such that for all $a, b \in L$*

- *η is order reversing: $a \leq b \rightarrow \eta(b) \leq \eta(a)$, and*
- *$\eta(0) = 1$ and $\eta(1) = 0$.*

(b) A negator such that $\eta(\eta(a)) = a$ is called **strong** *or* **involutive**.

From Prop. 4.33(1),(3) we see that negation as defined for residuated lattices is a negator in the above sense, though not an involutive one. We say that $\sqcap$ and $\sqcup$ are *dual with respect to the negator η* and denote this by $(\sqcap, \sqcup, \eta)$, called a **De Morgan triple**, iff

$$\eta(a \sqcup b) = \eta(a) \sqcap \eta(b) \quad \text{and} \quad \eta(a \sqcap b) = \eta(a) \sqcup \eta(b).$$

The following proposition gives some evidence that we are on a good way with the generalized concepts introduced so far.

Proposition 9.24. ([26], Theorem 3.20) *Let $(L, \wedge, \vee, 0, 1)$ be a bounded complete lattice. Given a t-norm $\sqcap$ and a strong negator η on L, the binary operation $\sqcup$ defined by*

$$a \sqcup b = \eta\big(\eta(a) \sqcap \eta(b)\big) \tag{9.5}$$

is a t-conorm on L such that $(\sqcap, \sqcup, \eta)$ is a De Morgan triple.

Proof. The proof proceeds by routine calculations in five steps.
(a) $a \sqcup 0 = \eta\big(\eta(a) \sqcap \eta(0)\big) = \eta\big(\eta(a) \sqcap 1\big) = \eta(\eta(a)) = a$.
(b) Assume $b \leq d$; then we have $\eta(b) \geq \eta(d)$, whence, $\eta(a) \sqcap \eta(b) \geq \eta(a) \sqcap \eta(d)$ by monotonicity of $\sqcap$. Thus $a \sqcup b = \eta\big(\eta(a) \sqcap \eta(b)\big) \leq \eta\big(\eta(a) \sqcap \eta(d)\big) = a \sqcup d$ which shows the monotonicity of $\sqcup$.
(c) $a \sqcup b = \eta\big(\eta(a) \sqcap \eta(b)\big) = \eta\big(\eta(b) \sqcap \eta(a)\big) = b \sqcup a$.
(d):

$$\begin{aligned} a \sqcup (b \sqcup c) &= \eta\big(\eta(a) \sqcap \eta(b \sqcup c)\big) \\ &= \eta\big(\eta(a) \sqcap \eta[\eta(\eta(b) \sqcap \eta(c))]\big) \\ &= \eta\big(\eta(a) \sqcap (\eta(b) \sqcap \eta(c))\big), \quad \text{since } \eta \text{ is involutive} \\ &= \eta\big((\eta(a) \sqcap \eta(b)) \sqcap \eta(c)\big), \quad \text{since } \sqcap \text{ is associative} \\ &= \eta\big(\eta[\eta(\eta(a) \sqcap \eta(b))] \sqcap \eta(c)\big), \quad \text{since } \eta \text{ is involutive} \\ &= \eta\big(\eta[a \sqcup b] \sqcap \eta(c)\big) \\ &= (a \sqcup b) \sqcup c. \end{aligned}$$

(e)

$$\begin{aligned} \eta[a \sqcup b] &= \eta[\eta\big(\eta(a) \sqcap \eta(b)\big)] = \eta(a) \sqcap \eta(b), \text{ and} \\ \eta[a \sqcap b] &= \eta[\eta(\eta(a)) \sqcap \eta(\eta(b))] = \eta(a) \sqcup \eta(b). \end{aligned}$$

□

Proposition 9.25. *Let $(L, \wedge, \vee, 0, 1)$ be a bounded complete lattice, let A, B, C denote L-fuzzy sets over a set X, and let $(\sqcap, \sqcup, \eta)$ be a De Morgan triple (on L). Then the following holds:*

(1) $\sqcap$ is commutative, associative, and idempotent:

$$A \sqcap B = B \sqcap A, \quad A \sqcap (B \sqcap C) = (A \sqcap B) \sqcap C, \quad A \sqcap A = A.$$

(2) $\sqcup$ is commutative, associative, and idempotent:

$$A \sqcup B = B \sqcup A, \quad A \sqcup (B \sqcup C) = (A \sqcup B) \sqcup C, \quad A \sqcup A = A.$$

(3) Monotonicity, identity laws:

$$A \sqcap B \subseteq A \subseteq A \sqcup B,$$
$$A \sqcap \emptyset = \emptyset, \quad A \sqcup \emptyset = A,$$
$$A \sqcap X = A, \quad A \sqcup X = X.$$

(4) Distributivity laws:

$$A \sqcap (B \sqcup C) = (A \sqcap B) \sqcup (A \sqcap C), \quad \text{if } \sqcap \text{ is distributive over } \sqcup,$$
$$A \sqcup (B \sqcap C) = (A \sqcup B) \sqcap (A \sqcup C), \quad \text{if } \sqcup \text{ is distributive over } \sqcap.$$

(5) De Morgan laws:

$$\eta(A \sqcap B) = \eta(A) \sqcup \eta(B), \quad \eta(A \sqcup B) = \eta(A) \sqcap \eta(B).$$

9.2.3 *Some Semantic Considerations*

We first turn to the operation of implication, together with a look on the inference rule of **modus ponens** in its two valued form (following [4, p. 26f] and [34, p. 1355]): In this inference rule of classical logic conjunction and implication play an important role. Modus ponens says that if P and $P \to Q$ are valid propositions, then we are allowed to infer that also Q is a valid proposition, i.e. the truth-value 1 is preserved by modus ponens, commonly written as $P, P \to Q \models Q$. Due to the *deduction theorem* for propositional logic we then have that $(P \wedge (P \to Q)) \to Q$ is a tautology (see [37, p. 21]). So the two-valued modus ponens rule is sound in the sense that the inferred proposition is at least as true as the conjunction of the antecedents.

In the many-valued case we want the same, viz. that inference be truth preserving, i.e. the truth value of the conclusion is not less than the truth value of the conjunction of the premises, which implies that the truer the premises are, the truer the conclusion must be.

Letting $\|P\| \in L$ denote the truth value of a proposition P, then - with $\|P \wedge Q\| = \|P\| \otimes \|Q\|$ and $\|P \to Q\| = \|P\| \curvearrowright \|Q\|$ - from Prop. 4.27(2) we get $\|P\| \otimes (\|P\| \curvearrowright \|Q\|) \leq \|Q\|$, which obviously meets the above requirements, since in a residuated lattice L $\otimes$ is a t-norm on L (it is commutative, associative and $x \otimes 1 = x$ by Def. 4.26, and non-decreasing in both arguments by Prop. 4.29). Thus, the higher the truth values of the premises P and $P \to Q$, the higher the truth values of their conjunction and of the conclusion. Moreover, by Prop. 4.27(3), from $\|P\| \otimes (\|P\| \curvearrowright \|Q\|) \leq \|Q\|$ we get

$\|(P \wedge (P \rightarrow Q)) \rightarrow Q\| = (\|P\| \otimes (\|P\| \frown \|Q\|)) \frown \|Q\| = 1$ as analogon to the classical tautology. So we got a many-valued version of the modus ponens which is sound.

In informal terms, a statement is called *stronger* if it implies more. Recall that '*Ex Falso Quodlibet*' is the medieval name for the rule of inference which allows that from a contradiction you may deduce anything whatsoever. Thus, the contradictory proposition, denoted by $\perp$, is the strongest, and the tautological proposition, denoted by $\top$, is the weakest formula.

With regard to propositions with valuations in a lattice L, "weaker than" means "higher in the partial ordering of L". It would, of course, be nice to know the weakest proposition W that that together with P implies Q. Applying the adjointness property Def. 4.26(3) we can see that $\|W\| \otimes \|P\| \leq \|Q\| \leftrightarrow \|W\| \leq \|P\| \frown \|Q\|$ for two propositions P and Q; thus $P \rightarrow Q$ gives us the weakest proposition W we are looking for.

Since I regard it an important aspect, let me continue with some arguments which form a strong indication that the lattice meet is a very suitable and plausible candidate for the multiplication $\otimes$ in a residuated lattice:

(1) *The universal quantifier semantics*: In connection with Def. 4.26 it has been said that the two additional operations $\otimes, \frown : L \times L \longrightarrow L$ in residuated lattices are intended to model conjunction and implication of the corresponding logical calculus, and supremum ($\bigvee$) and infimum ($\bigwedge$) are intended for modeling the existential and universal quantifier, respectively.
For a set $X = \{x_1, x_2, \ldots, x_n\}$ the common understanding is that the statements $\forall x \in X[P(x)]$ and $P(x_1) \wedge P(x_2) \wedge \ldots \wedge P(x_n)$ mean the same, whence one would expect them to have identical truth values. For the many-valued case this means that the infimum $\bigwedge$ and the conjunction $\otimes$ should be based on the same binary operation on L.

(2) *The idempotency aspect*: For every proposition φ we certainly want that its truth value $\|\varphi\|$ is the same as that of $\varphi \& \varphi$, i.e. $\|\varphi\| = \|\varphi \& \varphi\| = \|\varphi\| \otimes \|\varphi\|$. And, naturally, for any fuzzy set S we require $S = S \cap S$, which holds according to Def. 9.2(d): $\mu_S(x) = \mu_{S \cap S}(x) = \mu_S(x) \wedge \mu_S(x)$. So $\otimes$ should be be idempotent, and by Prop. 9.21(1) the lattice meet is the only idempotent t-norm.

(3) *The algebraic aspect*: By Prop. 9.22, lattice join and lattice meet are the only t-norm and t-conorm that distribute over each other.

So there are good reasons for focusing on residuated lattices in general and, given that we take $\otimes := \wedge$, on Heyting algebras in particular. Note that all results valid for residuated lattices in general do, of course, hold for Heyting algebras in particular. But despite of the above arguments, there are situations where idempotency simply does not work properly and which therefore need a separate treatment; see the remarks about the *Poincaré paradox* on page 302.

9.3 Fuzzy Relations: Basic Operations and Properties

The concept of fuzzy relations is one of the most fundamental notions in fuzzy set theory and a major mathematical tool in numerous application areas, where they usually appear in form of fuzzy relation equations or inequalities. I will try to present enough information to give a proper overview on both mathematical aspects and application possibilities of fuzzy relations as a solid base for further investigation.

In this section, not only the basic concepts around fuzzy relations in a technical sense will be introduced, but I will also take the opportunity to discuss some important semantical, by which I mean non-formal, aspects that automatically arise by going the crucial step from two-valued to many-valued conceptions: If we let $L^+ := \{a \in L : a > 0\}$ denote the *positive* elements of L, this step means a drastic change from $|L^+| = 1$ to $|L^+| > 1$, i.e. we now have more than only one way to say "*yes*", "*has*" or "*is*", allowing sharper classifications.

Of course, you may ask: Given the modeling power of (the usual) two-valued binary relations, why shall one bother with fuzzy relations? A confessing fuzzyist (as I am) would probably argue that, using fuzzy relations, you can build FAR BETTER models that describe various aspects of reality MUCH MORE ACCURATE because of abandoning the old tradition of black-or-white or all-or-nothing judgments and reasoning. There are many more examples of that kind: Take any suitable adjective and combine it with a linguistic modifier in an appropriate way, as e.g. MUCH HIGHER, A LITTLE BIT LOUDER, SLIGHTLY COOLER, CLOSE TO, VERY CLOSE TO, or CONSIDERABLY LARGER THAN etc.

As an example consider the $[0,1]$-fuzzy relation MUCH GREATER THAN on $\mathbb{R}$, $\gg \subseteq \mathbb{R}\times\mathbb{R}$, given by

$$\gg (x,y) := \begin{cases} (1+\frac{100}{(x-y)^2})^{-1}, & x > y \\ 0 & \text{otherwise.} \end{cases}$$

You can see from this example that membership functions usually can be adapted to a special context in an easy way: change the numerator from 100 to e.g. 300 or 653, or you may decide to choose some suitable linear function.

Let X and Y be two sets. In the classical two-valued theory, a relation from X to Y is a subset of the Cartesian product of the two sets, $\rho \subseteq X \times Y$. In order to achieve a seamless inclusion of the special two-valued case in the new general framework, we try to follow the classical way as close as possible. So, as a first approach, we proceed as follows: We define the **Cartesian product** of two fuzzy sets A and B of X and Y, respectively, as the fuzzy set $A \times B$ of $X \times Y$ with

$$A \times B(x,y) = A(x) \wedge B(y). \tag{9.6}$$

Following [38], let S be a crisp set and $T \subseteq S$. Then a relation ρ on S is also a relation on T if $x\rho y$ implies $x \in T$ and $y \in T$, for all x, y. Restated in terms of characteristic functions this gives: $R(x,y) = 1 \to T(x) = T(y) = 1$, or equivalently $R(x,y) \le T(x) \wedge T(y)$, which finally results in

Let A and B be two fuzzy sets over X and Y, respectively. A (binary) **fuzzy relation** *ρ from A to B is a fuzzy subset of $A \times B$.*

In other words, for ρ to be a fuzzy relation from A to B, it is required that the degree of membership of any pair of elements does not exceed the degree of membership of either of the elements in A and B, respectively. This concept originated in [38], called "fuzzy relations on fuzzy sets", and is also used in e.g. [7, 8, 31].

Though this definition would be a perfect twin of the bivalent variant (a fuzzy relation is a fuzzy subset of the Cartesian product of two fuzzy sets), it rules out some data from forming fuzzy relations. Consider the following example: Let $U = \{Adam, Eve\}$ be our universe of discourse, and suppose we are given two fuzzy sets over U as follows:

$$\text{BLOND} := \{\langle Adam, 0.2\rangle, \langle Eve, 0.1\rangle\}, \text{ and}$$

$$\text{LIGHT BROWN} := \{\langle Adam, 0.4\rangle, \langle Eve, 0.3\rangle\}.$$

Comparing the Cartesian product of BLOND and LIGHT BROWN (left matrix) with a possible description of the HAIR COLOR-SIMILARITY of Adam and Eve (right matrix) shows that the latter one fails to obey the restrictions for being a fuzzy relation from BLOND to LIGHT BROWN (using obvious abbreviations):

BL×LIBR	*Adam*	*Eve*
Adam	0.2	0.2
Eve	0.1	0.1

HC-SIM	*Adam*	*Eve*
Adam	1	0.8
Eve	0.8	1

Note that this can occur only in valuation spaces L with $|L^+| > 1$ and is, therefore, impossible in the bivalent case where all non-zero values are equal to 1. Be it out of deliberation or out of conceiving fuzzy relations as fuzzy sets over Cartesian products of crisp sets, the following definition has most often been used, at least outside fuzzy graph theory.

Definition 9.26. *Let X and Y denote two crisp sets, and $(L, \wedge, \vee, \otimes, \frown, 0, 1)$ be a complete residuated lattice. A (binary) L-***fuzzy relation** *ρ over $X \times Y$ (from X to Y) is an L-fuzzy subset of $X \times Y$.*

The requirement from above, viz. that the degree of membership of any pair of elements does not exceed the degree of membership of either of the elements in X and Y, respectively, is now fulfilled automatically.

Since fuzzy relations are fuzzy sets, they obey the same set-theoretic laws as shown in Prop. 9.25. The most prominent kinds of fuzzy relations clearly arise for $L = \mathbf{2}$ as the classical variant of (Boolean, two-valued) *relations*; for $L = [0, 1]$ as *fuzzy relations* as originally defined in [47], and for a more general lattice L as *L-fuzzy relations* in [18].

Written in form of generalized characteristic functions we get the sets of relations as $\mathbf{2}^{X \times Y}$, of fuzzy relations as $[0, 1]^{X \times Y}$, and of L-fuzzy relations as $L^{X \times Y}$, respectively. Analogous to $\mathcal{B}_{X \times Y}$ for classical two-valued relations, we will write $\mathcal{F}_{X \times Y}$ and $\mathcal{F}^L_{X \times Y}$ in the second and third case to denote the set of all fuzzy relations of the respective kind. If

$X = Y$ we simply write $\mathcal{F}_X^L$.

The set L of membership values is required to be a complete residuated lattice; so we can consider fuzzy relations on arbitrary, i.e. not necessarily finite sets. However, whenever we work with fuzzy subsets and relations on a finite set, the completeness assumption becomes needless and can be omitted.

As in the case of crisp binary relations, every fuzzy relation ρ on a set X can be represented by a directed fuzzy graph G_ρ with X as its set of nodes such that there is a weighted arc $\langle x, y\rangle$ connecting each pair of nodes x, y, with $\rho(x, y)$ being its associated weight. Following the common habit, arcs with weight 0 are not shown when drawing G_ρ. Basic concepts of fuzzy graphs are the topic of Chapter 13.

In most cases the translation of basic classical two-valued notions into corresponding many-valued ones can be performed schematically on the formal or syntactical level. As an example let us look at the definition of the *domain* of a relation $\rho \subseteq X \times Y$ as given in Def. 2.3:

$$\mathsf{dom}(\rho) := \{x \in X : \exists y \in Y[x\rho y]\} = \{x \in X : \exists y \in Y[\rho(x, y) = 1]\} \subseteq X.$$

In terms of membership functions this is equivalent to $\mu_{\mathsf{dom}(\rho)}(x) = \max\{y \in Y : \rho(x, y)\}$. Analogous considerations lead to

Definition 9.27. *Let ρ be a fuzzy relation from X to Y, and A be a fuzzy set over X, and $(L, \wedge, \vee, \otimes, \frown, 0, 1)$ be a complete residuated lattice.*

(a) *The* **afterset** *of $x \in X$ is the fuzzy set $x\rho$ over Y defined as*
$$x\rho(y) = \rho(x, y).$$
(b) *The* **foreset** *of $y \in Y$ is the fuzzy set ρy over X defined as*
$$\rho y(x) = \rho(x, y).$$
(c) *The* **image** *of A under ρ is the fuzzy set $\rho[A]$ over Y defined as*
$$\rho[A](y) = \bigvee_{x \in X} \big(A(x) \otimes \rho(x, y)\big).$$
(d) *The* **domain** *of ρ is the fuzzy set* $\mathsf{dom}(\rho)$ *over X defined as*
$$\mathsf{dom}(\rho)(x) = \bigvee_{y \in Y} \rho(x, y) = \mathsf{hgt}(x\rho).$$
(e) *The* **range** *of ρ is the fuzzy set* $\mathsf{rng}(\rho)$ *over Y defined as*
$$\mathsf{rng}(\rho)(y) = \bigvee_{x \in X} \rho(x, y) = \mathsf{hgt}(\rho y).$$

Conversion and composition of fuzzy relations can be defined in a nearly self-evident way:

Definition 9.28. *Let ρ and σ be two L-fuzzy relations from A to B and from B to C, respectively, and $(L, \wedge, \vee, \otimes, \frown, 0, 1)$ be a complete residuated lattice.*

(a) *The L-fuzzy relation $\rho^\smile$ defined as $\rho^\smile(a, b) = \rho(b, a)$ is called the* **converse** *of ρ and is an L-fuzzy relation from B to A.*
(b) *The* **product** *(or* **composition***) of ρ and σ is the L-fuzzy relation $\rho \circ \sigma$ from A to C*

defined by

$$\rho \circ \sigma(a,c) = \bigvee_b \big(\rho(a,b) \otimes \sigma(b,c)\big). \tag{9.7}$$

We simplify notation to remain close to the one used in the two-valued context and will write

$$\mu_\rho(a,b) = \rho(a,b) = a\rho b, \tag{9.8}$$

and use the second or third variant at will; so (9.7) will then become

$$a\rho\sigma c = \bigvee_b a\rho b \otimes b\sigma c.$$

Taking $\vee$ = max and $\otimes$ = min, (9.7) yields the so-called **standard composition**, since it is based on the standard t-conorm and t-norm. (Remember that max and min give the only pair of a t-norm and a t-conorm that distribute over each other; see Prop. 9.22).

Representation of fuzzy relations is analogous to the classical case as described in Def. 2.12.

Definition 9.29.

(a) *A $m \times n$ matrix with values from a lattice L is called an* L-**valued matrix**. *The set of all $m \times n$ L-valued matrices is denoted by $\mathfrak{M}^L_{m\times n}$.*

(b) *Let ρ be an L-fuzzy relation from $A = \{a_1, a_2, \ldots, a_m\}$ to $B = \{b_1, b_2, \ldots, b_n\}$. $M = (m_{ij}) \in \mathfrak{M}^L_{m\times n}$ is a* **representation of** ρ *in case $m_{ij} = \rho(a_i, b_j)$, and is then denoted by M_ρ.*

(c) *Let $M = (m_{ij}) \in \mathfrak{M}^L_{m\times n}$ and $P = (p_{ij}) \in \mathfrak{M}^L_{n\times p}$. The* **product of** L-**valued matrices** *M and P, denoted by MP is the matrix $Q = (q_{ij}) \in \mathfrak{M}^L_{m\times p}$ with*

$$q_{ij} = \bigvee_{1\le k\le n} (m_{ik} \otimes p_{kj}). \tag{9.9}$$

Proposition 9.30. *Let $\rho \subseteq A \times B$ and $\sigma \subseteq B \times C$ be fuzzy relations from A to B, and from B to C, respectively, with $A = \{a_1, a_2, \ldots, a_m\}$, $B = \{b_1, b_2, \ldots, b_n\}$, and $C = \{c_1, c_2, \ldots, c_p\}$. Furthermore, let $R = (r_{ij}) \in \mathfrak{M}^L_{m\times n}$ and $S = (s_{ij}) \in \mathfrak{M}^L_{n\times p}$ be representations of ρ and σ. Then $P = RS \in \mathfrak{M}^L_{m\times p}$ is a representation of $\rho\sigma$.*

Having transferred the most basic operations for relations to the many-valued context, one may now ask about properties of fuzzy relations and their operations. A few answers are given below, together with some sample proofs.

Proposition 9.31. *Let ρ and σ be two L-fuzzy relations from S to T.*

(1) *If $\rho \subseteq \sigma$, then $\mathsf{dom}(\rho) \subseteq \mathsf{dom}(\sigma)$ and $\mathsf{rng}(\rho) \subseteq \mathsf{rng}(\sigma)$.*
(2) $\mathsf{dom}(\rho \cup \sigma) = \mathsf{dom}(\rho) \cup \mathsf{dom}(\sigma)$.
(3) $\mathsf{dom}(\rho \cap \sigma) \subseteq \mathsf{dom}(\rho) \cap \mathsf{dom}(\sigma)$.
(4) $\mathsf{rng}(\rho \cup \sigma) = \mathsf{rng}(\rho) \cup \mathsf{rng}(\sigma)$.
(5) $\mathsf{rng}(\rho \cap \sigma) \subseteq \mathsf{rng}(\rho) \cap \mathsf{rng}(\sigma)$.

Proof.

$$
\begin{aligned}
(1):\ \rho \subseteq \sigma &\rightarrow \forall x \in S, y \in T[x\rho y \leq x\sigma y] \\
&\rightarrow \bigvee_{y\in T} x\rho y \leq \bigvee_{y\in T} x\sigma y \\
&\rightarrow \mathsf{dom}(\rho) \subseteq \mathsf{dom}(\sigma).
\end{aligned}
$$

$$
\begin{aligned}
(3):\ \mathsf{dom}(\rho \cap \sigma)(x) &= \bigvee_{y\in T} x\rho \cap \sigma y \\
&= \bigvee_{y\in T} (x\rho y \wedge x\sigma y) \\
&\leq \bigvee_{y\in T} x\rho y \wedge \bigvee_{y\in T} x\sigma y && \text{by Prop. 1.44(L4)} \\
&= [\mathsf{dom}(\rho)](x) \wedge [\mathsf{dom}(\sigma)](x) \\
&= [\mathsf{dom}(\rho) \cap \mathsf{dom}(\sigma)](x).
\end{aligned}
$$

□

We finish this section with a well known and often cited example from the area of fuzzy relations that, finally, initiated most of the investigations concerning fuzzy relation equations and inequalities, viz. the seminal contribution of Sanchez [39]. Suppose we are interested in what patients have what diseases. The medical knowledge can be structured by using three sets appropriately chosen for our considerations: a set P of patients, a set D of diseases, and a set S of symptoms of diseases.

Furthermore, we set up three fuzzy relations as follows: $\pi \subseteq P \times S$ with $\pi(p,s)$ indicating the strength of symptom s for patient p, $\sigma \subseteq S \times D$ with $\sigma(s,d)$ informing about the explanatory power of symptom s for disease d, and $\delta \subseteq P \times D$ with $\delta(p,d)$ as the degree to which patient p has disease d. Usually the data observed in this setup are about symptoms of patients (from medical examination) and about what symptoms are characteristic for what diseases (from medical expert knowledge). What we want to infer from this is information about the disease of patient p.

9.3.1 *Basic Operations on Fuzzy Relations*

This section summarizes some facts from what could be called *fuzzy relation calculus*. That means, we will list the most important relationships that hold for operations on fuzzy relations, and some more generalizations of theorems from the two-valued case. Though proofs are simple and do closely follow the ones for the two-valued case, we give some examples to get familiar with the notation we use in the many-valued context. We start with the many-valued version of Prop. 2.8 and Th. 2.9 and will see that most of the properties carry over.

If not stated otherwise, in the following we let L denote a complete residuated lattice $(L,\wedge,\vee,\otimes,\curvearrowright,0,1)$. If appropriate we will change our notation to differentiate between the lattice theoretical operators $\wedge,\vee$ and the logical operators $\&, \|$ in order to improve readability.

Proposition 9.32. *For any L-fuzzy relations ρ and σ we have*

(1) $\mathsf{dom}(\rho^{\smile}) = \mathsf{rng}(\rho)$.
(2) $\mathsf{rng}(\rho^{\smile}) = \mathsf{dom}(\rho)$.
(3) If $\rho \subseteq S \times T$*, then* $\rho^{\smile} \subseteq T \times S$.

(4) $\rho^{\smallsmile\smallsmile} = \rho$.
(5) $\overline{\rho}^{\smallsmile} = \overline{\rho^{\smallsmile}}$.
(6) *If* $\rho \subseteq \sigma$, *then* $\rho^{\smallsmile} \subseteq \sigma^{\smallsmile}$.
(7) $(\rho \cap \sigma)^{\smallsmile} = \rho^{\smallsmile} \cap \sigma^{\smallsmile}$.
(8) $(\rho \cup \sigma)^{\smallsmile} = \rho^{\smallsmile} \cup \sigma^{\smallsmile}$.
(9) $x\rho = \rho^{\smallsmile} x\,, \quad \rho y = y\rho^{\smallsmile}$.

Proof. (1): $\mathsf{dom}(\rho^{\smallsmile})(x) = \bigvee_y \rho^{\smallsmile}(x,y) = \bigvee_y \rho(y,x) = \mathsf{rng}(\rho)(x)$.

(5): $x\overline{\rho}^{\smallsmile} y = y\overline{\rho} x = y\rho x \frown 0 = x\rho^{\smallsmile} y \frown 0 = x\overline{\rho^{\smallsmile}} y$.

(7): $x(\rho \cap \sigma)^{\smallsmile} y = y(\rho \cap \sigma)x = y\rho x \wedge y\sigma x = x\rho^{\smallsmile} y \wedge x\sigma^{\smallsmile} y = x(\rho^{\smallsmile} \cap \sigma^{\smallsmile})y$.

(9): $x\rho(y) = \rho(x,y) = \rho^{\smallsmile}(y,x) = \rho^{\smallsmile} x(y)$. □

Theorem 9.33. *Let* ρ, σ , *and* τ *be L-fuzzy relations such that the operations involved are possible. Then the following holds:*

(1) $I_S\rho = \rho = \rho I_T$.
(2) $\emptyset\rho = \emptyset = \rho\emptyset$.
(3) *Associativity of relation product:* $\rho(\sigma\tau) = (\rho\sigma)\tau$.
(4) *Subdistributivity of relation product over intersection:*

$$\rho(\sigma \cap \tau) \subseteq (\rho\sigma) \cap (\rho\tau) \quad \text{and} \quad (\rho \cap \sigma)\tau \subseteq (\rho\tau) \cap (\sigma\tau).$$

(5) *Monotonicity of relation product: If* $\sigma \subseteq \tau$, *then*

$$\rho\sigma \subseteq \rho\tau \quad \text{and} \quad \sigma\rho \subseteq \tau\rho.$$

(6) *Distributivity of relation product over union:*

$$\rho(\sigma \cup \tau) = (\rho\sigma) \cup (\rho\tau) \quad \text{and} \quad (\rho \cup \sigma)\tau - (\rho\tau) \cup (\sigma\tau)$$

(7) *If* $\rho \subseteq \sigma$ *and* $\tau \subseteq \omega$, *then* $\rho\tau \subseteq \sigma\omega$.
(8) $(\rho\sigma)^{\smallsmile} = \sigma^{\smallsmile}\rho^{\smallsmile}$.
(9) $\rho\sigma = \bigcup_{y \in Y} (\rho y \otimes y\sigma)$.
(10) *If* $\otimes = \wedge$ *(as in Boolean and Heyting algebras) and* $\alpha > 0$, *then* $(\rho \circ \sigma)_\alpha = \rho_\alpha \circ \sigma_\alpha$.

Proof.
(The example proofs given below are routine applications of the laws that hold in residuated lattices; compare e.g. the proof of (3) with the corresponding proof in Th. 2.22. The proof of (5), in the compact notation introduced in (9.8) above, is actually an easy transfer of the proof for two-valued relations, and (7) is a mere copy; again see Th. 2.22.)

(1): $aI_S\rho c = \bigvee_b aI_S b \otimes b\rho c = aI_S a \otimes a\rho c = a\rho c$, since 1 is the unit for $\otimes$.

$$\begin{aligned}
(3):\ [\rho(\sigma\tau)](x,w) &= \bigvee_y [\rho(x,y) \otimes (\sigma\tau)(y,w)] \\
&= \bigvee_y [\rho(x,y) \otimes \bigvee_z \big(\sigma(y,z) \otimes \tau(z,w)\big)] \\
&= \bigvee_y \bigvee_z [\rho(x,y) \otimes \big(\sigma(y,z) \otimes \tau(z,w)\big)] \\
&= \bigvee_z [\bigvee_y \big(\rho(x,y) \otimes \sigma(y,z)\big) \otimes \tau(z,w))] \\
&= \bigvee_z [(\rho\sigma)(x,z) \otimes \tau(z,w)] \\
&= [(\rho\sigma)\tau](x,w).
\end{aligned}$$

(5): Assume $x\rho\sigma z$ (i.e. > 0). By (9.7) $x\rho\sigma z = \bigvee_y(x\rho y \otimes y\sigma z)$. By hypothesis we have $\sigma \subseteq \tau$, whence $y\sigma z \leq y\tau z$, for all y, z. Thus, by monotonicity of $\otimes$, $x\rho\sigma z \leq x\rho\tau z$, i.e. $\rho\sigma \subseteq \rho\tau$.

(6): Using Prop. 4.30(1) we get $x[\rho \circ (\bigcup_i \sigma_i)]y = x\rho z \otimes z(\bigcup_i \sigma_i)y = x\rho z \otimes \bigvee_i z\sigma_i y = \bigvee_i (x\rho z \otimes z\sigma_i y) = x \bigcup_i \rho\sigma_i\, y$.

(7): By (5) $\rho \subseteq \sigma \wedge \tau \subseteq \omega \rightarrow \rho\tau \subseteq \rho\omega \wedge \rho\omega \subseteq \sigma\omega$, whence $\rho\tau \subseteq \sigma\omega$.

(8): $x(\rho\sigma)\breve{}z = z(\rho\sigma)x = \bigvee_y(z\rho y \otimes y\sigma x) = \bigvee_y(x\breve{\sigma}y \otimes y\breve{\rho}z) = x\breve{\sigma}\breve{\rho}z$.

(9): $x\rho\sigma z = \bigvee_{y\in Y}(x\rho y \otimes y\sigma z) = \bigvee_{y\in Y}\big(\rho y(x) \otimes y\sigma(z)\big) = \bigvee_{y\in Y}\big[(\rho y \otimes y\sigma)(x, z)\big] = \bigcup_{y\in Y}(\rho y \otimes y\sigma)$.

(10):
$$
\begin{aligned}
\langle x, z\rangle \in (\rho\sigma)_\alpha &\leftrightarrow x\rho\sigma z \geq \alpha \\
&\leftrightarrow \bigvee_{y\in Y}(x\rho y \otimes y\sigma z) \geq \alpha \\
&\leftrightarrow \exists_{y\in Y}[(x\rho y \otimes y\sigma z) \geq \alpha] \\
&\leftrightarrow \exists_{y\in Y}[(x\rho y \wedge y\sigma z) \geq \alpha], \qquad \text{since } \otimes = \wedge \\
&\leftrightarrow \exists_{y\in Y}[x\rho y \geq \alpha \;\&\; y\sigma z \geq \alpha] \\
&\leftrightarrow \exists_{y\in Y}[\langle x, y\rangle \in \rho_\alpha \;\&\; \langle y, z\rangle \in \sigma_\alpha] \\
&\leftrightarrow \langle x, z\rangle \in \rho_\alpha\sigma_\alpha.
\end{aligned}
$$

□

Thus, analogous to the classical case (see the begin of Section 2.9), it holds that

- $(\mathcal{F}_X^L, \cdot, \breve{}, I_X)$ is an involution monoid,
- $(\mathcal{F}_X^L, \cdot)$ is a semigroup with zero element $\emptyset$, and
- $(\mathcal{F}_X^L, \cdot; \subseteq)$ is a partially ordered semigroup.

9.3.2 *Basic Properties: An Informal Discussion*

Roughly following the presentation for two-valued relations we now turn to properties a fuzzy relation may have, but we will have to confine ourselves to the most important ones, and also to a few selected aspects only that arise out of a many-valued approach. Why this is necessary you will shortly see from the examples and their discussion given below. These considerations are important, because they (1) should give you a feeling how the many-valued approach enriches the flexibility and modeling possibilities of your formal apparatus, but can also lead to new situations that have to be considered; and (2) should draw your attention to the fact that formal generalizations have to be done carefully, i.e. they also have to account for semantical aspects, if they should not degenerate to a mere mathematical playground.

In Def. 2.14 and Prop. 2.18 two equivalent definitions for reflexivity have been given as follows: $\forall x \in X[x\rho x]$ and $I_X \subseteq \rho$. This has also been done for symmetry and transitivity. The question is how to generalize the two-valued concepts appropriately. Some examples for reflexivity can be found in e.g. [13, Chapter II.3] and [8, 30, 46] where the following variants are listed (with different names):

(a) *reflexive* if $\forall x \in X$, $\rho(x,x) = 1$;
(b) *ϵ-reflexive* if $\forall x \in X$, $\rho(x,x) \geq \epsilon > 0$;
(c) *strictly ϵ-reflexive* if $\forall x \in X$, $\rho(x,x) = \epsilon > 0$;
(d) *weakly reflexive* if $\forall x \in X$, $\rho(x,x) > 0$;
(e) *strongly reflexive* if $\forall x \in X$, $\rho(x,x) \geq \bigvee_{y \in X} \rho(x,y) = \mathsf{dom}(\rho)(x)$.

(a) is a direct translation of the two-valued definition and can easily be justified. It just expresses identity: a thing is indisputably identical to itself, whence a two-valued classification suffices to describe this, and thus $I_X \subseteq \rho$ results as definition of reflexivity also in the many-valued context.

(b) - (d) can be seen as an innocent arithmetical generalizations (remember that $L = [0,1] \subset \mathbb{R}$ is the very first lattice used in FST). But unless we take $\epsilon = 1$, an ϵ-reflexive identity I_ϵ, i.e. $I_\epsilon(x,x) \geq \epsilon$ for all $x \in X$ and $I_\epsilon(x,y) = 0$ for $x \neq y$, will not be a unit for composition with an arbitrary fuzzy relation; in fact, $I_\epsilon \rho = \rho I_\epsilon = \rho$ does not hold as soon as $\mathsf{hgt}(\rho) > \epsilon$.

And concerning strong reflexivity (e), $\rho(x,x)$ becomes dependent on the elements $y \neq x$ to which x is ρ-related, and can no longer be interpreted as a 'degree of identity', but rather represents something like an upper bound for the degree of x's relatedness to the other elements of X.

As another example for the manifold extension possibilities within a many-valued approach we give, without further discussion, the following plausible variants of a generalized notion of symmetry. An L-fuzzy relation ρ on a set X, with $\epsilon > 0$, is called

- *symmetric*, if $\forall x, y \in X[x\rho y = y\rho x]$;
- *strongly ϵ-symmetric*, if $\forall x, y \in X[x\rho y \wedge y\rho x \geq \epsilon \rightarrow x\rho y = y\rho x]$;
- *ϵ-symmetric* if $\forall x, y \in X[x\rho y \geq \epsilon \rightarrow y\rho x \geq \epsilon]$;
- *weakly ϵ-symmetric* if $\forall x, y \in X[x\rho y \geq \epsilon \rightarrow y\rho x > 0]$;
- *weakly symmetric* if $\forall x, y \in X[x\rho y > 0 \rightarrow y\rho x > 0]$.

Note that in the Boolean two-valued context all these variants collapse - as is to be required from any meaningful generalization.

Let us consider two more examples: In [11, Definition 5.2(b)] and [15, Definition 9] a fuzzy relation ρ on a set X is called *ϵ-irreflexive* if $x\rho x \leq 1 - \epsilon$ for all $x \in X$. Now taking $x\rho x = 0.5$, for all $x \in X$, then we have that ρ is both 0.5-reflexive and also 0.5-irreflexive. In my opinion one should refuse this: Irreflexivity is of a negative character, stating the absence of a property. And as already mentioned at the begin of Section 9.3 fuzzy sets are intended to allow finer distinctions if an object posseses a certain property. This view is reflected in Def. 9.34(d) below.

And in [12, Definition 5] a fuzzy relation ρ is called *weakly antisymmetric* if $x\rho y = 1 \rightarrow y\rho x < 1$, for all $x, y \in X, x \neq y$. A reflexive fuzzy relation ρ with only 1's below the diagonal and only 0.99 above would then qualify as weakly antisymmetric. Do we really want such a terminology? From these examples you should already have got an idea of how many questions and aspects turn out to be worth considering in the context of transition from two-valued to many-valued concepts.

Another important aspect has to be mentioned before we proceed: We have defined symmetry in the same way as in the two-valued case, viz. by *equality* of the two values for *all* possible pairs of elements, or if you want, as a kind of *perfect* symmetry. Despite the many-valued approach used, symmetry of a fuzzy relation, and also equality of fuzzy sets (see Def. 9.2(b)), are sharp all-or-nothing concepts - thus, in a way, they have remained "un-fuzzy". In both cases one might think of more relaxed versions that emerge by a simple combinatorial process: On one side you either have usual, i.e. sharp, or APPROXIMATE EQUALITY of membership degrees, and on the other side you have *all* or NEARLY ALL, ALMOST ALL, or MOST of the elements possibly concerned. For whatever reason, a very weak form of symmetry may be sufficient for some purpose: It is okay for you, if for MOST pairs of elements (i.e. for e.g. 75%, 80%, ..., or 90%) the membership degrees in ρ and $\rho^{\smile}$ are VERY CLOSE TO each other. And you might eventually call a fuzzy relation ρ with this property *almost weakly symmetric* and, moreover, quantify this with a degree of symmetry, say $\mathrm{sym}(\rho) \in [0,1]$. Clearly, this kind of reasoning can also be applied to other properties. (Besides, you need not stop here: Why assign an ordinary real and not take $\mathrm{sym}(\rho)$ to be a fuzzy number in the interval $[0,1]$ or an interval-valued fuzzy set etc. Nevertheless, you will end up in a situation where ordinary real numbers are used to define a higher-level fuzzified concept.)

Do *not* smile at what you have just read. Taking the original semantic intentions of fuzzy set theory seriously, this is a typical rather than abstruse aspect of applying its principal concepts; see e.g. [20, Section 18.6] or [4, p. 82ff, Section 4.1.2], where graded properties are formally represented by many-valued predicates in the language of a many-valued logical system. For more recent contributions you may want to have a look into [5, e.g. Section 5.2] or [17]. But this approach is beyond the purposes of our text, and probably neither manageable nor necessary for practical purposes.

So, if you want to generalize from the collapsed two-valued to an appropriate many-valued definition of a concept, you will have to be careful in selecting a suitable way of "explosion", both with respect to its mathematical properties but also to its application possibilities, i.e. to both its syntactical and semantical aspects. Furthermore, a detailed analysis of the above arguments shows, that even the introduction of a higher level of flexibility, as in e.g. combining a high percentage of elements with sharp equality of membership grades, does not lead beyond two-valued mathematics and logic: They still serve as the metalanguage in which various models offering additional possibilities for finer grained descriptions are formulated.

9.3.3 *Basic Properties: A More Formal Discussion*

As one may suspect from the considerations above, fuzzy set theory offers a considerable variety of ways to extend definitions of properties of classical binary relations - a large playground one might say. A first fundamental task is to decide about working with fuzzy relations on fuzzy sets or on ordinary sets, with the latter one having evolved as the standard framework. As indicated above, another fundamental choice is the one between working with 'crisp' properties of fuzzy relations or with 'fuzzified' (graded) ones. (For fuzzy relations on fuzzy sets see e.g. [29, 38, 40] and for fuzzy relations with graded properties [4], [19, Chapter 2], or [20, Chapter 18].)

The considerations in this book will take place in the context of fuzzy relations on ordinary sets with crisp properties. Nonetheless, as will become clear below, there is still a wide range of generalization possibilities of well known properties. In this section we will present some commonly used properties of fuzzy relations in both point-based and algebraic form. See e.g. [6, 11, 12, 15, 26, 35] among many others. The definitions given below are, of course, only part of what has been proposed during the last few decades. For sake of brevity we write $\rho^{\Uparrow}$ for $\mathsf{supp}(\rho)$.

Definition 9.34. (Classical properties) *Let ρ be an L-fuzzy relation over X; then ρ is called*

(a) **reflexive**	*if* $\forall x \in X[x\rho x = 1]$	*or* $I_X \subseteq \rho$.
(b) **symmetric**	*if* $\forall x, y \in X[x\rho y = y\rho x]$	*or* $\rho = \rho^{\smile}$.
(c) **transitive**	*if* $\forall x, z \in X[\bigvee_{y\in X}(x\rho y \wedge y\rho z) \le x\rho z]$	*or* $\rho^2 \subseteq \rho$.
(d) **irreflexive**	*if* $\forall x \in X[x\rho x = 0]$	*or* $I_X \cap \rho = \emptyset$.
(e) **asymmetric**	*if* $\forall x, y \in X[x\rho y > 0 \rightarrow y\rho x = 0]$	*or* $\rho^{\Uparrow} \cap (\rho^{\smile})^{\Uparrow} = \emptyset$.
(f) **antisymmetric**	*if* $\forall x, y \in X[x\rho y \wedge y\rho x > 0 \rightarrow x = y]$	*or* $\rho^{\Uparrow} \cap (\rho^{\smile})^{\Uparrow} \subseteq I_X$.
(g) **connected**	*if* $\forall x, y \in X[x \ne y \rightarrow x\rho y \vee y\rho x = 1]$	*or* $\overline{I_X} \subseteq \rho \cup \rho^{\smile}$.
(h) **totally connected**	*if* $\forall x, y \in X[x\rho y \vee y\rho x = 1]$	*or* $\rho \cup \rho^{\smile} = U$.
(i) **idempotent**	*if* $\forall x, z \in X[\bigvee_{y\in X}(x\rho y \wedge y\rho z) = x\rho z]$	*or* $\rho^2 = \rho$.
(j) **functional**	*if* $\forall x, y, z[x\rho y \wedge x\rho z \rightarrow y = z]$	*or* $\rho^{\smile}\rho \subseteq I_X$.
(k) **injective**	*if* $\forall x, y, z[x\rho z \wedge y\rho z \rightarrow x = y]$	*or* $\rho\rho^{\smile} \subseteq I_X$.

In fact, properties (a) - (c) are the 'classical' properties: the first ones that have been defined for fuzzy relations in 1971 by Zadeh [48] to specify the concept of a fuzzy equivalence. We will have a closer look at these properties and some variants later on.

Before we proceed, let us pause for a moment and think about some algebraic characterizations listed in the definition above. The consideration below will involve the lattice shown on the right.

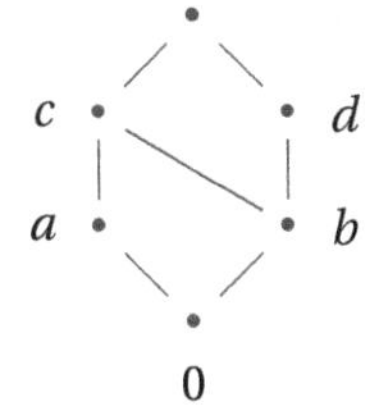

- Transitivity as defined above can more specifically be called meet-transitivity; and since $\wedge$ is the greatest t-norm on L, a meet-transitive relation ρ is T-transitive for every t-norm T on L, i.e. $\rho \circ_T \rho \subseteq \rho$ independent of which t-norm T else we use for relation composition.
- In Section 9.2.3 we have mentioned some arguments in favor of the meet operation. But as you all know, there are two sides to everything; so we need to

take a second look. Contrary to transitivity, the direct transfer of the algebraic criterion from the Boolean to general L-fuzzy relations fails for asymmetry. Let us see why:
Consider an irreflexive relation with all values below the main diagonal equal to a and all values above equal to d; then clearly, since $a \wedge d = 0$, $\rho \cap \rho^{\smile} = \emptyset$, and this holds for whatever t-norm is used for intersection. That clearly means, that the common viewpoint on asymmetry, viz."if there is a relationship from x to y, then there is none in the other direction" cannot be modeled this way. But there is another equivalent that helps: for common two-valued relations we have $\rho \cap \rho^{\smile} = \rho^{\Uparrow} \cap (\rho^{\smile})^{\Uparrow}$. And in fact, an L-fuzzy relation ρ is asymmetric iff $\rho^{\Uparrow} \cap (\rho^{\smile})^{\Uparrow} = \emptyset$. (Note, however, that for *chains* the original condition $\rho \cap \rho^{\smile} = \emptyset$ would still work, since for two non-zero elements a, b we always have $a \wedge b \neq 0$.)

- The problem just considered is due to the existence of zero divisors in our example lattice; it also occurs in connection to fuzzy conjunction operators and intersection of fuzzy sets, respectively: Let e.g. S and T be two L-fuzzy sets with, say, $\mu_S(x) = a$ and $\mu_T(x) = b$ such that $0 < a, b \in L$ are zero divisors of $\otimes$. Then we would have $\mu_{S \cap T}(x) = \mu_S(x) \otimes \mu_T(x) = 0$, most probably a result we did not intend to get.

(a) Reflexivity:

Our informal discussions in Section 9.3.2 did already indicate that the properties making up a fuzzy equivalence are the most often generalized ones. In the course of time, reflexivity has become the property with probably the greatest number of variants, whence we arrange some of them in a separate definition. After Zadeh [48] in 1971, the next generalization steps have been done by Yeh in [45] in 1973 and two years later in [46]. The notion of hgt-reflexivity emerges from [28, p. 123], G-reflexivity appears in [22], and the concept of locally reflexive relations has been introduced in [1].

Some additional notation will make the definitions and considerations below more convenient: For sake of brevity we write $\rho^{\vee} := \mathrm{hgt}(\rho)$ and $d_\rho^{\wedge} := \bigwedge_{z \in X} z\rho z$. We consider the following diagonals:

$$\Delta_\rho(x,x) := \rho(x,x), \quad \Delta_\epsilon(x,x) = \epsilon > 0, \quad \text{and} \quad \Delta_\rho^x(x,x) := \bigvee_{y \in X} x\rho y \vee y\rho x. \tag{9.10}$$

Of course, for all three diagonals we set $\Delta_\bullet(x,y) = 0$, for all x, y such that $x \neq y$. (Δ_ρ^x is called *local equality* in [1, 3.1 Definition], denoted with E_R.)

Definition 9.35. (Reflexivity variants) *Let L be a complete Heyting algebra, and $\rho \neq \emptyset$ be an L-fuzzy relation over X; then ρ is called*

(a) **reflexive**	*if* $\forall x \in X[\rho(x,x) = 1]$,	*or* $\Delta_\rho = I_X$.
(b) **strictly ϵ-reflexive**	*if* $\forall x \in X[\rho(x,x) = \epsilon > 0]$,	*or* $\Delta_\rho = \Delta_\epsilon$.
(c) **ϵ-reflexive**	*if* $\forall x \in X[\rho(x,x) \geq \epsilon > 0]$,	*or* $\Delta_\rho \supseteq \Delta_\epsilon$.
(d) **hgt-reflexive**	*if* $\forall x \in X[\rho(x,x) = \rho^{\vee}]$,	*or* $\Delta_\rho = \Delta_{\rho^{\vee}}$.
(e) **G-reflexive**	*if* $\forall x, y \in X[x\rho x > 0 \ \& \ x\rho y \leq d_\rho^{\wedge},\ x \neq y]$.	

(f) **weakly reflexive**	*if* $\forall x \in X[\rho(x,x) > 0]$,	*or* $(\Delta_\rho)^{\Uparrow} = I_X$.
(g) **locally reflexive**	*if* $\forall x \in X[x\rho x = \bigvee_{y\in X} x\rho y \vee y\rho x]$,	*or* $\Delta_\rho = \Delta^x_\rho$.
(h) **strong locally reflexive**	*if* $\forall x, y \in X[x\rho y \vee y\rho x > 0 \rightarrow x\rho x = 1]$.	
(i) **coreflexive**	*if* $\forall x, y \in X[x\rho y > 0 \rightarrow x = y]$,	*or* $\rho \subseteq I_X$.
(j) **ϵ-coreflexive**	*if* $\forall x, y \in X[x\rho y \geq \epsilon \rightarrow x = y]$,	*or* $(\rho_\epsilon)^{\Uparrow} \subseteq I_X$.

In addition to the above characterizations which are based on diagonals only, one can also use the more familiar ones in some cases: (a) $I_X \subseteq \rho$, (c) $\Delta_\epsilon \subseteq \rho$, (d) $\Delta_{\rho^\vee} \subseteq \rho$, (f) $I_X \subseteq \rho^{\Uparrow}$, and (g) $\Delta^x_\rho \subseteq \rho$. Moreover, the following property of Δ^x_ρ is worth mentioning:

Proposition 9.36. ([1, 3.3 Theorem]) *Δ^x_ρ is both a left and a right identity for ρ, and it is the smallest diagonal with this property*

Proof. $x\Delta^x_\rho\rho y = \bigvee_z(x\Delta^x_\rho z \wedge z\rho y) = x\Delta^x_\rho x \wedge x\rho y = x\rho y$, and similarly for $y\rho\Delta^x_\rho x$. Furthermore, $\forall x\exists y[x\Delta^x_\rho x = x\rho y \vee x\Delta^x_\rho x = y\rho x]$. Thus, for any $\delta \subset \Delta^x_\rho$, $x\delta\rho y < x\rho y$ or $y\rho\delta x < y\rho x$. □

Excluding $\emptyset$ from being a reflexive relations of any kind is rather obvious from semantical reasons and also avoids formal pathologies like $\emptyset$ being e.g. hgt-reflexive. A coreflexive relation is one that has nonzero values on the diagonal only, if it has any at all (so $\emptyset$ is coreflexive).

Though the above definition might appear as a kind of playground for formal symbol manipulation, in my opinion it is not. It rather reflects the historical development of the various attempts to generalize one of the most basic characteristics a binary relation can have. It is again an example demonstrating the increase of possibilities beyond two-valued concepts.

It is easy to see that the following relationships hold for the above mentioned reflexivity variants, thus providing a certain amount of consistency between them.

Corollary 9.37.

(1) reflexive $\rightarrow$ strictly ϵ-reflexive $(\epsilon = 1)$ $\rightarrow$ ϵ-reflexive $(0 < \epsilon \leq 1)$ $\rightarrow$ weakly reflexive.

(2) reflexive $\rightarrow$ hgt-reflexive $\rightarrow$ strictly ϵ-reflexive $(\epsilon = \rho^\vee)$ $\rightarrow$ ϵ-reflexive $(0 < \epsilon \leq \rho^\vee)$.

(3) reflexive $\rightarrow$ strong locally reflexive $\rightarrow$ locally reflexive.

(4) hgt-reflexive $\rightarrow$ G-reflexive $\rightarrow$ ϵ-reflexive $(0 < \epsilon \leq d^\wedge_\rho)$.

It is useful to summarize these relationships in a picture as shown below (with obvious abbreviations). Note that, as one would expect, classical reflexivity is in fact a special case of all the other properties mentioned (coreflexivity, of course, excluded).

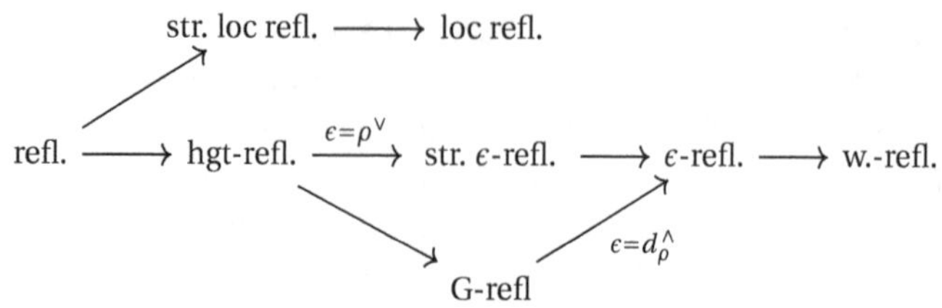

(b) Symmetry:

The second property relevant for the definition of equivalence relations is symmetry. The following generalizations have been conceived of:

Definition 9.38. (Symmetry variants) *Let L be a complete Heyting algebra, ρ be an L-fuzzy relation over X, and $\epsilon > 0$; then ρ is called*

(a) **symmetric**	*if* $\forall x, y \in X[x\rho y = y\rho x]$.
(b) **strongly ϵ-symmetric**	*if* $\forall x, y \in X[x\rho y \geq \epsilon \;\&\; y\rho x \geq \epsilon \rightarrow x\rho y = y\rho x]$.
(c) **ϵ-symmetric**	*if* $\forall x, y \in X[x\rho y \geq \epsilon \rightarrow y\rho x \geq \epsilon]$.
(d) **weakly symmetric**	*if* $\forall x, y \in X[x\rho y > 0 \rightarrow y\rho x > 0]$.
(e) **weakly ϵ-symmetric**	*if* $\forall x, y \in X[x\rho y \geq \epsilon \rightarrow y\rho x > 0]$.

Corollary 9.39. *Let L be a complete Heyting algebra, and ρ be an L-fuzzy relation over X; then the following consistency relationships hold for $\epsilon \in (0,1]$:*

(1) symmetric $\rightarrow$ strongly ϵ-symmetric.
(2) symmetric $\rightarrow$ ϵ-symmetric $\rightarrow$ weakly ϵ-symmetric.
(3) symmetric $\rightarrow$ weakly symmetric $\rightarrow$ weakly ϵ-symmetric.

As we have seen it for reflexivity, also (classical) symmetry is a special case of all the variants mentioned. Note that, contrary to reflexivity, the case $\rho = \emptyset$ is not excluded. We continue with some algebraic, i.e. point-free, characterizations of symmetry properties in

Corollary 9.40. *Let L be a complete Heyting algebra, and ρ be an L-fuzzy relation over X; then ρ is*

(1) symmetric	*if* $\rho = \rho\breve{}$.	
(2) strongly ϵ-symmetric,	*if* $\rho_{\vert\epsilon} = (\rho_{\vert\epsilon})\breve{}$,	*i.e. if $\rho_{\vert\epsilon}$ is symmetric.*
(3) ϵ-symmetric,	*if* $\rho_\epsilon = (\rho_\epsilon)\breve{}$,	*i.e. if ρ_ϵ is symmetric.*
(4) weakly symmetric,	*if* $\rho^\Uparrow = (\rho^\Uparrow)\breve{}$,	*i.e. if $\rho^\Uparrow$ is symmetric.*
(5) weakly ϵ-symmetric	*if* $\rho_\epsilon = (\rho^\Uparrow)\breve{}$.	

Proof. (2): Recall that by Def. 9.3(e) we have $\rho_{\vert\epsilon}(x,y) = \rho(x,y)$ if $\rho(x,y) \geq \epsilon$ and $\rho_{\vert\epsilon}(x,y) = 0$ otherwise.

(3): $x\rho_\epsilon y \leftrightarrow x\rho y \geq \epsilon \leftrightarrow y\rho x \geq \epsilon \leftrightarrow y\rho_\epsilon x$. □

(c) Transitivity:

Things become a little bit more subtle for transitivity. The commonly used classical two-valued definition $\forall x, y, z[x\rho y \,\&\, y\rho z \rightarrow x\rho z]$ can be read in two ways: For any z, if there is a connection from x to y and from y to z, then (i) there exists also a connection from x to z, and (ii) there is also a connection from x to z with strength $\rho(x,z)$ not less than the maximal value in $\{\min\{\rho(x,y), \rho(y,z)\}\}$. Contrary to the many-valued case these two interpretations coincide due to the collapsed situation for $L = \mathbf{2}$ and can conveniently be expressed by the well known compact algebraic criterion $\rho\rho \subseteq \rho$, equivalent to $\forall x, z \in X\left[\bigvee_{y\in X}(x\rho y \wedge y\rho z) \leq x\rho z\right]$. This is also the usual way how transitivity has originally been defined for L-fuzzy relations; it corresponds to interpretation (ii).

Definition 9.41. (Transitivity variants) *Let L be a complete Heyting algebra, ρ be an L-fuzzy relation over X, and $\epsilon > 0$; then ρ is called*

(a) **transitive** *if* $\forall x, y, z \in X[x\rho y \wedge y\rho z \leq x\rho z]$.
(b) **strongly ϵ-transitive** *if* $\forall x, y, z \in X[x\rho y \wedge y\rho z \geq \epsilon \rightarrow x\rho y \wedge y\rho z \leq x\rho z]$.
(c) **ϵ-transitive** *if* $\forall x, y, z \in X[x\rho y \wedge y\rho z \geq \epsilon \rightarrow x\rho z \geq \epsilon]$.
(d) **weakly transitive** *if* $\forall x, y, z \in X[x\rho y \wedge y\rho z > 0 \rightarrow x\rho z > 0]$.
(e) **weakly ϵ-transitive** *if* $\forall x, y, z \in X[x\rho y \wedge y\rho z \geq \epsilon \rightarrow x\rho z > 0]$.

A short look at some examples may be illustrative: Let $x\rho y = 0.7$, $y\rho z = 0.8$, and $x\rho z = 0.6$; then $x\rho z < x\rho y \wedge y\rho z - 0.7$. So ρ is neither transitive nor strongly ϵ-transitive for any $\epsilon > 0$, but it is an ϵ-transitive relation for all $\epsilon \in (0, 0.6]$. It is also weakly transitive, and weakly ϵ-transitive for all $\epsilon \in (0, 0.7]$. Now suppose we change ρ by setting $x\rho z = 0.9$. Then ρ becomes both transitive, and strongly ϵ-transitive and ϵ-transitive for all $\epsilon \in (0, 0.7]$, and remains weakly transitive, and also weakly ϵ-transitive for all $\epsilon \in (0, 0.7]$.

Let $X = \{x, y\}$ and $L = \{0 < \alpha < \epsilon < 1\}$. Consider two relations ρ, σ as follows:

$$\rho^2 = \begin{pmatrix} \epsilon & 1 \\ 1 & \epsilon \end{pmatrix} \cdot \begin{pmatrix} \epsilon & 1 \\ 1 & \epsilon \end{pmatrix} = \begin{pmatrix} 1 & \epsilon \\ \epsilon & 1 \end{pmatrix}, \text{ and } \sigma^2 = \begin{pmatrix} \epsilon & \alpha \\ \alpha & \epsilon \end{pmatrix} \cdot \begin{pmatrix} \epsilon & \alpha \\ \alpha & \epsilon \end{pmatrix} = \begin{pmatrix} \epsilon & \alpha \\ \alpha & \epsilon \end{pmatrix} = \sigma.$$

ρ is not strongly ϵ-transitive, since $x\rho y \wedge y\rho x = 1 > x\rho x$, but it is both α- and ϵ-transitive, since for all entries we have $\rho(s,t) \geq \epsilon > \alpha$. σ is idempotent, and hence transitive, and since $\sigma = \sigma_{|\alpha}$ it is strongly α-transitive too (see Cor. 9.43). σ clearly is α-transitive (for all entries $z \in \rho$ we have $z \geq \alpha$), but it is also ϵ-transitive ($x\sigma x \wedge x\sigma x = \epsilon = x\sigma x$, and analogous for y).

So the relationships between the transitivity notions are not as smooth as those for reflexivity and symmetry concepts:

Corollary 9.42. *Let L be a complete Heyting algebra, and ρ be an L-fuzzy relation over X, and $\epsilon > 0$; then the following holds:*

(1) *transitive $\rightarrow$ strongly ϵ-transitive.*
(2) *ϵ transitive $\rightarrow$ weakly ϵ-transitive.*
(3) *transitive $\rightarrow$ weakly transitive $\rightarrow$ weakly ϵ-transitive.*

We continue with some algebraic, i.e. point-free, characterizations of transitivity properties in

Corollary 9.43. *Let L be a complete Heyting algebra, and ρ be an L-fuzzy relation over X; then ρ is*

(1) transitive	*iff*	$\rho^2 \subseteq \rho$.
(2) strongly ϵ-transitive	*iff*	$(\rho_{\|\epsilon})^2 \subseteq \rho_{\|\epsilon}$.
(3) ϵ-transitive	*iff*	$(\rho_\epsilon)^2 \subseteq \rho_\epsilon$.
(4) weakly transitive	*iff*	$(\rho^{\Uparrow})^2 \subseteq \rho^{\Uparrow}$.
(5) weakly ϵ-transitive	*iff*	$(\rho_\epsilon)^2 \subseteq \rho^{\Uparrow}$.

Proof. The proofs are straightforward from the definitions; we have
(1): $\forall x,y,z \in X[x\rho y \wedge y\rho z \le x\rho z] \;\leftrightarrow\; \forall x,z \in X\left[\bigvee_{y\in X}(x\rho y \wedge y\rho z) \le x\rho z\right]$.

(2): Let ρ be strongly ϵ-transitive, and let $x\rho y \wedge y\rho z \ge \epsilon$. Then $x\rho y \ge \epsilon$ & $y\rho z \ge \epsilon$, so $x\rho_{|\epsilon} y$ & $y\rho_{|\epsilon} z$. But by assumption we also have $\epsilon \le x\rho y \wedge y\rho z \le x\rho z$, whence $x\rho_{|\epsilon} z$, i.e. $(\rho_{|\epsilon})^2 \subseteq \rho_{|\epsilon}$. Conversely, from $(\rho_{|\epsilon})^2 \subseteq \rho_{|\epsilon}$ and $x\rho_{|\epsilon} y$ & $y\rho_{|\epsilon} z$ we get $x\rho y \wedge y\rho z \ge \epsilon$ and $\bigvee_{y\in X} x\rho_{|\epsilon} y \wedge y\rho_{|\epsilon} z \le x\rho_{|\epsilon} z$. So $x\rho y \wedge y\rho z \le x\rho_{|\epsilon} z = x\rho z \ge \epsilon$, for all $y \in X$.

(3): Similar to (2).

(5): Similar to (2), let ρ be weakly ϵ-transitive, and $x\rho y \wedge y\rho z \ge \epsilon$. Then $x\rho y \ge \epsilon$ & $y\rho z \ge \epsilon$, so $x\rho_\epsilon y$ & $y\rho_\epsilon z$. And by assumption we have $x\rho z > 0$, whence $x\rho^{\Uparrow} z$, i.e. $(\rho_\epsilon)^2 \subseteq \rho^{\Uparrow}$. Conversely, from $(\rho_\epsilon)^2 \subseteq \rho^{\Uparrow}$ and $x\rho y \wedge y\rho z \ge \epsilon$ we get $x\rho_\epsilon y$ & $y\rho_\epsilon z$, and by assumption $x\rho^{\Uparrow} z$. Thus $x\rho z > 0$. □

(d) Other Properties:

We now turn to properties which are essential for the definition of various kinds of order relations:

Definition 9.44. (Property variants) *Let L be a complete Heyting algebra, ρ be an L-fuzzy relation over X, and $\epsilon > 0$; then ρ is called*

(a) **irreflexive**	*if* $\forall x \in X[x\rho x = 0]$,	*or* $I_X \cap \rho = \emptyset$.
(b) **asymmetric**	*if* $\forall x,y \in X[x\rho y > 0 \rightarrow y\rho x = 0]$,	*or* $\rho \cap \rho^{\smile} = \emptyset$.
(c) **ϵ-asymmetric**	*if* $\forall x,y \in X[x\rho y \ge \epsilon \rightarrow y\rho x = 0]$,	*or* $\rho_\epsilon \cap \rho_\epsilon^{\smile} = \emptyset$.
(d) **antisymmetric**	*if* $\forall x,y \in X[x\rho y > 0 \wedge y\rho x > 0 \rightarrow x = y]$,	*or* $\rho^{\Uparrow} \cap (\rho^{\smile})^{\Uparrow} \subseteq I_X$.
(e) **ϵ-antisymmetric**	*if* $\forall x,y \in X[x\rho y \ge \epsilon \wedge y\rho x \ge \epsilon \rightarrow x = y]$,	*or* $\rho_\epsilon \cap \rho_\epsilon^{\smile} \subseteq I_X$.
(f) **connected**	*if* $\forall x,y \in X[x \ne y \rightarrow x\rho y \vee y\rho x = 1]$	*or* $\overline{I_X} \subseteq \rho \cup \rho^{\smile}$.
(g) **ϵ-connected**	*if* $\forall x,y \in X[x \ne y \rightarrow x\rho y \vee y\rho x \ge \epsilon]$,	*or* $\overline{I_X} \subseteq \rho_\epsilon \cup \rho_\epsilon^{\smile}$.
(h) **weakly connected**	*if* $\forall x,y \in X[x \ne y \rightarrow x\rho y \vee y\rho x > 0]$,	*or* $\overline{I_X} \subseteq (\rho \cup \rho^{\smile})^{\Uparrow}$.
(i) **totally connected**	*if* $\forall x,y \in X[x\rho y \vee y\rho x = 1]$	*or* $\rho \cup \rho^{\smile} = U$.
(j) **totally ϵ-connected**	*if* $\forall x,y \in X[x\rho y \vee y\rho x \ge \epsilon]$,	*or* $\rho_\epsilon \cup \rho_\epsilon^{\smile} = U$.
(k) **totally weakly connected**	*if* $\forall x,y \in X[x\rho y \vee y\rho x > 0]$,	*or* $(\rho \cup \rho^{\smile})^{\Uparrow} = U$.

Corollary 9.45. *Let L be a complete Heyting algebra, and ρ be an L-fuzzy relation over X; then the following consistency relationships hold for properties of ρ, for each $0 < \epsilon \le 1$:*

(1) *asymmetric* $\to$ ϵ*-asymmetric.*
(2) *antisymmetric* $\to$ ϵ*-antisymmetric.*
(3) *connected* $\to$ ϵ*-connected* $\to$ *weakly connected.*
(4) *totally connected* $\to$ *totally* ϵ*-connected* $\to$ *totally weakly connected.*

Another common (and nearby) generalization originates from using t-norms and t-conorms instead of the meet and join operations. t-norms are usually denoted by T and t-conorms by S, where the latter are not always substituted for the sup-operation.

Definition 9.46. ([16,27]) *Let* $\rho, \sigma \in \mathcal{F}^X$ *be fuzzy relations.* ρ *is called*

(a) T-**asymmetric** *if* $T(x\rho y, y\rho x) = 0$ *for all* $x, y \in X$.
(b) T-**antisymmetric** *if* $T(x\rho y, y\rho x) = 0$ *for all* $x, y \in X, x \neq y$.
(c) T-**transitive** *if* $T(x\rho z, z\rho y) \leq x\rho y$ *for all* $x, y, z \in X$.
(d) S-**complete** *if* $S(x\rho y, y\rho x) = 1$ *for all* $x, y \in X$.
(e) T-**equivalence** *if it is reflexive, symmetric, and* T*-transitive.*
(f) *The* T-**composition** *of* ρ *and* σ *is given by* $x(\rho \circ_T \sigma)y = \sup_{z\in X} T(x\rho z, z\sigma y)$.

Though we have now dealt with a considerable number of properties and some of their variants, we feel urged to remind you of the fact that we did neither consider the topic of graded properties nor that of fuzzy relations on fuzzy sets. But you must not believe that even this list of concepts to be exhaustive. The 'relational playground' is a wide one: see e.g. [12,32], or [36].

9.3.4 *On Some Properties of Relation Properties*

Up to now we have mainly considered consistency relationships between generalized versions of properties of classical relations, but have barely analyzed how the various generalized properties interact with each other or how they behave with respect to union, intersection, and composition of fuzzy relations. This will be a main topic of this section, restricted, however, to the most relevant of the property variants mentioned above. But before turning to it, we include some short remarks on how properties of fuzzy relations are preserved by their cuts.

9.3.4.1 *Cutworthy Properties*

The results reported below are for $[0,1]$-fuzzy relations, since this is the context usually dealt with in literature and because also t-norms can be considered. It may be helpful to recall Def. 9.3 and Prop. 9.4. For more details you may want to consult e.g. [26, Section 2.1], or for a more advanced approach [4, Section 7.2.2].

On page 23 of the former it is stated that "Any property generalized from classical set theory into the domain of fuzzy set theory that is preserved in all α-cuts for $\alpha \in (0,1]$ in the classical sense is called a **cutworthy property**." It is significant that the standard operations of intersection and union of fuzzy sets (min and max operations) are cutworthy. As a consequence, other operations that are based solely on them are cutworthy as well, as, for example, the max–min composition of binary fuzzy relations; for a direct proof see Th. 9.33(10).

A fuzzy-set generalization of any classical property is, of course, required to reduce to its crisp counterpart in the special binary case (i.e. when membership grades are restricted to 0 and 1), but it is not required to be cutworthy. There are, in fact, many fuzzy-set generalizations of classical set properties that are not cutworthy. Examples are all operations of intersection and union of fuzzy sets (t-norms and t-conorms) except the standard ones (min and max), and all operations of complementation of fuzzy sets, even though all of them are, by definition, generalizations of classical complementation.

Corollary 9.47. ([16, Section 2.4], [44, Section 3.3]) *Let ρ be an $[0,1]$-fuzzy relation over X.*

(1) ρ is reflexive iff, for all $\alpha \in (0,1]$, ρ_α is reflexive.
(2) ρ is symmetric iff, for all $\alpha \in (0,1]$, ρ_α is symmetric.
(3) ρ is transitive iff, for all $\alpha \in (0,1]$, ρ_α is transitive.

Proof. (1): If ρ is reflexive, then $x\rho x = 1 \geq \alpha$, whence $x\rho x \in \rho_\alpha$, for all $\alpha \in (0,1]$. Conversely, assuming ρ_α to be reflexive for all $\alpha \in (0,1]$, it follows that ρ_1 is reflexive, i.e. for all $x \in X$ $x\rho x = 1$.

(2): If ρ is symmetric and $x\rho_\alpha y$, then $x\rho y = y\rho x \geq \alpha$. So $y\rho_\alpha x$ and ρ_α is symmetric. Conversely, assume that ρ_α is symmetric for all $\alpha \in (0,1]$. Let $x\rho y = \alpha$ for some $\alpha \in (0,1]$ and $x, y \in X$; then $x\rho_\alpha y$ and also $y\rho_\alpha x$, whence $y\rho x \geq \alpha = x\rho y$. By a similar argument one gets $x\rho y \geq y\rho x$, which demonstrates the symmetry of ρ.

(3): Assume that ρ is transitive, and $x\rho_\alpha y$ and $y\rho_\alpha z$ for a fixed $\alpha \in (0,1]$. Then $x\rho y \geq \alpha$ and $y\rho z \geq \alpha$. Since $x\rho z \geq x\rho y \wedge y\rho z$ by definition of transitivity, we get $x\rho z \geq \alpha$. So ρ_α is transitive. Conversely, assume that ρ_α is transitive or all $\alpha \in (0,1]$. If we choose $\alpha = x\rho y \wedge y\rho z$, then we get $x\rho_\alpha y$ and $y\rho_\alpha z$, whence $x\rho_\alpha z$ since ρ_α has been assumed transitive. So $x\rho z \geq \alpha = x\rho y \wedge y\rho z$, i.e. ρ is transitive. □

So we have shown that being a fuzzy preorder, tolerance, or equivalence is a cutworthy property. From Prop. 9.4 we know that the α-cuts of a fuzzy set form a collection of nested crisp sets, and that a crisp equivalence relation on a set X determines a partition of that set. So every ρ_α determines a partition $\pi(\rho_\alpha)$ of X if ρ is a fuzzy equivalence relation. It can easily be seen that two elements x and y belong to the same block of this partition if $x\rho y \geq \alpha$, and that these partitions are nested in the sense that $\pi(\rho_\alpha)$ is a refinement of $\pi(\rho_\beta)$ iff $\alpha \geq \beta$. Such a nested sequence of partitions may be represented diagrammatically in the form of a *partition tree* (An example can be found in [26, Fig. 5.7] or [48, Fig. 2]; see also [33, Section 7.2].)

Other cutworthy properties of fuzzy relations are (see [16, Section 2.5]): irreflexivity, antisymmetry, asymmetry, connectedness, and total connectedness.

9.3.4.2 *Some Interrelationships of Relation Properties*

It may be instructive to recall some of the classical results referred to in Prop. 2.16, Prop. 2.17 and Section 2.6. As you will notice, the proofs for the crisp are useful guidelines for the fuzzy context in many cases.

Proposition 9.48. *Some basic relationships involving irreflexivity, antisymmetry, and asymmetry are:*

(1) irreflexive & weakly transitive → asymmetric.
(2) asymmetric ↔ (irreflexive & antisymmetric).
(3) weakly symmetric & antisymmetric → weakly transitive.

Proof. (1): Let ρ be an irreflexive and weakly transitive fuzzy relation; assuming that it is also weakly symmetric yields $x\rho y > 0 \;\&\; y\rho x > 0 \;\rightarrow\; x\rho x > 0$ by the weak transitivity of ρ, a contradiction to its irreflexivity. So ρ must be asymmetric.

(2): Let ρ be weakly reflexive; then by asymmetry we get $x\rho x > 0 \;\rightarrow\; x\rho x = 0$, whence we get $x\rho x = 0$ by Th. 1.17(6), a contradiction. Thus, ρ is irreflexive. By asymmetry we have $x\rho y > 0 \;\rightarrow\; y\rho x = 0$, whence $x\rho y > 0 \;\&\; x \neq y \;\rightarrow\; y\rho x = 0$ by Th. 1.17(16). Thus, ρ is also antisymmetric. To prove the other direction, assume $x\rho y > 0$; then $x \neq y$ by the irreflexivity of ρ, whence $y\rho x = 0$ by antisymmetry. Thus ρ is asymmetric.

(3): Let $x\rho y > 0 \;\&\; y\rho z > 0$; then $y\rho x > 0$ by weak symmetry. But by the antisymmetry of ρ we get $x = y$; so from $y\rho z > 0$ we get $x\rho z > 0$, i.e. ρ is weakly transitive. □

If you consider the above proofs in the context of crisp relations, then we have $x\rho y > 0$ iff $x\rho y = 1$ iff $x\rho y$, with the latter one as the usual notation. Generalizing this to L-fuzzy relations we will often use it as a shorthand to denote the more cumbersome $x\rho y > 0$.

Proposition 9.49. *If an L-fuzzy relation ρ is reflexive, symmetric, transitive, irreflexive, asymmetric, antisymmetric, or connected, then also $\rho^{\smile}$ has this property.*

Proof. Transitivity: Assume $z\rho^{\smile}y \wedge y\rho^{\smile}x$; then $x\rho y \wedge y\rho z$, whence $x\rho z$ by transitivity of ρ. Thus, also $z\rho^{\smile}x$. (Or in the more elegant algebraic way: From $\rho\rho \subseteq \rho$ we get $(\rho\rho)^{\smile} = \rho^{\smile}\rho^{\smile} \subseteq \rho^{\smile}$ by Prop. 9.32(6) and Th. 9.33(8).)

Asymmetry: $\rho \cap \rho^{\smile} = \emptyset$ implies $(\rho \cap \rho^{\smile})^{\smile} = \rho^{\smile} \cap \rho^{\smile\smile} = \emptyset^{\smile} = \emptyset$. □

To show how much generalized reflexivity resembles the concept for classical two-valued relations, we check against some properties of the latter as proven in Section 2.5. From Th. 9.33 above we know that $(\mathcal{F}_X^L, \cdot, ^{\smile}, I_X; \subseteq)$ is a partially ordered involution monoid.

Lemma 9.50. *Let L be a complete Heyting algebra, and $\mathcal{F}_\epsilon^L := \{\rho \in \mathcal{F}_X^L : \rho^\vee \leq \epsilon\}$. Then the following holds:*

(1) $\rho I_X = \rho = I_X\rho$ for all $\rho \in \mathcal{F}_X^L$.
(2) $\rho\Delta_\epsilon = \rho = \Delta_\epsilon\rho$ for all $\rho \in \mathcal{F}_\epsilon^L$.

(3) $\rho\sigma, \sigma\rho \in \mathcal{F}_\epsilon^L$ *for all* $\rho, \sigma \in \mathcal{F}_\epsilon^L$.
(4) $(\mathcal{F}_\epsilon^L, \cdot, \breve{}, \Delta_\epsilon; \subseteq)$ *is a partially ordered involution monoid.*

Proof. (1): This is Th. 9.33(1).
(2): $x\rho\Delta_\epsilon y = \bigvee_z (x\rho z \wedge z\Delta_\epsilon y) = x\rho y \wedge y\Delta_\epsilon y = x\rho y \wedge \epsilon = x\rho y$.
(3): $x\rho\sigma y = \bigvee_z (x\rho z \wedge z\sigma y) \leq \bigvee_z (\epsilon \wedge \epsilon) = \epsilon$.
(4): Follows from (2) and (3). □

Proposition 9.51. *Let L be a complete Heyting algebra, ρ, σ be ϵ-reflexive L-fuzzy relations on X, and τ be arbitrary. Then*

(1) $\rho \cup \tau$ *is* ϵ*-reflexive.*
(2) $\rho \cap \sigma$ *is* ϵ*-reflexive.*
(3) $\rho\sigma$ *and* $\sigma\rho$ *are* ϵ*-reflexive.*

Proof. (Remember that intersection of fuzzy sets has been defined via the meet $\wedge$ operation in Def. 9.2(d).)
(1): $\Delta_\epsilon \subseteq \rho \;\rightarrow\; \Delta_\epsilon \subseteq \rho \cup \tau$.
(2): $\Delta_\epsilon \subseteq \rho \;\&\; \Delta_\epsilon \subseteq \sigma \;\rightarrow\; \Delta_\epsilon \subseteq \rho \cap \sigma$.
(3): Since $\bigvee_z (x\Delta_\epsilon z \wedge z\Delta_\epsilon y) \neq 0$ iff $x = y = z$, we get $\Delta_\epsilon = (\Delta_\epsilon)^2$. And $\Delta_\epsilon \subseteq \rho \;\&\; \Delta_\epsilon \subseteq \sigma$ yields $(\Delta_\epsilon)^2 \subseteq \rho\sigma$ by Th. 9.33(5). □

Proposition 9.52. *Let L be a complete Heyting algebra, and $\rho, \sigma \in \mathcal{F}_\epsilon^L$ be ϵ-reflexive L-fuzzy relations on X, and $\tau \in \mathcal{F}^L$ be arbitrary. Then*

(1) $\rho \cup \sigma \subseteq \rho\sigma \cap \sigma\rho$.
(2) *Let* ρ *also be transitive; then* ρ *is idempotent, i.e.* $\rho^2 = \rho$.
(3) $\tau \subseteq \tau\breve{\tau}\tau$.

Proof. (1): From $\Delta_\epsilon \subseteq \rho \;\&\; \Delta_\epsilon \subseteq \sigma$ we get $\rho = \rho\Delta_\epsilon \subseteq \rho\sigma$ and $\sigma = \Delta_\epsilon\sigma \subseteq \rho\sigma$, using Lemma 9.50(2). So $\rho \cup \sigma \subseteq \rho\sigma$. The remaining assertions follow in a similar way.
(2): By assumption $\Delta_\epsilon \subseteq \rho \;\&\; \rho^2 \subseteq \rho$; then $\rho = \rho\Delta_\epsilon \subseteq \rho^2$ by monotonicity of composition.
(3): $x\tau\breve{\tau}\tau y = \bigvee_{z\in X} \left[x\tau z \wedge \bigvee_{w\in X} \left(w\tau z \wedge w\tau y\right)\right] \geq x\tau y \wedge \bigvee_{w\in X} \left(w\tau y \wedge w\tau y\right) = x\tau y \wedge \bigvee_{w\in X} \left(w\tau y\right) \geq x\tau y \wedge x\tau y = x\tau y$ □

Comparing the results concerning reflexivity obtained in Section 2.6 with the ones just shown we see that the former can be transferred to $\mathcal{F}_\epsilon^L$ and ϵ-reflexive fuzzy relations. This does not really surprise, since the properties of ϵ-reflexive fuzzy relations in $\mathcal{F}_\epsilon^L$ are the same as that of reflexive two-valued relations: An ϵ-reflexive fuzzy relation $\rho \in \mathcal{F}_\epsilon^L$ clearly is strictly ϵ-reflexive, and $x\rho y \leq \epsilon$ for all x, y, $x \neq y$.

That the restriction to $\mathcal{F}_\epsilon^L$ is essential can be seen from the following example: Let $L = \{0 < \alpha < \epsilon < 1\}$. Then

$$\begin{pmatrix} \epsilon & 1 \\ 1 & \epsilon \end{pmatrix} \cdot \begin{pmatrix} \epsilon & 0 \\ 0 & \epsilon \end{pmatrix} = \begin{pmatrix} \epsilon & \epsilon \\ \epsilon & \epsilon \end{pmatrix}, \text{ i.e. } \rho\Delta_\epsilon \neq \rho; \text{ but } \quad \begin{pmatrix} \epsilon & \alpha \\ \alpha & \epsilon \end{pmatrix} \cdot \begin{pmatrix} \epsilon & 0 \\ 0 & \epsilon \end{pmatrix} = \begin{pmatrix} \epsilon & \alpha \\ \alpha & \epsilon \end{pmatrix}, \text{ i.e. } \sigma\Delta_\epsilon = \sigma.$$

In other words, for a diagonal being a neutral element becomes dependent on the context in which it is considered. We proceed with some results concerning irreflexivity, symmetry, and transitivity; note that the proofs are more or less identical to that of the crisp case (see Section 2.6).

Proposition 9.53. *Let ρ and σ be L-fuzzy relations. Then the following holds:*

(1) *ρ,σ irreflexive $\to$ $\rho\cap\sigma$ irreflexive $\wedge$ $\rho\cup\sigma$ irreflexive.*
(2) *ρ,σ symmetric $\to$ $\rho\cap\sigma$ symmetric $\wedge$ $\rho\cup\sigma$ symmetric.*
(3) *$\rho\cup\rho^{\smile},\rho\cap\rho^{\smile},\rho\rho^{\smile}$, and $\rho^{\smile}\rho$ are symmetric.*
(4) *ρ,σ transitive $\to$ $\rho\cap\sigma$ transitive.*

Proof. (1): We have $I\cap(\rho\cap\sigma)\subseteq I\cap\rho=\emptyset$, and $\emptyset=(I\cap\rho)\cup(I\cap\sigma)=I\cap(\rho\cup\sigma)$.
(2): By assumption $\rho^{\smile}=\rho$ and $\sigma^{\smile}=\sigma$. So $(\rho\cap\sigma)^{\smile}=\rho^{\smile}\cap\sigma^{\smile}=\rho\cap\sigma$, by Prop. 9.32(7).
(3): For union we have $(\rho\cup\rho^{\smile})^{\smile}=\rho^{\smile\smile}\cup\rho^{\smile}=\rho\cup\rho^{\smile}$ by Prop. 9.32(8); and for composition $(\rho\rho^{\smile})^{\smile}=\rho^{\smile\smile}\rho^{\smile}=\rho\rho^{\smile}$ by Th. 9.33(8).
(4): We have $\rho\cap\sigma\subseteq\rho$ and $\rho\cap\sigma\subseteq\sigma$, and by hypothesis $\rho^2\subseteq\rho$ and $\sigma^2\subseteq\sigma$. From this it follows that $(\rho\cap\sigma)^2\subseteq\rho^2\subseteq\rho$ by Th. 9.33(4) and (5). An analogous argument gives $(\rho\cap\sigma)^2\subseteq\sigma$, whence $(\rho\cap\sigma)^2\subseteq\rho\cap\sigma$. □

There are some more analogies to results obtained for crisp relations in Prop. 2.24, their proofs again being reusable:

Proposition 9.54.

(1) *Let ρ and σ be symmetric, and $\rho\sigma\subseteq\sigma\rho$; then $\rho\sigma=\sigma\rho$.*
(2) *Let ρ be coreflexive; then ρ is symmetric, transitive, and $\rho^2=\rho$.*
(3) *Let ρ and σ be coreflexive relations; then $\rho\sigma=\rho\cap\sigma$.*
(4) *Let ρ and σ be coreflexive relations; then $\rho\sigma=\sigma\rho$.*

Proof. (1): We have $(\rho\sigma)^{\smile}\subseteq(\sigma\rho)^{\smile}$ by Prop. 9.32(6); furthermore, by Th. 9.33(8), $(\rho\sigma)^{\smile}=\sigma^{\smile}\rho^{\smile}=\sigma\rho$, and $(\sigma\rho)^{\smile}=\rho^{\smile}\sigma^{\smile}=\rho\sigma$, i.e. $\sigma\rho\subseteq\rho\sigma$. □

9.4 Some Results from the Calculus of Relations

In Section 2.5, the SCHRÖDER equivalences and related results have been called the "most important results in the calculus of relations". The above considerations have shown that the operations with fuzzy sets and fuzzy relations have a lot of the properties that we know from their two-valued classical variants. So a natural question is for the validity of the SCHRÖDER equivalences $\rho\sigma\subseteq\tau \;\leftrightarrow\; \rho^{\smile}\overline{\tau}\subseteq\overline{\sigma} \;\leftrightarrow\; \overline{\tau}\sigma^{\smile}\subseteq\overline{\rho}$, the modular laws, and the DEDEKIND rule and their interrelationships for L-fuzzy relations.

(To improve readability we use different notations for the of logical negation in e.g. $\neg\exists[\ldots]$ and $\neg x\rho y := x\rho y \frown 0 \in L$, and simply write $x\overline{\rho}y$ for the latter one, since there is no danger of confusion with the standard complement. Remember that for the same reason we use & and ‖ for the logical *and* and *or*.)

Theorem 9.55. *Let L denote a complete* HEYTING *algebra, and ρ, σ, and τ L-fuzzy relations; as usual, we assume that the constructs are well formed. Then the following holds:*

(1) **SCHRÖDER *rule*** *for L-fuzzy relations:*
   (a) $\rho\sigma \subseteq \tau \;\rightarrow\; \rho^{\smile}\overline{\tau} \subseteq \overline{\overline{\sigma}} \;\rightarrow\; \rho\overline{\overline{\sigma}} \subseteq \overline{\overline{\tau}}$.
   (b) $\rho\sigma \subseteq \tau \;\rightarrow\; \overline{\tau}\sigma^{\smile} \subseteq \overline{\rho} \;\rightarrow\; \overline{\overline{\rho}}\sigma \subseteq \overline{\overline{\tau}}$.

(2) **DEDEKIND *rule*:** $\rho\sigma \cap \tau \subseteq (\rho \cap \tau\sigma^{\smile})(\sigma \cap \rho^{\smile}\tau)$.

(3) ***Modular laws*** *for L-fuzzy relations:*
   (a) $\rho\sigma \cap \tau \subseteq (\rho \cap \tau\sigma^{\smile})\sigma$.
   (b) $\rho\sigma \cap \tau \subseteq \rho(\sigma \cap \rho^{\smile}\tau)$.

Proof. *(1)* SCHRÖDER *rule* (a): Note that we need $\wedge$ instead of $\otimes$ for concluding line $(*)$ from line $(\dagger)$.

$$\begin{aligned}
\rho\sigma \subseteq \tau &\leftrightarrow \forall x, y[x\rho\sigma y \le x\tau y] \\
&\leftrightarrow \forall x, y[\textstyle\bigvee_z (x\rho z \wedge z\sigma y) \le x\tau y] \\
&\leftrightarrow \forall x, y[\forall z[x\rho z \wedge z\sigma y \le x\tau y]] && (\dagger) \\
&\leftrightarrow \forall x, y[\forall z[x\rho z \le x\tau y \;\|\; z\sigma y \le x\tau y]] && (*) \\
&\leftrightarrow \forall x, y[\forall z[z\rho^{\smile} x \le x\tau y \;\|\; x\overline{\tau} y \le z\overline{\sigma} y]] && \text{by Prop. 4.33(1).}
\end{aligned}$$

Using monotonicity and Prop. 4.33(4), from the first inequality we get: $z\rho^{\smile}x \wedge x\overline{\tau}y \le x\tau y \wedge x\overline{\tau}y = 0 \le z\overline{\sigma}y$. So $\bigvee_x(z\rho^{\smile}x \wedge x\overline{\tau}y) \le z\overline{\sigma}y$. In the second case $z\rho^{\smile}x \wedge x\overline{\tau}y \le x\overline{\tau}y \le z\overline{\sigma}y$; so again $\bigvee_x(z\rho^{\smile}x \wedge x\overline{\tau}y) \le z\overline{\sigma}y$. Together this yields $\rho^{\smile}\overline{\tau} \subseteq \overline{\sigma}$.

Furthermore, assume $\rho^{\smile}\overline{\tau} \subseteq \overline{\sigma}$. Then

$$\begin{aligned}
\rho^{\smile}\overline{\tau} \subseteq \overline{\sigma} &\leftrightarrow \forall y, z[\textstyle\bigvee_x (z\rho^{\smile}x \wedge x\overline{\tau}y) \le z\overline{\sigma}y] \\
&\leftrightarrow \forall y, z[\forall x[z\rho^{\smile}x \wedge x\overline{\tau}y \le z\overline{\sigma}y]] \\
&\leftrightarrow \forall x, y, z[x\rho z \le z\overline{\sigma}y \;\|\; x\overline{\tau}y \le z\overline{\sigma}y] \\
&\leftrightarrow \forall x, y, z[x\rho z \le z\overline{\sigma}y \;\|\; z\overline{\overline{\sigma}}y \le x\overline{\overline{\tau}}y].
\end{aligned}$$

From the first inequality we get $x\rho z \wedge z\overline{\overline{\sigma}}y \le z\overline{\sigma}y \wedge z\overline{\overline{\sigma}}y = 0 \le x\overline{\overline{\tau}}y$, using monotonicity and Prop. 4.33(4), so $\bigvee_z(x\rho z \wedge z\overline{\overline{\sigma}}y) \le x\overline{\overline{\tau}}y$, i.e. $\rho\overline{\overline{\sigma}} \subseteq \overline{\overline{\tau}}$. And the second inequality implies $x\rho z \wedge z\overline{\overline{\sigma}}y \le x\overline{\overline{\tau}}y$, whence again $\bigvee_z(x\rho z \wedge z\overline{\overline{\sigma}}y) \le x\overline{\overline{\tau}}y$, i.e. we get $\rho\overline{\overline{\sigma}} \subseteq \overline{\overline{\tau}}$.

(1): SCHRÖDER *rule* (b): The proof is similar to the one above, so we present only parts of it:

$$\begin{aligned}
\rho\sigma \subseteq \tau &\leftrightarrow \ldots \text{ (as above)} \\
&\leftrightarrow \forall x, y[\forall z[x\rho z \le x\tau y \;\|\; z\sigma y \le x\tau y]] && (*) \\
&\leftrightarrow \forall x, y[\forall z[x\overline{\tau}y \le x\overline{\rho}z \;\|\; y\sigma^{\smile}z \le x\tau y]].
\end{aligned}$$

Using monotonicity and Prop. 4.33(1), from the first inequality we get: $x\overline{\tau}y \wedge y\sigma^{\smile}z \le x\overline{\tau}y \le x\overline{\rho}z$. So $\bigvee_y(x\overline{\tau}y \wedge y\sigma^{\smile}z) \le x\overline{\rho}z$. In the second case $x\overline{\tau}y \wedge y\sigma^{\smile}z \le x\overline{\tau}y \wedge x\tau y = 0 \le x\overline{\rho}z$; so, again $\bigvee_y(x\overline{\tau}y \wedge y\sigma^{\smile}z) \le x\overline{\rho}z$, whence $\overline{\tau}\sigma^{\smile} \subseteq \overline{\rho}$.

Furthermore, assume $\rho^{\smile}\overline{\tau} \subseteq \overline{\sigma}$. Then

$$\begin{aligned}\overline{\tau}\breve{\sigma} \subseteq \overline{\rho} &\leftrightarrow \forall x,y[\bigvee_z(x\overline{\tau}z \wedge z\breve{\sigma}y) \leq x\overline{\rho}y] \\ &\leftrightarrow \forall x,y[\forall z[x\overline{\tau}z \wedge y\sigma z) \leq x\overline{\rho}y]] \\ &\leftrightarrow \forall x,y,z[x\overline{\tau}z \leq x\overline{\rho}y \parallel y\sigma z \leq x\overline{\rho}y] \\ &\leftrightarrow \forall x,y,z[x\overline{\overline{\rho}}y \leq x\overline{\overline{\tau}}z \parallel x\overline{\overline{\rho}}y \leq y\overline{\sigma}z].\end{aligned}$$

From the first inequality we get $x\overline{\overline{\rho}}y \wedge y\sigma z \leq x\overline{\overline{\tau}}z$, so $\overline{\overline{\rho}}\sigma \subseteq \overline{\overline{\tau}}$. And the second inequality implies $x\overline{\overline{\rho}}y \wedge y\sigma z \leq y\overline{\sigma}z \wedge y\sigma z = 0$, i.e. again we get $\overline{\overline{\rho}}\sigma \subseteq \overline{\overline{\tau}}$.

(2): DEDEKIND *rule*:

$$\begin{aligned}x(\rho\sigma \cap \tau)y &= \bigvee_z(x\rho z \wedge z\sigma y) \wedge x\tau y \\ &= \bigvee_z(x\rho z \wedge z\sigma y \wedge y\breve{\sigma}z \wedge z\breve{\rho}x) \wedge x\tau y \\ &= \bigvee_z(x\rho z \wedge (x\tau y \wedge y\breve{\sigma}z) \wedge (z\breve{\rho}x \wedge x\tau y) \wedge z\sigma y) \\ &\leq \bigvee_z(x\rho z \wedge \bigvee_y(x\tau y \wedge y\breve{\sigma}z) \wedge \bigvee_x(z\breve{\rho}x \wedge x\tau y) \wedge z\sigma y) \\ &= \bigvee_z[x(\rho \cap \tau\breve{\sigma})z \wedge z(\sigma \cap \breve{\rho}\tau)y] \\ &= x(\rho \cap \tau\breve{\sigma})(\sigma \cap \breve{\rho}\tau)y.\end{aligned}$$

(3): *Modular laws*: By the DEDEKIND rule (2) we have $\rho\sigma \cap \tau \subseteq (\rho \cap \tau\breve{\sigma})(\sigma \cap \breve{\rho}\tau)$. Applying the monotonicity property of relation composition to the right side yields the assertions: $(\rho \cap \tau\breve{\sigma})(\sigma \cap \breve{\rho}\tau) \subseteq (\rho \cap \tau\breve{\sigma})\sigma$ and $(\rho \cap \tau\breve{\sigma})(\sigma \cap \breve{\rho}\tau) \subseteq \rho(\sigma \cap \breve{\rho}\tau)$. □

Proposition 9.56. *Let L denote a* HEYTING *algebra, and* ρ, σ, *and* τ *L-fuzzy relations; then the following holds: The two modular laws together are equivalent to the* DEDEKIND *rule.*

Proof. The modular rules follow directly from the DEDEKIND rule by virtue of the monotonicity properties of relation composition:

$$\rho\sigma \cap \tau \subseteq (\rho \cap \tau\breve{\sigma})(\sigma \cap \breve{\rho}\tau) \subseteq \begin{cases}(\rho \cap \tau\breve{\sigma})\sigma \\ \rho(\sigma \cap \breve{\rho}\tau).\end{cases}$$

Conversely, assume that both modular rules hold; then we have

$\rho\sigma \cap \tau$	$\subseteq \rho(\sigma \cap \breve{\rho}\tau)$	Th. 9.55(3)(b); thus
$\rho\sigma \cap \tau \cap \tau$	$\subseteq \rho(\sigma \cap \breve{\rho}\tau) \cap \tau$	
	$\subseteq (\rho \cap \tau(\sigma \cap \breve{\rho}\tau)\breve{})(\sigma \cap \breve{\rho}\tau)$	Th. 9.55(3)(a)
	$= (\rho \cap \tau(\breve{\sigma} \cap \breve{\tau}\rho))(\sigma \cap \breve{\rho}\tau)$	
	$\subseteq (\rho \cap \tau\breve{\sigma})(\sigma \cap \breve{\rho}\tau)$	Monotonicity of product.

□

Some simple results concerning complements needed below are given in the next two lemmas. Note that the converse direction does not hold in Lemma 9.58(1).

Lemma 9.57. *Let L denote a* HEYTING *algebra; for the negation operation* $\neg a := a \curvearrowright 0$ *the following holds:*

(1) $a \wedge \neg a = 0$.
(2) $a \wedge b = 0 \leftrightarrow a \leq \neg b \leftrightarrow b \leq \neg a$.
(3) $a \leq b \rightarrow a \wedge \neg b = 0$.

Proof. (1): A HEYTING algebra is a residuated lattice with $\otimes = \wedge$; so the assertion follows from Prop. 4.33(4).

(2): The first equivalence follows from Prop. 4.33(7), and applying it to $b \wedge a = 0$ yields the second one since $b \wedge a = 0 \leftrightarrow a \wedge b = 0$.

(3): $a \le b$ implies $\neg b \le \neg a$ by Prop. 4.33(1); thus $a \wedge \neg b = 0$ by (2). □

Lemma 9.58. *Let L denote a* HEYTING *algebra, and let A, B be two L-fuzzy sets over X. Then*

(1) $A \subseteq B \rightarrow A \cap \overline{B} = \emptyset$.
(2) $A \cap B = \emptyset \leftrightarrow A \subseteq \overline{B}$.
(3) $A \subseteq \overline{\overline{A}}$.
(4) $A \cap \overline{A} = \emptyset$.

Proof. (1): By Lemma 9.57(3) we get $A(x) \le B(x) \rightarrow A(x) \wedge \neg B(x) = 0$.
(2): By Lemma 9.57(2) $A(x) \wedge B(x) = 0$ iff $A(x) \le \neg B(x)$.
(3): By Prop. 4.33(2) $A(x) \le \neg\neg A(x)$.
(4): Follows from (1), since $A \subseteq A$. □

Proposition 9.59. *Let L denote a* HEYTING *algebra, and ρ, σ, and τ L-fuzzy relations; then the following holds: The* DEDEKIND *implies the* SCHRÖDER *rule* Th. 9.55(1).

Proof. Assume that the DEDEKIND rule holds; then by Prop. 9.56 also the modular laws are valid.

Part (a) $\rho\sigma \subseteq \tau \rightarrow \rho^{\smile}\overline{\tau} \subseteq \overline{\sigma} \rightarrow \rho\overline{\overline{\sigma}} \subseteq \overline{\overline{\tau}}$: From $\rho\sigma \subseteq \tau$ we get $\rho\sigma \cap \overline{\tau} = \emptyset$ by Lemma 9.58(1). This result together with Th. 9.55(3)(b) yields $\rho^{\smile}\overline{\tau} \cap \sigma \subseteq \rho^{\smile}(\overline{\tau} \cap \rho\sigma) = \rho^{\smile}\emptyset = \emptyset$, i.e. $\rho^{\smile}\overline{\tau} \cap \sigma = \emptyset$. Thus $\rho^{\smile}\overline{\tau} \subseteq \overline{\sigma}$ by Lemma 9.58(2), whence the first implication has been shown. Applying it to $\rho^{\smile}\overline{\tau} \subseteq \overline{\sigma}$ immediately yields $\rho\overline{\overline{\sigma}} \subseteq \overline{\overline{\tau}}$.

Part (b) $\rho\sigma \subseteq \tau \rightarrow \overline{\tau}\sigma^{\smile} \subseteq \overline{\rho} \rightarrow \overline{\overline{\rho}}\sigma \subseteq \overline{\overline{\tau}}$: From $\rho\sigma \subseteq \tau$ we get $\rho\sigma \cap \overline{\tau} = \emptyset$ by Lemma 9.58(1), and by Th. 9.55(3)(a) we have $\overline{\tau}\sigma^{\smile} \cap \rho \subseteq (\overline{\tau} \cap \rho\sigma)\sigma^{\smile} = \emptyset$, so $\overline{\tau}\sigma^{\smile} \subseteq \overline{\rho}$. Applying the first implication to $\overline{\tau}\sigma^{\smile} \subseteq \overline{\rho}$ gives $\overline{\overline{\rho}}\sigma \subseteq \overline{\overline{\tau}}$. □

Note that, given the validity of the DEDEKIND rule, we can prove the SCHRÖDER rule in an element-free way, in contrast to the proof of Th. 9.55(1). From the above results we can also see that as soon as negation is involutive, i.e. $\alpha = \overline{\overline{\alpha}}$, we are back to the classical result as given in Th. 2.21; this is the case for $L = \mathbf{2}$ in particular, and for MV-algebras in general.

References

[1] Bandler, W., Kohout, L.J.: Special properties, closures and interiors of crisp and fuzzy relations, Fuzzy Sets and Systems **26**, 317-331 (1988).

[2] Bedregal, B., Beliakov, G. et al.: Negations Generated by Bounded Lattices t-Norms, in: Greco, S. et al.(Eds.), *Proceedings of IPMU 2012, Part III*, CCIS 299, pp. 326-335 (2012).

[3] Běhounek, L., Cintula, P., Hájek, P.: Introduction to Mathematical Fuzzy Logic, in: Cintula, P., Hájek, P., Noguera, C. (Eds), *Handbook of Mathematical Fuzzy Logic, Vol 1*, Individual author and College Publications, London, pp. 1-102 (2011).

[4] Bělohlávek, R.: *Fuzzy Relational Systems - Foundations and Principles*, Springer Science+ Business Media, New York (2002).

[5] Bělohlávek, R., Dauben, J.W., Klir, G.J.: *Fuzzy Logic and Mathematics - A Historical Perspective*, Oxford University Press (2017).

[6] Bentkowska, U., Pękala, B., Rząsa, W.: The stability of local properties of fuzzy relations under ordinal equivalence, Information Sciences **491**, 265-278 (2019).

[7] Chakraborty, M.K., Das, M.: Studies in fuzzy relations over fuzzy subsets, Fuzzy Sets and Systems **9**, 79-89 (1983).

[8] Chakraborty, M.K., Das, M.: On fuzzy equivalence I, Fuzzy Sets and Systems **11**, 185-193 (1983).

[9] De Baets, B., Mesiar, R.: Triangular norms on product lattices, Fuzzy Sets and Systems **104,** 61-75 (1999).

[10] De Cooman, G. Kerre E.E.: Order norms on bounded partially ordered sets, J. Fuzzy Math. **2**, 281-310 (1994).

[11] Drewniak, J.: *Fuzzy Relation Calculus*, Silesian University, Katowice (1989).

[12] Drewniak, J., Dudziak, U.: Aggregations preserving classes of fuzzy relations, Kybernetika **41**, 265-284 (2005).

[13] Dubois, D., Prade, H.: *Fuzzy Sets and Systems: Theory and Applications*, Academic Press, New York, (1980).

[14] Dubois, D., Prade, H.: A theorem on implication functions defined from triangular norms, Stochastica **8**, 267-279 (1984)

[15] Dudziak, U.: Weak and graded properties of fuzzy relations in the context of aggregation process, Fuzzy Sets and Systems **161**, 216-233 (2010).

[16] Fodor, J., Roubens, M.: *Fuzzy Preference Modeling and Multicriteria Decision Support*, Springer Science+Business Media, Dordrecht (1994).

[17] Gerla, G.: Vagueness and formal fuzzy fogic: Some criticism, Logic and Logical Philosophy **26**, 431-460 (2017).

[18] Goguen, J.: L-fuzzy sets, J. of Mathematical Analysis and Applications **18**, 145-174 (1967).

[19] Gottwald, S.: *Fuzzy Sets and Fuzzy Logic*, Friedr. Vieweg & Sohn Verlagsgesellschaft mbH, Braunschweig/Wiesbaden (1993).

[20] Gottwald, S.: *A Treatise on Many-Valued Logics*, Research Studies Press Ltd., Baldock, England (2001).

[21] Gottwald, S.: Mathematical fuzzy logics, The Bulletin of Symbolic Logic **14**, 210-239 (2008).

[22] Gupta, K.C., Gupta, R.K.: Fuzzy equivalence relation redefined, Fuzzy Sets and Systems **79**, 227-233 (1996).

[23] Hájek, P.: *Metamathematics of Fuzzy Logic*, Springer Science+Business Media, Dordrecht (1998).

[24] Klement, E.P., Mesiar, R., Pap, E.: *Triangular Norms*, Kluwer Academic Publishers, Dordrecht (2000).

[25] Klement, E.P., Mesiar, R., Pap, E.: Triangular norms: Basic notions and properties, in Klement, E.P., Mesiar, R. (Eds.), *Logical, Algebraic, Analytic and Probabilistic Aspects of Triangular Norms*, Elsevier B.V., Dordrecht, pp. 17-60 (2005).

[26] Klir, G., Yuan, B.: *Fuzzy Sets and fuzzy Logic: Theory and Applications*, Prentice Hall, New Jersey, (1995).

[27] Llamazares, B., De Baets, B.: Fuzzy strict preference relations compatible with fuzzy orderings , Int. J. of Uncertainty, Fuzziness and Knowledge-Based Systems **18**, 13-24 (2010).

[28] Makamba, B.B., Murali, V.: Normality and congruence in fuzzy subgroups, Information Sciences **59**, 121-129 (1992).

[29] Mathew, S., Mordeson, J.N., Malik, D.S.: *Fuzzy Graph Theory*, Springer International Publishing (2018)

[30] Montes, S., Jiměnez, J., Gil, P.: ϵ-Partitions and α-Equivalences, Mathware & Soft Computing **5**, 201-211 (1998).

[31] Mordeson, J.N., Nair, P.S.: *Fuzzy Graphs and Fuzzy Hypergraphs*, Physica Verlag, Heidelberg (2000).

[32] Murthy, N., Prasad, P.: Representation of L-fuzzy binary relations via a Galois connection, Tamkang J. of Mathematics **40**, 287-305 (2009).

[33] Nguyen, H.T., Walker, C.L., Walker, E.A.: *A First Course in Fuzzy Logic*, 4th ed., Taylor & Francis Group, Boca Raton (2019).

[34] Noguera, C., Esteva, F., Godo, L.: Generalized continuous and left-continuous t-norms arising from algebraic semantics for fuzzy logics, Information Sciences **180**, 1354-1372 (2010).

[35] Ovchinnikov, S.V.: Structure of fuzzy binary relations, Fuzzy Sets and Systems **6**, 169-195 (1981).

[36] Pękala, B. Bentkowska, U. et al.: General local properties of fuzzy relations and fuzzy multisets used to an algorithm for group decision making, FUZZ-IEEE 2020, 1-8 (2020).

[37] Rautenberg, W.: *A Concise Introduction to Mathematical Logic*, 3rd ed., Springer Science+Business Media, New York (2010).

[38] Rosenfeld, A.: Fuzzy Graphs, in: Zadeh, L.A., Fu, K.S., Tanaka, K., Shimura, M. (Eds), *Fuzzy Sets and their Application to Cognitive and Decision Processes*, Academic Press, New York, pp. 77-95 (1975).

[39] Sanchez, E.: Solutions in composite fuzzy relation equations: Applications to medical diagnosis in Brouwerian logic, in: Gupta, M.M., Saridis, M.M., Gaines, B.R. (Eds.), *Fuzzy Automata and Decision Processes*, Elsevier/North-Holland, New York, pp. 221-234 (1977).

[40] Šešelja, B., Tepavčević, A., Udovičić, M.: Fuzzy posets with fuzzy order applied to fuzzy ordered groups, Filomat **28**, 1835-1848 (2014).

[41] Toth, H.: Fuzziness: From epistemic considerations to terminological clarification, Int. J. of Uncertainty, Fuzziness and Knowledge-Based Systems **5**, 481-503 (1997).

[42] Toth, H.: Towards Fixing Some 'Fuzzy' Catchwords: A Terminological Primer, in: Zadeh, L.A., Kacprzyk, J. (Eds), *Computing with Words in Information/Intelligent Systems 1*, Springer-Verlag Berlin Heidelberg, pp. 154-179 (1999).

[43] Trillas, E.: Sobre funciones de negación en la teoría de conjuntos difusos, Stochastica **3**, 47–59 (1979). (Reprinted (English version) in: Barro S. et al. (Eds.), *Advances of Fuzzy Logic*, Press of the Universidad de Santiago de Compostela, pp. 31–43 (1998).)

[44] Wang, X., Ruan, D., Kerre, E.E..: *Mathematics of Fuzziness - Basic Issues*, Springer-Verlag Berlin Heidelberg (2009).

[45] Yeh, R.T.: Toward an algebraic theory of fuzzy relational systems, The Department of Computer Science, Univ. of Texas Austin, TR-25, 26 pp. (1973).

[46] Yeh, R.T., Bang, S.Y.: Fuzzy relations, fuzzy graphs, and their applications to clustering analysis, in: Zadeh, L.A., Fu, K.S., Tanaka, K., Shimura, M. (Eds), *Fuzzy Sets and their Application to Cognitive and Decision Processes,* Academic Press, New York, pp. 125-149 (1975).

[47] Zadeh, L.A.: Fuzzy Sets, Information and Control **8**, 338-353 (1965).

[48] Zadeh, L.A.: Similarity Relations and Fuzzy Orderings, Information Sciences **3**, 177-200 (1971).

[49] Zadeh, L.A.: Outline of a New Approach to the Analysis of Complex Systems and Decision Processes, IEEE Transactions on Systems, Man and Cybernetics **3**, 28-44 (1973).

CHAPTER 10

SOME BASIC KINDS OF FUZZY RELATIONS

There are two main kinds of relations which are of foremost importance in pure and applied mathematics, in computer science, and also, of course, in daily life: equivalence and order relations. Therefore, these two classes will be the main topic of this chapter. Remember the examples given at the beginning of [47, Chapter 1] (or the problems of fair partitioning of some resources, to pay appropriate prices etc.):

> (But be aware, that relations are such a common thing in everyday life that their usage usually goes unnoticed. That is to say: Long before relations have become an object of mathematical interest, they already had an important role in dealing with everyday situations. But associated epistemological considerations are not our topic here.)
>
> We can be sure of the fact that what we call relations nowadays played a significant, yet unconscious role in everyday life already very early in the development of mankind: It has always been important to be stronger, to associate with a companion more durable than most of the others, to own a larger area of land, to have available richer stocks for the winter period, to raise a greater number of cattle, to have more money at one's disposal than ... .

We will also consider generalized concepts of functions (in contrast to Chapter 1 now on the object level), and will also have a short look at fuzzy tolerance, regular and difunctional relations.

10.1 Fuzzy Tolerance and Equivalence Relations

10.1.1 *On the Epistemological Roots of Fuzziness*

It is hardly to be denied that much - if not most - of human knowledge is of qualitative nature. This kind of knowledge is cast in so-called *natural* languages, which are a highly economic tool for everyday communication since they include notions that allow for flexible interpretation. But, as it has been pointed out in [46, p. 488], this flexibility of natural language is not of an arbitrary nature: it rests upon the capability for generalization and abstraction as basic presuppositions for the evolution of language. Only on basis of these two processes we get a natural explanation for the emergence of fuzziness as follows:

- One of the main assumptions of *hypothetical realism*[1] – as well as of common sense and natural science – is the existence of a real world that is structured and is at least partially knowable by perception and experience.
- Structure means a certain kind of order, i.e. a temporal or spatial coincidence or succession of states or events. So living beings, and humans in particular, have the possibility for, and also the ability of *re*-cognizing objects and events: Only this makes the emergence of predicators and language as a whole both plausible and reasonable at all.
- *Re*-cognition in turn requires for comparisons, the result of which can be either 'discriminate' or 'make equal' (i.e. regard as equivalent).
- Fuzziness, finally, arises by the possibility of regarding as equal some attribute values (that are discriminable in principle and, hence, if necessary) in certain situations. In general, this can be done without having to fear any bad consequences or serious problems for the individual and, consequently, for the species too. (If you want e.g. to cross a road and you see a car coming along, you will neither be interested in its exact distance nor its exact speed; the only thing which is relevant for you is whether this distance is large enough and the speed is sufficiently low.)

Very similar points of view have been described already in 1772 long before fuzzy sets have been conceptualized, and also in 2004, when fuzzy sets had nearly reached the mature age of 40:

> In reality, all arguments from experience are founded on the similarity which we discover among natural objects, and by which we are induced to expect effects similar to those which we have found to follow from such objects. ... none but a fool or madman will ever pretend to dispute the authority of experience, or to reject that great guide of human life, ... From causes which appear similar we expect similar effects. This is the sum of all our experimental conclusions. ([29], chapter on *Cause and Effect*)

> Similarity is perhaps the most important phenomenon accompanying human perception and reasoning. On the one hand, similarity allows humans to think on an abstract level. Instead of storing individual elements (events, objects encountered in the past, etc.), people store classes of similar elements. Disregarding individual elements and the subsequent reasoning with similarity-based classes of these elements makes it possible to reason with complex evidence. On the other hand, similarity allows humans to act when they face a

[1] Described in Vollmer[48, pp. 188, 191]as follows: "All knowledge is hypothetical, i.e., conjectural, fallible, preliminary. There exists a real world, independent of our consciousness; it is structured, coherent, and quasi-continuous; it is at least partially knowable and explainable by perception, experience, and intersubjective science. ... Our forms of intuition and our categories of experience, the qualities of our perceptions, and the basic elements of inference, of classification, of abstraction, are, according to evolutionary epistemology, innate, ontogenetically a priori and constitutive for mesocosmic knowledge."

> situation not encountered previously. Usually, people try to act as in a known situation which is most similar to the actual one. ([3], first paragraph)

Approximate equality, similarity, indistinguishability, or nearness thus being at the epistemological roots of fuzziness, it is, of course, interesting to have a look at how these concepts are handled within FST as a kind of generalized equality. In the following considerations we will prefer the term '*similarity*'.

There are two principle approaches to be found in literature to formally describe similarity: A qualitative one by a crisp relation $\sigma \subseteq S \times S$, with $x\sigma y$ describing that x and y are similar, and a quantitative one using a mapping $S: S \times S \longrightarrow \mathbb{R}^+$ that associates a positive similarity index to x and y, with $S(x, y)$ the higher the more similar the two objects are. Since fuzzy sets can be regarded as a means to model (and calculate with) what might be called '*quantified qualities*' they should prove to be a suitable tool to handle similarity in an appropriate way.

Fuzzy equivalence relations have originally been introduced in Zadeh [56] under the name *similarity relations* as min-transitive fuzzy relations that assign to each pair$\langle x, y\rangle \in S \times S$ a truth degree $x\theta y \in [0, 1]$. We close this introductory section with the formal

Definition 10.1. *Let* $(L, \wedge, \vee, \otimes, \curvearrowright, 0, 1)$ *be a complete residuated lattice, and* S *a non-empty set.*

(a) *A* **fuzzy tolerance** *is a reflexive and symmetric L-fuzzy relation.*
(b) θ *is called a* **fuzzy equivalence**, *or* **similarity relation**, *on* S *if* θ *is a fuzzy tolerance that is transitive, i.e.* $x\theta y \otimes y\theta z \leq x\theta z$ *for all* $x, y, z \in S$. *The set of all L-fuzzy equivalence relations on* S *is denoted by* Eq_S^L.
(c) *A fuzzy equivalence is called a* **fuzzy equality** *if* $x\theta y = 1$ *implies* $x = y$.

10.1.2 *Fuzzy Tolerance Relations*

For reasons similar to that mentioned at the begin of Section 2.8.5.2, fuzzy tolerances are a popular generalization of fuzzy equivalences. In the literature, fuzzy tolerances are also called **proximity relations**. In the case of a finite (and small) underlying universe, fuzzy tolerances can be represented by symmetric matrices with all diagonal elements equal to 1. We begin with a simple result for fuzzy tolerances:

Corollary 10.2. *Let* ρ *and* σ *be two fuzzy tolerance relations on a set* X. *Then* $\rho \cap \sigma$, $\rho \cup \sigma$, *and* $\rho^{\smile}$ *are also fuzzy tolerances on* X

Proof. We use Prop. 9.53(2) and the fact that the intersection and union of reflexive fuzzy relations are reflexive. If ρ is reflexive and symmetric, then clearly $\rho^{\smile}$ also has these properties. □

The following proposition is from [49] where the authors give a point-based proof which you might perhaps want to compare with our algebraic one.

Proposition 10.3. ([49, Proposition 3.14]) *If ρ is a fuzzy tolerance on a set X; then ρ^k is a fuzzy tolerance for every $k \geq 1$.*

Proof. By definition, ρ is reflexive, i.e. $I \subseteq \rho$; thus, by monotonicity of composition, $\rho \subseteq \rho^2 \subseteq \rho^3 \subseteq \dots$. So ρ^k is reflexive for all $k \geq 1$. For symmetry we use proof by induction on k: Since ρ is also symmetric, we have $\rho = \rho^{\smile}$. Now assume that ρ^k is symmetric; then $(\rho^{k+1})^{\smile} = (\rho\rho^k)^{\smile} = (\rho^k)^{\smile}\rho^{\smile} = \rho^k\rho = \rho^{k+1}$, whence ρ^k is symmetric for all $k \geq 1$. □

Lemma 10.4. *Let L be a complete residuated lattice; then, for $\rho \in \mathbb{L}_n$, we have $\mathbf{t}(\rho) = \rho \cup \rho^2 \cup \dots \cup \rho^n$.*

Proof. From Cor. 4.31 we know that for a residuated lattice $(L, \wedge, \vee, \otimes, \curvearrowright, 0, 1)$ its reduct $(L, \vee, \otimes, 0, 1)$ is an incline. And by Cor. 8.10 the assertion follows. □

Proposition 10.5. ([49, Proposition 3.15]) *Let ρ be a fuzzy tolerance on a set X with $|X| = n$; then $\mathbf{t}(\rho) = \rho^n$, and it is the minimal equivalence relation containing ρ.*

Proof. Due to Lemma 10.4 $\mathbf{t}(\rho) = \rho \cup \rho^2 \cup \dots \cup \rho^n$, and from the proof of Prop. 10.3 we know that $\rho \subseteq \rho^2 \subseteq \rho^3 \subseteq \dots$. So $\mathbf{t}(\rho) = \rho^n$. By Cor. 10.2 ρ^n is a tolerance, i.e. reflexive and symmetric, and by definition $\mathbf{t}(\rho)$ is the minimal transitive relation τ such that $\rho \subseteq \tau$. □

In the same way as for ordinary relations in Def. 2.69, we introduce the following concepts for L-fuzzy relations (in [15, 41] the terms *pre-class* and *tolerance class* are used for (a) and (b) below):

Definition 10.6. *Let ρ be an arbitrary, and τ be a tolerance relation on a set S.*

(a) A **pre-block** *of ρ is a non-void fuzzy subset $B \subseteq S$, such that $B \times B \subseteq \rho$, i.e. if $B(x) \wedge B(y) \leq x\rho y$ for all $x, y \in X$ (see Def. 9.2). The set of pre-blocks of ρ is denoted by PBl_ρ.*

(b) A **block** *of ρ is a maximal pre-block with respect to inclusion: if $B \subseteq C$ and $C \times C \subseteq \rho$, then $B = C$. The set of blocks of ρ is denoted by Bl_ρ. We say that ρ has blocks provided $\mathsf{Bl}_\rho \neq \emptyset$.*

(c) The **tolerance class** *of an element $s \in S$ is the afterset $s\tau$, also denoted as s/τ, and sometimes also called the τ-***neighborhood of** *s. The set of all tolerance classes is called the* **quotient set** *of S modulo τ, denoted by S/τ.*

(d) A set S together with a fuzzy tolerance relation τ is called a **fuzzy tolerance space,** *denoted (S, τ); and Tol_S denotes the set of all tolerances on S.*

TABLE 10.1: Pre-blocks and blocks.

τ	x	y	z	$A \times A$	x	y	z	A	B	$B \times B$	x	y	z
x	0.8	0.3	0.6	x	0.6	0.3	0.6	0.6	0.6	x	0.6	0.3	0.6
y	0.3	0.8	0.7	y	0.3	0.3	0.3	0.3	0.3	y	0.3	0.3	0.3
z	0.6	0.7	1	z	0.6	0.3	0.7	0.7	1	z	0.6	0.3	0.7

We cite [15, Example 1]: Let $S = \{x, y, z\}$, and let τ be a fuzzy tolerance, and $A, B \subseteq S$ be two fuzzy sets as defined in Table 10.1. Then A is a pre-block but not a block since $A \subseteq B$, and B is also a pre-block. In fact, B is a block.

For sake of convenience we use the same compact notation for L-fuzzy relations as in the crisp case: $a\rho b$ should be read as $a\rho b = l > 0$, for some $l \in L$. We will now check for some properties of fuzzy tolerances if they hold in analogy to the crisp case (see e.g. [10], or [47, Section 4.3.2]):

Proposition 10.7. *Let ρ be an L-fuzzy relation on a finite set S, and $\emptyset \neq C \subseteq S$ be a fuzzy subset of S; then C is a pre-block of ρ iff there is a $B \in \mathsf{Bl}_\rho$ with $C \subseteq B$.*

Proof. Obviously, every fuzzy subset of a block of ρ is a pre-block. Conversely, every pre-block is contained in a block: For finite S this follows from the existence of maximal elements in every finite non-empty poset. □

Corollary 10.8. *Let ρ, σ be L-fuzzy relations on a finite set S; then*

(1) ρ has blocks iff $I \cap \rho \neq \emptyset$.
(2) If $\rho \subseteq \sigma$ and ρ has blocks, then also σ has blocks, and every block of ρ is contained in some block of σ.

Proof. (1): Take an x with $x(I \cap \rho)x$, and apply Prop. 10.7 to $C = \{x\}$.
(2): By Prop. 10.7 and (1). □

The following results are from [25, Section 4.1] and [41, Section 3] on which the former is based.

Definition 10.9. *Let A be a fuzzy set over a set $X \neq \emptyset$. A family $\mathscr{C} = \{A_i, i \in I\} \neq \emptyset$ of fuzzy sets A_i over X is called a* **fuzzy covering** *of A iff*

$$(1)\quad \forall i \in I[A_i \neq \emptyset], \qquad \text{and} \qquad (2)\quad \bigcup_{i \in I} A_i = A.$$

This definition is a natural generalization of Def. 1.3. Given a fuzzy covering $\mathscr{C} = \{C\}$ of A we define a fuzzy relation $\rho_{\mathscr{C}}$ by

$$\rho_{\mathscr{C}}(x, y) = \bigvee_{C \in \mathscr{C}} \big(C(x) \wedge C(y)\big), \tag{10.1}$$

called the fuzzy binary relation *associated with $\mathscr{C}$*.

Lemma 10.10. *The relation $\rho_{\mathscr{C}}$ defined in (10.1) satisfies the following properties:*

$$x\rho_{\mathscr{C}}y = y\rho_{\mathscr{C}}x, \tag{10.2}$$

$$x\rho_{\mathscr{C}}y \leq x\rho_{\mathscr{C}}x \wedge y\rho_{\mathscr{C}}y, \tag{10.3}$$

$$x\rho_{\mathscr{C}}x = A(x), \tag{10.4}$$

for all $x, y \in X$.

Proof. (10.2) and (10.4) are obvious; for (10.3) we get, using (10.4) and the fact that each $C \in \mathscr{C}$ is a fuzzy subset of A,

$$x\rho_{\mathscr{C}}y = \bigvee_{C \in \mathscr{C}} \big(C(x) \wedge C(y)\big) \leq A(x) \wedge A(y) = x\rho_{\mathscr{C}}x \wedge y\rho_{\mathscr{C}}y.$$

□

A relation obeying (10.2)-(10.4) is called a **proximity relation on** A in [41, Definition 3.3]. If $A = X$ then $\rho_{\mathscr{C}}$ is a reflexive and symmetric d-fuzzy relation on A, with (10.3) trivially satisfied; in other words $\rho_{\mathscr{C}}$ is a fuzzy tolerance on X. So there is a function $f : \mathscr{C} \longrightarrow \rho_{\mathscr{C}}$ from the set of fuzzy coverings to the fuzzy tolerances on a set X. As a special case of the following proposition we get that for any fuzzy tolerance $\rho \subseteq X$ there exists a fuzzy covering of X such that ρ is associated with that covering.

Proposition 10.11. ([41, Theorem 3.5]) *Let ρ be a proximity relation on a fuzzy set $A \subseteq X$. Then there exists a fuzzy covering $\mathscr{C}$ of A such that $\rho = \rho_{\mathscr{C}}$.*

Proof. We consider the family $\{K_{ab}\}$ of fuzzy sets of X given by $K_{ab}(x) = a\rho b$ if $x \in \{a, b\}$, and $K_{ab}(x) = 0$ otherwise, for all $a, b \in X$. Then the following holds:

(i) $K_{ab}(a) \wedge K_{ab}(a) = a\rho b \leq a\rho a$ by (10.3),
(ii) $K_{ab}(a) \wedge K_{ab}(b) = a\rho b$,
(iii) $K_{ab}(b) \wedge K_{ab}(a) = a\rho b = b\rho a$ by (10.2),
(iv) $K_{ab}(b) \wedge K_{ab}(b) = a\rho b \leq b\rho b$ by (10.3),
(v) $K_{ab}(x) \wedge K_{ab}(y) = 0$ if $x, y \notin \{a, b\}$.

So $K_{ab}(x) \wedge K_{ab}(y) \leq x\rho y$, for all $x, y \in X$, whence K_{ab} is a pre-block according to Def. 10.6(a), and by Prop. 10.7 there is a block $B \in \mathsf{Bl}_\rho$ with $K_{ab} \subseteq B$.

Let $\mathscr{C} = \{C\}$ be a set of blocks such that for all $a, b \in X$ we have $K_{ab} \subseteq C$ for some $C \in \mathscr{C}$. Then, for any $C \in \mathscr{C}$, $C(a) \wedge C(b) \leq a\rho b$ and, moreover, there is a C for which equality holds: Let $K_{ab} \subseteq C \in \mathscr{C}$, then $a\rho b = K_{ab}(a) \wedge K_{ab}(b) \leq C(a) \wedge C(b) \leq a\rho b$. Hence, $\bigvee_{C \in \mathscr{C}} \big(C(a) \wedge C(b)\big) = a\rho b$, for all $a, b \in X$. And since ρ is a proximity relation on U, we get $\bigvee_{C \in \mathscr{C}} C(a) = A(a)$, for all $a \in X$. So $\mathscr{C}$ is a cover of A, and $\rho = \rho_{\mathscr{C}}$. □

We have thus shown that the function $f : \mathscr{C} \longrightarrow \rho_{\mathscr{C}}$ from the set of fuzzy coverings to the fuzzy tolerances on a set X is onto. [41, Example 3.7] (or [25, p. 87]) shows that f is not a bijective mapping.

10.1.3 *Fuzzy Equivalence Relations*

10.1.3.1 *Some Algebraic Properties*

In case that L is a Heyting lattice then the following correspondence with crisp equivalence relations can be shown:

Proposition 10.12. *Let $(L, \wedge, \vee, \otimes, \frown, 0, 1)$ be a complete Heyting lattice, i.e. $\otimes = \wedge$. Then $\rho \in \mathsf{Eq}_S^L$ iff $\rho_\alpha \in \mathsf{Eq}_S$ for all $\alpha \in L^+$.*

Proof. We have to show that ρ is reflexive, symmetric, and transitive iff ρ_α is for all $\alpha \in L^+$. The proofs for reflexivity and symmetry are analogous to Cor. 9.47, so we prove the assertion for transitivity only. If ρ is $\wedge$-transitive and $\{\langle x,y\rangle,\langle y,z\rangle\} \subseteq \rho_\alpha$ for any fixed $\alpha \in L^+$, then $x\rho y \geq \alpha$ and $y\rho z \geq \alpha$, and by the assumption of $\wedge$-transitivity we have $x\rho z \geq x\rho y \wedge y\rho z \geq \alpha$. Thus $\langle x,z\rangle \in \rho_\alpha$. (It is here where we need $\otimes = \wedge$: Since $\wedge$ is the only idempotent t-norm by Prop. 9.21, and also the greatest one by Prop. 9.20, only $\wedge$ allows us to derive $x\rho z \geq \alpha$.) Conversely assume that ρ_α is an equivalence on S for all $\alpha \in L^+$, and let $x\rho y \wedge y\rho z \geq \alpha$. Then $\{\langle x,y\rangle,\langle y,z\rangle\} \subseteq \rho_\alpha$, whence $\langle x,z\rangle \in \rho_\alpha$ since ρ_α is transitive. Thus $x\rho z \geq \alpha$. □

In Prop. 2.66 and Prop. 2.67 some properties of crisp equivalence relations have been shown. We will now check if they can be generalized to fuzzy relations. (Recall that $\Delta_1 = I_X$ and $\mathcal{F}^L_\epsilon := \{\rho \in \mathcal{F}^L_X : \rho^\vee \leq \epsilon\}$; so $\mathcal{F}^L_1 = \mathcal{F}^L_X$.)

Proposition 10.13. *Let $\rho,\sigma \in \mathsf{Eq}^L_S$. Then the following holds:*

(1) $\rho \cap \sigma$ is an equivalence relation.
(2) $\rho\rho = \rho$.
(3) $\rho\sigma$ is an equivalence iff $\rho\sigma = \sigma\rho$.
(4) $\rho\rho^\smile\rho = \rho$.
(5) $\rho \subseteq \sigma$ implies $\rho\sigma = \sigma$.
(6) If $\rho \cup \sigma$ is an equivalence then $\rho \cup \sigma = \rho\sigma$.
(7) If $\rho \cup \sigma$ is an equivalence then ρ and σ commute.

Proof. (1): Reflexivity of intersection follows from Prop. 9.51(2). By Prop. 9.53 $\rho \cap \sigma$ is symmetric and transitive.

(2): By Prop. 9.52(2).

(3): If $\rho\sigma$ is an equivalence, then $\rho\sigma = (\rho\sigma)^\smile = \sigma^\smile\rho^\smile = \sigma\rho$, by the symmetry of each of $\rho\sigma, \rho$ and σ. Conversely, assume $\rho\sigma = \sigma\rho$. By Prop. 9.51(3) $\rho\sigma$ is reflexive; by the symmetry of ρ and σ we get $\rho\sigma = \sigma\rho = \sigma^\smile\rho^\smile = (\rho\sigma)^\smile$, whence $\rho\sigma$ is symmetric; and, finally, $\rho\sigma$ is transitive: $\rho\sigma\rho\sigma = \rho\sigma\sigma\rho = \rho\sigma\rho = \rho\rho\sigma = \rho\sigma$ by (2).

(4): Using symmetry of ρ and (2) we get $\rho\rho^\smile\rho = \rho^2 = \rho$.

(5): Obviously, $\sigma = I\sigma \subseteq \rho\sigma$; by assumption we also have $\rho\sigma \subseteq \sigma\sigma = \sigma$ by monotonicity of composition and (2).

(6): By Prop. 9.52(1) we have $\rho \cup \sigma \subseteq \rho\sigma$. Moreover, $\rho\sigma \subseteq (\rho \cup \sigma)^2 \subseteq \rho \cup \sigma$, using the assumption and monotonicity.

(7): By (6) and (3). □

The question about the structure of Eq^L_S can be answered using the same kind of arguments as for two-valued equivalences:

Proposition 10.14. *Let L be a complete Heyting lattice. Then $(\mathsf{Eq}^L_S, \subseteq)$ is a complete lattice.*

Proof. For every $S \neq \emptyset$ we have $\{I_S, U_S\} \subseteq \mathsf{Eq}_S^L$. Furthermore, $I_S \subseteq \rho \subseteq U_S$ for each $\rho \in \mathsf{Eq}_S^L$, whence Eq_S^L is a bounded poset. We have to show that any non-empty family $\{\rho_j : j \in J\} \subseteq \mathsf{Eq}_S^L$ has a supremum and infimum. From Prop. 10.13(1) we get that $\bigwedge\{\rho_j : j \in J\} = \bigcap_{j\in J} \rho_j \in \mathsf{Eq}_S^L$. The supremum $\bigvee\{\rho_j : j \in J\}$ of a familiy of equivalence relations on S is $\bigcap\{\rho \in \mathsf{Eq}_S^L : \rho \supseteq \rho_j$, for all $j \in J\}$. But, for each $j \in J$, $U_S \in \mathsf{Eq}_S^L$ and $\rho_j \subseteq U_S$, whence this intersection exists and is an equivalence on S due to what has just been shown. Moreover, it is the least such equivalence and therefore the supremum of $\{\rho_j : j \in J\} \subseteq \mathsf{Eq}_S^L$. □

Next we will have a look at the join operation in Eq_S^L.

Proposition 10.15. *Let L be a complete Heyting lattice and $\rho, \sigma \in \mathsf{Eq}_S^L$. Then the following holds:*

(1) $\rho \vee \sigma = \mathbf{t}(\rho \cup \sigma)$.
(2) $\rho \vee \sigma = \rho \cup \rho\sigma \cup \rho\sigma\rho \cup \ldots \cup \sigma \cup \sigma\rho \cup \sigma\rho\sigma \ldots$.
(3) ρ *and* σ *commute iff* $\rho \vee \sigma = \rho\sigma$ *iff* $\rho\sigma \in \mathsf{Eq}_S^L$.

Proof. (1): By Prop. 9.51(1) and Prop. 9.53(2) $\rho \cup \sigma$ is reflexive and symmetric, so it is a fuzzy tolerance. Thus, by Prop. 10.3, $(\rho \cup \sigma)^k$ is a fuzzy tolerance for every $k \geq 1$, and so is $\bigcup_{i\geq 1}(\rho \cup \sigma)^i$, again using Prop. 9.51(1) and Prop. 9.53(2). By definition, $\mathbf{t}(\rho \cup \sigma)$ is the smallest transitive relation τ with $\rho \subseteq \tau$ and $\sigma \subseteq \tau$. Thus, what remains to show is $\mathbf{t}(\rho \cup \sigma) = \bigcup_{i\geq 1}(\rho \cup \sigma)^i$. L is a bounded distributive lattice, hence an incline. So from Prop. 8.9 we get $A^{m+1} \leq \sum_{k=1}^{n} A^k$, for all $m \geq n$, which implies $\mathbf{t}(A) = \sum_{k\in\underline{m}} A^k$ by Prop. 8.9. Taking $A := \rho \cup \sigma$ yields the desired result.

(2): Same proof as for Prop. 2.67(3).

(3): Assume $\rho\sigma = \sigma\rho$; then $\rho \vee \sigma$ reduces to $\rho \cup \rho\sigma \cup \sigma$. But by reflexivity we get $\rho = \rho I_S \subseteq \rho\sigma$ and $\sigma = I_S\sigma \subseteq \rho\sigma$, so $\rho \vee \sigma = \rho\sigma$. By (1), $\rho \vee \sigma \in \mathsf{Eq}_S^L$; thus $\rho\sigma \in \mathsf{Eq}_S^L$ and by Prop. 10.13(3) ρ and σ commute. □

10.1.3.2 *Fuzzy Equivalence Classes*

It has been shown in Section 2.8.5.1 that a crisp equivalence relation groups the elements of its universe into disjoint classes, and that there is a one-to-one correspondence between partitions and crisp equivalence relations on a set S. A fuzzy equivalence ρ on a set S induces a crisp partition π_{ρ_α} of S, for each $\alpha \in L^+$. This clearly means that a fuzzy equivalence groups elements into crisp sets whose members are similar to each other at least to degree $\alpha \in L^+$: two elements $x, y \in S$ belong to the same block of the partition π_{ρ_α} if $x\rho y \geq \alpha$. If $\alpha \geq \beta$, then $\pi_{\rho_\alpha} \leq \pi_{\rho_\beta}$, i.e. π_{ρ_α} is a refinement of π_{ρ_β}. Such a nested family of partitions generated by α-cuts of S can be represented in the form of partition tree (see [56, p. 187]).

Besides this nice connection to crisp concepts, fuzzy equivalences and their properties are of interest on their own. So we now turn to proper many-valued generalizations of some other crisp concepts connected to equivalence relations. As a first straightforward result we are led to

Definition 10.16. *Let θ be a fuzzy equivalence relation on a non-empty set S.*

(a) *The aftersets $x\theta$, denoted as $[x]_\theta \in \mathcal{F}^L(S)$ or simply by $[x]$ if no confusion about the equivalence involved can occur, is called the* **fuzzy equivalence class** *of x induced by θ, and is given by $[x]_\theta(y) = x\theta y$.*

(b) *The set of all fuzzy equivalence classes is called the* **quotient set** *of S modulo θ, denoted by S/θ.*

For crisp equivalence relations, by Prop. 2.64 we have $[x] = [y]$ iff $x\rho y$; the fuzzy counterpart to this is shown in

Proposition 10.17. ([40, Section 3]) *Let θ be a fuzzy equivalence relation on a set S. Then the following holds:*

(1) $[x]_\theta = [y]_\theta$ *iff* $x\theta y = 1$.

(2) *If* $[x]_\theta \neq [y]_\theta \;\rightarrow\; \mathsf{hgt}([x]_\theta \cap [y]_\theta) < 1$.

(3) $x\theta y = \mathsf{hgt}([x]_\theta \otimes [y]_\theta)$.

Proof. (1): If $[x]_\theta = [y]_\theta$ then $x\theta y = [x]_\theta(y) = [y]_\theta(y) = y\theta y = 1$. Conversely, from $x\theta y = 1$ we get by transitivity of θ that, for each $s \in S$,

$$[x]_\theta(s) = x\theta s \geq x\theta y \otimes y\theta s = y\theta s = [y]_\theta(s).$$

Similarly, $[y]_\theta(s) \geq [x]_\theta(s)$, whence $[x]_\theta = [y]_\theta$.

(2): Assume $[x]_\theta \neq [y]_\theta$ and $\mathsf{hgt}([x]_\theta \cap [y]_\theta) = 1$. From the last equation we conclude that $\exists s \in S[[x]_\theta(s) = 1 = [y]_\theta(s)]$. Then, by transitivity and symmetry of θ,

$$x\theta y \geq \textstyle\bigvee_{s\in S}(x\theta s \otimes s\theta y) = \bigvee_{s\in S}\big([x]_\theta(s) \otimes [y]_\theta(s)\big) = 1,$$

a contradiction to (1).

(3): $\mathsf{hgt}([x]_\theta \otimes [y]_\theta) = \bigvee_{s\in S}\big([x]_\theta(s) \otimes [y]_\theta(s)\big) = \bigvee_{s\in S}(x\theta s \otimes s\theta y) = x\theta y$. □

Similarly to crisp equivalence relations, also fuzzy equivalences are a very important concept. So, before proceeding on the formal level, some excursion into semantical aspects is advisable, providing another example for problems that eventually remain hidden at a first look on the generalization from the two-vlaued to a many-valued formalism. There have been some debates of whether the properties of reflexivity, symmetry, and transitivity are appropriate for a concept of similarity. The first two are commonly agreed upon: Reflexivity expresses the fact that every object is completely indistinguishable from, or similar to, itself. Symmetry says that the degree in which x is indistinguishable from y coincides with the degree in which y is indistinguishable from x; so it models "*similar to each other*" which is different from "*the son resembles his father*" which is sometimes used as a counterexample (and then implicitly incorporates a temporal aspect).

From an epistemological point of view, transitivity is the most interesting property and has been a point of disagreement long before fuzzy sets appeared. Already at the beginning of the twentieth century Poincaré (in [42], Chapter II on *Mathematical Magnitude and Experiment*) discussed the problems that arise if one follows the common intuition that indistinguishability be transitive. The so-called **Poincaré paradox**, usually symbolized as $A = B \wedge B = C \wedge A \neq C$, which is the name given to Poincaré's observation that indistinguishability might behave non-transitive in real life, can be illustrated by the following example: A microscope or a telescope has a certain resolution so that we can only distinguish between two points, if their distance exceeds a certain value, say d. So it is possible that we are not able to distinguish v_1 from v_2 , v_2 from v_3 , . . . , v_{m-1} from v_m, but can easily do so between v_1 and v_m.

Now, if we try model such a situation e.g. in the context of height in cm by a fuzzy equivalence, then we get $151 \approx 152 = 1, \dots 199 \approx 200 = 1$ and by transitivity $151 \approx 200 = 1$, which is of course absurd. But if we take a value $\alpha \neq 1$ but close to it, how could we justify this choice for perfectly indistinguishable pairs? Besides, assuming $v_i \approx v_{i+1} = \alpha$, we conclude from transitivity that $v_1 \approx v_m \geq \alpha^{m-1}$. If $\otimes = \wedge$ this does not diminish the value of $v_1 \approx v_m$, so L-fuzzy equivalences with L a Heyting lattice will not be suitable. Further discussions on this topic are beyond the current scope, whence we have to refer the reader to [26, 43] for details. See also [16] and related papers in this issue.

To become well prepared for the considerations to follow below, we need some more definitions. We follow mainly the presentations in [12, 31, 40, 43].

Definition 10.18. *Let* $\mu \in \mathcal{F}^L(X)$ *and* $\theta \in \mathsf{Eq}^L_X$.

(a) μ *is called* **extensional** *(or* **observable***) w.r.t.* θ *if*

$$\mu(x) \otimes \theta(x, y) \leq \mu(y) \text{ for all } x, y \in X.$$

(b) μ *is called* **indistinguishable** *(or* **indiscernible***) w.r.t.* θ *if*

$$\mu(x) \otimes \mu(y) \leq \theta(x, y) \text{ for all } x, y \in X.$$

(c) *If* μ *is normal, extensional and indistinguishable w.r.t.* θ*, then it is called a* **fuzzy point** *with respect to* θ.

Some comments on this definition are in order: Interpreting an ordinary equivalence $\approx$ on X in the sense that equivalent elements cannot be distinguished or may be identified, then only those subsets $A \subseteq X$ satisfying $x \in A \;\&\; x \approx y \;\rightarrow\; y \in A$ "behave well" with respect to $\approx$. Extensionality generalizes this condition, and thus is the most important property a fuzzy set can fulfill with respect to a fuzzy equivalence. Similarly, indistinguishability generalizes $x, y \in A \rightarrow x \approx y$.

Lemma 10.19. *Let* $\mu \in \mathcal{F}^L(X)$ *and* $\theta \in \mathsf{Eq}^L_X$.

(1) μ *is extensional w.r.t.* θ *iff* $\mu(x) \leftrightarrows \mu(y) \geq \theta(x, y)$ *for all* $x, y \in X$.
(2) *If* μ *is normal, then it is indistinguishable w.r.t.* θ *iff it is contained in an equivalence class of* θ.
(3) μ *is a fuzzy point w.r.t.* θ *iff it is an equivalence class of* θ.

Proof. (1): Let μ be extensional w.r.t. θ; then the extensionality condition $\mu(x)\otimes\theta(x,y)\le\mu(y)$ can be rewritten as $\theta(x,y)\le\mu(x)\curvearrowright\mu(y)$. Therefore,

$$\theta(x,y)=\theta(x,y)\wedge\theta(y,x)\le(\mu(x)\curvearrowright\mu(y))\wedge(\mu(y)\curvearrowright\mu(x))=\mu(x)\leftrightharpoons\mu(y).$$

For the other direction assume $\theta(x,y)\le\mu(x)\leftrightharpoons\mu(y)$. Then $\theta(x,y)\le\mu(x)\curvearrowright\mu(y)$, and by adjointness $\mu(x)\otimes\theta(x,y)\le\mu(y)$.

(2): Let $a\in X$ be such that $\mu(a)=1$. If μ is indistinguishable with respect to θ, then $\mu(x)=\mu(a)\otimes\mu(x)\le\theta(a,x)=[a]_\theta(x)$; so $\mu\subseteq[a]_\theta$, for every $x\in X$. Conversely, if $\mu\subseteq[a]_\theta$ for some $a\in X$, then $\mu(x)\le\theta(a,x)$ and $\mu(y)\le\theta(a,y)$ for every $x,y\in X$, whence $\mu(x)\otimes\mu(y)\le\theta(a,x)\otimes\theta(a,y)\le\theta(x,y)$, using symmetry and transitivity of θ.

(3): If μ is a fuzzy point w.r.t. θ, then by (2) $\mu\subseteq[a]_\theta$, with $a\in X$ such that $\mu(a)=1$. And the extensionality of μ w.r.t. θ yields $[a]_\theta(x)=\theta(a,x)=\mu(a)\otimes\theta(a,x)\le\mu(x)$. Therefore, $[a]_\theta\subseteq\mu$, whence $\mu=[a]_\theta$.
Conversely, let $\mu=[a]_\theta$ for some $a\in X$; then $\mu(a)=1$ and μ is normal. From the $\otimes$-transitivity of θ we get $\mu(x)\otimes\theta(x,y)=\theta(a,x)\otimes\theta(x,y)\le\theta(a,y)=\mu(y)$ and $\mu(x)\otimes\mu(y)=\theta(x,a)\otimes\theta(a,y)\le\theta(x,y)$. for any $x,y\in X$. Thus, μ is a fuzzy point. □

Definition 10.20. *Let $\mu\in\mathcal{F}^L(X)$, and $\mathcal{C}\subseteq\mathcal{F}^L(X)$ for a set X. Following* [12, p. 1299] *we define fuzzy relations as follows:*

(a) $S_\mu(x,y)=1$ *if* $x=y$, *and* $S_\mu(x,y)=\mu(x)\otimes\mu(y)$ *if* $x\ne y$;
(b) $T_\mu(x,y)=1$ *if* $\mu(x)=\mu(y)$, *and* $T_\mu(x,y)=\mu(x)\otimes\mu(y)$ *if* $\mu(x)\ne\mu(y)$;
(c) $E_\mu(x,y)=\mu(x)\leftrightharpoons\mu(y)$;
(d) $E_\mathcal{C}=\bigcap_{\mu\in\mathcal{C}}E_\mu$, *i.e.* $E_\mathcal{C}(x,y)=\bigwedge_{\mu\in\mathcal{C}}(\mu(x)\leftrightharpoons\mu(y))$;
(e) $S_\mathcal{C}$ *denotes the least* $\theta\in\mathsf{Eq}_X^L$ *such that* $S_\mu\subseteq\theta$ *for all* $\mu\in\mathcal{C}$ *(such an equivalence exists since* $U_X\in\mathsf{Eq}_X^L$*).*

Lemma 10.21. *All of* S_μ, T_μ, E_μ, *and* $E_\mathcal{C}$ *are fuzzy equivalence relations.*

Proof. Obviously, S_μ and T_μ are both reflexive and symmetric, and E_μ is so by Cor. 4.32(1) and (2). As an example we prove the transitivity of S_μ: We have to show $S_\mu(x,y)\otimes S_\mu(y,z)\le S_\mu(x,z)$. If $x=z$ then $S_\mu(x,z)=1$, and the inequality trivially holds. Now assume $x\ne z$. The product $S_\mu(x,y)\otimes S_\mu(y,z)$ equals (a) $S_\mu(y,z)=S_\mu(x,z)$ if $x=y$; (b) $S_\mu(x,y)=S_\mu(x,z)$ if $y=z$; and (c) $\mu(x)\otimes\mu(y)\otimes\mu(z)\le\mu(x)\otimes\mu(z)=S_\mu(x,z)$ if $x\ne y$ and $y\ne z$ (using Prop. 4.27(6)). By Prop. 10.13(1), the intersection of fuzzy equivalences is again a fuzzy equivalence. □

The next proposition characterizes all fuzzy equivalence relations having a given fuzzy set as an equivalence class.

Proposition 10.22. ([12, Theorem 3.1]) *$\mu\in\mathcal{F}^L(X)$ is an equivalence class of some $\theta\in\mathsf{Eq}_X^L$ iff it is normal. In this case, the closed interval $[S_\mu,E_\mu]$ is the set of all $\theta\in\mathsf{Eq}_X^L$ such that $\mu\in X/\theta$.*

Proof. Let $\mu = [x]_\theta$; then by reflexivity of θ we have $x\theta x = 1$ for all $x \in X$, whence $\mathsf{hgt}([x]_\theta) = 1$. Conversely, let $\mu \in \mathcal{F}^L(X)$ be normal; then it is an equivalence class of any of the fuzzy equivalence relations S_μ, T_μ, and E_μ defined above. Assume $\mu(a) = 1$; then for e.g. S_μ we get $[a]_{S_\mu}(x) = 1 = \mu(a)$ for $x = a$, and $[a]_{S_\mu}(x) = \mu(a) \otimes \mu(x) = \mu(x)$ otherwise.

By Lemma 10.19(1) we have that μ is extensional w.r.t. θ iff $\theta(x,y) \le \mu(x) \leftrightharpoons \mu(y) = E_\mu(x,y)$ for all $x,y \in X$. Thus, the set of all $\theta \in \mathsf{Eq}^L_X$ such that μ is extensional w.r.t. θ is the principal ideal $(E_\mu]$ of Eq^L_X generated by E_μ. Furthermore, for any $\theta \in \mathsf{Eq}^L_X$ and $x,y \in X, x \ne y$ we have that μ is indistinguishable with respect to θ iff $S_\mu \subseteq \theta$, whence the set of all $\theta \in \mathsf{Eq}^L_X$ such that μ is indistinguishable w.r.t. θ is the principal filter $[S_\mu)$ of Eq^L_X generated by E_μ. By Lemma 10.19(3), the set of all $\theta \in \mathsf{Eq}^L_X$ such that μ is an equivalence class of θ is the intersection $[S_\mu) \cap (E_\mu] = [S_\mu, E_\mu]$ (note that by Cor. 4.32(6) $S_\mu \subseteq E_\mu$). □

Now we give a partial fuzzy counterpart of Th. 2.65, in which properties of the set of equivalence classes have been shown for the crisp case.

Proposition 10.23. *Let $\theta \in \mathsf{Eq}^L_X$ for a set X. Then the following holds:*

(1) $[x]_\theta \ne \emptyset$ and $\mathsf{hgt}([x]_\theta) = 1$.
(2) For $x,y \in X$, either $[x]_\theta = [y]_\theta$ or $\mathsf{hgt}([x]_\theta \cap [y]_\theta) < 1$.
(3) $\bigcup\{[x]_\theta : x \in X\} = X$.

Proof. (1): Since θ is reflexive, $[x]_\theta(x) = 1$ for each $x \in X$.
(2): By Prop. 10.17(1),(2).
(3): Since for each $x \in X$ we have $[x]_\theta(x) = 1$, $x \in \mathsf{core}([x]_\theta)$, and we get $X \subseteq \bigcup\{[x]_\theta : x \in X\}$. On the other hand, $[x]_\theta \subseteq X$ gives $\bigcup\{[x]_\theta : x \in X\} \subseteq X$. □

As you can see, from the three characteristic properties of crisp partitions, the one concerning disjointness of fuzzy equivalence classes is missing: the main distinction between fuzzy and crisp equivalence classes is that the former may have nonempty intersections. So Th. 2.65(4), i.e. X/θ is a partition of X, cannot be transferred. But one can in fact proceed in a "disjointness-free" way as e.g. in [12, 40]. It is well-known that $\mathcal{A} \subseteq \mathbf{2}^X$ is a partition of a crisp set X if and only if it is the set of all equivalence classes of some equivalence relation on X. Likewise one can define a **fuzzy partition** as a family $\mathcal{B}$ of normal fuzzy subsets of a set X if it is the set of all distinct equivalence classes of some fuzzy equivalence relation on X. A family $\mathcal{C} \subseteq \mathcal{B}$ is said to be a **fuzzy semi-partition** of X. Following [12], fuzzy semi-partitions can be characterized in several ways.

To this end two more notions are introduced in the following definition, which can often be encountered explicitly (e.g. as *natural T-indistinguishability operator* in [43]) or implicitly ([12, 31]) in connection with fuzzy equivalences and their discussion. The subsequent presentation of results follows the one of [12].

Definition 10.24. *Let $\mu, \nu \in \mathcal{F}^L(X)$.*

*(a) The **degree of compatibility** of μ and ν is $\mathsf{cmp}(\mu,\nu) = \bigvee_{x\in X}(\mu(x) \otimes \nu(x))$.*
*(b) The **degree of equality** of μ and ν is $\mathsf{equ}(\mu,\nu) = \bigwedge_{x\in X}(\mu(x) \leftrightharpoons \nu(x))$.*

Proposition 10.25. ([12, Theorem 3.2]) (Recall Def. 10.20.) *Let $\mathcal{C} \subseteq \mathcal{F}^L(X)$ be a set of normal fuzzy subsets of X; then the following conditions are equivalent:*

(1) $\mathcal{C}$ is a fuzzy semi-partition of X.
(2) for all $\mu, \nu \in \mathcal{C}$, $\mathsf{cmp}(\mu,\nu) \le \mathsf{equ}(\mu,\nu)$.
(3) for all $x, y \in X$, $\bigvee_{\mu\in\mathcal{C}}(\mu(x) \otimes \mu(y)) \le \bigwedge_{\nu\in\mathcal{C}}(\nu(x) \leftrightharpoons \nu(y))$.
(4) $S_{\mathcal{C}} \subseteq E_{\mathcal{C}}$.

Moreover, if any of these conditions holds, then the set of all fuzzy equivalence relations on X having $\mathcal{C}$ as fuzzy semi-partition is given by the interval $[S_{\mathcal{C}}, E_{\mathcal{C}}] \subseteq \mathsf{Eq}^L_X$.

Proof. (1) → (2): Let $\mathcal{C} \subseteq X/\theta$ for some $\theta \in \mathsf{Eq}^L_X$, and $\mu, \nu \in \mathcal{C}$. Then $\mu = [a]_\theta$ and $\nu = [b]_\theta$, for some $a, b \in X$. Using Prop. 10.17(3) we have

$$\mathsf{cmp}(\mu,\nu) = \bigvee_{x\in X}(\mu(x) \otimes \nu(x)) = \mathsf{hgt}(\mu\otimes\nu) = \mathsf{hgt}([a]_\theta \otimes [b]_\theta) = \theta(a,b).$$

Now take an arbitrary $y \in X$; then by Prop. 10.22

$$\theta(a,b) \le E_{[y]_\theta}(a,b) = [y]_\theta(a) \leftrightharpoons [y]_\theta(b) = [a]_\theta(y) \leftrightharpoons [b]_\theta(y) = \mu(y) \leftrightharpoons \nu(y)$$

Therefore, $\theta(a,b) \le \bigwedge_{y\in X}(\mu(y) \leftrightharpoons \nu(y)) = \mathsf{equ}(\mu,\nu)$.

(2) → (3): Note that (2) can be stated as

$$\text{for all } \mu, \nu \in \mathcal{C} \text{ and all } x, y \in X,\ \mu(x) \otimes \nu(x) \le \mu(y) \leftrightharpoons \nu(y).$$

For arbitrary $\mu, \nu \in \mathcal{C}$ and $x, y \in X$ this yields

$$\mu(x) \otimes \nu(x) \le \mu(y) \curvearrowright \nu(y) \quad \text{and} \quad \mu(y) \otimes \nu(y) \le \mu(x) \curvearrowright \nu(x).$$

By the adjunction property we get from the first inequality $\mu(x) \otimes \nu(x) \le \mu(y) \curvearrowright \nu(y) \leftrightarrow \mu(x) \otimes \nu(x) \otimes \mu(y) \le \nu(y) \leftrightarrow \mu(x) \otimes \mu(y) \otimes \nu(x) \le \nu(y) \leftrightarrow \mu(x) \otimes \mu(y) \le \nu(x) \curvearrowright \nu(y)$; then, by a similar argumentation for the second inequality, in sum we have

$$\mu(x) \otimes \mu(y) \le \nu(x) \curvearrowright \nu(y) \quad \text{and} \quad \mu(x) \otimes \mu(y) \le \nu(y) \curvearrowright \nu(x).$$

Therefore, for all $\mu, \nu \in \mathcal{C}$ and all $x, y \in X$, $\mu(x) \otimes \mu(y) \le \nu(x) \leftrightharpoons \nu(y)$.

(3) → (4): For $\mu \in \mathcal{C}$, and $x, y \in X$ with $x \neq y$, (3) implies $\mu(x) \otimes \mu(y) \le \bigwedge_{\nu\in\mathcal{C}}(\nu(x) \leftrightharpoons \nu(y)) = E_{\mathcal{C}}(x,y)$; thus $S_\mu \subseteq E_{\mathcal{C}}$, for every $\mu \in \mathcal{C}$, whence $S_{\mathcal{C}} \subseteq E_{\mathcal{C}}$.

(4) → (1): For any $\theta \in [S_{\mathcal{C}}, E_{\mathcal{C}}]$ we have $\theta \in [S_\mu, E_\mu]$, for every $\mu \in \mathcal{C}$. By Prop. 10.22, any $\mu \in \mathcal{C}$ is an equivalence class of θ, and $\mathcal{C}$ therefore a fuzzy semi-partition of X.

Finally, if $\mathcal{C}$ is a fuzzy semi-partition of X, i.e. $S_{\mathcal{C}} \subseteq E_{\mathcal{C}}$, then $[S_{\mathcal{C}}, E_{\mathcal{C}}]$ is the intersection $\bigcap_{\mu\in\mathcal{C}}[S_\mu, E_\mu]$, and by Prop. 10.22 this is the set of all fuzzy equivalence relations on X having every $\mu \in \mathcal{C}$ as one of its equivalence classes. □

Corollary 10.26. ([12, Corollaries 3.1 and 3.2]) *Let $\mathcal{C} \subseteq \mathcal{F}^L(X)$ be a fuzzy semi-partition of X. Then*

(1) for all $\mu, \nu \in \mathcal{C}$, $\mathsf{cmp}(\mu,\nu) = \mathsf{equ}(\mu,\nu)$.
(2) for all $\mu, \nu \in \mathcal{C}$ with $\mu \neq \nu$, the following two assertions hold:
 (A1) $\mathsf{hgt}(\mu\otimes\nu) < 1$, *and*
 (A2) $\nu(x) = \mathsf{hgt}(\mu\otimes\nu)$ *for every* $x \in \mathsf{core}(\mu)$.

(3) *$\mathcal{C}$ does not contain two different members with the same core.*

Proof. (1): Let $\mu, \nu \in \mathcal{C}$; so each of μ, ν is normal and thus has a non-empty core. For $a \in \mathsf{core}(\mu)$, by Prop. 4.22(3) and (5) we have

$$\bigwedge_{x\in X}(\mu(x) \leftrightharpoons \nu(x)) \leq \mu(a) \leftrightharpoons \nu(a) = \nu(a) = \mu(a) \otimes \nu(a) \leq \bigvee_{x\in X}(\mu(x) \otimes \nu(x)),$$

i.e. $\mathsf{equ}(\mu,\nu) \leq \mathsf{cmp}(\mu,\nu)$. Together with Prop. 10.25(2) this proves the assertion.

(2): Noting that $\mathsf{hgt}(\mu \otimes \nu) = \mathsf{cmp}(\mu,\nu)$, (A1) follows from (1) and Cor. 4.32(4), and (A2) is an immediate consequence of the proof of (1).

(3): (2)(A1) is equivalent to $\bigvee_{x,y\in X}(\mu(x) \otimes \nu(y)) < 1$, i.e. there are no $x, y \in X$ such that $\mu(x) = \nu(y) = 1$. □

Proposition 10.27. ([12, Theorem 3.3]) *Let $\mathcal{C}$ be a fuzzy semi-partition of X, and $x, y \in X, x \neq y$. Then for $S_{\mathcal{C}}$ the following holds: $S_{\mathcal{C}}(x,x) = 1$ and $S_{\mathcal{C}}(x,y) = \bigvee_{\mu\in\mathcal{C}}(\mu(x) \otimes \mu(y))$ for any two different $x, y \in X$.*

Proof. Assume $S \in \mathcal{F}^L_X$ to fulfill the two conditions mentioned in the assertion, viz. $S(x,x) = 1$ and $S(x,y) = \bigvee_{\mu\in\mathcal{C}}(\mu(x) \otimes \mu(y))$ for $x \neq y \in X$: By Def. 10.20 $S_\mu \subseteq S_{\mathcal{C}}$, for every $\mu \in S_{\mathcal{C}}$, and by the definition of S_μ and S one gets $S \leq S_{\mathcal{C}}$. Furthermore, due to the properties assumed for S, $S_\mu \leq S$, for every $\mu \in \mathcal{C}$, and since S_μ is reflexive so is S. Clearly, S is also symmetric. We will show that S is also transitive, i.e. a fuzzy equivalence relation. But then $S_{\mathcal{C}} \leq S$, since by Def. 10.20 it is the smallest $\theta \in \mathsf{Eq}^L_X$ such that $S_\mu \leq \theta$ for all $\mu \in \mathcal{C}$.

Since $\mathcal{C}$ is a fuzzy semi-partition of X, $\mathcal{C} = \{[a]_\theta\}_{a\in A}$ for some $\theta \in \mathsf{Eq}^L_X$ and $A \subseteq X$. Then, $\bigvee_{\mu\in\mathcal{C}}(\mu(x) \otimes \mu(y)) = \bigvee_{\mu\in\mathcal{C}}(\theta(x,a) \otimes \theta(a,y))$, and for arbitrary $x, y, z \in X$, we have

$$\begin{aligned}
S(x,y) \otimes S(y,z) &= \left[\textstyle\bigvee_{a\in A}(\theta(x,a) \otimes \theta(a,y))\right] \otimes \left[\textstyle\bigvee_{b\in A}(\theta(y,b) \otimes \theta(b,z))\right] \\
&\leq \left[\textstyle\bigvee_{a\in A}(\theta(x,a) \otimes \theta(a,y))\right] \otimes \theta(y,z) \\
&= \textstyle\bigvee_{a\in A}(\theta(x,a) \otimes \theta(a,y) \otimes \theta(y,z)) \\
&\quad \text{(by transitivity of } \theta \text{ and Prop. 4.30(1))} \\
&\leq \textstyle\bigvee_{a\in A}(\theta(x,a) \otimes \theta(a,z)) \\
&= S(x,z).
\end{aligned}$$

□

As has been shown in Prop. 10.25 a fuzzy equivalence relation corresponding to a fuzzy semi-partition need not necessarily be uniquely determined, but in some cases it is as shown in

Corollary 10.28. ([12, Corollary 3.3]) *Let $\mathcal{C} \subseteq \mathcal{F}^L(X)$ be a set of normal fuzzy subsets of X; then the following conditions are equivalent:*

(1) *There exists a unique $\theta \in \mathsf{Eq}^L_X$ having $\mathcal{C}$ as fuzzy semi-partition.*
(2) *for any two different $x, y \in X$: $\bigvee_{\mu\in\mathcal{C}}(\mu(x) \otimes \mu(y)) = \bigwedge_{\nu\in\mathcal{C}}(\nu(x) \leftrightharpoons \nu(y))$.*
(3) *$S_{\mathcal{C}} = E_{\mathcal{C}}$.*

Proof. (1) and (3) are equivalent by Prop. 10.25(1),(4). (2) and (3) are equivalent by Prop. 10.27, since by Def. 10.20(d) $\bigwedge_{\nu\in\mathcal{C}}(\nu(x) \leftrightharpoons \nu(y)) = E_{\mathcal{C}}$. □

Now we are ready to characterize fuzzy partitions as follows (recall Cor. 10.26 above for (A1) and (A2)):

Theorem 10.29. ([12, Theorem 3.4]) *Let $C \subseteq \mathcal{F}^L(X)$ be a set of normal fuzzy subsets of a set X; then the following conditions are equivalent:*

(1) C is a fuzzy partition of X.
(2) C is a fuzzy semi-partition of X, and satisfies the following condition:
  (A3) *for every $x \in X$ there is an $\mu \in C$ with $\mu(x) = 1$.*
(3) C satisfies the following two conditions:
  (A4) *for every $x \in X$ there is a unique $\mu_x \in C$ with $\mu_x(x) = 1$, and*
  (A5) *$\mu_x(y) = \mathsf{hgt}(\mu_x \otimes \mu_y)$ for all $x, y \in X$.*

Moreover, if any of the above five conditions is satisfied, then $S_C = E_C$ and it is the unique fuzzy equivalence relation on X such that C is its fuzzy partition.

Proof. (1) $\to$ (2): If C is a fuzzy partition of some $\theta \in \mathsf{Eq}^L_X$, then for every $x \in X$ $[x]_\theta \in C$ and $[x]_\theta = 1$.

(2) $\to$ (3): By (A3) for every $x \in X$ there is an $\mu \in C$ with $\mu(x) = 1$. If there is another $\nu \in C$ with $\nu(x) = 1$, then $\mathsf{hgt}(\mu \otimes \nu) \geq \mu(x) \otimes \nu(x) = 1$, which contradicts Cor. 10.26(A1). Therefore, (A4) holds. (A5) is just a reformulation of Cor. 10.26(A2).

(3) $\to$ (1): Using (A4) and (A5) we define a fuzzy relation $\theta \in \mathcal{F}^L_X$ by $\theta(x, y) = \mathsf{hgt}(\mu_x \otimes \mu_y)$. Clearly, θ is reflexive and symmetric. To prove transitivity we use (A5) and get $\theta(x, y) \otimes \theta(y, z) = \mathsf{hgt}(\mu_x \otimes \mu_y) \otimes \mathsf{hgt}(\mu_y \otimes \mu_z) = \mu_x(y) \otimes \mu_z(y) \leq \mathsf{hgt}(\mu_x \otimes \mu_z) = \theta(x, z)$. Hence $\theta \in \mathsf{Eq}^L_X$ and by (A5) $\theta(x, y) = [x]_\theta(y)$ for all $x, y \in X$, whence $C = X/\theta$.

Now assume that any of the conditions (1)-(3) above holds. Then by (A4), (A5) and Cor. 10.26(1) we have

$$\begin{aligned} \bigvee_{\mu \in C}(\mu(x) \otimes \mu(y)) &= \bigvee_{a \in A}(\mu_a(x) \otimes \mu_a(y)) = \bigvee_{a \in A}(\mu_x(a) \otimes \mu_y(a)) \\ &= \bigwedge_{a \in A}(\mu_x(a) \leftrightarrows \mu_y(a)) = \bigwedge_{a \in A}(\mu_a(x) \leftrightarrows \mu_a(y)) \\ &= \bigwedge_{\nu \in C}(\nu(x) \leftrightarrows \nu(y)). \end{aligned}$$

By Cor. 10.28 we conclude that $S_C = E_C$, and that it is the unique $\theta \in \mathsf{Eq}^L_X$ with $C = X/\theta$. □

Observe that conditions (A3) and (A4) are, respectively, equivalent to
(A3)' $\{\mathsf{core}(\mu)\}_{\mu \in C}$ is a covering of X; and
(A4)' $\{\mathsf{core}(\mu)\}_{\mu \in C}$ is a partition of X.

10.2 Fuzzy Functions

Functions are one of the most fundamental concepts in whole mathematics and simultaneously, in their usual context of classical set theory, also a very simple one. As we will see below, this changes significantly when looking at the research on fuzzified versions of this notion and the results obtained: It is a little bit cumbersome to find one's way, so I will try to provide a kind of systematic overview on the various concepts.

First recall Def. 1.11 and Def. 2.36. It is clear from these definitions that, in the classical two-valued context, a function is a special kind of relation with some characteristic properties. As may be expected on account of the detailed considerations on fuzzy relation properties in Chapter 9, things do considerably change in the many-valued case. Consequently, various formal concepts have been proposed as possible designata for the term "*fuzzy function*", beginning as early as 1975 in [37, Section 1-2], where the following variants are mentioned as "*fuzzy function of type i*" (below abbreviated as *fF-i*):

Let $A \in \mathcal{F}(X), B \in \mathcal{F}(Y)$, and let $A \times B$ be the Cartesian product of A and B, i.e. $A \times B(x, y) = A(x) \wedge B(y)$; see (9.6).

fF-5 A fuzzy function f from X in Y is $f: \mathcal{F}(X) \longrightarrow \mathcal{F}(Y)$. i.e. f associates to each fuzzy subset of X a fuzzy subset of Y. Example: "*Good books have high prices.*"

fF-6 A fuzzy function (fuzzy relation) f from X in Y is $f: X \times Y \longrightarrow [0,1]$. One can consider $f(x,y)$ as the "degree of membership of y at the image of x by f" or, alternatively, as the "intensity of the relation between x and y". Example: "*x and y are approximately equal.*"

fF-8 A fuzzy function f from A in B is $f: X \longrightarrow Y$ such that $B \circ f = A$.

fF-9 A fuzzy function f from A in B is $f: X \longrightarrow Y$ such that $B \circ f \geq A$.

fF-11 A fuzzy function f from A in B is $f \subseteq A \times B$.

One can easily uncover two entirely different groups of "*fuzzy functions*": *fF-6* and *fF-11* are based on fuzzy relations and d-fuzzy relations (to be introduced in Section 13.1), respectively, but *without* any additional function-specific conditions. The other alternatives use ordinary functions f which in some way involve fuzzy sets, *with* certain additional characteristics. However, none of these concepts works as an intuitive and nearby generalization of the classical notion of a function.

As one might suspect from this fact and the diversity of concepts mentioned in the list given above, a number of different approaches to the notion of fuzzy function has been proposed in the course of time. So just below we will continue our considerations with another early (point-based) generalization which considers a fuzzy function as a special kind of fuzzy relation.

And later on, in Section 10.2.2, we will learn about more complex notions of fuzzy function which do not only depend on the fuzzy relation under consideration, but also on fuzzy equivalences on the universes for the independent and the dependent variables involved. Fuzzy functions of that kind turned out to be very useful in approximate reasoning, fuzzy control, vague algebra and other fields. Moreover, by abandoning crisp in favor of approximate equality is much more in the original spirit of fuzzy sets.

10.2.1 *Fuzzy Functions - Syntactical Approaches*

In [54],[55, p. 130] (see also [23, p. 73]) fuzzy functions are regarded as fuzzy relations ρ *with* specific properties as shown below (the authors use the terms ϵ-**determinate** and

ϵ-**productive** for properties (a) and (b)).

Definition 10.30. *Let X, Y be two ordinary (finite) sets, and $\rho \in \mathcal{F}(X \times Y)$ be a fuzzy relation on $X \times Y$. ρ is called*

(a) ϵ-**functional** *iff* $\forall x \in X[x\rho y \geq \epsilon \wedge x\rho z \geq \epsilon \rightarrow y = z]$.
(b) ϵ-**total** *iff* $\forall x \in X \exists y \in Y[x\rho y \geq \epsilon]$.
(c) ϵ-**function** *iff it is both ϵ-functional and ϵ-total.*
(d) ϵ-**surjective** *iff* $\forall y \in Y \exists x \in X[x\rho y \geq \epsilon]$.
(e) ϵ-**injective** *iff ρ is an ϵ-function and $\rho\breve{}$ is ϵ-functional.*
(f) ϵ-**bijective** *iff ρ and $\rho\breve{}$ are both ϵ-functions.*

That means, an ϵ-function is a fuzzy relation that has the properties of a function on all those pairs (x, y) for which the strength of the relationship modeled by ρ exceeds or is equal to a certain threshold ϵ: In other words, if ρ is an ϵ-function, then its ϵ-cut ρ_ϵ is an ordinary function, i.e. it never does assign two or more elements from the codomain to an element of the domain.

We continue with generalizations of the algebraic characterization of functions, following the approaches of [38] and [40, p. 173], first having a look at the very short remarks in the latter one: Let X, Y be sets. An *L-diagonal* is a reflexive L-fuzzy relation D_X on X such that $D(x, y) < 1$ for all $x \neq y$ in X. An *L-mapping* of X in a set Y is an L-relation μ on $X \times Y$ such that (i) $\mu\breve{}\mu \subseteq D_Y$, and (ii) $D_X \subseteq \mu\mu\breve{}$ where D_X, D_Y are L-diagonals over X and Y, respectively. So by (i) there is not more that one $y \in Y$ with $\mu(x, y) = 1$, and by (ii) for each $x \in X$ there is such a $y \in Y$. But, for $x \neq y$, one can have $\mu\breve{}\mu(x, y) \leq D_Y(x, y) > 0$, i.e. there is a $z \in X$ such that $\mu\breve{}(y, z) \wedge \mu(z, x) > 0$, whence two or more different elements can be assigned to z. Thus, μ need not have the constituent function property. Accordingly, this concept has practically not found attention.

Nemitz's [38, Definition 2.6] *L-valued fuzzy functions* are much nearer to the classical situation. Its definition and important properties are shown below in their original form.

Definition 10.31. *Let L be a complete bounded lattice, and let X, Y be two ordinary sets and $f \in \mathcal{F}^L(X \times Y)$ be a fuzzy relation on $X \times Y$. f is called an L-valued* **fuzzy function** *from X onto Y if*

(1) $f\breve{}f = I_Y$,
(2) $ff\breve{}$ *is a fuzzy equivalence on X.*

Lemma 10.32. ([38, Lemma 2.7]) *Let X, Y be two ordinary sets, and $f \in \mathcal{F}^L(X \times Y)$ be a fuzzy relation on $X \times Y$. Then $f\breve{}f = I_Y$ is equivalent to the two conditions*

(a) $f(x, y) \wedge f(x, y') = 0$ *for all* $x \in X$, $y, y' \in Y$ *and* $y \neq y'$,
(b) $\bigvee_{x \in X} f(x, y) = 1$, *for any* $y \in Y$.

Proof. The assertion follows from

$$f\breve{}f(y, y') = \bigvee_{x \in X} (f\breve{}(y, x) \wedge f(x, y')) = \bigvee_{x \in X} (f(x, y) \wedge f(x, y')) = I(y, y') = \delta_{yy'}. \quad (10.5)$$

Clearly, (10.5) implies (a) and (b). Conversely, from (a) we get $f(x,y) \wedge f(x,y') = 0$, for $y, y' \in Y$ and $y \neq y'$. So, if at all, $f(x,y) \wedge f(x,y) = f(x,y) > 0$. But by (b), for any $y \in Y$ there is an $x \in X$ with $f(x,y) = 1$. Now assume $x \neq y$; then by (a) $f(x,x) \wedge f(x,y) = 0$. So we must have $x = y$, and $f(y,y) = 1 = \delta_{yy}$. □

Lemma 10.33. ([38, Lemma 2.8]) *Let X, Y be two ordinary sets, and $f \in \mathcal{F}^L(X \times Y)$ be a fuzzy relation on $X \times Y$, and $f^\smile f = I_Y$. Then $ff^\smile$ is a fuzzy equivalence iff $\bigvee_{y\in Y} f(x,y) = 1$, for any $x \in X$.*

Proof. We have $ff^\smile(x,x') = \bigvee_{y\in Y}(f(x,y) \wedge f(x',y))$. If $ff^\smile$ is an equivalence, then $ff^\smile(x,x) = 1$, i.e. $\bigvee_{y\in Y} f(x,y) = 1$. Conversely, assume that $\bigvee_{y\in Y} f(x,y) = 1$, for any $x \in X$. Let $f(x,z) = 1$; then $f^\smile(z,x) = 1$, whence $ff^\smile$ is reflexive. By Th. 9.33(8) we get $(ff^\smile)^\smile = f^{\smile\smile}f^\smile = ff^\smile$, so $ff^\smile$ is symmetric. Moreover, using the assumption $f^\smile f = I_Y$, $(ff^\smile)(ff^\smile) = f(f^\smile f)f^\smile = ff^\smile$, so $ff^\smile$ is transitive. □

An essential property which each meaningful concept for fuzzy functions must clearly guarantee is that they are closed with respect to composition.

Proposition 10.34. ([38, Theorem 2.9]) *Let X, Y, and Z be sets, and f, g be L-valued fuzzy functions from X onto Y and from Y onto Z, respectively. Then fg is an L-valued fuzzy function from X onto Z.*

Proof. We have $(fg)^\smile(fg) = g^\smile(f^\smile f)g = g^\smile I_Y g = g^\smile g = I_Z$. So we can apply Lemma 10.33 to fg and calculate as follows:

$$\begin{aligned}
\bigvee_{z\in Z} fg(x,z) &= \bigvee_{z\in Z}\left[\bigvee_{y\in Y}(f(x,y) \wedge g(y,z))\right] = \bigvee_{y\in Y}\left[\bigvee_{z\in Z}(f(x,y) \wedge g(y,z))\right] \\
&= \bigvee_{y\in Y}\left[f(x,y) \wedge \bigvee_{z\in Z} g(y,z)\right] = \bigvee_{y\in Y}[f(x,y) \wedge 1] \\
&= \bigvee_{y\in Y} f(x,y) \\
&= 1.
\end{aligned}$$

□

It is interesting to note that Prop. 10.34 does no longer hold if condition Lemma 10.32(b) is changed to $\bigvee_{x\in X} f(x,y) > 0$; see the example in [38, p. 181].

10.2.2 *Fuzzy Functions - Semantical Approaches*

We have used the term "*syntactical*" in the header of the foregoing section because the generalizations are based on formal considerations only, and are mainly based on crisp equality. Equality on sets is the paradigmatic equivalence relation: Take a set X, and define

$$E_X(x,y) = \begin{cases} 1 & \text{if } x = y \\ 0 & \text{otherwise;} \end{cases} \tag{10.6}$$

then reflexivity, symmetry, and transitivity of $=$ on X can clearly be expressed for the ordinary case by

(Eq1) $E_X(x,x) = 1$,
(Eq2) $E_X(x,y) = E_X(y,x)$, and
(Eq3) $E_X(x,y) \wedge E_X(y,z) \leq E_X(x,z)$,

for all $x,y,z \in X$. (Note that this is just how fuzzy equivalences are defined in Def. 10.1.)

Next we consider an ordinary function $f : X \longrightarrow Y$. It can be represented as a relation $f_\rho \subseteq X \times Y$ satisfying the conditions

(1a) $\forall a \in X \exists c \in Y[f_\rho(a,c) = 1]$,
(2a) $\forall a,b \in X \forall c,d \in Y[f_\rho(a,c) = 1 \,\&\, f_\rho(b,d) = 1 \,\&\, a = b \rightarrow c = d]$.

Taking (10.6) into account, (1a) and (2a) can obviously be written in the form

(1a') $\forall a \in X \exists c \in Y[f_\rho(a,c) > 0]$,
(2a') $\forall a,b \in X \forall c,d \in Y[f_\rho(a,c) \wedge f_\rho(b,d) \wedge E_X(a,b) \leq E_Y(c,d)]$.

These motivating considerations lead us to the first semantically inspired notion (recall from Def. 10.1 that a fuzzy equality E_X on a set X is a fuzzy equivalence such that $E_X(x,y) = 1 \iff x = y$, for all $x,y \in X$):

Definition 10.35. ([17, p. 240], [33, Definition 3.2]) *An L-fuzzy relation $f \subseteq X \times Y$ is called a* **weak fuzzy function** *with respect to the fuzzy equalities E_X and E_Y if*

(W1) $\forall x \in X \exists y \in Y[f(x,y) > 0]$,
(W2) $\forall x,x' \in X \forall y,y' \in Y[f(x,y) \wedge f(x',y') \wedge E_X(x,x') \leq E_Y(y,y')]$.

(This is called a *fuzzy function* and *fuzzy mapping*, respectively, in the papers mentioned, but in view of Def. 10.30 and Def. 10.36(d) property (W1) may be called *weakly total*, whence our choice of terminology.)

According to [31, p. 434] fuzzy equivalences are intended to model gradual indistinguishability of elements. In the crisp case we can substitute equal elements in any formula or predicate without affecting its truth value. In the context of fuzzy sets this substitution property is called *extensionality*. In Def. 10.18 we have already learned that a fuzzy set $\mu \in \mathcal{F}^L(X)$ is called extensional with respect to a fuzzy equivalence $\theta \in \mathsf{Eq}_X^L$ if $\mu(x) \otimes \theta(x,y) \leq \mu(y)$ for all $x,y \in X$. The translation of this condition is "*If x belongs to μ and y equals x, then also y belongs to μ.*" In view of this we read condition (W2) of the last definition as: "*If $f(x) = y$, x equals x', and $f(x') = y'$ then y equals y'.*"

Along this way of argumentation the following notions play an important role in the considerations on fuzzy functions; see e.g. [31, p. 434] or [9, 18]. The first two conditions in (a) simply say that we can replace equal elements in both places of a fuzzy relation; and extensionality of a partial fuzzy function means that it maps indistinguishable elements to indistinguishable elements.

Definition 10.36. ([18, Section 3]) *Let $\rho \subseteq X \times Y$ be a weak fuzzy function with respect to the fuzzy equalities $\overset{X}{\sim}$ and $\overset{Y}{\sim}$; ρ is called*

(a) **partial fuzzy function** *with respect to* $\overset{X}{\sim}$ *and* $\overset{Y}{\sim}$ *if it satisfies the following conditions:*

(E1) $x\rho y \otimes x \overset{X}{\sim} x' \leq x'\rho y$ *for all* $x, x' \in X$ *and all* $y \in Y$ *(*ρ *is called* **extensional** *with respect to* $\overset{X}{\sim}$*.);*

(E2) $x\rho y \otimes y \overset{Y}{\sim} y' \leq x\rho y'$ *for all* $x \in X$ *and all* $y, y' \in Y$ *(*ρ *is called* **extensional** *with respect to* $\overset{Y}{\sim}$*);*

(PF) $x\rho y \otimes x\rho y' \leq y \overset{Y}{\sim} y'$ *for all* $x \in X$ *and all* $y, y' \in Y$*.*

(b) **fuzzy function** *if it is a partial fuzzy function such that*

(FF) $\forall x \in X[\bigvee_{y \in Y} x\rho y = 1]$.

(c) **perfect fuzzy function** *if it is a partial fuzzy function such that*

(F1) $\forall x \in X \exists y \in Y[x\rho y = 1]$.

(d) **strong fuzzy function** *with respect to* $\overset{X}{\sim}$ *and* $\overset{Y}{\sim}$ *if it satisfies conditions* (F1) *and*

(F2) $x\rho y \otimes x'\rho y' \otimes x \overset{X}{\sim} x' \leq y \overset{Y}{\sim} y'$ *for all* $x, x' \in X$ *and all* $y, y' \in Y$*.*

Note that a perfect fuzzy function is nothing else than 1-total in the terminology of Def. 10.30; so there is no real need to introduce a new notion, whence, in accordance with the classical case, we use the term **total fuzzy function** instead. It is obvious that a total fuzzy function is also a fuzzy function.

Note: *In the following L is always meant to denote a complete residuated lattice if not specified otherwise, possibly with* $\otimes = \wedge$*, and* ρ, σ *will always denote L-fuzzy relations.*

We do already know from Prop. 10.34 that the first test a generalized notion of a fuzzy function has to pass is that the composition of two fuzzy functions of a certain kind must give a function of that kind again. The next proposition shows that this holds true for partial fuzzy functions.

(For sake of simplicity, below we will often write $x \sim x'$ etc. if it is clear from the context which set is the basis for the equivalence, and (E1/ρ) etc to denote the relation for which we consider the condition mentioned. And in the proofs we will use Prop. 4.27(1): $y \leq y' \rightarrow x \otimes y \leq x \otimes y'$ and Prop. 4.30(1): $x \otimes \bigvee_{i \in I} y_i = \bigvee_{i \in I}(x \otimes y_i)$, as well as commutativity of $\otimes$, possibly without explicit referencing them.)

Proposition 10.37. *Let* $\rho \subseteq X \times Y$ *and* $\sigma \subseteq Y \times Z$ *be partial fuzzy functions with respect to* $\overset{X}{\sim}$ *and* $\overset{Y}{\sim}$*, and* $\overset{Y}{\sim}$ *and* $\overset{Z}{\sim}$*, respectively. Then* $\rho\sigma \subseteq X \times Z$ *is a partial fuzzy function with respect to* $\overset{X}{\sim}$ *and* $\overset{Z}{\sim}$*.*

Proof. (E1): We have $x\rho y \otimes x \sim x' \leq x'\rho y$ for all $y \in Y$ by (E1/ρ). Thus $x\rho\sigma z \otimes x \sim x' = (\bigvee_{y \in Y}(x\rho y \otimes y\sigma z)) \otimes x \sim x' = \bigvee_{y \in Y}(x\rho y \otimes y\sigma z \otimes x \sim x') \leq \bigvee_{y \in Y}(x'\rho y \otimes y\sigma z) = x'\rho\sigma z$.

(E2): Similar, using $y\sigma z \otimes z \sim z' \leq y\sigma z'$ by (E2/σ).

(PF): We have to show that $x\rho\sigma z \otimes x\rho\sigma z' \leq z \sim z'$, for all $x \in X, z, z' \in Z$ and conclude as follows:

$$\begin{aligned}
x\rho\sigma z\otimes x\rho\sigma z' &= \textstyle\bigvee_{t\in Y}(x\rho t\otimes t\sigma z)\otimes\bigvee_{u\in Y}(x\rho u\otimes u\sigma z') \\
&= \textstyle\bigvee_{t\in Y}\left[x\rho t\otimes t\sigma z\otimes\bigvee_{u\in Y}(x\rho u\otimes u\sigma z')\right] && \text{by Prop. 4.30(1)} \\
&= \textstyle\bigvee_{t\in Y}\bigvee_{u\in Y}\left[x\rho t\otimes t\sigma z\otimes x\rho u\otimes u\sigma z'\right] && \text{by Prop. 4.30(1)} \\
&\le \textstyle\bigvee_{t\in Y}\bigvee_{u\in Y}\left[t\overset{Y}{\sim}u\otimes t\sigma z\otimes u\sigma z'\right] && \text{by (PF}/\rho) \\
&\le \textstyle\bigvee_{t\in Y}\bigvee_{u\in Y}\left[u\sigma z\otimes u\sigma z'\right] && \text{by (E1}/\sigma) \\
&\le \textstyle\bigvee_{t\in Y}\bigvee_{u\in Y}\left[z\sim z'\right] && \text{by (PF}/\sigma) \\
&= z\sim z'.
\end{aligned}$$

□

Proposition 10.38. *Let $\rho\subseteq X\times Y$ and $\sigma\subseteq Y\times Z$ be total (strong) fuzzy functions with respect to $\overset{X}{\sim}$ and $\overset{Y}{\sim}$, and $\overset{Y}{\sim}$ and $\overset{Z}{\sim}$, respectively. Then $\rho\sigma\subseteq X\times Z$ is a total (strong) fuzzy function with respect to $\overset{X}{\sim}$ and $\overset{Z}{\sim}$.*

Proof. Assume that ρ and σ are total. Then for each $x\in X$ there is a $y\in Y$ such that $x\rho y=1$, say y_0, and for each $y\in Y$ there is a $z\in Z$ such that $y\sigma z=1$. So $y_0\sigma z_0=1$ for some $z_0\in Z$, and $x\rho y_0\otimes y_0\sigma z_0=1$, i.e. for each $x\in X$ there is a $z\in Z$ such that $x\rho\sigma z=1$.

Now let ρ and σ be strong. We have to show that $x\rho\sigma z\otimes x'\rho\sigma z'\otimes x\overset{X}{\sim}x'\le z\overset{Z}{\sim}z'$ for all $x,x'\in X$ and all $z,z'\in Z$. Similar to the proof of (PF) in Prop. 10.37 we have

$$\begin{aligned}
&x\rho\sigma z\otimes x'\rho\sigma z'\otimes x\sim x' \\
&\quad= \textstyle\bigvee_{t\in Y}(x\rho t\otimes t\sigma z)\otimes\bigvee_{u\in Y}(x'\rho u\otimes u\sigma z')\otimes x\sim x' \\
&\quad= \textstyle\bigvee_{t\in Y}\bigvee_{u\in Y}\left[x\rho t\otimes t\sigma z\otimes x'\rho u\otimes u\sigma z'\right]\otimes x\sim x' \\
&\quad= \textstyle\bigvee_{t\in Y}\bigvee_{u\in Y}\left[x\rho t\otimes x'\rho u\otimes x\sim x'\otimes t\sigma z\otimes u\sigma z'\right] \\
&\quad\le \textstyle\bigvee_{t\in Y}\bigvee_{u\in Y}\left[t\overset{Y}{\sim}u\otimes t\sigma z\otimes u\sigma z'\right] && \text{by (F2}/\rho) \\
&\quad\le \textstyle\bigvee_{t\in Y}\bigvee_{u\in Y}\left[z\sim z'\right] && \text{by (F2}/\sigma) \\
&\quad= z\sim z'.
\end{aligned}$$

□

The following results are adapted from [13, Section 3]. Let $\rho\subseteq X\times Y$ be a partial fuzzy function with respect to $\overset{X}{\sim}$ and $\overset{Y}{\sim}$. Due to adjointness and commutativity of $\otimes$, for condition (E1) in Def. 10.36 we get

$$x\sim x'\otimes x\rho y\le x'\rho y \;\leftrightarrow\; x\sim x'\le x\rho y\curvearrowright x'\rho y \text{ for all } x,x'\in X \text{ and all } y\in Y.$$

Thus, because of the symmetry of $\sim$, also $x'\sim x\otimes x'\rho y\le x\rho y \;\leftrightarrow\; x\sim x'\le x'\rho y\curvearrowright x\rho y$. So the conditions (E1) and (E2) for extensionality of an L-fuzzy relation can be rewritten as

(E1') $x\overset{X}{\sim}x'\le x\rho y\leftrightarrows x'\rho y$ for all $x,x'\in X$ and all $y\in Y$,

(E2') $y\overset{Y}{\sim}y'\le x\rho y\leftrightarrows x\rho y'$ for all $x\in X$ and all $y,y'\in Y$.

For any L-fuzzy relation $\rho\subseteq X\times Y$ we define L-fuzzy relations E^X_ρ and E^Y_ρ on X and Y, respectively, by

$$xE^X_\rho x'=\bigwedge_{y\in Y}x\rho y\leftrightarrows x'\rho y \qquad\text{and}\qquad yE^Y_\rho y'=\bigwedge_{x\in X}x\rho y\leftrightarrows x\rho y'. \tag{10.7}$$

Since (E1') holds for all $y \in Y$ we get $x \overset{X}{\sim} x' \leq \bigwedge_{y\in Y} x\rho y \leftrightharpoons x'\rho y = x E_\rho^X x'$; similar for $y \overset{Y}{\sim} y'$. In the next proposition we show that E_ρ^X and E_ρ^Y are indeed L-fuzzy equivalences on X and Y, and due to (E1') and (E2') they are the greatest L-fuzzy equivalences such that ρ is extensional with respect to them.

Proposition 10.39. *For any L-fuzzy relation $\rho \subseteq X \times Y$, E_ρ^X and E_ρ^Y are L-fuzzy equivalences on X and Y, respectively.*

Proof. Obviously, E_ρ^X is reflexive and symmetric by Cor. 4.32(1) and (2). And with regard to transitivity we have

$$\begin{aligned}
&\bigvee_{t\in X} [x E_\rho^X t \otimes t E_\rho^X x'] \\
&= \bigvee_{t\in X} [(\bigwedge_{y\in Y} x\rho y \leftrightharpoons t\rho y) \otimes (\bigwedge_{y\in Y} t\rho y \leftrightharpoons x'\rho y)] \\
&\leq \bigvee_{t\in X} [(x\rho y \leftrightharpoons t\rho y) \otimes (t\rho y \leftrightharpoons x'\rho y)] && \text{monotonicity of } \otimes\text{, for any } y \in Y \\
&\leq \bigvee_{t\in X} [(x\rho y \leftrightharpoons x'\rho y)] && \text{by Cor. 4.32(5), for any } y \in Y \\
&= x\rho y \leftrightharpoons x'\rho y && \text{for any } y \in Y.
\end{aligned}$$

So $\bigvee_{t\in X} [x E_\rho^X t \otimes t E_\rho^X x'] \leq \bigwedge_{y\in Y} (x\rho y \leftrightharpoons x'\rho y) = x E_\rho^X x'$. □

Proposition 10.40. ([13, Theorem 3.1], [14, Theorem 4.1]) *Let X, Y be non-empty sets and $\rho \subseteq X \times Y$ be an L-fuzzy relation. Then the following conditions are equivalent:*

(1) *ρ is a partial fuzzy function with respect to E_ρ^X and E_ρ^Y.*
(2) *$\rho^{\smile}$ is a partial fuzzy function with respect to E_ρ^X and E_ρ^Y.*
(3) *$\rho^{\smile}\rho \subseteq E_\rho^Y$.*
(4) *$\rho\rho^{\smile} \subseteq E_\rho^X$.*
(5) *$\rho\rho^{\smile}\rho \subseteq \rho$ (i.e. ρ is difunctional).*

Proof. (1) ⇒ (3): By Def. 10.36 ρ is extensional with respect to some equivalences $\overset{X}{\sim}$ and $\overset{Y}{\sim}$; moreover we have argued just before that E_ρ^Y is the greatest L-fuzzy equivalence on Y such that ρ is extensional with respect to it. Thus, using condition Def. 10.36(PF) we obtain $x\rho y \otimes x\rho y' = y\rho^{\smile} x \otimes x\rho y' \leq y \overset{Y}{\sim} y' \leq y E_\rho^Y y'$ for all $x \in X$ and all $y, y' \in Y$, whence $\bigvee_{x\in X} y\rho^{\smile} x \otimes x\rho y' \leq y E_\rho^Y y'$.

(3) ⇒ (1): Obvious, since ρ is extensional with respect to E_ρ^X and E_ρ^Y.

(3) ⇒ (5): By assumption, for arbitrary $x, x' \in X$ and $y, y' \in Y$ we have $y\rho^{\smile} x \otimes x\rho y' = x\rho y \otimes x\rho y' \leq y E_\rho^Y y'$, and from (E2') and the definition of $\leftrightharpoons$ we get $y E_\rho^Y y' \leq x'\rho y \leftrightharpoons x'\rho y' \leq x'\rho y \rightharpoonup x'\rho y'$. Thus, $x\rho y \otimes x\rho y' \leq x'\rho y \rightharpoonup x'\rho y'$; so by adjointness and commutativity of $\otimes$ we get $x'\rho y \otimes y\rho^{\smile} x \otimes x\rho y' \leq x'\rho y'$.

(5) ⇒ (3): Given (5), then for all $x, x' \in X$ and $y, y' \in Y$ we get $y\rho^{\smile} x \otimes x\rho y' \otimes x'\rho y \leq x'\rho y'$, and changing the roles of y and y' by symmetry of argument, also $y'\rho^{\smile} x \otimes x\rho y \otimes x'\rho y' = y\rho^{\smile} x \otimes x\rho y' \otimes x'\rho y' \leq x'\rho y$. Using adjointness yields $y\rho^{\smile} x \otimes x\rho y' \leq x'\rho y \leftrightharpoons x'\rho y'$. Thus, for all $x \in X$ and $y, y' \in Y$, $y\rho^{\smile} x \otimes x\rho y' \leq \bigwedge_{x'\in X} (x'\rho y \leftrightharpoons x'\rho y') = y E_\rho^Y y'$.

(4) ⇒ (5), (5) ⇒ (4): Similar to the above.

(1) $\iff$ (2): By the equivalence of (1) and (5) we have that $\rho^{\smile}$ is a partial fuzzy function iff $\rho^{\smile}\rho^{\smile\smile}\rho^{\smile} \subseteq \rho^{\smile}$ iff (by Prop. 9.32(6)) $\rho\rho^{\smile}\rho \subseteq \rho$ iff ρ is a partial fuzzy function. □

Similar characterizations for total fuzzy functions and uniform relations (total fuzzy functions which are surjective in the sense of Def. 10.42(b)) are given in [13, Theorems 3.2 and 3.3] and [14, Theorem 4.2].

10.2.3 *Properties of Fuzzy Functions*

Proposition 10.41. ([18, Proposition 3.9]) *Let $\overset{X}{\sim}$ and $\overset{Y}{\sim}$ be fuzzy equalities on X and Y, respectively, and $\rho \subseteq X \times Y$ an L-fuzzy relation.*

(1) If ρ is extensional with respect to $\overset{X}{\sim}$ and fulfills conditions (F1) and (PF) then it is a stong fuzzy function.
(2) A total fuzzy function is a strong fuzzy function.
(3) A strong fuzzy function is a weak fuzzy function.

Proof. (1): By (PF) we have $x'\rho y \otimes x'\rho y' \leq y \overset{Y}{\sim} y'$ and by (E1) $x\rho y \otimes x \overset{X}{\sim} x' \leq x'\rho y$ which, for sake of convenience, we abbreviate as $ab \leq c$ and $de \leq a$, respectively. Applying commutativity of $\otimes$ to Prop. 4.29(1) we get $y \leq y' \rightarrow y \otimes x \leq y' \otimes x$. So $deb \leq ab \leq c$, i.e. $x\rho y \otimes x \overset{X}{\sim} x' \otimes x'\rho y' \leq y \overset{Y}{\sim} y'$, whence we have shown (F2).
(2): A total is a partial fuzzy function for which (F1), and both (E1) and (PF) hold, by the respective definitions. Thus we get (F2) by (1).
(3): Obvious from Def. 10.35 and Def. 10.36(d), since (F1) implies (W1) and (F2) is equivalent to (W2). □

We now turn to another group of elementary and well known properties of ordinary functions and will learn how they may be generalized in a suitable way to fuzzy functions. The presentation below is a combination of [17, 18] and [33]. The terminology we use differs slightly from the one used in the papers mentioned, but, hopefully, is a little bit more systematic.

Definition 10.42. *Let $\rho \subseteq X \times Y$ be a weak fuzzy function with respect to the fuzzy equalities $\overset{X}{\sim}$ and $\overset{Y}{\sim}$; f is called*

(a) **weakly surjective** *if* $\forall y \in Y \exists x \in X[x\rho y > 0]$.
(b) **surjective** *if* $\forall y \in Y \exists x \in X[x\rho y = 1]$.
(c) **injective** *if* $\forall x, x' \in X \forall y, y' \in Y[x\rho y \otimes x'\rho y' \otimes y \overset{Y}{\sim} y' \leq x \overset{X}{\sim} x']$. (IWF)
(d) **weakly bijective** *if it is weakly surjective and injective.*
(e) **bijective** *if it is surjective and injective.*

(IWF), injectivity according to Def. 10.42(c), is also attributed to strong fuzzy functions in [18, Definition 4.2], where in addition a separate notion of injectivity for total fuzzy function is presented:

$$x\rho y \otimes x'\rho y \leq x \overset{X}{\sim} x', \text{ for all } x, x' \in X, y \in Y. \qquad \text{(IPF)}$$

Since we have just learned above that strong fuzzy functions are also weak, (IWF) can be applied consistently. Perfect are also weak fuzzy functions, so both (IPF) and (IWF) apply, and thus there is need to clarify their logical relationship.

Lemma 10.43. *Let* $\overset{X}{\sim}$ *and* $\overset{Y}{\sim}$ *be fuzzy equalities on* X *and* Y*, respectively, and* $\rho \subseteq X \times Y$ *be a weak fuzzy function.*

(1) *(IWF) implies (IPF).*
(2) *If* ρ *is total then (IPF) implies (IWF).*
(3) *For total fuzzy functions (IPF) and (IWF) are equivalent.*

Proof. (1): $x\rho y \otimes x'\rho y = x\rho y \otimes x'\rho y \otimes 1 = x\rho y \otimes x'\rho y \otimes y \overset{Y}{\sim} y \leq x \overset{X}{\sim} x'$, since in a residuated lattice $(L, \otimes, 1)$ is a monoid, and with the inequality following from (IWF).

(2): Let ρ be a total fuzzy function. Then by (E2/ρ) we have $x'\rho y' \otimes y' \sim y \leq x'\rho y$. And since $y' \sim y = y \sim y'$ we get $x'\rho y' \otimes y \sim y' \leq x'\rho y$. From (IPF) we have $x\rho y \otimes x'\rho y \leq x \overset{X}{\sim} x'$, whence $x\rho y \otimes x'\rho y' \otimes y \sim y' \leq x\rho y \otimes x'\rho y \leq x \overset{X}{\sim} x'$, i.e. (IWF) holds.

(3): Follows from (1) and (2). □

We continue with results and examples borrowed from [33, Section 3], where the real unit interval is used as underlying valuation set: Let $X = \{a, b, c\}$ and $Y = \{x, y\}$; then E_X and E_Y are fuzzy equalities, and the fuzzy relation f is a weak fuzzy function with respect to E_X and E_Y. f is not a perfect, but a weakly surjective fuzzy function. Moreover, f is not injective, since $f(a, y) \wedge f(b, y) \wedge E_Y(y, y) = 0.7 \not\leq 0.3 = E_X(a, b)$.

E_X	a	b	c
a	1	0.3	0.3
b	0.3	1	0.8
c	0.3	0.8	1

E_Y	x	y
x	1	0.7
y	0.7	1

f	x	y
a	0.5	1
b	0.4	0.7
c	1	0

For the fuzzy relations $g \subseteq X \times Y$ and $h, k \subseteq Y \times X$ shown below the following holds: g is total and surjective with respect to E_X and E_Y, but it is not injective. h is not a weak fuzzy function with respect to E_X and E_Y, since $h(x, a) \wedge h(y, b) \wedge E_Y(x, y) = 0.5 \not\leq 0.3 = E_X(a, b)$ (see Def. 10.35). And k is total and injective with respect to E_X and E_Y.

g	x	y
a	0.5	1
b	1	0.7
c	1	0

h	a	b	c
x	0.5	0.4	1
y	1	0.7	0

k	a	b	c
x	1	0.3	0.2
y	1	0	0.3

We also like to mention [33, Remark 3.4]: $I_X : X \longrightarrow X$ is bijective with respect to any fuzzy equality on X, and is itself a fuzzy equality. Any function $f : X \longrightarrow Y$ is a total fuzzy function with respect to I_X and I_Y; if f is injective (surjective, bijective) then it is injective (surjecive, bijective) with respect to I_X and I_Y.

In Prop. 10.37 and Prop. 10.38 above it has been shown that the partial and strong fuzzy functions fulfill the minimal requirement that the composition gives again a fuzzy function of the same kind. The following results give counterparts to some more properties of ordinary mappings (see Cor. 2.38). (Note that, as we did above, we keep the relational way of notation instead of the more common one for functions which would read as follows: *Let* $f : X \longrightarrow Y$ *and* $g : Y \longrightarrow Z$ *be weak fuzzy functions with respect to* $\overset{X}{\sim}$ *and* $\overset{Y}{\sim}$, *and* $\overset{Y}{\sim}$ *and* $\overset{Z}{\sim}$, *respectively. If* f *and* g *are weakly surjective, surjective, injective, weakly bijective, or bijective, then so is* $gf : X \longrightarrow Z$.)

Proposition 10.44. ([33, Proposition 3.5]) *Let* $\rho \subseteq X \times Y$ *and* $\sigma \subseteq Y \times Z$ *be weak fuzzy functions with respect to* $\overset{X}{\sim}$ *and* $\overset{Y}{\sim}$, *and* $\overset{Y}{\sim}$ *and* $\overset{Z}{\sim}$, *respectively. If* ρ *and* σ *are weakly surjective, surjective, injective, weakly bijective, or bijective, then so is* $\rho\sigma \subseteq X \times Z$.

Proof. Suppose ρ and σ are surjective, and $z \in Z$. Then there exist $y' \in Y$ and $x' \in X$ such that $y'\sigma z = 1$ and $x'\rho y' = 1$. Thus $x'\rho\sigma z = \bigvee_{y \in Y}(x'\rho y \wedge y\sigma z) \geq x'\rho y' \wedge y'\sigma z = 1$.

Suppose ρ and σ are injective, $x, x' \in X$, $y, y' \in Y$, and $z, z' \in Z$. Since ρ is injective, $x\rho y \wedge x'\rho y' \wedge y \overset{Y}{\sim} y' \leq x \overset{X}{\sim} x'$, and similarly $y\sigma z \wedge y'\sigma z' \wedge z \overset{Z}{\sim} z' \leq y \overset{Y}{\sim} y'$. So

$$(x\rho y \wedge y\sigma z) \wedge (x'\rho y' \wedge y'\sigma z') \wedge z \overset{Z}{\sim} z' \leq x \overset{X}{\sim} x',$$

for all $y, y' \in Y$. Therefore,

$$\left(\bigvee_{y \in Y}(x\rho y \wedge y\sigma z)\right) \wedge \left(\bigvee_{y \in Y}(x'\rho y \wedge y\sigma z')\right) \wedge z \overset{Z}{\sim} z' \leq x \overset{X}{\sim} x'.$$

Thus, $x\rho\sigma z \wedge x'\rho\sigma z' \wedge z \overset{Z}{\sim} z' \leq x \overset{X}{\sim} x'$, i.e. $\rho\sigma$ is injective. □

Proposition 10.45. ([33, Proposition 3.6]) *Let* $\rho \subseteq X \times Y$ *and* $\sigma \subseteq Y \times Z$ *be weak fuzzy functions with respect to* $\overset{X}{\sim}$ *and* $\overset{Y}{\sim}$, *and* $\overset{Y}{\sim}$ *and* $\overset{Z}{\sim}$, *respectively.*

(1) If $\rho\sigma$ *is total, then so is* ρ.
(2) If $\rho\sigma$ *is (weakly) surjective, then so is* σ.

Proof. (1): Suppose $\rho\sigma$ is total and $x \in X$. Then there exists $z \in Z$ such that $x\rho\sigma z = 1$. Thus $x\rho\sigma z = \bigvee_{y \in Y}(x\rho y \wedge y\sigma z) = 1$, whence $x\rho y' \wedge y'\sigma z = 1$ for some $y' \in Y$. Therefore, $x\rho y' = 1$.

(2): Suppose $\rho\sigma$ is weakly surjective and $z \in Z$. Then there exists $x \in X$ such that $x\rho\sigma z > 0$. Thus $x\rho\sigma z = \bigvee_{y \in Y}(x\rho y \wedge y\sigma z) > 0$, whence $x\rho y' \wedge y'\sigma z > 0$ for some $y' \in Y$. Therefore, $y'\sigma z > 0$, i.e. σ is weakly surjective. □

For two ordinary mappings $f : X \longrightarrow Y$ and $g : Y \longrightarrow Z$ it is well known that if $gf : X \longrightarrow Z$ is injective, then so is f. However, [33, Example 3.6] shows that this does not hold for weak fuzzy functions. For more information about various properties of fuzzy functions have a look at e.g. [17] and [33].

10.3 Regular and Difunctional Fuzzy Relations

Remember that we did already learn a few things about g-inverses and regularity in Section 2.11 and Section 6.6. Here we shall add some facts on regular relations and matrices, and we will also consider the special case of difunctional relations, for both crisp and fuzzy relations. For the latter ones we take fuzzy algebras $\mathbb{F} = (F \subseteq [0,1] \subseteq \mathbb{R}, \max, \min)$ as valuation space (see Section 4.1). Remember that fuzzy algebras are commutative path algebras. We first give some basic facts in

Lemma 10.46. *Let ρ, σ be fuzzy relations over a set X, and $\alpha \in F \subseteq [0,1] \subseteq \mathbb{R}$. Then*

(1) $(\alpha\rho)^{\smile} = \alpha\rho^{\smile}$.
(2) $\alpha(\rho\sigma) = (\alpha\rho)(\alpha\sigma)$.

Proof. (2): $x\,\alpha(\rho\sigma)\,z = \alpha \wedge (x\rho\sigma z) = \alpha \wedge \left(\bigvee_{y\in Y}(x\rho y \wedge y\sigma z)\right) = \bigvee_{y\in Y}(\alpha \wedge x\rho y) \wedge (\alpha \wedge y\sigma z) = x\,(\alpha\rho)(\alpha\sigma)\,z$. □

In [11, Theorem 2.3] it is shown that for regular square matrices $A \in \mathbb{F}_n$ the row rank, the column rank, and the factor rank are all equal. A more general version, which we give without proof, is

Theorem 10.47. ([36, p. 81, Theorem 3.1.1]) *Let $A \in \mathbb{F}_{m\times n}$ be a regular matrix. Then the row rank, the column rank, and the factor rank of A are all equal.*

Some basic properties of g-inverses of regular fuzzy matrices are given in

Proposition 10.48. ([36, Theorem 3.1.2]) *Let $A \in \mathbb{F}_{m\times n}$ be a regular matrix, X be a g-inverse of A, and $\alpha \in F$. Then*

(1) $X^T \in A^T\{1\}$.
(2) $\alpha X \in (\alpha A)\{1\}$.
(3) AX and XA are idempotent and regular.
(4) $\mathsf{fr}(A) \le \mathsf{fr}(X)$.
(5) For permutation matrices P, Q we have $Q^T X P^T \in PAQ\{1\}$.

Proof. (1): By Prop. 6.60(1).

(2): Since $\alpha A = [\min\{\alpha, a_{ij}\}] = A\alpha$ and $\min\{\alpha,\alpha\} = \alpha$, we get $(\alpha A)(\alpha X)(\alpha A) = \alpha A$.

(3): By Prop. 6.58(2).

(4): By Lemma 6.47(2) we get $\mathsf{fr}(A) \le \min\{\mathsf{fr}(A), \mathsf{fr}(XA)\} \le \min\left\{\mathsf{fr}(A), \min\{\mathsf{fr}(X), \mathsf{fr}(A)\}\right\} = \min\{\mathsf{fr}(A), \mathsf{fr}(X)\}$. So the assertion follows.

(5): From Section 2.10 we know that permutation matrices are invertible and that $P^{-1} = P^T$ and $Q^{-1} = Q^T$. Assume $AXA = A$; then $(PAQ)(Q^T X P^T)(PAQ) = PA(QQ^T)X(P^T P)AQ = P(AXA)Q = PAQ$. □

10.3.1 *Fuzzy Difunctional Relations*

Regular relations are defined by $\exists\sigma \in \mathcal{B}_S[\rho\sigma\rho = \rho]$, and **difunctional relations** by $\rho\rho^\smile\rho \subseteq \rho$. Since $\rho \subseteq \rho\rho^\smile\rho$ by Prop. 9.52(3) for any fuzzy relation ρ, difunctional are a special kind of regular fuzzy relations, whence they got this place.

Proposition 10.49. ([39, Proposition 2]) *Let $\alpha \in F$; if ρ is a difunctional fuzzy relation, then also $\alpha\rho$.*

Proof. Using Lemma 10.46, from $\rho = \rho\rho^\smile\rho$ we get that $\alpha\rho = \alpha(\rho\rho^\smile\rho) = (\alpha\rho)(\alpha\rho^\smile)(\alpha\rho) = (\alpha\rho)(\alpha\rho)^\smile(\alpha\rho)$. □

Proposition 10.50. ([39, Propositions 6-10]) *Let ρ and σ be fuzzy relations; then the following holds:*

(1) *If ρ is difunctional, then $\rho\rho^\smile$ and $\rho^\smile\rho$ are transitive.*
(2) *If ρ is symmetric and transitive, then it is difunctional.*
(3) *If ρ and σ are difunctional, then also $\rho\cap\sigma$.*
(4) *ρ is difunctional iff $\rho^\smile$ is.*
(5) *If ρ is functional, then it is difunctional.*

Proof. (1): By assumption, $\rho\rho^\smile\rho \subseteq \rho$, so $\rho^\smile(\rho\rho^\smile\rho) \subseteq \rho^\smile\rho$ by monotonicity of composition.

(2): $\rho\rho^\smile\rho = (\rho\rho)\rho \subseteq \rho\rho \subseteq \rho$.

(3): Using monotonicity, $(\rho\cap\sigma)(\rho\cap\sigma)^\smile(\rho\cap\sigma) = (\rho\cap\sigma)(\rho^\smile\cap\sigma^\smile)(\rho\cap\sigma) \subseteq (\rho\rho^\smile\rho)\cap(\sigma\sigma^\smile\sigma)$. But $(\rho\rho^\smile\rho)\cap(\sigma\sigma^\smile\sigma) \subseteq \rho\cap\sigma$, since ρ and σ are difunctional.

(4): Using Prop. 9.32(6) and Th. 9.33(8), from $\rho\rho^\smile\rho \subseteq \rho$ we get $\rho^\smile \supseteq (\rho\rho^\smile\rho)^\smile = \rho^\smile\rho^{\smile\smile}\rho^\smile = \rho^\smile\rho\rho^\smile$, so $\rho^\smile$ is difunctional. Conversely, $\rho^\smile\rho\rho^\smile \subseteq \rho^\smile$ implies $(\rho^\smile\rho\rho^\smile)^\smile \subseteq \rho^{\smile\smile}$, i.e. $\rho\rho^\smile\rho \subseteq \rho$.

(5): $\rho^\smile\rho \subseteq I$ implies $\rho\rho^\smile\rho \subseteq \rho I = \rho$. □

Proposition 10.51. ([39, Proposition 5]) *ρ is a difunctional fuzzy relation iff ρ_α is difunctional for all $\alpha \in \Lambda(\rho)$* (see Def. 9.3(b)).

Proof. Let ρ be difunctional; using Lemma 10.46(2) and the fact that $\alpha\rho_\alpha \subseteq \rho$ we get $\alpha(\rho_\alpha\rho_\alpha{}^\smile\rho_\alpha) = (\alpha\rho_\alpha)(\alpha\rho_\alpha{}^\smile)(\alpha\rho_\alpha) \subseteq \rho\rho^\smile\rho \subseteq \rho$. Note that the α-cut ρ_α is a crisp relation with $\rho_\alpha(x,y) = 1$ iff $\rho(x,y) \geq \alpha$. Consider any crisp relation τ with $\alpha\tau \subseteq \rho$; then $\alpha\tau(x,y) \in \{0,\alpha\}$, and $\alpha\tau(x,y) = \alpha$ implies $\rho(x,y) \geq \alpha$. Let $[\rho\alpha] = \{\tau \text{ is crisp}, \alpha\tau \subseteq \rho\}$. Then $\rho_\alpha = \bigcup[\rho\alpha]$. By what we have just shown above, $\alpha(\rho_\alpha\rho_\alpha{}^\smile\rho_\alpha) \subseteq \rho$; and since $\rho_\alpha\rho_\alpha{}^\smile\rho_\alpha \in [\rho\alpha]$, $\rho_\alpha\rho_\alpha{}^\smile\rho_\alpha \subseteq \rho_\alpha$ follows.

Conversely, assume that ρ_α is difunctional for all $\alpha \in \Lambda(\rho)$. Then

$$\begin{aligned}
\rho\rho^\smile\rho &= \textstyle\bigcup_{\alpha,\beta,\gamma}(\alpha\rho_\alpha)(\beta\rho_\beta)^\smile(\gamma\rho_\gamma) \\
&\subseteq \textstyle\bigcup_\alpha(\alpha\rho_\alpha)(\alpha\rho_\alpha)^\smile(\alpha\rho_\alpha) && \text{assuming } \alpha \leq \beta\wedge\gamma, \text{ so } \rho_\beta,\rho_\gamma \subseteq \rho_\alpha \text{ by Prop. 9.4(3);} \\
&= \textstyle\bigcup_\alpha \alpha(\rho_\alpha\rho_\alpha{}^\smile\rho_\alpha) && \text{by Lemma 10.46(2)} \\
&\subseteq \textstyle\bigcup_\alpha \alpha\rho_\alpha && \text{since } \rho_\alpha \text{ is difunctional} \\
&= \rho && \text{by decomposition Prop. 9.5.}
\end{aligned}$$

□

Proposition 10.52. ([39, Proposition 12]) *The difunctional closure of a fuzzy relation ρ is given by* $\mathbf{d}(\rho) = \bigcup_{k\geq 0} \rho(\rho\breve{}\rho)^k$ (recall Def. 2.29).

Proof. (a) Clearly, $\rho \subseteq \mathbf{d}(\rho)$, since $\rho = \rho(\rho\breve{}\rho)^0 = \rho I$.

(b) $\mathbf{d}(\rho)$ is difunctional: First we will concentrate on $\mathbf{d}(\rho)\breve{}$. Note that $((\sigma\tau)^m)\breve{} = (\tau\breve{}\sigma\breve{})^m$, so $(\rho(\rho\breve{}\rho)^m)\breve{} = ((\rho\breve{}\rho)^m)\breve{}\rho\breve{} = (\rho\breve{}\rho\breve{}\breve{})^m\rho\breve{} = (\rho\breve{}\rho)^m\rho\breve{} = \rho\breve{}(\rho\rho\breve{})^m$, whence by Prop. 9.32(8)

$$\mathbf{d}(\rho)\breve{} = \left(\bigcup_{j\geq 0} \rho(\rho\breve{}\rho)^j\right)\breve{} = \bigcup_{j\geq 0}\left(\rho(\rho\breve{}\rho)^j\right)\breve{} = \bigcup_{j\geq 0} \rho\breve{}(\rho\rho\breve{})^j.$$

Therefore, by straightforward calculation we get

$$\begin{aligned}
\mathbf{d}(\rho)\mathbf{d}(\rho)\breve{}\mathbf{d}(\rho) &= \left(\bigcup_{i\geq 0} \rho(\rho\breve{}\rho)^i\right)\mathbf{d}(\rho)\breve{}\mathbf{d}(\rho)\\
&= \left(\bigcup_{i\geq 0} \rho(\rho\breve{}\rho)^i\mathbf{d}(\rho)\breve{}\right)\mathbf{d}(\rho)\\
&= \left(\left[\rho\bigcup_{j\geq 0} \rho\breve{}(\rho\rho\breve{})^j\right] \cup \left[\rho\rho\breve{}\rho\bigcup_{j\geq 0} \rho\breve{}(\rho\rho\breve{})^j\right] \cup \ldots\right)\mathbf{d}(\rho)\\
&= \left(\left[\bigcup_{j\geq 1} (\rho\rho\breve{})^j\right] \cup \left[\bigcup_{j\geq 2} (\rho\rho\breve{})^j\right] \cup \ldots\right)\mathbf{d}(\rho)\\
&= \left(\bigcup_{j\geq 1} (\rho\rho\breve{})^j\right)\mathbf{d}(\rho)\\
&= \left(\left[\bigcup_{j\geq 1} (\rho\rho\breve{})^j\rho\right] \cup \left[\bigcup_{j\geq 1} (\rho\rho\breve{})^j\rho\rho\breve{}\rho\right] \cup \ldots\right)\\
&= \bigcup_{m\geq 1} \rho(\rho\breve{}\rho)^m\\
&\subseteq \mathbf{d}(\rho).
\end{aligned}$$

(c): For all difunctional fuzzy relations ρ' we have $\rho \subseteq \rho' \rightarrow \mathbf{d}(\rho) \subseteq \rho'$: Let ρ' be difunctional with $\rho \subseteq \rho'$. To show that $\mathbf{d}(\rho) \subseteq \rho'$ we prove $\rho(\rho\breve{}\rho)^m \subseteq \rho'$ for each $m \geq 0$. The case $m = 0$ follows from the assumption $\rho = \rho(\rho\breve{}\rho)^0 = \rho I \subseteq \rho'$. If $\rho(\rho\breve{}\rho)^k \subseteq \rho'$, then $\rho(\rho\breve{}\rho)^{k+1} = \rho(\rho\breve{}\rho)^k\rho\breve{}\rho \subseteq \rho'\rho'\breve{}\rho' \subseteq \rho'$ by monotonicity of composition and since by assumption $\rho \subseteq \rho'$ and ρ' is difunctional. Thus we have shown that indeed $\rho(\rho\breve{}\rho)^m \subseteq \rho'$ for each $m \geq 0$, whence $\mathbf{d}(\rho) \subseteq \rho'$, i.e. $\mathbf{d}(\rho)$ is the smallest difunctional fuzzy relation containing ρ. □

Proposition 10.53. ([45, Theorem 2]) *If ρ is functional, then both $\rho\rho\breve{}$ and $\rho\breve{}\rho$ are fuzzy difunctional relations.*

Proof. We have $(\rho\rho\breve{})(\rho\rho\breve{})\breve{}(\rho\rho\breve{}) = (\rho\rho\breve{})(\rho\rho\breve{})(\rho\rho\breve{}) = \rho(\rho\breve{}\rho)(\rho\breve{}\rho)\rho\breve{} \subseteq \rho\rho\breve{}$ since ρ is assumed to be functional, i.e. $\rho\breve{}\rho \subseteq I$. □

Proposition 10.54. ([45, Theorem 3]) *If ρ, σ are functional, then both $\rho\sigma$ and $\sigma\rho$ are fuzzy difunctional relations.*

Proof. $(\rho\sigma)(\rho\sigma)\breve{}(\rho\sigma) = (\rho\sigma)(\sigma\breve{}\rho\breve{})(\rho\sigma) = \rho(\sigma\sigma\breve{})(\rho\breve{}\rho)\sigma \subseteq \rho(\sigma\sigma\breve{})\sigma = \rho\sigma(\sigma\breve{}\sigma) \subseteq \rho\sigma$. □

Proposition 10.55. ([45, Theorem 4]) *If ρ is reflexive and difunctional, then it is a fuzzy equivalence.*

Proof. Note that $\rho\breve{}$ is reflexive, and by Prop. 10.50(4) also difunctional. So we have $\rho\breve{} = I\rho\breve{}I \subseteq \rho\rho\breve{}\rho = \rho$, and likewise $\rho = I\rho I \subseteq \rho\breve{}\rho\rho\breve{} = \rho\breve{}$. Therefore, $\rho\breve{} = \rho$, i.e. ρ is symmetric. And for transitivity we have $\rho\rho = \rho\breve{}\rho = I\rho\breve{}\rho \subseteq \rho\rho\breve{}\rho = \rho$. (This is an opportunity to compare the above algebraic with the corresponding point-based proof in [45].) □

Corollary 10.56. ([45, Theorem 6]) *If ρ is reflexive and difunctional, then it is idempotent, and $\rho^{\smile}\rho = \rho\rho^{\smile} = \rho$.*

Proof. By Prop. 10.55 ρ is a fuzzy equivalence and thus idempotent by Prop. 10.13(2). The rest follows from the symmetry of ρ. □

10.4 Fuzzy Order Relations

The most obvious idea to define fuzzy orderings clearly is to replace the classical axioms reflexivity, antisymmetry, and transitivity of the crisp case with suitable generalizations. For the sake of historical appreciation we have to mention that the first definition of this kind (also for fuzzy equivalences) has been given as early as 1971 by Zadeh [56] himself: A fuzzy relation ρ on a set X is a **fuzzy partial order** if it is reflexive, transitive, and antisymmetric, where the properties involved are meant to be those from Def. 9.34. Since then different kinds of fuzzy orders have been introduced and studied, generalizing the original concept for both formal and intuitive reasons as we will see below by way of a short sketch of its historical development (see e.g. [53]).

Later on two groups of concepts crystallize in literature (see [8, p. 118]): The first follows the above mentioned paper of Zadeh and uses crisp equality in the antisymmetry condition (e.g. [32, p. 130], [40, 50]); the other, more recent one, a fuzzy equivalence relation or the specific case of fuzzy equality (e.g. [2], [19, Section 3], [28, 34, 35]). A still more general concept considers order relations on fuzzy sets (later on called d-fuzzy relations in Section 13.1) like a group of TALL people or a bag of INTERESTING books (see e.g. [20]).

Crisp partial orders exhibit two well known algebraic properties that may serve as criteria for checking whether a new concept is a suitable generalization (see e.g. [7]): The Szpilrajn theorem is the "best known connection between partial orders and linear orders" and states that every partial order can be extended to a linear order on the same set. The theorem by Dushnik and Miller states, moreover, that every partial order is the intersection of its linear extensions, and has a much simpler, nearly trivial, proof than the former (so we include it).

Theorem 10.57. (Szpilrajn theorem; [27, p. 53, Theorem 2.3.2], [44, Section 10.1]) *Let $(P, \leq)$ be a partially ordered set, and let $a, b \in P$ be incomparable. Then there exists a linear order $\leq^*$ on P, called a* **linear extension** *of $\leq$, such that $\leq$ is contained in $\leq^*$ and $a \leq^* b$.*

Zadeh [56, pp. 195-199] has proven an extended version of the Szpilrajn theorem valid for his (the classical) version of a fuzzy partial order.

Theorem 10.58. ([27, p. 206, Theorem 7.2.1]) *Let $(P, \leq)$ be a partially ordered set. Then there exists a set of linear orders $(P, \leq_i)$, $i \in I$, such that $\leq$ is the intersections of these order relations: $\leq = \bigcap_{i \in I} \leq_i$.*

Proof. For every pair of incomparable elements $a, b \in P$ we choose two linear extensions $\leq_{ab}$ and $\leq_{ba}$ which exist due to the Szpilrajn theorem. Then the intersection of all linear orderings $\leq_{ab}$ and $\leq_{ba}$, with $a, b \in P$ all the pairs of incomprable elements, equals $\leq$. □

Before we proceed we have to warn the reader that, unfortunately, there is a plethora of both names and slightly different concepts one has to come to grips with (see e.g. [19, pp. 440, 445]). Below we will present a few of all these variants together with some of their relevant properties. Let us start with a look at transitivity and antisymmetry, using as valuation space the real unit interval. For both properties the expression $x\rho y \wedge y\rho z$ is used for their definitions. A first step towards a more general concept is by replacing the minimum t-norm $\wedge$ by an arbitrary t-norm T.

Definition 10.59. *Let ρ be a fuzzy relation over X; then ρ is called*

(a) T-**transitive** *if* $\forall x, z \in X[\bigvee_{y\in X}\big(T(x\rho y, y\rho z)\big) \leq x\rho z]$.
(b) T-**antisymmetric** *if* $\forall x, y \in X[T(x\rho y, y\rho x) > 0 \rightarrow x = y]$.

Definition 10.60. *Let ρ be a fuzzy relation over X; then ρ is called a*

(a) T-**preorder**, *if it is reflexive and T-transitive.*
(b) **partial** T-**order**, *if it is a T-antisymmetric T-preorder.*
(c) T-**equivalence**, *if it is a symmetric T-preorder.*

T-antisymmetry is a rather strong and, moreover, a hybrid requirement: Recalling the classical axiom of antisymmetry: $x \leq y \wedge y \leq x \rightarrow x = y$, we see that only $\leq$ and $\wedge$ have been replaced by a fuzzy relation ρ and a t-norm T, respectively, whereas crisp equality on the right hand side has been retained. So the definition of T-antisymmetry is a kind of "half-way fuzzification" ([5, p. 600]. There are also some other semantical aspects discussed in this paper (see also [4, Chapter 3] or [6, Section 4]) which give rise to the next straightforward generalization step in Def. 10.62 below.

As an aside, let us consider implication from Boolean propositional logic: Let φ and ψ be two arbitrary well-formed formulas. Defining $\varphi \preceq \psi$ iff $\models \varphi \rightarrow \psi$ (i.e. $\varphi \rightarrow \psi$ is a tautology) makes $\preceq$ a preorder, and if two formulas are considered to be equal in case $\models \varphi \equiv \psi$, then $\preceq$ becomes a partial order on the set of well-formed formulas. Now let T be a left-continuous t-norm and $\curvearrowright_T$ be the associated residual implication (see Def. 9.7). The question whether this fundamental correspondence still holds for fuzzy logical systems based on left-continuous triangular norms and corresponding residual implications, gets a negative answer. To see this we need

Lemma 10.61. ([6, Lemma 2.9], [25, Proposition 1.10] for (2)) *For a left-continuous t-norm T the following holds for all $x, y, z \in [0, 1]$:*

(1) $x \leq y$ *iff* $\curvearrowright_T(x, y) = 1$.
(2) $T\big(\curvearrowright_T(x, y), \curvearrowright_T(y, z)\big) \leq \curvearrowright_T(x, z)$.
(3) $\curvearrowright_T(1, y) = y$.

By the well known facts (1) and (2) of the foregoing lemma $\curvearrowright_T$ is reflexive and T-transitive. But the T-antisymmetry of $\curvearrowright_T$ would imply $T\big(\curvearrowright_T(x,y), \curvearrowright_T(y,x)\big) = 0$ for any $x \neq y$ which cannot hold for $0 < y \neq 1 = x$ due to Lemma 10.61(3). Thus, the residual implication $\curvearrowright_T$ for a left-continuous t-norm T cannot be a partial T-order.

Definition 10.62. *Let ρ be a T-transitive fuzzy relation, and E be a T-equivalence on X; then ρ is called*

*(a) E-**reflexive***	*if $\forall x,y \in X[xEy \leq x\rho y]$.*
*(b) T-E-**antisymmetric***	*if $\forall x,y \in X[T(x\rho y, y\rho x) \leq xEy]$.*
*(c) **partial** T-E-**order***	*if it is E-reflexive, T-transitive, and T-E-antisymmetric.*

At first glance the requirement of E-reflexivity may seem surprising. But note that it is nothing else than a reformulation of reflexivity which is based on crisp identity: For a reflexive relation ρ we clearly have $I_X \subseteq \rho$, i.e.$I_X(x,y) \leq \rho(x,y)$. It is easy to check that any partial T-order is a T-I_X-order, so the concept of T-E-orders does indeed generalize the one of partial T-orders.

We now turn to residual biimplication given as $\overset{T}{\leftrightarrow}(x,y) := T\big(\curvearrowright_T(x,y), \curvearrowright_T(y,x)\big)$. By Lemma 10.61(1) we get that $\overset{T}{\leftrightarrow}(x,x) = T\big(\curvearrowright_T(x,x), \curvearrowright_T(x,x)\big) = T(1,1) = 1$, so $\overset{T}{\leftrightarrow}$ is reflexive, and by (2) that it is T-transitive. Moreover, by its definition, $\curvearrowright_T$ clearly is T-$\overset{T}{\leftrightarrow}$-antisymmetric. Therefore, the residual implication of a left-continuous t-norm T is a partial T-$\overset{T}{\leftrightarrow}$-order which is in perfect accordance to the Boolean case. This example demonstrates that sometimes we can, and need to, do better than with the straightforward and nearby generalization only.

Another kind of generalization we het by replacing the real unit interval by a (complete residuated) lattice. There are two definitions of this kind which have, however, been shown to be equivalent.

Definition 10.63. ([51, Section 3]) *Let L be a complete residuated lattice and $\preceq$ be a L-fuzzy relation on a set X; then $\preceq$ is called an*

*(a) L-**fuzzy preorder** if $\preceq$ is reflexive and transitive, and X an L-**fuzzy preordered set**.*

*(b) L-**fuzzy partial order** (in the sense of Fan) if $\preceq$ is an L-fuzzy preorder for which condition*

(F) $x \preceq y = y \preceq x = 1 \rightarrow x = y$

*holds. X is said to be an L-**fuzzy partially ordered set**.*

*(c) L-**fuzzy partial order** (in the sense of Bělohlávek) if $\approx$ is an L-equality and $\preceq$ an L-fuzzy preorder on X which is **compatible with** $\approx$, i.e.*

(C) $(x \approx y) \otimes (x' \approx y') \otimes (x \preceq x') \leq (y \preceq y')$

satisfying the condition

(B) $(x \preceq y) \wedge (y \preceq x) \leq (x \approx y)$.

*Again, X is said to be an L-**fuzzy partially ordered set**. (In [1, Definition 4.42] $\langle\langle X, \approx\rangle, \preceq\rangle$ is called an L-ordered set.)*

(Note that the antisymmetry condition (F) required for L-fuzzy partial orders is stronger than the ones introduced in Def. 9.44: it corresponds to being 1-antisymmetric, and may perhaps better be called *strongly antisymmetric.*)

We will now show that L-fuzzy partial orders in the sense of Fan (Fpo for short) are equivalent to those in the sense of Bělohlávek (Bpo for short). We need

Lemma 10.64. ([1, Lemma 4.45]) *Let L be a complete residuated lattice. In an L-ordered set $\langle\langle X,\approx\rangle,\preceq\rangle$ we have $x\preceq y \wedge y\preceq x = x\approx y$.*

Proof. The $\leq$-direction is requirement (B), the other we get from the compatibility of $\preceq$ with $\approx$: Replacing both x', y' with x in requirement (C) gives $(x\approx y)\otimes(x\approx x)\otimes(x\preceq x) = (x\approx y)\otimes 1\otimes 1\leq (y\preceq x)$, so $(x\approx y)\leq(y\preceq x)$. Similar for $(x\approx y)\leq(x\preceq y)$ (with $x', y\leftarrow x$ and $y'\leftarrow y$), whence $(x\approx y)\leq(x\preceq y)\wedge(y\preceq x)$ as desired. □

Proposition 10.65. ([51, Proposition 3.4]) *L-fuzzy partial orders in the sense of Fan and in the sense of Bělohlávek are equivalent concepts.*

Proof. To show that a Bpo is an Fpo we have to prove that (B) implies (F): Assume $(x\preceq y)\wedge(y\preceq x) = 1\leq(x\approx y)$; then $x\approx y = 1$ and by Def. 10.1(c) this implies $x = y$.

For the converse direction let ρ be an Fpo over X. We define the L-fuzzy relation $x\approx_\rho y := x\rho y\wedge y\rho x$. Then, as a preorder, ρ clearly is reflexive and transitive, and by definition $(x\rho y)\wedge(y\rho x)\leq(x\approx_\rho y)$, so (B) is satisfied. It remains to show (a) that $\approx_\rho$ is an L-equality and (b) that ρ is a Bpo compatible with it.

(a) Obviously, $\approx_\rho$ is reflexive and symmetric. Moreover we have

$$(x\approx_\rho y)\otimes(y\approx_\rho z) = \big((x\rho y)\wedge(y\rho x)\big)\otimes\big((y\rho z)\wedge(z\rho y)\big).$$

Using the shorthand $(a\wedge b)\otimes(c\wedge d)$ for the right hand side, by twice applying Prop. 4.30(2) we get

$$(a\wedge b)\otimes(c\wedge d)\leq(a\otimes c)\wedge(b\otimes c)\wedge(a\otimes d)\wedge(b\otimes d)\leq(a\otimes c)\wedge(d\otimes b).$$

Resubstituting in the last expression we get

$$\big((x\rho y)\otimes(y\rho z)\big)\wedge\big((z\rho y)\otimes(y\rho x)\big)\leq(x\rho z)\wedge(z\rho y) = x\approx_\rho z$$

by the transitivity of ρ. So we could show that $\approx_\rho$ is transitive. Finally assume $x\approx_\rho y = 1$; then $x\rho y = y\rho x = 1$, and from (F) we get $x = y$. Thus $\approx_\rho$ is an L-equality

(b) ρ is compatible with $\approx_\rho$ since, similar to (a), we have

$$(x\approx_\rho y)\otimes(x'\approx_\rho y')\otimes(x\rho x')\leq(y\rho x)\otimes(x'\rho y')\otimes(x\rho x')\leq(y\rho y').$$

□

We continue with a short look at some examples which confirm the suitability of the concept:

Corollary 10.66. ([1, Theorem 2.17]) *Let L be a residuated lattice; then $\curvearrowright$ is a fuzzy partial order on L.*

Proof. (i) $x\curvearrowright x = 1$: By Prop. 4.27(4).
(ii) $(x\curvearrowright y)\otimes(y\curvearrowright z)\leq(x\curvearrowright z)$: By Prop. 4.27(2) we have both $x\otimes(x\curvearrowright y)\leq y$ and $y\otimes$

$(y \curvearrowright z) \leq z$, whence $x \otimes (x \curvearrowright y) \otimes (y \curvearrowright z) \leq z$. Then, using commutativity of $\otimes$, by the adjointness property $a \otimes b \leq c \;\leftrightarrow\; a \leq b \curvearrowright c$ (see Def. 4.26) the assertion follows.
(iii) $(x \curvearrowright y) = 1 = (y \curvearrowright x) \;\rightarrow\; x = y$: By Prop. 4.27(3). □

Let ρ be a fuzzy partial order on X; then $\leq_\rho := \{\langle x, y\rangle : x\rho y = 1\}$ is a crisp partial order on X. And if $\leq$ is a crisp partial order on X, then $\rho_\leq : X \times Y \longrightarrow L$, with $\rho_\leq(x, y) = 1$ if $x \leq y$ and 0 otherwise, is a fuzzy partial order. ([51, Example 3.6(3)]).

Our next example also serves to mention paradigmatic instances of graded concepts proposed in fuzzy set theory (which otherwise are neglected in this text), since they also attest the plausibility of the concept of fuzzy partial order. We follow the presentation in [1, p. 82ff], which is based on its (3.3) and (3.12), and Theorems 3.12 and 3.14.

The relation $\subseteq$ on L^X is crisp, i.e. for any two L-sets either $A \subseteq B$ or not, but from the viewpoint of a graded approach (as it is the essence of fuzzy set theory) to consider a kind of graded subsethood between L-sets is a nearby idea: For two L-sets A, B the **subsethood degree of** A **in** B is defined as

$$S(A,B) = \bigwedge_{x\in X}(A(x) \curvearrowright B(x)).$$

The L-fuzzy set $S(A,B)$ may then be taken as the truth value of "*each element of A is also in B*", and the relation $\subseteq$ can be conistently described since $S(A,B) = 1$ iff $A \subseteq B$ (recall Prop. 4.27(3)). A concept of graded equality then naturally arises as

$$A \approx B = \bigwedge_{x\in X}(A(x) \leftrightharpoons B(x)),$$

called the **degree of equality** of A and B. Intuitively, $A \approx B$ expresses the truth value of "*A and B contain the same elements*". $=$ can be defined by $\approx$ since $A \approx B = 1$ iff $A = B$ (recall Cor. 4.32(4)). (We will not need graded equality in the corollary below, since it resorts to the simpler Fpo version, but thought it worthwhile to be presented anyway.) Now we are ready to turn to

Corollary 10.67. ([1, Example 4.47]) *Let L be a residuated lattice; then* $S(A,B)$ *is a fuzzy partial order on* L^X *for any set* $X \neq \emptyset$*, any subset* $\emptyset \neq M \subseteq L^X$*, and* $A, B \in M$.

Proof. (i) $S(A, A) = 1$: By Cor. 4.32(2).
(ii) $S(A,B) \otimes S(B,C) \leq S(A,C)$: By Cor. 4.32(5).
(iii) $S(A,B) = 1 = S(B,A) \;\rightarrow\; A = B$: By Prop. 4.27(4). □

Though there are a lot of various other results concerning the current topic, we have to stop our considerations on fuzzy orders here. For first steps beyond what has been offered above, the interested reader may want to continue with [52], or with the more advanced [19, 35], and the references given there to learn about joins and meets, minimal and maximal elements, fuzzy complete lattices, fuzzy Galois connections etc. And, of course, there are fuzzy counterparts to Ferrers, semi- and interval orders: see e.g. [21, 22, 30].

References

[1] Bělohlávek, R.: *Fuzzy Relational Systems - Foundations and Principles*, Springer Science + Business Media, New York (2002).

[2] Bělohlávek, R.: Concept lattices and order in fuzzy logic, Annals of Pure and Applied Logic **128**, 277-298 (2004).

[3] Bělohlávek, R., Funioková, T.: Similarity and Fuzzy Tolerance Spaces, J. Logic Computat. **14**, 827-855 (2004).

[4] Bodenhofer, U.: A Similarity-Based Generalization of Fuzzy Orderings, PhD Thesis, Johannes Kepler Universität, Linz, Austria (1998).

[5] Bodenhofer, U.: A similarity-based generalization of fuzzy orderings preserving the classical axioms. Int. J. of Uncertainty, Fuzziness and Knowledge-Based Systems **8**, 593-610 (2000).

[6] Bodenhofer, U.: Representations and constructions of similarity-based fuzzy orderings, Fuzzy Sets and Systems **137**, 113-136 (2003).

[7] Bonnet, R., Pouzet, M.: Linear extensions of ordered sets, in: Rival, I. (Ed.), *Ordered Sets*, D. Reidel Publishing Company, Dordrecht/Boston/London, pp. 125-170 (1982).

[8] Borzooei, R.A., Dvurečenskij, A., Zahiri, O.: L-ordered and L-lattice ordered groups. Information Sciences **314**, 118-134 (2015).

[9] Cabrera, I.P., Cordero, P., Ojeda-Aciego, M.: On fuzzy relations, adjunctions, and functional fuzzy relations, 2016 IEEE Symposium Series on Computational Intelligence (SSCI), pp. 1-7 (2016).

[10] Chajda, I., Duda, J.: Blocks of binary relations, Annales Univ. Sci. Budapest, sectio Math **22**, 3-9 (1979).

[11] Cho, H.H.: Regular fuzzy matrices and fuzzy equations, Fuzzy Sets and Systems **105**, 445-451 (1999).

[12] Ćirić, M., Ignjatović, J., Bogdanović, S.: Fuzzy equivalence relations and their equivalence classes, Fuzzy Sets and Systems **158**, 1295-1313 (2007).

[13] Ćirić, M., Ignjatović, J., Bogdanović, S.: Uniform fuzzy relations and fuzzy functions, Fuzzy Sets and Systems **160**, 1054-1081 (2009).

[14] Ćirić, M., Ignjatović, J., Damljanović, N., Bašić, M.: Bisimulations for fuzzy automata, Fuzzy Sets and Systems **186**, 100-139 (2012).

[15] Das, M., Chakraborty, M.K., Ghoshal, T.K.: Fuzzy tolerance relation, fuzzy tolerance space and basis, Fuzzy Sets and Systems **97**, 361-369 (1998).

[16] De Cock, M., Kerre, E.E.: On (un)suitable fuzzy relations to model approximate equality, Fuzzy Sets and Systems **133**, 137-153 (2003).

[17] Demirci, M.: Fuzzy functions and their fundamental properties, Fuzzy Sets and Systems **106**, 239-246 (1999).

[18] Demirci, M.: Fuzzy functions and their applications, J. of Mathematical Analysis and Applications **252**, 495-517 (2000).

[19] Demirci, M.: A theory of vague lattices based on many-valued equivalence relations I: general representation results, Fuzzy Sets and Systems **151**, 437-472 (2005).

[20] Demirci, M.: On the representations of L-equivalence relations on L-fuzzy sets with applications to locally vague environments, Int.J. of General Systems **49**, 334-354 (2020).

[21] Díaz, S., De Baets, B., Montes, S.: On the Ferrers property of valued interval orders, Top **19**, 421-447 (2011).

[22] Díaz, S., Montes, S., Campus, V., De Baets, B.: A comparative study of T-interval orders, T-Ferrers and T-biorders, in: Magdalena, L. et al. (Eds), *Proceedings of IPMU'08*, pp. 109-116 (2008).

[23] Dubois, D., Prade, H.: *Fuzzy Sets and Systems: Theory and Applications*, Academic Press, New York, (1980).

[24] Dubois, D., Prade, H.: A theorem on implication functions defined from triangular norms, Stochastica **8**, 267-279 (1984)

[25] Fodor, J., Roubens, M.: *Fuzzy Preference Modeling and Multicriteria Decision Support*, Springer Science+Business Media, Dordrecht (1994).

[26] Gerla, G.: Approximate Similarities and Poincaré Paradox, Notre Dame J. of Formal Logic **49**, 203-226 (2008).

[27] Harzheim, E.: *Ordered Sets*, Springer Science+Business Media (2005).

[28] Höhle, U., Blanchard, N.: Partial ordering in L-underdeterminate sets, Information Sciences **35**, 133-144 (1985).

[29] Hume, D.: *An Enquiry Concerning Human Understanding*, Hackett Publ Co., Cambridge, MA 1993 (1772).

[30] Induráin, E., Martinetti, D. et al.: On the preservation of semiorders from the fuzzy to the crisp setting, Int. J. of Uncertainty, Fuzziness and Knowledge-Based Systems **19**, 899-920 (2011).

[31] Klawonn, F.: Fuzzy points, fuzzy relations and fuzzy functions, in: Novák, V., Perfilieva, I. (Eds.), *Discovering the World with Fuzzy Logic*, Physica-Verlag, Heidelberg, pp. 431-453 (2000).

[32] Klir, G., Yuan, B.: *Fuzzy Sets and fuzzy Logic: Theory and Applications*, Prentice Hall, New Jersey, (1995).

[33] Lim, P.K., Choi, G.H., Hur, K.: Fuzzy Mappings and Fuzzy Equivalence Relations, Int J. of Fuzzy Logic and Intelligent Systems **11**, 153-164 (2011).

[34] Llamazares, B., De Baets, B.: Fuzzy strict preference relations compatible with fuzzy orderings, Int J. of Uncertainty, Fuzziness and Knowledge-Based Systems **18**, 13-24 (2010).

[35] Martinek, P.: Completely lattice L-ordered sets with and without L-equality, Fuzzy Sets and Systems **166**, 44-55 (2011).

[36] Meenakshi, A.R.: *Fuzzy Matrix: Theory and Applications*, MJP Publishers, Chennai, India (2008).

[37] Negoiţă, C.V., Ralescu, D.: *Applications of Fuzzy Sets to System Analysis*, Springer Basel AG (1975).

[38] Nemitz, W.C.: Fuzzy relations and fuzzy functions, Fuzzy Sets and Systems **19**, 177-191 (1986).

[39] Ounalli, H., Jaoua, A.: On fuzzy difunctional relations, Information Sciences **95**, 219-232 (1996).

[40] Ovchinnikov, S.V.: Structure of fuzzy binary relations, Fuzzy Sets and Systems **6**, 169-195 (1981).

[41] Ovchinnikov, S.V.: Similarity relations, fuzzy partitions, and fuzzy orderings, Fuzzy Sets and Systems **40**, 107-126 (1991).

[42] Poincaré, H.: *La Science et l'Hypothèse*, Flammarion, Paris (1902).

[43] Recasens, J.: *Indistinguishability Operators - Modelling Fuzzy Equalities and Fuzzy Equivalence Relations*, Springer-Verlag Berlin Heidelberg (2010).

[44] Schröder, B.: *Ordered Sets*, 2nd edition, Springer International Publishing (2016).

[45] Seo, C., Han, K., Sung, Y., Eun, H.: On the relationships between fuzzy equivalence relations and fuzzy difunctional relations, and their properties, Fuzzy Sets and Systems **109**, 459-462 (2000).

[46] Toth, H.: Fuzziness: From epistemic considerations to terminological clarification, Int. J. of Uncertainty, Fuzziness and Knowledge-Based Systems **5**, 481-503 (1997).

[47] Toth, H.: *Relations: Concrete, Abstract and Applied - An Introduction*, World Scientific Publishing, Singapore (2020).

[48] Vollmer, G.: On supposed circularities in an empirically oriented epistemology, in: Radnitzky , G., Bartley III, W.W. (Eds.), *Evolutionary Epistemology, Theory of Rationality, and the Sociology of Knowledge*, Open Court, La Salle, Illinois, pp. 163-200 (1987).

[49] Wang, X., Ruan, D., Kerre, E.E.: *Mathematics of Fuzziness - Basic Issues*, Springer-Verlag Berlin Heidelberg (2009).

[50] Wang, K., Zhao, B.: Join-completions of L-ordered sets, Fuzzy Sets and Systems **199**, 92-107 (2012).

[51] Yao, W.: Quantitative domains via fuzzy sets: Part I: Continuity of fuzzy directed complete posets, Fuzzy Sets and Systems **161**, 973-987 (2010).

[52] Yao, W., Lu, L.X.: Fuzzy Galois connections on fuzzy posets, Math. Log. Quart. **55**, 105-112 (2009).

[53] Yao, W., Zhao, B.: Kernel systems on L-ordered sets, Fuzzy Sets and Systems **182**, 101-109 (2011).

[54] Yeh, R.T.: Toward an algebraic theory of fuzzy relational systems, TR-25, Univ. of Texas, Austin, Dpt. of Computer Science (1973).

[55] Yeh, R.T., Bang, S.Y.: Fuzzy relations, fuzzy graphs, and their applications to clustering analysis, in: Zadeh, L.A., Fu, K.S., Tanaka, K., Shimura, M. (Eds.), *Fuzzy Sets and their Application to Cognitive and Decision Processes*, Academic Press, New York, pp. 125-149 (1975).

[56] Zadeh, L.A.: Similarity relations and fuzzy orderings, Information Science **3**, 177-200 (1971).

CHAPTER 11

FUZZY PREFERENCE STRUCTURES AND SOCIAL CHOICE

Obviously, as stated in [22], to develop preferences is an inevitable step in a lot of human activities and is necessary for making one's decisions in daily life. This is, of course, not only true for individuals but also for collectives, varying in size and characteristic properties, from small teams to whole democracies. A group choice, or collective action, should be based on the desires or preferences of the individuals in that society. Therefore, there is a long history of analysis, modeling, and measurement of individual preferences and of preference aggregation.

But it is not only because preference modeling is an important application area for crisp and fuzzy relations that this short chapter has been included: to be honest, the main reason is that preference structures provide us with a nice example where the naive fuzzification of a familiar concept leads to remarkable and surprising results.

11.1 Crisp Preference Structures

We begin with a look at the concept of a *preference structure* as introduced on basis of two-valued relations to provide a reference starting point for the subsequent considerations. This concept has been used for the original definition of a semiorder in Luce [12]. (Let us note that the concept of a semiorder is usually attributed to this paper, but, as a matter of fact, it is older and dates back under a different name to [25] of which a "translation" to modern notation can be found in [3].)

In a sense, the conception of preference structures starts out from a behavioral point of view by considering the reaction possibilities for decision makers (individuals, agents) when they have to rate elements from a set of alternative actions, candidates, objects, etc..

Following [1,15,22], let us consider a set A of alternatives. When asked about a pair of distinct elements in A, individuals are supposed to respond with one of the two answers "*yes*" or "*no*" to the question "*Is object a at least as good as object b?*". Asking this question for all ordered pairs of $A \times A$ results in the definition of a binary relation $\preceq$ on

A, and we say $a \preceq b$ iff a is confirmed to be at least as good as b by a "*yes*" to the above question. You may think of $a \preceq b$ meaning "*a comes before b*" or "*a is better than b*", because it will not be worse then b. It is then natural to consider $\preceq$ as reflexive. Moreover, we suppose that individuals are consistent in their judgements, so that we presuppose $\preceq$ to be transitive too, i.e. a preorder.

Based on two "*yes*" |"*no*" answers for every two objects $a, b \in A$ one of the following four patterns applies:

(1) $a \preceq b \wedge \neg b \preceq a$, denoted by aPb, and interpreted as "*a is strictly preferred to b*";
(2) $\neg a \preceq b \wedge b \preceq a$, denoted by bPa, and interpreted as "*b is strictly preferred to a*";
(3) $a \preceq b \wedge b \preceq a$, denoted by aIb, and interpreted as "*a is indifferent to b*";
(4) $\neg a \preceq b \wedge \neg b \preceq a$, denoted by aJb, and interpreted as "*a is incomparable to b*".

So one, and only one, of the following three situations occurs: An individual

- clearly prefers an object a over another object b, or
- feels indifferent about them, or
- is unable to compare the two alternatives under consideration.

(Remark concerning the notation: Our usage of P, I, J to denote relations involved follows the usual habit in the field. Therefore, in order to avoid misunderstandings concerning I, we will use Δ_X for the diagonal $\{\langle x, x\rangle : x \in X\}$ if necessary.)

It is thus a triplet of binary relations that contains all the possible answers of any decision maker in a pairwise comparison experiment. According to these cases, the following offers itself as a suitable framework since, by construction, P is asymmetric, I and J are symmetric, and from the reflexivity of $\preceq$ we get that I is also reflexive, and that J is irreflexive. Given that all pairs are presented to a decision maker, we get three subsets of the set of all ordered pairs in $A \times A$ as follows:

(1) $P := \{\langle a, b\rangle : a \textit{ is preferred to } b\}$,
(2) $I := \{\langle a, b\rangle : a \textit{ is indifferent to } b\}$, and
(3) $J := \{\langle a, b\rangle : a \textit{ and } b \textit{ cannot be compared}\}$.

Taking I as the complement of $P \cup P^{\smile}$ with respect to $A \times A - J$, gives us the following

Definition 11.1. ([4, Section 3.1]) *A* **preference structure** *on a set A is a triple* $\langle P, I, J\rangle$ *of relations on A such that:*

(1) P is asymmetric;
(2) I is reflexive and symmetric, i.e. a tolerance;
(3) J is irreflexive and symmetric;
(4) $P \cup P^{\smile} \cup I \cup J = A^2$;
(5) P, I, and J are pairwise disjoint.

Note that, due to the last two requirements, $P, P^{\smile}, I$ and J form a partition of $A \times A$: The asymmetry of P implies $P \cap P^{\smile} = \emptyset$, and since I, J are symmetric, using Prop. 2.8(7) we get $P \cap I = \emptyset \rightarrow (P \cap I)^{\smile} = P^{\smile} \cap I^{\smile} = P^{\smile} \cap I = \emptyset^{\smile} = \emptyset$, and likewise $P^{\smile} \cap J = \emptyset$.

Any preference structure (P, I, J) on a set A can be completely characterized by a single relation, called its **large preference relation**, defined as $R := P \cup I$. R clearly is reflexive, since $\Delta_A \subseteq I$, and one has $R^d := \overline{R^{\smile}} = (\overline{R})^{\smile} = P \cup J$ (see Def. 2.42). Note that $R^{\smile} = P^{\smile} \cup I$, since I is symmetric. Moreover, using elementary calculations, it is not hard to check that the following holds:

$$P = R \cap R^d, \quad I = R \cap R^{\smile}, \text{ and } \quad J = \overline{R} \cap R^d. \tag{11.1}$$

As an example we give a proof of the second equation:

$$\begin{aligned} R \cap R^{\smile} &= (P \cup I) \cap (P \cup I)^{\smile} \\ &= (P \cup I) \cap (P^{\smile} \cup I^{\smile}) \\ &= [(P \cup I) \cap P^{\smile}] \cup [(P \cup I) \cap I^{\smile})] \\ &= [(P \cap P^{\smile}) \cup (I \cap P^{\smile})] \cup [(P \cap I^{\smile}) \cup (I \cap I^{\smile})] \\ &= [\emptyset \cup \emptyset] \cup [\emptyset \cup I] \\ &= I. \end{aligned}$$

The definitions of preference structures which one can find in literature differ along three dimensions:

- They are defined by the properties of its characteristic relation, or by the properties of its constituting relations P, I, and J.
- They differ depending on whether individuals are assumed (urged in an experient/able) to judge for every two distinct elements $a, b \in A$ which one they prefer, or not. In the former case we have $J = \emptyset$ and will write the resulting preference structure as (P, I).
- They may differ in the assumptions made about the basic relation observed from an individual's judgements being irreflexive or reflexive, and which is usually assumed to fulfill the consistency requirement of transitivity.

As an example result from the area of classical preference structures (which makes intelligible both their theoretical and practical concern) we present without proof the following:

Theorem 11.2. ([21, Theorem 3.4]) *Let R be the characteristic relation of a preference structure* (P, I) *on a finite set A. Then the following conditions are equivalent:*

(1) R is a total preorder; i.e. reflexive, complete and transitive.

(2) $\begin{cases} P \text{ is negatively transitive,} \\ R \text{ is reflexive and complete.} \end{cases}$

(3) $\begin{cases} P \text{ is transitive,} \\ I \text{ is transitive,} \\ R \text{ is reflexive and complete.} \end{cases}$

(4) $\begin{cases} P \text{ is transitive,} \\ PI \subseteq P \text{ and } IP \subseteq P, \\ R \text{ is reflexive and complete.} \end{cases}$

(5) *There exists a function $g : A \longrightarrow \mathbb{R}$ such that, for all $a, b \in A$,*

$$\begin{cases} aPb \leftrightarrow g(a) > g(b), \\ aIb \leftrightarrow g(a) = g(b). \end{cases}$$

11.2 Fuzzy Preference Structures

There is a good reason why the heading above has not been taken as 'L-Fuzzy Preference Structures': Neaarly all research has been done for $[0,1]$-valued preference relations with extensive usage of t-norms and t-conorms, i.e. in a classical and well-known fuzzy setting. The only exception known to the author is [8], a paper on "Poset-valued preference relations".

The investigations on fuzzy preference relations and structures have a tradition lasting for more than four decades, starting with [17, 18]. The results we are going to report below have been found mainly in the late nineties of the last century, and, in our opinion, are a very nice and typical example for what may hide behind seemingly reasonable and cautious generalizations of well-known classical concepts. However, we will not be very detailed (this would require much more advanced results on triangular (co)norms), but instead will give some results only, often without proof (but providing the relevant references), among them primarily the monograph [4] or the overview paper [20]. Of course, fuzzy preferences have remained a vivid research topic also during the first twenty years of the current century as can be seen from the books [7, 14], the more recent overviews [2,23], or e.g. [10, 11, 13].

In addition to the information presented in Section 9.2 we need some more concepts and results on t-(co)norms for our considerations on fuzzy preference structures (our notation follows the common one):

Definition 11.3.

(a) *A continuous t-norm T is called* **Archimedean** *if $T(x,x) < x$ for all $x \in (0,1)$.*

(b) *A t-norm T is said to have* **zero divisors** *if there exist $x, y \in (0,1]$ such that $T(x,y) = 0$.*

(c) *A t-norm T is called* **positive** *if it has no zero divisors.*

Let us recall from Def. 11.1 that a classical *preference structure* on a set A is a triple $\langle P, I, J\rangle$ of relations on A with the following properties:

(1) P is asymmetric, i.e. $P \cap P^{\smile} = \emptyset$;

(2) I is reflexive and symmetric, i.e. $\Delta_A \subseteq I$ and $I = I^{\smile}$;

(3) J is irreflexive and symmetric, i.e. $\Delta_A \cap J = \emptyset$ and $J = J^{\smile}$;

(4) P, I, and J are pairwise disjoint, i.e. $P \cap I = P \cap J = I \cap J = \emptyset$;

(5) $P \cup P^{\smile} \cup I \cup J = A^2$.

Note that the asymmetry of P implies its irreflexivity. Furthermore, remember that these relations can be reconstructed from the large preference relation $R = P \cup I$ as shown in

(11.1). A self-evident idea is to substitute a De Morgan triple (T,S,N) (see Def. 9.13) for the classical set theoretical operations of union, intersection and complement which gives

$$P = R \cap R^d: \quad P(a,b) = T(R(a,b),\ N(R(b,a))), \tag{11.2}$$

$$I = R \cap R^{\smile}: \quad I(a,b) = T(R(a,b),\ R(b,a)), \tag{11.3}$$

$$J = \overline{R} \cap R^d: \quad J(a,b) = T(N(R(a,b)),\ N(R(b,a))). \tag{11.4}$$

At a first glance, the generalization of the classical definition of a preference structure to a fuzzy version thus seems to be a straightforward formal exercise. However, as will immediately be shown, this leads to rather surprising outcomes.

11.2.1 *The Definitional Approach*

A possibility is to generalize the definition of a classical preference structure in an obvious way as follows:

Definition 11.4. ([24, Definition 4.3]) *Consider a De Morgan triple $M = (T,S,N)$. A* **fuzzy preference structure** *with respect to M (M-FPS) on a set of alternatives A is a triple (P,I,J) of binary fuzzy relations on A such that*

(1) P is T-asymmetric and irreflexive;
(2) I is reflexive and symmetric;
(3) J is irreflexive and symmetric;
(4) $P \cap_T I = P \cap_T J = I \cap_T J = \emptyset$;
(5) $P \cup_S P^{\smile} \cup_S I \cup_S J = A^2$.

(Note that for a t-norm T the irreflexivity of P does not necessarily follow from T-asymmetry, since T may have zero-divisors. Thus it has to be required explicitly.)

Theorem 11.5. ([24, Theorem 4.1]) *Let $M = (T,S,N)$ be a De Morgan triple, with T a positive t-norm. If a triple of binary fuzzy relations (P,I,J) is an M-FPS on A, then the fuzzy relations P, I and J are all crisp.*

Proof. Suppose there exist $a,b \in A$ such that $0 < P(a,b) < 1$. Then, by the T-asymmetry of P we get $T(P(a,b),P(b.a)) = 0$. Since $P(a,b) > 0$ and T has no zero divisors, it follows that $P(b,a) = 0$. Def. 11.4(4) implies $T(P(a,b),I(a,b)) = 0$, whence $I(a,b) = 0$, and similarly we get $J(a,b) = 0$. Finally, from Def. 11.4(5) we get $P(a,b) \cup_S P^{\smile}(a,b) \cup_S I(a,b) \cup_S J(a,b) = P(a,b) \cup_S 0 = P(a,b) = 1$, a contradiction. The proofs for I and J are similar. □

So, after having deliberately applied a standard fuzzification process to a well known classical concept, we end up where we have started from: crisp relations. Therefore, we have to use t-norms having zero divisors to obtain proper fuzzy preference structures. That means that neither $\min(a,b)$ nor $a \cdot b$ (usual product) can be used. And, due to the existence of zero divisors, we may have $T(P(a,b),P(b,a)) = 0$ although $P(a,b) > 0$

and $P(b,a) > 0$, i.e. asymmetry in the sense that at most one of the two directions gets a positive value is not guaranteed.

Recall from Def. 9.13 that Ł_φ denotes the triple $(W_\varphi, W'_\varphi, N_\varphi)$, called a Łukasiewicz triple. For better orientation we list some results valid for t-norms:

Proposition 11.6.

(1) ([19, Theorem 3.1]) *A t-norm T is a continuous Archimedean t-norm with zero divisors if and only if there exists an automorphism φ of the unit interval $[0,1]$ such that $T(x,y) = W_\varphi(x,y) = \varphi^{-1}(W(\varphi(x),\varphi(y)))$.*

(2) (quoted from [23, Theorem 6], [24, Theorem 4.2]) *Let $M = (T,S,N)$ be a De Morgan triple, with T a non-Archimedean t-norm having zero divisors. If a triple of binary fuzzy relations (P,I,J) is an M-FPS on A, then there exists $\beta \in (0,1]$ such that the relations P,I,J cannot have values in the interval $[\beta,1]$.*

Due to these results there are two possibilities to carry on with continuous t-norms with zero divisors: If T is Archimedian then it must be equal to a φ-transform of the Łukasiewicz t-norm; and if T is a non-Archimedian t-norm then the possible values of (P,I,J) are restricted, hence the expressive power of the strict preference, indifference and incomparability relations that make up a fuzzy preference structure is limited. These findings are thus a strong evidence to use continuous Archimedean t-norms having zero divisors (or equivalently, Łukasiewicz triples) in the definition of fuzzy preference structures.

Definition 11.7. ([24, Definition 5.1]) *A* **Łukasiewicz fuzzy preference structure** *with respect to Ł_φ (Ł_φ-FPS) on a set of alternatives A is a triple (P,I,J) of binary fuzzy relations on A such that*

(1) *P is W_φ-asymmetric;*
(2) *I is reflexive and symmetric;*
(3) *J is irreflexive and symmetric;*
(4) *$P \cap_{W_\varphi} I = P \cap_{W_\varphi} J = I \cap_{W_\varphi} J = \emptyset$;*
(5) *$P \cup_{W'_\varphi} P^{\smile} \cup_{W'_\varphi} I \cup_{W'_\varphi} J = A^2$.*

Since a Łukasiewicz triple is completely determined by the choice of the underlying $[0,1]$-automorphism φ, we can simplify notation and use the much more convenient $\cup_\varphi$ instead of the cumbersome $\cup_{W'_\varphi}$ etc.

Since W_φ is a t-norm with zero divisors, the W_φ-asymmetry of a binary fuzzy relation does not imply its irreflexivity. But as observed at the end of [24], the irreflexivity of P need not be mentioned explicitly in the context of an Ł_φ-FPS, since it follows from the W_φ-asymmetry of P together with the reflexivity of I and the requirement $P \cap_\varphi I = \emptyset$. Recalling from the context of Def. 9.9 that $T_\varphi(x,y) := \varphi^{-1}(T(\varphi(x),\varphi(y)))$, and that φ^{-1} is an order automorphism if φ is, for all $a,b \in A$ we have

$$\text{(i) } P \cap_\varphi I = \emptyset \;\rightarrow\; \varphi^{-1}\big(\max\{\varphi(P(a,b)) + \varphi(I(a,b)) - 1, 0\}\big) = 0 = \varphi^{-1}(0)$$
$$\rightarrow\; \varphi(P(a,b)) + \varphi(I(a,b)) \le 1;$$

(ii) $I(a,a) = 1 = \varphi(1)$.

From (i) and (ii) we get $\varphi(P(a,a)) + \varphi(I(a,a)) \leq 1$, so $\varphi(P(a,a)) = 0$ follows.

Moreover, an $Ł_\varphi$-FPS does not suffer from the limited freedom of choice reported in Prop. 11.6(2) above: for a φ-transforms of Łukasiewicz t-norm we get $\beta = 1$.

11.2.2 *The Axiomatic Approach*

Proposition 11.8. ([4, Proposition 3.1]) *There is no De Morgan triple (T,S,N) with strong negation such that $R = P \cup I$ with P, I defined by* (11.2) *and* (11.3).

Proof. In order to make writing and reading more convenient, we let $x = R(a,b)$ and $y = R(b,a)$. Then $R = P \cup I$ becomes

$$S(T(x,N(y)),\ T(x,y)) = x. \tag{11.5}$$

Taking $x = 1$, (11.5) gives $S(N(y),\ y) = 1$, or equivalently, $T(z,N(z)) = 0$ for all $z \in [0,1]$. (Since in De Morgan triples $S(x,y) = N[T(N(x),N(y))]$, we get

$$1 = S(N(y),\ y) = N[T(N(N(y)),\ N(y))] = N[T(y,\ N(y))].$$

So $N(1) = 0 = T(y,\ N(y))$, with y be abitrary in $[0,1]$.)

Choosing $y = x$ such that $0 < x \leq N(x)$, (11.5) therefore yields

$$x = S(T(x,N(x)),\ T(x,x)) = S(0,T(x,x)) = T(x,x) \leq T(x,N(x)) = 0,$$

contradicting the assumption $0 < x$. □

Motivated by this negative result, in [4,5] the following axiomatic approach has been proposed (see also [23]): Assume R to be a reflexive fuzzy binary relation on a set A of alternatives, and let (T,S,N) be a continuous De Morgan triple. The axioms are

- **(IA) Independence of Irrelevant Alternatives:**
  For any two alternatives a,b the values of $P(a,b), I(a,b)$ and $J(a,b)$ depend only on the values $R(a,b)$ and $R(b,a)$.
- **(PA) Positive Association Principle:**
  The functions $p(x,N(y)), i(x,y), j(N(x),N(y))$ are nondecreasing with respect to both arguments.
- **(S) Symmetry:**
  i and j are symmetric functions.
- **(LP) Preserving Large Preference:**
  $P \cup_S I = R$ *and* $P \cup_S J = R^d$.

According to these requirements there are functions $p, i, j : [0,1] \times [0,1] \longrightarrow [0,1]$ such that

$$P(a,b) = p(R(a,b),R(b,a)), \qquad I(a,b) = i(R(a,b),R(b,a)),$$
$$J(a,b) = j(R(a,b),R(b,a)).$$

(PA) is a property that holds in the classical case, and (S) is a natural and self-evident assumption, also valid in the classical case. Note that asymmetry of P has not been assumed as an axiom. Do also note that in the classical context both $P \cup I = R$ and $P \cup J = R^d$ are valid, and that (LP) can be rewritten as

$$S(P(a,b), I(a,b)) = R(a,b) \quad \text{and} \quad S(P(a,b), J(a,b)) = R^d(a,b),$$

for all $a, b \in A$, and hence, due to (IA), as a system of two functional equations in the following form:

$$S(p(x,y),\ i(x,y)) = x, \tag{11.6}$$

$$S(p(x,y),\ j(x,y)) = N(y), \tag{11.7}$$

for all $x, y \in [0,1]$, and with $x = R(a,b)$ and $y = R(b,a)$ as above.

Based upon the chosen axioms the following has been shown in [4, Section 3.3] and [5] with (11.6) and (11.7) as the key requirements (recall Def. 9.13):

Theorem 11.9. *Suppose that the $\langle p, i, j, T, S, N\rangle$ satisfies axioms (IA), (PA), (S), and (LP), i.e.* (11.6) *and* (11.7)*; then there exists an order automorphism φ of $[0,1]$ such that $(T,S,N) = (W_\varphi, W'_\varphi, N_\varphi)$, i.e. (T,S,N) is a Łukasiewicz triple.*

Finally we give a few connections among the functions p, i, j without proof that will let us conclude some facts on the relations P, I, J.

Proposition 11.10. ([5, Theorem 4.2]) *For any solution $\langle p, i, j, T, S, N\rangle$ of* (11.6) *and* (11.7) *according to* Th. 11.9 *the following equalities hold for all $x, y \in [0,1]$:*

(1) $T(p(x,y), p(y,x)) = 0.$
(2) $T(p(x,y), i(x,y)) = 0.$
(3) $T(p(x,y), j(x,y)) = 0.$
(4) $T(i(x,y), j(x,y)) = 0.$
(5) $(p(x,y), p(y,x), i(x,y), j(y,x)) = 1.$

Corollary 11.11. ([5, Corollary 4.1]) *For any solution $\langle p, i, j, T, S, N\rangle$ of* (11.6) *and* (11.7) *according to* Th. 11.9 *we have (with $(T,S,N) = \mathbf{Ł}_\varphi$ and using the simplified notation introduced above):*

(1) P is φ-asymmetric.
(2) $P \cap_\varphi I = \emptyset.$
(3) $P \cap_\varphi J = \emptyset.$
(4) $I \cap_\varphi J = \emptyset.$
(5) $P \cup_\varphi P^{\smile} \cup_\varphi I \cup_\varphi J = A \times A.$

Since I and J are clearly symmetric, following the axiomatic approach as outlined above also leads to Łukasiewicz fuzzy preference structures as defined in Def. 11.7.

11.3 Social Choice Theory

Of course, as many other (crisp, classical) concepts, also social choice theory has not gone unnoticed by the fuzzy community, which is the reason why we insert this very short section. Only a general orientation and informal characterization of Social Choice Theory is included, followed by some hints for further study of detailed concepts and results.

We characterize the topic under consideration - one that is at the basis of living together and cooperation - with the very first sentences of two valuable sources of information:

> "Social choice theory is an analysis of collective decision making. The theory of social choice starts out from the articulated opinions or values of the members of a given community or the citizens of a given society and attempts to derive a collective verdict or statement. Such a situation can be called direct democracy, where public actions are determined directly by the members of society." ([6])
>
> "The term *preferences* or *system of preferences* relates to tastes defined on a set of alternatives. The term *choice* refers to a pattern of behavior that takes the form of choice from a set of alternatives. In the context of collective choice, that is, choice by a number of individuals, the alternatives can be possible decisions of a group of judges or of a jury, the possible modes of action faced by a board of directors, different policies considered by a government, local community, committee of experts, a political party that wishes to win an election, or a group of civil servants. The use of the terms *social preferences* or *social choice* is common in social contexts in which the chosen alternatives affect several individuals (some or all members of society). ([16])

A lot of books and papers have been devoted not only to formal models of the subject area, but also to its philosophical, political, and economic aspects, the whole discussions by far outreaching what can be touched here.

Even the minimal amount of conceptual and formal concepts needed to discuss one of the most famous results as an example, the so-called Arrow's Impossibility Theorem, is too special and exceeds the aims of this chapter. We suggest [6, 16] or [9] as starting points for the classical, and [7, 14] for the fuzzy version.

References

[1] Bouyssou, D., Vincke, P.: Binary relations and preference modeling, in: Bouyssou, D., Dubois, D., Pirlot, M., Prade, H. (Eds.), *Decision-Making Process: Concepts and Methods*, ISTE Ltd and John Wiley & Sons, Inc., pp. 49-84 (2009).

[2] De Baets, B., Fodor, J.: Preference modelling, a matter of degree, in: Ehrgott, M. et al. (Eds.), *Trends in Multiple Criteria Decision Analysis*, Springer Science+Business Media, pp. 141-176 (2010).

[3] Fishburn, P.C., Monjardet, B.: Norbert Wiener on the theory of measurement (1914, 1915, 1921), J. of Mathematical Psychology **36**, 165-184 (1992).

[4] Fodor, J., Roubens, M.: *Fuzzy Preference Modeling and Multicriteria Decision Support*, Springer Science+Business Media, Dordrecht (1994).

[5] Fodor, J., Roubens, M.: Valued preference structures, European J. of Operational Research **79**, 277-286 (1994).

[6] Gaertner, W.: *A Primer in Social Choice Theory*, Revised Edition, Oxford University Press (2009).

[7] Gibilisco, M.B., Gowen, A.M. et al.: *Fuzzy Social Choice Theory*, Springer International Publishing Switzerland (2014).

[8] Janiš, V., Montes, S., Šešelja, B., Tepavčević, A.: Poset-valued preference relations, Kybernetika **51**, 747-764 (2015).

[9] List, C.: Social Choice Theory, Stanford Encyclopedia of Philosophy, `https://plato.stanford.edu/archives/win2022/entries/social-choice/` (2022).

[10] Llamazares, B.: Factorization of fuzzy preferences, Social Choice and Welfare **24**, 475-496 (2005).

[11] Llamazars, B., De Baets, B.: Fuzzy strict preference relations compatible with fuzzy orderings, Int. J. of Uncertainty, Fuzziness and Knowledge-Based Systems **18**, 13-24 (2010).

[12] Luce, R. D.: Semiorders and a theory of utility discrimination, Econometrica **24**, 178-191 (1956).

[13] Montes, I., Diaz, S., Montes, S.: On complete fuzzy preorders and their characterizations, Soft Computing **15**, 1999-2011 (2011).

[14] Mordeson, J.N., Malik, D.S., Clark, T.D.: *Application of Fuzzy Logic to Social Choice Theory*, Taylor & Francis Group (2015).

[15] Moretti, S., Öztürk, M., Tsoukiàs, A.: Preference Modelling, in: Greco, S., Figueira, J., Ehrgott, M. (Eds.), *Multiple Criteria Decision Analysis: State of the Art Surveys*, 2nd ed., Springer Science+Business Media New York, pp. 43-95 (2016).

[16] Nitzan, S.: *Collective Preference and Choice*, Cambridge University Press (2010).

[17] Orlovsky, S.A.: Decision-making with a fuzzy preference relation, Fuzzy Sets and Systems **1**, 155-167 (1978).

[18] Ovchinnikov, S.V.: Structure of fuzzy binary relations, Fuzzy Sets and Systems **6**, 169-195 (1981).

[19] Ovchinnikov, S.V., Roubens, M.: On strict preference relations, Fuzzy Sets and Systems **43**, 319-326 (1991).

[20] Perny, P., Roubens, M.: Fuzzy preference modeling, in: R. Słowinski (Ed), *Fuzzy Sets in Decision Analysis, Operations Research and Statistics*, Springer Science+Business Media New York, pp. 3-30 (1998).

[21] Pirlot, M., Vincke, P.: *Semiorders - Properties, Representations, Applications*, Springer Science+Business Media Dordrecht (1997).

[22] Roubens, M., Vincke, P.: *Preference Modelling*, Springer-Verlag Berlin Heidelberg (1985).

[23] Rudas, I.J., Fodor, J.: Fundamentals of Fuzzy Preference Modeling, in: *Proc. of the 8th WSEAS International Conference on Applied Informatics and Communications (AIC'08)*, pp. 134-145 (2008).

[24] Van de Walle, B., De Baets, B., Kerre, E.: A plea for the use of Łukasiewicz triplets in the definition of fuzzy preference structures, (I). General argumentation, Fuzzy Sets and Systems **97**, 349-359 (1998).

[25] Wiener, N.: Contribution to the theory of relative position, Mathematical Proc. of the Cambridge Philosophical Soc. **17**, 441-449 (1914).

CHAPTER 12

FUZZY RELATION EQUATIONS AND INEQUALITIES

Fuzzy relation equations (FREs) and inequalities (FRIs) have a long tradition in Fuzzy Set Theory, starting in the mid-seventies with two seminal contributions by Sanchez [50, 51] in the context of medical research. Since then, fuzzy relation equations have been investigated throughout and have become one of the most intensively studied problems in the field of fuzzy sets and fuzzy systems.

As a result a good deal more contributions of theoretical nature have been published than a single person can work through with reasonable effort, if this is possible at all. Not to mention the vast number of application oriented papers. But, fortunately, there exist a few excellent overviews: [27, Chapter 6] and [45, Chapters 2,3] may serve as a first orientation on the basic aspects of FREs on the real unit interval. For more details on both logical operators and further types of FREs you may best start with [35] which, though in principle based mainly on the older work [11] and [72], gives a comprehensive and well readable overview with a lot of references. And in [33] the problem of solving a finite system of sup-T equations with a general continuous triangular norm T is investigated in detail.

From the sources mentioned above one can see that there is no account of the newest research work and its achievements during the last almost fifteen years. It is the author's hope that, due to its special organization, the information provided in Section 12.4 below can be of some help in this respect.

Apart from pushing forward research in many-valued logic, fuzzy relation equations are the second area where a vast progress in theory has been achieved under the roof of Fuzzy Set Theory during the last four decades. During this period, fuzzy relations have also found a wide field of applications and nowadays are used in e.g. fuzzy control, approximate reasoning, decision-making, fuzzy information retrieval, and fuzzy pattern recognition, to name a few.

The considerations in this chapter are focused to the questions of solvability of sup-$\sqcap$ fuzzy relation equations and inequalities ($\sqcap$ being some t-norm on L) and also to the structure of their solution sets on complete BROUWERian lattices (often called "*rel-*

atively pseudocomplemented lattice", which you must not forget; see Def. 4.20). FREs of this kind are in line with what has been initiated in Sanchez [50, 51], have been thoroughly analyzed over the years, and are a solid basis for various applications. Algorithms and their complexity, or properties of FREs on more general or more special algebraic structures, or other kinds of fuzzy relation equations and inequalities will not be our concern here. For these topics we have to refer interested readers to e.g. [1–3, 11, 45].

12.1 Types of Fuzzy Relation Equations

Various types of FREs naturally originate from the variants of relation composition. Besides the commonly used circle product $\rho \circ \sigma$, the so-called subdirect $\rho \lhd \sigma$, and the superdirect product $\rho \rhd \sigma$ for classical two-valued relations are also used. Their definitions are as follows:

(a) $\rho \circ \sigma = \{\langle x, z\rangle : \exists y \in Y[x\rho y \wedge y\sigma z\}$,
(b) $\rho \lhd \sigma = \{\langle x, z\rangle : \forall y \in Y[x\rho y \rightarrow y\sigma z\}$,
(c) $\rho \rhd \sigma = \{\langle x, z\rangle : \forall y \in Y[y\sigma z \rightarrow x\rho y\}$,

Recall Def. 9.14 for the concepts of conjunctor, disjunctor, and implicator as generalizations of the classical logical operations of conjunction, disjunction, and implication. If you, furthermore, remember that the supremum and infimum of L model the existential and universal quantifier, respectively, then we get the many-valued versions of the above compositions as follows:

(a) $\rho \circ \sigma(x, z) = \sup_{y\in Y} \mathcal{C}(\rho(x, y), \sigma(y, z))$,
(b) $\rho \lhd \sigma(x, z) = \inf_{y\in Y} \mathcal{I}(\rho(x, y), \sigma(y, z))$,
(c) $\rho \rhd \sigma(x, z) = \inf_{y\in Y} \mathcal{I}(\sigma(y, z), \rho(x, y))$.

Since very often a t-norm $\mathcal{T}$ is used as a conjunctor we get the two basic groups of sup-$\mathcal{C}$ or sup-$\mathcal{T}$, and inf-$\mathcal{I}$ equations.

12.2 The Lattice Based Approach: General Results

In this section we will focus on the first group, for which (12.1) is an example, and we will outline the principle strategies that have been applied to tackle the resolution problem for FREs as well as some general and special results, often providing no proofs but links only.

12.2.1 *Some Preliminaries*

For sake of simplicity, in what follows let X, Y and Z denote three non-empty finite sets with $|X| = m, |Y| = n$ and $|Z| = r$; remember that, for any $n \in \mathbb{N}$, we denote by $\underline{n}$ the set $\{1, 2, \ldots, n\}$. Let L a complete BROUWERian lattice (in the terminology of Sanchez [50]), i.e. a complete residuated lattice $(L, \wedge, \vee, \otimes, \frown, 0, 1)$ where $\otimes = \wedge$, or HEYTING lattice, in our terminology.

Assume that $\rho \in \mathcal{F}^L_{X\times Y}$, $\sigma \in \mathcal{F}^L_{Y\times Z}$, and $\tau \in \mathcal{F}^L_{X\times Z}$. Let $R = (r_{ij}) \in \mathfrak{M}^L_{m\times n}$, $S = (s_{jk}) \in \mathfrak{M}^L_{n\times r}$ and $T = (t_{mr}) \in \mathfrak{M}^L_{m\times r}$ be representations of ρ, σ, and τ, respectively. Furthermore, let the three relations be constrained by

$$\rho \circ \sigma = \tau, \qquad i.e.\ t_{ik} = \bigvee_j (r_{ij} \wedge s_{jk}) \tag{12.1}$$

If only two of the three relations are given, there is a "problem": The *direct resolution problem,* if we know both ρ and σ; we can get τ by direct computation. If however the result τ and only one of the relations on the left side are known, then we have an *inverse resolution problem,* viz.

$$\text{(e1)}\ \chi \circ \sigma = \tau \quad \text{or} \quad \text{(e2)}\ \rho \circ \chi = \tau,$$

with χ unknown. Whereas the direct resolution problem can simply be solved by calculation and has a unique solution, inverse problems are far from being trivial and a solution is guaranteed neither to exist nor to be unique.

Before we are going to present results known about solutions for equation (12.1), note the following decomposition possibilities to reduce the complexity of the problem and its structure ([11,27,35,72]):

(1) If we have a method for solving (e1) $\chi \circ \sigma = \tau$, we can also use it for solving (e2) $\rho \circ \chi = \tau$: Rewrite the last equation as $\chi^{\smile} \circ \rho^{\smile} = \tau^{\smile}$ and try to solve this to get $\chi^{\smile}$ by the solution method for (e1); if there is such a solution then clearly $\chi^{\smile\smile} = \chi$ is a solution for (e2).

(2) Let $\mathbb{S}(\sigma, \tau) := \{\rho : \rho\sigma = \tau\}$ or, equivalently, $\mathbb{S}(S, T) = \{R : R \circ S = T\}$ (correspondingly $\mathbb{S}(\rho, \tau) := \{\sigma : \rho\sigma = \tau\}$) denote the set of solutions for equation (12.1). Then it is easy to see that (12.1) can be partitioned into a set of simpler equations (called *matrix-by-vector* equations in the terminology of [1]) as follows: Let $S \in \mathbb{S}(R, T)$, i.e. $R \circ S = T$; then we obviously have,

$$\begin{bmatrix} r_{11} & r_{12} & \cdots & r_{1n} \\ r_{21} & r_{22} & \cdots & r_{2n} \\ & \vdots & & \\ r_{m1} & r_{22} & \cdots & r_{mn} \end{bmatrix} \circ \begin{bmatrix} s_{1k} \\ s_{2k} \\ \vdots \\ s_{nk} \end{bmatrix} = \begin{bmatrix} t_{1k} \\ t_{2k} \\ \vdots \\ t_{mk} \end{bmatrix} \tag{12.2}$$

for all $k \in \underline{r}$.

As an equivalent vector formulation we have, for all $k \in \underline{r}$, with $\boldsymbol{s}_k = (s_{ik})_{i\in\underline{n}}$, and $\boldsymbol{t}_k = (t_{ik})_{i\in\underline{m}}$

$$R \circ \boldsymbol{s}_k = \boldsymbol{t}_k. \tag{12.3}$$

If $\mathbb{S}_k(R, \boldsymbol{t}_k) = \{\boldsymbol{s}_k : R \circ \boldsymbol{s}_k = \boldsymbol{t}_k\}$ then we get the solutions S as one-row matrices

$$S = \begin{bmatrix} \boldsymbol{s}_1 & \boldsymbol{s}_2 & \cdots & \boldsymbol{s}_r \end{bmatrix},$$

where $\boldsymbol{s}_k \in \mathbb{S}_k(R, \boldsymbol{t}_k)$ for all $k \in \underline{r}$.

(3) Clearly, (12.2) is a system of m independent (called *vector-by vector* in [1], and *fuzzy elementary* in [60]) equations in n variables s_j over L of the form

$$\begin{bmatrix} r_{i1} \; r_{i2} \; \cdots \; r_{in} \end{bmatrix} \circ \begin{bmatrix} s_1 \\ s_2 \\ \vdots \\ s_n \end{bmatrix} = t_i, \qquad \text{or } \bigvee_{j \in \underline{n}} (r_{ij} \wedge s_j) = t_i. \tag{12.4}$$

(This system is also called a "*fuzzy (max-min) linear system of equations*" (FLSE) in [43, 45].)

(4) And this, in turn, means that we finally deal with n elementary (in [1] called *scalar-by scalar*) equations of the form $r_j \wedge s_j = t_{i_j}$ or, in a more simple formulation: with n equations $r \wedge s = t$, or, equivalently, with n pairs of inequalities $r \wedge s \leq t$ and $r \wedge s \geq t$.

12.2.2 *Starting with the Begin: The Maximal Solution*

As already mentioned, investigation on fuzzy relational equations started in the mid-seventies with Sanchez's groundbreaking papers [50, 51] in which he has presented results that remained an important cornerstone throughout all the time since then. We begin with some definitions necessary for the subsequent presentation of these results, for which we do only partially adhere to the notation that originated from the papers just mentioned and which has commonly been used in early contributions as e.g. [13]. Therefore, it makes sense to insert the following

CAVEAT: In [50, 51], and often also in related papers, $S \circ R$ is written for $\rho\sigma$, R^{-1} for $\rho^{\smile}$, and $a\alpha b$ instead of $a \frown b$. For @-composition, however, the order is not changed.

Let L be a complete Brouwerian lattice, and $\rho \in \mathcal{F}^L_{X\times Y}$, $\sigma \in \mathcal{F}^L_{Y\times Z}$, and $\tau \in \mathcal{F}^L_{X\times Z}$ denote L-fuzzy relations.

Definition 12.1. *The* **@-composition of relations** ρ *and* σ *is defined by*

$$\rho\, @\sigma(x,z) = \bigwedge\nolimits_y \left(x\rho y \frown y\sigma z\right) \tag{12.5}$$

The following results are due to [50]; see also [45, Chapter 2.2] for an elaborate exposition (with a notation that avoids the change of factors in relation composition).

Proposition 12.2.

(1) $\sigma \subseteq \rho^{\smile} @(\rho\sigma)$;
(2) $\rho \subseteq (\sigma\, @(\rho\sigma)^{\smile})^{\smile}$;
(3) $\rho(\rho^{\smile} @\tau) \subseteq \tau$;
(4) $(\sigma\, @\tau^{\smile})^{\smile}\sigma \subseteq \tau$.

Proof. Noting that $\otimes = \wedge$, and using monotonicity properties of $\frown$ we get

(1): $$(\rho^{\smile} @(\rho\sigma))(y,z) = \bigwedge\nolimits_x \left[x\rho y \frown (x\rho\sigma z)\right] = \bigwedge\nolimits_x \left[x\rho y \frown \left(\bigvee\nolimits_t (x\rho t \wedge t\sigma z)\right)\right]$$

$$
\begin{aligned}
&= \bigwedge_x \left[x\rho y \frown \left((x\rho y \wedge y\sigma z) \vee \bigvee_{t \neq y} (x\rho t \wedge t\sigma z) \right) \right] \\
&\geq \bigwedge_x \left[x\rho y \frown (x\rho y \wedge y\sigma z) \right] \qquad \text{by Prop. 4.29(1)} \\
&\geq \bigwedge_x \left[y\sigma z \right] \qquad \text{by Prop. 4.27(2), second inequality} \\
&= y\sigma z
\end{aligned}
$$

$$
\begin{aligned}
(3): \quad \rho(\rho\breve{}\,@\tau)(x,z) &= \bigvee_y \left[x\rho y \wedge y(\rho\breve{}\,@\tau)z \right] \\
&= \bigvee_y \left[x\rho y \wedge \bigwedge_t (t\rho y \frown t\tau z) \right] \\
&= \bigvee_y \left[x\rho y \wedge (x\rho y \frown x\tau z) \wedge \bigwedge_{t \neq x} (t\rho y \frown t\tau z) \right] \\
&\leq \bigvee_y \left[x\rho y \wedge (x\rho y \frown x\tau z) \right] \\
&\leq \bigvee_y \left[x\tau z \right] \qquad \text{by Prop. 4.27(2), first inequality} \\
&= x\tau z
\end{aligned}
$$

□

Based upon this lemma the fundamental results on the solution sets of fuzzy relation equations are given in

Theorem 12.3. ([50, Theorems 5 and 6]) *Let $S_\sigma := \mathbb{S}(\rho,\tau)$; $S_\rho := \mathbb{S}(\sigma,\tau)$; then*

(1) *$S_\sigma \neq \emptyset$ iff $\hat{\sigma} := \rho\breve{}\,@\tau \in S_\sigma$; $\hat{\sigma}$ is the greatest element in S_σ.*
(2) *$S_\rho \neq \emptyset$ iff $\hat{\rho} := (\sigma\,@\tau\breve{})\breve{} \in S_\rho$, and $\hat{\rho}$ is the greatest element in S_ρ.*

Proof. (1): Assume $S_\sigma \neq \emptyset$; then there is a relation σ' with $\rho\sigma' = \tau$. We get

$$\sigma' \subseteq \rho\breve{}\,@(\rho\sigma') = \rho\breve{}\,@\tau = \hat{\sigma} \tag{12.6}$$

from Prop. 12.2(1). Thus $\tau = \rho\sigma' \subseteq \rho(\rho\breve{}\,@\tau)$, which, together with Prop. 12.2(3), gives $\rho(\rho\breve{}\,@\tau) = \tau$, i.e. $\hat{\sigma} \in S_\sigma$. And (12.6) shows that $\hat{\sigma}$ is the greatest solution.
(2): The proof is analogous to (1) using Prop. 12.2(2) and (4). □

Before we have a closer look at the solution sets of fuzzy relation equations in Section 12.2.4, a few words on fuzzy relation inequalities are in order.

12.2.3 *Fuzzy Relation Inequalities*

Lemma 12.4. *Let $\rho \in \mathcal{F}^L_{X\times Y}$ and $\sigma_1,\sigma_2 \in \mathcal{F}^L_{Y\times Z}$ denote L-fuzzy relations such that $\sigma_1 \subseteq \sigma_2$. Then $\rho\,@\sigma_1 \subseteq \rho\,@\sigma_2$.*

Proof. By Prop. 4.29(2) we have $x\rho y \frown y\sigma_1 z \leq x\rho y \frown y\sigma_2 z$ for each $y \in Y$, so $\rho\,@\sigma_1(x,z) \leq \rho\,@\sigma_2(x,z)$. □

Theorem 12.5. ([51, p. 225]) *Let $\rho \in \mathcal{F}^L_{X\times Y}$, $\sigma \in \mathcal{F}^L_{Y\times Z}$, and $\tau \in \mathcal{F}^L_{X\times Z}$ denote L-fuzzy relations. Then the following holds*

(1) *Let $S^{\subseteq}_\sigma := \{\sigma : \rho\sigma \subseteq \tau\}$; then $\hat{\sigma} := \rho\breve{}\,@\tau$ is the greatest element in $S^{\subseteq}_\sigma$.*
(2) *Let $S^{\subseteq}_\rho := \{\rho : \rho\sigma \subseteq \tau\}$; then $\hat{\rho} := (\sigma\,@\tau\breve{})\breve{}$ is the greatest element in $S^{\subseteq}_\rho$.*

Proof. (1): $S_\sigma^\subseteq \neq \emptyset$ since $\emptyset \in S_\rho^\subseteq$. Suppose $\sigma \in S_\sigma^\subseteq$, then $\rho\sigma \subseteq \tau$, whence $\rho^\smile @(\rho\sigma) \subseteq \rho^\smile @\tau$ by Lemma 12.4. Therefore, by Prop. 12.2(1), we get $\sigma \subseteq \rho^\smile @\tau$, and Prop. 12.2(3) gives $\rho(\rho^\smile @\tau) \subseteq \tau$. Thus, $\hat{\sigma} \in S_\sigma^\subseteq$, and $\sigma \subseteq \hat{\sigma}$ for all $\sigma \in S_\sigma^\subseteq$.

(2): $S_\rho^\subseteq \neq \emptyset$ since $\emptyset \in S_\rho^\subseteq$. Note that Prop. 9.32(6) implies $\rho \subseteq \sigma \leftrightarrow \rho^\smile \subseteq \sigma^\smile$. Suppose $\rho \in S_\rho^\subseteq$, then $\rho\sigma \subseteq \tau$. But $\rho\sigma \subseteq \tau \leftrightarrow (\rho\sigma)^\smile \subseteq \tau^\smile$, whence $\sigma @(\rho\sigma)^\smile \subseteq \sigma @\tau^\smile$ by Lemma 12.4. This implies $(\sigma @(\rho\sigma)^\smile)^\smile \subseteq (\sigma @\tau^\smile)^\smile$, so $\rho \subseteq (\sigma @\tau^\smile)^\smile$ by Prop. 12.2(2). Finally, Prop. 12.2(4) gives $(\sigma @\tau^\smile)^\smile\sigma \subseteq \tau$; so $\hat{\rho} \in S_\rho^\subseteq$ and $\rho \subseteq \hat{\rho}$ for all $\rho \in S_\rho^\subseteq$. □

12.2.4 *Minimal Solutions*

Now that in Th. 12.3 we have given necessary and sufficient conditions for the existence of solutions of equation (12.1), what more can be said about the solution set? Some trivial but important facts are listed in

Proposition 12.6. *Assume $S_\sigma := \mathbb{S}(\rho,\tau) \neq \emptyset$, $|S_\rho| \geq 2$, and $\rho_1, \rho_2 \in S_\rho$. Then*

(1) $\rho \in [\rho_1, \rho_2]$ (i.e. $\rho_1 \subseteq \rho \subseteq \rho_2$) implies $\rho \in S_\rho$.
(2) $\rho_1 \cup \rho_2 \in S_\rho$.

Proof. (1): $\tau = \rho_1\sigma \subseteq \rho\sigma \subseteq \rho_2\sigma = \tau$.
(2): $(\rho_1 \cup \rho_2)\sigma = \rho_1\sigma \cup \rho_2\sigma = \tau \cup \tau = \tau$. □

Let $\check{S}_\rho$ denote the set of minimal elements in $S_\rho \neq \emptyset$. If we know the minimal solutions of (12.1) then, by what has just been shown, the set of all solutions is given by

$$\mathbb{S}(\rho,\tau) = \bigcup_{\check{\rho}\in\check{S}_\rho} [\check{\rho}, \hat{\rho}].$$

So minimal solutions play a crucial role in describing all solutions of relational equations, and we are therefore left with the task to find a general method for obtaining them. Unfortunately, the problem of finding minimal solutions is by far more complicated than that of finding the greatest one as will become clear from the considerations below: Neither the decision about their existence for a given class of fuzzy relations, nor the construction of efficient algorithms for their computation is a trivial task. Accordingly, throughout the last four decades, a great number of research papers have been devoted to this topic. The top-down problem decomposition from (12.4) suggests to search the problem solution in a bottom-up way, and this has really been done starting with Zhao [72], followed by De Baets [9–11] and others (see e.g. [35,60] for an overview).

Before we can proceed we need some more concepts that play an essential role in the analysis of the minimal solution problem. Looking back to equation (12.4) it is clear that any solution $\mathbf{s}_k = (s_{ik})_{i\in\underline{n}}$ means a join decomposition $\bigvee_{j\in\underline{n}}(r_{ij} \wedge s_j)$ of t_i. Therefore, many researchers discuss conditions for the existence of minimal solutions of (12.4) via considering decompositions of the form $\bigvee_{j\in\underline{n}}(r_j \wedge s_j) = t$.

Definition 12.7. *Let L be a complete lattice.*

(a) An element $0 \neq a \in L$ is called **join-irreducible** *if, for all $b, c \in L$,*

$$a = b \vee c \rightarrow a \in \{b, c\}.$$

(b) *A set $A \subseteq L$ of join-irreducible elements of L, with $|A| \geq 2$, such that $b = \bigvee_{a \in A} a$, is called a* **join decomposition** *of b, and $b \in L$ is called* **join-decomposable**.

(c) *A join decomposition A of $b \in L$ is called* **irredundant** *if there is no join decomposition A' of b such that $A' \subsetneq A$.*

All nonzero elements of a chain are join-irreducible and all nonzero elements of a finite distributive lattice or a direct product of a finite number of chains are either join-irreducible or join-decomposable. On the other hand, it is known that every element in a finite lattice is a join of join-irreducible elements with the convention that $\bigvee \emptyset = 0$.

Definition 12.8. *Let L be a complete lattice, and f be a mapping $f : L \longrightarrow L$;*

(a) *L is called* **completely join-distributive** *if it is distributive and*

$$b \wedge (\bigvee_{a \in A} a) = \bigvee_{a \in A} (b \wedge a).$$

(b) *f is called a* **sup-morphism** *(an* **inf-morphism***) if for any non-empty $A \subseteq L$ we have*

$$f(\bigvee_{a \in A} a) = \bigvee_{a \in A} f(a) \qquad \big(f(\bigwedge_{a \in A} a) = \bigwedge_{a \in A} f(a)\big).$$

(c) *f is called a* **homomorphism** *if it is both a sup-morphism and inf-morphism.*

Lemma 12.9. *The following are well known results from lattice theory:*

(1) *Let L be a distributive lattice and $b \in L$ be a join-irreducible element. Then $b \leq \bigvee_{i \in \underline{n}} a_i$ iff $\exists i \in \underline{n}\, [b \leq a_i]$.*

(2) *In a distributive lattice L, the irredundant join-decomposition of an element $b \in L$ is unique.*

(3) *A Brouwerian lattice is distributive.*

(4) *A complete lattice is Brouwerian if and only if it is completely join-distributive.*

Proof. (1): By the assumption we have $p = p \wedge \bigvee_{i \in \underline{n}} a_i$, and by distributivity $p = \bigvee_{i \in \underline{n}} (p \wedge a_i)$. Since p is join-irreducible, $p = p \wedge a_i$ for some i, i.e. $p \leq a_i$.

(2): Let L be a distributive lattice; assume that an element $b \in L$ has two irredundant join-decompositions, i.e.

$$b = x_1 \vee \cdots \vee x_r = y_1 \vee \cdots \vee y_s.$$

Then, by (1) we get for a given x_i that, for some $j \in \underline{s}$, $y_j \geq x_i$ and similarly $x_k \geq y_j$. The irredundancy of the join-decompositions implies that the x_i and y_i are equal in pairs, and $r = s$.

(3): By Prop. 4.21(1).

(4): See [5, p. 128, THEOREM 24]. □

Our presentation follows De Baets [9–11] which generalizes the results of Sanchez reported above by replacing the meet operation by t-norms on L with certain properties. Note that these results are therefore valid in residuated lattices L since, since $\otimes$ is a t-norm on L.

Definition 12.10. ([9, 11]) *Let $(L, \leq)$ be a poset; a subset $R \subseteq L$ is called a* **root system** *of L if there is an element $m \in R$ and an antichain $O \subseteq m^{\downarrow}$, the principal ideal generated by m, such that $R = \bigcup_{o \in O}[o, m]$. m is called the* **stem**, *and the elements $o \in O$ the* **offshoots** *of R.*

A root system is called *finitely generated* if its set of offshoots is finite. In a root system R with stem m and set of offshoots O, m is the greatest element, and the elements of O are the minimal elements of R. Hence, a root system clearly is an order convex subset that is completely determined by its greatest element and its minimal elements. Below we will see that the solution set of sup-$\sqcap$ FREs, if not empty, is a root system under some rather general restrictions. In [9, 10] the following results have been stated without proof:

Lemma 12.11.

(1) (Intersection of finitely generated root systems)
Let $\{R_i : i \in I\}$ be a family of finitely generated root systems of a complete lattice L with stem m_i and set of offshoots O_i. If $\bigcap_{i \in I} R_i \neq \emptyset$, then this intersection is a root system with stem $m = \inf_{i \in I} m_i$ and with offshoots the minimal elements of $\{\sup_{i \in I} o_i : \forall i \in I[o_i \in O_i \wedge o_i \leq m]\}$.

(2) (Finite intersection of finitely generated root systems)
If the intersection of a finite family of finitely generated root systems of a complete lattice is non-empty, then it is a finitely generated root system.

(a) Single Equations:

Let $\sqcap$ be a t-norm on a bounded poset; then for the map $\sqcap_a$ defined by $\sqcap_a(x) = a \sqcap x$ we have $\mathrm{rng}(\sqcap_a) \subseteq [0, a]$: for all $x \in L$, $a \sqcap x \leq a \wedge x$ by Prop. 9.20(1), and clearly $a \wedge x \in [0, a]$ for all $x \in L$. The partial mapping $\sqcap_a$ is called **maximally surjective** if $\mathrm{rng}(\sqcap_a) = [0, a]$. $\wedge_a$ is an example of a maximally surjective map: for $b \in [0, a]$ we clearly have $a \wedge b = b$, so $b \in \mathrm{rng}(\wedge_a)$. The t-norm $\sqcap$ is called **maximally surjective** iff all its partial mappings $\sqcap_a$ are.

The most basic equation we consider is the equation $a \sqcap x = b$. In order to describe its solution set, two operators are associated to a t-norm $\sqcap$ as follows:

$$a \check{\sqcap} b = \bigvee\{c \in L : a \sqcap c \leq b\},$$
$$a \hat{\sqcap} b = \bigwedge\{c \in L : a \sqcap c \geq b\}.$$

Proposition 12.12. ([9, Propositions 12-14]) *Let L be a complete lattice, $\sqcap$ be a t-norm on L, and $a, b \in L$.*

(1) *If $\sqcap_a$ is a sup-morphism, then the solution set of the inequality $a \sqcap x \leq b$ is $[0, a \check{\sqcap} b]$.*
(2) *If $\sqcap_a$ is a inf-morphism, then the solution set of the inequality $a \sqcap x \geq b$ is $[a \hat{\sqcap} b, 1]$ if $a \geq b$, and $\emptyset$ otherwise.*
(3) *If $\sqcap_a$ is a homomorphism, then the solution set of the equation $a \sqcap x = b$ is $[a \hat{\sqcap} b, a \check{\sqcap} b]$ if $b \in \mathrm{rng}(\sqcap_a)$, and $\emptyset$ otherwise.*

Proof. (1): Due to the monotonicity of $\sqcap_a$ and the definition of $\check{\sqcap}$ it suffices to show that $a \check{\sqcap} b$ is a solution of $a \sqcap x \leq b$. The set $\{c \in L : a \sqcap c \leq b\}$ is not empty since 0 belongs to it. Thus, since $\sqcap_a$ is a sup-morphism,

$$a \sqcap (a \check{\sqcap} b) = a \sqcap \bigvee\{c \in L : a \sqcap c \leq b\} = \bigvee\{a \sqcap c : c \in L,\ a \sqcap c \leq b\} \leq b.$$

(2): Assume $a \geq b$; then due to the monotonicity of $\sqcap_a$ and the definition of $\hat{\sqcap}$ it suffices to show that $a \hat{\sqcap} b$ is a solution of $a \sqcap x \geq b$. The set $\{c \in L : a \sqcap c \geq b\}$ is not empty since 1 belongs to it. Thus, since $\sqcap_a$ is an inf-morphism,

$$a \sqcap (a \hat{\sqcap} b) = a \sqcap \bigwedge\{c \in L : a \sqcap c \geq b\} = \bigwedge\{a \sqcap c : c \in L,\ a \sqcap c \geq b\} \geq b.$$

(3): Immediate consequence of (1) and (2): If $b \notin \mathsf{rng}(\sqcap_a)$, then there is no $c \in L$ such that $a \sqcap c = b$. Otherwise, we have $a \sqcap c = b$ for at least one $c \in L$. So $c \in [0, a \check{\sqcap} b] \cap [a \hat{\sqcap} b, 1]$, and by construction $a \hat{\sqcap} b \leq c \leq a \check{\sqcap} b$, whence $[a \hat{\sqcap} b, a \check{\sqcap} b]$ is the set of solutions. □

Corollary 12.13. *Let L be a complete lattice, $\sqcap$ be a t-norm on L, and $a, b \in L$. If $\sqcap_a$ is a maximally surjective homomorphism, then the solution set of the equation $a \sqcap x = b$ is $[a \hat{\sqcap} b, a \check{\sqcap} b]$ if $a \geq b$, and $\emptyset$ otherwise.*

(b) Equations of n variables:

We now turn to equations of the form (12.4): Given $b \in L$ and $\boldsymbol{a} = (a_1, a_2, \ldots, a_n)$, determine $\boldsymbol{x}^T = (x_1, x_2, \ldots, x_n)^T$ such that $\boldsymbol{a} \circ \boldsymbol{x} = b$, i.e. $\bigvee_{i \in \underline{n}} (a_i \sqcap x_i) = b$.

Note that both $\boldsymbol{a}$ and $\boldsymbol{x}$ can be regarded as fuzzy sets over $\underline{n}$, i.e. $\boldsymbol{a}, \boldsymbol{x} \in \mathcal{F}^L(\underline{n})$. It is well known that $(\mathcal{F}^L(\underline{n}), \subseteq)$ is a complete lattice if L is; thus $\mathbb{S}(\boldsymbol{a}, \boldsymbol{b}) \subseteq \mathcal{F}^L(\underline{n})$.

Proposition 12.14. ([9, Propositions 16-18]) *Let L be a complete lattice, $\sqcap$ be a t-norm on L, $b \in L$, and $\boldsymbol{a} \in \mathcal{F}^L(\underline{n})$.*

(1) *If the mappings $\sqcap_a$ are sup-morphisms then the solution set of the inequality $\sup_{i \in \underline{n}} (a_i \sqcap x_i) \leq b$ is given by $[\emptyset, \mathbf{g}]$, with $\mathbf{g} \subset \mathcal{F}^L(\underline{n})$ defined by $g_i = a_i \check{\sqcap} b$.*

(2) *Let L be distributive and b be a join-irreducible element of L. If the mappings $\sqcap_a$ are inf-morphisms then the solution set of the inequality $\sup_{i \in \underline{n}} (a_i \sqcap x_i) \geq b$ is non-empty iff $b \leq \sup_{i \in \underline{n}} a_i$. If it is non-empty then it is a finitely generated root system with stem $\mathbf{1}_n := \chi_{\underline{n}}$ and with offshoots the elements of*

$$O = \{\boldsymbol{o}_k \in \mathcal{F}^L(\underline{n}) : b \leq a_k\},$$

where $\boldsymbol{o}_k(i) = a_k \hat{\sqcap} b$ for $i = k$, and $\boldsymbol{o}_k(i) = 0$ otherwise.

(3) *Let b be a join-decomposable element of L with $b = \sup_{j \in J} b_j$. If the mappings $\sqcap_a$ are inf-morphisms and the solution set of the inequality $\sup_{i \in \underline{n}} (a_i \sqcap x_i) \geq b$ is non-empty then it is a root system with stem $\mathbf{1}_n$ and with offshoots the elements of*

$$O = \{\textstyle\bigcup_{j \in J} \boldsymbol{o}_k^j \in \mathcal{F}^L(\underline{n}) : b_j \leq a_k\},$$

where $\boldsymbol{o}_k^j(i) = a_k \hat{\sqcap} b_j$ for $i = k$, and $\boldsymbol{o}_k^j(i) = 0$ otherwise. If the join-decomposition is finite, then this root system is also finitely generated.

Proof. (1): By Prop. 12.12(1), the solution set of the inequality $a \sqcap x \leq b$ is $[0, a \check{\sqcap} b]$. Observing the equivalences

$$\sup_{i \in \underline{n}}(a_i \sqcap x_i) \leq b \;\leftrightarrow\; \forall i \in \underline{n}\,[a_i \sqcap x_i \leq b] \;\leftrightarrow\; \forall i \in \underline{n}\,[0 \leq x_i \leq a_i \check{\sqcap} b],$$

the assertion follows immediately.

(2): By Lemma 12.9(1),

$$\sup_{i \in \underline{n}}(a_i \sqcap x_i) \geq b \leftrightarrow \exists k \in \underline{n}[a_k \sqcap x_k \geq b],$$

and from Prop. 12.12(2) we get

$$\sup_{i \in \underline{n}}(a_i \sqcap x_i) \geq b \leftrightarrow \exists k \in \underline{n}[b \leq a_k \;\&\; a_k \hat{\sqcap} b \leq x_k].$$

Again with Lemma 12.9(1) it follows that $\sup_{i \in \underline{n}}(a_i \sqcap x_i) \geq b$ is solvable iff $b \leq \sup_{i \in \underline{n}} a_i$, whence the solution set is given by $\bigcup_{\mathbf{o}_k \in O}[\mathbf{o}_k, \mathbf{1}_n]$. For the solution set to be a root system, O must be an antichain: If $b = 0$ then obviously $O = \{\emptyset\}$ (since $a \hat{\sqcap} 0 = \bigwedge\{c \in L : a \sqcap c \geq b\} = 0$). For $b > 0$ it follows (since $\forall a \in L\, \forall b \in L - \{0\}[a \hat{\sqcap} b > 0]$) that $\mathbf{o}_k(k) > 0$, and thus the elements of O form an antichain.

(3): Note that

$$\sup_{i \in \underline{n}}(a_i \sqcap x_i) \geq b = \sup_{j \in J} b_j \;\leftrightarrow\; \forall j \in J[\sup_{i \in \underline{n}}(a_i \sqcap x_i) \geq b_j].$$

We complete the proof by applying (2) (to the join-irreducible elements of b's join-decomposition) and Lemma 12.11. □

Combining the results obtained for the inequalities we get the solution sets for the corresponding equations $\bigvee_{i \in \underline{n}}(a_i \sqcap x_i) = b$ (which clearly is the intersection of the solution sets of the two corresponding inequalities) as stated in the following proposition.

Proposition 12.15. ([9, Theorems 2 and 3]) *Let L be a complete lattice, $\sqcap$ be a t-norm on L, $b \in L$, and $\mathbf{a} \in \mathcal{F}^L(\underline{n})$.*

(1) *Let b be a join-irreducible element of L. If the mappings $\sqcap_a$ are homomorphisms, and the solution set of the equation $\bigvee_{i \in \underline{n}}(a_i \sqcap x_i) = b$ is non-empty then it is a finitely generated root system with stem $\mathbf{g} \in \mathcal{F}^L(\underline{n})$ defined by $g_i = a_i \check{\sqcap} b$ and with offshoots the elements of*

$$O = \{\mathbf{o}_k \in \mathcal{F}^L(\underline{n}) : b \leq a_k \wedge \mathbf{o}_k \subseteq \mathbf{g}\},$$

where $\mathbf{o}_k()i = a_k \hat{\sqcap} b$ for $i = k$, and $\mathbf{o}_k(i) = 0$ otherwise.
If $\sqcap$ is also maximally surjective, then the set of offshoots can be written as

$$O = \{\mathbf{o}_k \in \mathcal{F}^L(\underline{n}) : b \leq a_k\}.$$

A necessary and sufficient solvability condition therefore is that $\mathbf{g}$ is a solution.

(2) *Let b be a join-decomposable element of L with $b = \sup_{j \in J} b_j$. If the mappings $\sqcap_a$ are homomorphisms and the solution set of the inequality $\bigvee_{i \in \underline{n}}(a_i \sqcap x_i) \geq b$ is non-empty then it is a root system with stem $\mathbf{g} \in \mathcal{F}^L(\underline{n})$ defined by $g_i = a_i \check{\sqcap} b$ and with offshoots the elements of*

$$S = \{\bigcup_{j \in J} \mathbf{o}_k^j \in \mathcal{F}^L(\underline{n}) : b_j \leq a_k \wedge \mathbf{o}_k^j \subseteq \mathbf{g}\},$$

where $\mathbf{o}_k^j(i) = a_k \hat{\sqcap} b_j$ for $i = k$, and $\mathbf{o}_k(i) = 0$ otherwise. If the join-decomposition *is finite, then this root system is also finitely generated, and if $x \hat{\sqcap} y \leq x \check{\sqcap} y$, then the offshoots are the minimal elements of the*

$$T = \{\bigcup_{j \in J} \boldsymbol{o}_k^j \in \mathcal{F}^L(\underline{n}) : b_j \leq a_k\}.$$

A necessary and sufficient solvability condition therefore is that $\mathbf{g}$ *is a solution.*

Proof. (1): Let $O = \{\bigcup_{j \in J} \boldsymbol{o}_k^j \in \mathcal{F}^L(\underline{n}) : b_j \leq a_k\}$, and assume that the solution set is non-empty. Then, by Prop. 12.14(1) and (2), it is given by $[\emptyset, \mathbf{g}] \cap \bigcup_{\boldsymbol{o}_k \in O} [\boldsymbol{o}_k, \mathbf{1}_n]$, which can be written as

$$\bigcup_{\boldsymbol{o}_k \in O} [\boldsymbol{o}_k, \mathbf{g}] \text{ and also as } \bigcup_{\boldsymbol{o}_k \in O'} [\boldsymbol{o}_k, \mathbf{g}],$$

with $O' = \{\boldsymbol{o}_k : \boldsymbol{o}_k \in O \wedge \boldsymbol{o}_k \subseteq \mathbf{g}\}$. If $\sqcap$ is maximally surjective, then $O = O'$, since $a_k \hat{\sqcap} b \leq a_k \check{\sqcap} b$ for $a_k \geq b$ by Cor. 12.13.

(2): Let O denote the minimal elements of T, and assume that the solution set is non-empty. Then, by Prop. 12.14(1) and (3), it is given by

$$[\emptyset, \mathbf{g}] \cap \bigcup_{\boldsymbol{o} \in O} [\boldsymbol{o}, \mathbf{1}_n] = \bigcup_{\boldsymbol{o} \in O'} [\boldsymbol{o}, \mathbf{g}]$$

with $O' = \{\boldsymbol{o} : \boldsymbol{o} \in O \wedge \boldsymbol{o} \subseteq \mathbf{g}\}$. Let $\boldsymbol{o} = \bigcup_{j \in J} \boldsymbol{o}_k^j \in O$, then $\boldsymbol{o} \subseteq \mathbf{g}$ iff $\forall j \in J[\boldsymbol{o}_k^j \subseteq \mathbf{g}]$. The elements of O' are the minimal elements of S. If $\hat{\sqcap} \leq \check{\sqcap}$, then $\boldsymbol{o}_k^j \subseteq \mathbf{g}$, since $b_j \leq b$ implies $a_k \hat{\sqcap} b_j \leq a_k \hat{\sqcap} b \leq a_k \check{\sqcap} b$ by the monotoncity of $\sqcap$. □

(c) System of equations:

Let a system $SE_m = \{\mathrm{EQ}_j : j \in \underline{m}\}$ of m equations be given as

$$\mathrm{EQ}_j : \quad \bigvee a_{ji} \sqcap x_i = b_j, \; j \in \underline{m}.$$

For the solution set of SE_m we have

Proposition 12.16. ([11, Section 6.4.4]) *Let L be a complete lattice $\sqcap$ be a t-norm on L, $\mathbf{b} \in \mathcal{F}^L(\underline{m})$, and $\boldsymbol{a}_j \in \mathcal{F}^L(\underline{n})$, such that each b_j is either a join-irreducible element of L or has a finite join decomposition. If the mappings $\sqcap_a$ are homomorphisms, and the solution set of SE_m is non-empty then it is a root system of $(\mathcal{F}^L(\underline{n}), \subseteq)$. Moreover, if the index set J is finite, then this root system is finitely generated.*

Suppose the solution set of SE_m is non-empty, and that the EQ_j *have solution sets with stem G_j and offshoots O_j; then the solution set of SE_m is the root system with stem and offshoots given by*

$$G = \bigcap_{j \in \underline{m}} G_j \text{ and the minimal elements of } \left\{\bigcup_{j \in \underline{m}} N_j \in O_j : N_j \subseteq G\right\}.$$

A necessary and sufficient solvability condition therefore is that G is the greatest solution.

12.3 Some Particular Aspects and Results

In the foregoing section we learned about (abstract and elegant) results on the existence conditions for solutions of FREs and about the structure of its solution set for the wide class of complete lattices, hence also for complete Brouwerian lattices which in fact are (together with the special case $[0,1] \subseteq \mathbb{R}$) the predominant underlying structure in the investigation of FREs from the very beginning. (Once again we would like to remind you that, in the context of fuzzy relation equations, *Brouwerian* always means *relatively pseudocomplemented.*) To have this knowledge is very fine, but does not itself provide us with a single lower solution in practice, nor does it give us an idea of the many different approaches, or the detailed and often cumbersome arguments and techniques needed to tackle the problem. To get a more complete overview on the whole scenario, in this section we will provide information on related facets by taking a short look on some selected examples.

Taking into account the (rather rough) classification and the number of the items listed in Section 12.4 as well as the many diverse aspects they deal with, you can certainly imagine that it has not been an easy task to select results (and some proofs) which can reasonably be presented within the framework of the current text. Looking into some of these papers you will immediately recognize that, from the beginning of the FRE story, there are so many disjoint strands using different techniques with a lot of various notions and notations that one can hardly find one's way: It really needs a lot of effort to compare, translate etc. to spot existing relationships.

So, to keep things manageable within the framework of this text, we have to limit ourselves to a few important and well analyzed aspects. Hence, in what follows below, readers will get just a glimpse of small parts of the different approaches as well as of who contributed what and when - or if you want: something like a small potpourri from the history of fuzzy relation equations. We will put our focus on the two nearby branches of investigation: On linear lattices (as the direct generalization of the real unit interval) and on complete Brouwerian lattices (which gave the starting signal to the study of fuzzy relation equations as one of most vivid subareas in fuzzy set theory).

Note: In the following we will use a slightly sloppy instead of a fully consistent notation in cases when a proper matrix and a vector are involved. That means we will write $A\boldsymbol{b} = \boldsymbol{c}$ and also $\boldsymbol{a}B = \boldsymbol{c}$, since it is clear from the context if we deal with a column or row vector. This will avoid overloading formulas with superscripts for transposition without causing confusion.

Before we proceed we would like to recall a few well known or notable facts:

(R1) Be careful abut the definition of relation composition: In older papers (from the pioneers, mainly dating back to the eighties) our $\rho \circ \sigma$ is often written as $S \circ R$, i.e. with reverse order of factors.

(R2) Do also observe the various types of relation equations (we use uppercase letters for matrices, and bold lower case for vectors; $U, \boldsymbol{u}$ stand for the unknown component): Most often one of $R \circ U = T$, $U \circ S = T$, $R \circ \boldsymbol{u} = \boldsymbol{t}$, or $\boldsymbol{u} \circ S = \boldsymbol{t}$ is used. (An exception is [12] which considers also equations of the form $\boldsymbol{r} \circ U = \boldsymbol{t}$.) (We will use the following obvious abbreviations to indicate the type of FREs: RU, Ru, rU and US, uS, Us.)

(R3) Due to remark (1) in Section 12.2.1, these six schemes can be reduced to three, since $U \circ S = T$ corresponds to $S^{\smile} \circ U^{\smile} = T^{\smile}$, and similar if vectors are involved.

(R4) The Big Bang of research on FREs has already been mentioned above in Section 12.2.2: Sanchez' seminal papers [50, 51] and the results obtained therein can be used as starting point for analyzing the solvability status of any FRE, since Th. 12.3 possibly provides the greatest solution for both the RU and US type.

(R5) Let X, Y and Z be three non-empty finite sets with $|X| = m, |Y| = n$ and $|Z| = r$, and $R = (r_{ij}) \in \mathbb{L}_{m\times n}$, $S = (s_{jk}) \in \mathbb{L}_{n\times r}$ and $T = (t_{ik}) \in \mathbb{L}_{m\times r}$ be three relations. Analogous to the decomposition of (12.1) into matrix-by-vector equations indicated above in Section 12.2.1, the relation equation $R \circ S = T$ is equivalent to a system of m fuzzy equations of the form $\boldsymbol{r}_i \circ S = \boldsymbol{t}_i$, $i \in \underline{m}$, where $\boldsymbol{r}_i = (r_{ij})_{j\in\underline{n}}$, and $\boldsymbol{t}_i = (t_{ik})_{k\in\underline{r}}$. Let $\mathbb{S}_k(\boldsymbol{r}_i, T) = \{S : \boldsymbol{r}_i \circ S = \boldsymbol{t}_i\}$ then we get the solution set $\mathbb{S}(R, T)$ as $\bigcap_k \mathbb{S}_k(\boldsymbol{r}_i, T)$.

(R6) Finally, a few words about the general structure of the meta-algorithm for consistency analysis of a fuzzy relation equation are in order:

(1) Check if a given FRE, say $US = T$, has any solutions using one of the following criteria:

(a) If $RS = T$, then by definition we have $\max_j \min(r_{ij}, s_{jk}) = t_{ik}$. So, if $\max_j s_{jk} < t_{ik}$ for some k, then $S_R = \emptyset$. If there is no such k, then:

(b) By Th. 12.3 we have, with $S_R := \mathbb{S}(S, T)$: $S_R \neq \emptyset$ iff $\hat{R} := (S @ T^{\smile})^{\smile} \in S_R$. Then $\hat{R}$ is the greatest element in S_R.

(2) If $US = T$ has solutions, then according to your needs calculate them:

(a) If not yet done, compute $\hat{R}$.

(b) In case you are interested in other solutions, calculate members of $\check{S}_R$, the set of minimal elements in $S_R \neq \emptyset$ by a suitable algorithm.

Some concrete instances of solution algorithms (in chronological order) can be found in e.g. [12, 65] or [27, p. 160ff]. A more recent overview on algorithms for various kinds of matrix composition, including complexity considerations, is given in [44, 45].

(R7) If an equation has no exact solutions (i.e. is *inconsistent*), or such solutions are not necessarily required, then (in the spirit of fuzzy set theory) you might be interested in approximate ones. Various approaches of this kind can be found in [16, Chapter 10], or in e.g. [24, 40, 46, 58]. A more recent contribution is [68].

In the outline of the meta-algorithm given above in (R6) it has been suggested to look for a "suitable algorithm" to find minimal solutions. Doing so you will soon have to cope with the fact that the investigations in the relevant literature not only are usually reported for two different kinds of FREs, viz. either the uS/US or the Ru/RU type. But having a closer look at the various contributions you will immediately also recognize that all the concepts and arguments, though they are considerably different on the one hand, on the other they have in common to be intrinsically very detailed and sophisticated.

This has caused very much deliberation on the author's side if to abandon this section completely, or if not, what to include, finally resulting in the following conclusion: We will first consider some general aspects and strategies involved in finding the solutions for fuzzy relation equations, with the focus on finite sets with valuations from a bounded linear lattice (maybe sometimes specialized to $[0,1] \subseteq \mathbb{R}$). Also some contributions concerning minimal or unique solutions will be mentioned. But because of the many details and subtleties mentioned above that come along with the definitions and proofs, we can only show some typical examples from various periods together with hints to related references. Thus, our exposition is more of a small collection of snapshots on a few corners of research on fuzzy relation equations than a systematic treatment of even a small part of it: Hopefully just enough to provide an insight into the kind of arguments applied.

12.3.1 *Solutions of Finite FREs in Linear Lattices*

It is, of course, very helpful to get some practice in the transition between uS and Ru results at hand, at least for basic notions and results. Therefore, at a few occasions, we include both variants for comparison purposes (see the insertion at the end of Section 12.3.1.3), but otherwise will freely use the same type as in the literature on which the respective considerations are based. Thus additional and artificial obstacles for further studies can be avoided. (Besides, it is far beyond the modest aims of the current chapter to strive for a unifying and systematic overview that can supplement the comprehensive contributions [33, 35] of Li and Fang with more technical details and update them with newer results. Such an undertaking needs a lot more of space and time to get done properly than is available here and now.)

TABLE 12.1: Some Basic Facts on Fuzzy Relation Equations.

Ru-**type**	uS-**type**
$Ru = t$	$uS = t$
$\bigvee_j (r_{ij} \wedge u_j) = t_i$	$\bigvee_i (u_i \wedge s_{ij}) = t_j$
$[R^{\smile} @ t]_j = \bigwedge_i (r_{ij} \curvearrowright t_i)$	$[S @ t^{\smile}]_i := \bigwedge_j (s_{ij} \curvearrowright t_j)$
$S_s \neq \emptyset$ iff $\hat{s} := R^{\smile} @ t \in S_s$	$S_r \neq \emptyset$ iff $\hat{r} := (S @ t^{\smile})^{\smile} \in S_r$

Table 12.1 does not offer news but shows some facts which we know already from Section 12.2.2, emphasizing that there is a definite anchor point for further investigation of both Ru-type and uS-type fuzzy relation equations. We continue with a characterization of consistent systems in

Proposition 12.17. ([65, Lemma 2.2]) *Let $A = [a_{ij}]_{m\times n}$ and $\mathbb{S} := \{x : xA = b\}$. then $\mathbb{S} \neq \emptyset$ iff for each column a_{*j} of A, $j \in \underline{n}$, there exists an $i \in \underline{m}$ such that (i) $x_i \wedge a_{ij} = b_j$, and (ii) $x_i \wedge a_{ij'} \leq b_{j'}$ for all $j' \neq j$.*

Proof. [⇒] : Follows trivially since $\bigvee_i(x_i \wedge a_{ij}) = b_j$ for any $\boldsymbol{x} \in \mathbb{S}$.

[⇐] : Conversely, assume that for each column $\boldsymbol{a}_{*j}$ of A, $j \in \underline{n}$, there exists an $i \in \underline{m}$ such that (i) $x_i \wedge a_{ij} = b_j$, and (ii) $x_i \wedge a_{ij'} \leq b_{j'}$ for all $j' \neq j$, and that $\mathbb{S} = \emptyset$. But $\mathbb{S} = \emptyset$ means that $\boldsymbol{x}A \neq \boldsymbol{b}$ for all $\boldsymbol{x}$. Thus, there exists a column $\boldsymbol{a}_{*j_0}$ such that for all $\boldsymbol{x} = [x_i], i \in \underline{m}, x_i \wedge x_{ij_0} \neq b_{j_0}$, or for some $k \in \underline{n}, k \neq j_0$, we have $x_i \wedge a_{ik} > b_k$. This means, that for column $\boldsymbol{a}_{*j_0}$ there is no x_i satisfying conditions (i) and (ii), a contradiction to our assumption. □

12.3.1.1 *Some General Mechanisms and Strategies*

Next we introduce the concept of *equivalence* of FREs and *solution-set invariance* of matrices (see [42, p. 341] or [66]: Two systems $\boldsymbol{x} \cdot A = \boldsymbol{b}$ and $\boldsymbol{x} \cdot A' = \boldsymbol{b}$ with solutions sets $\mathbb{S}$ and $\mathbb{S}'$, respectively, are called **equivalent** if $\mathbb{S} = \mathbb{S}'$, and the matrices A and A' are then said to be **solution set invariant** with respect to $\boldsymbol{b}$. The concept of solution-set invariance is important because it is an often used strategy to code various 'qualitative′ properties of a system $\boldsymbol{x} \cdot A = \boldsymbol{b}$ in some derived matrix, called **characteristic** in e.g. [39], *augmented* and *help* matrix in [42], or *quasi-characteristic* in [65]. Clearly, such a derived matrix A' is useful only if it is solution set invariant to the original matrix A with respect to $\boldsymbol{b}$ and if it makes calculation of solutions easier.

As a simple example we mention the transformation of an FLSE into its so-called **normal form**: Let $E: \bigvee_i(u_i \wedge s_{ij}) = t_j, j \in \underline{m}$ be a system of fuzzy linear equations, and $\pi((1,2,\ldots,m)) = (j_1, j_2, \ldots, j_m)$ be a permutation such that $t_{j_1} \geq t_{j_2} \geq \cdots \geq t_{j_m}$. Then the system $E': \bigvee_i(u_i \wedge s_{i\pi(j)j}) = t_{\pi(j)}, j \in \underline{m}$ is said to be the normal form of E, and clearly is equivalent to E. There are algorithms which use normal forms to reduce computational complexity as in [52, 65].

By [18, Lemma 3.1], we get another way for complexity reduction that arises from the possibility to get rid of zeros on the right-hand side: Let $B_0 = \{i \mid b_i = 0\}$ and $A_0 = \{j \mid \exists i \in B_0[a_{ij} > 0]\}$. Then clearly for any solution of $A \cdot \boldsymbol{x} = \boldsymbol{b}$ we must have $\max_j(\min(a_{ij}, x_j)) = 0$, whence $x_j = 0$ for all $j \in A_0$ follows. Therefore, one can omit the equations with indices from B_0 and the columns of A with indices in A_0. The solutions of the original and the reduced system correspond to each other: One simply extends the solutions of the reduced equation by inserting zeros at the locations that have been eliminated in the reduction step.

For demontration purposes we quote [18, Example 1]: We take $[0,5] \subseteq \mathbb{R}$ as our linear lattice, and let

$$A = \begin{bmatrix} 3 & 1 & 2 & 0 \\ 2 & 0 & 0 & 1 \\ 1 & 2 & 0 & 3 \end{bmatrix}, \quad \text{and} \quad \boldsymbol{b} = \begin{bmatrix} 2 \\ 0 \\ 1 \end{bmatrix}$$

The second equation in $A \cdot \boldsymbol{x} = \boldsymbol{b}$ is

$$(2 \wedge x_1) \vee (0 \wedge x_2) \vee (0 \wedge x_3) \vee (1 \wedge x_4) = 0,$$

so we get $x_1 = 0$, x_2, x_3 arbitrary and $x_4 = 0$. Deleting row 2 and columns 1 and 4 yields

$$A' = \begin{bmatrix} 1 & 2 \\ 2 & 0 \end{bmatrix}, \quad \boldsymbol{x}' = (x_2, x_3), \qquad \text{and} \quad \boldsymbol{b}' = \begin{bmatrix} 2 \\ 1 \end{bmatrix}$$

From the first equation $(1 \wedge x_2) \vee (2 \wedge x_3) = 2$ we get x_2 arbitrary and $x_3 \geq 2$; the second equation $(2 \wedge x_2) \vee (0 \wedge x_3) = 1$ implies $x_2 = 1$ and x_3 arbitrary. The original equation thus has the solutions $x_1 = 0, x_2 = 1, 2 \leq x_3 \leq 5, x_4 = 0$.

Further inspecting existing literature you will soon become aware of a second strategy applied to cope with the problem of finding (all) solutions of FREs: To get a first basic information on a given equation, a natural and often used method is grouping the coefficients involved as shown in Table 12.2; there are also more specific sets of indices used at various occasions. For easier accessibility of the original literature, we will consider both Ru- and uS-type equations, and slightly change notation if appropriate.

Our account of the just mentioned classification of the matrix coefficients involved in an FRE follows the one given in [42, 43]. Similar approaches can also be found in e.g. [26, p. 67], [31, p. 319], or [52, p. 20].

TABLE 12.2: Coefficient Classification of a Fuzzy Relation Equation.

Ru-type	uS-type
$A\boldsymbol{x} = \boldsymbol{b}$	$\boldsymbol{x}A = \boldsymbol{b}$
$\bigvee_j (a_{ij} \wedge x_j) = b_i$	$\bigvee_i (x_i \wedge a_{ij}) = b_j$
Types of coefficients	
Peeva(1985) [41, p. 197]	Higashi & Klir(1984) [26, p. 67]
$S_i = \{j \in J : a_{ij} < b_i\} =: J_i^<$	${}^3I_j = \{i \in I : a_{ij} < b_j\} =: I_j^<$
$E_i = \{j \in J : a_{ij} = b_i\} =: J_i^=$	${}^2I_j = \{i \in I : a_{ij} = b_j\} =: I_j^=$
$G_i = \{j \in J : a_{ij} > b_i\} =: J_i^>$	${}^1I_j = \{i \in I : a_{ij} > b_j\} =: I_j^>$

The coefficients in the sets S_i, E_i and G_i mentioned in Table 12.2 have been called S-type, E-type, and G-type coefficients, respectively (from: Smaller, Equal, and Greater) in [43]. The importance of this fundamental classification for the investigation of fuzzy relation equations becomes immediately clear from a look back to the proof of Prop. 12.17, or to (12.7): In fact the latter one can (using an obvious adapted notation) be rephrased as

$$a\sigma b := 0, \text{ if } a < b, \text{ i.e. if } a \in S_b, \quad \text{and} \quad a\sigma b := b, \text{ if } a \geq b, \text{ i.e. } a \in E_b \cup G_b.$$

Furthermore, recall Def. 12.1 introduced in [50]: $\rho @ \sigma(x,z) = \bigwedge_y (x\rho y \curvearrowright y\sigma z)$, with $a \curvearrowright b$ denoting the relative pseudocomplement of a in b, which is the greatest element x such that $a \wedge x \leq b$. In the context of linear lattices we have $a \curvearrowright b = 1$ whenever $a \leq b$, i.e. $a \in S_b \cup E_b$, and $a \curvearrowright b = b$ otherwise, i.e. if $a \in G_b$. So the trichotomy of S-, E-, and G-type coefficients has, albeit only implicitly, been used already from the very beginning. And it is indeed a fundamental information for the analysis of fuzzy relation equations.

12.3.1.2 *Minimal Solutions*

We now turn to the problem of finding and characterizing minimal (also: *lower*) solutions, and to this end have to introduce another kind of composition: Let $a, b \in L$ (by assumption for this section a bounded linear lattice); then define

$$a\sigma b := 0, \text{ if } a < b, \quad \text{and} \quad a\sigma b := b, \text{ if } a \geq b. \tag{12.7}$$

For two vectors $\boldsymbol{x} \in \mathbb{L}_{m\times 1}$ and $\boldsymbol{y} \in \mathbb{L}_{1\times n}$ we define their **σ-composition** $\boldsymbol{x} @ \boldsymbol{y}$ as the L-fuzzy relation $W \in \mathbb{L}_{m\times n}$ with $w_{ij} = x_i \sigma y_j$.

For the remainder of this section we assume that $X = \{x_1, x_2, \ldots, x_n\}$, $Y = \{y_1, y_2, \ldots, y_m\}$, and $Z = \{z_1, z_2, \ldots, z_p\}$ are non-empty finite sets, with indices from sets I, J and K, respectively; and that $R \subseteq X \times Y$, $S \subseteq Y \times Z$, and $T \subseteq X \times Z$ are three finite fuzzy relations with values in a totally ordered lattice.

Consider an equation of RU-type, say $RS = T$ with S unknown and R, T given, with non-empty solution set $\mathbb{S} := \mathbb{S}(R, T) = \{S : RS = T\}$. Then, as already mentioned in (R5) above, this equation is equivalent to a system of fuzzy equations $\boldsymbol{r}_{i*} \circ S = \boldsymbol{t}_{i*}$, with solution sets $\mathbb{S}_i$, $i \in I$. For the solution set $\mathbb{S}$ we then have $\bigcap_{i\in I} \mathbb{S}_i$.

Theorem 12.18. ([12, Theorem 3]) *If $\mathbb{S}_i \neq \emptyset$, $i \in I$, then there exists a lower solution $L_i \leq S$ for any $S \in \mathbb{S}_i$.*

Proof. For any $S \in \mathbb{S}_i$ and $k \in K$ we have $\bigvee_j (r_{ij} \wedge s_{jk}) = t_{ik}$. Thus, for any $k \in K$, there is a $j_k \in J$ such that $r_{ij_k} \geq r_{ij_k} \wedge s_{j_k k} = t_{ik}$. Therefore, $t_{ik} \leq s_{j_k k}$, and $r_{ij_k} \sigma t_{ik} = t_{ik}$. Then, defining $L_i(j, k) = 0$ for $j \in J - j_k$ and $L_i(j_k, k) = t_{ik}$ makes $L_i \subseteq Y \times Z$ a lower solution in $\mathbb{S}_i$: We clearly have $\bigvee_j \big(r_{ij} \wedge L_i(j, k)\big) = r_{ij_k} \wedge L_i(j_k, k) = r_{ij_k} \wedge t_{ik} = t_{ik}$. Furthermore, there is at most one non-zero value $L_i(j_k, k)$ in column k of L_i, as the minimal requirement for satisfying the equality, so $L_i \subseteq S$ for any $S \in \mathbb{S}_i$. □

Proposition 12.19. ([12, Theorems 4, 5]) *Assume $\mathbb{S} := \mathbb{S}(R, T) \neq \emptyset$, and let $\hat{S} := R^{\smile} @ T$. Then the following holds:*

(1) *$\Lambda_i := \{L_i \in \mathbb{S}_i : L_i \leq \hat{S}\} \neq \emptyset$, for any $i \in I$.*
(2) *Let the fuzzy relation L be given as $L := \bigvee_{i\in I} L_i \subseteq Y \times Z$, where $L_i \in \Lambda_i$, for $i \in I$; then $L \in \mathbb{S}$.*

Proof. (1): By Th. 12.3(1), $\hat{S}$ is the greatest solution, so $\hat{S} \in \mathbb{S}$, and therefore $\hat{S} \in \mathbb{S}_i$ for any $i \in I$.

(2): By (1), $\Lambda_i \neq \emptyset$, and by definition of L, $L_i \leq L \leq \hat{S}$, for any $i \in I$. So $T = R \circ L_i \leq R \circ L \leq R \circ \hat{S} = T$. □

A well known result from lattice theory is

Lemma 12.20. ([5, p. 4, Theorem 3]) *Any finite non-empty subset X of a poset has minimal and maximal elements.*

Theorem 12.21. ([12, Theorem 9]) *If $\mathbb{S} \neq \emptyset$, then the minimal elements of $\Lambda := \{L \in \mathbb{S} : L = \bigvee_{i\in I} L_i, L_i \in \Lambda_i, i \in I\}$ are the lower solutions in $\mathbb{S}$ and vice versa.*

Proof. If $\mathbb{S} \neq \emptyset$, then by Prop. 12.19 $\Lambda \neq \emptyset$ follows. Since $\mathbb{S}$ is a partially ordered and Λ a finite set, Λ has a minimal element M by Lemma 12.20. Let $S \in \mathbb{S}$ such that $S \leq M$. Then $S \in \mathbb{S}_i$ for each $i \in I$, and by Th. 12.18 there is a lower solution $L_i \in \mathbb{S}_i$ such that $L_i \leq S$ for each $i \in I$. So $L_i \leq S \leq \hat{S}$, whence $L_i \in \Lambda_i$ for each $i \in I$. Therefore, $L = \bigvee_{i\in I} L_i \in \Lambda$ with $L \leq S \leq M$. But since M is minimal in $\mathbb{S}$, $L = M$ follows.

Conversely, let $R \in \mathbb{S}$ be a lower solution. As above we can find an element $L \in \Lambda$ such that $L \leq R$. Then we get $L = R$ since R is minimal in $\mathbb{S}$. So $R \in \Lambda$, being clearly minimal in Λ too. □

You may also have a look at [70] for a few additional characterizations of minimal solutions. Let me also add a hint to [59, Example 2.1] which shows that a minimal solution may not exist at all. Besides, there are two more questions that one can ask in connection with minimal solutions of a fuzzy relation equation: What are the conditions for the existence of a unique minimal solution $\check{S}$? And when do we have $\check{S} = \hat{S}$, i.e. only one solution? These questions will be considered below.

12.3.1.3 *Least Solution*

Our considerations on a unique minimal solution of equations $\boldsymbol{x}\cdot A = \boldsymbol{b}$ over the real unit interval $[0,1]$ are based on [31, Section 3]. So we again consider a system of equations $\bigvee_j (x_i \wedge a_{ij}) = b_j$, $j \in J$. As above we denote the set of solutions by $\mathbb{S}$, and for each $\boldsymbol{x} \in \mathbb{S}$ we set $H_i^{\boldsymbol{x}} = \{j \in J : x_i \wedge a_{ij} = b_j\}$ for any $i \in I$ and any $j \in J$.

Proposition 12.22. ([31, Theorem 3.1]) *Let $\mathbb{S} \neq \emptyset$ and $\check{\boldsymbol{x}} = (\check{x}_1, \check{x}_2, \dots, \check{x}_n)$ be minimal in $\mathbb{S}$. Then for any $i \in I$ we have*

(1) If $H_i^{\check{\boldsymbol{x}}} = \emptyset$, then $\check{x}_i = 0$.

(2) If $H_i^{\check{\boldsymbol{x}}} \neq \emptyset$, then there exists $j \in J$ such that $\check{x}_i \wedge a_{ij} = b_j$ and $\check{x}_i = b_j$.

Proof. Substituting $\check{\boldsymbol{x}}$ in equation $\boldsymbol{x}\cdot A = \boldsymbol{b}$ gives a system

$$b_j = (\check{x}_1 \wedge a_{1j}) \vee \cdots \vee (\check{x}_i \wedge a_{ij}) \vee \cdots \vee (\check{x}_n \wedge a_{nj}), \quad j \in J. \tag{12.8}$$

(1): $H_i^{\check{\boldsymbol{x}}} = \emptyset$ implies $\check{x}_i \wedge a_{ij} < b_j$ for each $j \in J$. Thus, replacing $\check{x}_i$ by 0 in (12.8) and keeping all $\check{x}_k, k \neq i$ unchanged, all equations in (12.8) remain valid. Therefore, $\boldsymbol{x}' = (\check{x}_1, \dots, \check{x}_{i-1}, 0, \check{x}_{i+1}, \dots, \check{x}_m) \in \mathbb{S}$, so the assumption $\check{x}_i > 0$ would give a contradiction to the hypothesis that $\check{\boldsymbol{x}}$ is a minimal.

(2): Let $p := |J|$, $H_i^{\check{\boldsymbol{x}}} = \{j_1, j_2, \dots, j_m\}$, $m \in \underline{p}$, and $1 \leq j_1 < j_2 < \cdots < j_m \leq p$. Therefore

$$\check{x}_i \wedge a_{ij'} = b_{j'} \quad \text{for each} \quad j' \in \{j_1, j_2, \dots, j_m\} \tag{12.9}$$

and

$$\check{x}_i \wedge a_{ik} < b_k \quad \text{for each} \quad k \in \underline{p} - \{j_1, j_2, \dots, j_m\}. \tag{12.10}$$

From (12.9) we get $\check{x}_i \geq b_{j'}$ for each $j' \in \{j_1, j_2, \dots, j_m\}$. We now have to show that there exists an index $j_k \in \{j_1, j_2, \dots, j_m\}$ such that $\check{x}_i = b_{j_k}$. Assume that

$$\check{x}_i > b_{j'} \quad \text{for each} \quad j' \in \{j_1, j_2, \ldots, j_m\}. \tag{12.11}$$

By (12.11) and (12.9) we then get

$$a_{ij'} = b_{j'} \quad \text{for each} \quad j' \in \{j_1, j_2, \ldots, j_m\}. \tag{12.12}$$

Let $x_i^* = \max\{b_{j_1}, b_{j_2}, \ldots, b_{j_m}\}$. From (12.11) it follows that

$$b_{j'} \leq x_i^* < \check{x}_i \quad \text{for each} \quad j' \in \{j_1, j_2, \ldots, j_m\}. \tag{12.13}$$

By (12.12) and (12.13) we have

$$x_i^* \wedge a_{ij'} = b_{j'} \quad \text{for each} \quad j' \in \{j_1, j_2, \ldots, j_m\}, \tag{12.14}$$

and by (12.10) and (12.13)

$$x_i^* \wedge a_{ij'} < b_k \quad \text{for each} \quad k \in \underline{p} - \{j_1, j_2, \ldots, j_m\}. \tag{12.15}$$

Therefore, replacing $\check{x}_i$ by x_i^* in (12.8) and keeping all $\check{x}_k, k \neq i$ unchanged, all equations in (12.8) can be shown to remain valid using (12.9), (12.14) and (12.10), (12.15). So, $\boldsymbol{x}^* = (\check{x}_1, \ldots, \check{x}_{i-1}, x_i^*, \check{x}_{i+1}, \ldots, \check{x}_m) \in \mathbb{S}$, a contradiction to the hypothesis that $\check{\boldsymbol{x}}$ is minimal (since by (12.13) $x_i^* < \check{x}_i$). □

Corollary 12.23. ([31, Corollary 3.2]) *Let $\boldsymbol{x} = (x_1, x_2, \ldots, x_n)$ be a minimal element of $\mathbb{S}$. Then, for $i \in \underline{n}$, either $x_i = 0$ or there exists a $j \in \underline{n}$ such that $x_i = b_j$.*

Lemma 12.24. ([26, Theorem 4]) *$\boldsymbol{x} \in \mathbb{S}$ iff $\mathbb{S}$ has a minimal element $\check{\boldsymbol{x}}$ such that $\check{\boldsymbol{x}} \leq \boldsymbol{x} \leq \hat{\boldsymbol{x}}$.*

Theorem 12.25. ([31, Theorem 3.3]) *Let $I_j^{\boldsymbol{x}} = \{i \in I : x_i \wedge a_{ij} = b_j\}$. $\mathbb{S}$ has a unique minimal element iff $\mathbb{S}$ has a minimal element $\boldsymbol{x}^*$ such that for any $\boldsymbol{x} \in \mathbb{S}$, $I_j^{\boldsymbol{x}^*} \subseteq I_j^{\boldsymbol{x}}$ for any $j \in J$.*

Proof. Assume that $\boldsymbol{x}^* = (x_1^*, x_2^*, \ldots, x_n^*) \in \mathbb{S}$ is the unique minimal element, and that for some $\boldsymbol{x} = (x_1, x_2, \ldots, x_n) \in \mathbb{S}$ we have $I_j^{\boldsymbol{x}^*} \not\subseteq I_j^{\boldsymbol{x}}$ for some $j \in J$. Then there is an $i \in I_j^{\boldsymbol{x}^*}$ with $i \notin I_j^{\boldsymbol{x}}$, whence

$$x_i^* \wedge a_{ij} = b_j \tag{12.16}$$

and

$$x_i \wedge a_{ij} < b_j \tag{12.17}$$

From (12.16) we get that x_i^*, $a_{ij} \geq b_j$, whence from (12.17) it follows that $x_i < b_j$. Thus, $x_i^* \geq b_j > x_i$. Lemma 12.24 implies that $\mathbb{S}$ has a minimal element $\check{\boldsymbol{x}}$ such that $\check{\boldsymbol{x}} \leq \boldsymbol{x}$, whence $\check{x}_i \leq x_i < x_i^*$. So $\boldsymbol{x}^* \neq \check{\boldsymbol{x}}$, which contradicts our assumption that $\boldsymbol{x}^*$ is the unique minimal element of $\mathbb{S}$.

Conversely, let $\boldsymbol{x}^* = (x_1^*, x_2^*, \ldots, x_n^*) \in \mathbb{S}$ be a minimal element such that for any $\boldsymbol{x} \in \mathbb{S}$, $I_j^{\boldsymbol{x}^*} \subseteq I_j^{\boldsymbol{x}}$ for any $j \in J$, and let $\boldsymbol{x} = (x_1, x_2, \ldots, x_n)$ be an arbitrary minimal element of $\mathbb{S}$. We will now show that the assumption $x_i^* > x_i$, for some i, leads to a contradiction, whence $x_i^* \leq x_i$ as desired. $x_i^* > x_i \geq 0$ clearly implies $x_i^* > 0$. whence by Prop. 12.22(1) we get $H_i^{\boldsymbol{x}^*} \neq \emptyset$. So by Prop. 12.22(2) there exists an index $j \in J$ such that $x_i^* \wedge a_{ij} = b_j$ and and $x_i^* = b_j$. This implies $i \in I_j^{\boldsymbol{x}^*}$, and since $I_j^{\boldsymbol{x}^*} \subseteq I_j^{\boldsymbol{x}}$ by assumption, $i \in I_j^{\boldsymbol{x}}$ follows. Hence $x_i \wedge a_{ij} = b_j$, and therefore $x_i \geq x_i^*$, since $\boldsymbol{x}^*$ is minimal in $\mathbb{S}$, a contradiction to the assumption $x_i^* > x_i$. Furthermore, because $\boldsymbol{x}$ has been assumed minimal in $\mathbb{S}$, we obtain $\boldsymbol{x} = \boldsymbol{x}^*$. So $\boldsymbol{x}^*$ is the unique minimal element of $\mathbb{S}$. □

An example for which no least solution exists is given in [8, Example 1]: Let

$$A = \begin{bmatrix} 0.1 & 0.2 \\ 0.4 & 0.1 \end{bmatrix}, \qquad \text{and} \qquad \boldsymbol{b} = [0.1, 0.1].$$

The set of solutions for $\boldsymbol{x} \cdot A = \boldsymbol{b}$ is $\mathbb{S} = \{v(t) := [0.1, t] : 0 \leq t \leq 0.1\} \cup \{w(t) := [t, 0.1] : 0 \leq t \leq 0.1\}$. Thus we have two chains of solutions, and no pair $v(t), w(s)$ is comparable for $t, s \in [0, 0.1)$, in particular the two minimal solutions $v(0)$ and $w(0)$ are incomparable. Clearly, $v(0.1) = w(0.1)$ is the greatest solution.

For ease of reference we repeat an adapted version of Table 12.1.

Ru-**type**	uS-**type**
$A\boldsymbol{x} = \boldsymbol{b}$	$\boldsymbol{x}A = \boldsymbol{b}$
$\bigvee_j (a_{ij} \wedge x_j) = b_i$	$\bigvee_i (x_i \wedge a_{ij}) = b_j$
$[A^{\check{}} @ \boldsymbol{b}]_j = \bigwedge_i (a_{ij} \curvearrowright b_i)$	$[A @ \boldsymbol{b}^{\check{}}]_i := \bigwedge_j (a_{ij} \curvearrowright b_j)$
$\mathbb{S} \neq \emptyset$ iff $\hat{\boldsymbol{x}} := A^{\check{}} @ \boldsymbol{b} \in \mathbb{S}$	$\mathbb{S} \neq \emptyset$ iff $\hat{\boldsymbol{x}} := (A @ \boldsymbol{b}^{\check{}})^{\check{}} \in \mathbb{S}$

Recall that in the context of linear lattices we have $a \curvearrowright b = 1$ whenever $a \leq b$, i.e. $a \in S_b \cup E_b$ (or $a \in I_b^{\leq}$) and $a \curvearrowright b = b$ otherwise, i.e. if $a \in G_b$ (or $a \in I_b^{>}$); you might want to have a look at Table 12.2. So

- $a_{ij} \curvearrowright b_i = 1$ whenever $a_{ij} \leq b_i$, i.e. $j \in J_i^{\leq}$, and $a_{ij} \curvearrowright b_i = b_i$ otherwise, i.e. $j \in J_i^{>}$, whence $\bigwedge_i (a_{ij} \curvearrowright b_i) = 1$ iff $j \in J_i^{\leq}$ for all i, and $\bigwedge_i (a_{ij} \curvearrowright b_i) = \bigwedge_i b_i$, $j \in J_i^{>}$, otherwise;
- $a_{ij} \curvearrowright b_j = 1$ whenever $a_{ij} \leq b_j$, i.e. $i \in I_j^{\leq}$, and $a_{ij} \curvearrowright b_j = b_j$ otherwise, i.e. $i \in I_j^{>}$, whence $\bigwedge_j (a_{ij} \curvearrowright b_j) = 1$ iff $i \in I_j^{\leq}$ for all j, and $\bigwedge_j (a_{ij} \curvearrowright b_j) = \bigwedge_j b_j$, $i \in I_j^{>}$, otherwise.

For sake of completeness we next transform the results about least solutions from their original uS-form to the corresponding Ru-type counterpart:

- Prop. 12.22 reads as: Let $\mathbb{S} \neq \emptyset$ and $\check{\boldsymbol{x}} = (\check{x}_1, \check{x}_2, \ldots, \check{x}_n)$ be minimal in $\mathbb{S}$. Then for any $i \in I$ we have
  (1) If $H_j^{\check{\boldsymbol{x}}} = \emptyset$, then $\check{x}_j = 0$.
  (2) If $H_j^{\check{\boldsymbol{x}}} \neq \emptyset$, then there exists $i \in I$ such that $a_{ij} \wedge \check{x}_j = b_i$ and $\check{x}_j = b_i$.
- Cor. 12.23 remains unchanged, i.e. for a minimal element $\boldsymbol{x} = (x_1, x_2, \ldots, x_n) \in \mathbb{S}$ we get that for any $i \in \underline{n}$, either $x_i = 0$ or there exists a $j \in \underline{n}$ such that $x_i = b_j$.
- And Th. 12.25 changes to: *$\mathbb{S}$ has a unique minimal element iff $\mathbb{S}$ has a minimal element $\boldsymbol{x}^*$ such that for any $\boldsymbol{x} \in \mathbb{S}$, $J_i^{\boldsymbol{x}^*} \subseteq J_i^{\boldsymbol{x}}$ for any $i \in I$, where* $J_i^{\boldsymbol{x}} = \{j \in J : a_{ij} \wedge x_j = b_i\}$. Since we are in the context of underlying linear lattices , $a_{ij} \wedge x_j = b_i$ implies that at least one of a_{ij} and x_j is equal to b_i. Thus $j \in J_i^{\boldsymbol{x}}$ if $a_{ij} = b_i$ or $x_j = b_i$.

12.3.1.4 *Unique Solution*

From what we have learned up to here, if a fuzzy relation equation $A\cdot\boldsymbol{x}=\boldsymbol{b}$ has only a single solution $\boldsymbol{s}$, then we know that $\boldsymbol{s}=\hat{\boldsymbol{x}}$: $|\mathbb{S}|=1$ implies $\mathbb{S}=\{\hat{\boldsymbol{x}}\}$. Clearly, $\boldsymbol{s}$ is also a unique minimal solution, and so it must have the characteristic properties of both the greatest and a minimal solution. Moreover, it must fulfill the conditions for a unique minimal solution as given above in Th. 12.25.

The interesting question then is for the characterization of when fuzzy relation equations have only one solution, and not how this solution looks like. Our presentation in this section follows the work of Cechlárová [6] and Gavalec [18]. Other approaches to the problem of uniquely solvable fuzzy relation equations over linear lattices can be found in e.g. [14, 29, 30, 36]. To simplify matters for the following considerations, let a matrix $A\in\mathbb{L}_{m\times n}$ and a vector $\boldsymbol{b}\in\mathbb{L}_{m\times 1}$ be fixed, and $\boldsymbol{x}\in\mathbb{L}_{n\times 1}$.

Lemma 12.26. ([18, Lemma 3.2]) *Set $M_j^> := \{i\in\underline{m} : a_{ij}>b_i\}$ and $\hat{x}_j := \bigwedge\{b_k : k\in M_j^>\}$, where $\hat{x}_j := 1$ if $M_j^>=\emptyset$. Let $\boldsymbol{x}\in\mathbb{S}$; then $\boldsymbol{x}\leq\hat{\boldsymbol{x}}$ and $\hat{\boldsymbol{x}}\in\mathbb{S}$.*

Proof. Let $i\in\underline{m}$ and $\boldsymbol{x}\in\mathbb{S}$. Then $\bigvee_{j\in\underline{n}}(a_{ij}\wedge x_j)=b_i$, so $a_{ij}\wedge x_j\leq b_i$ for all $j\in\underline{n}$ and, for all $i\in M_j^>$, $x_j\leq b_i$ follows. Thus, $x_j\leq\hat{x}_j$, and the inequality $\boldsymbol{x}\leq\hat{\boldsymbol{x}}$ is proved, and by monotonicity of matrix multiplication we get $\boldsymbol{b}=A\cdot\boldsymbol{x}\leq A\cdot\hat{\boldsymbol{x}}$. For the converse inequality note that

$$a_{ij}\wedge\hat{x}_j=\begin{cases} a_{ij}\wedge 1\leq b_i \text{ for all } i\in\underline{m}, & \text{if } M_j^>=\emptyset,\\ a_{ij}\wedge(\bigwedge\{b_k : a_{kj}>b_k\})\leq b_i, \text{ for all } k\in M_j^>, & \text{otherwise.}\end{cases}$$

Therefore, if $i\in M_j^>$ then clearly $a_{ij}\wedge\hat{x}_j\leq b_i$, and since $a_{ij}\leq b_i$ for all other $i\in\underline{m}$, we finally get $\bigvee_{j\in\underline{n}}(a_{ij}\wedge\hat{x}_j)\leq b_i$, i.e. $A\cdot\hat{\boldsymbol{x}}\leq\boldsymbol{b}$. □

Let $\hat{S}=\{\boldsymbol{x}\in\mathbb{L}_{n\times 1} : \boldsymbol{x}\leq\hat{\boldsymbol{x}}\}$ and $F_{ij}=\{\boldsymbol{x}\in\hat{S} : a_{ij}\wedge x_j=b_i\}$. If $\boldsymbol{x}\in F_{ij}$ then we say that x_j fulfills the ith equation in $A\cdot\boldsymbol{x}=\boldsymbol{b}$. This does, of course, not mean that $\boldsymbol{x}\in\mathbb{S}$.

Lemma 12.27. ([18, Lemmas 4.1, 4.2]) *Let $\boldsymbol{x}\in\hat{S}$.*

(1) For $i\in\underline{m}$, $j\in\underline{n}$, if $\boldsymbol{x}\in\hat{S}-F_{ij}$ then $a_{ij}\wedge x_j<b_i$.
(2) $\boldsymbol{x}\in\mathbb{S}$ iff there is a mapping $\varphi:\underline{m}\longrightarrow\underline{n}$ such that $\forall i\in\underline{m}[\boldsymbol{x}\in F_{i\varphi(i)}]$.

Proof. (1): For $\boldsymbol{x}\in\hat{S}-F_{ij}$ we have $a_{ij}\wedge x_j\leq b_i$, and since $\boldsymbol{x}\notin F_{ij}$, equality cannot hold.
(2): Assume that $\boldsymbol{x}\leq\hat{\boldsymbol{x}}$ is a solution of $A\cdot\boldsymbol{x}=\boldsymbol{b}$. Then $\bigvee_{j\in\underline{n}}(a_{ij}\wedge x_j)=b_i$, for all $i\in\underline{m}$. So, for all $i\in\underline{m}$, there exists an index j_i such that $a_{ij_i}\wedge x_{j_i}=b_i$. Now take $\varphi(i):=j_i$ to get $\boldsymbol{x}\in F_{i\varphi(i)}$.

Conversely, assume that there is a mapping $\varphi:\underline{m}\longrightarrow\underline{n}$ such that $\boldsymbol{x}\in F_{i\varphi(i)}$ for all $i\in\underline{m}$, i.e. $\boldsymbol{x}\in\hat{S}$ and $a_{i\varphi(i)}\wedge x_{\varphi(i)}=b_i$. From (1) we get $a_{ij}\wedge x_j\leq b_i$ for all $j\in\underline{n}$, whence $\bigvee_{j\in\underline{n}}(a_{ij}\wedge x_j)=b_i$ for all $i\in\underline{m}$, i.e. $\boldsymbol{x}\in\mathbb{S}$. □

We need some more notation for the characterization of unique solvability of fuzzy relation equations $A\cdot\boldsymbol{x}=\boldsymbol{b}$. For each $j\in\underline{n}$ we define

$$I_j := \{i \in \underline{m} : a_{ij} \geq b_i = \hat{x}_j\}, \qquad \mathcal{I} := \{I_j : j \in \underline{n}\}, \quad \text{and}$$
$$K_j := \{i \in \underline{m} : a_{ij} = b_i < \hat{x}_j\}, \qquad \mathcal{K} := \{K_j : j \in \underline{n}\}.$$

Lemma 12.28. ([18, Lemma 4.3]) *Let $i \in \underline{m}$, $j \in \underline{n}$. Then*

(1) $F_{ij} = \{\boldsymbol{x} \in \hat{S} : x_j = \hat{x}_j\}$, *for every* $i \in I_j$;
(2) $F_{ij} = \{\boldsymbol{x} \in \hat{S} : b_i \leq x_j \leq \hat{x}_j\}$, *for every* $i \in K_j$;
(3) $F_{ij} = \emptyset$, *for every* $i \in \underline{m} - (I_j \cup K_j)$.

Proof. (1): Let $i \in I_j$; then $a_{ij} \geq b_i = \hat{x}_j$. Since for any $\boldsymbol{x} \in \hat{S}$ we have $x_j \leq \hat{x}_j$, $a_{ij} \wedge x_j = b_i$ iff $x_j = b_i = \hat{x}_j$ follows.
(2): Let $i \in K_j$; then $a_{ij} = b_i < \hat{x}_j$. So, for any $\boldsymbol{x} \in \hat{S}$, $a_{ij} \wedge x_j = b_i$ iff $b_i \leq x_j \leq \hat{x}_j$.
(3): Let $i \in \underline{m} - (I_j \cup K_j)$. Thus, $i \notin I_j$ and $i \notin K_j$, which by basic propositional logic gives

$$\big((a_{ij} < b_i \text{ or } b_i \neq \hat{x}_j) \;\&\; (a_{ij} \neq b_i \text{ or } b_i \geq \hat{x}_j)\big).$$

If $a_{ij} < b_i$, then for any $\boldsymbol{x} \in \hat{S}$ we have $a_{ij} \wedge x_j < b_i$, and $F_{ij} = \emptyset$. For $a_{ij} \geq b_i$ it remains to consider

$$\big((b_i \neq \hat{x}_j) \;\&\; (a_{ij} \neq b_i \text{ or } b_i \geq \hat{x}_j)\big) \;\equiv\; \big((a_{ij} > b_i \neq \hat{x}_j) \;\text{ or }\; (b_i > \hat{x}_j)\big).$$

If $a_{ij} > b_i \neq \hat{x}_j$, then $i \in M_j^{>}$, and $b_i > \hat{x}_j$ by definition of $\hat{x}_j$. Conversely, $b_i > \hat{x}_j$ implies $b_i > \hat{x}_j \geq x_j \geq a_{ij} \wedge x_j$, for any $\boldsymbol{x} \in \hat{S}$. Therefore, $F_{ij} = \emptyset$. □

In [6], a necessary condition for unique solvability is given, using the notion of minimal covering: A covering $\mathscr{A}$ of a set A (see Def. 1.3) is called **minimal** if $\bigcup(\mathscr{A} - C) \neq A$ for each $C \in \mathscr{A}$. For the (much longer) proof of the second statement of the following lemma we refer to the original paper. Do also note that the underlying linear lattice is supposed to be bounded.

Lemma 12.29. ([18, Theorem 4.6]) *Let $A \in \mathbb{L}_{m\times n}$ and $\boldsymbol{b} \in \mathbb{L}_{m\times 1}$. Then*

(1) $|\mathbb{S}| \geq 1$ *iff* $\underline{m} = \bigcup(\mathcal{I} \cup \mathcal{K})$;
(2) $|\mathbb{S}| \leq 1$ *iff* $\underline{m} \neq \bigcup\big((\mathcal{I} - I_j) \cup \mathcal{K})\big)$.

Proof. (1): Let $|\mathbb{S}| \geq 1$ and let $\boldsymbol{x} \in \mathbb{S}$. Then, by Lemma 12.27(2), there is a mapping $\varphi : \underline{m} \longrightarrow \underline{n}$, such that $\boldsymbol{x} \in F_{i\varphi(i)}$ for every $i \in \underline{m}$. By Lemma 12.28(3) this implies $i \in I_{\varphi(i)} \cup K_{\varphi(i)}$. So we get $\underline{m} = \bigcup(\mathcal{I} \cup \mathcal{K})$.

Conversely, assume $\underline{m} = \bigcup(\mathcal{I} \cup \mathcal{K})$, i.e. $\forall i \in \underline{m} \exists j \in \underline{n} [a_{ij} \geq b_i = \hat{x}_j \text{ or } a_{ij} = b_i < \hat{x}_j]$. In both cases we have $a_{ij} \wedge \hat{x}_j = b_i$. So there is a mapping $\varphi : \underline{m} \longrightarrow \underline{n}$, such that $\boldsymbol{x} \in F_{i\varphi(i)}$ for every $i \in \underline{m}$. $F_\varphi := \bigcap\{F_{i\varphi(i)} : i \in \underline{m}\} \neq \emptyset$ sinc for the vector $\boldsymbol{x} \in \hat{S}$ as $x_j := \hat{x}_j$ we get $\boldsymbol{x} \in F_\varphi$ due to Lemma 12.28(1),(2), and by Lemma 12.27(2) $\boldsymbol{x} \in \mathbb{S}$. □

Theorem 12.30. ([18, Theorem 4.5]) *Let $A \in \mathbb{L}_{m\times n}$ and $\boldsymbol{b} \in \mathbb{L}_{m\times 1}$. The equation $A \cdot \boldsymbol{x} = \boldsymbol{b}$ has a unique solution $\boldsymbol{x} \in \mathbb{L}_{n\times 1}$ if and only if the system $\mathcal{I}$ is a minimal covering of $\underline{m} - \bigcup\mathcal{K}$.*

Proof. Follows from the assertions in Lemma 12.29. □

12.3.1.5 *Approximate Solutions*

It has already been mentioned at the end of Section 9.3.2 in connection with properties of fuzzy relations that sharp all-or-nothing concepts do not fit well to the proper spirit of fuzzy set theory, and we have also presented some ideas which have been proposed to overcome this situation. It has also been stated that, whatever method is applied, one always ends up using real numbers and/or crisp versions of the $<, \leq$ or $=$-relations, even in defining higher-level fuzzy concepts. In other words, the ordinary and well-known mathematical concepts are maintained as the meta-language used to *model* approximation, flexibility or tolerance etc..

In the context of fuzzy relation equations Gottwald proposes solvability degrees whose calculations are based on formulas in a many-valued logical calculus; see [19, 20, 22] and [21, Chapter 3]. In addition recall the references already mentioned in (R7) on page 351. Do also note the contributions on equations of interval-valued fuzzy relations in [32, 62, 69] and the group of papers [34, 63, 64].

As argued in [4, Chapter 6.2], considering approximate solutions of fuzzy relational equations is quite natural in the context of fuzzy relational modeling: Exact solutions may not exist at all, and even if they do, very good approximate solutions may be cheaper or easier to implement. Besides, disregarding good approximations would be somewhat artificial with regard to the fuzzy spirit, especially in situations where the equation results from an expert's description. So the following approach is suggested: Given relations R, S and T and an equation $R * S = T$ over the sets X, Y, and Z, we like to have a suitable measure of equality $\approx$ such that $(R * S \approx T) = 1$ iff $R * S = T$. According to Cor. 4.32(4) we have $x = y$ iff $x \leftrightharpoons y = 1$, so $R * S \approx T := \bigwedge_{\langle x,z\rangle \in X \times Z} \left(R * S(x,z) \leftrightharpoons T(x,z)\right)$ fulfills our needs.

Still another approach has been taken in [57–59] from the late eighties: Instead of searching for exact solutions of an equation $\boldsymbol{r}S = \boldsymbol{t}$ the authors consider tolerance regions such that $\boldsymbol{l} \subseteq \boldsymbol{r}S \subseteq \boldsymbol{u}$, where either $\boldsymbol{r}$ or S are unknown, and $\boldsymbol{l}$ and $\boldsymbol{u}$ are given, where $[\boldsymbol{l}, \boldsymbol{u}]$ give the lower and upper bound of tolerable outputs of the system S for input $\boldsymbol{r}$. If $\boldsymbol{r}$ is given, then we are asking for system configurations obeying the inequalities, and for given S we look for suitable inputs.

This approach is appealing since it allows the user to directly contribute information and his demands about acceptable regions for a solution. Moreover, in case of finite input and output spaces the structure of the solution set can be completely described. The limitation to finite basic spaces does not seem to be too restrictive, since practical problems can usually be modeled in this way.

The remainder of this section follows [58, Sections 2 and 4] which in turn are based on ideas put forward in [26, Section 3].

Lemma 12.31. ([26, Lemma 1], [5, p. 4, Theorem 3]) *Let $(P, \leq)$ be a finite poset. Then, for each $p \in P$ there is a minimal element $p' \in P$ such that $p' \leq p$. Equivalently: Any finite nonempty subset X of P has minimal elements.*

Proof. Let $X = \{x_1, x_2, \dots, x_n\}$. Define $m_1 := x_1$, and $m_k := x_k$ if $x_k < m_{k-1}$ and $m_k := x_{k-1}$ otherwise. Then m_n will be minimal. □

Henceforth the set of all the minimal elements of P will be denoted with P_0.

Lemma 12.32. ([26, Lemma 2]) *Let $(P, \leq)$ be a poset and $P' \subseteq P$. If for each $p \in P$ there is a $p' \in P'$ with $p' \leq p$ and $P_0 \subseteq P'$, then $P_0 = P'_0$.*

Proof. Let $p \in P_0 \subseteq P'$ and $p' \in P' \subseteq P$ such that $p' \leq p$. Then $p' \leq p \in P_0$ implies $p' = p$, whence $p \in P'_0$, i.e. $P_0 \subseteq P'_0$. Conversely, let $p \in P'_0 \subseteq P' \subseteq P$ and $p' \in P$ such that $p' \leq p$. Then, by the assumption, there is a $p'' \in P'$ with $p'' \leq p'$, and since $p \in P'_0$ and $p'' \leq p' \leq p$ we get $p'' = p$ which implies $p = p'$. Therefore, since we have taken $p' \in P$, $p \in P_0$, i.e. $P'_0 \subseteq P_0$. □

Before we can proceed we need to insert a few results about posets for later usage. We now consider a point-to-set mapping F with the following properties:

(F1) $F : P \longrightarrow \mathcal{P}(P)$
(F2) For $p \in P$ and $q \in F(p)$ we have $q \leq p$.
(F3) Let $p, p' \in P$ with $p \leq p'$; then $F(p) \subseteq F(p')$.
(F4) For all $p \in P$ the set $F(p)$ has minimal elements.

Proposition 12.33. ([58, Theorem 2.1]) *Let P be a poset and F be a mapping obeying (F1) - (F4). Then $p \in P_0$ iff $F(p) = \{p\}$.*

Proof. Let $p \in P_0$, then (F2) implies $q \leq p$ for any $q \in F(p)$. So $q = p$ and $F(p) = \{p\}$. Conversely, let $F(p) = \{p\}$, and $p' \leq p$. From (F3) we get $F(p') \subseteq F(p)$, i.e. $F(p') = \{p\}$, whence $q \in F(p')$ implies $q = p$. Since from (F2) we get $q \leq p' \leq p$, $p' = p$ follows, i.e. $p \in P_0$. □

Proposition 12.34. ([58, Theorem 2.2]) *Let P be a poset, $p \in P$ and F be a mapping obeying (F1) - (F4). Then there exists a $p' \in P_0$ such that $p' \leq p$.*

Proof. Let $p \in P$; then by (F4) there is a $p' \in F(p)_0$, and by (F2) we have $p' \leq p$. Assuming $p' \notin P_0$ Prop. 12.33 implies that there is a $q' \in F(p')$ with $q' \neq p'$, and by (F2) $q' < p'$. But from $q' \in F(p')$ and $p' \leq p$ by (F3) we get $q' < p' \leq p$. So $q' \in F(p)$, whence $p' \notin F(p)_0$, a contradiction. □

Proposition 12.35. ([58, Theorems 2.3, 2.4]) *Let $\hat{p}$ be the greatest element of a poset P; then the following holds:*

(1) $P_0 \subseteq F(\hat{p})$.
(2) $F(\hat{p})_0 = P_0$.

Proof. (1): Let $p \in P_0$; then $F(p) \subseteq F(\hat{p})$ by (F3), and $F(p) = \{p\}$ by Prop. 12.33. Therefore, $p \in F(\hat{p})$.

(2): Let $P' = F(\hat{p})$. By Prop. 12.34 there is a $\check{p} \in P_0$ with $\check{p} \le p$ for $p \in P$, and by (1) $\check{p} \in P'$. Thus $P_0 \subseteq P'$, and equality of the two sets follows from Lemma 12.32. □

The following considerations are less general than those in [58, Section 4] in that we restrict them to the sup-inf / max-min composition instead of any continuous t-norm h. This makes argumentation a little bit easier, but still conveys the idea lying behind: The definition $h^{-1}(a,b) = \{z \in [0,1] : h(a,z) = b\}$ on p. 96 reduces to $h^{-1}(a,b) = \{b\}$ if $a \ge b$ and $h^{-1}(a,b) = \emptyset$ otherwise, if we choose $h = \wedge$, whence (12.18) and the proof of Prop. 12.36 below become slightly simpler.

Let $X = \{x_1, x_2, \dots, x_m\}$ and $Y = \{y_1, y_2, \dots, y_n\}$ be two finite sets; let $\boldsymbol{x} \in \mathcal{F}(X)$, $\boldsymbol{l}, \boldsymbol{u} \in \mathcal{F}(Y)$, and $R \in \mathcal{F}(X \times Y)$. We will consider a system $\boldsymbol{l}_i \subseteq \boldsymbol{x}R_i \subseteq \boldsymbol{u}_i$, $i \in \underline{N}$, of fuzzy relation inequalities, with $\mathscr{V}$ denoting the set of its solutions. (Note that the special case $\boldsymbol{l}_i = \boldsymbol{u}_i$ gives a system of fuzzy relation equalities $\boldsymbol{x}R_i = \boldsymbol{y}_i$.) In order to make use of the results on posets mentioned above, we must construct a mapping from the poset $\mathscr{V}$ to its power set $\mathcal{P}(\mathscr{V})$ with the properties required.

To simplify notation of what we will need below we use $l^i_k = \boldsymbol{l}_i(y_k)$, $u^i_k = \boldsymbol{u}_i(y_k)$, and $r^i_{jk} = R_i(x_j, y_k)$, all three denoting elements in the real unit interval $[0,1]$. We define, for each $\boldsymbol{l}_i \subseteq \boldsymbol{v}R_i \subseteq \boldsymbol{u}_i$, $\boldsymbol{v} \in \mathscr{V}$:

$$I^i_k(\boldsymbol{v}) := \{j \in \underline{m} : v_j \wedge r^i_{jk} \ge l^i_k\} \quad \text{and} \quad I^i(\boldsymbol{v}) := \times_{k \in \underline{n}} I^i_k(\boldsymbol{v}), \quad i \in \underline{N}.$$

So $I^i_k(\boldsymbol{v})$ contains the information in which pairs (v_j, r^i_{jk}) both values are not less than the k-th component of the lower tolerance vector $\boldsymbol{l}_i$. Thus, $I^i(\boldsymbol{v})$ is the set of all n-tuples $f^i_v = (f^i_v(1), f^i_v(2), \dots, f^i_v(n)) = (j_1, j_2, \dots, j_n) \in \underline{m}^n$ such that $v_{j_k} \wedge r^i_{j_k k} \ge l^i_k$, for all $k \in \underline{n}$, i.e. it gives us all (the index n-tuples of) the solutions of the inequality $\boldsymbol{l}_i \subseteq \boldsymbol{x}R_i$.

We set $I(\boldsymbol{v}) := \times_{i \in \underline{N}} I^i(\boldsymbol{v})$. An element $f_v \in I(\boldsymbol{v})$ thus contains N n-tuples f^i_v, $i \in \underline{N}$, each of the latter representing a solution of inequality $\boldsymbol{l}_i \subseteq \boldsymbol{x}R_i$, whence each $f_v = (f^{i_1}_v, f^{i_2}_v, \dots, f^{i_N}_v) \in I(\boldsymbol{v})$ stands for a solution of the system of inequalities $\boldsymbol{l}_i \subseteq \boldsymbol{x}R_i$, $i \in N$. (Note that for $\boldsymbol{v} \in \mathscr{V}$ we have: for each $i \in \underline{N}$ and each $k \in \underline{n}$ there is a $j \in \underline{m}$ such that $v_j \wedge r^i_{jk} \ge l^i_k$, since by definition $\bigvee_{j \in \underline{m}} (v_j \wedge r^i_{jk}) \ge l^i_k$. So $I^i_k(\boldsymbol{v}) \ne \emptyset$ for all $k \in \underline{n}$, whence also $I^i(\boldsymbol{v}) \ne \emptyset$, for all $i \in \underline{N}$, and consequently $I(\boldsymbol{v}) \ne \emptyset$.)

Then, for each $f^i_v \in I^i(\boldsymbol{v})$, $\boldsymbol{v} \in \mathscr{V}$, define the index set

$$J_{f^i_v}(j) := \{k \in \underline{n} : f^i_v(k) = j\}, \text{ i.e. } (v_j \wedge r^i_{jk}) \ge l^i_k,$$

and let $h^i_v = (h^i_v(1), h^i_v(2), \dots, h^i_v(n))$ where

$$h^i_v(j) = \begin{cases} \bigvee\limits_{k \in J_{f^i_v}(j)} l^i_k & \text{if } J_{f^i_v}(j) \ne \emptyset \\ 0 & \text{if } J_{f^i_v}(j) = \emptyset \end{cases}, \text{ for } j \in \underline{m}. \tag{12.18}$$

Finally we define two mappings as follows:

$$\check{f} : \mathscr{V} \longrightarrow \mathcal{F}(X) \text{ by } [\check{f}(\boldsymbol{v})]_j = \bigvee_{i \in \underline{N}} h^i_v(j), \text{ and}$$

$$F : \mathscr{V} \longrightarrow \mathcal{P}(\mathscr{V}) \text{ by } F(\boldsymbol{v}) = \{\check{f}(\boldsymbol{v}) : f_v \in I(\boldsymbol{v})\}. \tag{12.19}$$

Describing the case $J_{f_v^i}(j) \neq \emptyset$ in detail we get:

$$[\check{f}(\boldsymbol{v})]_j = \bigvee_{i\in\underline{N}} h_v^i(j) = \bigvee_{i\in\underline{N}} \bigvee_{\substack{k\in\underline{n}:f_v^i(k)=j\ \&\\ (v_j\wedge r_{jk}^i)\geq l_k^i}} l_k^i.$$

So, fixing j and a solution $\boldsymbol{v}$, the j-th position in the vector $\check{f}(\boldsymbol{v})$ gets the maximal value l_k^i in *all* the lower bounds $\boldsymbol{l}_i$ such that $(v_j \wedge r_{jk}^i) \geq l_k^i$.

Proposition 12.36. ([58, Theorem 4.3(a)]) *Let $\boldsymbol{v}\in\mathcal{V}$; then $F(\boldsymbol{v})\subseteq\mathcal{V}$, i.e. $F:\mathcal{V}\longrightarrow\mathcal{P}(\mathcal{V})$.*

Proof. Assume that $\boldsymbol{l}_i \subseteq \check{f}(\boldsymbol{v})\cdot R_i$ does not hold for some $i\in\underline{N}$. Then there are $k'\in\underline{n}$ and $i'\in\underline{N}$ such that $[\check{f}(\boldsymbol{v})\cdot R_{i'}]_{k'} < l_{k'}^{i'}$, whence $v_j\wedge r_{jk'}^{i'} < l_{k'}^{i'}$, for all $j\in\underline{m}$. But $\boldsymbol{v}\in\mathcal{V}$ implies $v_{j'}\wedge r_{j'k'}^{i'} \geq l_{k'}^{i'}$, for some $j'\in\underline{m}$, which gives a contradiction. □

Proposition 12.37. ([58, Theorem 4.3(b)]) *Let $\boldsymbol{v}\in\mathcal{V}$ and $f_v\in I(\boldsymbol{v})$; then $\check{f}(\boldsymbol{v})\subseteq\boldsymbol{v}$.*

Proof. Let $\boldsymbol{v}\in\mathcal{V}$ and suppose $\check{f}(\boldsymbol{v})\not\subseteq\boldsymbol{v}$. Then there is a $j'\in\underline{n}$ such that $[\check{f}(\boldsymbol{v})]_{j'} > v_{j'}$, and thus $h_v^{i'}(j') > v_{j'}$ for some $i'\in\underline{n}$. Therefore, from $\boldsymbol{v}\in\mathcal{V}$ and (12.18) we get that there is a $k'\in\underline{n}$ such that $\bigvee_{j\in\underline{m}}(v_j\wedge r_{jk'}^{i'}) \geq l_{k'}^{i'} > v_{j'}$. By the first inequality we get $k'\in J_{f_v^{i'}}(j') := \{k\in\underline{n} : f_v^{i'}(k) = j'\ \&\ (v_{j'}\wedge r_{j'k}^{i'}) \geq l_k^{i'}\}$, and the second implies $(v_{j'}\wedge r_{j'k'}^{i'}) < l_{k'}^{i'}$ which is a contradiction to $k'\in J_{f_v^{i'}}(j')$. □

Proposition 12.38. ([58, Theorem 4.3(c)]) *Let $\boldsymbol{v},\boldsymbol{w}\in\mathcal{V}$ and $\boldsymbol{v}\subseteq\boldsymbol{w}$. If $\check{f}(\boldsymbol{v})\in F(\boldsymbol{v})$ then $\check{f}(\boldsymbol{w})\in F(\boldsymbol{w})$ and $\check{f}(\boldsymbol{v}) = \check{f}(\boldsymbol{w})$.*

Proof. Suppose $\boldsymbol{v},\boldsymbol{w}\in\mathcal{V}$, $\boldsymbol{v}\subseteq\boldsymbol{w}$, and $\check{f}(\boldsymbol{v})\in F(\boldsymbol{v}) = \{\check{f}(\boldsymbol{v}) : f_v\in I(\boldsymbol{v})\}$. Then, recalling the construction steps from above, we have

$$f_v\in I(\boldsymbol{v}) = \times_{i\in\underline{N}} I^i(\boldsymbol{v}) = \times_{i\in\underline{N}}\times_{k\in\underline{n}} I_k^i(\boldsymbol{v}) \quad\text{and}\quad I_k^i(\boldsymbol{v}) := \{j\in\underline{m} : v_j\wedge r_{jk}^i \geq l_k^i\}.$$

Then for some $f_v^i\in I^i(\boldsymbol{v})$, $i\in\underline{N}$, from $v_j\leq w_j$ for all $j\in\underline{m}$, we get $f_v^i\in I^i(\boldsymbol{w})$. Therefore, $h_v^i = h_w^i$ and $\check{f}(\boldsymbol{v}) = \check{f}(\boldsymbol{w})$. □

The last three propositions show that conditions (F1) - (F3) hold for the mapping F as constructed in (12.19). And since $F(p)$ is a finite set, it always has a minimal element due to Lemma 12.31, whence also (F4) holds. So, using Prop. 12.35 we can conclude

Theorem 12.39. ([58, Theorem 4.4]) *$\mathcal{V}_0 = (F(\hat{p}))_0$, i.e. $\mathcal{V}$ and $F(\hat{p})$ have the same minimal elements.*

12.3.2 *Solutions of FREs in Brouwerian Lattices*

As already mentioned at the begin of Section 12.3, complete Brouwerian lattices have been a main player from the beginning of the investigation of fuzzy relation equations with [50, 51] and are still in the focus of many research efforts.

To begin with, we would once more like to emphasize that Brouwerian lattices in the sense of [50] are called relatively pseudocomplemented lattices in Def. 4.20 and elsewhere. But to follow the common naming within the area of fuzzy relation equations, we adhere to the term *'Brouwerian'*. Furthermore, recall from Section 4.4 that every chain becomes a Brouwerian lattice by defining $a \curvearrowright b = 1$ whenever $a \le b$, and $a \curvearrowright b = b$ otherwise.

Moreover, recall from Lemma 12.9 that Brouwerian lattices are distributive (so the results obtained for distributive also hold for Brouwerian lattices), that a complete lattice is Brouwerian if and only if it is completely join-distributive, and that finite distributive lattices are Brouwerian. Every finite distributive lattice, complete chains, and especially the real unit interval are all examples of complete and completely distributive lattices where each element has an irredundant finite decomposition into join- and meet-irreducible elements.

Though the starting point and its basic results concerning fuzzy relation equations based on complete Brouwerian lattices have already been reported in Section 12.2.2 and Section 12.2.3, for sake of a more convenient look up we repeat a few things here in a form adapted to our current needs:

As usual, let $\mathbb{L}_{m\times n} := \{[a_{ij}] : a_{ij} \in L, i \in \underline{m}, j \in \underline{n}\}$ the set of $m \times n$ matrices over L. In the following we consider equations of Ru-type

$$A\boldsymbol{x} = \boldsymbol{b}, \quad \text{with} \quad \bigvee_{j\in\underline{n}}(a_{ij} \wedge x_j) = b_i, \forall i \in \underline{m}, \tag{12.20}$$

where $A \in \mathbb{L}_{m\times n}$, $\boldsymbol{x} \in \mathbb{L}_{n\times 1}$, and $\boldsymbol{b} \in \mathbb{L}_{m\times 1}$, and will again use $\mathbb{S} := \mathbb{S}(A, \boldsymbol{b})$ to denote their solution sets. Recall that the @-composition of A and $\boldsymbol{b}$ is defined as $(A^{\smile} @\boldsymbol{b})_j = \bigwedge_{i\in\underline{m}}(a_{ij} \curvearrowright b_i)$. Then in analogy to Th. 12.3(1) we get (see Table 12.1)

Corollary 12.40. ([25, Theorem 2.1]) *Let $A \in \mathbb{L}_{m\times n}$, $\boldsymbol{x} \in \mathbb{L}_{n\times 1}$, and $\boldsymbol{b} \in \mathbb{L}_{m\times 1}$; then $\mathbb{S} \neq \emptyset$ iff $\hat{\boldsymbol{x}} := A^{\smile} @\boldsymbol{b} \in \mathbb{S}$, and $\hat{\boldsymbol{x}}$ is the greatest element in $\mathbb{S}$.*

Proof. Let $\boldsymbol{x} \in \mathbb{S}$. For all $i \in \underline{m}, j \in \underline{n}$ we have $a_{ij} \wedge x_j \le b_i$, so $x_j \le a_{ij} \curvearrowright b_i$. Therefore, $x_j \le \bigwedge_{i\in\underline{m}}(a_{ij} \curvearrowright b_i) =: \hat{x}_j$, so $\boldsymbol{x} \le \hat{\boldsymbol{x}}$ for all $\boldsymbol{x} \in \mathbb{S}$. It remains to show that $\hat{\boldsymbol{x}}$ itself is a solution. We have $\bigvee_{j\in\underline{n}}(a_{ij} \wedge \hat{x}_j) \ge \bigvee_{j\in\underline{n}}(a_{ij} \wedge x_j) = b_i$. On the other hand, $a_{ij} \wedge \hat{x}_j = a_{ij} \wedge \bigwedge_{k\in\underline{m}}(a_{kj} \curvearrowright b_k) \le a_{ij} \wedge (a_{ij} \curvearrowright b_i) \le b_i$, whence $\bigvee_{j\in\underline{n}}(a_{ij} \wedge \hat{x}_j) \le b_i$. Thus, $A\hat{\boldsymbol{x}} = \boldsymbol{b}$. □

Let L be an arbitrary Brouwerian lattice, and let $[a]_b$ denote the solution set of $a \wedge x = b$, i.e. $[a]_b = \{c \in L : a \wedge c = b\}$.

Lemma 12.41. ([25, Lemma 3.1]) *Over a Brouwerian lattice L equation $a \wedge x = b$ is solvable iff $a \ge b$ and $[a]_b = [b, a \curvearrowright b]$.*

Proof. $a \ge b$ implies $b \in [a]_b$. Now suppose that $a \wedge x = b$ is solvable and $x \in [a]_b$; then $b = a \wedge x \le a$. Furthermore, $b = a \wedge x \le x \le a \curvearrowright b$. So for an element $y \in [b, a \curvearrowright b]$, i.e. $b \le y \le a \curvearrowright b$, using monotonicity we get $b = a \wedge b \le a \wedge y \le a \wedge (a \curvearrowright b) \le b$. Therefore, $b = a \wedge y$, whence $[a]_b = [b, a \curvearrowright b]$. □

Proposition 12.42. ([25, Theorem 3.1]) *For the system of equations $a_i \wedge x = b_i$, $i \in \underline{m}$ and $m \geq 2$, the following conditions are equivalent:*

(1) *The system of equations $a_i \wedge x = b_i$ is solvable;*
(2) *$a_i \geq b_i$ and $a_j \wedge b_i = b_j \wedge b_i$, for all $i, j \in \underline{m}$;*
(3) *$\bigvee_{i \in \underline{m}} b_i$ is a solution;*
(4) *$\bigwedge_{i \in \underline{m}} (a_i \curvearrowright b_i)$ is a solution.*

The solution set is $\left[\bigvee_{i \in \underline{m}} b_i, \bigwedge_{i \in \underline{m}} (a_i \curvearrowright b_i)\right]$.

Proof. (1) $\leftrightarrow$ (4): Follows from Cor. 12.40.
(1) $\rightarrow$ (2): For a solution x one has $x \geq b_i$ and $a_i \geq b_i$, for all $i \in \underline{m}$. Thus $a_j \wedge b_i = a_j \wedge (x \wedge b_i) = (a_j \wedge x) \wedge b_i = b_j \wedge b_i$ for any $j \in \underline{m}$.
(2) $\rightarrow$ (3): Since $a_j \wedge b_i = b_j \wedge b_i \leq b_j$, for any $i, j \in \underline{m}$, we get $\bigvee_{i \in \underline{m}} (a_j \wedge b_i) = a_j \wedge \left(\bigvee_{i \in \underline{m}} b_i\right) \leq b_j$. And, from $a_j \geq b_j$, $a_j \wedge \left(\bigvee_{i \in \underline{m}} b_i\right) \geq b_j \wedge \left(\bigvee_{i \in \underline{m}} b_i\right) = b_j$ follows.
(3) $\rightarrow$ (1): Obvious.
Finally, by Lemma 12.41, the solution set of the system $a_i \wedge x = b_i$ is $\bigcap_{i \in \underline{m}} [b_i, a_i \curvearrowright b_i] = \left[\bigvee_{i \in \underline{m}} b_i, \bigwedge_{i \in \underline{m}} (a_i \curvearrowright b_i)\right]$. □

Note that conditions (1) - (3) are already equivalent when the system is assigned on a distributive lattice. A further characterization of solvability is given in

Corollary 12.43. ([55, Corollary 2.7]) *The system of equations $a_i \wedge x = b_i$, $i \in \underline{m}$, over L is solvable iff $b_i \leq a_i$ for all $i \in \underline{m}$ and $\bigvee_{i \in \underline{m}} b_i \leq \bigwedge_{i \in \underline{m}} (a_i \curvearrowright b_i)$.*

Proof. If system of equations $a_i \wedge x = b_i$ is solvable the assertion is immediate from Prop. 12.42. Conversely, assume that $b_i \leq a_i$ for all $i \in \underline{m}$ and $\bigvee_{k \in \underline{m}} b_i \leq \bigwedge_{k \in \underline{m}} (a_k \curvearrowright b_k)$. Then

$$a_i \wedge \bigvee_{k \in \underline{m}} b_k = (a_i \wedge b_i) \vee \left(a_i \wedge \bigvee_{k \in \underline{m} - \{i\}} b_k\right) = b_i \vee \left(a_i \wedge \bigvee_{k \in \underline{m} - \{i\}} b_k\right) \geq b_i,$$

and, by Prop. 4.27(2) and remembering that a Brouwerian is a residuated lattice with $a \otimes b := a \wedge b$,

$$a_i \wedge \bigwedge_{k \in \underline{m}} (a_k \curvearrowright b_k) = a_i \wedge (a_i \curvearrowright b_i) \wedge \bigwedge_{k \in \underline{m} - \{i\}} (a_k \curvearrowright b_k) \leq b_i \wedge \bigwedge_{k \in \underline{m} - \{i\}} (a_k \curvearrowright b_k) \leq b_i.$$

So, from the assumption $\bigvee_{i \in \underline{m}} b_i \leq \bigwedge_{i \in \underline{m}} (a_i \curvearrowright b_i)$ and the two inequalities above we get

$$b_i \leq a_i \wedge \bigvee_{k \in \underline{m}} b_k \leq a_i \wedge \bigwedge_{k \in \underline{m}} (a_k \curvearrowright b_k) \leq b_i,$$

for all $i \in \underline{m}$, i.e. $a_i \wedge \bigvee_{k \in \underline{m}} b_k = a_i \wedge \bigwedge_{k \in \underline{m}} (a_k \curvearrowright b_k) = b_i$ holds for all $i \in \underline{m}$, which means that the system is solvable. □

12.3.2.1 *The Pre-Solution Matrix*

Another characterization for the solvability (i.e. not using the greatest solution) has been given in [16, Theorem 2.7] (see also [13, Theorem 3]). We take a closer look at this approach, based on recent investigations in [55].

Setting $\mathscr{Q}_j := \bigcap_{i\in I}[a_{ij}]_{q_{ij}}$, for any $j \in J$, we define

$$\mathscr{Q} := \{Q \in \mathbb{L}_{I\times J} : \bigvee_{j\in J} q_{ij} = b_i, \forall i \in I \quad \& \quad \mathscr{Q}_j \neq \emptyset, \forall j \in J\}.$$

$Q \in \mathscr{Q}$ is called a **pre-solution matrix** in [55]. Clearly, an equation $a \wedge x = b$ is solvable in a lattice L iff $a \geq b$. Therefore, since $[a_{ij}]_{q_{ij}} = \{c \in L : a_{ij} \wedge c = q_{ij}\}$, we need to consider relations $Q \subseteq A$ only.

Theorem 12.44. (adapted from [13, Theorem 3]) *Let $\mathbb{S}$ be the solution set of $A\boldsymbol{x} = \boldsymbol{b}$ and $\mathscr{Q}$ as defined above. Then $\mathbb{S} \neq \emptyset$ iff $\mathscr{Q} \neq \emptyset$.*

Proof. Let $\mathbb{S} \neq \emptyset$ and $\boldsymbol{c} \in \mathbb{S}$. Then $\bigvee_{j\in J}(a_{ij} \wedge c_j) = b_i$, for all $i \in I$. We define relations $Q_i = [q_{ij}]$, where $q_{ij} = a_{ij} \wedge c_j \leq a_{ij}$, for every $i \in I, j \in J$; so clearly $\bigvee_{j\in J} q_{ij} = b_i$, for all $i \in I$. Furthermore, we get $c_j \in [a_{ij}]_{q_{ij}}$ for all $i \in I$, whence $\mathscr{Q}_j \neq \emptyset$. Therefore, $Q_i \in \mathscr{Q}$.

Conversely, let $Q_i = [q_{ij}] \in \mathscr{Q}$; then $Q_i \subseteq A$, $\bigvee_{j\in J} q_{ij} = b_i$ and $\mathscr{Q}_j \neq \emptyset$, for all $i \in I, j \in J$. For an element $c_j \in \mathscr{Q}_j$ we have $a_{ij} \wedge c_j = q_{ij}$, and therefore $\bigvee_{j\in J}(a_{ij} \wedge c_j) = \bigvee_{j\in J} q_{ij} = b_i$, for each $i \in I$. Therefore, $\boldsymbol{c}$ is a solution of $A\boldsymbol{x} = \boldsymbol{b}$, i.e. $\boldsymbol{c} \in \mathbb{S}$. □

Before we turn to solutions of $A\boldsymbol{x} = \boldsymbol{b}$ themselves we have a look at some of their relationships to pre-solution matrices following [55, Section 3].

Proposition 12.45. ([55, Theorem 3.1]) *Let $\boldsymbol{x} \in \mathbb{S}$. Then there is a $Q \in \mathscr{Q}$ with $q_{ij} = a_{ij} \wedge x_j$, for all $i \in I, j \in J$.*

Proof. Define $Q \in \mathbb{L}_{I\times J}$ by $q_{ij} = a_{ij} \wedge x_j$ for all $i \in I, j \in J$. The assertion follows by arguments analogous to the first half of the proof of Th. 12.44. □

The above proposition shows how one can get a relation $Q \in \mathscr{Q}$ from a solution $\boldsymbol{x}$ of $A\boldsymbol{x} = \boldsymbol{b}$. Conversely, one can get a set of solutions (i.e. possibly more than one) from any $Q \in \mathscr{Q}$, as shown below. We let vectors $\boldsymbol{l}^Q, \boldsymbol{u}^Q \in \mathbb{L}_{J\times 1}$ be defined as

$$\boldsymbol{l}^Q_j = \bigvee_{i\in I} q_{ij} \quad \text{and} \quad \boldsymbol{u}^Q_j = \bigwedge_{i\in I}(a_{ij} \curvearrowright q_{ij}), \text{ for all } j \in J,$$

and for $\boldsymbol{x} \in \mathbb{L}_{I\times 1}$ and $A \in \mathbb{L}_{I\times J}$ we define $A^{\boldsymbol{x}} = [a^{\boldsymbol{x}}_{ij}] \in \mathbb{L}_{I\times J}$ by $a^{\boldsymbol{x}}_{ij} = a_{ij} \wedge x_j, \forall i \in I, j \in J$.

Proposition 12.46. ([55, Theorem 3.2]) *Let $\mathbb{S}$ be the solution set of $A\boldsymbol{x} = \boldsymbol{b}$. For each $Q \in \mathscr{Q}$ we have $S_Q := [\boldsymbol{l}^Q, \boldsymbol{u}^Q] \subseteq \mathbb{S}$. Moreover, $A^{\boldsymbol{x}} = Q$ for any $\boldsymbol{x} \in S_Q$.*

Proof. Suppose $Q \in \mathscr{Q}$; then from Cor. 12.43 we get $\boldsymbol{l}^Q \leq \boldsymbol{u}^Q$, and for any $\boldsymbol{x} \in S_Q$ we have $\bigvee_{i\in I} q_{ij} \leq x_j \leq \bigwedge_{i\in I}(a_{ij} \curvearrowright q_{ij})$, for all $j \in J$. Using Prop. 12.42(3) and (4) we get $a_{ij} \wedge x_j \geq a_{ij} \wedge \bigvee_{i\in I} q_{ij} = q_{ij}$, and $a_{ij} \wedge x_j \leq a_{ij} \wedge \bigwedge_{i\in I}(a_{ij} \curvearrowright q_{ij}) = q_{ij}$. Thus $a_{ij} \wedge x_j = q_{ij}$, for all $i \in I, j \in J$. This means $\bigvee_{j\in J}(a_{ij} \wedge x_j) = \bigvee_{j\in J} q_{ij} = b_i$. for all $i \in I$, i.e. $\boldsymbol{x} \in \mathbb{S}$. Thus $S_Q \subseteq \mathbb{S}$. Finally, for $\boldsymbol{x} \in S_Q$ we get $A^{\boldsymbol{x}} = [a^{\boldsymbol{x}}_{ij}] = [a_{ij} \wedge x_j] = [q_{ij}]$. □

The last proposition shows that a pre-solution matrix Q can induce a whole set of solutions S_Q with $\boldsymbol{l}^Q$ as least and $\boldsymbol{u}^Q$ as greatest element. It has also been shown that, for any $\boldsymbol{x} \in S_Q$, the pre-solution matrix $A^{\boldsymbol{x}}$ is Q, i.e. different solutions may induce the same pre-solution matrix. A natural question then is if different pre-solution matrices can give the same solution, answered in

Proposition 12.47. ([55, Theorem 3.3]) *Let $Q, Q' \in \mathcal{Q}$ and $Q \neq Q'$. Then $S_Q \cap S_{Q'} = \emptyset$.*

Proof. Assume $\boldsymbol{x} \in S_Q \cap S_{Q'}$ for some different $Q, Q' \in \mathcal{Q}$. From Prop. 12.46 we get $Q = A^{\boldsymbol{x}} = Q'$, a contradiction. □

The next proposition demonstrates that when a pre-solution matrix Q is induced from the solution $\boldsymbol{x}$, then $\boldsymbol{x}$ belongs to the set of solutions S_Q induced from Q:

Proposition 12.48. ([55, Theorem 3.4]) *Let $\boldsymbol{x} \in \mathbb{S}$ and $Q = A^{\boldsymbol{x}}$. Then $Q \in \mathcal{Q}$ and $\boldsymbol{x} \in S_Q$.*

Proof. Suppose $\boldsymbol{x} \in \mathbb{S}$ and $Q = A^{\boldsymbol{x}}$. Prop. 12.45 and Prop. 12.46 imply $Q \in \mathcal{Q}$ and $S_Q \subseteq \mathbb{S}$. Moreover, for all $j \in J$, $\boldsymbol{l}_j^Q = \bigvee_{j \in J}(a_{ij} \wedge x_j) \leq x_j \leq \bigwedge_{i \in I}(a_{ij} \curvearrowright (a_{ij} \wedge x_j)) = \bigwedge_{i \in I}(a_{ij} \curvearrowright q_{ij}) = \boldsymbol{u}_j^Q$, applying Cor. 4.24(2) for the second inequality. Therefore, $\boldsymbol{x} \in S_Q$. □

The results in Prop. 12.45 and Prop. 12.46 allow us to characterize $\mathcal{Q}$ and $\mathbb{S}$ from each other as follows:

$$\mathcal{Q} = \{A^{\boldsymbol{x}} : \boldsymbol{x} \in \mathbb{S}\} \quad \text{and} \quad \mathbb{S} = \bigcup_{Q \in \mathcal{Q}} S_Q. \tag{12.21}$$

This relationship suggests that determining $\mathcal{Q}$ could be an alternative and maybe more efficient way to solve a system of equations $A\boldsymbol{x} = \boldsymbol{b}$ since a pre-solution matrix Q can induce a whole set of solutions S_Q as shown in Prop. 12.46.

We conclude this section with a few results on properties of pre-solution matrices, omitting their proofs (compare with Prop. 12.6 and Cor. 12.40):

Proposition 12.49. ([55, Section 4])

(1) *Let $\hat{Q} \in \mathbb{L}_{I \times J}$ by $\hat{q}_{ij} = a_{ij} \wedge \bigvee_{k \in I}(a_{kj} \curvearrowright b_k)$, for all $i \in I, j \in J$. Then $\mathcal{Q} \neq \emptyset$ iff $\hat{Q} \in \mathcal{Q}$. Furthermore, $\hat{Q}$ is the greatest element in $\mathcal{Q}$.*

(2) *If $\hat{Q}$ is the greatest element in $\mathcal{Q}$, then $\boldsymbol{u}^{\hat{Q}}$ is the greatest solution $\hat{\boldsymbol{x}}$.*

(3) *For any $Q, Q' \in \mathcal{Q}$, $Q \vee Q' \in \mathcal{Q}$.*

(4) *$\mathcal{Q}$ is not convex, i.e. for some $Q, Q' \in \mathcal{Q}$ with $Q' \subseteq Q$, there is a Q'' with $Q' \subseteq Q'' \subseteq Q$ and $Q'' \notin \mathcal{Q}$.*

12.3.2.2 *Minimal Solutions*

Recall that by Prop. 12.6 solutions sets are not only upper semilattices but also convex sets. Thus, as already noted, minimal solutions play a crucial role in describing all solutions of relation equations since $S_\rho := \mathbb{S}(\sigma, \tau) = \bigcup_{\check{\rho} \in \check{S}_\rho} [\check{\rho}, \hat{\rho}]$ with $\check{S}_\rho$ denoting the set of minimal elements in $S_\rho \neq \emptyset$. However, despite their importance for determining the solution set of fuzzy relational equations, we do not generally know if S_ρ has or does not have minimal elements. (That minimal solutions do not always exist has been shown in e.g. [54, Example 3.1] or [59, Examples 2.1, 2.2].)

The basic investigations on minimal (lower) solutions of fuzzy relation equations over complete Brouwerian lattices have been contributed more than three decades ago in [13] and [16, Section 3.5]. In the following we use these concepts and ideas presented in a form adapted later on to the Ru-type of equations by various authors; see e.g. [54, 55, 67] or [25].

Let $\mathbb{S} \neq \emptyset$. For any $\boldsymbol{x} \in \mathbb{S}$ and any $j' \in J$, let

$$\gamma_{j'}(\boldsymbol{x}) = \{l \in L : (a_{ij'} \wedge l) \vee [\bigvee_{j \in J - \{j'\}} (a_{ij} \wedge x_j)] = b_i, \forall i \in I\}. \tag{12.22}$$

Then the minimal solutions can be characterized as follows:

Theorem 12.50. (adapted from [13, Theorem 5] , [16, Theorem 3.21]) *$\boldsymbol{x} \in \mathbb{S}$ is a minimal solution iff $x_j = \min \gamma_j(\boldsymbol{x})$ for every $j \in J$.*

Proof. Let $\boldsymbol{x} \in \mathbb{S}$ be a minimal solution. From $\bigvee_{j \in J}(a_{ij} \wedge x_j) = b_i, \forall i \in I$, we get $\gamma_j(\boldsymbol{x}) \neq \emptyset$ because $x_j \in \gamma_j(\boldsymbol{x})$, for every $j \in J$. Furthermore, for every $l \in \gamma_j(\boldsymbol{x})$ we have $(x_j \wedge l) \in \gamma_j(\boldsymbol{x})$:

$$\begin{aligned}
&(a_{im} \wedge (x_j \wedge l)) \vee [\bigvee_{j \in J-\{m\}}(a_{ij} \wedge x_j)] \\
&\quad = [(a_{im} \wedge l) \wedge (a_{im} \wedge x_j)] \vee [\bigvee_{j \in J-\{m\}}(a_{ij} \wedge x_j)] \\
&\quad = [(a_{im} \wedge l) \vee (\bigvee_{j \in J-\{m\}}(a_{ij} \wedge x_j))] \wedge [(a_{im} \wedge x_j) \vee (\bigvee_{j \in J-\{m\}}(a_{ij} \wedge x_j))] \\
&\quad = b_i \wedge b_i = b_i.
\end{aligned}$$

Then the fuzzy vector $\boldsymbol{x}'$ defined as

$$x'_j = \begin{cases} x_j & j \in J - \{m\} \\ x_j \wedge l & j = m \end{cases}$$

is a solution such that $\boldsymbol{x}' < \boldsymbol{x}$. But the minimality of $\boldsymbol{x}$ implies $x_j \wedge l = x_j$, so $x_j \leq l$ for every $l \in \gamma_j(\boldsymbol{x})$. Thus $x_j = \min \gamma_j(\boldsymbol{x})$.

Conversely, assume that $x_j = \min \gamma_j(\boldsymbol{x})$, for every $j \in J$, and let $\boldsymbol{x}' \in \mathbb{S}$ such that $\boldsymbol{x}' \leq \boldsymbol{x}$. Then for the fuzzy vector $\boldsymbol{s}$ defined as

$$s_j = \begin{cases} x_j & j \in J - \{m\} \\ x'_j & j = m \end{cases}$$

we get $\boldsymbol{x}' \leq \boldsymbol{s} \leq \boldsymbol{x}$, so $\boldsymbol{s} \in \mathbb{S}$. Thus $(a_{im} \wedge s_m) \vee [\bigvee_{j \in J-\{m\}}(a_{ij} \wedge x_j)] = b_i$, i.e. $s_m = x'_m \in \gamma_j(\boldsymbol{x})$. But by assumption $x_m = \min \gamma_m(\boldsymbol{x})$, so $x_m \leq x'_m$. From the hypothesis $\boldsymbol{x}' \leq \boldsymbol{x}$, and since m is arbitrary, $\boldsymbol{x}' = \boldsymbol{x}$ follows. □

As shown in e.g. [54, Example 3.1] or [67, Example 3.1], minimal solutions do not always exist in case of I a finite and J an infinite set. Under certain circumstances, however, it is possible to find conditions for their existence. It is beyond the scope of this text to include details in addition to Th. 12.53 below, whence we refer the interested reader to [54, Theorem 4.3] and [67, Section 3] for more results.

From the considerations above we see that there are two problems concerning the kinds of solutions of a fuzzy relation equation: (P1) Give conditions which characterize the minimal / least / unique solution(s), i.e. provide the means that allow to check a known solution for these properties; and (P2) give conditions which characterize the existence of each of these kinds of solutions, i.e. provide the means that allow to check for a given system of fuzzy relation equations if a solution of a certain kind does exist at all. With Th. 12.50 we have an answer to (P1) concerning minimal solutions, so we turn to (P2) as the last subject of this section.

A join decomposition (see Def. 12.7) $0 \neq p = \bigvee_{k\in K} q_k$, where $q_k \neq 0$ for each $k \in K$, in a complete lattice L is called **minimal** if $p \neq \left(\bigvee_{k\in K-\{k'\}} q_k\right) \vee t_{k'}$ for each $k' \in K$ and for any $t_{k'} < q_{k'}$. It is obvious that minimal join decompositions are irredundant.

Proposition 12.51. ([56, Theorem 3.4]) *Let $\bigvee_{k\in K} q_k$ be a minimal join decomposition of p in a complete Brouwerian lattice, and $\{K_t : t \in T\}$ be a partition of K. Then $\bigvee_{t\in T}\left(\bigvee_{k\in K_t} q_k\right)$ is a minimal join decomposition of p.*

Proof. If $\bigvee_{t\in T}\left(\bigvee_{k\in K_t} q_k\right)$ is not a minimal join decomposition of p, then there exists some

$$c < \bigvee_{k\in K_{t'}} q_k \text{ for some } t' \in T \tag{12.23}$$

such that $\bigvee_{t\in T-\{t'\}}\left(\bigvee_{k\in K_t} q_k\right) \vee c = p$. Thus, using the fact that Brouwerian latices are completely join distributive, we conclude as follows:

$$\begin{aligned} p &= \left[\bigvee_{t\in T-\{t'\}}\left(\bigvee_{k\in K_t} q_k\right)\right] \vee c \\ &= \left[\bigvee_{t\in T-\{t'\}}\left(\bigvee_{k\in K_t} q_k\right)\right] \vee \left[c \wedge \left(\bigvee_{k\in K_{t'}} q_k\right)\right] \\ &= \left[\bigvee_{t\in T-\{t'\}}\left(\bigvee_{k\in K_t} q_k\right)\right] \vee \left[\bigvee_{k\in K_{t'}} (c \wedge q_k)\right] \\ &= \left[\bigvee_{k\in K-K_{t'}} q_k\right] \vee \left[\bigvee_{k\in K_{t'}} (c \wedge q_k)\right]. \end{aligned}$$

Clearly, $c \wedge q_k \leq q_k$ for all $k \in K_{t'}$, Indeed we have $c \wedge q_k = q_k$ for all $k \in K_{t'}$, since otherwise there is some $k' \in K_{t'}$ such that $c \wedge q_{k'} < q_{k'}$; but then $\left(\bigvee_{k\in K-\{k'\}} q_k\right) \vee (c \wedge q_{k'}) = p$, contrary to the hypothesis that $\bigvee_{k\in K} q_k$ is a minimal join decomposition of p. So $c \wedge q_k = q_k$, i.e. $q_k \leq c$, for all $k \in K_{t'}$, whence $\bigvee_{k\in K_{t'}} q_k \leq c$, a contradiction to (12.23). Therefore, we have shown that $\bigvee_{t\in T}\left(\bigvee_{k\in K_t} q_k\right)$ is a minimal join decomposition of p. □

For the results below the following context is used: Let L be a complete Brouwerian lattice, $\boldsymbol{a}$ a row, and $\boldsymbol{x}$ a column vector over L. We investigate equations of the form $\boldsymbol{ax} = b$, $b \in L$, equivalently written as $\bigvee_{j\in J}(a_j \wedge x_j) = b$. The solution set is denoted by $\mathbb{S}$, and the set of minimal solutions by $\mathbb{S}_0$, and $N_{\boldsymbol{x}} = \{j \in J : x_j \neq 0\}$ for any $\boldsymbol{x}$. Note that $b = 0$ implies that $\mathbf{0}$ is the least solution; so we assume $b \neq 0$.

Proposition 12.52. ([56, Theorem 4.1]) *If $\boldsymbol{x} \in \mathbb{S}_0$ then $\bigvee_{j\in N_{\boldsymbol{x}}} x_j$ is a minimal join decomposition of b.*

Proof. From $b = \bigvee_{j\in J}(a_j \wedge x_j) = \bigvee_{j\in J}[a_j \wedge (a_j \wedge x_j)]$ we see that $\boldsymbol{x}' \in \mathbb{S}$, where $x'_j = a_j \wedge x_j$. Since $\boldsymbol{x}$ is minimal, $a_j \wedge x_j = x_j$ follows for any $j \in J$. Therefore, $\bigvee_{j\in N_{\boldsymbol{x}}} x_j = b$, i.e. $\bigvee_{j\in N_{\boldsymbol{x}}} x_j$ is a minimal join decomposition of b. □

Theorem 12.53. ([56, Theorem 4.2]) *$\mathbb{S}_0 \neq \emptyset$ if and only if b has a minimal join decomposition $b = \bigvee_{k\in K} p_k$ and for each p_k, $k \in K$, there is an index $j \in J$ such that $p_k \leq a_j$. Moreover, for any partition $\{K_t : t \in T\}$ of K such that $F_{\{K_t:t\in T\}}$, denoting set of one to one mappings f from T onto some subset $J' \subseteq J$ with $\bigvee_{k\in K} p_k \leq a_{f(t)}$ for any $t \in T$, is nonempty, the (row) vector $\boldsymbol{x}$ with*

$$x_j = \bigvee_{k\in K_{f^{-1}(j)}} p_k,\ j \in \{f(t) : t \in T\} \quad \textit{and} \quad x_j = 0, \textit{otherwise},$$

for any $f \in F_{\{K_t:t\in T\}}$, is a minimal solution of $\boldsymbol{ax} = b$.

Proof. If $\boldsymbol{x} \in \mathbb{S}_0$, then $\bigvee_{j \in N_{\boldsymbol{x}}} x_j$ is a minimal join decomposition of b and $x_j \leq a_j$ for any $j \in N_{\boldsymbol{x}}$ by Prop. 12.52.

Conversely, if b has a minimal join decomposition $\bigvee_{k \in K} p_k$ and for each p_k, $k \in K$, there is an index $j \in J$ such that $p_k \leq a_j$, then there exists a partition $\{K_t : t \in T\}$ of K such that $F_{\{K_t:t\in T\}} \neq \emptyset$. Since for any $f \in F_{\{K_t:t\in T\}}$ it is obvious that $x_j = \bigvee_{k \in K_{f^{-1}(j)}} p_k \leq a_{f(f^{-1}(j))} = a_j$, for any $j \in \{f(t) : t \in T\}$, we get

$$\bigvee_{j\in J}(a_j \wedge x_j) = \bigvee_{j\in J} x_j = \bigvee_{j\in\{f(t):t\in T\}} \Big(\bigvee_{k\in K_{f^{-1}(j)}} p_k\Big) = \bigvee_{t\in T}\Big(\bigvee_{k\in K_t} p_k\Big) = \bigvee_{k\in K} p_k = b.$$

Therefore, $\boldsymbol{x} \in \mathbb{S}$. By assumption, $\bigvee_{k \in K} p_k$ is a minimal join decomposition of b, so by Prop. 12.51, $\bigvee_{j\in\{f(t):t\in T\}}\left(\bigvee_{k\in K_{f^{-1}(j)}} p_k\right) = \bigvee_{j\in N_{\boldsymbol{x}}} x_j$ is a minimal join decomposition of b, whence $\boldsymbol{x} \in \mathbb{S}_0$. □

Note that Th. 12.53 suggests a way to construct a minimal solution from a minimal join decomposition of b, and every minimal solution can be obtained in this way.

12.3.2.3 *Least and Unique Solution*

Proposition 12.54. ([67, Theorem 3.2]) *If* $\boldsymbol{x} \in \mathbb{S}$, *then* $x_j \in \gamma_j(\boldsymbol{x}')$ *for any* $\boldsymbol{x}' \in \mathbb{S}$ *with* $\boldsymbol{x} \leq \boldsymbol{x}'$ (see (12.22)).

Proof. For any $\boldsymbol{x} \in \mathbb{S}$, and $j' \in J$, we take $\overline{\boldsymbol{x}}$ as $\overline{x}_j = x_{j'}$ if $j = j'$, and $\overline{x}_j = x'_j$, otherwise. Then $\boldsymbol{x} \leq \overline{\boldsymbol{x}} \leq \boldsymbol{x}'$ and $\overline{\boldsymbol{x}} \in \mathbb{S}$. Therefore, $(a_{ij'} \wedge x_{j'}) \vee \left[\bigvee_{j\in J-\{j'\}}(a_{ij} \wedge x'_j)\right] = b_i$, i.e. $j' \in \gamma_j(\boldsymbol{x}')$. □

Proposition 12.55. ([54, Theorem 4.1]) *Let* $\mathbb{S} \neq \emptyset$. *Then* $\{x_{j'} : \boldsymbol{x} \in \mathbb{S}\} = \gamma_{j'}(\hat{\boldsymbol{x}})$ *for any* $j' \in J$, *where* $\hat{\boldsymbol{x}}$ *is the greatest solution.*

Proof. By Prop. 12.54 it is obvious that $\{x_{j'} : \boldsymbol{x} \in \mathbb{S}\} \subseteq \gamma_{j'}(\hat{\boldsymbol{x}})$. Conversely, let $l \in \gamma_{j'}(\hat{\boldsymbol{x}})$ and define $\boldsymbol{x}'$ as $x'_j = \hat{x}_j$ if $j \neq j'$, and $x'_j = l$, otherwise. Then $\boldsymbol{x}' \in \mathbb{S}$ and $l \in \{x_{j'} : \boldsymbol{x} \in \mathbb{S}\}$, i.e. $\gamma_{j'}(\hat{\boldsymbol{x}}) \subseteq \{x_{j'} : \boldsymbol{x} \in \mathbb{S}\}$. □

Proposition 12.56. ([54, Corollary 4.1]) *Equation* $A\boldsymbol{x} = \boldsymbol{b}$ *has the least solution iff* $\overline{\boldsymbol{x}} \in \mathbb{S}$, *with* $\overline{x}_j = \min \gamma_j(\hat{\boldsymbol{x}})$ *for any* $j \in J$, *where* $\hat{\boldsymbol{x}}$ *is the greatest solution. Moreover,* $\overline{\boldsymbol{x}}$ *is the least solution.*

Proof. Let $\boldsymbol{x}^*$ be the least element in $\mathbb{S}$. Then from Prop. 12.55 we have $x^*_j \in \{x_{j'} : \boldsymbol{x} \in \mathbb{S}\} = \gamma_{j'}(\hat{\boldsymbol{x}})$, and $x^*_j \leq x_j$ since $\boldsymbol{x}^*$ is the least solution. Conversely, if $\overline{\boldsymbol{x}} \in \mathbb{S}$, with $\overline{x}_j = \min \gamma_j(\hat{\boldsymbol{x}})$ for any $j \in J$, then from Prop. 12.55 one can easily see that $\overline{\boldsymbol{x}}$ is the least solution. □

Proposition 12.57. ([54, Corollary 4.2]) *Equation* $A\boldsymbol{x} = \boldsymbol{b}$ *has a unique solution iff* $\min \gamma_j(\hat{\boldsymbol{x}}) = \{\hat{x}_j\}$ *for any* $j \in J$, *where* $\hat{\boldsymbol{x}}$ *is the greatest solution.*

Proof. Straightforward from Prop. 12.56 (or also from Th. 12.50). □

12.4 A Classified and Chronological Bibliography on FREs

For the reader's convenience, this section provides a (though very long, but certainly not complete) list of papers dealing with theoretical aspects of fuzzy relation equations and inequalities, which to my best knowledge is the most comprehensive available at the time of its compilation. The entries collected in this list are (1) tentatively classified with respect to the underlying kind of valuation structure, and (2) in chronological order to facilitate a reconstruction of the research progress.

The big number of publications indicates a long and difficult period of intensive research in order to determine the complete solution set of a fuzzy relation equation. The first and most essential are the results of Sanchez already discussed above in sections 12.2.2 and 12.2.3 for the max-min (join-meet) composition of fuzzy relations: They provide a check if there are solutions at all, simultaneously yielding the greatest one if $S_\rho \neq \emptyset$. And if this is the case, then by Prop. 12.6 we know that we "only" have to find all minimal solutions to get the whole solution set. So, since the mid-seventies a journey, as interesting as laborious, looking for minimal solutions and solution algorithms takes place.

Let me insert some preliminary remarks on the list of contributions given below:

- Obviously, fuzzy relation inequalities are of interest of their own, but also because common solutions of $a \leq b$ and $b \leq a$ give solutions of $a = b$.
- Concerning the vast number of applications of FREs only papers on topics of formal theoretical nature have been included, such as graph and automata theory.
- Recall that Brouwerian lattices in the sense of [50] are called relatively pseudo-complemented lattices in Def. 4.20 and elsewhere. But to follow the common naming within the area of fuzzy relations, the term '*Brouwerian*' will be used in this section. Furthermore, recall from Section 4.4 that every chain becomes a Brouwerian lattice by defining $a \frown b = 1$ whenever $a \leq b$, and $a \frown b = b$ otherwise. Moreover, by Prop. 4.21, Brouwerian lattices are distributive (so the results obtained for distributive also hold for Brouwerian lattices), and finite distributive lattices are Brouwerian.

A short remark about the first names of researchers from China seems in order. They often consist of two parts and are written in one of the following forms: Firstsecond, First-Second, or First-second. And their abbreviations in references vary from F., F.-S., or F.-s. to F.S.. Below, and also in the reference sections, we use the latter form.

12.4.1 *Books and Chapters on Fuzzy Relation Equations*

1. Bělohlávek, R.: *Fuzzy Relational Systems-Foundations and Principles*, Springer Science+ Business Media, New York (2002); Chapter 6.
2. DiNola, A., Sessa, S., Pedrycz, W., Sanchez, E.: *Fuzzy Relation Equations and Their Applications to Knowledge Engineering*, Kluwer Academic Publishers, Dordrecht (1989).

3. Gottwald, S.: *Fuzzy Sets and Fuzzy Logic*, Friedr. Vieweg & Sohn, Braunschweig/Wiesbaden (1993); Chapter 3.
4. Klir, G., Yuan, B.: *Fuzzy Sets and fuzzy Logic: Theory and Applications*, Prentice Hall, New Jersey (1995); Chapter 6.
5. Peeva, K., Kyosev, Y.: *Fuzzy Relational Calculus. Theory, Applications and Software*, World Scientific, Singapore (2004).

12.4.2 *Brouwerian and Residuated Lattices*

1. Sanchez, E.: Resolution of composite fuzzy relation equations. Information and Control **30**, 38–48 (1976).
2. Sanchez, E.: Solutions in composite fuzzy relation equations-Applications to medical diagnosis in Brouwerian logic, in Gupta, M.M., Saridis, G.N., Gaines, B. (Eds): *Fuzzy Automata and Decision Processes*, North-Holland, pp. 221-234 (1977). (Reprinted in Dubois, D., Prade, H., Yager, R.R. (Eds): *Readings in Fuzzy Sets for Intelligent Systems*, Morgan Kaufmann, pp. 159-165 (1993).)
3. Di Nola, A., Lettieri, A.: Relation equations in residuated lattices. Rendiconti del Circolo matematico di Palermo **38**(2), 246-256 (1989).
4. Sessa, S.: Finite fuzzy relation equations with unique solution in complete Brouwerian lattices, Fuzzy Sets and Systems **29**, 103-113 (1989).
5. Di Nola, A.: On solving relational equations in Brouwerian lattices, Fuzzy Sets and Systems **34**, 365-376 (1990).
6. De Baets, B.: Analytical solution methods for fuzzy relation equations. In Dubois, D., Prade, H. (Eds): *Fundamentals of Fuzzy Sets*, Springer Science+Business Media New York, pp 291-340 (2000).
7. Wang, X.P.: Method of solution to fuzzy relational equations in a complete Brouwerian lattice, Fuzzy Sets and Systems **120**, 409-414 (2001).
8. Wang, X.P.: Infinite fuzzy relational equations in a complete Brouwerian lattice, Indian J. Pure Appl. Math. **33**, 87-95 (2002).
9. Wang, X.P.: Conditions under which a fuzzy relational equation has minimal solution in a complete Brouwerian lattice, Adv. in Math. **31** 220–228 (2002), (in Chinese).
10. Gottwald, S., Novák, V., Perfilieva, I.: Fuzzy Control and Pseudo-Solutions of Fuzzy Relation Equations, in *Betrachtungen zur Systemtheorie*, Hochschule Zittau/Görlitz, pp. 217-225 (2003).
11. Li, Y.M., Wang, X.P.: The solution sets of @-fuzzy relational equations in finite domains and on a complete Brouwerian lattice, Indian J. Pure Appl. Math. **34**, 1249-1257 (2003).
12. Perfilieva, I.: Solvability of a system of fuzzy relation equations: easy to check conditions, Neural Network World **13**, 571-579 (2003).
13. Perfilieva, I., Gottwald, S.: Solvability and approximate solvability of fuzzy relation equations, Int. J. of General Systems **32**, 361-372 (2003).
14. Wang, X.P.: Infinite fuzzy relational equations on a complete Brouwerian lattice, Fuzzy Sets and Systems **138**, 657-666 (2003).
15. Perfilieva, I.: Fuzzy function as an approximate solution to a system of fuzzy relation equations, Fuzzy Sets and Systems **147**, 363-383 (2004).
16. Wang, X.P., Xiong, Q.Q.: The solution set of a fuzzy relational equation with sup-conjunctor composition in a complete Brouwerian lattice, Fuzzy Sets and Systems **153**, 249-260 (2005).
17. Nosková, L.: System of Fuzzy Relation equations-Criteria of solvability, IRAFM Research Rep. 95 (2006).

18. Wang, X.P., Qu, X.B.: Continuous join-irreducible elements and their applications to describe the solution set of fuzzy relational equations, Acta Math. Sinica, Chinese Series **49**, 1171-1180 (2006), (in Chinese).
19. Nosková, L., Perfilieva, I.: System of fuzzy relation equations with sup-$*$ composition in semi-linear spaces: minimal solutions, in *Proc. Fuzz-IEEE Conf. on Fuzzy Systems*, London, pp. 1520-1525 (2007).
20. Perfilieva, I., Novák, V.: System of fuzzy relation equations as a continuous model of IF-THEN rules. Information Sciences **177**, 3218-3227 (2007).
21. Qu, X.B., Wang, X.P.: Some properties of infinite fuzzy relational equations on complete Brouwerian lattices, Fuzzy Sets and Systems **158**, 1327-1339 (2007).
22. Xiong, Q.Q., Wang, X.P.: Solution sets of inf-α_T fuzzy relational equations on complete Brouwerian lattices, Information Sciences **177**, 4757-4767 (2007).
23. Han, S.C., Li, H.X., Wang, J.Y.: Resolution of matrix equations over arbitrary Brouwerian lattices, Fuzzy Sets and Systems **159**, 40-46 (2008).
24. Li, Y.M., Wang, X.P.: Necessary and sufficient conditions for existence of maximal solutions for inf-α composite fuzzy relational equations, Computers and Mathematics with Applications **55**, 1961-1973 (2008).
25. Perfilieva, I.: Systems of Fuzzy Relation Equations in a Space with Fuzzy Preorder, in: *IFSA-EUSFLAT 2009 Proc.*, pp. 1601-1605 (2009)
26. Wang, X.P., Sun, F.: Resolution of Fuzzy Relation Equations with Sup-Inf Composition over Complete Brouwerian Lattices-a review, in: *IFSA-EUSFLAT 2009 Proc.*, pp. 82-87 (2009).
27. Ćirić, M., Stamenković, A., Ignjatović, J., Petković, T.: Fuzzy relation equations and reduction of fuzzy automata, J. of Computer and System Sciences **76**, 609-633 (2010).
28. Ignjatović, J., Ćirić, M., Bogdanović, S.: On the greatest solutions to weakly linear systems of fuzzy relation inequalities and equations, Fuzzy Sets and Systems **161**, 3081-3113 (2010).
29. Perfilieva, I.: Fuzzy Relation Equations in Semilinear Spaces, in: Hüllermeier, E., Kruse, R., Hoffmann, F. (Eds.), *IPMU 2010, Part I*, Springer-Verlag Berlin Heidelberg, pp. 545-552 (2010).
30. Perfilieva, I.: Semilinear Space, Galois Connections and Fuzzy Relation Equations, in: Cao, B.-Y. et al. (Eds.), *Quantitative Logic and Soft Computing 2010*, Springer-Verlag Berlin Heidelberg, pp. 71-79 (2010).
31. Qu, X.B., Wang, X.P., Xiong, Q.Q.: The lattice properties of the solution set of fuzzy relational equations, in: Da Ruan, Li, T. et. al. (Eds.), *Computational Intelligence - Foundations and Applications*, FLINS 2010, World Scientific, pp. 124-129 (2010).
32. Zhou, J., Li, Q.: The largest solution of linear equation over the complete Heyting algebra, Acta Mathematica Scientia **30B**, 810-818 (2010).
33. Alcalde, C., Burusco, A., Fuentes-González, R.: On the solutions of L-fuzzy relational equations with the L-Fuzzy Concept Theory, *Proc. EUSFLAT-LFA 2011*, Atlantis Press, 986-991 (2011).
34. Alcalde, C., Burusco, A., Fuentes-Gonzalez, R.: Analysis of certain L-fuzzy relational equations and the study of its solutions by means of the L-fuzzy concept theory. Int. J. of Uncertainty, Fuzziness and Knowledge-Based Systems **20**, 21-40 (2012).
35. Ignjatović, J., Ćirić, M.: Weakly linear systems of fuzzy relation inequalities and their applications: A brief survey, Filomat **26**, 207-241 (2012).
36. Ignjatović, J., Ćirić, M., Damljanović, N., Jančić, I.: Weakly linear systems of fuzzy relation inequalities: The heterogeneous case, Fuzzy Sets and Systems **199**, 64-91 (2012).
37. Xiong, Q.Q., Lei, M.H., Wang, X.P.: Solution sets of fuzzy relation equations in complete Brouwerian lattices, in: Wang, G., Zhao, B., Li, Y. (Eds.), *Quantitative Logic and Soft Computing*, World Scientific, pp. 273-278 (2012).

38. Qu, X.B., Wang, X.P., Sun, F.: A method for solving fuzzy relation equations in complete Brouwerian lattices, in: Kahraman, C., Kerre, E.E., Bozbura, F. (Eds.), *Uncertainty Modeling in Knowledge Engineering and Decision Making*, FLINS 2012, World Scientific, pp. 608-614 (2012).
39. Sun, F.: Conditions for the existence of the least solution and minimal solutions to fuzzy relation equations over complete Brouwerian lattices. Information Sciences **205**, 86–92 (2012).
40. Xiong, Q.Q., Wang, X.P.: Fuzzy relational equations on complete Brouwerian lattices, Information Sciences **193**, 141-152 (2012).
41. Xiong, Q.Q., Wang, X.P.: Interval-valued fuzzy relational equations with min-implication, in: Kahraman, C., Kerre, E.E., Bozbura, F. (Eds.), *Uncertainty Modeling in Knowledge Engineering and Decision Making*, FLINS 2012, World Scientific, pp. 627-632 (2012).
42. Kim, Y.C.: Fuzzy relation equations and Galois connections, Mathematica Aeterna **3**, 25-27 (2013).
43. Perfilieva, I.: Finitary solvability conditions for systems of fuzzy relation equations, Information Sciences **234**, 29–43 (2013).
44. Sun, F., Wang, X.P., Qu, X.B.: Minimal join decompositions and their applications to fuzzy relation equations over complete Brouwerian lattices. Information Sciences **224**, 143–151 (2013).
45. Wang, X.P., Zhao, S.: Solution sets of finite fuzzy relation equations with sup-inf composition over bounded Brouwerian lattices, Information Sciences **234**, 80-85 (2013).
46. Xiong, Q.Q., Wang, X.P.: Some Results of Interval-valued fuzzy relational equations with sup-conjunctor composition, in: *Proc. IFSA/NAFIPS 2013*, IEEE, pp. 333-337 (2013).
47. Kim, Y.C.: Solutions of fuzzy relation equations in generalized residuated lattices, Int. J. of Pure and Applied Mathematics **94**, 101-118 (2014).
48. Qu, X.B., Wang, X.P., Lei, M.H.: Conditions under which the solution sets of fuzzy relational equations over complete Brouwerian lattices form lattices, Fuzzy Sets and Systems **234**, 34-45 (2014).
49. Xiong, Q.Q., Wang, X.P.: Infinite fuzzy relational equations with sup-conjunctor on complete Brouwerian lattices, Information Sciences **279**, 691-701 (2014).
50. Ignjatović, J., Ćirić, M., Šešelja, B., & Tepavčević, A.: Fuzzy relational inequalities and equations, fuzzy quasi-orders, closures and openings of fuzzy sets, Fuzzy Sets and Systems **260**, 1-24 (2015).
51. Qu, X.B., Sun, F., Wang, T.F.: Matrix elementary transformations in solving systems of fuzzy relation equations, Applied Soft Computing **31**, 25-29 (2015).
52. Medina, J.: Notes on "Solution sets of inf-αT fuzzy relational equations on complete Brouwerian lattices" and "Fuzzy relational equations on complete Brouwerian lattices". Information Sciences **402**, 82-90 (2017).
53. Stanković, I., Ćirić, M., Ignjatović, J.: Fuzzy relation equations and inequalities with two unknowns and their applications, Fuzzy Sets and Systems **322**, 86-105 (2017).
54. Sun, F., Qu, X.B., Wang, X.P., Zhu, L.: On pre-solution matrices of fuzzy relation equations over complete Brouwerian lattices, Fuzzy Sets and Systems **384**, 34-53 (2020).
55. Stanimirović, S., Micić, I.: On the solvability of weakly linear systems of fuzzy relation equations, Information Sciences **607**, 670-687 (2022).

12.4.3 *The Real Unit Interval*

1. Sanchez, E.: Resolution of Eigen fuzzy sets equations, Fuzzy Sets and Systems **1**, 69-74 (1978).
2. Czogala, E., Drewniak, J., Pedrycz, W.: Fuzzy relation equations on a finite set. Fuzzy Sets and Systems **7**, 89-101 (1982).
3. Di Nola, A., Sessa, S.: On the fuzziness of solutions of σ-fuzzy relation equations on finite spaces, Fuzzy Sets and Systems **11**, 65-77 (1983).

4. Pedrycz, W.: Fuzzy relational equations with generalized connectives and their applications, Fuzzy Sets and Systems **10**, 185-201 (1983).
5. Pedrycz, W.: Numerical and applicational aspects of fuzzy relational equations, Fuzzy Sets and Systems **11**, 1-18 (1983).
6. Di Nola, A., Pedrycz, W., Sessa, S.: Some theoretical aspects of fuzzy-relation equations describing fuzzy systems, Information Sciences **34**, 242-264 (1984).
7. Di Nola, A., Pedrycz, W., Sessa, S., Wang, P.Z.: Fuzzy relation equations under triangular norms-A survey and new results, Stochastica **VIII**, 99-145 (1984).
8. Higashi, M., Klir, G.J.: Resolution of finite fuzzy relation equations. Fuzzy Sets and Systems **13**, 65-82 (1984).
9. Luo, C.Z.: Reachable solution set of a fuzzy relation equation, J. of Math. Analysis and Applications **103**, 524-532 (1984).
10. Wang, P.Z., Sessa, S.. Di Nola, A., Pedrycz, W.: How many lower solutions does a fuzzy relation equation have? BUSEFAL **18**, 67-74 (1984).
11. Gottwald, S.: Generalized solvability criteria for fuzzy equations, Fuzzy Sets and Systems **17**, 285-296 (1985).
12. Miyakoshi, M., Shimbo, M.: Solutions of composite fuzzy relational equations with triangular norms, Fuzzy Sets and Systems **16**, 53-63 (1985).
13. Pappis C., Sugeno M.: Fuzzy relational equations and the inverse problem, Fuzzy Sets and Systems **15**, 79-90 (1985).
14. Pedrycz, W.: On generalized fuzzy relational equations and their applications, J. of Math. Analysis and Applications **107**, 520-536 (1985).
15. Gottwald, S.: Characterizations of the solvability of fuzzy equations, Elektronische Informationsverarbeitung und Kybernetik **22**, 67-91 (1986).
16. Gottwald, S., Pedrycz, W.: Solvability of fuzzy relational equations and manipulation of fuzzy data, Fuzzy Sets and Systems **18**, 45-65 (1986).
17. Gottwald, S., Pedrycz, W.: On the suitability of fuzzy models: an evaluation through fuzzy integrals, Int. J. Man-Machine Studies **24**, 141-151 (1986).
18. Miyakoshi, M., Shimbo, M.: Lower solutions of systems of fuzzy equations. Fuzzy Sets and Systems **19**, 37-46 (1986).
19. Wu, W.: Fuzzy reasoning and fuzzy relational equations, Fuzzy Sets and Systems **20**, 67-78 (1986).
20. Wagenknecht, M., Hartmann, K.: On the construction of fuzzy eigen solutions in given regions, Fuzzy Sets and Systems **20**, 55-65 (1986).
21. Wagenknecht, M., Hartmann, K.: Fuzzy modeling with tolerances, Fuzzy Sets and Systems **20**, 325-332 (1986).
22. Miyakoshi, M., Shimbo, M.: Sets of solution-set-invariant coefficient matrices of simple fuzzy relation equations, Fuzzy Sets and Systems **21**, 59-83 (1987).
23. Wagenknecht, M., Hartmann, K.: On direct and inverse problems for fuzzy equation systems with tolerances, Fuzzy Sets and Systems **24**, 93-102 (1987).
24. Cheng, L., Peng, B.: The fuzzy relation equation with union or intersection preserving operator, Fuzzy Sets and Systems **25**, 191-204 (1988).
25. Guo, S.Z., Wang, P.Z., Di Nola, A., Sessa, S.: Further contributions to the study of finite fuzzy relation equations, Fuzzy Sets and Systems **26**, 93-104 (1988).
26. Pedrycz, W.: Approximate solutions of fuzzy relational equations. Fuzzy Sets and Systems **28**, 183-202 (1988).
27. Tang, F.: Fuzzy bilinear equations, Fuzzy Sets and Systems **28**, 217-226 (1988).

28. Di Nola, A., Sessa, S., Pedrycz, W.: On some finite fuzzy relation equations, Information Sciences **50**, 93-109 (1990).

29. Hsu, H.M., Wang, H.F.: Sensitivity analysis on generalized fuzzy relation equations, *Proc. First International Symposium on Uncertainty Modeling and Analysis*, IEEE, pp. 312-317 (1990).

30. Miyakoshi, M., Shimbo, M.: Sets of solution-set equivalent coefficient matrices of fuzzy relation equations, Fuzzy Sets and Systems **35**, 357-387 (1990)

31. Pedrycz, W.: Inverse problem in fuzzy relational equations, Fuzzy Sets and Systems **36**, 277-291 (1990).

32. Wagenknecht, M., Hartmann, K.: On the existence of minimal solutions for fuzzy equations with tolerances, Fuzzy Sets and Systems **34**, 237-244 (1990).

33. Li, J.X.: The smallest solution of max-min fuzzy equations, Fuzzy Sets and Systems **41**, 317-327 (1990).

34. Pedrycz, W.: Processing in relational structures: Fuzzy relational equations, Fuzzy Sets and Systems **40**, 77-106 (1991).

35. Wang, H.F., Chang, Y.C.: Resolution of composite interval-valued fuzzy relation equations, Fuzzy Sets and Systems **44**, 227-240 (1991).

36. Wang, H.F., Hsu, H.M.: Sensitivity analysis of fuzzy relation equations, Int. J. of General Systems **9**, 155-169 (1991).

37. Wang, P.Z., Zhang, D.Z., Sanchez, E., Lee, E.S.: Latticized linear programming and fuzzy relation inequalities, J. of Math. Analysis and Applications **159**, 72-87 (1991).

38. Di Nola, A., Sessa, S., Pedrycz, W.: A study on approximate reasoning mechanisms via fuzzy relation equations, Int. J. of Approximate Reasoning **6**, 33-44 (1992).

39. Han, S.R., Sekiguchi, T.: Solution of Fuzzy relational equations using Sign Matrices, Japanese J. of Fuzzy Theory and Systems **4**, 13-27 (1992).

40. Wang, H.F., Hsu, H.M.: An alternative approach to the resolution of fuzzy relation equations, Fuzzy Sets and Systems **45**, 203-213 (1992).

41. Adamopoulos, G.I., Pappis, C.P.: Some results on the resolution of fuzzy relation equations, Fuzzy Sets and Systems **60**, 83-88 (1993).

42. Arnould T., Tano S.: A rule-based method to calculate exactly the widest solution sets of a max-min fuzzy relational inequality, Fuzzy Sets and Systems **64**, 39-58 (1994).

43. Arnould T., Tano S.: A rule-based method to calculate the widest solution sets of a max-min fuzzy relational equation, Int. J. of Uncertainty, Fuzziness and Knowledge-Based Systems **2**, 247-256 (1994).

44. De Baets, B., Kerre, E.E.: A primer on solving fuzzy relational equations on the unit interval, Int. J. of Uncertainty, Fuzziness and Knowledge-Based Systems **2**, 205-225 (1994).

45. De Baets, B., Kerre, E.E.: A representation of solution sets of fuzzy relational equations, in: Trappl, R. (Ed.), *Cybernetics and Systems'94*, World Scientific Publishing, Singapore, pp. 287-294 (1994).

46. Gottwald, S.: Approximately solving fuzzy relation equations: Some mathematical results and some heuristic proposals. Fuzzy Sets and Systems **66**, 175-193 (1994).

47. Gottwald, S., Pedrycz, W.: Revealing logical dependencies in fuzzy data with the aid of multi-level fuzzy relational equations, Information Sciences **78**, 311-324 (1994).

48. Liu, X.: The Structure of Fuzzy Matrices, The J. of Fuzzy Mathematics **2**, 311-325 (1994).

49. Liu, X.: The Solution of Fuzzy Equations, The J. of Fuzzy Mathematics **2**, 327-333 (1994).

50. Mohanty, B.K., Singh, N.: Fuzzy relational equations in analytical hierarchy process, Fuzzy Sets and Systems **63**, 11-19 (1994).

51. Cechlárová, K.: Unique solvability of Max-min fuzzy equations and strong regularity of matrices over fuzzy algebra, Fuzzy Sets and Systems **75**, 165-177 (1995).

52. Drewniak, J.: Equations in classes of fuzzy relations, Fuzzy Sets and Systems **75**, 215-228 (1995).
53. Gottwald, S.: Approximate solutions of fuzzy relational equations and a characterization of t-norms that define metrics for fuzzy sets, Fuzzy Sets and Systems **75**, 189-201 (1995).
54. Pappis, C.P., Karacapilidis, N.I.: Application of a similarity measure of fuzzy sets to fuzzy relational equations, Fuzzy Sets and Systems **75**, 135-142 (1995).
55. Wagenknecht, M.: On transitive solutions of fuzzy equations, inequalities and lower approximations of fuzzy relations, Fuzzy Sets and Systems **75**, 229-240 (1995).
56. Zhang, K.L.: On fuzzy bilinear equations, Fuzzy Sets and Systems **76**, 91-96 (1995).
57. Gottwald, S.: Fundamentals of fuzzy relation calculus, in: Pedrycz, W. (Ed.), *Fuzzy Modelling*, Springer, Boston, MA, pp. 25-47 (1996).
58. Miyagi, H., Noborikawa, M., Fukumura, H., Yamashita, K.: Solution of fuzzy relation equations using the rule of exclusion, in: *Proc. 1996 IEEE International Conference on Systems, Man and Cybernetics*, pp. 2517-2521 (1996).
59. Neundorf, D., Böhm, R.: Solvability criteria for systems of fuzzy relation equations, Fuzzy Sets and Systems **80**, 345-352 (1996).
60. Chung, F.L., Lee, T.: A new look at solving a system of fuzzy relational equations, Fuzzy Sets and Systems **88**, 343-353 (1997).
61. Imai, H., Miyakoshi, M., Da-te, T.: Some properties of minimal solutions for a fuzzy relation equation, Fuzzy Sets and Systems **90**, 335-340 (1997).
62. Miyagi, H., Fukumura, H., Kamiya, K., Yamashita, K.: Algorithm for solution of fuzzy relation equations, in: *Proc. 1997 IEEE International Conference on Systems, Man and Cybernetics*, pp. 4005-4009 (1997).
63. Bourke, M.M., Fisher, D.G.: Solution algorithms for fuzzy relational equations with max-product composition, Fuzzy Sets and Systems **94**, 61-69 (1998).
64. Imai, H., Kikuchi, K., Miyakoshi, M.: Unattainable solutions of a fuzzy relation equation, Fuzzy Sets and Systems **99**, 193-196 (1998).
65. Kawaguchi, M.F., Miyakoshi, M. (1998): Composite fuzzy relational equations with non-commutative conjunctions. Information Sciences **110**, 113-125.
66. Li, G.Z., Fang, S.C.: Solving interval-valued fuzzy relation equations, IEEE Transactions on Fuzzy Systems **6**, 321-324 (1998).
67. Miyagi, H., Fukumura, H., Yamashita, K.: Solution of Fuzzy Relation Equations Using the Rule of Exclusion, Japanese J. of Fuzzy Theory and Systems **10**, 113-125 (1998).
68. Tzafestas, S.G., Stamou, G.B.: A new solution methodology for fuzzy relation equation, in: *Proc. IEA/AIE*, LNCS 1415, Springer Verlag, pp. 210–219 (1998).
69. Fan, Y., Miyagi, H.: Studying on Solution of Fuzzy Relation Equations, *Third International Conference on Knowledge-Based Intelligent Information Engineering Systems*, pp. 357-360 (1999).
70. Stamou, G.B., Tzafestas, S.G.: Fuzzy relation equations and fuzzy inference systems: An Inside Approach, IEEE Trans. SMC-Part B **29**, 694-702 (1999).
71. Cechlárová, K.: A note on unsolvable systems of max-min (fuzzy) equations, Linear Algebra and its Applications **310**, 123-128 (2000).
72. Gottwald, S.: Generalised solvability behaviour for systems of fuzzy equations, in: Novák, V., Perfilieva, I. (Eds.), *Discovering the World With Fuzzy Logic*, Physica Verlag, Springer, pp. 401-430 (2000).
73. Perfilieva, I., Tonis, A.: Compatibility of systems of fuzzy relation equations. Int. J. of General Systems **29**, 511-528 (2000).
74. Pedrycz, W.: Fuzzy relational equations: Bridging theory, methodology and practice, Int. J. of General Systems **29**, 529-554 (2000).

75. Tang, F.C.: Perturbation techniques for fuzzy matrix equations, Fuzzy Sets and Systems **109**, 363-369 (2000).
76. Zimmermann, K.: Generalized fuzzy relational inequalities and related problems, Int. J. of General Systems **29**, 637-644 (2000).
77. Stamou, G.B., Tzafestas, S.G.: Resolution of composite fuzzy relation equations based on Archimedean triangular norms, Fuzzy Sets and Systems **120**, 395-407 (2001).
78. Chen, L., Wang, P.P.: Fuzzy relation equations (I): the general and specialized solving algorithms, Soft Computing **6**, 428-435 (2002).
79. Imai, H., Miyakoshi, M., Sato, Y.: Properties of the solution set of a fuzzy relation equation from the viewpoint of attainability, Soft Computing **6**, 87-91 (2002).
80. Luoh, L., Wang, W.J., Liaw, Y.K.: New algorithms for solving fuzzy relation equations, Mathematics and Computers in Simulation **59**, 329-333 (2002).
81. Gottwald, S.: Fuzzy relation equations and approximate solutions. Some recent results, in *Proc. 3rd Conference of the EUSFLAT*, pp. 6-11 (2003).
82. Luoh, L., Wang, W.J., Liaw, Y.K.: Matrix-pattern-based computer algorithm for solving fuzzy relation equations, IEEE Transactions on Fuzzy Systems **11**, 100-108 (2003).
83. Wang, S., Fang, S.C., Nuttle, H.L.W.: Solution sets of interval-valued fuzzy relational equations, Fuzzy Optimization and Decision Making **2**, 41-60 (2003).
84. Wang, S., Fang, S.C., Nuttle, H.L.W.: Errata, Fuzzy Optimization and Decision Making **6**, 297 (2007).
85. Fernández, M.J., Gil, P.: Some specific types of fuzzy relation equations, Information Sciences **164**, 189-195 (2004).
86. Gottwald, S.: Fuzzy control and fuzzy relation equations. A unified view as interpolation problem, in: *Proc. NAFIPS 2004*, pp. 270-275 (2004).
87. Xia, Z.Q., Guo, F.F.: An ABS-FRE algorithm for solving systems of fuzzy relation equations, J. Appl. Math. & Computing **15**, 285-297 (2004).
88. Zhang, C., Li, D., Dang, P.: On perturbation properties of fuzzy relation equations with max-product composition, Int. J. Pure and Applied Mathematics **17**, 63-69 (2004).
89. Gottwald, S.: Mathematical Fuzzy Control. A Survey of Some Recent Results, Logic Journal of the IGPL **13**, 525-541 (2005).
90. Xiong, Q.Q., Wang, X.P.: Some properties of sup-min fuzzy relational equations on infinite domains, Fuzzy Sets and Systems **151**, 393-402 (2005).
91. Gottwald, S.: On a Generalization of Fuzzy Relation Equations, in: *Proc. 11th IPMU Conf. 2006*, pp. 2572-2577 (2006).
92. Han, S.C., Li, H.X., Wang, J.Y.: Resolution of finite fuzzy relation equations based on strong pseudo-t-norms, Applied Mathematics Letters **19**, 752-757 (2006).
93. Peeva, K.: Universal algorithm for solving fuzzy relational equations, Italian J. of Pure and Applied Mathematics **19**, 169-188 (2006).
94. Perfilieva, I.: Logical foundations of rule-based systems, Fuzzy Sets and Systems **157**, 615-621 (2006).
95. Chen, L., Wang, P.P.: Fuzzy relation equations (II): the branch-point-solutions, Soft Computing **11**, 33-40 (2007).
96. Peeva, K., Kyosev, Y.: Algorithm for solving max-product fuzzy relational equations. Soft Computing **11**, 593-605 (2007).
97. Shieh, B.S.: Solutions of fuzzy relation equations based on continuous t-norms, Information Sciences **177**, 4208-4215 (2007).

98. Gottwald, S.: Calculi of Information Granules: Fuzzy Relational Equations, in: Pedrycz, W., Skowron, A., Kreinovich, V. (Eds.), *Handbook of Granular Computing*, John Wiley & Sons, pp. 225-248 (2008).
99. Li, J.X.: A new algorithm for minimizing a linear objective function with fuzzy relation equation constraints, Fuzzy Sets and Systems **159**, 2278-2298 (2008).
100. Li, P., Fang, S.C.: On the resolution and optimization of a system of fuzzy relational equations with sup-T composition. Fuzzy Optimization and Decision Making, **7**, 169-214 (2008).
101. Lin, J.L.: Finding the Complete Set of Minimal Solutions for Fuzzy Max-Archimedean t-Norm Relational Equations, *Proc. Eighth International Conference on Intelligent Systems Design and Applications, ISDA 2008*, IEEE, pp. 501-506 (2008).
102. Molai, A.A., Khorram, E.: An algorithm for solving fuzzy relation equations with max-T composition operator, Information Sciences **178**, 1293-1308 (2008).
103. Shieh, B.S.: New resolution of finite fuzzy relation equations with max-min composition. Int. J. of Uncertainty, Fuzziness and Knowledge-Based Systems **16**, 19-33 (2008).
104. Wu, Y.K., Guu, S.M.: An efficient procedure for solving a fuzzy relational equation with max-Archimedean t-norm composition. IEEE Transactions on Fuzzy Systems **16**, 73-84 (2008).
105. Yeh, C.T.: On the minimal solutions of max-min fuzzy relational equations. Fuzzy Sets and Systems **159**, 23-39 (2008).
106. Li, P.: Fuzzy Relational Equations: Resolution and Optimization, PhD dissertation, North Carolina State University, Industrial Engineering (2009).
107. Li, P., Fang, S.C.: A note on solution sets of interval-valued fuzzy relational equations, Fuzzy Optimization and Decision Making, **8**, 115-121 (2009).
108. Li, P., Fang, S.C.: Latticized Linear Optimization on the Unit Interval, IEEE Transactions on Fuzzy Systems **17**, 1353-1365 (2009).
109. Lin, J.L.: On the relation between fuzzy max-Archimedean t-norm relational equations and the covering problem, Fuzzy Sets and Systems **160**, 2328-2344 (2009).
110. Drewniak, J., Matusiewicz, Z.: Properties of max-$*$ fuzzy relation equations, Soft Computing **14**, 1037-1041 (2010).
111. Luoh, L., Liaw, Y.K.: Novel approximate solving algorithm for fuzzy relational equations, Mathematical and Computer Modelling **52**, 303-308 (2010).
112. Wu, Y.K., Liu, C.C., Lur, Y.Y., Guu, S.M.: One Simple Procedure to Finding the Best Approximate Solution for a Particular Fuzzy Relational Equation with Max-min Composition, in: *Proc. Third International Joint Conference on Computational Science and Optimization 2010*, IEEE, pp. 141-144 (2010).
113. Xiong, Q.Q., Wang, X.P.: Some properties of infinite fuzzy relational equations $A \odot X = B$ in a lattice $[0,1]$, in: *Proc. FLINS 2010*, pp. 130-135 (2010)
114. Li, P., Fang, S.C.: On the unique solvability of fuzzy relational equations. Fuzzy Optimization and Decision Making **10**, 115-124 (2011).
115. Li, P., Jin, Q.: Fuzzy relational equations with min-biimplication composition. Fuzzy Optimization and Decision Making **11**, 227-240 (2012).
116. Lin, J.L., Wu, Y.K., Guu, S.M.: On fuzzy relational equations and the covering problem, Information Sciences **181**, 2951–2963 (2011).
117. Cheng, D., Feng, J., Lv, H.: Solving fuzzy relational equations via semitensor product, IEEE Transactions on Fuzzy Systems **20**, 390-396 (2012).
118. Drewniak, J., Matusiewicz, Z.: Fuzzy equations max-$*$ with conditionally cancellative operations, Information Sciences **206**, 18–29 (2012).
119. Fan, H., Feng, J., Zhang, L.: Solving a class of fuzzy relation inequalities via semi-tensor product, in: *Proc. 10th World Congress on Intelligent Control and Automation*, IEEE, pp. 1465-1470 (2012).

120. Ghodousian, A., Khorram, E.: Linear optimization with an arbitrary fuzzy relational inequality. Fuzzy Sets and Systems **206**, 89-102 (2012).

121. Li, P.: A note on fuzzy relational equations with min-implication composition. IEEE Transactions on Fuzzy Systems **21**, 983-986 (2013).

122. Matusiewicz, Z., Drewniak, J.: Increasing continuous operations in fuzzy max-$*$ equations and inequalities, Fuzzy Sets and Systems **232**, 120-133 (2013).

123. Peeva, K.: Resolution of fuzzy relational equations–method, algorithm and software with applications, Information Sciences **234**, 44-63 (2013).

124. Yang, Y., Liu, Q.Y., Wang, X.P.: Decomposition of realizable fuzzy relations, Soft Computing **17**, 363-367 (2013).

125. Xiong, Q.Q., Wang, X.P.: Some properties of infinite fuzzy relational equations with sup-inf composition, Information Sciences **252**, 32-41 (2013).

126. Zahariev, Z.: Solving max-min fuzzy linear systems of equations-algorithm and software, Annual of "Informatics" Section, Union of Scientists in Bulgaria **6**, 1-16 (2013)

127. Medina, J., Turunen, E., Bartl, E., Díaz-Moreno, J.C.: Minimal Solutions of Fuzzy Relation Equations with General Operators on the Unit Interval, in: Laurent, A. et al. (Eds.), *IPMU 2014, Part III, Communications in Computer and Information Science 444*, Springer, pp. 81-90 (2014).

128. Molai, A.A.: An algorithm for solving fuzzy relation programming with the max-t composition operator, TWMS J. App. Eng. Math. **5**, 21-29 (2015).

129. Qu, X.B., Sun,F., Wang, T., Xiong, Q.Q.: The structure of solution sets of fuzzy relation equations, in: *Proc. of the 2015 Conference IFSA-EUSFLAT*, Atlantis Press, pp. 28-33 (2015).

130. Xiong, Q.Q., Shu, Q.Y.: Fuzzy relational equations and the covering problem, in: *Proc. 2015 Conference IFSA-EUSFLAT*, Atlantis Press, pp. 77-84 (2015).

131. Kouchakinejad, F., Mashinchi, M., Mesiar, R.: Solution-set invariant matrices and vectors in fuzzy relation inequalities based on max-aggregation function composition, Iranian J. of Fuzzy Systems **13**, 91-100 (2016).

132. Yang, X.B., Yang, X.P., Hayat, K.: A New Characterization of the Minimal Solution Set to Max-min Fuzzy Relation Inequalitites, Fuzzy Inf. Eng. **9**, 423-435 (2017).

133. Lin, H., Yang, X.: Optimal Strong Solution of the Weighted Minimax Problem With Fuzzy Relation Equation Constraints, IEEE Access **6**, 27593-27603 (2018).

134. Wang, X.P., Liu, L.J., Zhang, X.: Sets of solution-set-invariant coefficient matrices of fuzzy relation equations with max-min composition. J. of Intelligent & Fuzzy Systems **34**, 4067-4078 (2018).

135. Chai, Z.L.: Using max-continuous T fuzzy relation equations to determine the vector, Soft Computing **23**, 10477-10484 (2019).

136. Wang, S., Li, H.: Column stacking approach to resolution of systems of fuzzy relational inequalities, J. of the Franklin Institute **356**, 3314-3332 (2019).

137. Xiao, G., Zhu, T., Chen, Y., Yang, X.: Linear searching method for solving approximate solution to system of max-min fuzzy relation equations with application in the instructional information resources allocation, IEEE Access **7**, 65019–65028, (2019).

138. Fan, H., Feng, J., Meng, M., Wang, B.: General decomposition of fuzzy relations - Semi-tensor product approach, Fuzzy Sets and Systems **384**, 75-90 (2020).

139. Ma, Y., Yang, X., Cao, B.Y.: Fuzzy-Relation-Based Lexicographic Minimum Solution to the P2P Network System, IEEE Access **8**, 195447-195458 (2020).

140. Wang, S., Li, H.: Resolution of Fuzzy Relational Inequalities with Boolean Semi-Tensor Product Composition, Mathematics **9**, 937 (2021).

141. Chen, S., Hayat, K., Yang, X.: Upper Bounded Minimal Solution of the Max-Min Fuzzy Relation Inequality System, IEEE Access **10**, 84384-84397 (2022).
142. Wen, C.F., Wu, Y.K., Li, Z.: Algebraic formulae for solving systems of max-min inverse fuzzy relational equations, Information Sciences **622**, 1162–1183 (2023).
143. Zahariev, Z., Zaharieva, G., Peeva, K.: Fuzzy Relational Equations - Min-Goguen Implication, *AIP Conference Proceedings **2505***, 120004 (2022).
144. Guo, F.F., Fu, R., Shen, J.: Inverses of fuzzy relation matrices with addition-min composition, Fuzzy Sets and Systems **490**, 109037 (2024).

12.4.4 *Linear Lattices*

1. Di Nola, A., Sessa, S.: On the set of solutions of composite fuzzy relation equations, Fuzzy Sets and Systems **9**, 275-285 (1983).
2. Di Nola, A.: An algorithm of calculation of lower solutions of fuzzy relation equation, Stochastica **VIII**, 33-40 (1984).
3. Sessa, S.: Some results in the setting of fuzzy relation equations theory, Fuzzy Sets and Systems **14**, 281-297 (1984).
4. Di Nola, A.: Relational equations in totally ordered lattices and their complete resolution, J. of Math. Analysis and Applications **107**, 148-155 (1985).
5. Di Nola, A., Sessa, S., Pedrycz, W., Higashi, M.: Minimal and maximal solutions of a decomposition problem of fuzzy relations, Int. J. of General Systems **11**, 103-116 (1985).
6. Lettieri, A., Liguori, F.: Some results on fuzzy relation equations provided with one solution, Fuzzy Sets and Systems **17**, 199-209 (1985).
7. Peeva, K.: Systems of linear equations over a bounded chain, Acta Cybernetica **7**, 195-202 (1985).
8. Drewniak, J.: Maximal Number of Lower Solutions Fuzzy Relation Equation, in: Trappl, R.(Ed.), *Cybernetics and Systems' 86*, Springer, Dordrecht, pp. 585-590 (1986).
9. Di Nola, A., Sessa, S.: Finite Fuzzy Relation Equations with a Unique Solution in Linear Lattices, J. Math. Anal. Appl. **132**, 39-49 (1988).
10. Di Nola, A., Kolodziejczyk, W., Sessa, S.: Transitive solutions of relational equations on finite sets and linear lattices, in: Bouchon-Meunier, B., Yager, R.R., Zadeh, L.A. (Eds.), *Proc IPMU 1990.* LNCS 521, Springer, Berlin, pp. 173-182 (1991).
11. Peeva, K.: Fuzzy linear systems, Annales Univ Sci Budapest, Sect. Comp **12**, 201-207 (1991).
12. Vrba, J.: General decomposition problem of fuzzy relations, Fuzzy Sets and Systems **54**, 69-79 (1991).
13. Peeva, K.: Fuzzy linear systems, Fuzzy Sets and Systems **49**, 339-355 (1992).
14. Liguori, F., Martini, G., Sessa, S.: Idempotent and compact matrices on linear lattices-A survey of some lattice results and related solutions of finite relation equations, Int. J. Math. & Math. Sci. **16**, 301-310 (1993).
15. Cechlárová, K.: Unique solvability of max-min fuzzy equations and strong regularity of matrices over fuzzy algebra, Fuzzy Sets and Systems **75**, 165-177 (1995).
16. Wagenknecht, M.: On transitive solutions of fuzzy equations, inequalities and lower approximations of fuzzy relations, Fuzzy Sets and Systems **75**, 229-240 (1995).
17. Gavalec, M.: Solvability and unique solvability of max–min fuzzy equations, Fuzzy Sets and Systems, **124**, 385-393 (2001).
18. Gavalec, M., Plávka, J.: Strong regularity of matrices in general max–min algebra, Linear Algebra and its Applications **371**, 241-254(2003).

19. Cechlárová, K.: On max-min linear inequalities and Coalitional Resource Games with sharable resources, Linear Algebra and its Applications **433**, 127-135 (2010).
20. Myšková, H.: Max-min interval systems of linear equations with bounded solution, Kybernetika **48**, 299-308 (2012)
21. Plávka, J.: On the weak robustness of fuzzy matrices, Kybernetika **49**, 128-140 (2013).
22. Sun, F., Wang, X.P., Qu, X.B., Wang, T.F.: Square roots of matrices over a complete lattice, in: *Proc. 2015 Conference IFSA-EUSFLAT*, Atlantis Press, pp. 21-27 (2015).
23. Díaz-Moreno, J.C., Medina, J., Turunen, E.: Computing the minimal solutions of finite fuzzy relation equations on linear carriers, Annals of Computer Science and Information Systems **9**, 19-23 (2016).
24. Díaz-Moreno, J.C., Medina, J., Turunen, E.: Minimal solutions of general fuzzy relation equations on linear carriers, Fuzzy Sets and Systems, **311**, 112-123 (2017).
25. Myšková, H., Plávka, J.: On the solvability of interval max-min matrix equations, Linear Algebra and its Applications **590**, 85-96 (2019).

12.4.5 *(Complete / Distributive / Residuated) Lattices*

1. Drewniak, J.: Fuzzy relation equations and inequalities, Fuzzy Sets and Systems **14**, 237-247 (1984).
2. Turunen, E.: On generalized fuzzy relation equations necessary and sufficient conditions for the existence of solutions, Acta Universitatis Carolinae. Mathematica et Physica **28**, 33-37 (1987).
3. Zhao, C.K.: On matrix equations in a class of complete and completely distributive lattice, Fuzzy Sets and Systems **22**, 303-320 (1987).
4. Liu, W.J.: On some systems of simultaneous equations in a completely distributive lattice, Information Sciences **50**, 185-196 (1990).
5. Zhang, K. L.: Matrix equations in distributive lattices, Acta Scientiarum Naturalium Universitatis Intramongolicas **22**, 311-317 (1991).
6. Zhang, K.L.: The inverse problem for distributive lattices, Fuzzy Sets and Systems **59**, 109-113 (1993).
7. De Baets, B.: An Order-Theoretic Approach to Solving Sup-T Equations, in: Ruan, D. (Ed.), *Fuzzy Set Theory and Advanced Mathematical Applications*, Springer Science+Business Media, New York, pp. 67-87 (1995).
8. De Baets, B.: Sup-T Equations: State of the Art, in: Kaynak, O., Zadeh, L.A. et al.(Eds.), *Computational Intelligence: Soft Computing and Fuzzy-Neural Integration with Applications*, Springer Verlag, Berlin Heidelberg, pp. 80-93 (1998).
9. De Baets, B.: Analytical solution methods for fuzzy relational equations, in: Dubois, D., Prade, H. (Eds.), *Fundamentals of fuzzy sets*, Springer Science+Business Media, New York, pp. 291-340 (2000).
10. Wang, X.P., Xiong, Q.: The solution set of a fuzzy relational equation with sup-conjunctor composition in a complete lattice, Fuzzy Sets and Systems **153**, 249-260 (2005).
11. Hosseinyazdi, M.: The optimization problem over a distributive lattice, J. of Global Optimization **41**, 283-298 (2008).
12. Hosseinyazdi, M.: An Application of Linear Algebra over Lattices, J. of Mathematical Extension **3**, 71-86 (2008).
13. Perfilieva, I.: Systems of Fuzzy Relation Equations in a Space with Fuzzy Preorder, in: *Proc. IFSA/EUSFLAT Conf. 2009*, pp. 1601-1605 (2009).

14. Wang, X.P., Xia, C.: The solution sets of infinite fuzzy relational equations with sup-conjunctor composition on complete distributive lattices, Fuzzy Sets and Systems **160**, 2989-3006 (2009).
15. Zhang, K.L., Li, D.H., Song, L.X.: On finite relation equations with sup-conjunctor composition over a complete lattice, Fuzzy Sets and Systems **160**, 119-128 (2009).
16. Ignjatović, J., Ćirić, M., Bogdanović, S.: On the greatest solutions to weakly linear systems of fuzzy relation inequalities and equations, Fuzzy Sets and Systems **161**, 3081-3113 (2010).
17. Bartl, E., Bělohlávek, R.: Sup-t-norm and inf-residuum are a single type of relational equations, Int. J. of General Systems **40**, 599-609 (2011).
18. Díaz, J.C., Medina, J.: Concept lattices in fuzzy relation equations, in: *Proc. 8th Conf. on Concept Lattices and Their Applications (CLA 2011)*, pp. 75-86 (2011).
19. Jiménez, J., Montes, S., Šešelja, B., Tepavčević, A.: Fuzzy Relational Inequations and Equations in the Framework of Control Problems, in: Liu, W. (Ed.), *Proc ECSQARU 2011*, LNAI 6717, Springer-Verlag Berlin Heidelberg, pp. 606-615 (2011).
20. Bartl, E.: Fuzzy Relational Equations, Ph.D. Dissertation, Dpt. of Computer Science, Palacky University of Olomouc (2012).
21. Bělohlávek, R.: Sup-t-norm and inf-residuum are one type of relational product: Unifying framework and consequences. Fuzzy Sets and Systems **197**, 45-58 (2012).
22. Díaz, J.C., Medina, J.: Solving systems of fuzzy relation equations by fuzzy property-oriented concepts, Information Sciences **222**, 405-412 (2013).
23. Perfilieva, I.: Finitary solvability conditions for systems of fuzzy relation equations, Information Sciences **234**, 29-43 (2013).
24. Bartl, E., Klir G.J.: Fuzzy relational equations in general framework, Int. J. of General Systems **43**, 1-18 (2014).
25. Jiménez, J., Montes, S., Šešelja, B., Tepavčević, A.: Fuzzy correspondence inequations and equations, Fuzzy Sets and Systems **239**, 81-90 (2014).
26. Sun, F., Qu, X.B., Wang, X.P., Zhu, L.: Remarks on minimal solutions of a system of fuzzy relation equations over complete infinitely distributive lattices, Soft Computing **20**, 423-428 (2016).
27. Medina, J.: Minimal solutions of generalized fuzzy relational equations: Clarifications and corrections towards a more flexible setting, Int. J. of Approximate Reasoning **84**, 33-38 (2017).
28. Zuo, K., Wang, X.P., Zhang, X.: Solutions of fuzzy correspondence inequations with sup-conjunctor composition, in: *Proc. 3rd Int. Conf. on Computational Intelligence & Communication Technology (CICT 2017)*, IEEE, pp. 1-6 (2017).
29. Stepanović, V.: Fuzzy set inequations and equations with a meet-continuous codomain lattice, J. of Intelligent & Fuzzy Systems **34**, 4009-4021 (2018).
30. Cornejo, M.E., Lobo, D., Medina, J.: Solving Generalized Equations with Bounded Variables and Multiple Residuated Operators, Mathematics **8**, 1992, 1-22 (2020).
31. Qiao, S., Zhu, P.: Extremal solutions to fuzzy relation equations and inequalities with three unknowns, J. of Intelligent & Fuzzy Systems **38**, 5055-5076 (2020).
32. Turunen, E.: Necessary and sufficient conditions for the existence of solution of generalized fuzzy relation equations $A \leftrightarrow X = B$, Information Sciences **536**, 351-357 (2020).
33. Bartl, E., Trnecka, M.: Covering of minimal solutions to fuzzy relational equations, Int. J. of General Systems **50**, 117-138 (2021).
34. Sun, F., Qu, X. B.: Resolution of fuzzy relation equations with increasing operations over complete lattices, Information Sciences **570**, 451-467 (2021).
35. Cao, N., Štěpnička, M.: On solvability of systems of partial fuzzy relational equations, Fuzzy Sets and Systems **450**, 87-117(2022).

36. Micić, I., Jančić, Z., Stanimirović, S.: Computation of solutions to certain nonlinear systems of fuzzy relation inequations, in: Poulakis, D., Rahonis, G. (Eds.), *Algebraic Informatics. Proc. CAI 2022*, LNCS 13706. Springer, pp. 192-202 (2022).
37. Stepanović, V., Tepavčevič, A.: Fuzzy sets (in)equations with a complete codomain lattice, Kybernetika **58**, 145-162 (2022).

12.4.6 *Other Variants and General Contributions*

1. Cheng, L., Peng, B.: The fuzzy relation equation with union or intersection preserving operator, Fuzzy Sets and Systems **25**, 191-204 (1988).
2. Zhao, C.K.: Inverses of L-fuzzy matrices, Fuzzy Sets and Systems **34**, 103-116 (1990).
3. Fernández, M.J., Suárez, F., Gil, P.: Equations of fuzzy relations defined on fuzzy subsets, Fuzzy Sets and Systems **52**, 319-336 (1992).
4. Cho, H.H.: Regular fuzzy matrices and fuzzy equations, Fuzzy Sets and Systems **105**, 445-451 (1999).
5. Fang, S.C., Li, G.: Solving fuzzy relation equations with a linear objective function, Fuzzy Sets and Systems **103**, 107-113 (1999).
6. Wang, S.: Interval computations for fuzzy relational equations and cooperative game theory, PhD Dissertation, North Carolina State University (2002).
7. Kohout, L.J.: Equations and Inequalities With BK-Products of Relations, in: Nikravesh, M. et al. (Eds.), *Fuzzy Partial Differential Equations and Relational Equations*, Springer-Verlag Berlin Heidelberg, pp. 225-250 (2004).
8. Kohout, L.J.: Theory of Fuzzy Generalized Morphisms and Relational Inequalities, Int. J. of General Systems **33**, 339-360 (2004).
9. Han, S., Li, H.: Solvability of L-fuzzy Relation Equations in Generalized One-sided Residuated Groupoids, Southeast Asian Bulletin of Mathematics **30**, 671-683 (2006).
10. Perfilieva,I., Nosková, L.: System of fuzzy relation equations with inf $\rightarrow$ composition: Complete set of solutions, Fuzzy Sets and Systems **159**, 2256-2271 (2008).
11. Li, P., Fang, S.C.: A survey on fuzzy relational equations, part I: classification and solvability. Fuzzy Optimization and Decision Making **8**, 179-229 (2009).
12. Perfilieva, I.: Fuzzy relation equations in semilinear spaces, in: Hüllermeier, E. et al. (Eds.), *Proc. IPMU 2010, Part I*, Springer-Verlag Berlin Heidelberg, pp. 545–552 (2010).
13. Díaz, J. C., Medina, J.: Multi-adjoint relation equations: Definition, properties and solutions using concept lattices, Information Sciences **253**, 100-109 (2013).
14. Kim, Y.C.: Fuzzy relation equations in pseudo BL-algebras, Int. J. of Fuzzy Logic and Intelligent Systems **13**, 208-214 (2013).
15. Tan, Y.J.: On invertible matrices over commutative semirings, Linear and Multilinear Algebra **61**, 310-324 (2013).
16. Bartl, E.: Minimal solutions of generalized fuzzy relational equations: Probabilistic algorithm based on greedy approach, Fuzzy Sets and Systems **260**, 25-42 (2015).
17. Liu, C.C., Lur, Y.Y., Wu, Y.K.: Some properties of bipolar max-min fuzzy relational equations, in: *Proc. of the Int. MultiConference of Engineers and Computer Scientists 2015, Vol II*, pp. 955–969 (2015).
18. Li, P., Jin, Q.: On the resolution of bipolar max-min equations, Kybernetika **52**, 514-530 (2016).
19. Bartl, E., Prochazka, P.: Do We Need Minimal Solutions of Fuzzy Relational Equations in Advance, IEEE Transactions on Fuzzy Systems **25**, 1356-1363 (2017).

20. Medina, J.: Minimal solutions of generalized fuzzy relational equations: Clarifications and corrections towards a more flexible setting, Int. J. of Approximate Reasoning **84**, 33-38 (2017).
21. Cornejo, M.E., Lobo, D., Medina, J., De Baets, B.: Bipolar equations on complete distributive symmetric residuated lattices: The case of a join-irreducible right-hand side, Fuzzy Sets and Systems **442**, 92-108 (2022).

12.5 Applications of Fuzzy Relation Equations

At the risk of being accused of another of already many repetitions, I deliberately include this short section for sake of completeness to present one of the most popular running examples for the application of fuzzy relation equations.

At the beginning of Section 10.1.3 above it has been asserted that large parts of human knowledge are qualitative and cast in natural languages expressions and sentences. As a consequence, there can be no doubt that human reasoning proceeds in such qualitative terms using suitable mechanisms. Formal models within FST for describing that kind of information processing run under the heading of **approximate reasoning**. Following the historical development we will outline the main concepts of approximate reasoning and their usage by presenting two classical concepts of FST: the generalized modus ponens conceived in [71], and fuzzy rule based systems initiated by the idea of fuzzy control as presented in [37].

12.5.1 *Inference Rules*

An *argument* is a sequence of statements (or propositions), of which all but the final one are called *premises*, and the final proposition is called the *conclusion*. An argument is called *valid* if the truth of all its premises implies that also the conclusion is true.

A **rule of inference** is a template for building valid arguments and is usually given in one of the two forms shown below. The rule is called **valid** with respect to the semantics of (classical) logic if true premises yield a conclusion that is also true (under a certain interpretation). In other words such a rule states that whenever in the course of some logical derivation the given premises have been obtained, the specified conclusion can be taken for granted as well. That is what we are usually interested in when searching for a proof. For some more details about the intricate relationship between the truth of statements and the validity of arguments see the section on 'Validity and Truth' in [7, p. 27ff].

Premise p_1 $\wedge$ *Premise* p_2 $\wedge$ ...
Premise p_n

Conclusion q $\therefore$

In a more compact notation this can be written as

$$p_1, p_2, \ldots, p_n \vdash q.$$

So an inference rule is valid if and only if $p_1 \wedge p_2 \wedge \ldots \wedge p_n \rightarrow q$ is a tautology.

In what follows we will focus on the so-called **hypothetical syllogisms** which contain one hypothetical proposition (or conditional) as premise (mixed), or only conditionals (pure):

- **modus ponens:** $p, p \to q \vdash q$, (MP)
- **modus tollens:** $p \to q, \neg q \vdash \neg p$,
- Pure hypothetical syllogism: $p \to q, q \to r \vdash p \to r$.

Within the area of approximate reasoning and its application in fuzzy control, decision-making etc., generalized versions of modus ponens are of utmost significance. The original extension of inference from two-valued logic to fuzzy logic was given by Zadeh's generalized modus ponens (GMP). In [71], another seminal work of Zadeh, he introduced the idea of formulating control algorithms by logical rules of qualitative nature. Such rules have the structure "*If A then B*", where A and B denote fuzzy sets over corresponding universes, whence they are also called fuzzy conditionals.

As an example consider *linguistic variables* (i.e variables whose possible values are defined in linguistic terms) like TEMPERATURE or SPEED with possible values say low, medium, and high, for sake of simplicity. An elementary piece of information is e.g. "TEMPERATURE *is* high", and typical fuzzy conditionals are "*If* SPEED *is* low *then* TEMPERATURE *is* high'. These are the elementary building blocks out of which fuzzy rule bases and fuzzy inference rules are built. If we have suitable mechanisms that can process qualitative information of this kind, then we can use the knowledge and experience of human experts in solving certain problems, instead of working with e.g. complex systems of differential equations.

Let us now try to find a formulation of the two-valued form of MP that is suitable for generalization to the fuzzy case by following the proven standard strategy of identifying sets with their characteristic functions (see [27, Section 8.6]): Let x and y be two variables that take their values from sets U and V, respectively, and assume that for all values $u \in U, v \in V$ the variables are related by a function $y = f(x)$. If the value of x is in a given set $A \subseteq U$, then one can infer that $f(x) \in \{v \in V : v = f(x), x \in A\}$.

We generalize this by assuming that the variables are related by an arbitrary relation $\rho \subseteq U \times V$, and let furthermore $A \subseteq U$ and $B \subseteq V$. Then, given $x = u$ we can infer that $v \in u\rho$, and given $u \in A$ we have $v \in B = \bigcup_{u \in A} u\rho = \{v \in V : u \in A \wedge u\rho v\}$. Thus, in terms of characteristic functions we get, for all $v \in V$,

$$B(v) = \bigvee_{u \in A} \big(A(u) \wedge \rho_{\to}(u, v)\big), \quad i.e. \quad B = A \circ \rho_{\to}, \tag{12.24}$$

where $\rho_{\to}$ models the IF-THEN-rule, and A models the fact (obtained by a measurement or observation).

Generalizing the above further by letting A be a fuzzy set over U and ρ be a fuzzy relation over $U \times V$, (12.24) is called the **compositional rule of inference** (CRI), and can clearly be written as the relation equation $B = A \circ \rho_{\to}$. We have finally arrived at a formal model for a generalized version of MP which allows us to handle e.g. inferences like

If temperature is high then fan speed is medium.
Temperature is high.

Fan speed is medium. $\therefore$

The relation $\rho_\rightarrow$ which is implicit in fuzzy conditionals can be modeled by using a conjunctor $\mathcal{C}$ and an implicator $\mathcal{I}$ (see Def. 9.14) so that the general form of the CRI appears as

$$B(y) = \bigvee_{x \in U} \mathcal{C}\Big(A(x), \mathcal{I}\big(A(x), B(y)\big)\Big). \tag{12.25}$$

An entry $\rho_\rightarrow(u, v)$ is thought of the strength of the rule 'IF u THEN v', or the degree to which v is related to u.

12.5.2 *Fuzzy Rule Based Systems*

Among all the techniques developed using fuzzy sets, fuzzy IF-THEN rules are by far the most visible due to their wide range of successful applications in many disciplines such as control systems, decision making, pattern recognition, etc. Among these fuzzy control has been the most popular application of rule based approximate reasoning, since it has been used around the mid nineties as a marketing argument for e.g. washing machines, cameras, or cars. Fuzzy control[1] consists of a set of techniques for transforming such qualitative rules into precise control actions. To describe its main concepts and strategies and their properties is beyond the current scope, but we will give a short overview on how relations and relation equations contribute to the evaluation of rule bases in general. For a short introduction to fuzzy control see e.g [28] [17, Chapter 1], or [27, Section 12.2], and for more details [17, 38] or [61].

Fuzzy IF-THEN statements can be regarded in two ways: as *fuzzy mapping rules* or as *fuzzy implication rules.* A fuzzy mapping rule describes a functional relationship between inputs and an output using linguistic terms. Each rule approximates a small segment of the function, and the entire function is approximated by a set of rules. Such a collection of fuzzy mapping rules is sometimes called a *fuzzy rule-based model,* and is basic to the so-called Mamdani-approach to fuzzy control to which we now turn.

From the example above it can be seen that GMP is very near to classical MP, since its formal structure is $A, A \rightarrow B \vdash B$, i.e. it still assumes an exact formal matching of the fact with the antecedent of the implication. In the spirit of fuzzy set theory (compare the quotations on page 294), more can be achieved than only this mere symbolic matching: Fuzzy rules allow to draw a conclusion for a given input that was possibly never encountered before. Using the following running example in approximate reasoning literature,

---

[1]Quite generally spoken, **control theory** is concerned with influencing systems in order to achieve that certain output quantities take a desired course. The nature of the system does not matter, only its dynamic behavior is of interest. In classical control theory, which is mostly concerned with technical systems, the system that will be influenced is called the *plant.* But systems could as well be social, chemical or biological, as, for instance, the system of a national economy with the rate of inflation as output quantity. ([38])

If a tomato is red then it is ripe.
This tomato is red.

This tomato is ripe. $\therefore$

what can be said of a tomato that is slightly red? Put more formally, we want to make inferences of the form

If x is A then z is C.
x is A'.

---

z is C'. $\therefore$

So, the formal structure of GMP thus becomes $A', A \to C \vdash C'$, where A and A' are fuzzy sets over the same universe describing 'similar' meanings.

Now we will take a short look on the role and structure of fuzzy rule bases in fuzzy control. In fuzzy control terminology, a rule is a fuzzy conditional in which the antecedent is a restriction in its application domain and the consequent is some control action for the system under control. In cases when rules need a more complex structure than the one we have considered until now, one can use

$$\text{If } x_1 \text{ is } A_1 \text{ and ... and } x_m \text{ is } A_m \text{ then } y \text{ is } C\,.$$

for the description of a multidimensional antecedent. If we take the Cartesian product $A = A_1 \times \cdots \times A_m$, which gives $A(x_1,\ldots,x_m) = \bigwedge_{i \in \underline{m}} A_i(x_i)$ by (9.6), then we have again the simple form of a fuzzy conditional that we have used above in (12.25).

It has been said above that the rule base of the most common approach to fuzzy controllers ([37]) is regarded as collection of fuzzy mapping rules, i.e. its IF-THEN statements do not represent a causal relationship nor an implication in the usual sense. They convey the idea that for the rule's antecedent, the consequent is related to it in a "reasonable" way such that the rule contributes to the piecewise definition of a function (see [38]):

$$f(x_1, x_2, \ldots, x_n) := \begin{cases} C_1 & \text{if } x_1 \text{ is } A_{11} \text{ and...and } x_m \text{ is } A_{1m}, \\ \vdots & \\ C_n & \text{if } x_1 \text{ is } A_{n1} \text{ and...and } x_m \text{ is } A_{nm}. \end{cases}$$

Each elementary piece of expert knowledge, i.e. each fuzzy conditional, contributes to the overall conclusion to the amount its antecedent matches with a given fact or observation.

Henceforth, in order to achieve a clear representation of the main concepts, we will restrict the subsequent considerations to simple rules. Thus, a rule base $\mathcal{R} = \{R_1, \ldots, R_n\}$ of a Mamdani controller with rules of the form R_k: 'If x is A_k then y is C_k' comprises information about n of possible value pairs for a functional relationship:

$$A_1 \text{ and } C_1 \text{ OR } \cdots \text{ OR } A_n \text{ and } C_n. \tag{12.26}$$

This means that we have to use a disjunctor for aggregating the single conclusions of all rules in $\mathcal{R}$, and how much the conclusion from each rule does contribute to the overall result must be influenced by how much its antecedent matches with the fact.

Given the observation 'x is A'' and the rule R_i: 'If x is A_i then y is C_i' we first calculate the *compatibility* (or *matching*) *degree* between antecedent A_i and A' as

$$\gamma_i^{A'} = \bigvee_{x \in U} \big(A'(x) \wedge A_i(x)\big) \tag{12.27}$$

and subsequently apply it to the fuzzy set C_i describing the rule's consequent to get the output fuzzy set of the rule as

$$C_i'(y) = \gamma_i^{A'} \wedge C_i(y). \tag{12.28}$$

This means clipping or cutting off C_i. Note that using product instead of minimum would result in some kind of scaling C_i down. $\gamma_i^{A'}$ is also called *activation* or *firing degree* of a rule. (12.28) is called the *CM approach* (for **C**ompatibility or **M**atching). According to (12.26) the overall result is given by

$$C'(y) = \bigvee_{i \in \underline{n}} C_i'(y) \tag{12.29}$$

Recalling that in the Mamdani view a rule 'If x is A then y is C' has been rephrased as 'A and C', i.e. $\rho_{\rightarrow}(a, c) = a \wedge c$, the following result holds:

Proposition 12.58. ([27, Section 11.2]) *For every fuzzy input A', the CM approach* (12.29) *is equivalent to the CRI approach* (12.24).

Proof. Let 'If x is A then y is C' be a fuzzy conditional, and let A' represent an observed value. Using (12.28) and (12.29), with the CM approach the membership degree of an element y in C' is calculated as

$$\begin{aligned}
C'(y) &= \textstyle\bigvee_{i \in \underline{n}} C_i'(y) \\
&= \textstyle\bigvee_{i \in \underline{n}} \Big(\bigvee_{x \in U} \big[A'(x) \wedge A_i(x)\big] \wedge C_i(y)\Big) \\
&= \textstyle\bigvee_{i \in \underline{n}} \bigvee_{x \in U} \big[A'(x) \wedge A_i(x) \wedge C_i(y)\big] \\
&= \textstyle\bigvee_{x \in U} \bigvee_{i \in \underline{n}} \big[A'(x) \wedge \big(A_i(x) \wedge C_i(y)\big)\big] \\
&= \textstyle\bigvee_{x \in U} \big[A'(x) \wedge \bigvee_{i \in \underline{n}} \big(A_i(x) \wedge C_i(y)\big)\big] \\
&= \textstyle\bigvee_{x \in U} \big[A'(x) \wedge \rho_{\rightarrow}(x, y)\big] \\
&= (A' \circ \rho_{\rightarrow})(y).
\end{aligned}$$

□ □

We have now demonstrated how, in the context of the Mamdani interpretation of fuzzy rules, fuzzy relations arise in a natural way. But given a rule base $\mathcal{R} = \{R_1, \ldots, R_n\}$, what we really like to know is a fuzzy relation ρ which is a solution for the system of fuzzy relation equations

$$A_i \circ \rho = C_i, \quad i \in \underline{n}. \tag{12.30}$$

In other words, we are interested in a function that is in agreement with the n data points (A_i, C_i) given by $\mathcal{R}$. A relation ρ with the property (12.30) is also called a *correct model* or *interpolation* of $\mathcal{R}$. In general, such a correct model may not exist; so it is interesting to know necessary and sufficient conditions for when it does.

Let $\rho \in \mathcal{F}^L_{X\times Y}$, $\sigma \in \mathcal{F}^L_{Y\times Z}$, and $\tau \in \mathcal{F}^L_{X\times Z}$ denote three L-fuzzy relations. From Th. 12.3(1) we know that $S_\sigma := \mathbb{S}(\rho,\tau) \neq \emptyset$ iff $\hat{\sigma} := \rho^{\smile} @\tau \in S_\sigma$, and that $\hat{\sigma}$ then is the greatest solution of $\rho\sigma = \tau$. Now, by Def. 12.1 we have

$$\rho @\sigma(x,z) = \bigwedge_y \big(x\rho y \frown y\sigma z\big).$$

Adapting this to the rule base configuration we get: $A_i \in \mathcal{F}^L(U)$, $\rho \in \mathcal{F}^L_{U\times V}$, and $C_i \in \mathcal{F}^L(V)$. Furthermore, the greatest solution $\hat{\rho}_i \in \mathcal{F}^L_{U\times V}$ of $A_i \circ \rho = C_i$ is $A_i^{\smile} @C_i = A_i @C_i$, i.e. $\hat{\rho}_i(u,v) = A_i(u) \frown C_i(v)$.

Proposition 12.59. *A system of fuzzy equations* (12.30) *has a solution* $\hat{\rho}$ *iff*

$$A_i \circ \hat{\rho} = C_i,\ i \in \underline{n} \tag{12.31}$$

with $\hat{\rho} = \bigcap_{i\in\underline{n}}(A_i @C_i)$. *In case a solution exists,* $\hat{\rho}$ *is the greatest one.*

Proof. Obviously, if (12.31) holds true, then the system (12.30) has $\hat{\rho}$ as a solution. Conversely, assume that ρ' is a solution of (12.30); then $\rho' \subseteq A_i @C_i$ for each $i \in \underline{n}$, whence $\rho' \subseteq \hat{\rho}$. Using monotonicity and subdistributivity of relation composition we get for each $j \in \underline{n}$

$$\begin{aligned} C_j &= A_j \circ \rho' \\ &\subseteq A_j \circ \hat{\rho} \\ &= A_j \circ \textstyle\bigcap_{i\in\underline{n}}(A_i @C_i) \\ &\subseteq \textstyle\bigcap_{i\in\underline{n}}\big(A_j \circ (A_i @C_i)\big) \\ &= C_j \cap \textstyle\bigcap_{j\neq i\in\underline{n}}\big(A_j \circ (A_i @C_i)\big) \\ &\subseteq C_j. \end{aligned}$$

Hence, $A_j \circ \hat{\rho} = C_j$. Thus, given any solution ρ' of the system (12.30), (12.31) holds and $\rho' \subseteq \hat{\rho}$, and $\hat{\rho}(u,v) = \bigwedge_{i\in\underline{n}}\big(A_i(u) \frown C_i(v)\big)$. □ ⊔

The proposition above gives us a result about the solvability of certain systems of fuzzy relation equations in connection to the Mamdani approach to fuzzy control in terms of sharp/exact equality; but, as we have noted already earlier at the end of Section 9.3.2 on properties of fuzzy relations, this is not fully in the spirit of fuzzy set theory. Accordingly, there have also been investigations about approximate solutions in case exact ones do not exist; see e.g. [23, 46]. There are also other necessary and sufficient conditions that avoid the check via $\hat{\rho}$ ([48]), among them ones that allow to get a decision directly in terms of the input-output fuzzy sets only ([49]).

The considerations above provide a tiny glimpse only onto the rich literature on approximate reasoning and fuzzy control: they comprise the classical starting points of a huge amount of subsequent investigations dealing with various focal subjects. Let me finally add a few more hints indicating the broad spectrum of ideas in connection with the current topic: [15, 47, 53].

References

[1] Bartl, E.: Minimal solutions of generalized fuzzy relational equations: Probabilistic algorithm based on greedy approach. Fuzzy Sets and Systems **260**, 25-42 (2015).

[2] Bartl, E., Bělohlàvek, R.: Hardness of Solving Relational Equations, IEEE Transactions on Fuzzy Systems **23**, 2435-2438 (2015).

[3] Bartl, E., Klir, G.J.: Fuzzy relational equations in general framework, Int. J. of General Systems **43**, 1-18 (2014).

[4] Bělohlàvek, R.: *Fuzzy Relational Systems - Foundations and Principles*, Springer Science + Business Media, New York (2002).

[5] Birkhoff, G.: *Lattice Theory*, 3rd ed., American Mathematical Society, Providence, Rhode Island (1973).

[6] Cechlárová, K.: Unique solvability of Max-min fuzzy equations and strong regularity of matrices over fuzzy algebra, Fuzzy sets and systems **75**, 165-177 (1995).

[7] Copi, I.M., Cohen, C., McMahon, K.: *Introduction to Logic*, 14th ed., Pearson Education Limited, Harlow, Essex (2014).

[8] Czogala, E., Drewniak, J., Pedrycz, W.: Fuzzy relation equations on a finite set. Fuzzy Sets and systems **7**, 89-101 (1982).

[9] De Baets, B.: An Order-Theoretic Approach to Solving Sup-T Equations, in: Ruan, D. (Ed), *Fuzzy Set Theory and Advanced Mathematical Applications*, Springer Science+Business Media, New York, pp. 67-87 (1995).

[10] De Baets, B.: Sup-T Equations: State of the Art, in: Kaynak, O., Zadeh, L.A. et al. (Eds), *Computational Intelligence: Soft Computing and Fuzzy-Neural Integration with Applications*, Springer Verlag, Berlin Heidelberg, pp. 80-93 (1998).

[11] De Baets, B.: Analytical solution methods for fuzzy relational equations, in: Dubois, D., Prade, H. (Eds.), *Fundamentals of Fuzzy Sets*, Springer Science+Business Media, New York, pp. 291-340 (2000).

[12] Di Nola, A.: Relational equations in totally ordered lattices and their complete resolution, J. of Mathematical Analysis and Applications **107**, 148-155 (1985).

[13] Di Nola, A.: On solving relational equations in Brouwerian lattices, Fuzzy Sets and Systems **34**, 365-376 (1990).

[14] Di Nola, A., Sessa, S.: Finite Fuzzy Relation Equations with a Unique Solution in Linear Lattices, J. of Mathematical Analysis and Applications **132**, 39-49 (1988).

[15] Di Nola, A., Sessa, S., Pedrycz, W.: A study on approximate reasoning mechanisms via fuzzy relation equations, Int. J. of Approximate Reasoning **6**, 33-44 (1992).

[16] Di Nola, A., Sessa, S., Pedrycz, W., Sanchez, E.: *Fuzzy Relation Equations and Their Applications to Knowledge Engineering*, Kluwer Academic Publishers, Dordrecht (1989).

[17] Feng, G.: *Analysis and Synthesis of Fuzzy Control Systems*, CRC Press, Boca Raton (2010).

[18] Gavalec, M.: Solvability and unique solvability of max–min fuzzy equations, Fuzzy Sets and Systems **124**, 385-393 (2001).

[19] Gottwald, S.: Generalized solvability criteria for fuzzy equations, Fuzzy Sets and Systems **17**, 285-296 (1985).

[20] Gottwald, S.: Characterizations of the solvability of fuzzy equations, Elektronische Informationsverarbeitung und Kybernetik **22**, 67-91 (1986).

[21] Gottwald, S.: *Fuzzy Sets and Fuzzy Logic*, Vieweg Verlag, Braunschweig/Wiesbaden (1993).

[22] Gottwald, S.: Approximately solving fuzzy relation equations: Some mathematical results and some heuristic proposals, Fuzzy Sets and Systems **66**, 175-193 (1994).

[23] Gottwald, S.: Fuzzy control and fuzzy relation equations. A unified view as interpolation problem, in: *NAFIPS 2004*, IEEE Comp. Soc., Piscataway/NJ, pp. 270-275 (2004).

[24] Gottwald, S.: Calculi of Information Granules: Fuzzy Relational Equations, in: Pedrycz, W., Skowron, A., Kreinovich, V. (Eds), *Handbook of Granular Computing*, John Wiley & Sons, pp. 225-248 (2008).

[25] Han, S.C., Li, H.X., Wang, J.Y.: Resolution of matrix equations over arbitrary Brouwerian lattices, Fuzzy Sets and Systems **159**, 40-46 (2008).

[26] Higashi, M., Klir, G.J.: Resolution of finite fuzzy relation equations, Fuzzy Sets and Systems **13**, 65-82 (1984).

[27] Klir, G., Yuan, B.: *Fuzzy Sets and Fuzzy Logic: Theory and Applications*, Prentice Hall, New Jersey (1995).

[28] Lee, C.C.: Fuzzy logic in control systems: fuzzy logic controller. I, IEEE Transactions on Systems, Man and Cybernetics **20**, 404-418 (1990).

[29] Lettieri, A., Liguori, F.: Characterization of some fuzzy relation equations provided with one solution on a finite set, Fuzzy Sets and Systems **13**, 83-94 (1984).

[30] Lettieri, A., Liguori, F.: Some results on fuzzy relation equations provided with one solution, Fuzzy Sets and Systems **17**, 199-209 (1985)

[31] Li, J.X.: The smallest solution of max-min fuzzy equations, Fuzzy Sets and Systems **41**, 317-327 (1990).

[32] Li, G.Z., Fang, S.C.: Solving Interval-Valued Fuzzy Relation Equations, IEEE Transactions on Fuzzy Systems **6**, 321-324 (1998).

[33] Li, P., Fang, S.C.: On the resolution and optimization of a system of fuzzy relational equations with sup-T composition, Fuzzy Optimization and Decision Making, **7**, 169-214 (2008).

[34] Li, P., Fang, S.C.: A note on solution sets of interval-valued fuzzy relational equations, Fuzzy Optimization and Decision Making, **8**, 115-121 (2009).

[35] Li, P., Fang, S.C.: A survey on fuzzy relational equations, part I: classification and solvability, Fuzzy Optimization and Decision Making **8**, 179-229 (2009).

[36] Li, P., Fang, S.C.: On the unique solvability of fuzzy relational equations. Fuzzy Optimization and Decision Making, **10**, 115-124 (2011).

[37] Mamdani, E.H.: Advances in the linguistic synthesis of fuzzy controllers, Int. J. Man-Machine Studies **8**, 669-678 (1976).

[38] Michels, K., Klawonn, F., Kruse, R., Nürnberger, A.: *Fuzzy Control-Fundamentals, Stability and Design of Fuzzy Controllers*, Springer-Verlag Berlin Heidelberg (2006).

[39] Miyakoshi, M., Shimbo, M.: Sets of solution-set-invariant coefficient matrices of simple fuzzy relation equations, Fuzzy Sets and Systems **21**, 59-83 (1987).

[40] Pedrycz, W.: Approximate solutions of fuzzy relational equations, Fuzzy Sets and Systems **28**, 183-202 (1988).

[41] Peeva, K.: Systems of linear equations over a bounded chain, Acta Cybernetica **7**, 195-202 (1985).

[42] Peeva, K.: Fuzzy linear systems, Fuzzy Sets and Systems **49**, 339-355 (1992).

[43] Peeva, K.: Universal algorithm for solving fuzzy relational equations, Italian J. of Pure and Applied Mathematics **19**, 169-188 (2006).

[44] Peeva, K.: Resolution of fuzzy relational equations - Method, algorithm and software with applications. Information Sciences **234**, 44-63 (2013).

[45] Peeva, K., Kyosev, Y.: *Fuzzy Relational Calculus. Theory, Applications and Software*, World Scientific, Singapore (2004).

[46] Perfilieva, I.: Fuzzy function as an approximate solution to a system of fuzzy relation equations, Fuzzy Sets and Systems **147**, 363-383 (2004).

[47] Perfilieva, I.: Logical foundations of rule-based systems, Fuzzy Sets and Systems **157**, 615-621 (2006).

[48] Perfilieva, I., Novák, V.: System of fuzzy relation equations as a continuous model of IF-THEN rules, Information Sciences **177**, 3218-3227 (2007).

[49] Perfilieva, I., Tonis, A.: Compatibility of systems of fuzzy relation equations, Int. J. of General Systems **29**, 511-528 (2000).

[50] Sanchez, E.: Resolution of composite fuzzy relation equations, Information and Control **30**, 38-48 (1976).

[51] Sanchez, E.: Solutions in composite fuzzy relation equations: Applications to medical diagnosis in Brouwerian logic, in: Gupta, M.M., Saridis, M.M., Gaines, B.R. (Eds.), *Fuzzy Automata and Decision Processes*, Elsevier/North-Holland, New York, pp. 221-234 (1977).

[52] Shieh, B.S.: New resolution of finite fuzzy relation equations with max-min composition, Int. J. of Uncertainty, Fuzziness and Knowledge-Based Systems **16**, 19-33 (2008).

[53] Štěpnička, M., Mandal, S.: Fuzzy inference systems preserving Moser–Navara axioms, Fuzzy Sets and Systems **338**, 97-116 (2018).

[54] Sun, F.: Conditions for the existence of the least solution and minimal solutions to fuzzy relation equations over complete Brouwerian lattices, Information Sciences **205**, 86-92 (2012).

[55] Sun, F., Qu, X.B., Wang, X.P., Zhu, L.: On pre-solution matrices of fuzzy relation equations over complete Brouwerian lattices, Fuzzy Sets and Systems **384**, 34-53 (2020).

[56] Sun, F., Wang, X.P., Qu, X.B.: Minimal join decompositions and their applications to fuzzy relation equations over complete Brouwerian lattices, Information Sciences **224**, 143-151 (2013).

[57] Wagenknecht, M., Hartmann, K.: Fuzzy modeling with tolerances, Fuzzy Sets and Systems **20**, 325-332 (1986).

[58] Wagenknecht, M., Hartmann, K.: On direct and inverse problems for fuzzy equation systems with tolerances, Fuzzy Sets and Systems **24**, 93-102 (1987).

[59] Wagenknecht, M., Hartmann, K.: On the existence of minimal solutions for fuzzy equations with tolerances, Fuzzy Sets and Systems **34**, 237-244 (1990).

[60] Wang, X.P.: Method of solution to fuzzy relation equations in a complete Brouwerian lattice, Fuzzy Sets and Systems **120**, 409-414 (2001).

[61] Wang, L.X.: *A Course in Fuzzy Systems and Control*, Prentice Hall, Upper Saddle River, New Jersey (1996).

[62] Wang, H.F., Chang, Y.C.: Resolution of composite interval-valued fuzzy relation equations, Fuzzy Sets and Systems **44**, 227-240 (1991).

[63] Wang, S., Fang, S.C., Nuttle, H.L.W.: Solution sets of interval-valued fuzzy relational equations, Fuzzy Optimization and Decision Making **2**, 41-60 (2003).

[64] Wang, S., Fang, S.C., Nuttle, H.L.W.: Errata, Fuzzy Optimization and Decision Making **6**, 297 (2007).

[65] Wang, H.F., Hsu, H.M.: An alternative approach to the resolution of fuzzy relation equations, Fuzzy Sets and Systems **45**, 203-213 (1992).

[66] Wang, X.P., Liu, L.J., Zhang, X.: Sets of solution-set-invariant coefficient matrices of fuzzy relation equations with max-min composition, Journal of Intelligent & Fuzzy Systems **34**, 4067-4078 (2018).

[67] Wang, X.P., Sun, F.: Resolution of Fuzzy Relation Equations with Sup-Inf Composition over Complete Brouwerian Lattices-a review, in: *IFSA-EUSFLAT 2009 Proc.*, pp. 82-87 (2009).

[68] Xiao, G., Zhu, T., Chen, Y., Yang, X.: Linear searching method for solving approximate solution to system of max-min fuzzy relation equations with application in the instructional information resources allocation, IEEE Access **7**, 65019–65028 (2019).

[69] Xiong, Q.Q., Wang, X.P.: Some Results of Interval-valued fuzzy relational equations with sup-conjunctor composition, in: *IFSA/NAFIPS 2013 Proc.*, pp. 333-337 (2013).

[70] Yeh, C.T.: On the minimal solutions of max–min fuzzy relational equations, Fuzzy Sets and Systems **159**, 23-39 (2008).

[71] Zadeh, L.A.: Outline of a New Approach to the Analysis of Complex Systems and Decision Processes, IEEE Transactions on Systems, Man and Cybernetics **3**, 28-44 (1973).

[72] Zhao, C.K.: On matrix equations in a class of complete and completely distributive lattice, Fuzzy Sets and Systems **22**, 303-320 (1987).

CHAPTER 13

FUZZY GRAPHS

The situation in the fuzzy context is quite analogous to the classical one: A topic that must not be missing in connection to fuzzy relations are fuzzy graphs, whence we include this short introduction to the subject quasi as an appetizer. Let us start with a quotation from [33, page 5]:

> "The most simple and least strict **definition of a graph** is the following: a graph is a set of points and lines connecting some pairs of the points. Mathematicians name and number everything: in graph theory, points are called **vertices**, and lines are called **edges**."

Reading these two sentences, probably nobody would expect that there exist dozens of books on graph theory, often comprising several hundred pages and about rather sophisticated things. Our ambitions in this chapter, however, are very modest ones: Only a few rather basic concepts and some of their properties will be discussed to give the reader a feeling how naturally fuzzy sets fit into the graph theoretical world. A lot more of information than we can include below, and many hints to journal papers, can be found in the four monographs [16, 19, 21, 23].

The close relationship between (both crisp and fuzzy) relations and graphs is but one reason for the current chapter; the other, from a technical point of view more interesting one, is the opportunity to illustrate once more the surprising subtleties behind a seemingly nearby and simple generalization of a well known classical concept. This needs a detailed discussion, which (despite its already mentioned introductory character) results in a chapter of considerable length, since a number of intricacies pop up caused by changing the valuation space from $\{0,1\}$ to $[0,1] \subseteq \mathbb{R}$.

Fuzzy graphs have been conceived of rather early in the history of fuzzy sets in [13] (the world's first textbook on fuzzy sets), the real start taking place two years later with its English translation and the two papers [24] and [35]. Rosenfeld [24] introduced fuzzy analogs of several basic graph-theoretic concepts, including subgraphs, paths and connectedness, bridges and cut-nodes, and forests and trees, using fuzzy node sets and undirected graphs. Contrarily and independently, Yeh and Bang [35] use crisp vertex sets and deal with reachability and (edge and vertex) connectivity in both directed and undirected fuzzy graphs and apply these concepts to cluster analysis.

In order to provide access to the main stream concepts of the existing literature, in this chapter we will deviate from considering fuzzy relations on crisp sets: Thus, as it is done in all of [16, 19, 21, 23] fuzzy graphs will be defined as structures which do not only have valued arcs/edges, but also valued nodes/vertices. This is, of course, a reasonable approach: Think of a graph representing a network with a number of cities together with highways connecting them, and assume you want to drive from A to D and have to pass B or C; then, if B has a bypass and C has not, you probably will assign a greater value to B in order to indicate the higher driving comfort because there is no need to cross straight through it. But, of course, your preference may be the other way round, in case you like e.g. the challenges of urban traffic, or the possibility of sightseeing or shopping.

13.1 Fuzzy Relations on Fuzzy Sets

In order to recall the general set-up: We know that a (crisp) relation ρ on a set S is a subset of $S\times S$. Similarly, the concept of a fuzzy relation on a fuzzy set σ of S as a fuzzy subset of $\sigma\times\sigma$ has been mentioned and shortly discussed at the beginning of Section 9.3. Since fuzzy relations of this kind are at the basis of the notion of a fuzzy graph we present an official definition and call them **d-fuzzy relations** (short for double-fuzzy) to distinguish them from the kind of fuzzy relations used in the rest of the book.

Definition 13.1. *Let S and T be two sets, and σ and τ be fuzzy sets on S and T, respectively. A fuzzy set ρ on $S\times T$ is called a* **d-fuzzy relation from σ to τ** *if*

$$\rho(x,y)\le\sigma(x)\wedge\tau(y) \tag{13.1}$$

for all $x\in S$ and $y\in T$. If $\sigma=\tau$ then ρ is called a **d-fuzzy relation on σ**.

The rationale behind this definition is obvious: From Section 9.1 we know that the membership function of a fuzzy set is a generalization of the characteristic function of classical (naive) set theory, the latter taking values from $\{0,1\}\subseteq[0,1]$ only. Let $T\subseteq S$ and ρ be a relation on S. Then ρ becomes a relation on T only if $\langle x,y\rangle\in\rho$ implies $x\in T$ and $y\in T$. If t and r are the characteristic functions of T and ρ, respectively, then this condition becomes $r(x,y)=1\to t(x)=t(y)=1$ for all $x,y\in T$ so $r(x,y)\le t(x)\wedge t(y)$.

On page 5 of [16] the authors state that inequality (13.1) allows "to organize the flow through an edge of a fuzzy graph in such a way that it never exceeds the capacities of its end vertices". So the concept of a d-fuzzy relation perfectly fits to this kind of problem. But remember our example from Section 9.3 concerning the similarity of hair colors for which d-fuzzy relations are improper. This gives room and need for interesting semantical considerations of when which kind of concept can serve as an appropriate model. A few additional remarks will be given in connection with the notion of connectedness just after Def. 13.20 below.

Definition 13.2. *Let S and T be two sets, σ and τ be fuzzy sets on S and T, respectively, and ρ be a d-fuzzy relation.*

(a) $\rho\subseteq\sigma\times\tau$ *is called* **strong** *if $\rho(x,y)=\sigma(x)\wedge\tau(y)$, for all x,y.*

(b) *The d-fuzzy relation* $\rho^{\smile} \subseteq \tau \times \sigma$ *given by* $\rho^{\smile}(x, y) = \rho(y, x)$ *is called the* **converse** *of* ρ.

(c) $\rho \subseteq \sigma \times \sigma$ *is called* **d-reflexive** *if* $\rho(x, x) = \sigma(x)$, *and* **symmetric** *if* $\rho(x, y) = \rho(y, x)$ *for all* $x, y \in S$.

(d) $\rho \subseteq \sigma \times \sigma$ *is called* **transitive** *if* $\rho^2 \subseteq \rho$.

(e) $\rho \subseteq \sigma \times \sigma$ *is called a* **d-fuzzy equivalence** *if it is d-reflexive, symmetric and transitive.*

To avoid possible misunderstandings and to emphasize the kind of relation we use the term *d-reflexive* if appropriate. The concept itself is due to [24, p. 82]. For any set S, a reflexive relation ρ on S is d-reflexive on χ_S, the characteristic function of S, whence formal consistency with the two-vlued case is given. Moreover, if ρ is d-reflexive, then we know $\rho(x, y) \leq \sigma(x) \wedge \sigma(y) \leq \sigma(x) = \rho(x, x)$, for all $x, y \in S$. So d-reflexivity is a plausible generalization since it preserves a number of characteristic properties of the classical notion, as e.g. that the representation matrix of ρ (see Def. 9.29) is diagonally dominant (see Def. 7.5). Some more such properties are given below.

Technically speaking, fuzzy relations are built upon the Cartesian product of crisp sets, and d-fuzzy relations on the Cartesian product of fuzzy sets; see (9.6). So we can define the composition of d-fuzzy relations as follows:

Definition 13.3. *Let* $\rho \subseteq \sigma \times \tau$ *and* $\rho' \subseteq \tau \times \nu$ *be d-fuzzy relations, with* σ, τ, ν *fuzzy sets on* S, T *and* U *respectively. The* **max-min composition** *of* ρ *and* ρ' *is the fuzzy relation* $\rho \circ \rho' \subseteq \sigma \times \nu$ *given as* $\rho \circ \rho'(x, z) = \bigvee_{y \in T} \big(\rho(x, y) \wedge \rho'(y, z)\big)$ *for all* $x \in S$ *and* $z \in U$.

Some straightforward consequences of the above definitions are listed in

Proposition 13.4. *Let* $\rho, \rho' \subseteq \sigma \times \tau$ *and* $\pi \subseteq \tau \times \nu$ *be d-fuzzy relations, with* σ, τ, ν *fuzzy sets on* S, T *and* U *respectively.*

(1) $\rho^{\smile}$ *is a d-fuzzy relation on* $\tau \times \sigma$.

(2) $\rho \circ \pi \subseteq \sigma \times \nu$, *i.e. it is a d-fuzzy relation from* σ *to* ν.

(3) *Both* $\rho \cup \rho'$ *and* $\rho \cap \rho'$ *are d-fuzzy relations.*

Proof. (1): Trivial, since $\rho(y, x) = \rho^{\smile}(x, y) \leq \sigma(x) \wedge \tau(y) = \tau(y) \wedge \sigma(x)$.

(2): $\rho(x, y) \leq \sigma(x) \wedge \tau(y)$ and $\pi(y, z) \leq \tau(y) \wedge \nu(z)$ implies $\rho(x, y) \wedge \pi(y, z) \leq \sigma(x) \wedge \tau(y) \wedge \nu(z) \leq \sigma(x) \wedge \nu(z)$. Thus, $\bigvee_{y \in T} \big(\rho(x, y) \wedge \pi(y, z)\big) = \rho \circ \pi(x, z) \leq \sigma(x) \wedge \nu(z)$ for all $x \in S$ and $z \in U$. □

The following properties are well known to hold for fuzzy relations; they are also valid for d-fuzzy relations. The example shows that proofs are really easy.

Proposition 13.5. ([21, Theorem 1.2], [16, Theorem 1.2.11]) *Let* $\nu, \pi, \rho, \tau \subseteq \sigma \times \sigma$ *be d-fuzzy relations on a fuzzy set* σ *of a set* S. *Then the following properties hold.*

(1) $\rho \cup \tau = \tau \cup \rho$.

(2) $\rho \cap \tau = \tau \cap \rho$.

(3) $\overline{\overline{\rho}} = \rho$.
(4) $\rho \cup (\tau \cup \pi) = (\rho \cup \tau) \cup \pi$.
(5) $\rho \cap (\tau \cap \pi) = (\rho \cap \tau) \cap \pi$.
(6) $\rho \circ (\tau \circ \pi) = (\rho \circ \tau) \circ \pi$.
(7) $\rho \cap (\tau \cup \pi) = (\rho \cap \tau) \cup (\rho \cap \pi)$.
(8) $\rho \cup (\tau \cap \pi) = (\rho \cup \tau) \cap (\rho \cup \pi)$.
(9) $\overline{(\rho \cup \tau)} = \overline{\rho} \cap \overline{\tau}$.
(10) $\overline{(\rho \cap \tau)} = \overline{\rho} \cup \overline{\tau}$.
(11) $\nu \subseteq \rho \wedge \pi \subseteq \tau \rightarrow \nu \cup \pi \subseteq \rho \cup \tau$.
(12) $\nu \subseteq \rho \wedge \pi \subseteq \tau \rightarrow \nu \cap \pi \subseteq \rho \cap \tau$.
(13) $\nu \subseteq \rho \wedge \pi \subseteq \tau \rightarrow \nu \circ \pi \subseteq \rho \circ \tau$.

Proof. (13): We have $\nu \circ \pi(x,y) = \bigvee_{z\in S}\big(\nu(x,z) \wedge \pi(z,y)\big) \leq \bigvee_{z\in S}\big(\rho(x,z) \wedge \tau(z,y)\big) = \rho \circ \tau(x,y)$ for all $x,y \in S$. □

The last two propositions show that d-fuzzy relations are closed with respect to conversion, intersection, union, and composition, and, moreover, fulfill many relationships familiar from two-valued and fuzzy relations. This does, however, not hold for the standard complement $\overline{\rho}$ as defined above in the same way as for fuzzy relations; see Section 13.1.1 below.

Proposition 13.6. ([21, Theorem 1.3], [16, Theorem 1.2.13]) *Let ρ, τ be d-fuzzy relations on a fuzzy set σ of a set S, with ρ d-reflexive. Then $\tau \subseteq \rho \circ \tau$ and $\tau \subseteq \tau \circ \rho$.*

Proof. For all $x, z \in S$ we have $\rho \circ \tau(x,z) = \bigvee_{y\in S}\big(\rho(x,y) \wedge \tau(y,z)\big) \geq \rho(x,x) \wedge \tau(x,z) = \sigma(x) \wedge \tau(x,z)$. But $\tau(x,z) \leq \sigma(x) \wedge \sigma(z)$, whence $\sigma(x) \wedge \tau(x,z) = \tau(x,z)$. Therefore, $\tau \subseteq \rho \circ \tau$. □

Proposition 13.7. ([21, Theorem 1.3], [16, Corollary 1.2.14 ff]) *Let ρ, τ be d-reflexive fuzzy relations on a fuzzy set $\sigma \subseteq S$.*

(1) $\rho \subseteq \rho^2$.
(2) $\Delta_\rho =: \rho^0 \subseteq \rho \subseteq \rho^2 \subseteq \rho^3 \subseteq \cdots \subseteq \rho^\infty := \bigcup_{k\geq 1} \rho^k$ *(see (9.10)).*
(3) $\rho^0(x,x) = \rho(x,x) = \rho^2(x,x) = \rho^3(x,x) = \cdots = \rho^\infty(x,x) = \sigma(x)$.
(4) *Both $\rho \circ \tau$ and $\tau \circ \rho$ are d-reflexive.*
(5) *ρ_α is reflexive on σ_α for all $\alpha \in [0,1]$.*

Proof. (1), (2): Take $\rho, \rho^2, \ldots$ for τ in the foregoing proposition.

(3): The proof runs by induction on n. We have $\rho(x,x) = \sigma(x)$ by definition of d-reflexivity, and assume that $\rho^n(x,x) = \sigma(x)$, for all $x \in S$. Then $\rho^{n+1}(x,x) = \bigvee_{y\in S}\big(\rho^n(x,y) \wedge \rho(y,x)\big) \leq \bigvee_{y\in S}\big(\sigma(x) \wedge \sigma(x)\big) = \sigma(x)$. But we also have $\rho^{n+1}(x,x) = \bigvee_{y\in S}\big(\rho^n(x,y) \wedge \rho(y,x)\big) \geq \rho^n(x,x) \wedge \rho(x,x) = \sigma(x)$.

(4): We have $\rho \circ \tau(x,x) = \bigvee_{y\in S}\big(\rho(x,y) \wedge \tau(y,x)\big) \leq \bigvee_{y\in S}\big(\sigma(x) \wedge \sigma(x)\big) = \sigma(x)$, but also $\rho \circ \tau(x,x) = \bigvee_{y\in S}\big(\rho(x,y) \wedge \tau(y,x)\big) \geq \rho(x,x) \wedge \tau(x,x) = \sigma(x)$.

(5): If $x \in \sigma_\alpha$, then $\rho(x,x) = \sigma(x) \geq \alpha$, whence $\rho_\alpha(x,x)$ (ρ_α being crisp). □

Completing the overview on the properties of d-fuzzy relations we finally list the following results:

Proposition 13.8. ([21, Theorem 1.4], [16, Corollary 1.2.20]) *Let ρ, τ be symmetric d-fuzzy relations on a fuzzy set $\sigma \subseteq S$.*

(1) *$\rho \circ \tau$ is symmetric iff $\rho \circ \tau = \tau \circ \rho$.*
(2) *ρ^k is symmetric for all $k \geq 1$.*
(3) *ρ_α is symmetric on σ_α for all $\alpha \in [0,1]$.*

Proof.

$$\begin{aligned}(1)\colon \rho \circ \tau(x,y) &= \rho \circ \tau(y,x)\\ &\leftrightarrow \textstyle\bigvee_{z \in S}\big(\rho(x,z) \wedge \tau(z,y)\big) = \bigvee_{z \in S}\big(\rho(y,z) \wedge \tau(z,x)\big)\\ &\leftrightarrow \textstyle\bigvee_{z \in S}\big(\rho(x,z) \wedge \tau(z,y)\big) = \bigvee_{z \in S}\big(\tau(x,z) \wedge \rho(z,y)\big)\\ &\leftrightarrow \rho \circ \tau(x,y) = \tau \circ \rho(x,y).\end{aligned}$$

□

Proposition 13.9. ([21, Theorem 1.5], [16, Corollary 1.2.23]) *Let ρ, τ be transitive d-fuzzy relations on a fuzzy set $\sigma \subseteq S$.*

(1) *If $\rho \circ \tau = \tau \circ \rho$ then $\rho \circ \tau$ is transitive.*
(2) *If ρ is also symmetric then $\rho(x,y) = \rho(y,x) \leq \rho(x,x)$.*
(3) *ρ_α is transitive on σ_α for all $\alpha \in [0,1]$.*

Proof. (2): We have $\rho^2(x,x) = \bigvee_{y \in S}\big(\rho(x,y) \wedge \rho(y,x)\big) \leq \rho(x,x)$. Using symmetry this implies $\bigvee_{y \in S}\big(\rho(x,y) \wedge \rho(x,y)\big) \leq \rho(x,x)$; thus $\rho(x,y) \leq \rho(x,x)$. Again using symmetry gives $\rho(y,x) \leq \rho(x,x)$. □

13.1.1 *Complement of d-Fuzzy Relations*

Note that all the operations on d-fuzzy relations we have considered above conform to (in fact: are the same as) those defined for (usual) fuzzy relations. And they should be, since the latter are a special case of d-fuzzy relations, with $\sigma(x) = 1$ for all $x \in S$. Turning to complement, things change: If we use standard negation $\overline{\rho}(x,y) = 1 - \rho(x,y)$ then in general we do not get $\overline{\rho}(x,y) \leq \sigma(x) \wedge \sigma(y)$: take e.g. $\rho(x,y) = 0.4$ and $\sigma(x) = \sigma(y) = 0.5$. Thus, the set of d-fuzzy relations is not closed with respect to standard complementation. But clearly $\overline{\overline{\rho}} = \rho$.

In [22, Definition 3.9] another concept has shortly been mentioned: Let ρ be a d-fuzzy relation on a fuzzy set $\sigma \subseteq S$. Set $\sigma^c = \sigma$ and $\rho^c(x,y) = 0$ if $\rho(x,y) > 0$, and $\rho^c(x,y) = \sigma(x) \wedge \sigma(y)$ otherwise. Then ρ^c is always a d-fuzzy relation on σ, even if ρ is not; but it is missing the property that $(\rho^c)^c = \rho$: you can take the same example as above. (This property holds if and only if ρ is strong which, of course, holds true for two-valued relations.) The definition is very close to the crisp complement in the sense that $\langle x,y\rangle \in \rho \to \langle x,y\rangle \notin \rho^c$ and $\langle x,y\rangle \notin \rho \to \langle x,y\rangle \in \rho^c$.

Still another concept has been presented in [29] which has become the prevalent one in the context of d-fuzzy relations: Again $\sigma^c = \sigma$, but $\rho^c(x,y) = (\sigma(x) \wedge \sigma(y)) - \rho(x,y)$. ρ^c is then always a d-fuzzy relation as desired. Moreover, we have $(\sigma^c)^c = \sigma^c = \sigma$, and $(\rho^c)^c(x,y) = (\sigma^c(x) \wedge \sigma^c(y)) - \rho^c(x,y) = (\sigma(x) \wedge \sigma(y)) - [(\sigma(x) \wedge \sigma(y)) - \rho(x,y)] = \rho(x,y)$. Note that for usual fuzzy relations $\rho^c = \overline{\rho}$. We will discuss the operation of complement again below in connection with fuzzy graphs.

13.2 Preliminaries about Crisp Graphs

As usual we start with a short look at some classical concepts to provide a reference for the subsequent generalizations. Some helpful introductory sources of information are e.g. [2, 33, 34]. More advanced texts are e.g. the classic [5] or [10, 11] and [26]. For a comprehensive discussion of directed graphs and related algorithms (on a level much more advanced than our introductory remarks given below) see [3] or [4, Chapter 1].

The short list of intuitive and easily comprehensible definitions given below might possibly suggest that graph theory is a collection of rather trivial concepts and relationships between them. To see that this is only partly true, but to a very large extent it is not, you need to take a closer look into the literature just mentioned.

Definition 13.10.

(a) *A* **directed graph** *(or* **digraph** *for short) is a pair $G = (V, E)$, where $E \subseteq V \times V$. An element $v \in V = \{v_1, v_2, \ldots, v_n\}$ is called a* **vertex** *(or* **node***), and $e \in E$ an* **edge** *(or* **arc***) of G. If $e = \langle a, b\rangle$ then a is called the* **initial vertex** *or* **tail***, and b the* **target vertex** *or* **head** *of e.*

(b) *The Boolean matrix representing E is called the* **adjacency matrix** *of G: $A_G \in \mathbb{B}_n$ defined by $A_G = [a_{ij}]$, $i, j \in \underline{n}$, with $a_{ij} = 1$ iff $\langle v_i, v_j\rangle \in E$.*

Since arcs are defined to be ordered pairs, directed graphs provide a natural means to represent and visualize relations. Conversely, it is clear that to every Boolean matrix $A \in \mathbb{B}_n$ there is an associated digraph of order n, denoted by G_A, such that $A_{G_A} = A$.

We have to introduce some additional commonly used vocabulary indispensable for talking about graphs: The terms "*head*" and "*tail*" clearly come from the arrows used to draw directed graphs. An edge $e = \langle a, a\rangle$ is called a **loop**. Note that the above definition does not allow multiple edges, i.e. edges having the same endpoints. Such a digraph is called **simple** if it has no loops. A digraph of the form $G = (\{v\}, \emptyset)$ is called **trivial.**

Vertex $v \in V$ and edge $e \in E$ are said to be **incident** if v is the tail or head of e. The two vertices incident with an edge e are called its **endnodes**, and e is said to be an *edge from its tail to its head.* Two vertices $v \neq w$ of G are called **adjacent**, or **neighbors**, if they are endnodes of an edge e. Two edges $e \neq f$ are **adjacent** if they have an endnode in common.

The properties of a graph G are very often determined by the presence or absence of some specific structures as a part of it. We begin with

Definition 13.11. *Let $G = (V, E)$ and $G' = (V', E')$ be two graphs.*

(a) *G' is called a* **subgraph** *of G, written $G' \subseteq G$, if $V' \subseteq V$ and $E' \subseteq E$; G' is called a* **proper subgraph**, *$G \subset G'$, if $V' \subset V$ or $E' \subset E$, and a* **spanning subgraph** *of G if $V' = V$.*

(b) *Let $V' \subseteq V$; if E' consists of all those edges of G joining pairs of vertices of V', then G' is called the* **vertex induced subgraphvertex!induced subgraph** *of G, often simply denoted by $G' = \langle V' \rangle$ since E' is completely determined by V'.*

(c) *Similarly, if $E' \subseteq E$, and V' is the set of all end vertices of the edges of E', then G' is called the* **edge induced subgraph** *of G, often simply denoted by $G' = \langle E' \rangle$.*

The subgraph relation clearly emerges by removing/adding something from/to a given graph. For later purposes we take a look at deletion of vertices and edges, two operations on graphs that will be needed in connection with the concepts of cut-nodes and bridges:

Definition 13.12. *Let $G = (V, E)$ be a graph, $v \in V$ and $e \in E$.*

(a) **Vertex removal**: *$G - v$ is the induced subgraph of G on the vertex set $V - \{v\}$, i.e. the graph obtained from G by removing the vertex v and all the edges incident with it.*

(b) **Edge removal**: *$G - e$ is the subgraph of G that results after removing edge e from G. (Since the end vertices of e remain in this subgraph, it is not the induced subgraph on $E - \{e\}$.)*

The removal of a set of vertices or edges from a graph can be performed by a successive removal of single vertices or edges in an arbitrary order.

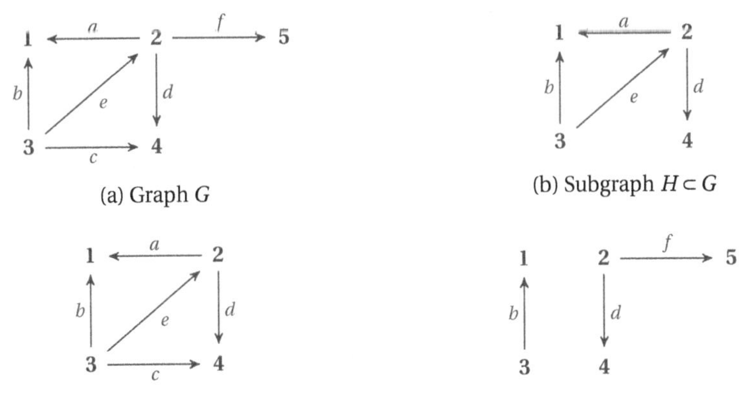

(a) Graph G

(b) Subgraph $H \subset G$

(c) $\langle\{1,2,3,4\}\rangle \subset G$

(d) $\langle\{b,d,f\}\rangle \subset G$

FIG. 13.1: Various subgraphs.

Definition 13.13. *Let $G = (V, E)$ be a digraph.*

(a) *A* **path** *p in G from vertex u to vertex v of length k is a sequence of $k+1$ distinct nodes starting with u and ending with v, i.e. $p := u = v_0, v_1, \ldots, v_k = v$ such that $\langle v_m, v_{m+1}\rangle \in E$, for $0 \leq m < k$. v is said to be* **reachable** *from u via path p.*
(b) *If $u = v_0 = v_k = v$ and the vertices $v_1, \ldots, v_{k-1}$ are distinct, then the sequence $v_0, v_1, \ldots, v_k$ is said to form a* **cycle**. *G is called* **acyclic** *if it has no cycles.*
(c) *Two distinct vertices u and v are called* **strongly connected** *if there is a path from u to v and a path from v to u. G is* **strongly connected** *if every two distinct vertices are. (A single vertex may be regarded as trivially strongly connected to itself.)*

Obviously, strong connectivity between nodes is a reflexive, symmetric, and transitive relation, and hence an equivalence on V, say γ; thus the elements of the quotient set V/γ, called the **strong components** of G, form a partition of V. Clearly, G is a strongly connected graph if $|V/\gamma| = 1$.

To simplify the subsequent discussions we will from now on consider undirected graphs, i.e. graphs with an underlying symmetric relation. It is not hard to see that the generalization of definitions and results is straightforward, using slightly adapted and a little bit more complex concepts.

Most of the notions introduced so far do apply unchanged or without substantial modification. E.g. edges have of course endnodes, but it does not make sense to tag one of them as head or tail; and *strong* looses its meaning: two vertices are connected if there is a path between them, and a graph is **connected** if any two of its vertices are connected by some path. Otherwise it is called **disconnected**, i.e. there exists a pair of vertices having no path connecting them. Any disconnected graph G is a union of two or more connected graphs, called the **connected components** of G. As above we can easily see that the relation $\gamma \subseteq V \times V$, with $\gamma(x, y)$ iff there is a path between x and y, is an equivalence. So for a connected graph we have $|V/\gamma| = 1$, and for a disconnected graph $|V/\gamma| > 1$.

Connectivity of a graph $G = (V, E)$ may be weak, so that the removal of a single vertex or edge will result in in a disconnected subgraph. This leads to

Definition 13.14. *Let $G = (V, E)$ be a connected graph, and let $cp(G)$ denote the number of components of G.*

(a) *A subset $W \subseteq V$ is called a* **vertex cut** *if $G - W$ is disconnected, i.e. $cp(G - W) > cp(G)$. W is a k-***vertex cut** *if $|W| = k$. A vertex v is called a* **cut vertex** *(or* **cut node***) of G if $\{v\}$ is a vertex cut.*
(b) *A subset $F \subseteq E$ is called an* **edge cut** *if $G - F$ is disconnected, i.e. $cp(G - F) > cp(G)$. F is a k-***edge cut** *if $|F| = k$. An edge e is called a* **bridge** *of G if $\{e\}$ is an edge cut.*

As one can see from Fig. 13.2(a), an isolated node such as **6** cannot be a cut node of a graph, whereas, as illustrated in (b), a single edge such as h is a bridge, since its removal

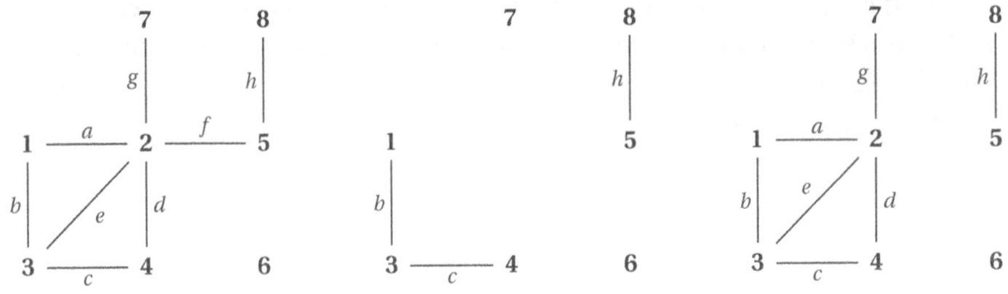

(a) Graph G (b) Graph $G-\{2\}$ (c) Graph $G-\{f\}$

FIG. 13.2: Cut-nodes and bridges.

would leave two isolated nodes. $\{b,d,e\}$ is a 3-edge cut, and $\{2,3\}$ is a 2-vertex-cut leaving h as the only edge in $G-\{2,3\}$. Note that $G-\{2,3\} = \langle\{5,6,7,8\}\rangle \subset G-\{f\}$. Moreover, from (c) we can also see that the removal of a bridge increases $\mathrm{cp}(G)$ by 1, whereas by removing a cut node v we can get $\mathrm{cp}(G-v) > \mathrm{cp}(G)+1$. In fact, the deletion of a cut node from a connected graph may yield a graph with any number of components. As an example consider the star $K_{1,n}$ with $n+1$ vertices; deleting the central vertex yields n components, all of them isolated nodes.

A graph is called **nonseparable** if it is connected, nontrivial and contains no cut nodes. All other graphs are **separable**. A graph that is not connected is thus separable. A **block** of a separable graph G is a maximal nonseparable subgraph of G, that means it is a nonseparable graph that is not properly contained in any other nonseparable subgraph of G. The graphs usually denoted by K_1 : • and K_2 : • —— • are blocks. Note that K_2 is the only block containing a bridge. Graphs in which there is an edge between every pair of n vertices are called **complete** graphs and are denoted K_n. So K_3 is a triangle, simultaneously also the cycle C_3 containing three edges.

We close this introductory section with a few characteristic results and an example proof to demonstrate that graph theory is more than just a list of definitions. Nevertheless, we still remain on a more or less intuitive and elementary level.

Proposition 13.15.

(1) *v is a cut node of a connected graph G with at least three vertices if and only if there exist two vertices u and w distinct from v such that v is on every u-w path.*

(2) *A connected graph G with at least two vertices contains at least two vertices that are not cut vertices.*

(3) *An edge e of a connected graph G is a bridge if and only if it belongs to no cycle of G.*

(4) *An edge e of a connected graph G is a bridge if and only if it there exist vertices u and v such that e belongs to every u-v path in G.*

(5) *The deletion of a bridge from a connected graph yields a graph with exactly two components.*

Proof. (3): If $e = \langle u, v\rangle$ is a bridge of G, its ends u and v belong to different components of $G - \{e\}$, and so are not connected by a path in $G - \{e\}$; equivalently, e is not part in a cycle of G. Conversely, if e is not a bridge of G, then u and v are in the same component of $G - \{e\}$ and there is a path P between them. Thus, adding e to P gives a cycle in G through e. □

13.3 Fuzzy Graphs

In this section we will provide generalizations along the guidelines given by the few basic concepts which have been introduced for crisp graphs above. On the basis of a detailed analysis we will see that the commonly used concept of fuzzy graphs predominant in literature has some questionable properties.

And although we will confine our considerations to some basic concepts and results only in this introductory overview (viz. forests, trees, and cycles; cutnodes, blocks and bridges), the material offered is sufficient to, once again, give the reader a feeling of how much more possibilities and complexities arise by the transition from the crisp $\{0,1\}$-valued to the fuzzy $[0,1]$-valued graphs. We will also give hints at where crisp and fuzzy graphs have different properties, and will emphasize them Δ *by using a different font and enclosing the text by Greek upper case deltas* Δ.

Most of the material included below is adapted from small parts of the monographs [16, 19, 21]. In these books, as well as in numerous papers, a lot more of concepts and corresponding results have been given. You may also want to have a look at [23]. Short, but comprehensive, overviews on definitions and results (without any proofs) are provided in [28] or [19, Section 1.2]. In addition, by clearly indicating by whom and when a contribution has been made, these overviews also give historical insights into the development of fuzzy graph theory.

13.3.1 *The Basic Concepts*

As introduced in Def. 13.10 earlier in this chapter, a *directed graph* is a pair (V, E) with V the set of *vertices* (or *nodes*) and E a relation on V, the set of *edges* (or *arcs*). Similarly, any (d-)fuzzy relation ρ on a (fuzzy subset σ of a) set V can be regarded as defining a directed fuzzy graph, where the edge $\langle x, y\rangle \in V \times V$ has weight or strength $\rho(x, y) \in [0,1]$. The common understanding of a fuzzy graph is that of a triple $G = (V, \sigma, \mu)$ consisting of an underlying finite non-empty set V of nodes, a fuzzy set σ on V, and a fuzzy set $\mu \subseteq V \times V$ such that $\mu(x, y) \leq \sigma(x) \wedge \sigma(y)$, i.e. $\mu \subseteq \sigma \times \sigma$. The formal definition as adapted from [16, 19] is as follows:

Definition 13.16. *A* **fuzzy graph** $G = (V, \sigma, \mu)$ *is a triple consisting of a finite set* $V \neq \emptyset$, *a fuzzy set* $\sigma \subseteq V$, *and a d-fuzzy relation* μ *on* σ. *If not stated otherwise,* μ *is assumed to be d-reflexive and symmetric.* σ *is called the* **fuzzy vertex set** *and* μ *the* **fuzzy edge set** *of* G.

In the way how Def. 13.16 is stated, μ is symmetric, i.e. as we have done it for crisp graphs, in what follows below we consider undirected fuzzy graphs for sake of simplicity. It is in fact rather obvious how to adapt the various notions in a direction specific way. (Directed fuzzy networks are dealt with in e.g. [15, 18].) A digraph $G = (V, E)$ according to Def. 13.10 is a fuzzy graph with $\sigma(x) = 1$, and $\mu(x, y) \in \{0, 1\}$ for all $x, y \in V$.

We continue with a discussion of the seemingly obvious and simple generalization of the concept of a subgraph to the context of fuzzy graphs. And as we will see below by a critical examination of existing literature, even for such a fundamental notion things turn out to be a little bit ambiguous or not fully developed. We put all this in an aside which is not necessary for the understanding of the main text. Thus, readers not interested in such detailed considerations may skip them and continue with our official definition. (But, of course, I strongly recommend reading these remarks: they really present a very good example illustrating how a syntactically appropriate and 'obvious' generalization of a well known 'simple' concept can cause severe semantical problems.)

Both for the aside just below, and also later on we need some more notation: Let $G = (V, \sigma, \mu)$ be a fuzzy graph. Using $\mu_G^* := \mathsf{supp}(\mu)$ and $\sigma_G^* := \mathsf{supp}(\sigma)$, or simply μ^* and σ^* if G is clear from the context, the crisp graph $G^* = (V, \sigma^*, \mu^*)$, is called the **underlying graph** of $G = (V, \sigma, \mu)$. In the following, an (undirected) edge e between two nodes u and v will simply be denoted as an unordered pair $\underline{uv}$, and we will also write $e \in \mu^*$.

Aside 1: On syntactical and semantical aspects of V and σ

Some remarks are in order concerning the role of V and σ in the definition of subgraphs: The nodes of crisp graphs $G = (V, E)$, symbolizing entities relevant for the current problem context, come into existence by their introduction in the definition as the underlying set V. The implicit valuation $\sigma(x) = 1$, for all $x \in V$, only says that x is an entity under consideration, or simply that x *exists* as a node in the graph G.

In the context of d-fuzzy relations, or equivalently fuzzy graphs, however, the situation is different and a little bit of a hybrid character. In fact there are the following two variants:

(1) A valuation $\sigma(x) = 0$ indicates that node x does not cause costs (e.g. time or money) or does not have any benefits (e.g. comfort, good reachability, low traffic noise), but x still remains a node of G. Hence it is important to notice that σ^* only singles out those nodes in G which do have a certain property under consideration (to a certain extent) and does not modify the underlying node set V of G.

(2) On the other hand, sometimes fuzzy graphs are defined on the basis of σ^* instead of V as in e.g. [19, Definition 1.2.2], sometimes with the additional restriction "Here we assume $\sigma^* = V$" as e.g. in [32, p. 2], but without it in the former reference. Moreover, in [32] V is called the vertex set of G, whereas [19, top of p. 5] only states that (note that now V comes into play) "*The elements of $\sigma^* = V$ are called* ***vertices*** *of the fuzzy graph and the elements of μ^* are called the* ***edges***." Then what about $x \notin \sigma^*$? If there are such vertices x, y at all, do they belong to V, with possibly $V \neq \sigma^*$? In any case $\mu(x, y) = 0$, so $\underline{xy}$ is not an edge of G.

Looking into publications on fuzzy graphs will make you note that most examples carry weights on their edges only. This is often described by phrases like "In all the examples σ is chosen suitably." ([14, p. 220]). The reason for it is that the valuations of nodes have indirect influence on a fuzzy graph because they restrict the weights of their incident edges, and that almost all derived concepts are based on the valuations of the edges alone.

We now turn to fuzzy subgraphs and start our journey with the forefathers of all subsequent definitions, viz. the ones in Rosenfeld's paper from 1975 [24, p. 86]:

> (FSG-1) Let $G = (V, \sigma, \mu)$ be a fuzzy graph. A fuzzy graph $H = (V, \tau, \nu)$ is called a **(partial) fuzzy subgraph** of G if $\tau \subseteq \sigma$ and $\nu \subseteq \mu$.
>
> (FSG-2) H is the fuzzy subgraph of G **induced by** τ if it is the maximal fuzzy subgraph of G that has fuzzy node set τ, which is the case if $\nu(x, y) = \tau(x) \wedge \tau(y) \wedge \mu(x, y)$, for all $x, y \in V$.

Note that it is already here where problems or questions (denoted as(PQ-i)) arise:

(PQ-1) To compare two fuzzy sets with respect to $\subseteq$, both must have the same underlying universe. Thus, both G and H have V as their node set, so there is no formal (or syntactical) problem. But (FSG-1) does not give us any additional information, e.g. if $\sigma^* = V$, and if not, does an $x \notin \sigma^*$ still count as node in G or not. In other words, as mentioned above, the question is: Does σ evaluate a 'degree of existence' of nodes, or does $\sigma(x) = 0$ mean that x does not have e.g. any benefits or cause any troubles in the context under consideration.

Our next station is the first monograph on fuzzy graphs, viz. [21, pp. 19/20], which inherits the above definitions (the first under the name *partial fuzzy subgraph*), adds a new one and characterizes it as a special case of a partial fuzzy subgraph:

> (FSG-3) Let $G = (V, \sigma, \mu)$ be a fuzzy graph. A fuzzy graph $H = (W, \tau, \nu)$ is called the **fuzzy subgraph of** G **induced by** $W \subseteq V$, denoted $\langle G_W \rangle$ (since it clearly is unique), if $\tau(u) = \sigma(u)$ and $\nu(u, v) = \mu(u, v)$, for all $u, v \in W$ (i.e. $\tau = \sigma_{|W}$, and $\nu = \mu_{|W \times W}$).
>
> ...
>
> The authors note that a fuzzy subgraph (W, τ, ν) of (V, σ, μ) induced by W is a special case of a partial fuzzy subgraph by setting
>
> $$\tau(x) = \sigma(x) \text{ for } x \in W, \text{ and } \tau(x) = 0 \text{ for } x \in V - W;$$
> $$\nu(x, y) = \mu(x, y) \text{ for } x, y \in W^2, \text{ and } \nu(x, y) = 0 \text{ for } \langle x, y \rangle \in V^2 - W^2.$$

(PQ-2) The last assertion (still claimed at the bottom of [16, p. 14]) is invalid: As soon as $W \subset V$, $\langle G_W \rangle$ is no longer a (partial) fuzzy subgraph of G, since in (FSG-1)-fuzzy subgraphs are required to have the same node set as their original graph.

Graphs have been invented as a natural tool for visualization of certain configurations, relationships etc. of interest. So we continue with some examples in form of drawings:

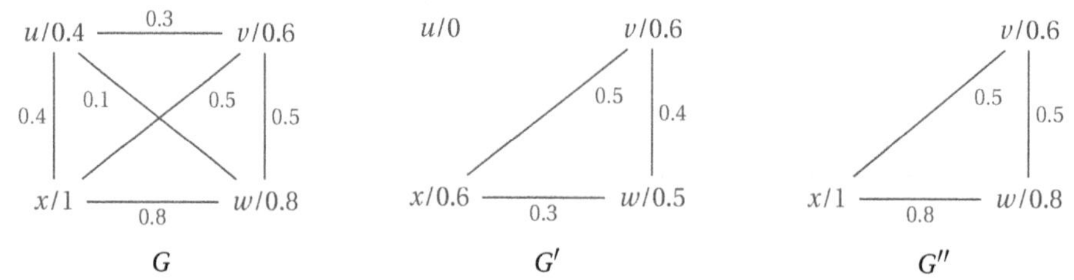

FIG. 13.3: Various subgraphs.

G'' in Fig. 13.3 is the fuzzy subgraph of $G = (V, \sigma, \tau)$ induced by $W = \{v, w, x\}$, and $G' = (V, \sigma', \tau')$ is a partial fuzzy subgraph of G, since $V' = V, \sigma' \subseteq \sigma$ and $\tau' \subseteq \tau$. Adding the node $u/0.4$ to G'' makes the resulting graph a spanning (i.e. partial) fuzzy subgraph of G.

Summarizing the above considerations we must accept the fact that the first approaches to the notion of a fuzzy subgraph, dating back to the mid-seventies, are not without their problems. We thus have a look at what has been offered more then four decades later.

[16, Definition 2.1.3] corresponds to (FSG-1) and (FSG-3), and the first half of [19, Definition 1.2.2] is as follows

> (FSG-1a/b) "A fuzzy graph $H = (V, \tau, \nu)$ is called a **partial fuzzy subgraph** of $G = (V, \sigma, \mu)$ if $\tau(u) \le \sigma(u)$, for all $u \in \sigma^*$ and $\nu(u, v) \le \mu(u, v)$ for each $\underline{uv} \in \mu^*$. In particular, we call $H = (V, \tau, \nu)$ a **fuzzy subgraph** of $G = (V, \sigma, \mu)$ if $\tau(u) = \sigma(u)$ for all $u \in \tau^*$ and $\nu(u, v) = \mu(u, v)$ for every $\underline{uv} \in \nu^*$."

A fuzzy subgraph is indeed a special case of a partial fuzzy subgraph, since we obviously have $0 < \tau(u) = \sigma(u)$ or $0 = \tau(u) \le \sigma(u)$; similar for μ and ν. But:

(PQ-3) The definition of a partial fuzzy subgraph in (FSG-1a/b) suffers from the same deficiency as (FSG-3): The values $\tau(u)$ for $u \in V - \sigma^*$ remain undefined, so one could have $\tau(u) > \sigma(u) = 0$ for such a node, and similar for μ^*: Let $G = (V, \sigma, \mu)$ a fuzzy graph such that $u, v \notin \sigma^*$. Then $\sigma(u) = \sigma(v) = 0$, whence $\underline{uv} \notin \mu^*$. Further assume that $\tau(u) = \tau(v) = a > 0$, whence $\nu(u, v) = a > 0$ would be a feasible value for $H = (V, \tau, \nu)$, contrary to $\mu(u, v) = 0$.

TABLE 13.1: Fuzzy subgraph concepts.

FSG Variants	[21, pp. 19/20], [24, p. 86]	[19, Def. 1.2.2]
partial fsg	(FSG-1): $\tau \subseteq \sigma$, $\nu \subseteq \mu$	(FSG-1a): $\tau(x) \le \sigma(x)$ for all $x \in \sigma^*$, $\nu(x, y) \le \mu(x, y)$ for all $\underline{xy} \in \mu^*$
fsg	---	(FSG-1b): $\tau(x) = \sigma(x)$ for all $x \in \tau^*$, $\nu(x, y) = \mu(x, y)$ for all $\underline{xy} \in \nu^*$
spanning	$\tau = \sigma$	$\tau = \sigma$
τ-induced	(FSG-2): $\tau \subseteq \sigma$, $\nu(x, y) = \tau(x) \wedge \tau(y) \wedge \mu(x, y)$	---
W-induced	(FSG-3): $H = (W, \tau, \nu)$, $W \subseteq V$, $\tau(x) = \sigma(x)$, $\nu(x, y) = \mu(x, y)$ for all $x, y \in W$	(FSG-3a): $H = (W, \tau, \nu)$, $W \subseteq \sigma^*$, $\tau(x) = \sigma(x)$, $\nu(x, y) = \mu(x, y)$ for all $x, y \in W$

Table 13.1 offers a compact overview of the various concepts mentioned above ('fsg' clearly stands for "*fuzzy subgraph*"). As already done before, we let $G = (V, \sigma, \mu)$ denote a fuzzy graph and, if not indicated otherwise, $H = (V, \tau, \nu)$ a subgraph of G.

As one can see from the detailed considerations above, the concept of a subgraph of a fuzzy graph $G = (V, \sigma, \mu)$ depends on two aspects: the underlying structure (set of entities) under consideration, given by the node set V, and the valuations μ and σ. Given the inconsistencies we have mentioned above, these two aspects seem to interfere with one another in an unwanted, or at least unexpected, way. Taking into account how a fuzzy subset is defined, one would certainly want (FSG-1) to describe a certain kind of fuzzy subgraph, and, under certain restrictions, likewise for (FSG-3). Another guideline is

(Struct) If a node x does belong to a fuzzy graph $G = (V, \sigma, \mu)$ only depends on whether $x \in V$ or not, but not on the value $\sigma(x)$. Changes in the structure of a fuzzy graph occur only by means of changes in the node or edge set (compare with Def. 13.12).

Based on the above considerations on fuzzy subgraphs we propose the following

Definition 13.17. *Let $G = (V, \sigma, \mu)$ be a fuzzy graph. A fuzzy graph $H = (W, \tau, \nu)$ is called*

(a) *a* **fuzzy v-subgraph** *of G if $W = V$, $\tau \subseteq \sigma$ and $\nu \subseteq \mu$;*
(b) *a* **(weak) spanning fuzzy subgraph** *of G if it is a fuzzy v-subgraph such that $\tau = \sigma$ ($\tau^* = \sigma^*$);*
(c) *the fuzzy subgraph of G* **node-induced by** *$\tau \subseteq \sigma$, denoted $\langle G^n_\tau \rangle$ if it is the maximal fuzzy v-subgraph of G with fuzzy node set τ, i.e. $\nu(x, y) = \tau(x) \wedge \tau(y) \wedge \mu(x, y)$, for all $x, y \in V$;*
(d) *the fuzzy subgraph of G* **edge-induced by** *$\nu \subseteq \mu$, denoted $\langle G^e_\nu \rangle$, if $W = \{x, y : \underline{xy} \in \nu^*\}$ and $\tau = \sigma_{|W}$;*
(e) *a* **fuzzy s-subgraph** *of G if $W \subseteq V$, $\tau(x) \leq \sigma(x)$, and $\nu(x, y) \leq \mu(x, y)$ for all $x, y \in W$;*
(f) *the fuzzy s-subgraph of G* **induced by** *$W \subseteq V$, denoted $\langle G_W \rangle$, if $\tau = \sigma_{|W}$, and $\nu = \mu_{|W \times W}$, (i.e. $\tau(x) = \sigma(x)$ and $\nu(x, y) = \mu(x, y)$ for all $x, y \in W$).*

(The prefixes "v-" and "s-" are meant to indicate that the subgraph property depends on the values assigned to nodes and edges, or on the structure given by the node set itself, respectively.)

A few consistency checks of these notions are in order:

(1) As a special case $G = (V, \sigma, \mu)$ itself should fit in each of the six categories of Def. 13.17, and it is indeed easy to check that in fact it does: Cases (a), (b) and (e) are trivial; moreover we have $G = \langle G^n_\sigma \rangle = \langle G^e_\mu \rangle = \langle G_V \rangle$.
(2) Each fuzzy v-subgraph is a fuzzy s-subgraph, but only fuzzy s-subgraphs with $W = V$ qualify as fuzzy v-subgraphs. This is in accordance with requirement (Struct). So the problems related to (PQ-1) are eliminated.
(3) From (1) and (2) it follows that the only edge-induced subgraph which is also node-induced is G itself.
(4) Let $W_\nu = \{x, y : \underline{xy} \in \nu^*\}$ and $\tau_\nu = \sigma_{|W_\nu}$; then $\langle G^e_\nu \rangle = (W_\nu, \tau_\nu, \nu) = \langle G_{W_\nu} \rangle$.
(5) A spanning fuzzy subgraph is a special case of a weak spanning fuzzy subgraph.
(6) $G = (V, \sigma, \mu)$ is a weak spanning fuzzy subgraph of its underlying graph $G^* = (V, \sigma^*, \mu^*)$.

In Section 13.2 it has already been mentioned that the concept of a subgraph emerges by removing/adding vertices or edges from/to a given graph. This is also true for fuzzy graphs, but additionally we can decrease/increase the valuations on nodes or edges as we have seen in Def. 13.17. Analogous to Def. 13.12 we give

Definition 13.18. *Let $G = (V, \sigma, \mu)$ be a fuzzy graph, $u \in V$ and $e = \underline{vw} \in \mu^*$.*

(a) **Vertex removal:** *$\langle G - u\rangle$ is the fuzzy s-subgraph induced by $W = V - \{u\}$.*
(b) **Edge removal:** *$\langle G - e\rangle$ is the spanning fuzzy subgraph $H = (V, \sigma, \nu)$ of G with $\nu(v, w) = \nu(w, v) = 0$ and $\nu(x, y) = \mu(x, y)$ for all other pairs $x, y \in V$.*

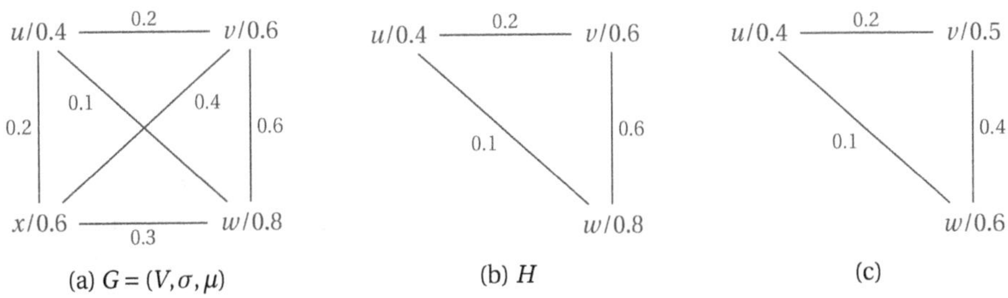

FIG. 13.4: Fuzzy subgraphs.

As an example consider the fuzzy graph $G = (V, \sigma, \mu)$ shown in Fig. 13.4(a): Taking away the diagonals and leaving the rest unchanged results in a spanning fuzzy subgraph, which clearly is different from $\langle G^n_\sigma\rangle = \langle G_V\rangle = G$. However, the triangle H shown in (b), as well as the other three (analog) triangles of G form fuzzy s-subgraphs induced by the three nodes they have. The triangle in (c) is a weak spanning fuzzy subgraph of H but not a fuzzy v-subgraph of G (since it misses node x). Moreover, $H = \langle G - x\rangle = \langle G_{V-\{x\}}\rangle$.

There is another concept, where valuations play an important role: As we have seen in Section 13.1.1, complementation of d-fuzzy relations always is - in terms of fuzzy graphs - about the valuation of its edges, the set of nodes and their valuations remaining untouched. This means that, in one way or the other, the values associated to nodes and edges seem to have different interpretations or even to belong to different semantical categories. Do also note that defining complementation this way makes it fully conforming to the classical concept.

Fig. 13.5 illustrates complements of fuzzy graphs by means of fuzzy cycles with 3 and 4 nodes and is borrowed from [25]. For sake of simplicity the example graphs are assumed to have crisp node sets, that is they are fuzzy graphs in the sense of [35]. Note that $C_3 = C_3^c$, and that C_4^c is again a fuzzy cycle. Δ *Do also compare this to the crisp case, where the complements would consist of 3 isolated nodes and the cross, respectively, none of them being a cycle.* Δ

13.3.2 *Paths and Connectedness*

One of the most important questions when graphs are used as models in certain problem areas is for the existence of paths between two vertices, and about their benefits or

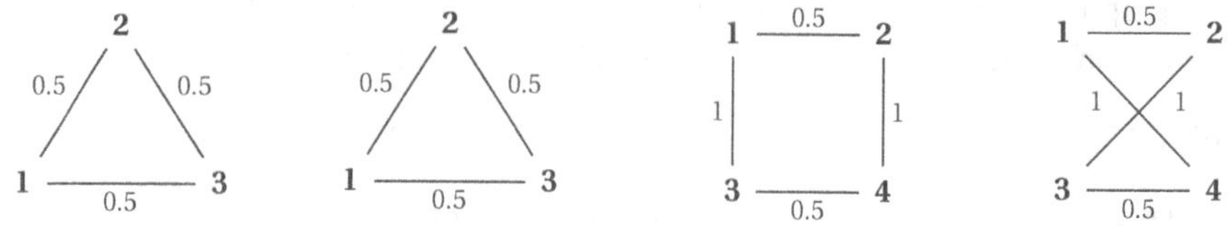

(a) Graph C_3 (b) Graph C_3^c (c) Graph C_4 (d) Graph C_4^c

FIG. 13.5: Complement of fuzzy graphs.

costs. Being connected is one of the most elementary properties a graph or parts of it can exhibit.

Definition 13.19. *Let $G = (V, \sigma, \rho)$ be a fuzzy graph.*

(a) *A* **path** *in G between two vertices u and v of length k is a sequence of k+1 distinct nodes starting with u and ending with v: $P_{uv} = (v_0 v_1 \dots v_k)$, with $u = v_0$ and $v_k = v$, such that $\sigma(v_j) > 0$ for $0 \le j \le k$, and $\rho(v_m, v_{m+1}) > 0$ for $0 \le m < k$. u and v are said to be* **connected** *via path P_{uv}, also called a* **uv-path**.

(b) *If $u = v_0 = v_k = v$ and the vertices $v_1, \dots, v_{k-1}$ are distinct, then $(v_0 v_1 \dots v_k)$ is said to form a* **cycle**. *G is* **acyclic** *if it has no cycles.*

Definition 13.20. *Let $G = (V, \sigma, \rho)$ be a fuzzy graph.*

(a) *Let $P_{uv} = (v_0, v_1, \dots, v_k)$ be a path in G. Then $\mathsf{str}(P_{uv}) = \bigwedge_{0 \le m < k} \rho(v_m, v_{m+1})$ is called the* **strength** *of P_{uv}.*

(b) $\mathsf{conn}_G(u, v) := \bigvee_{P_{uv}} \mathsf{str}(P_{uv})$ *is called the* **strength of connectedness** *between two distinct nodes $u, v \in V$ (often denoted with $\mu^\infty(u, v)$), and* $\mathsf{conn}_G(u, u) := 1$.

(c) *A path P_{uv} is called a* **strongest path** *if* $\mathsf{str}(P_{uv}) = \mathsf{conn}_G(u, v)$.

(d) *G is* **connected** *if G^* is connected, i.e.* $\mathsf{conn}_G(u, v) > 0$ *for all pairs of distinct nodes $u, v \in V$.*

Aside 2: On the semantical and syntactical role of V and σ, again

Whereas a single node in the context of crisp graphs is regarded as trivially connected to itself (i.e. a graph with just one vertex is connected), problems arise in the case of fuzzy graphs $G = (V, \sigma, \rho)$: If we want a single vertex to be connected to itself, then for consistency reasons we must have $\mathsf{conn}_G(u, u) > 0$. A plausible solution would be to require ρ to be d-reflexive on σ, so that possibly $\mathsf{conn}_G(u, u) = \sigma(u) > 0$. But the definition of d-reflexivity as introduced in Def. 13.2(c) does not require $\sigma(x) > 0$, for all $x \in V$. Had we really defined d-reflexivity in that way, then σ would be converted into something like the "degree of existence" of a node $x \in V$ in G. This, in turn, means that σ is not free for characterizing/evaluating e.g. the attractiveness of cities, the importance of persons in a project, or capacities of a store etc. So the only possibility that keeps σ free from an 'existential' nature and also conforms to the crisp case is to define conn_G as we did above.

Another difficulty with d-fuzzy relations is their restricted suitability for modeling certain situations, as already indicated with the example given above. Things become still more subtle when dealing with fuzzy graphs: Consider the case that you plan a traveling route with the goal of minimizing your accommodation costs. Then, if appropriate, you will certainly want to visit a friend (an additional benefit) living in say x. Then you set $\sigma(x) = 0$ because you will not have to pay anything. But this will make x an isolated node in your modeling graph $G = (V, \sigma, \rho)$ since, by definition, $\sigma(x) \wedge \sigma(y) = 0 = \rho(x, y)$, for all $y \in V$. Certainly not an effect you are looking for.

Using the generic terms costs/benefits for negatively/positively connoted properties then fuzzy graphs can be used for the latter ones: If I do not expect any benefit from using a node $x \in V$, then setting $\sigma(x) = 0$ is okay, since I do not want to visit x under normal circumstances.

Evidently, the strength of a path is defined to be the weight of its weakest edge, and the strength of connectedness between two nodes u, v is defined as the maximal strength of the paths between u and v. Following [7] an edge $\underline{uv}$ with $\rho(u, v) = \sigma(u) \wedge \sigma(v)$ is called **M-strong**. For all $x, y \in V$, $\sigma(u) \wedge \sigma(v)$ is the maximal value that can be chosen to make ρ a d-fuzzy relation on σ. Note that for M-strong edges we have $\rho^c(u, v) = 0$. A fuzzy graph $G = (V, \sigma, \rho)$ with all edges being M-strong has been named a **strong graph** in Mordeson and Peng [22]. M-strong fuzzy graphs are also called **complete** in e.g. [6, 21]. A slightly more general concept has been introduced in [8] and further refined in [17]:

Definition 13.21. *Let $G = (V, \sigma, \rho)$ be a fuzzy graph; an edge $e = \underline{uv} \in \rho^*$ is called*

(a) **strong** *if* $\rho(u, v) \geq \mathsf{conn}_{G-e}(u, v)$;
(b) α**-strong** *if* $\rho(u, v) > \mathsf{conn}_{G-e}(u, v)$;
(c) β**-strong** *if* $\rho(u, v) = \mathsf{conn}_{G-e}(u, v)$;
(d) δ**-edge** *if* $\rho(u, v) < \mathsf{conn}_{G-e}(u, v)$;
(e) δ^***-edge** *if it is a δ-edge such that $\rho(u, v) > w$, where w is the weight of the weakest edge in G.*

So an edge $e = \underline{uv} \in \rho^*$ is strong if its weight is at least as great as the connectedness of its end nodes when it is deleted. Obviously, an M-strong edge is strong; but the converse need not be true: In a cycle which has more than one arc with minimum weight every arc is strong.

A path in a fuzzy graph is called **strong** if it contains strong arcs only. A strongest path need not be a strong path and a strong path need not be strongest. A path is called an α**-strong path** if all its edges are α-strong, and it is called β**-strong** if all its edges are β-strong.

Proposition 13.22. ([8, Proposition 2]) *Let $G = (V, \sigma, \rho)$ be a fuzzy graph and $e = \underline{uv} \in \rho^*$. e is strong iff* $\mathsf{conn}_G(u, v) = \rho(u, v)$.

Proof. Obviously, $\mathsf{conn}_G(u, v) := \bigvee_{P_{uv}} \mathsf{str}(P_{uv}) \geq \rho(u, v)$. If a path P_{uv} contains e, then $\mathsf{str}P_{uv} \leq \rho(u, v)$. If it does not contain e it lies in $G - e$ and $\mathsf{str}P_{uv} \leq \mathsf{conn}_{G-e}(u, v) \leq \rho(u, v)$ since e is strong. Hence in either case the strength of a path from u to v is at most $\rho(u, v)$, whence $\mathsf{conn}_G(u, v) \leq \rho(u, v)$. Conversely, if $\rho(u, v) = \mathsf{conn}_G(u, v)$ then $\rho(u, v) \geq \mathsf{conn}_{G-e}(u, v)$, so e is strong. □

Fig. 13.6 shows (a) Example 5 from [31] and (b) Example 1 from [17]. Edges $\underline{vx}$ and $\underline{vw}$ are α-strong, $\underline{uv}$ and $\underline{ux}$ are β-strong, and $\underline{wu}$ and $\underline{wx}$ are δ-edges. The path $(uxvw)$

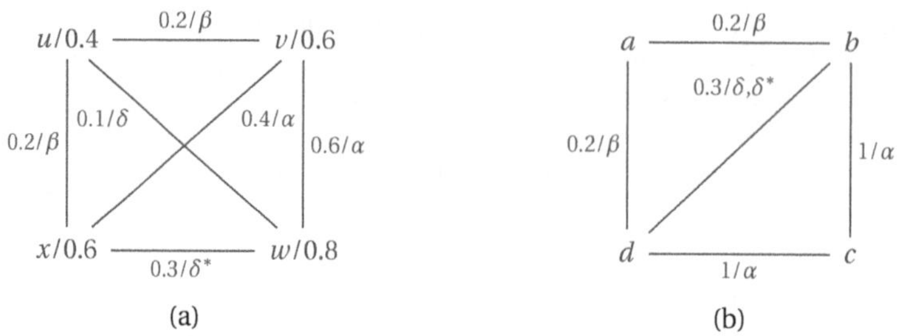

FIG. 13.6: Types of fuzzy edges.

is a strongest as well as a strong path, and (uxw) is a strongest path but not a strong path since $\underline{xw}$ is a δ^*-arc. If we increase the weight of edge $\underline{wx}$ by setting $\rho(w,x) = 0.4$ then it will become β-strong and the path (vxw) will become a strong but not a strongest path since $\text{conn}_G(v,w) = 0.6$.

In Fig. 13.6(b) $\underline{bc}$ and $\underline{cd}$ are α-strong, $\underline{ab}$ and $\underline{ad}$ are β-strong, and $\underline{bd}$ is δ^*. (dcb) is an α-strong, and (dab) a β-strong path. Note that in fuzzy graphs the types of edges cannot be determined by simply examining their weights (for details see [17, Section 3]): The weight of a δ-arc can exceed those of α-strong and β-strong arcs; and the weight of a β-strong edge can be greater than that of an α-strong edge. In Fig. 13.6(b) we have $\rho(b,d) = 0.3 > 0.2 = \rho(a,b)$, where $\underline{bd}$ is a δ-edge and $\underline{ab}$ is β-strong.

13.3.3 *Forests, Trees, and Cycles*

In the theory of crisp graphs a connected graph without any cycle is called a **tree**. Two or more trees form a **forest**. So, removing any edge or node from a crisp tree makes it disconnected, i.e. the number of components will increase. Thus, every edge (node) of a crisp tree is a bridge (cut node). Δ *But as we will see below, the structure of fuzzy trees is significantly different from that of their crisp counterparts: a fuzzy tree can contain cycles in the classical sense.* Δ

The generalization of trees in fuzzy graph theory appeared already in the very first contribution, viz. the classical paper [24] by Rosenfeld which also contains assertions and proofs of some fundamental properties of fuzzy trees. Corresponding concepts and results for cycles have been given in [20].

Definition 13.23. ([24, Section 10], [21, p. 22]) *Let $G = (V,\sigma,\rho)$ be a fuzzy graph. Then G is called*

(a) a **forest** *if the underlying graph G^* is a forest in the classical sense, and a* **tree** *if it is a forest which is connected;*

(b) a **fuzzy forest** *if there exists a spanning fuzzy subgraph $F = (V,\sigma,\nu)$ which is a forest such that for all edges $e = \underline{uv} \in \rho^* - \nu^*$ $\rho(u,v) < \text{conn}_F(u,v)$, i.e. there exists a path from u to v in F, whose strength is greater than $\rho(u,v)$. A fuzzy forest which is connected is called a* **fuzzy tree**.

The (counter)example graphs for fuzzy trees shown in Fig. 13.7 and Fig. 13.8 are the original ones appearing in [24, p. 92] and also in [21, p. 23]. Note that if $<$ had been replaced by $\leq$ in Def. 13.23, then even a triangle as in Fig. 13.7 with edge weights all equal to 1 (i.e. the crisp circle C_3) would qualify as a fuzzy tree since any two of its arcs would give a spanning fuzzy tree.

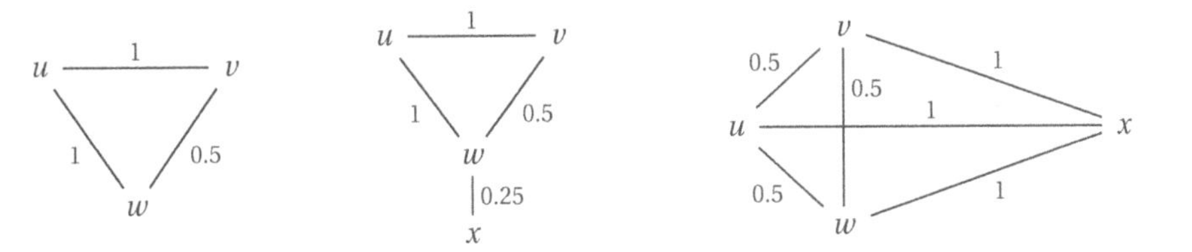

FIG. 13.7: Fuzzy trees.

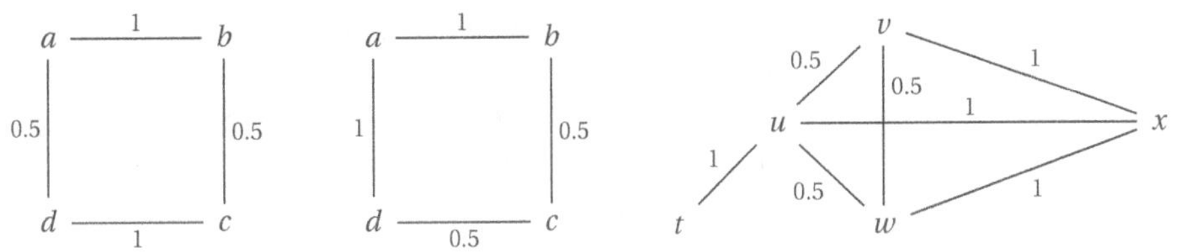

FIG. 13.8: Fuzzy graphs, but not fuzzy trees.

Definition 13.24. ([20, DEFINITION 3], [21, Definition 2.2]) *Let $G = (V, \sigma, \rho)$ be a fuzzy graph. Then G is called*

(a) a **cycle** *if the underlying graph G^* is a crisp cycle;*
(b) a **fuzzy cycle** *if it is a cycle and there is no unique $\underline{uv} \in \rho^*$ such that $\rho(u,v) = \bigwedge_{\underline{xy} \in \rho^*} \rho(x,y)$, i.e. G has more than one weakest edge.*

Our example graphs in connection with cycles and fuzzy cycles shown in Fig. 13.9 are from [16, Examples 2.3.7, 2.3.8] (see also [21, Examples 2.1, 2.2]). F is a fuzzy tree, but not a tree ($G - \underline{cd}$ is the spanning fuzzy subgraph required) and not a fuzzy cycle ($\underline{ab}$ is the unique weakest edge), while G is a fuzzy cycle, but not a fuzzy tree; and H is neither a fuzzy cycle nor a fuzzy tree.

The next two results appear already in [24], and may therefore be regarded as the classical ones on fuzzy trees. Although their proofs have also been (literally) reproduced at the places additionally mentioned below, we include them (also almost literally) since later on we have to resort to the construction used in

Theorem 13.25. ([21, Theorem 2.5], [16, Theorem 2.3.1]) *$G = (V, \sigma, \rho)$ is a fuzzy forest if and only if in any cycle of G there is an edge $\underline{uv}$ such that $\rho(u,v) < \mathrm{conn}_{G-\underline{uv}}(u,v)$.*

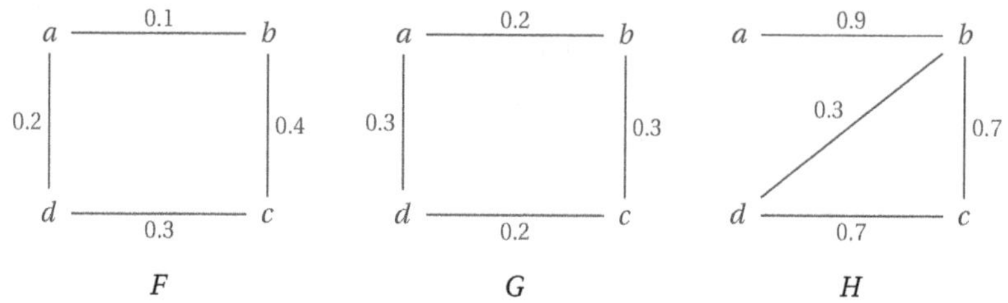

FIG. 13.9: Fuzzy cycles and fuzzy trees.

Proof. Assume that G has at least one cycle, that $e = \underline{uv}$ with $\rho(u,v) < \mathsf{conn}_{G-e}(u,v)$ belongs to one of them, say C, and that $\rho(u,v)$ is smallest, i.e. e is a weakest arc in C. (Note that if there is another weakest edge $\underline{st}$ in C, then there must be a st-path with strength greater than $\rho(s,t)$.) If we delete e, the resulting fuzzy subgraph satisfies the path property of a fuzzy forest. As long as there are still cycles in the resulting subgraph, we can repeat the process, noting that at each stage, no previously deleted arc is stronger than the arc being currently deleted. That means, that the path guaranteed by the property of the theorem involves only edges that have not yet been deleted. When no more cycle remains, the resulting fuzzy subgraph is a forest F. Let $\underline{st}$ be one of the arcs that have been deleted in the process of constructing F; then there is a path P from s to t that is stronger than $\rho(s,t)$ and that does not involve $\underline{st}$ itself or any of the arcs deleted prior to it. If P involves arcs that were deleted later, it can be redirected around them using a path of still stronger arcs. If any of these arcs were deleted later, the path can be further redirected, and so on. This process eventually stops with a path consisting entirely of arcs of F. Thus G is a fuzzy forest.

Conversely, if G is a fuzzy forest and C is any cycle, then some edge $\underline{uv}$ of C is not in $F = (V, \sigma, \nu)$, a fuzzy spanning tree of G existing by definition of a fuzzy forest. Thus, we have $\rho(u,v) < \mathsf{conn}_F(u,v) \le \mathsf{conn}_{G-\underline{uv}}(u,v)$. □

Proposition 13.26. ([21, Proposition 2.6], [16, Proposition 2.3.2]) *If there is at most one strongest path between any two vertices of a fuzzy graph G, then G is a fuzzy forest.*

Proof. Assuming that G is not a fuzzy forest, Th. 13.25 implies that there is a cycle C in G such that $\rho(u,v) \ge \mathsf{conn}_{G-\underline{uv}}(u,v)$ for all edges $\underline{uv} \in C$. Thus, $\underline{uv}$ is a strongest path from u to v. If we choose $\underline{uv}$ to be a weakest edge of C, it follows that the rest of C also is a strongest path from u to v, i.e. there is more than one strongest path from u to v. The assertion then follows by Th. 1.17(8). □

The converse of the previous proposition does not hold: A fuzzy forest can still have multiple strongest paths between nodes. The fuzzy graph G shown in Fig. 13.10 ([16, Example 2.3.3]) is a fuzzy forest, and $F := G - \underline{cd}$ is the spanning tree of G constructed as described in the proof of Th. 13.25. But both $(edcb)$ and $(edab)$ are strongest eb-paths in G.

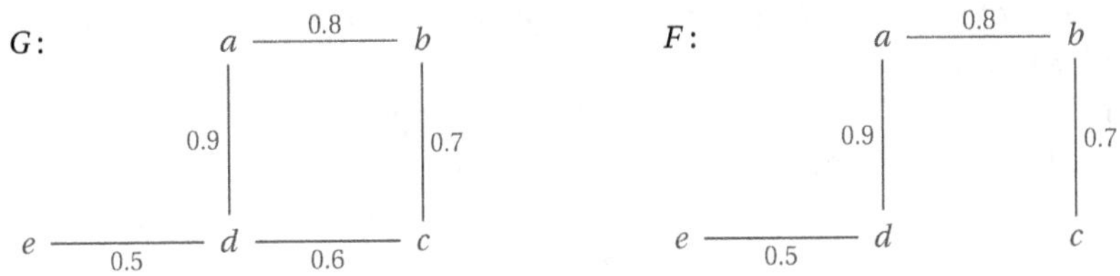

FIG. 13.10: A fuzzy tree with its spanning tree.

Proposition 13.27. ([21, Theorem 2.8], [16, Theorem 2.3.9]) *Let $G = (V,\sigma,\rho)$ be a cycle. Then G is a fuzzy cycle if and only if it is not a fuzzy tree.*

Proof. If G is a fuzzy cycle, then there are edges $\underline{st}, \underline{uv} \in \rho^*$ such that $\rho(s,t) = \rho(u,v) = \bigwedge_{xy \in \rho^*} \rho(x,y)$. If $H = (V,\sigma,\nu)$ is any spanning tree of G, then $\rho^* - \nu^* = \{\underline{xy}\}$ for some $x, y \in V$ because G is a cycle, whence there is no path P_{xy} in H with a strength $\mathsf{str}(P_{xy}) > \rho(x,y)$. Therefore, G is not a fuzzy tree.

Conversely, suppose that G is not a fuzzy tree. Since G is a cycle, for all $\underline{uv} \in \rho^*$, $H = (V,\sigma,\rho^* - \{\underline{uv}\})$ is a fuzzy spanning subgraph of G which is a tree, and $\mathsf{conn}_H(u,v) \leq \rho(u,v)$, where $\nu(u,v) = 0$ and $\nu(x,y) = \rho(x,y)$ for all $\underline{xy} \in \rho^* - \{\underline{uv}\}$. So, ρ does not attain $\bigwedge_{\underline{xy} \in \rho^*} \rho(x,y)$ uniquely, whence G is a fuzzy cycle. □

Proposition 13.28. ([16, Theorem 2.3.12]) *Let $G - (V,\sigma,\mu)$ be a fuzzy tree such that (σ^*,μ^*) is not a tree; then there exists at least one edge $\underline{uv}$ in μ^* for which $\mu(u,v) <$ $\mathsf{conn}_G(u,v)$.*

Proof. If G is a fuzzy tree, then by definition there exists a fuzzy spanning subgraph $H = (V,\sigma,\nu)$, which is a tree and $\mu(u,v) < \mathsf{conn}_H(u,v)$ for all edges $\underline{uv}$ not in H. Since also $\mathsf{conn}_H(u,v) \leq \mathsf{conn}_G(u,v)$, $\mu(u,v) < \mathsf{conn}_G(u,v)$ for all $\underline{uv}$ not in H, and by hypothesis there exists at least one edge $\underline{uv}$ not in H. □

Proposition 13.29. ([16, Theorem 2.3.13]) *Let $G = (V,\sigma,\rho)$ be a connected fuzzy graph without a fuzzy cycle. Then G is a fuzzy tree.*

Proof. If there is no cycle in G^*, then it is a tree and G is a fuzzy tree. Now assume that G has cycles, none of them being a fuzzy cycle by hypothesis. So each cycle C in G has exactly one weakest edge, say e. If, after removing e from G, there are still cycles in the resulting fuzzy graph, repeat this process. This will eventually result in a fuzzy subgraph of G which is a tree and the spanning fuzzy subgraph required. □

13.3.4 *Cutnodes and Bridges*

In this section, we will confine our considerations to the concepts of f-bridges and f-cutnodes in their classical sense as already introduced in [24], together with some of their properties and characterizations. As one might expect, some more fine-grained variants of both concepts have been proposed since then, described in [16, Section 2.5], called weak, partial and full fuzzy bridges, cutvertices and blocks. We also omit the more general notions of a *vertex cut* $S \subseteq V$ and an *edge cut* $T \subseteq V \times V$, or the related concepts of *edge* and *vertex connectivity*. See e.g. [9,27] or [21, p. 31].

Looking back at Def. 13.14 we see that in crisp graphs cut nodes and bridges are characterized by the fact that removing them increases the number of components. So in the resulting graph there are at least two vertices which are not connected any more, and a connected graph becomes disconnected. Recasting "*not connected any more*" in the spirit of fuzzy graphs gives "the strength of connectedness between some pair of nodes has been reduced" (from 1 to 0).

Obviously, "*being connected*" is an equivalence relation $\gamma \subseteq V \times V$ on the set of nodes of a fuzzy graph, with the equivalence classes called **connected components.** We have $\gamma(x,y)$ iff there is a path between x and y; thus, like in the crisp case, for a connected graph we have $|V/\gamma| = 1$, and $|V/\gamma| > 1$ for a disconnected one.

Definition 13.30. *Let $G = (V,\sigma,\rho)$ be a fuzzy graph, $u \in \sigma^*$, and $e = \underline{vw} \in \rho^*$.*

(a) *e is called a* **fuzzy bridge** *(short:* **f-bridge***) of G, if $\mathrm{conn}_{G-e}(v,w) < \mathrm{conn}_G(v,w)$ for some nodes $v, w \in \sigma^*$.*
(b) *u is called a* **fuzzy cutnode** *(short:* **f-cutnode***) of G, if $\mathrm{conn}_{G-u}(v,w) < \mathrm{conn}_G(v,w)$ for some nodes $v, w \in \sigma^*$ other than u.*
(c) *A fuzzy graph is said to be a* **fuzzy block** *(short:* **f-block***) if it has no fuzzy cutnodes.*

Thus, an edge e is an f-bridge if and only if there exist vertices u, v such that it is an edge of every strongest path from u to v. Likewise, $w \in V$ is an f-cutnode if and only if there exist two vertices $u, v \in V$ distinct from w such that it is a node on every strongest uv-path.

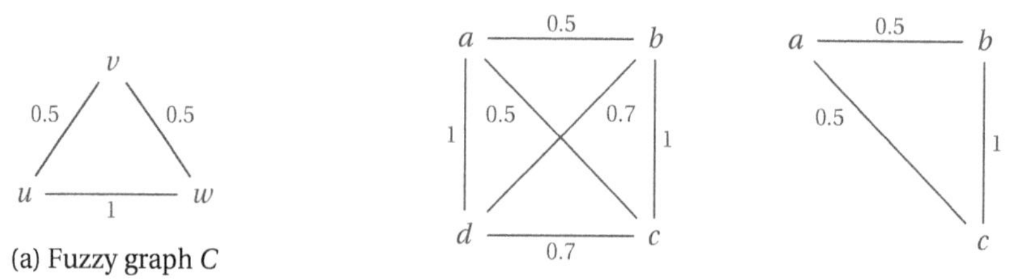

(a) Fuzzy graph C (b) Fuzzy graph G (c) Fuzzy graph G-d

FIG. 13.11: Fuzzy cutnodes and bridges.

We give examples inspired by [23, p. 119] in Fig. 13.11: $\underline{uw}$ is an f-bridge since $\mathrm{conn}_{C-\underline{uw}}(u,w) = 0.5 < \mathrm{conn}_C(u,w) = 1$, and d is an f-cutnode since $\mathrm{conn}_{G-d}(a,c) = 0.5 < \mathrm{conn}_G(a,c) = 0.7$. Δ *Note the difference to crisp graphs: Neither is $\underline{uw}$ a bridge in C^*, nor d a cutnode in G^*. The other way round, the removal of f-bridges or f-cutnodes from connected fuzzy graphs does not necessarily make them disconnected.* Δ

The fuzzy graph shown in Fig. 13.12 has several f-cutnodes (the ones with boxes around them) as well as several f-bridges (all edges except $\underline{ab}$ and $\underline{bc}$). E.g.

$\mathsf{conn}_{G-e}(d,f) = 0.4 < \mathsf{conn}_G(d,f) = 0.8$, or $\mathsf{conn}_{G-\underline{eb}}(f,b) = 0.4 < \mathsf{conn}_G(f,b) = 0.6$. None of the possible removals of an f-cutnode or an f-bridge in G would result in a disconnected fuzzy graph.

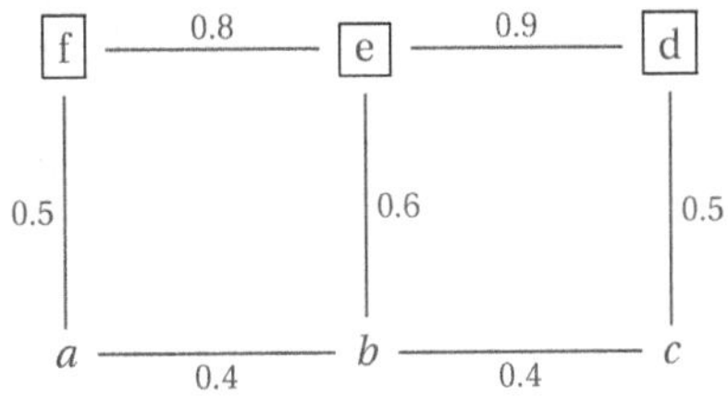

FIG. 13.12: Fuzzy cutnodes and bridges.

We now present characterizations and properties of f-bridges and f-cutnodes similar to those well known for their crisp counterparts; see e.g. [12, Chapter 3]. The following theorem first appeared already as [24, Theorem 9.1].

Theorem 13.31. ([21, Theorem 2.4], [16, Theorem 2.2.1]) *Let $G = (V,\sigma,\rho)$ be a fuzzy graph and $e = \underline{uv} \in \rho^*$. The following statements are equivalent:*

(1) *e is a fuzzy bridge.*
(2) $\mathsf{conn}_{G-e}(u,v) < \rho(u,v)$.
(3) *e is not the weakest arc of any cycle.*

Proof. (2) → (1): If e is not an f-bridge, there are no $u,v \in V$ such that $\mathsf{conn}_{G-e}(u,v) < \mathsf{conn}_G(u,v)$. So $\mathsf{conn}_{G-e}(u,v) = \mathsf{conn}_G(u,v) \geq \rho(u,v)$. We have thus shown ¬(1) → ¬(2), whence (2) → (1) by Th. 1.17(8).

(1) → (3): If e is the weakest edge of a cycle, then any path P_{uv} involving it can be converted into a path P'_{uv} not containing e but at least as strong as P_{uv} by using the rest of the cycle as a path from u to v. Thus, e cannot be an f-bridge.

(3) → (2): If $\mathsf{conn}_{G-e}(u,v) \geq \mathsf{conn}_G(u,v)$, then there exists a path P_{xy} not containing e such that $\mathsf{str}(P_{uv}) \geq \rho(u,v)$ and which together with e forms a cycle of G with e as a weakest edge. □

Proposition 13.32. ([8, Proposition 1]) *Every bridge in a fuzzy graph $G = (V,\sigma,\rho)$ is strong, but a strong edge need not be a bridge.*

Proof. Assume that $e = \underline{uv}$ is not strong, then $\rho(u,v) < \mathsf{conn}_{G-e}(u,v)$. Let P be a strongest path from u to v in $G-e$; then $\mathsf{str}(P) = \mathsf{conn}_{G-e}(u,v)$. If we adjoin e to P we get a cycle with e as its weakest edge. Hence, e is not a fuzzy bridge of G by Th. 13.31. So a fuzzy bridge must be a strong edge. That the converse does not hold can be seen from a triangle in which all the edges have the same weight. □

The following proposition presents a necessary and sufficient condition for an edge to be a fuzzy bridge:

Proposition 13.33. ([17, Theorem 1]) *Let $G = (V,\sigma,\rho)$ be a fuzzy graph. An edge $e = \underline{uv} \in \rho^*$ is an f-bridge iff it is α-strong.*

Proof. Let e be an f-bridge in G. Then by definition $\mathsf{conn}_{G-e}(u,v) < \mathsf{conn}_G(u,v)$, and by Prop. 13.22 we have $\mathsf{conn}_G(u,v) = \rho(u,v)$. So $\mathsf{conn}_{G-e}(u,v) < \rho(u,v)$, whence e is α-strong. Conversely, suppose that e is α-strong. Then by Def. 13.21(b) it follows that $\rho(u,v)$ is the unique strongest path from u to v and the removal of e will reduce $\mathsf{conn}_G(u,v)$. Thus e is an f-bridge. □

Theorem 13.34. ([21, Theorem 2.14], [16, Theorem 2.2.14]) *If $\underline{uv}$ is an f-bridge in a fuzzy graph $G = (V,\sigma,\rho)$, then $\mathsf{conn}_{(G)}(u,v) = \rho(u,v)$.*

Proof. Suppoe that $\underline{uv}$ is a fuzzy bridge and $\mathsf{conn}_G(u,v) > \rho(u,v)$. Then there exists a strongest uv-path P_{uv} with strength greater than $\rho(u,v)$, i.e. all edges P_{uv} have strength greater than $\rho(u,v)$. But P_{uv} together with $\underline{uv}$ forms a cycle with $\underline{uv}$ its weakest edge, contradicting our assumption that $\underline{uv}$ is a fuzzy bridge. □

Proposition 13.35. ([21, Proposition 2.7], [16, Proposition 2.3.4]) *If $G = (V,\sigma,\rho)$ is a fuzzy forest, then the edges of its spanning tree $F = (V,\tau,\nu)$ are precisely the bridges of G.*

Proof. An edge $e = \underline{uv} \in G^* - F^*$ cannot be an f-bridge because $\rho(u,v) < \mathsf{conn}_F(u,v) \le \mathsf{conn}_G(u,v)$. Now suppose an edge $e \in F^*$ not to be an f-bridge. Then there would exist a path P between u and v not involving e, with $\mathsf{str}(P) \ge \rho(u,v)$. P must involve edges in $G^* - F^*$ because F is a forest and thus has no cycles. However, by definition, any such edge $e_i = \underline{u_i v_i}$ can be replaced by a path P_i in F of strength greater than $\rho(u_i,v_i)$. e is not part of P_i because all its edges are strictly stronger than $\rho(u_i,v_i) \ge \rho(u,v)$. But by replacing each e_i by a suitable path P_i, one can construct a path in F between u and v which does not contain $\underline{uv}$, resulting in a cycle in F, a contradiction. □

A **maximum spanning tree** of a connected fuzzy graph $G = (V,\sigma,\rho)$ is a spanning fuzzy subgraph $T = (V,\sigma,\nu)$ which is a tree such that $\mathsf{conn}_G(u,v)$ is the strength of the unique strongest uv-path in T for all $u,v \in V$.

Theorem 13.36. ([16, Theorem 2.2.2]) *A node w of a fuzzy graph $G = (V,\sigma,\rho)$ is a fuzzy cutnode if and only if w is an internal vertex of every maximum spanning tree of G.*

Corollary 13.37. ([16, Corollary 2.2.4]) *An edge $\underline{uv}$ of a fuzzy graph $G = (V,\sigma,\rho)$ is a fuzzy bridge if and only if $\underline{uv}$ is in every maximum spanning tree of G.*

Proposition 13.38. ([16, Theorem 2.3.14]) *In a fuzzy tree $G = (V,\sigma,\rho)$ the removal of any fuzzy bridge $\underline{uv}$ reduces the strength of connectedness between its end nodes and also between some other pair of vertices.*

Proof. By Prop. 13.35 $\underline{uv}$ is an edge of the maximum spanning tree T of G, and T contains unique strongest paths joining every pair of nodes. So removing $\underline{uv}$ reduces the strength of connectedness between some other pair of vertices $x, y \in V$, where $\underline{xu}, \underline{yv} \in \rho^*$ if $\underline{uv}$ is an internal edge of T, and $u = x$ or $v = y$ otherwise. □

Theorem 13.39. ([21, p. 29, Theorem 2.20], [16, Theorem 2.3.18]) *A fuzzy graph $G = (V, \sigma, \rho)$ is a fuzzy tree if and only if the following two conditions are equivalent for all $u, v \in V$:*

(1) $\underline{uv}$ is a fuzzy bridge.
(2) $\mathsf{conn}_G(u, v) = \rho(u, v)$.

Theorem 13.40. ([21, Theorem 2.21], [16, Theorem 2.3.19]) *A fuzzy graph is a fuzzy tree if and only if it has a unique maximum fuzzy spanning tree.*

Next we have a look on f-cutnodes. and continue with some of their connectivity properties in fuzzy graphs.

Theorem 13.41. ([21, Theorem 2.10], [16, Theorem 2.2.10]) *Let $G = (V, \sigma, \rho)$ be a cycle. Then a node of G is an f-cutnode if and only if it is a common vertex of two f-bridges.*

Proof. Let w be a fuzzy cutnode of G. Then there are $u \neq w \neq v \in V$ such that w is on every strongest uv-path. Because G^* is a cycle, there exists only one strongest uv-path containing w, and all its edges are f-bridges. Conversely, if w is a common vertex of two f-bridges , then neither $\underline{uw}$ not $\underline{wv}$ is a weakest edge of G. The uv-path not containing $\underline{uw}$ and $\underline{wv}$ has strength less than $\rho(u, w) \wedge \rho(w, v)$, whence the strongest uv-path is (uwv), and $\mathsf{conn}_G(u, v) = \rho(u, w) \wedge \rho(w, v)$. Thus, w is an f-cutnode. □

Theorem 13.42. ([21, Theorem 2.11], [16, Theorem 2.2.11]) *If $w \in V$ is a common vertex of at least two fuzzy bridges, then it is a fuzzy cutnode.*

Proof. Let $\underline{u_1 w}$ and $\underline{w u_2}$ be two fuzzy bridges. Then there exist nodes u, v such that $\underline{u_1 w}$ is on every strongest uv-path. If w is distinct from u, v, then w is a fuzzy cutnode. Next suppose that $u = w$ or $v = w$, so that $\underline{u_1 w}$ is on every strongest uw-path or $\underline{wu_2}$ is on every strongest wv-path. Assume that w is not a fuzzy cutnode; then between every two vertices there exists at least one strongest path not containing w. In particular, there exists at least one strongest $u_1 u_2$-path, say P, not containing w which together with $\underline{u_1 w}$ and $\underline{wu_2}$ forms a cycle.

We consider two cases: (i) If $(u_1 w u_2)$ is not a strongest path, then clearly at least one of $\underline{u_1 w}$ and $\underline{wu_2}$ is a weakest edge of a cycle. But, due to Th. 13.31, this contradicts the assumption that $\underline{u_1 w}$ and $\underline{wu_2}$ are fuzzy bridges. (ii) If $(u_1 w u_2)$ is a strongest path, then $\mathsf{conn}_G(u_1, u_2) = \rho(u_1, w) \wedge \rho(w, u_2)$. Therefore, the edges of P are at least as strong as $\rho(u_1, w)$ and $\rho(w, u_2)$, which implies that $\underline{u_1 w}$ and $\underline{wu_2}$ are both weakest edges of a cycle, again a contradiction. □

The condition in Th. 13.42 is not necessary, as can be seen from graph G in Fig. 13.13 ([16, Example 2.2.12]): b is a fuzzy cutnode, but $\underline{ab}$ and $\underline{cd}$ are the only fuzzy bridges. Δ

In graph H $\underline{ab}$ and $\underline{cd}$ are fuzzy bridges, but there is no fuzzy cutnode. This is a significant difference to crisp graph theory. Δ

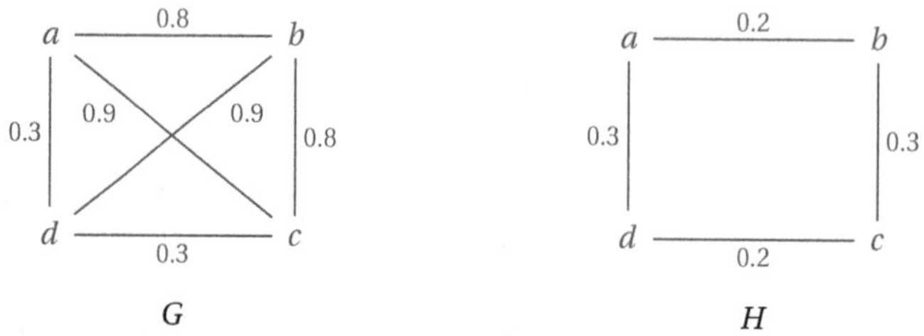

FIG. 13.13: Fuzzy bridges and cutnodes.

Proposition 13.43. ([21, Theorem 2.18], [16, Theorem 2.3.16]) *In a fuzzy tree $G = (V, \sigma, \rho)$ the internal nodes of a fuzzy spanning subgraph F are fuzzy cutnodes of G.*

Proof. Let w be any vertex in G, which is not an end vertex of F. Then w is the common node of at least two edges in F, which are fuzzy bridges of G and by Th. 13.42, w is a fuzzy cutnode. Also, if w is an end node of F, then it is not a fuzzy cutnode, because otherwise there would exist nodes u, v distinct from w such that w is on every uv- path and one such path certainly lies in F. But because w is assumed to be an end vertex of F, this is not possible. □

13.3.5 *Blocks*

In crisp graph theory blocks are maximal connected subgraphs without cutvertices. This concept has been generalized to fuzzy graphs in a straightforward way in [24, p. 91]: A maximal connected subgraph of $G = (V, \sigma, \rho)$ induced by a subset of vertices $P \subseteq V$ without f-cutnodes is said to be a **fuzzy block** or simply an **f-block** of G. Obviously, a complete fuzzy graph is an f-block. For more details see e.g. [16, Section 2.7], which together with [1] is the main source for the presentation below.

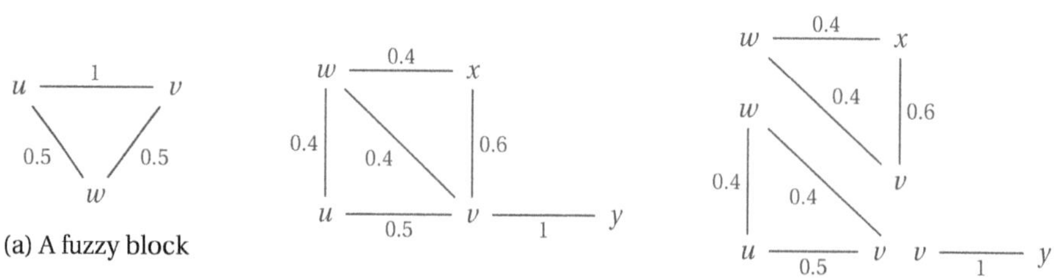

(a) A fuzzy block

(b) A fuzzy graph with 3 blocks

FIG. 13.14: Blocks in fuzzy graphs.

Δ *The properties of f-blocks in fuzzy graphs are different from those of blocks in crisp graphs:*

(1) Contrary to crisp graphs, an f-block may have one or more f-bridges as illustrated in Fig. 13.14(a). *Edge $\underline{uv}$ is an f-bridge of the triangle graph. (2) Two blocks of a graph share at most one vertex which then is a cutnode of the graph. But two f-blocks may share one or more vertices which need not be f-cutnodes of the fuzzy graph.* Fig. 13.14(b) *(this is Example 2 from [1]) shows a fuzzy graph which itself is not a block but which has three blocks shown on the right, each of them having an f-bridge. The two triangles have the nodes v and w in common, but only the former is an f-cutnode.* Δ

As has been stated in Rosenfeld [24, p. 91], if between any two vertices $u, v \in G$ there exist two strongest paths that are disjoint (except for u, v) then, obviously, G is an f-block. This corresponds to the classical property of crisp graphs that G is block with at least three vertices if every two of them lie on a common cycle. Δ *For crisp graphs also the converse direction holds: In a block every two points lie on a common cycle ([12, Theorem 3.3]). As the f-block shown in* Fig. 13.14(a) *demonstrates, the corresponding assertion does not hold for fuzzy graphs: Nodes u and v do not lie on two disjoint strongest paths, since* $\mathrm{conn}_G(uv) = 1$ *but* $\mathrm{str}\big((uwv)\big) = 0.5$. Δ

In [30] it has been shown that the converse of Rosenfeld's observation above is true only for blocks having no fuzzy bridges. So we have

Theorem 13.44. ([16, Theorem 2.7.4]) *The following statements are equivalent for a fuzzy graph $G = (V, \sigma, \rho)$:*

(1) G is a block.
(2) Any two vertices u and v such that $\underline{uv}$ is not a fuzzy bridge are joined by two internally disjoint strongest paths.
(3) For every three distinct vertices of G, there is a strongest path joining any two of them not containing the third.

Theorem 13.45. ([16, Theorem 2.7.2]) *In a block with at least one fuzzy bridge the removal of any fuzzy bridge reduces the strength of connectedness between its end nodes only.*

Proof. Let $G = (V, \sigma, \rho)$ be a block and $e = \underline{uv}$ be an f-bridge of G. Assume that $\mathrm{conn}_{G-e}(w, w') < \mathrm{conn}_G(w, w')$ for some pair of nodes w, w' other than u, v.

We first consider the case that $\{w, w'\} \cap \{u, v\} = \emptyset$, i.e. both w and w' are distinct from u and v. By assumption every strongest ww'-path contains e, whence the removal of either u or v obviously reduces the strength of connectedness between w an w'. So u and v are f-cutnodes in G, contradicting the assumption of G being a block. And if one of $u, v \in \{w, w'\}$, say $w = u$ and $w' \neq v$, then as before $\mathrm{conn}_{G-u}(w, w') < \mathrm{conn}_G(w, w')$, whence u is f-cutnode of G; and similarly for $w \neq u$ and $w' = v$. Both cases contradict the hypothesis that G is a block. We thus have to conclude that $w = u$ and $w' = v$. □

We close our section about fuzzy blocks with the following characterization, but without reproducing its proof of three and a half pages:

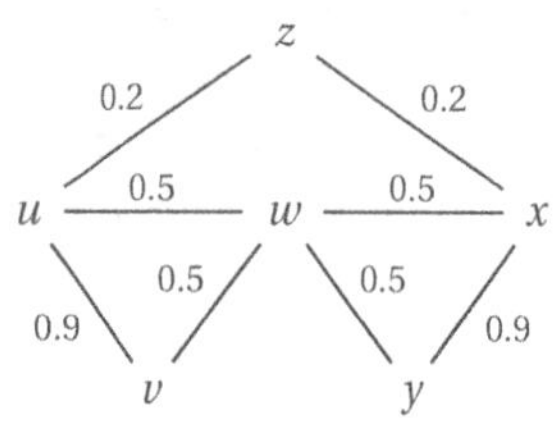

That the converse of Th. 13.45 does not hold can be seen from the example on the left ([16, Example 2.7.3]): Both $\underline{uv}$ and $\underline{xy}$ are fuzzy bridges and their removal does not reduce the strength of connectedness between any pair of nodes other than their end nodes. But since w is a fuzzy cutnode, G is not a block.

Theorem 13.46. ([16, Theorem 2.7.9]) *The following statements are equivalent for a fuzzy graph $G = (V, \sigma, \rho)$ with at least three vertices and without fuzzy bridges:*

(1) G is a block.
(2) For any two nodes $u, v \in V$ there exists a cycle containing them which is formed by two strongest strong paths.
(3) For each vertex u and each strong edge $\underline{vw}$ of G, there exists a cycle containing u and $\underline{vw}$ which is formed by two strongest strong uv-paths or uw-paths.
(4) For any two strong edges $\underline{uv}$ and $\underline{wx}$ of G, there exists a cycle containing them which is formed by two strongest strong uw-paths or vw-paths.
(5) For every three distinct vertices of G, there is a strongest strong path joining any two of them not containing the third.

A slightly more general version, together with some further results on fuzzy blocks, is given in Section 2 of [14]; see also [16, Section 4.3].

References

[1] Anjali, N., Mathew, S.: On blocks and stars in fuzzy graphs, J. of Intelligent & Fuzzy Systems **28**, 1659-1665 (2015).

[2] Balakrishnan, R., Ranganathan, K.: *A Textbook of Graph Theory*, 2nd ed., Springer Science +Business Media New York (2012).

[3] Bang-Jensen, J., Gutin, G.Z.: *Digraphs - Theory, Algorithms and Applications*, 2nd ed., Springer-Verlag London Limited (2009).

[4] Bang-Jensen, J., Gutin, G.Z. (Eds): *Classes of Directed Graphs*, Springer International Publishing AG (2018).

[5] Berge, C.: *Graphs and Hypergraphs*, North-Holland Publishing Company (1973).

[6] Bhutani, K.R.: On automorphisms of fuzzy graphs, Pattern Recognition Letters **9**, 159-162 (1989).

[7] Bhutani, K.R., Battou, A.: On M-strong fuzzy graphs, Information Sciences **155**, 103-109 (2003).

[8] Bhutani, K.R., Rosenfeld, A.: Strong arcs in fuzzy graphs, Information Sciences **152**, 319-322 (2003).

[9] Binu, M., Mathew, S., Mordeson, J.N.: Connectivity index of a fuzzy graph and its application to human trafficking, Fuzzy Sets and Systems **360**, 117-136 (2019).

[10] Bondy, J.A., Murty, U.S.R.: *Graph Theory*, Springer (2008).

[11] Diestel, R.: *Graph Theory*, 5th ed., Springer-Verlag GmbH Germany (2017).

[12] Harary, F.: *Graph Theory*, Taylor & Francis Group, Boca Raton, (1969).

[13] Kaufmann, A.: *Introduction a la Theorie des Sous-Ensembles Flous* (Vol. 1), Masson, Paris (1973). English translation: *Introduction to the Theory of Fuzzy Subsets* (Vol. 1), Academic Press, New York (1975).

[14] Mathew, S., Anjali, N., Mordeson, J.N.: Transitive blocks and their applications in fuzzy interconnection networks, Fuzzy Sets and Systems **352**, 142-160 (2018).

[15] Mathew, S., Mordeson, J.N.: Directed Fuzzy Networks as a Model to Illicit Flows and Max Flow Min Cut Theorem, New Mathematics and Natural Computation **13**, 219-229 (2017).

[16] Mathew, S., Mordeson, J.N., Malik, D.S.: *Fuzzy Graph Theory*, Springer International Publishing AG (2018).

[17] Mathew, S., Sunitha, M.S.: Types of arcs in fuzzy graphs, Information Sciences **179**, 1760-1768 (2009).

[18] Mordeson, J.N., Mathew, S.: Non-Deterministic Flow in Fuzzy Networks and Its Application in Identification of Human Trafficking Chains, New Mathematics and Natural Computation **13**, 231-243 (2017).

[19] Mordeson, J.N., Mathew, S.: *Advanced Topics in Fuzzy Graph Theory*, Springer Nature Switzerland AG (2019).

[20] Mordeson, J.N., Nair, P.S.: Cycles and cocycles of fuzzy graphs, Information Sciences **90**, 39-49 (1996).

[21] Mordeson, J.N., Nair, P.S.: *Fuzzy Graphs and Fuzzy Hypergraphs*, Physica-Verlag, Heidelberg (2000).

[22] Mordeson, J.N., Peng, C.S.: Operations on fuzzy graphs, Information Sciences **79**, 159-170 (1994).

[23] Pal, M., Samanta, S., Ghorai, G.: *Modern Trends in Fuzzy Graph Theory*, Springer Nature Singapore (2020).

[24] Rosenfeld, A.: Fuzzy graphs, in: Zadeh, L.A., Fu, K.S., Shimura, M. (Eds.), *Fuzzy Sets and Their Applications*, Academic Press, New York, pp. 77-95 (1975).

[25] Sandeep Narayan, K.R., Sunitha, M.S.: Connectivity in a Fuzzy Graph and its Complement, General Mathematics Notes **9**, 38-43(2012).

[26] Schmidt, G., Ströhlein, T.: *Relations and Graphs*, Springer-Verlag Berlin Heidelberg (1993).

[27] Sebastian, A., Mordeson, J.N., Mathew, S.: Generalized Fuzzy Graph Connectivity Parameters with Application to Human Trafficking, Mathematics **2020**, 8, 424 (2020).

[28] Sunitha, M.S., Mathew, S.: Fuzzy Graph Theory - A Survey, Annals of Pure and Applied Mathematics **4**, 92-110 (2013).

[29] Sunitha, M.S., Vijayakumar, A.: Complement of a fuzzy graph, Indian Journal of Pure Applied Mathematics **33**, 1451-1464 (2002).

[30] Sunitha, M.S., Vijayakumar, A.: Blocks in fuzzy graphs, The Journal of Fuzzy Mathematics **13**, 13-23 (2005).

[31] Tom, M., Sunitha, M.S.: On Strongest Paths, Delta Arcs and Blocks in Fuzzy Graphs, World Applied Sciences Journal **22**, 10-17 (2013).

[32] Tom, M., Sunitha, M.S.: Strong sum distance in fuzzy graphs, SpringerPlus **4**:214 (2015).

[33] Voloshin, V.I.: *Introduction to Graph and Hypergraph Theory*, Nova Science Publishers, New York (2009).

[34] Wallis, W.D.: *A Beginner's Guide to Graph Theory*, 2nd ed., Birkhauser Boston (2007).

[35] Yeh, R.T., Bang, S.Y.: Fuzzy graphs, fuzzy relations, and their applications to cluster analysis, in: Zadeh, L.A., Fu, K.S., Shimura, M. (Eds.), *Fuzzy Sets and Their Applications*, Academic Press, New York, pp. 125-149 (1975).

PART 4

Some Abstract Algebra of Fuzzy Relations

Chapter 14

Fuzzy Relation Algebras

14.1 Some Historical Remarks

The begin of our considerations on the concept of a *relation algebra* is a good place to give a very sketchy outline of the historical milestones in the development of relations in becoming a subject of mathematical research. More information can be found in [10, Introduction] or [31, Chapter 1]. Compared to e.g. geometry (Euclid, about 300 BCE), or calculus (Gottfried Wilhelm Leibniz (1646-1716), Isaac Newton (1643-1727); for more details see e.g [28, Chapter 9]), it took a rather long time until relations have become a major subject of mathematical research beginning around the middle of the 19th century, most notably by the work of Augustus De Morgan(1806-1871) and Charles Sanders Peirce (1839-1914).

During the second half of the 19th century two crucial development steps in the history of modern logic took place: George Boole (1815-1864) initiated the algebraic treatment of logic, and Augustus De Morgan started with the formulation of the logic of relations. We have to be aware that around 1850 the main object of logical investigations have still been the syllogisms created and analyzed by Aristotle (384-322 BCE). The main questions being: How to formalize? How to enrich? What are the mathematical properties and structure obtained? How brilliant and admirable Aristotle's syllogistic system doubtlessly is, it has serious limitations: Grattan-Guinness [12, p. 32] calls it "a curious feature of the history of philosophy that ... nobody seems to have taken seriously the fact that relational propositions, such as 'John is taller than Jeremy', cannot fall within the compass of syllogistic logic".

Or, as Alfred Tarski (1901-1983), the father of the axiomatic treatment of the calculus of relations, puts it in ([29, p. 73]):

> The first beginnings of the contemporary theory of relations are to be found in the writings of A. De Morgan, who carried out extensive investigations in this domain in the fifties of the Nineteenth Century. De Morgan clearly realized the inadequacy of traditional logic for the expression and justification, not merely of the more intricate arguments of mathematics and the sciences,

> but even of simple arguments occurring in every-day life; witness his famous aphorism, that all the logic of Aristotle does not permit us, from the fact that a horse is an animal, to conclude that the head of a horse is the head of an animal. In his effort to break the bonds of traditional logic and to expand the limits of logical inquiry, he directed his attention to the general concept of relations and fully recognized its significance. Nevertheless, De Morgan cannot be regarded as the creator of the modern theory of relations, ...
>
> The title of creator of the theory of relations was reserved for C. S. Peirce. In several papers published between 1870 and 1882, he introduced and made precise all the fundamental concepts of the theory of relations and formulated and established its fundamental laws. Thus Peirce laid the foundation for the theory of relations as a deductive discipline; ...

The development of a mathematical theory of binary relations began in 1864 with [3] by AUGUSTUS DE MORGAN, followed approximately ten years later by CHARLES SANDERS PEIRCE ([23]). Their work has been highly extended by ERNST SCHRÖDER in 1895 with the 649 pages of [27]. The basic objects of these investigations are binary relations between elements of some given base set S, using as fundamental operations the Boolean operations of set theory, where the complement $\overline{\rho}$ of a relation ρ is taken with respect to the universal (greatest) relation $U_S := S \times S$. Also the operations of composition and conversion, specific to binary relations, have been considered.

14.2 Relation Algebras

The element-free style of operations with relations in relational calculus makes it a very useful tool for the study of mathematics, logics, and theoretical computer science, but also for various applications. A further step towards a more abstract level has been the introduction of relation algebras in [15] as abstract algebraic models for the study of binary relations, gathering into a single system set-theoretical and relational aspects, i.e. union, intersection, and complement, as well as composition, converse and diagonal.

TARSKI's axiomatic algebraic approach to the calculus of relations has two important features that distinguish it from the older viewpoint ([19, p. 443]):

- First, PEIRCE and SCHRÖDER worked with the set of *all* binary relations on a universe of discourse U, i.e. with $\mathcal{B}_U$, whereas TARSKI's axioms may hold in models consisting of only *some* of them: any set U of relations containing the identity relation on U and closed under union, complementation with respect to $U \times U$, relative multiplication, and conversion will comply with.
- The second distinguishing feature is that there are relation algebras of binary relations in which the universal relation need not relate every element of the universe of discourse to every other.

As it is customary in the context of relation algebras, and to emphasize their character as an abstract algebraic structure, we will use ';' to denote composition of $x, y \in A$ explicitly, and we will sometimes take the freedom to still call elements of the carrier A

of a relation algebra "*relations*". But we must keep in mind that they can, but need not, represent relations in the common mathematical sense as we will see in examples given below. When concepts from the theory of Boolean algebras are applied to a relation algebra $\mathfrak{R}$, it is understood that they refer to the Boolean part of $\mathfrak{R}$.

The following definition is due to Tarski and Givant [30, §8.2]. (Remark: Axioms (R1)-(R10) are equations, so relation algebras form what is called a **variety** in universal algebra; that is, the class of relation algebras is closed under the formation of subalgebras, homomorphic images, and direct products. Therefore, one can construct other examples of relation algebras by applying these operations to given ones.)

Definition 14.1. *A* **relation algebra** *is a structure* $\mathfrak{R} = (A, +, {}^{-}, ;, {}^{\smile}, \mathbb{1})$ *of type* $\langle 2,1,2,1,0\rangle$ *such that for any* $x, y, z \in A$*:*

(R1)	$x + y = y + x,$	*(+-commutativity)*
(R2)	$x + (y + z) = (x + y) + z,$	*(+-associativity)*
(R3)	$\overline{\overline{x} + y} + \overline{\overline{x} + \overline{y}} = x,$	*(Huntington's axiom)*
(R4)	$x; (y; z) = (x; y); z,$	*(;-associativity)*
(R5)	$(x + y); z = x; z + y; z,$	*(right ;-distributivity)*
(R6)	$x; \mathbb{1} = x,$	*(right identity law)*
(R7)	$x^{\smile\smile} = x,$	*($^{\smile}$-involution)*
(R8)	$(x + y)^{\smile} = x^{\smile} + y^{\smile},$	*($^{\smile}$-distributivity)*
(R9)	$(x; y)^{\smile} = y^{\smile}; x^{\smile},$	*($^{\smile}$-involution product)*
(R10)	$x^{\smile}; \overline{x; y} + \overline{y} = \overline{y}.$	*(TARSKI / DE MORGAN axiom)*

The operation ; is referred to as *relative multiplication*, the operation $^{\smile}$ as *conversion*, and the element $\mathbb{1}$ as the *identity element*. $\mathbb{0} := \overline{\mathbb{1}}$ is called the **diversity** element. Equations (R1) - (R3), due to Huntington [14], assert that $(A, +, {}^{-})$ is a Boolean algebra, so that $\mathfrak{R}$ is a Boolean algebra with operators. (R10) is the link between the Boolean and relative parts of $\mathfrak{R}$, and is equivalent to $x^{\smile}; \overline{x; y} \leq \overline{y}$, since $\leq$ is defined as $x \leq y :\iff x + y = y$. It can also be proven equivalent to $x^{\smile}; \overline{x; y} \cdot y = 0$ using (R1) - (R3).

We give two examples of relation algebras, where the base set of the second does not consist of binary relations. More examples with detailed descriptions can be found in [10, Chapter 3].

(1) The full algebra of relations $(\mathcal{B}_S, \cup, {}^{-}, \circ, {}^{\smile}, I_S)$ (recall Th. 2.2 and definitions 2.5 and 2.6) is the standard example for relation algebras, where

$$\rho \cap \sigma = \overline{\overline{\rho} \cup \overline{\sigma}},$$
$$\rho \subseteq \sigma \text{ iff } \rho \cup \sigma = \sigma,$$
$$U_S = I_S \cup \overline{I_S},$$
$$\emptyset = \overline{U_S}.$$

(2) Let $(G, \cdot, {}^{-1}, e)$ be a group, and let $X, Y \subseteq G$. Define

$$X; Y = \{x \cdot y : x \in X \wedge y \in Y\},$$
$$X^{\smile} = \{x^{-1} : x \in X\},$$
$$1' = \{e\},$$

and let $^{-G}$ denote the complementation relative to G, i.e. $\overline{X}^G = G - X$. Then $(\mathbf{2}^G, \cup, ^{-G}, ;, ^{\smile}, 1')$ is a relation algebra.

Looking through the relevant literature one soon encounters various logically equivalent axiomatizations of relation algebras that differ mainly in which axioms are used to describe the Boolean algebra reduct (above done via (R1) - (R3)), and in how they relate the Boolean with the relational part. Maddux [19] gives an overview on some of them, including remarks on their historical development which show that Tarski himself contributed to the topic over more than three decades. Def. 14.1 is the last one Tarski can be said to have been involved in (he died in 1983), and it is parsimonious in that it uses none of the definable constants and operations.

A definition that is often used and which exhibits all the relevant operations and constants explicitly and also clearly keeps apart the set-theoretical and relational component is the following (see e.g. [18]; its equivalence to Def. 14.1 is shown in e.g. [31, Theorem 10.35]):

Definition 14.2. *A structure* $\mathfrak{R} = (A, +, \cdot, ^{-}, 0, 1, ;, ^{\smile}, \mathbb{1})$ *of type* $\langle 2,2,1,0,0,2,1,0\rangle$ *is called a* **relation algebra** *iff the following holds:*

(1) $(A, +, \cdot, ^{-}, 0, 1)$ *is a Boolean algebra,*
(2) $(A, ;, ^{\smile}, \mathbb{1})$ *is an involution monoid, and*
(3) $\forall x, y, z \in A[(x;y)\cdot z = 0 \leftrightarrow (x^{\smile};z)\cdot y = 0 \leftrightarrow (z;y^{\smile})\cdot x = 0]$.

14.2.1 *Some Properties of Relation Algebras*

In order to convince ourselves that the concept of relation algebras is a suitable abstraction, we have to "check" it somehow. It has already been mentioned in example (1) above that full algebras of relations are indeed models of (R1) - (R10). So we are on a right way, but this is only a conditio sine qua non. Additionally we must check further results and important theorems that hold for relations. This means that we have to prove them anew from the axioms postulated for relation algebras, and *must not* reuse the argumentations from Chapter 2.

Without giving their proofs, below we repeat some results from [31, Section 10.3.1] that further confirm the concept of a relation algebra to be an appropriate abstraction.

Proposition 14.3. *In a relation algebra* $\mathfrak{R}$ *the following holds:*

(1) $x \le y \rightarrow x^{\smile} \le y^{\smile}$.
(2) $x \le y \leftrightarrow x^{\smile} \le y^{\smile}$.
(3) $x \le y^{\smile} \leftrightarrow x^{\smile} \le y$.
(4) $1^{\smile} = 1$.
(5) $x + y^{\smile} = 1 \leftrightarrow x^{\smile} + y = 1$.
(6) $\overline{x}^{\smile} = \overline{x^{\smile}}$.
(7) $0^{\smile} = 0$, $1^{\smile} = 1$.
(8) $(x\cdot y)^{\smile} = x^{\smile}\cdot y^{\smile}$.

(9) $x^{\smile} = y^{\smile} \leftrightarrow x = y$.
(10) $x^{\smile} \cdot y = 0 \leftrightarrow x \cdot y^{\smile} = 0$.
(11) $x = \mathbb{1}; x$.

Proposition 14.4. *Let $\mathfrak{R}$ be a relation algebra; then the following holds:*

(1) $x; (y + z) = (x; y) + (x; z)$ *(distributivity of ; over +)*
(2) $x \leq y \rightarrow z; x \leq z; y \wedge x; z \leq y; z$ *(monotonicity of ;)*
(3) $\overline{y; \overline{x}; x^{\smile}} \leq \overline{y}$
(4) $x; y \leq z \leftrightarrow x^{\smile}; \overline{z} \leq \overline{y} \leftrightarrow \overline{z}; y^{\smile} \leq \overline{x}$ *(Schröder rule)*
(5) $1; 1 = 1$
(6) $a \leq 1; a$ *and* $a \leq a; 1$

Proposition 14.5. *Let $\mathfrak{R}$ be a relation algebra; then the following holds:*

(1) $x; 0 = 0; x = 0$
(2) $x; (y \cdot z) \leq (x; y) \cdot (x; z)$ *subdistributivity of ; over ·*
$(x \cdot y); z \leq (x; z) \cdot (y; z)$
(3) $x; y \cdot z \leq (x \cdot z; y^{\smile}); y$ *modular law,* Th. 2.20(2)(a)
(4) $x; y \cdot z \leq x; (y \cdot x^{\smile}; z)$ *modular law,* Th. 2.20(2)(b)
(5) $x; y \cdot z \leq (x \cdot z; y^{\smile}); (y \cdot x^{\smile}; z)$ *Dedekind rule,* Th. 2.20(3)
(6) $r \leq s \wedge x \leq y \rightarrow r; x \leq s; y$.

In [31, Section 10.3.2] on "Special Kinds of Elements" properties of symmetric, transitive, and equivalence elements are shown which furthermore confirm the appropriateness of relation algebras as an abstract algebraic framework to model and investigate concrete binary relations.

14.3 Fuzzy Relation Algebras

Contrary to the classical crisp case, where there exist different but equivalent definitions for the concept of a relation algebra, in the fuzzy (i.e. more generally, the many-valued) context some different concepts, generalizing the classical two-valued one, have been proposed. We will present three of them in detail and will give hints to three other ones at the end of this chapter.

14.3.1 *Furusawa's Approach: [0,1]-Fuzzy Relation Algebras*

The historically first approach to fuzzy relation algebras has been initiated in the mid nineties of the last century resulting in e.g. [9,16]; see also [4, Section 14.7]. These will be the primary sources for this section, where we mainly follow the structuring and naming of the last one.

Looking back to Def. 14.2 it can clearly be seen that the definition of classical relation algebras consists of three parts: the Boolean component which models the set-theoretical structure of relations, the purely relational part describing their algebraic structure as an involution monoid, and as third ingredient a rule how these two parts interact. Fuzzy relation algebras are constructed in an analogous way, but the fuzzy component comes along from outside as a set of scalars $k \in [0,1] \subseteq \mathbb{R}$, acting upon relations.

Definition 14.6. ([4, p. 340]) *A structure* $(R,+,\cdot,0,1,;,\breve{\ },\mathbb{1})$ *of type* $\langle 2,2,0,0,2,1,0\rangle$ *is called a* **generalized relation algebra** *iff the following holds:*

(1) $(R,+,\cdot,0,1)$ *is a complete distributive lattice,*
(2) $(R,;,\breve{\ },\mathbb{1})$ *is an involution monoid such that* $x \le y$ *implies* $x\breve{\ } \le y\breve{\ }$,
(3) $x;\left(\sum_i y_i\right) = \sum_i (x;y_i)$ *and* $\left(\sum_i x_i\right);y = \sum_i (x_i;y)$, *and*
(4) $(x;y)\cdot z \le x;(y\cdot(x\breve{\ };z))$.

Condition (4) is referred to as the Dedekind formula in [9, Definition 2.3.2]; it is one of the modular laws, given in Th. 9.55(3)(b).

We let Z denote the structure $([0,1]\subseteq\mathbb{R},\max,\min,\frown_Z,0,1)$, i.e. the complete lattice originally used by Zadeh [35]. The relative pseudocomplement is defined by $x\frown_Z y = 1$ whenever $x \le y$ and $x \frown_Z y = y$ otherwise. In fact, Z can be interpreted as residuated lattice if we take $a\otimes b = a\wedge b$ and $a\frown_Z b = \bigvee\{c : a\wedge c \le b\}$; see Section 4.4. It can easily be concluded from what has been shown in Section 9.3.1 that for $\mathcal{F}^Z_X$, the set of all L-fuzzy relations on X, the following holds:

Proposition 14.7. ([4, Lemma 14.7.3])

(1) $(\mathcal{F}^Z_X,\cup,\cap,1-.,\emptyset,U_X)$ *is a complete and bounded distributive lattice.*
(2) $(\mathcal{F}^Z_X,\circ,\breve{\ },I_X)$ *is an involution monoid.*
(3) $R\le S$ *implies* $R\breve{\ }\le S\breve{\ }$ *for all* $R,S\in\mathcal{F}^Z_X$.
(4) $(\mathcal{F}^Z_X,\cup,\cap,\emptyset,U_X,\circ,\breve{\ },I_X)$ *is a generalized relation algebra.*

Proof. (1): We clearly have $\emptyset\subseteq R\subseteq U_X$, for all $R\in\mathcal{F}^Z_X$. Moreover, defining $(\cup_i R_i)(x,y) = \bigvee_i R_i(x,y)$ and $(\cap_i R_i)(x,y) = \bigwedge_i R_i(x,y)$ for all $x,y\in X$, it is clear that $\cup_i R_i$ and $\cap_i R_i$ are the least upper and the greatest lower bound, respectively, of the family $\{R_i\}$, with respect to $\subseteq$.

(4): What is to be shown in addition to (1) - (3) is the validity of the Dedekind formula for $R,S,T\in\mathcal{F}^Z_X$:

$$\begin{aligned}
(RS)\cap T(x,y) &= \textstyle\bigvee_{z\in X}\big(R(x,z)\wedge S(z,y)\big)\wedge T(x,y)\\
&= \textstyle\bigvee_{z\in X}\big(R(x,z)\wedge S(z,y)\wedge T(x,y)\big)\\
&= \textstyle\bigvee_{z\in X}\big(R(x,z)\wedge S(z,y)\wedge R\breve{\ }(z,x)\wedge T(x,y)\big)\\
&\le \textstyle\bigvee_{z\in X}\big(R(x,z)\wedge S(z,y)\wedge R\breve{\ }T(z,y)\big)\\
&= \textstyle\bigvee_{z\in X}\big(R(x,z)\wedge (S\cap R\breve{\ }T)(z,y)\big)\\
&= [R(S\cap R\breve{\ }T)](x,y)
\end{aligned}$$

□

Definition 14.8. ([4, p. 338], [9, Definition 3.1.1]) *A structure $(F,\sqcup,\sqcap,\odot,\bot,\top)$ of type $\langle 2,2,2,0,0\rangle$ is called a* **fuzzy algebra** *iff the following holds:*

(1) *$(F,\sqcup,\sqcap,\bot,\top)$ is a complete and bounded distributive lattice with $\bot$ as least and $\top$ as greatest element;*

(2) *the mapping $\odot : [0,1]\times W \longrightarrow W$ is called* **semiscalar product**, *satisfying the following conditions for all $k,k'\in[0,1]$ and all $x,y\in F$:*

 (2a) *$0\odot x=\bot$ and $1\odot x = x$;*

 (2b) *$k\odot(k'\odot x)=(kk')\odot x$, using ordinary multiplication for k,k';*

 (2c) *$k\odot(\bigsqcup_i x_i)=\bigsqcup_i(k\odot x_i)$ and $k\odot(\bigsqcap_i x_i)=\bigsqcap_i(k\odot x_i)$;*

 (2d) *$(\bigwedge_i k_i)\odot x=\bigsqcap_i(k_i\odot x)$;*

 (2e) *$k\odot x\sqsubseteq k\odot y, 0<k$ implies $x\sqsubseteq y$, where $\sqsubseteq$ is the usual lattice ordering on F;*

(3) *if $x\sqcap(k\odot\top)=k\odot x$ for all $k\in[0,1]$, then there exists $y\in F$ such that $x\sqcup y=\top$ and $x\sqcap y=\bot$.*

Proposition 14.9. ([4, Lemma 14.7.1] *Let $(F,\sqcup,\sqcap,\odot,\bot,\top)$ be a fuzzy algebra; then for all $x,y\in F$ and all $k,k'\in[0,1]$ we have*

(1) *If $x\sqsubseteq y$, then $k\odot x\sqsubseteq k\odot y$;*

(2) *If $k\le k'$, then $k\odot x\sqsubseteq k'\odot x$;*

(3) *$k\odot x\sqsubseteq x$ and $k\odot\bot=\bot$;*

(4) *$k\odot x\sqsubset x\sqcap(k\odot\top)$;*

(5) *$(k\odot x)\sqcup(k'\odot x)=\max\{k,k'\}\odot x$.*

Proof. (1): $k\odot x=k\odot(x\sqcap y)=(k\odot x)\sqcap(k\odot y)$.
(2): $k\odot x=\min\{k,k'\}\odot x=(k\odot x)\sqcap(k'\odot x)\sqsubseteq k'\odot x$, using Def. 14.8(2d).
(3): From $k\le 1$ and (2) we get $k\odot x\sqsubseteq 1\odot x=x$. From this $\bot\sqsubseteq k\odot\bot\sqsubseteq\bot$ follows.
(4): From $x\sqsubseteq\top$ one gets $k\odot x\sqsubseteq k\odot\top$ by (1), and from (3) we have $k\odot x\sqsubseteq x$. So $k\odot x\sqsubseteq x\sqcap(k\odot\top)$.
(5): Assuming $k\le k'$, from (2) we get $(k\odot x)\sqcup(k'\odot x)=(k'\odot x)=\max\{k,k'\}\odot x$. Similar for $k'\le k$. □

An element of a fuzzy algebra is called **crisp** if $x\sqcap(k\odot\top)=k\odot x$ for alle $k\in[0,1]$.

Lemma 14.10. ([4, Lemma 14.7.2], [9, Proposition 3.1.2]) *Let $(F,\sqcup,\sqcap,\odot,\bot,\top)$ be a fuzzy algebra, and $x,y\in F$.*

(1) *If x and y are crisp, then also $x\sqcup y$ and $x\sqcap y$.*

(2) *The set of crisp elements of every fuzzy algebra is a Boolean algebra.*

Proof. (1): Using distributivity we get $(x\sqcup y)\sqcap k\odot\top=(x\sqcap k\odot\top)\sqcup(y\sqcap k\odot\top)=(k\odot x)\sqcup(k\odot y)=k\odot(x\sqcup y)$, applying Def. 14.8(2c) for the last step. Similarly, $(x\sqcap y)\sqcap k\odot\top=(x\sqcap k\odot\top)\sqcap(y\sqcap k\odot\top)=(k\odot x)\sqcap(k\odot y)=k\odot(x\sqcap y)$.

(2): Obviously, both $\bot$ and $\top$ are crisp, and the existence of complements for crisp elements is guaranteed by Def. 14.8(3). Together with (1) this proves the assertion. □

Definition 14.11. ([4, p. 341], [9, Definition 3.2.1]) *A structure* $(F, \sqcup, \sqcap, \odot, \bot, \top, ;, \breve{}, \mathbb{1})$ *of type* $\langle 2,2,2,0,0\rangle$ *is called a* **fuzzy relation algebra** *iff the following holds:*

(1) $(F, \sqcup, \sqcap, \odot, \bot, \top)$ *is a fuzzy algebra* (Def. 14.8);
(2) $(F, \sqcup, \sqcap, \bot, \top, ;, \breve{}, \mathbb{1})$ *is a generalized relation algebra* (Def. 14.6);
(3) $k \odot (x; y) = (k \odot x); (k \odot y)$;
(4) $(k \odot x); y = (k \odot x); (y \sqcap (k \odot \top))$,
(5) $(k \odot x)\breve{} = k \odot x\breve{}$.

Proposition 14.12. ([4, Lemma 14.7.4])

(1) $\mathfrak{F} := (\mathcal{F}^Z_X, \cup, \cap, \odot, \emptyset, U_X, \circ, \breve{}, I_X)$ *becomes a fuzzy relation algebra by defining the semiscalar product* $\odot : [0,1] \times \mathcal{F}^Z_X \longrightarrow \mathcal{F}^Z_X$ *as* $(k \odot R)(x, y) = kR(x, y)$.
(2) *A relation* R *in* $\mathfrak{F}$ *is crisp iff* $R(x, y) = 1$ *or* $R(x, y) = 0$ *for all* $x, y \in X$.

Proof. (1): Obviously, $(\mathcal{F}^Z_X, \cup, \cap, \cdot, \emptyset, U_X)$ is a fuzzy algebra. And by Prop. 14.7(4) $(\mathcal{F}^Z_X, \cup, \cap, \emptyset, U_X, \circ, \breve{}, I_X)$ is a generalized relation algebra. We still have to show that the requirements (3) to (5) of Def. 14.11 are valid, with (5) being trivial.

Using elementary arithmetic, viz. $k\min\{a, b\} = \min\{ka, kb\}$ and $k\max\{a, b\} = \max\{ka, kb\}$, we get $k \odot (R; S)(x, y) = k\big(\bigvee_z R(x, z) \wedge S(z, y)\big) = \bigvee_z k(R(x, z) \wedge S(z, y)) = \bigvee_z (kR(x, z) \wedge kS(z, y)) = [(k \odot R); (k \odot S)](x, y)$.

Moreover we have $[(k \odot R) \circ (S \cap k \odot U_X)](x, y) = \bigvee_z \big(kR(x, z) \wedge (S \cap k \odot U_X)(z, y)\big) = \bigvee_z \big(kR(x, z) \wedge S(z, y) \wedge k\big) = \bigvee_z \big((kR(x, z) \wedge k) \wedge S(z, y)\big) = \bigvee_z \big(kR(x, z) \wedge S(z, y)\big) = [(k \odot R) \circ S](x, y)$.

(2): We have $[k \odot U_X](x, y) = k$, for all $x, y \in X$. Furthermore, $R \cap (k \odot U_X) = k \odot R$ iff $\forall k \in [0,1][\min\{R(x, y), k\} = kR(x, y)$ iff $R(x, y) \in \{0,1\}$. So, for relations in $\mathfrak{F}$, the abstract characterization of crispness corresponds adequately to its intuitive meaning. □

Lemma 14.13. ([4, Lemmas 14.7.5, 14.7.6]) *Let* x, y *be relations in a fuzzy relation algebra* $(F, \sqcup, \sqcap, \odot, \bot, \top, ;, \breve{}, \mathbb{1})$, *and* $k \in [0,1]$. *Then the following holds:*

(1) $\bot\breve{} = \bot$, $\top\breve{} = \top$, *and* $\mathbb{1}\breve{} = \mathbb{1}$;
(2) $x\breve{} \sqcup y\breve{} = (x \sqcup y)\breve{}$, $x\breve{} \sqcap y\breve{} = (x \sqcap y)\breve{}$;
(3) *if* $x \sqsubseteq y$ *then* $z; x \sqsubseteq z; y$ *and* $x; z \sqsubseteq x; y$;
(4) $(k \odot x); y \sqsubseteq k \odot (x; \top)$;
(5) $x; (k \odot y) \sqsubseteq k \odot (\top; y)$;
(6) $x; (k \odot y) \sqsupseteq k \odot (x; y)$;
(7) $(k \odot x); y \sqsupseteq k \odot (x; y)$;
(8) *if* y *is crisp, then* $(k \odot x); y = k \odot (x; y)$;
(9) *if* x *is crisp, then* $x; (k \odot y) = k \odot (x; y)$.

Proof. (1): We have $\bot \sqsubseteq \bot\breve{}$; so $\bot\breve{} \sqsubseteq \bot\breve{}\breve{} = \bot$ by Def. 14.6(2). Similarly, $\mathbb{1}\breve{} = \mathbb{1}\breve{}; \mathbb{1} = \mathbb{1}\breve{}; \mathbb{1}\breve{}\breve{} = (\mathbb{1}\breve{}; \mathbb{1})\breve{} = (\mathbb{1}\breve{})\breve{} = \mathbb{1}$.

(2): Again using Def. 14.6(2) we get $x\breve{} \sqcup y\breve{} \sqsubseteq (x \sqcup y)\breve{}$. Applying this result to $x\breve{}, y\breve{}$ gives $x \sqcup y = x\breve{}\breve{} \sqcup y\breve{}\breve{} \sqsubseteq (x\breve{} \sqcup y\breve{})\breve{}$, whence $(x \sqcup y)\breve{} \sqsubseteq (x\breve{} \sqcup y\breve{})\breve{}\breve{} = x\breve{} \sqcup y\breve{}$. Similarly, $(x \sqcap y)\breve{} \sqsubseteq$

$x^{\smile} \sqcap y^{\smile}$. So $(x^{\smile} \sqcap y^{\smile})^{\smile} \sqsubseteq x^{\smile\smile} \sqcap y^{\smile\smile} = x \sqcap y$, whence $(x^{\smile} \sqcap y^{\smile})^{\smile\smile} = x^{\smile} \sqcap y^{\smile} \sqsubseteq (x \sqcap y)^{\smile}$.

(3): $z; x \sqsubseteq z; x \sqcup z; y = z; (x \sqcup y) = z; y$, using Def. 14.6(3) and the assumption $x \sqsubseteq y$.

(4):
$$\begin{aligned}(k \odot x); y &= (k \odot x); (y \sqcap (k \odot \top)) && \text{, by Def. 14.11(4)}\\ &\sqsubseteq (k \odot x); (k \odot \top) && \text{, by (3) since } y \sqcap (k \odot \top) \sqsubseteq k \odot \top\\ &= k \odot (x; \top) && \text{, by Def. 14.11(3).}\end{aligned}$$

(5): By the involution properties and Def. 14.11(5) we get
$$x; (k \odot y) = \big(x; (k \odot y)\big)^{\smile\smile} = \big((k \odot y)^{\smile}; x^{\smile}\big)^{\smile} = \big((k \odot y^{\smile}); x^{\smile}\big)^{\smile}.$$
But by (4) we have $(k \odot y^{\smile}); x^{\smile} \sqsubseteq k \odot (y^{\smile}; \top)$, so by Def. 14.11(5) and (1) we get
$$\big((k \odot y^{\smile}); x^{\smile}\big)^{\smile} \sqsubseteq \big(k \odot (y^{\smile}; \top)\big)^{\smile} = k \odot (y^{\smile}; \top)^{\smile} = k \odot (\top; y).$$

(6): $k \odot x \sqsubseteq x$ by Prop. 14.9(3). So $k \odot (x; y) = (k \odot x)(k \odot y) \sqsubseteq x; (k \odot y)$ using (3).

(8):
$$\begin{aligned}(k \odot x); y &= (k \odot x); (y \sqcap (k \odot \top)) && \text{, by Def. 14.11(4)}\\ &= (k \odot x); (k \odot y) && \text{, since } y \text{ is crisp}\\ &= k \odot (x; y) && \text{, by Def. 14.11(3).}\end{aligned}$$

(9):
$$\begin{aligned}x; (k \odot y) &= \big((k \odot y^{\smile}); x^{\smile}\big)^{\smile} && \text{, see proof of (5)}\\ &= \big((k \odot y^{\smile}); (x^{\smile} \sqcap k \odot \top)\big)^{\smile} && \text{, by Def. 14.11(4)}\\ &= (x^{\smile} \sqcap k \odot \top)^{\smile}; (k \odot y^{\smile})^{\smile} && \text{, by involution}\\ &= (x \sqcap (k \odot \top)^{\smile}); (k \odot y^{\smile})^{\smile} && \text{, by (2)}\\ &= (x \sqcap k \odot \top); (k \odot y) && \text{, by (1), Def. 14.11(5)}\\ &= (k \odot x); (k \odot y) && \text{, since } x \text{ is crisp}\\ &= k \odot (x; y) && \text{, by Def. 14.11(3).}\end{aligned}$$

□

Proposition 14.14. ([4, Lemma 14.7.7], [9, Proposition 3.2.2]) *Let x, y be relations in a fuzzy relation algebra $(F, \sqcup, \sqcap, \odot, \bot, \top, ;, ^{\smile}, \mathbb{1})$, and $k \in [0,1]$. Then the following holds:*

(1) *If x and y are crisp, then also $x^{\smile}$ and $x; y$.*
(2) *The identity relation $\mathbb{1}$ is crisp.*

Proof. (1): Assume that x and y are crisp. Concerning $x^{\smile}$ we have
$$\begin{aligned}x^{\smile} \sqcap (k \odot \top) = x^{\smile} \sqcap (k \odot \top^{\smile}) = x^{\smile} \sqcap (k \odot \top)^{\smile} && \text{, by Lemma 14.13(1), Def. 14.11(5)}\\ = (x \sqcap (k \odot \top))^{\smile} && \text{, by Lemma 14.13(2)}\\ = (k \odot x)^{\smile} && \text{, since } x \text{ is crisp}\\ = k \odot x^{\smile} && \text{, by Def. 14.11(5).}\end{aligned}$$

Furthermore, for the product $x; y$ we have
$$\begin{aligned}x; y \sqcap (k \odot \top) &\sqsubseteq x; [y \sqcap (x^{\smile}; (k \odot \top))] && \text{, by the modular law Def. 14.6(4)}\\ &\sqsubseteq x; [y \sqcap (k \odot (\top; \top))] && \text{, by Lemma 14.13(5)}\\ &= x; [y \sqcap (k \odot \top)] && \text{, since: } \top^2 \sqsubseteq \top \text{ and } \top = \mathbb{1}; \top \sqsubseteq \top^2\\ &= x; [k \odot y] && \text{, since } y \text{ is crisp}\\ &= k \odot (x; y) && \text{, by Lemma 14.13(9).}\end{aligned}$$

By Prop. 14.9(4) the reverse inequality $k \odot (x; y) \sqsubseteq x; y \sqcap (k \odot \top)$ holds.

$$
\begin{aligned}
(2):\ k \odot \mathbb{1} &\sqsubseteq \mathbb{1} \sqcap k \odot \top && \text{, by Prop. 14.9(4)} \\
&= \mathbb{1} \sqcap k \odot (\mathbb{1};\top) \\
&= \mathbb{1} \sqcap (k \odot \mathbb{1});\top && \text{, by Lemma 14.13(8), since } \top \text{ is crisp} \\
&\sqsubseteq (k \odot \mathbb{1});[(k \odot \mathbb{1})^{\smile};\mathbb{1} \sqcap \top] && \text{, by Def. 14.6(4): modular law} \\
&= (k \odot \mathbb{1});[(k \odot \mathbb{1}^{\smile});\mathbb{1}] && \text{, by Def. 14.11(5)} \\
&= (k \odot \mathbb{1});[k \odot \mathbb{1}] && \text{, by Lemma 14.13(1) } \mathbb{1}^{\smile} = \mathbb{1} \\
&\sqsubseteq (k \odot \mathbb{1});\mathbb{1} && \text{, by Prop. 14.9(3)} \\
&= k \odot \mathbb{1}.
\end{aligned}
$$

□

Proposition 14.15. ([4, Lemma 14.7.7]) *The set of crisp relations of a fuzzy relation algebra is a Boolean algebra.*

Proof. Let $\mathfrak{F} = (F, +, \cdot, \odot, \bot, \top, ;, ^{\smile}, \mathbb{1})$ be a fuzzy relation algebra. By Lemma 14.10 and Prop. 14.14, the set of crisp elements in $\mathfrak{F}$ is closed under $+, \cdot, ;$, and conversion, and includes all of $\bot, \top$and $\mathbb{1}$. □

14.3.2 *MV-Algebras and MV-Relation Algebras*

14.3.2.1 *MV-Algebras*

The notion of an MV-algebra has already, but only parenthetically, been mentioned at the end of Section 4.4 as a special case of BL-algebras, which in turn are residuated lattices. For the following considerations we need a little bit more information about these algebraic structures. There are two equivalent definitions of MV-algebras which can be found in literature: The original version from [1] which is deliberately redundant and, therefore, also very intuitive, and a more parsimonious one, given in

Definition 14.16. ([2, Def. 1.1.1], [11, Def. 9.2.1]) *An algebraic structure* $(A, \oplus, \neg, 0)$ *of type* $(2, 1, 0)$ *is called an* **MV-algebra** *iff, for all* $x, y \in A$*, the following holds:*

(MV1) $(A, \oplus, 0)$ *is an Abelian monoid,*
(MV2) $\neg\neg x = x$,
(MV3) $x \oplus \neg 0 = \neg 0$.
(MV4) $\neg(\neg x \oplus y) \oplus y = \neg(\neg y \oplus x) \oplus x$.

Suppose that $(A, \oplus, \neg, 0)$ is an MV-algebra. Then we define for any $x, y \in A$:

$$x \odot y := \neg(\neg x \oplus \neg y), \tag{14.1}$$

$$x \vee y := (x \odot \neg y) \oplus y, \tag{14.2}$$

$$x \wedge y := (x \oplus \neg y) \odot y, \tag{14.3}$$

$$1 := \neg 0, \tag{14.4}$$

$$x \leq y \leftrightarrow \exists z \in A[x \oplus z = y]. \tag{14.5}$$

Definition 14.17. ([1]) *An algebraic structure* $(A, \oplus, \odot, \neg, 0, 1)$ *of type* $(2, 2, 1, 0, 0)$ *is called an* **MV-algebra** *iff for all* $x, y, z \in A$*, the following holds:*

(C1) $x \oplus y = y \oplus x$,	(C1') $x \odot y = y \odot x$,
(C2) $x \oplus (y \oplus z) = (x \oplus y) \oplus z$,	(C2') $x \odot (y \odot z) = (x \odot y) \odot z$,
(C3) $x \oplus \neg x = 1$,	(C3') $x \odot \neg x = 0$,
(C4) $x \oplus 1 = 1$	(C4') $x \odot 0 = 0$,
(C5) $x \oplus 0 = x$.	(C5') $x \odot 1 = x$,
(C6) $\neg(x \oplus y) = \neg x \odot \neg y$.	(C6') $\neg(x \odot y) = \neg x \oplus \neg y$,
(C7) $\neg\neg x = x$,	
(C8) $\neg 0 = 1$,	
(C9) $x \vee y = y \vee x$,	(C9') $x \wedge y = y \wedge x$,
(C10) $x \vee (y \vee z) = (x \vee y) \vee z$,	(C10') $x \wedge (y \wedge z) = (x \wedge y) \wedge z$,
(C11) $x \oplus (y \wedge z) = (x \oplus y) \wedge (x \oplus z)$.	(C11') $x \odot (y \vee z) = (x \odot y) \vee (x \odot z)$.

The standard example for MV-algebras is given by $\mathfrak{L} = ([0,1], \oplus, \odot, \neg, 0, 1)$, where $\oplus := \min\{a+b, 1\}$ is the Łukasiewicz t-conorm, $\odot := \max\{a+b-1, 0\}$ is the Łukasiewicz t-norm, and $\neg a := 1 - a$ is the standard complement on $[0,1]$. Recall Def. 9.6.

It is important to insert a clarifying remark on the proof structures of Prop. 14.18 and Prop. 14.19: There is no circularity in the proof of the last four equalities, since the chain of implications used is as follows:

Def. 14.16 → Def. 14.17(C1) - (C9) → Def. 14.17(C1') - (C6'), (C9')
→ Prop. 14.19 → Prop. 14.20(5) - (7)
→ Def. 14.17(C10), (C11), (C10'), (C11').

Proposition 14.18. Def. 14.16 *and* Def. 14.17 *are equivalent.*

Proof. [⇐] Clearly, (C1), (C2) and (C5) make $(A, \oplus, 0)$ an Abelian monoid, so (MV1) holds. (C7) is (MV2), and (C4) together with (C8) gives (MV3). To prove (MV4) note that using (C6) and (C7) we get $\neg(\neg x \oplus y) \oplus y = (x \odot \neg y) \oplus y = x \vee y = y \vee x = \neg(\neg y \oplus x) \oplus x$. So, (C1) - (C8) allow us to derive the conditions of Def. 14.16.

[⇒] Clearly, (MV1) implies (C1), (C2) and (C5). (MV2) is (C7), (MV3) together with (14.2) yields (C4). Using (MV4) with $y = \neg 0$ we get $\neg(\neg x \oplus \neg 0) \oplus \neg 0 = \neg(0 \oplus x) \oplus x$; applying (MV2) yields the equation $\neg(\neg 0) \oplus \neg 0 = \neg 0 = 1 = \neg x \oplus x$, giving (C3). For (C6) we get $\neg x \odot \neg y = \neg(\neg\neg x \oplus \neg\neg y) = \neg(x \oplus y)$ using (14.1) and (C7). (C8) is (14.4). Concerning (C9) we have with the above definitions $x \vee y = (x \odot \neg y) \oplus y = \neg(\neg x \oplus y) \oplus y$. We can now apply (MV4) to the last expression and get $\neg(\neg y \oplus x) \oplus x = y \vee x$.

So far we have shown that Def. 14.16 implies each of (C1) - (C9) which can now also be used to prove the remaining conditions. We get

(C1'): $x \odot y = \neg(\neg x \oplus \neg y) = \neg(\neg y \oplus \neg x) = y \odot x$.

(C2'): $(x \odot y) \odot z = [\neg(\neg x \oplus \neg y)] \odot z = \neg[(\neg x \oplus \neg y) \oplus \neg z] = \neg[\neg x \oplus (\neg y \oplus \neg z)] = x \odot [\neg(\neg y \oplus \neg z)] = x \odot (y \odot z)$.

(C3'): By (C3) $x \oplus \neg x = 1$, so $\neg(\neg x \oplus x) = \neg 1$, and therefore $x \odot \neg x = \neg\neg 0 = 0$.

(C4'): By (C4) $\neg x \oplus 1 = 1$, i.e. $\neg x \oplus \neg 0 = \neg 0$; so using (C6) $x \odot 0 = \neg(\neg x \oplus \neg 0) = 0$.

(C5'): $a \odot 1 = \neg(\neg a \oplus \neg 1) = \neg\neg a = a$.

(C6'): $\neg(x \odot y) = \neg(\neg(\neg x \oplus \neg y))$ from (14.1).
(C9'): $x \wedge y = (x \oplus \neg y) \odot y = \neg[\neg(x \oplus \neg y) \oplus \neg y] = \neg[\neg(y \oplus \neg x) \oplus \neg x] = (y \oplus \neg x) \odot x = y \wedge x$, using (MV4) and (14.3).
(C10),(C10'): Using Prop. 14.20(5) below we can argue in a way analogous to that used in the proof of Prop. 1.42(SL1).
(C11), (C11'): See Prop. 14.20(9),(10) below. □

Proposition 14.19. ([11, Corollary 9.2.2]) *In any MV-algebra* $(A, \oplus, \odot, \neg, 0, 1)$ *for all* $x, y, z \in A$ *we have* $x \leq y \leftrightarrow \neg x \oplus y = 1 \leftrightarrow x \odot \neg y = 0 \leftrightarrow x \oplus (y \odot \neg x) = y \leftrightarrow \neg y \leq \neg x$.

Proof. Assume $x \leq y$; then there exists a $z \in A$ such that $\neg x \oplus y = \neg x \oplus (x \oplus z) = 1 \oplus z = 1$, applying Def. 14.17(C1),C2) and (C4). / Moreover, $\neg x \oplus y = 1$ gives $x \odot \neg y = \neg(\neg x \oplus y) = \neg 1 = 0$. / Then, using Def. 14.16(MV4) and Def. 14.17(C5), $x \odot \neg y = 0$ implies $(y \odot \neg x) \oplus x = \neg(\neg y \oplus x) \oplus x = \neg(\neg x \oplus y) \oplus y = (x \odot \neg y) \oplus y = y$. / But from $\neg(\neg y \oplus x) \oplus x = y$ we get $x \leq y$ due to (14.5), whence we have shown the equivalence of the first four conditions. / To complete the proof, using these results, it follows that $x \leq y \leftrightarrow \neg x \oplus y = 1 \leftrightarrow 0 = \neg(\neg x \oplus y) = \neg y \odot \neg\neg x \leftrightarrow \neg y \leq \neg x$. □

Proposition 14.20. ([2, Proposition 1.1.6], [11, Corollary 9.2.3]) *In any MV-algebra* $(A, \oplus, \odot, \neg, 0, 1)$ *the following holds for all* $x, y, z \in A$:

(1) $x \leq y \rightarrow x \oplus z \leq y \oplus z$.
(2) $x \leq y \rightarrow x \odot z \leq y \odot z$.
(3) $z \leq x \oplus y \leftrightarrow z \odot \neg x \leq y$.
(4) $x \odot y \leq z \leftrightarrow x \leq \neg y \oplus z$.
(5) $x \wedge y \leq x \leq x \vee y$.
(6) $0 \leq x \leq 1$.
(7) $x \odot y \leq x \leq x \oplus y$.
(8) $x \odot y \leq x \wedge y \leq x \leq x \vee y \leq x \oplus y$.
(9) $x \odot (y \vee z) = (x \odot y) \vee (x \odot z)$.
(10) $x \oplus (y \wedge z) = (x \oplus y) \wedge (x \oplus z)$.

Proof. (2): By (1) $x \leq y \rightarrow \neg y \leq \neg x \rightarrow \neg y \oplus \neg z \leq \neg x \oplus \neg z \rightarrow \neg(\neg x \oplus \neg z) \leq \neg(\neg y \oplus \neg z)$, i.e. $x \odot z \leq y \odot z$ by applying Def. 14.17(C6), (C6') and (C7) to the last inequality.

(3): Prop. 14.19 implies $z \leq x \oplus y \leftrightarrow 0 = z \odot \neg(x \oplus y) = z \odot \neg x \odot \neg y \leftrightarrow z \odot \neg x \leq y$.

(4): Prop. 14.19 implies $x \odot y \leq z \leftrightarrow 0 = (x \odot y) \odot \neg z = x \odot (y \odot \neg z) \leftrightarrow x \leq \neg(y \odot \neg z) = \neg y \oplus z$.

(5): By Prop. 14.19 we have $x \leq x \vee y \leftrightarrow x \odot \neg(x \vee y) = 0$. But $\neg(x \vee y) = \neg((x \odot \neg y) \oplus y) = (\neg(x \odot \neg y) \odot \neg y)$, whence we get $x \odot \neg(x \vee y) = x \odot (\neg(x \odot \neg y) \odot \neg y) = (x \odot \neg y) \odot \neg(x \odot \neg y) = 0$, using Def. 14.17(C3').

(6): Using Def. 14.17(C4),(C4') and Prop. 14.19 we have $x \oplus 1 = 1$ and $x \odot 0 = 0$, so $\neg 0 \oplus x = 1$, i.e. $0 \leq x$, and $\neg 1 \odot x = 0$, i.e. $x \leq 1$.

(7): $x \odot y \leq x \odot 1 = x$; $x = x \oplus 0 \leq x \oplus y$, for all $y \in A$ by monotonicity.

(8): Due to (14.2) and (14.3) we have $x \vee y = (x \odot \neg y) \oplus y$ and $x \wedge y = (x \oplus \neg y) \odot y$. So, using

(7) and monotonicity we get $x \odot y \le x \odot (\neg x \oplus y) = x \wedge y$ and $x \vee y = (x \odot \neg y) \oplus y \le x \oplus y$, whence the assertion follows by (5).

(9): By (2) we have $x \odot y \le x \odot (y \vee z)$ and $x \odot z \le x \odot (y \vee z)$. So $(x \odot y) \vee (x \odot z) \le x \odot (y \vee z)$. Conversely, suppose $x \odot y \le t$ and $x \odot z \le t$; then by (4) $y \le \neg x \oplus t$ and $z \le \neg x \oplus t$, whence $y \vee z \le \neg x \oplus t$. Again using (4) yields $y \vee z \odot x \le t$, i.e. $x \odot (y \vee z) \le t$ by Def. 14.17(C1'). In particular this holds for $t = x \odot (y \vee z)$.

(10): By (1) we have $x \oplus (y \wedge z) \le x \oplus y$ and $x \oplus (y \wedge z) \le x \oplus z$. So $x \oplus (y \wedge z) \le (x \oplus y) \wedge (x \oplus z)$. Conversely, suppose $t \le x \oplus y$ and $t \le x \oplus z$; then by (3) $t \odot \neg x \le y$ and $t \odot \neg x \le z$, whence $t \odot \neg x \le y \wedge z$. Again using (3) yields $t \le x \oplus (y \wedge z)$. In particular this holds for $(x \oplus y) \wedge (x \oplus z)$. □

It is worth noticing that $\oplus$, $\odot$, and $\wedge$ distribute over any existing join and, analogously, $\oplus$, $\odot$ and $\vee$ distribute over any existing meet, in any MV-algebra A (see [5, p. 137]). In other words, for any family $\{b_i\}_{i \in I}$ of elements of A for which there exists $\bigvee_{i \in I} b_i$, for any family $\{c_i\}_{i \in I}$ for which there exists $\bigwedge_{i \in I} c_i$, and for any $a \in A$,

$$\begin{aligned} a \bullet \bigvee_{i \in I} b_i &= \bigvee_{i \in I} a \bullet b_i, \text{ for } \bullet \in \{\oplus, \odot, \wedge\}, \\ a \bullet \bigwedge_{i \in I} c_i &= \bigwedge_{i \in I} a \bullet c_i, \text{ for } \bullet \in \{\oplus, \odot, \vee\}. \end{aligned} \tag{14.6}$$

Proposition 14.21. ([11, Proposition 9.2.1]) *For an MV-algebra $(A, \oplus, \odot, \neg, 0, 1)$ the following holds:*

(1) *The relation $\le$ defined by (14.5) is a lattice ordering and has $\wedge$ and $\vee$ as the corresponding lattice operations.*
(2) $x \le y \;\leftrightarrow\; x \wedge y = x \;\leftrightarrow\; x \vee y = y$ *for all $x, y \in A$.*

Proof. (1): It is easy to show that $\le$ is reflexive and transitive. $x \le y$ together with $y \le x$ implies $x \odot \neg y = 0$ and $x = y \oplus (x \odot \neg y)$ by Prop. 14.19, whence $x = y \oplus 0 = y$. So, $\le$ is also antisymmetric and thus a partial order.

By Prop. 14.20(5) $x \le x \vee y$ and $y \le x \vee y$, so $\sup\{x, y\} \le x \vee y$. Conversely we must show $x \vee y \le z$ for each upper bound z of x, y. $x \le z$ and $y \le z$ imply $\neg x \oplus z = 1$ and $y \oplus (z \odot \neg y) = z$ by Prop. 14.19. By (14.2) we have $x \vee y = (x \odot \neg y) \oplus y$, so

$$\begin{aligned} \neg(x \vee y) \oplus z &= \neg[(x \odot \neg y) \oplus y] \oplus [y \oplus (z \odot \neg y)] \\ &= [\neg(x \odot \neg y) \odot \neg y] \oplus [y \oplus (z \odot \neg y)] \\ &= [(\neg(x \odot \neg y) \odot \neg y) \oplus y] \oplus (z \odot \neg y)] \\ &= [\neg(x \odot \neg y) \vee y] \oplus (z \odot \neg y) && \text{by (14.2)} \\ &= [(\neg x \oplus y) \vee y] \oplus (z \odot \neg y) \\ &\ge [\neg x \oplus y] \oplus (z \odot \neg y) && \text{by Prop. 14.20(5),(1)} \\ &= \neg x \oplus (y \oplus (z \odot \neg y)) \\ &= \neg x \oplus z = 1. && \text{by Prop. 14.19 and assumption } x, y \le z. \end{aligned}$$

And since $0 \le x \le 1$ for all $x \in A$ by Prop. 14.20(6), we get $\neg(x \vee y) \oplus z = 1$ as desired.

(2): Due to (1) and Prop. 1.44(L2). □

14.3.2.2 *MV-Relation Algebras*

A second approach to relation algebras based on MV-algebras originated in [24, 25] and is much more tailored along the lines of the classical concept of Def. 14.2. The definition in [25] extends the one of [24] with the two additional requirements (MR5) and (MR6) that allow to establish results on some special kinds of elements in a relation algebra; see [25, Section 5] for details.

Definition 14.22. *A structure* $\mathfrak{M} = (A, \oplus, \odot, \neg, 0, 1, ;, \breve{}, \mathbb{1})$ *of type* $\langle 2,2,1,0,0,2,1,0\rangle$ *is called an* **MV-relation algebra** *(MVRA for short) iff the following holds:*

(MR1) $(A, \oplus, \odot, \neg, 0, 1)$ *is an MV-algebra,*
(MR2) $(A, ;, \mathbb{1})$ *is a monoid,*
(MR3) $(x;y) \odot z = 0 \leftrightarrow (x^{\smile};z) \odot y = 0 \leftrightarrow (z;y^{\smile}) \odot x = 0.$
(MR4) $(x \oplus y)^{\smile} = x^{\smile} \oplus y^{\smile}$,
(MR5) $(a \odot x);(b \odot y) \leq (a;b) \odot (x;y)$,
(MR6) $\mathbb{1} \odot \mathbb{1} = \mathbb{1}$.

The paradigmatic example is based on the standard MV-algebra $\mathfrak{L}$ as described just before Prop. 14.18: $([0,1]^{X\times X}, \oplus, \odot, \neg, 0, 1, ;, \breve{}, \mathbb{1})$ is an MV-relation algebra where the operations $\oplus, \odot, \neg$ defined component-wise on $\mathfrak{L}$, 0 and 1 are the constant functions, $\mathbb{1}(x,y) = \delta_{xy}$. It is straightforward to show that ; is associative, and that $([0,1]^{X\times X}, ;, \mathbb{1})$ is a monoid: $P;\mathbb{1}(x,y) = \bigvee_{z\in X} P(x,z) \odot \mathbb{1}(z,y) = P(x,y) \odot \mathbb{1}(y,y) = P(x,y)$. To show (MR3) note that $(P;Q);R = 0$ means that $P(x,y) \odot Q(y,z) \odot R(x,z) = 0$ for all $x,y,z \in X$; and considering that $P^{\smile}(x,y) = P(y,x)$ and that $\odot$ is commutative one immediately gets $P;Q(x,z) \odot R(x,z) = 0$ iff $P^{\smile};R(y,z) \odot Q(y,z) = 0$ iff $R;Q^{\smile}(x,y) \odot P(x,y) = 0$. (MR4) and (MR6) are trivial. Using Prop. 14.20 and (14.6) we finally show (MR5):

$$\begin{aligned}
[(P \odot Q);(S \odot T)](x,y) &= \textstyle\bigvee_{u\in X}[P(x,u) \odot Q(x,u) \odot S(u,y) \odot T(u,y)] \\
&= \textstyle\bigvee_{u\in X}[P(x,u) \odot S(u,y) \odot Q(x,u) \odot T(u,y)] \\
&\leq \textstyle\bigvee_{u\in X}\left[P(x,u) \odot S(u,y) \odot \bigvee_{u'\in X}[Q(x,u') \odot T(u',y)]\right] \\
&= \left[\textstyle\bigvee_{u\in X} P(x,u) \odot S(u,y)\right] \odot \left[\textstyle\bigvee_{u'\in X} Q(x,u') \odot T(u',y)\right] \\
&= (P;S)(x,y) \odot (Q;T)(x,y) \\
&= [(P;S) \odot (Q;T)](x,y).
\end{aligned}$$

As mentioned at the beginning of this section, the current approach resembles very much the classical one. So it is interesting to compare the properties of MVRAs with those of classical relation algebras as listed in Def. 14.1 and Prop. 14.3 - Prop. 14.5. We get analogous results below which are grouped accordingly; see also [24, Proposition 2.21].

Proposition 14.23. *In an MV-relation algebra* $(A, \oplus, \odot, \neg, 0, 1, ;, \breve{}, \mathbb{1})$ *the following holds for any* $x, y, z \in A$ *(in analogy to* Def. 14.1*):*

(1) $x \oplus y = y \oplus x$,
(2) $x \oplus (y \oplus z) = (x \oplus y) \oplus z$,
(3) $x;(y;z) = (x;y);z$,

(4) $x;\mathbb{1} = x$,
(5) $x^{\smile\smile} = x$,
(6) $(x \oplus y)^{\smile} = x^{\smile} \oplus y^{\smile}$,
(7) $x^{\smile} \odot y = 0 \leftrightarrow x \odot y^{\smile} = 0$,
(8) $(x;y)^{\smile} = y^{\smile};x^{\smile}$,
(9) $x^{\smile};\neg(x;y) \leq \neg y$.

Proof. (1),(2) and (3),(4) follow from the fact that $(A,\oplus,0)$ is an Abelian monoid and $(A,;,\mathbb{1})$ is a monoid by definition.

(5): Using Prop. 14.19, (4), and (MR3) twice in that order we get the following equivalences: $x \leq t \leftrightarrow x \odot \neg t = 0 \leftrightarrow (x;\mathbb{1}) \odot \neg t = 0 \leftrightarrow (x^{\smile};\neg t) \odot \mathbb{1} = 0 \leftrightarrow (x^{\smile\smile};\mathbb{1}) \odot \neg t = 0 \leftrightarrow x^{\smile\smile} \odot \neg t = 0 \leftrightarrow x^{\smile\smile} \leq t$. This gives $x \leq x \leftrightarrow x^{\smile\smile} \leq x$ for $t = x$, and the converse inequality for $t = x^{\smile\smile}$.

(6) is (MR4).

(7): Using (MR3) yields $x \odot y^{\smile} = 0 \leftrightarrow (x;\mathbb{1}) \odot y^{\smile} = 0 \leftrightarrow (x^{\smile};y^{\smile}) \odot \mathbb{1} = 0 \leftrightarrow (\mathbb{1};y^{\smile\smile}) \odot x^{\smile} = 0 \leftrightarrow y \odot x^{\smile} = 0$.

(8): By Prop. 14.19, (7) and several times (MR3) we get $(x;y)^{\smile} \leq t \leftrightarrow (x;y)^{\smile} \odot \neg t = 0 \leftrightarrow (x;y) \odot \neg t^{\smile} = 0 \leftrightarrow (x^{\smile};\neg t^{\smile}) \odot y = 0 \leftrightarrow (y;\neg t^{\smile\smile}) \odot x^{\smile} = 0 \leftrightarrow (y;\neg t) \odot x^{\smile} = 0 \leftrightarrow (y^{\smile};x^{\smile}) \odot \neg t = 0 \leftrightarrow y^{\smile};x^{\smile} \leq t$. Therefore, $(x;y)^{\smile} \leq t \leftrightarrow y^{\smile};x^{\smile} \leq t$, for any $t \in A$. Choosing $t = (x;y)^{\smile}$ and $t = y^{\smile};x^{\smile}$ gives the two inequalities required to prove $(x;y)^{\smile} = y^{\smile};x^{\smile}$.

(9): Applying Prop. 14.19 and (MR3) one gets $x^{\smile};\neg(x;y) \leq \neg y \leftrightarrow (x^{\smile};\neg(x;y)) \odot y = 0 \leftrightarrow (x;y) \odot \neg(x;y) = 0$ which holds due to Def. 14.17(C3'). □

Proposition 14.24. *In an MV-relation algebra $(A,\oplus,\odot,\neg,0,1,;,{}^{\smile},\mathbb{1})$ the following holds for any $x,y \in A$:*

(1) $(\neg x)^{\smile} \leq \neg(x^{\smile})$.
(2) $x \leq y \leftrightarrow x^{\smile} \leq y^{\smile}$.
(3) $x \leq y^{\smile} \leftrightarrow x^{\smile} \leq y$.
(4) $0^{\smile} = 0$, $1^{\smile} = 1$.
(5) $x \oplus y^{\smile} = 1 \leftrightarrow x^{\smile} \oplus y = 1$.
(6) $(x \odot y)^{\smile} = x^{\smile} \odot y^{\smile}$.
(7) $x^{\smile} = y^{\smile} \leftrightarrow x = y$.
(8) $\mathbb{1}^{\smile} = \mathbb{1}$.
(9) $\mathbb{1};x = x$.
(10) $(\neg x)^{\smile} = \neg(x^{\smile})$.

Proof. (1): From $\neg x \odot x = 0$ we get $\neg x \odot x^{\smile\smile} = 0$, which by Prop. 14.23(7) implies $(\neg x)^{\smile} \odot x^{\smile} = 0$. Thus $(\neg x)^{\smile} \leq \neg(x^{\smile})$ by (C7) and Prop. 14.19.

(2): If $x^{\smile} \leq y^{\smile}$, then $x^{\smile} \odot \neg(y^{\smile}) = 0$ by Prop. 14.19. But by (1) we have $(\neg y)^{\smile} \leq \neg(y^{\smile})$, so by Prop. 14.20(2) $x^{\smile} \odot (\neg y)^{\smile} = 0$, whence $x^{\smile\smile} \odot \neg y = x \odot \neg y = 0$ by Prop. 14.23(7). So $x \leq y$. Conversely, $x \leq y \rightarrow x^{\smile\smile} \leq y^{\smile\smile}$, so $x^{\smile} \leq y^{\smile}$ by what we just have proved.

(3): Take $y := y^{\smile}$ in (2).

(4): From (3) we get $0 \leq 0^{\smile} \leftrightarrow 0^{\smile} \leq 0$, taking $x = y = 0$; now apply Prop. 14.20(6). Similar for the second assertion.

(5): Assume $x \oplus y^{\smile} = 1$; then $x^{\smile} \oplus y = x^{\smile} \oplus y^{\smile\smile} = (x \oplus y^{\smile})^{\smile} = 1^{\smile} = 1$ by (4).

(6): Using (10) for the second and fourth equality we get $(x \odot y)^{\smile} = (\neg(\neg x \oplus \neg y))^{\smile} = \neg((\neg x \oplus \neg y)^{\smile}) = \neg((\neg x)^{\smile} \oplus (\neg y)^{\smile}) = \neg(\neg(x^{\smile}) \oplus \neg(y^{\smile})) = x^{\smile} \odot y^{\smile}$.

(7): Obvious from (2).

(8): Since $(A, ;, \mathbb{1})$ is a monoid by definition, $\mathbb{1}^{\smile} = \mathbb{1}; \mathbb{1}^{\smile}$, so $\mathbb{1}^{\smile\smile} = (\mathbb{1}; \mathbb{1}^{\smile})^{\smile} = \mathbb{1}^{\smile\smile}; \mathbb{1}^{\smile} = \mathbb{1}; \mathbb{1}^{\smile} = \mathbb{1}^{\smile}$.

(9): By (MR2).

(10): We take $x := x^{\smile}$ in (1) to get $[\neg(x^{\smile})]^{\smile} \leq \neg(x^{\smile\smile}) = \neg x$. So by (3) $\neg(x^{\smile}) \leq (\neg x)^{\smile}$, which together with (1) yields equality. □

Proposition 14.25. *In an MV-relation algebra* $(A, \oplus, \odot, \neg, 0, 1, ;, {}^{\smile}, \mathbb{1})$ *the following holds for any* $x, x', y, y', z \in A$:

(1) $(x \vee x'); y = (x; y) \vee (x'; y)$ *and* $x; (y \vee y') = (x; y) \vee (x; y')$.
(2) $x \leq y \rightarrow z; x \leq z; y \;\&\; x; z \leq y; z$.
(3) $\neg(y; x); x^{\smile} \leq \neg y$.
(4) $x; y \leq z \leftrightarrow x^{\smile}; \neg z \leq \neg y \leftrightarrow \neg z; y^{\smile} \leq \neg x$. *(Schröder rule)*
(5) $1; 1 = 1$.
(6) $a \leq 1; a$ *and* $a \leq a; 1$.

Proof.

(1):
$$\begin{aligned}
(x \vee x'); y \leq t &\leftrightarrow \big((x \vee x'); y\big) \odot \neg t = 0 && \text{Prop. 14.19}\\
&\leftrightarrow (\neg t; y^{\smile}) \odot (x \vee x') = 0 && \text{(MR3)}\\
&\leftrightarrow ((\neg t; y^{\smile}) \odot x) \vee ((\neg t; y^{\smile}) \odot x') = 0 && \text{Prop. 14.20(9)}\\
&\leftrightarrow (\neg t; y^{\smile}) \odot x = 0 \;\&\; (\neg t; y^{\smile}) \odot x' = 0 &&\\
&\leftrightarrow (x; y^{\smile\smile}) \odot \neg t = 0 \;\&\; (x'; y^{\smile\smile}) \odot \neg t = 0 && \text{(MR3)}\\
&\leftrightarrow (x; y) \odot \neg t = 0 \;\&\; (x'; y) \odot \neg t = 0 &&\\
&\leftrightarrow x; y \leq t \;\&\; x'; y \leq t && \text{Prop. 14.19}\\
&\leftrightarrow (x; y) \vee (x'; y) \leq t. &&
\end{aligned}$$

t was chosen arbitrarily, so the equivalences hold in particular for $t = (x; y) \vee (x'; y)$ and $t = (x \vee x'); y$, together yielding the equality as asserted.

(2): By Prop. 14.20(5) $x \leq x \vee y$, and from Prop. 14.21(2) we have $x \leq y \leftrightarrow x \vee y = y$, for all $x, y \in A$. Thus, assuming $x \leq y$ we get $x; z \leq x; z \vee y; z = (x \vee y); z = y; z$ by (1).

(3):
$$\begin{aligned}
\neg(y; x); x^{\smile} \leq \neg y &\leftrightarrow (\neg(y; x); x^{\smile}) \odot y = 0 && \text{Prop. 14.19}\\
&\leftrightarrow y; x \odot \neg(y; x) = 0 && \text{(M3)}\\
&\leftrightarrow \text{Def. 14.17(C3')}. &&
\end{aligned}$$

(4): Using (14.5) and (MR3) we get $x; y \leq z \leftrightarrow (x; z) \odot \neg z = 0 \leftrightarrow (x^{\smile}; \neg z) \odot y = 0 \leftrightarrow x^{\smile}; \neg z \leq y$. Similar for the second equivalence.

(5): Applying Prop. 14.20(6) yields $1; 1 \leq 1$, and from $\mathbb{1} \leq 1$ we get $1 = \mathbb{1}; 1 \leq 1; 1$ using (2).

(6): Obvious: $a = \mathbb{1}; a \le 1; a$. □

Proposition 14.26. *In an MV-relation algebra* $(A, \oplus, \odot, \neg, 0, 1, ;, \breve{\ }, \mathbb{1})$ *the following holds for any* $x, y, z \in A$*:*

(1) $x;0 = 0;x = 0$.
(2) $x;(y \odot z) \le x;(y \wedge z) \le (x;y) \wedge (x;z)$.

Proof. (1): By Def. 14.17(C5') we have $x;0 = (x;0) \odot 1$, so by Def. 14.22(MR3) $0 = x;0 = (x;0) \odot 1 \leftrightarrow (x\breve{\ };1) \odot 0 = 0$, with the last equality being true by Def. 14.17(C4'). Similarly, $0 = 0;x = (0;x) \odot 1 \leftrightarrow (1;x\breve{\ }) \odot 0 = 0$.
(2): By monotonicity properties Prop. 14.20(8) and Prop. 14.25(2). □

Though we have been able to reproduce a lot of results valid in crisp relation algebras also for MV-relation algebras, a careful look will make you miss distributivity properties analogous to those valid for classical relation algebras, viz. Def. 14.1(R5) $(x + y);z = x;z + y;z$ (right ;-distributivity) and Prop. 14.4(1) $x;(y + z) = (x;y) + (x;z)$ (left ;-distributivity), where addition $+$ is based on an underlying Boolean algebra $(A, +, \cdot, \bar{\ }, 0, 1)$ (see Def. 14.2(1)). Note that the distributivity properties shown in Prop. 14.25(1) are not what we are looking for.

Before we can provide a result in connection with this problem below, we need a new notion and have to show some of its properties, analogous to arguments used for Boolean algebras in e.g. [20, Theorem 18] or [31, Lemma 10.27].

Definition 14.27. *Let f and g be two functions on an MV-algebra* $(A, \oplus, \odot, \neg, 0, 1)$.

(a) *f is called* **additive** *if* $f(x \oplus y) = f(x) \oplus f(y)$, *for all* $x, y \in A$.
(b) *f is called* **completely additive** *if, for every indexed set* $\{x_i : i \in I\} \subseteq A$, *if* $\bigoplus_{i\in I} x_i$ *exists and* $I \ne \emptyset$ *then also* $\bigoplus_{i\in I} f(x_i)$ *exists and* $f(\bigoplus_{i\in I} x_i) = \bigoplus_{i\in I} f(x_i)$.
(c) *f is called a* **normal function** *if* $f(0) = 0$.
(d) *g is called a* **conjugate of** *f if* $f(x) \odot y = 0 \leftrightarrow x \odot g(y) = 0$, *for all* $x, y \in A$.

Lemma 14.28. (analogous to a part of [20, Theorem 18]) *Let $f : A \longrightarrow A$ be a function on an MV-algebra* $(A, \oplus, \odot, \neg, 0, 1)$. *If f has a conjugate then it is normal and completely additive.*

Proof. Let g be a conjugate to f. Then, since $0 = 0 \odot g(1) = f(0) \odot 1 = f(0)$ by definition, we get $f(0) = 0$, i.e. f is normal. Now assume that $I \ne \emptyset$, $\{x_i : i \in I\} \subseteq A$, and that $\bigoplus_{i\in I} x_i$ exists. Then the following equivalences hold:

	$f(\bigoplus_{i\in I} x_i) \le y$	
$\leftrightarrow$	$f(\bigoplus_{i\in I} x_i) \odot \neg y = 0$	by Prop. 14.19
$\leftrightarrow$	$\bigoplus_{i\in I} x_i \odot g(\neg y) = 0$	since f and g are conjugate
$\leftrightarrow$	$\bigoplus_{i\in I} x_i \le \neg g(\neg y)$	by Prop. 14.19
$\leftrightarrow$	$x_i \le \neg g(\neg y)$ for every $i \in I$	by Prop. 14.20(7)

$$\begin{array}{lll} \leftrightarrow & x_i \odot g(\neg y) = 0 \text{ for every } i \in I & \text{by Prop. 14.19} \\ \leftrightarrow & f(x_i) \odot \neg y = 0 \text{ for every } i \in I & \text{since } f \text{ and } g \text{ are conjugate} \\ \leftrightarrow & f(x_i) \le y \text{ for every } i \in I & \text{by Prop. 14.19} \\ \leftrightarrow & y \text{ is an upper bound of } \{f(x_i) : i \in I\} & \end{array}$$

With $y = f(\bigoplus_{i\in I} x_i)$ the first statement becomes true, so $f(\bigoplus_{i\in I} x_i)$ is an upper bound of $F := \{f(x_i) : i \in I\}$. Reading bottom up gives: y is an upper bound of F iff $f(\bigoplus_{i\in I} x_i) \le y$, for any upper bound y, so $f(\bigoplus_{i\in I} x_i)$ is the least one. Thus $f(\bigoplus_{i\in I} x_i) = \bigoplus_{i\in I} f(x_i)$, whence f is completely additive. □

Proposition 14.29. *In an MV-relation algebra $\mathfrak{A} = (A, \oplus, \odot, \neg, 0, 1, ;, \breve{}, \mathbb{1})$ the following holds for any $x, y, z \in A$:*

(1) $x;(y \oplus z) = (x; y) \oplus (x; z)$.
(2) $(x \oplus y); z = x; z \oplus y; z$.

Proof. Assume $\mathfrak{A}$ to be relation algebra according to Def. 14.22; then its condition (M3),

$$\forall x, y, z \in A[(x; y) \odot z = 0 \leftrightarrow (x\breve{}; z) \odot y = 0 \leftrightarrow (z; y\breve{}) \odot x = 0],$$

asserts that, for a fixed element a, the maps $x \mapsto a; x$ and $y \mapsto a\breve{}; y$ are conjugate, and also the maps $x \mapsto x; a$ and $y \mapsto y; a\breve{}$.
(1): Let $f(x) = a; x$ and $g(y) = a\breve{}; y$, then $0 = f(x) \odot z = a; x \odot z \leftrightarrow a\breve{}; z \odot x = g(z) \odot x = 0$ from the first equivalence. By Lemma 14.28, f preserves existing sums, so $a;(x \oplus y) = f(x \oplus y) = f(x) \oplus f(y) = a; x \oplus a; y$.

(2): Let $f(x) = x; a$ and $g(y) = y; a\breve{}$, then $0 = f(x) \odot z = x; a \odot z \leftrightarrow z; a\breve{} \odot x = g(z) \odot x = 0$ from the second equivalence. By Lemma 14.28, f preserves existing sums, so $(x \oplus y); a = f(x \oplus y) = f(x) \oplus f(y) = x; a \oplus y; a$. □

Insertion: An open problem

In the case of (Boolean, crisp) relation algebras conjugate functions have been be used to show the validity of the modular laws and the Dedekind rule (see e.g. [31, Section 10.2.5 and Proposition 10.38], or [10, Lemma 4.19, Corollary 4.20, Lemma 4.25]):

$$\begin{array}{ll} x; y \odot z \le (x \odot z; y\breve{}); y, & \text{modular law, Prop. 14.5(3)} \\ x; y \odot z \le x; (y \odot x\breve{}; z), & \text{modular law, Prop. 14.5(4)} \\ x; y \odot z \le (x \odot z; y\breve{}); (y \odot x\breve{}; z). & \text{Dedekind rule, Prop. 14.5(5)} \end{array}$$

Contrary to distributivity shown above in Prop. 14.29, using conjugate functions analogous to the crisp case does not seem promising to find a proof for these three inequalities (at least I could not succeed). So the problem remains open. Nonetheless, there are a few results that hold in analogy to those valid in the context of Boolean algebras (see [31, Proposition 10.28]) - but not the one that would be needed for a proof quite analogous to the Boolean case, viz. (1) ⇒ (3)(*a*), or (2) ⇒ (3) - as shown in

Lemma 14.30. *Let f and g be functions on an MV-algebra $(A, \oplus, \odot, \neg, 0, 1)$. For the statements below the following relationships hold:* (1) ⇔ (2), (3) ⇒ (1), *so also* (3) ⇒ (2).

(1) f and g are conjugates of each other.

(2) *for all* $x, y \in A$

(a) $f(x \wedge \neg g(y)) \leq f(x), \neg y$,
(b) $g(y \wedge \neg f(x)) \leq g(y), \neg x$.

(3) *for all* $x, y \in A$,

(a) $f(x) \odot y \leq f(x \odot g(y))$,
(b) $g(y) \odot x \leq g(y \odot f(x))$.
(c) $f(0) = 0, g(0) = 0$

Proof.
(1) $\Leftrightarrow$ (2): Suppose f and g are conjugate. Then, by Lemma 14.28, f is completely additive. Obviously, a completely additive function is additive and additive functions are monotone, so $f(x \wedge \neg g(y)) \leq f(x)$, since $x \wedge y \leq x$ by Prop. 14.20(5). From $(x \wedge \neg g(y)) \odot g(y) = 0$, since f and g are assumed to be conjugates, we get $f(x \wedge \neg g(y)) \odot y = 0$. Hence $f(x \wedge \neg g(y)) \leq \neg y$ by Prop. 14.19. Therefore, $f(x \wedge \neg g(y)) \leq f(x), \neg y$. And by a symmetric argument we get $g(y \wedge \neg f(x)) \leq g(y), \neg x$, and we are done with (1) $\rightarrow$ (2).

Conversely, assume (2) and $x \odot g(y) = 0$. By Prop. 14.19 we get $x \leq \neg g(y)$, so $f(x) = f(x \wedge \neg g(y)) \leq \neg y$, using Prop. 14.21(2) and assumption (2)(a). Thus, $f(x) \odot y = 0$. If $f(x) \odot y = 0$, then $y \leq \neg f(x)$, whence $g(y) = g(y \wedge \neg f(x)) \leq \neg x$ by assumption (2)(b). Thus $x \odot g(y) = 0$, and (2) $\Rightarrow$ (1) is shown.

(1) $\Rightarrow$ (3c): Let g be a conjugate to f. Then, since by definition $0 = 0 \odot g(1) = f(0) \odot 1 = f(0)$, we get $f(0) = 0$, i.e. f is normal; and similar for g.

(3) $\Rightarrow$ (1): Assume that (3) and $f(y) \odot z = 0$ hold. Then we get $g(z) \odot y \leq g(z \odot f(y)) = g(0) = 0$ by (3)(b) and the normality of g. Conversely, if $g(z) \odot y = 0$ then $f(y) \odot z \leq f(y \odot g(z)) = f(0) = 0$ by (3)(a) and the normality of f. So f and g are conjugate. □

There is, however, a group of equivalences that carry over to MV-relation algebras, given in one of the two equivalent forms (see Prop. 14.19): $(a;b) \odot c = 0 \leftrightarrow a;b \leq \neg c$. These equivalences appear under various names as can be seen below.

Some of them do already appear in Def. 14.22(MR3) as

$$(x;y) \odot z = 0 \leftrightarrow (x^{\smile};z) \odot y = 0 \leftrightarrow (z;y^{\smile}) \odot x = 0.$$

Another group runs under the name **Schröder equivalences** or **Schröder rule** ([26, 2.3.4 Proposition]), also called **De Morgan equivalences** in [21, Theorem 290]:

$$x;y \leq z \leftrightarrow x^{\smile};\neg z \leq \neg y \leftrightarrow \neg z;y^{\smile} \leq \neg x,$$

Their validity for MV-relation algebras has been shown in Prop. 14.25(4).

A third group we encounter as **De Morgan-Tarski laws** in [10, Corollary 4.9], or as **Tarski equivalences** in [21, Theorem 291]:

Proposition 14.31. ([10, Lemma 4.8, Corollary 4.9]) *In an MV-relation algebra* $(A, \oplus, \odot, \neg, 0, 1, ;, {}^{\smile}, \mathbb{1})$ *the following inequalities are equivalent for any* $x, y, z \in A$:

(1) $x;y \leq \neg z$.
(2) $x^{\smile};z \leq \neg y$.
(3) $z;y^{\smile} \leq \neg x$.
(4) $z^{\smile};x \leq \neg(y^{\smile})$.

(5) $y;z^{\smile} \leq \neg(x^{\smile})$.
(6) $y^{\smile};x^{\smile} \leq \neg(z^{\smile})$.

Proof. (1) implies $x;y \odot z = 0 = z \odot x;y$ iff $z \leq \neg(x;y)$, whence $x^{\smile};z \leq x^{\smile};\neg(x;y)$. By Prop. 14.23(9) we have $x^{\smile};\neg(x;y) \leq \neg y$, so $x^{\smile};z \leq \neg y$. For the reverse implication assume (2) and apply (1) $\rightarrow$ (2) with $x^{\smile}$, z, and y in place of x, y, and z, respectively.

The equivalence of (2) and (3) is given by the second one in Def. 14.22(MR3). We have $x^{\smile};z \leq \neg y \leftrightarrow (x^{\smile};z)^{\smile} \leq (\neg y)^{\smile} \leftrightarrow z^{\smile};x \leq \neg(y^{\smile})$, applying Prop. 14.24(1),(2), whence (2) is equivalent to (4). The equivalences of (3) and (5), and (1) and (6) can be shown in the same way. □

Note that, due to Prop. 14.19, it is evident that the De Morgan-Tarski laws are the same as *The cycle law* ([21, Theorem 294]):

Corollary 14.32. ([24, Proposition 2.30]) *In an MV-relation algebra the following equivalences hold for any $x, y, z \in A$:*
$(x;y) \odot z = 0 \leftrightarrow (x^{\smile};z) \odot y = 0 \leftrightarrow (z;y^{\smile}) \odot x = 0 \leftrightarrow (z^{\smile};x) \odot y^{\smile} = 0 \leftrightarrow (y;z^{\smile}) \odot x^{\smile} = 0 \leftrightarrow (y^{\smile};x^{\smile}) \odot z^{\smile} = 0$.

14.3.3 *Lattice Implication Relation Algebras*

(Note: Instead of the author's original rather unspecific "*lattice-valued relation algebra*" in [22], which serves as guideline for this section, we will use the name as shown in the header hinting to the underlying structure of *lattice implication algebras*.)

14.3.3.1 *Lattice Implication Algebras*

Definition 14.33. ([22, Definition 1], [33, Section 2.1]) *Let $(L, \wedge, \vee, 0, 1)$ be a bounded lattice with order-reversing involution $'$ and a binary operation $\rightarrow$ on L; then $(L, \wedge, \vee, ', \rightarrow, 0, 1)$ is called a* **lattice implication algebra** *if it satisfies the following axioms for all $x, y, z \in L$:*

(1) $x \rightarrow (y \rightarrow z) = y \rightarrow (x \rightarrow z)$,
(2) $x \rightarrow x = 1$,
(3) $x \rightarrow y = y' \rightarrow x'$,
(4) $x \rightarrow y = y \rightarrow x = 1$ *implies* $x = y$,
(5) $(x \rightarrow y) \rightarrow y = (y \rightarrow x) \rightarrow x$,
(6) $(x \vee y) \rightarrow z = (x \rightarrow z) \wedge (y \rightarrow z)$,
(7) $(x \wedge y) \rightarrow z = (x \rightarrow z) \vee (y \rightarrow z)$.

There are many familiar examples of lattice implication algebras (for more examples, and a long list of properties of implication algebras see [33, Section 2.1]):

- The *Łukasiewicz implication algebra* $([0,1], \min, \max, \neg, \curvearrowright_Ł, 0, 1)$ with $\neg a = 1-a$ and $a \curvearrowright_Ł b = \min\{1, (1-a+b)\}$.

- The *Łukasiewicz implication algebra on finite chains*: Let $L = \{a_1 < a_2 < \ldots < a_n\}$. For $j,k \in \underline{n}$ one defines $a_j \vee a_k = a_{\max\{j,k\}}$, $a_j \wedge a_k = a_{\min\{j,k\}}$, $(a_j)' = a_{n-j+1}$, and $a_j \rightarrow a_k = a_{\min\{n-j+k,n\}}$.
- Let $(B, \wedge, \vee, ^{-}, 0, 1)$ be a Boolean algebra, and define $a \rightarrow b := \overline{a} \vee b$. Then $(B, \wedge, \vee, ^{-}, \rightarrow, 0, 1)$ is an implication algebra.

Lemma 14.34. *In a lattice implication algebra the following holds for any $x \in L$:*

(1) $x \rightarrow y = 1$ *iff* $x \leq y$.
(2) $1' = 0, \qquad 0' = 1.$

Proof. (1): Assume $x \leq y$; then $x \rightarrow y = (x \wedge y) \rightarrow y = (x \rightarrow y) \vee (y \rightarrow y) = (x \rightarrow y) \vee 1 = 1$, applying Def. 14.33(7),(1). Conversely, if $x \rightarrow y = 1$ then $(x \vee y) \rightarrow y = (x \rightarrow y) \wedge (y \rightarrow y) = 1$, and since $y \rightarrow (x \vee y) = (y \rightarrow x) \vee (y \rightarrow y) = (y \rightarrow x) \vee 1 = 1$, by Def. 14.33(4) we get $x \vee y = y$, i.e. $x \leq y$.

(2): By definition, 1 is the greatest element in L, so $x \rightarrow 1 = 1$ for all $x \in L$ by (1). Therefore, $1 = x' \rightarrow 1 = 1' \rightarrow x'' = 1' \rightarrow x$, whence $1' \leq x$ by (1), for all $x \in L$, i.e. $1' = 0$. □

Lemma 14.35. ([33, Theorem 2.1.2]) *In a lattice implication algebra the following holds for any $x \in L$:*

(1) $1 \rightarrow x = 1$ *implies* $x = 1$.
(2) $1 \rightarrow x = x, \qquad x \rightarrow 0 = x'$.
(3) $0 \rightarrow x = 1, \qquad x \rightarrow 1 = 1$.

Proof. (1): Assume $1 \rightarrow x = 1$; then $x \rightarrow 1 = x \rightarrow (1 \rightarrow x) = 1 \rightarrow (x \rightarrow x) = 1 \rightarrow 1 = 1$. So $x = 1$ due to Def. 14.33(4).

(2): By Def. 14.33(1), $1 \rightarrow ((1 \rightarrow x) \rightarrow x) = (1 \rightarrow x) \rightarrow (1 \rightarrow x) = 1$, so by (1) $(1 \rightarrow x) \rightarrow x = 1$. Since $x \rightarrow (1 \rightarrow x) = 1 \rightarrow (x \rightarrow x) = 1 \rightarrow 1 = 1$, we get $1 \rightarrow x = x$ by Def. 14.33(4). Therefore, using Lemma 14.34(2), $x' = 1 \rightarrow x' = 0' \rightarrow x' = x \rightarrow 0$.

(3): We have $x \rightarrow 1 = x \rightarrow (0 \rightarrow 0) = 0 \rightarrow (x \rightarrow 0) = 0 \rightarrow ((x' \rightarrow 0) \rightarrow 0)$ (take $x := x'$ in (2)). Applying Def. 14.33(5),(1) to the last expression we get $0 \rightarrow ((0 \rightarrow x') \rightarrow x') = (0 \rightarrow x') \rightarrow (0 \rightarrow x') = 1$. Therefore, $0 \rightarrow x = x' \rightarrow 0' = x' \rightarrow 1 = 1$. □

Proposition 14.36. ([34, Lemma 2.4]) *Let $(L, \wedge, \vee, ', \rightarrow, 0, 1)$ be a lattice implication algebra; then, for all $x, y, z \in L$:*

(1) $x \leq y \rightarrow z$ *implies* $y \leq x \rightarrow z$.
(2) $(x \rightarrow y) \leq (y \rightarrow z) \rightarrow (x \rightarrow z)$.
(3) $(x \rightarrow y) \leq (z \rightarrow x) \rightarrow (z \rightarrow y)$.
(4) $x \leq y$ *implies* $(y \rightarrow z) \leq (x \rightarrow z)$.
(5) $x \leq y$ *implies* $(z \rightarrow x) \leq (z \rightarrow y)$.
(6) $x \leq y$ *implies* $y' \leq x'$.
(7) $y \leq x \rightarrow y$.
(8) $x \leq (x \rightarrow y) \rightarrow y$.

Proof. (1): From $x \leq (y \to z)$ we get $y \to (x \to z) = x \to (y \to z) = 1$; so $y \leq (x \to z)$.

(2):

$$
\begin{aligned}
& (x \to y) \to ((y \to z) \to (x \to z)) & \\
= \; & (x \to y) \to (x \to ((y \to z) \to z)) & \text{Def. 14.33(1)} \\
= \; & (x \to y) \to (x \to ((z \to y) \to y)) & \text{Def. 14.33(5)} \\
= \; & (x \to y) \to ((z \to y) \to (x \to y)) & \text{Def. 14.33(1)} \\
= \; & (z \to y) \to ((x \to y) \to (x \to y)) & \text{Def. 14.33(1)} \\
= \; & (z \to y) \to 1 & \text{Def. 14.33(2)} \\
= \; & 1 & \text{Lemma 14.34(1)}
\end{aligned}
$$

(3): Taking $x := z, y := x$, and $z := y$, (2) implies $z \to x \leq (x \to y) \to (z \to y)$. Therefore, by (1) we get $x \to y \leq (z \to x) \to (z \to y)$.

(4): From $x \leq y$ and (2) we get $1 = x \to y \leq (y \to z) \to (x \to z)$. Therefore, $(y \to z) \to (x \to z) = 1$, whence $(y \to z) \leq (x \to z)$.

(6): $x \leq y$ implies $1 = x \to y = y' \to x'$; so $y' \leq x'$.

(7): $y \to (x \to y) = x \to (y \to y) = x \to 1 = 1$. By Lemma 14.34(1) the assertion follows.

(8): Using Def. 14.33(1),(2) we get $x \to ((x \to y) \to y) = (x \to y) \to (x \to y) = 1$. □

Proposition 14.37. ([33, Theorems 2.1.4, 2.1.5], [34, Theorem 2.5]) *In a lattice implication algebra the following holds for any $x \in L$:*

(1) $((x \to y) \to y) \to y = x \to y$
(2) $x \vee y = (x \to y) \to y$.
(3) $x \wedge y = (x' \vee y')'$.
(4) $x \wedge y = ((x \to y) \to x')'$.
(5) $x \vee y = (x' \wedge y')'$.

Proof. (1): Recall that by Lemma 14.34(1) $x \to y = 1$ iff $x \leq y$. We have

$$
\begin{aligned}
& [((x \to y) \to y) \to y] \to (x \to y) & \\
= \; & [(y \to ((x \to y) \to (x \to y))) \to y] \to (x \to y) & \text{Def. 14.33(5)} \\
= \; & [(y \to 1) \to y] \to (x \to y) & \text{Def. 14.33(2)} \\
= \; & [1 \to y] \to (x \to y) & \text{Lemma 14.35(3)} \\
= \; & y \to (x \to y) & \text{Lemma 14.35(2)} \\
= \; & x \to (y \to y) & \text{Def. 14.33(1)} \\
= \; & x \to 1 & \text{Def. 14.33(2)} \\
= \; & 1 & \text{Lemma 14.35(3)}
\end{aligned}
$$

So we have shown $[((x \to y) \to y) \to y] \leq (x \to y)$. Conversely we get

$$
\begin{aligned}
& (x \to y) \to [((x \to y) \to y) \to y] & \\
= \; & ((x \to y) \to y) \to ((x \to y) \to y) & \text{Def. 14.33(1)} \\
= \; & 1 & \text{Def. 14.33(2)}
\end{aligned}
$$

(2): We have $x \to ((x \to y) \to y) = (x \to y) \to (x \to y) = 1$, and $y \to ((x \to y) \to y) = (x \to y) \to (y \to y) = (x \to y) \to 1 = 1$. Thus $x \vee y \leq (x \to y) \to y$, since $x \vee y$ is the least upper bound.

Conversely, for any $a \in L$ such that $x \le a$ and $y \le a$, $x \to a = 1$ and $y \to a = 1$. holds. Therefore,

$$\begin{aligned}
& ((x \to y) \to y) \to a & \\
= \ & ((x \to y) \to y) \to (1 \to a) & \text{Lemma 14.35(2)} \\
= \ & ((x \to y) \to y) \to ((y \to a) \to a) & \text{by hypothesis } y \le a \\
= \ & ((x \to y) \to y) \to ((a \to y) \to y) & \text{Def. 14.33(5)} \\
= \ & (a \to y) \to (((x \to y) \to y) \to y) & \text{Def. 14.33(1)} \\
= \ & (a \to y) \to (x \to y) & \text{by (1)} \\
= \ & x \to ((a \to y) \to y) & \text{Def. 14.33(1)} \\
= \ & x \to ((y \to a) \to a) & \text{Def. 14.33(5)} \\
= \ & 1. & \text{Lemma 14.35(2), hypothesis}
\end{aligned}$$

Therefore, $(x \to y) \to y \le a$, in particular this holds for $a = x \vee y$. So the assertion follows.

(3): Clearly $x', y' \le x' \vee y'$, so $(x' \vee y')' \le x'', y''$ since involution is order-reversing. Hence $(x' \vee y')'$ is a lower bound of x and y. Suppose that l is another lower bound of x and y; then $x' \le l'$ and $y' \le l'$ by Prop. 14.36(6). So $x' \vee y' \le l'$, whence $l'' = l \le (x' \vee y')'$. Therefore, $(x' \vee y')'$ is the greatest lower bound of x and y, so $x \wedge y = (x' \vee y')'$.

(4): Using (3) and (2) we get $x \wedge y = (x' \vee y')' = (y' \vee x')' = ((y' \to x') \to x')' = ((x \to y) \to x')'$, the last equality resulting from Def. 14.33(3).

(5): Using (4) gives $(x' \wedge y')' = ((x' \to y') \to x)'' = (y \to x) \to x = y \vee x = x \vee y$. □

In a lattice implication algebra L we define two additional binary operations

$$\otimes \text{ and } \oplus : x \otimes y = (x \to y')' \quad \text{and} \quad x \oplus y = x' \to y. \tag{14.7}$$

The next two propositions give some of their properties.

Proposition 14.38. ([33, Theorem 2.1.21]) *Let L be a lattice implication algebra, then for any $x, y, z \in L$:*

(1) $x \otimes y = y \otimes x, \qquad x \oplus y = y \oplus x.$
(2) $(x \otimes y) \otimes z = x \otimes (y \otimes z), \qquad (x \oplus y) \oplus z = x \oplus (y \oplus z).$
(3) $x \le y \to z$ *iff* $x \otimes y \le z.$
(4) $x \otimes y \le x, y \le x \oplus y.$
(5) $x \le a \ \& \ y \le b \to x \otimes y \le a \otimes b \ \& \ x \oplus y \le a \oplus b.$
(6) *(a)* $x \otimes (y \vee z) = (x \otimes y) \vee (x \otimes z), \qquad$ *(b)* $x \otimes (y \wedge z) = (x \otimes y) \wedge (x \otimes z).$
(7) *(a)* $x \oplus (y \vee z) = (x \oplus y) \vee (x \oplus z), \qquad$ *(b)* $x \oplus (y \wedge z) = (x \oplus y) \wedge (x \oplus z).$

Proof. (1): Using Def. 14.33(3) we have $x \otimes y = (x \to y')' = (y \to x')' = y \otimes x$ and $x \oplus y = x' \to y = y' \to x = y \oplus x$.

(2):
$$\begin{aligned}
(x \otimes y) \otimes z &= ((x \to y')' \to z')' & \\
&= (z \to (x \to y'))' & \text{Def. 14.33(3)} \\
&= (x \to (z \to y'))' & \text{Def. 14.33(1)} \\
&= (x \to (y \to z'))' & \text{Def. 14.33(3)} \\
&= ((y \to z')' \to x')' & \text{Def. 14.33(3)}
\end{aligned}$$

$$= (y \otimes z) \otimes x$$
$$= x \otimes (y \otimes z). \qquad \text{by (1)}$$

(3): By (1) $x \otimes y = y \otimes x$, so $x \otimes y \leq z$ iff $y \otimes x \leq z$ iff $(y \to x')' \to z = 1$. But

$$(y \to x')' \to z = z' \to (y \to x') = y \to (z' \to x') = y \to (x \to z) = x \to (y \to z)$$

Thus, $x \otimes y \leq z$ iff $x \to (y \to z) = 1$ iff $x \leq y \to z$.

(4): By Prop. 14.36(7) we have $x \leq y \to x$, and by (3) $x \leq y \to x$ iff $x \otimes y \leq x$. Similarly, $x \leq y' \to x = y \oplus x = x \oplus y$ using (1).

(5): From the assumption we get $b' \leq y'$ by Prop. 14.36(6). So $(a \to b') \leq (a \to y') \leq (x \to y')$ by Prop. 14.36(4),(5). Therefore, again using Prop. 14.36(6), $(x \to y')' \leq (a \to y')' \leq (a \to b')'$ so $x \otimes y \leq a \otimes y \leq a \otimes b$.
Since $b' \leq y'$, $y' \to x \leq b' \to x$ by Prop. 14.36(4), i.e. $x \oplus y \leq x \oplus b$. Moreover, $x \leq a$ implies $b' \to x \leq b' \to a$ using Prop. 14.36(5), i.e. $x \oplus b \leq a \oplus b$

(6): (a) ([17, Lemma 3]) Since $x \leq x$ and $y, z \leq (y \vee z)$, from (5) we have $x \otimes y \leq x \otimes (y \vee z)$ and $x \otimes z \leq x \otimes (y \vee z)$, so $(x \otimes y) \vee (x \otimes z) \leq x \otimes (y \vee z)$. Let $u \in L$ be another upper bound, i.e. $x \otimes y, x \otimes z \leq u$. Then, by (1), $y \otimes x, z \otimes x \leq u$, whence $y \leq x \to u$ and $z \leq x \to u$ from (3), whence $y \vee z \leq x \to u$, and so $x \otimes (y \vee z) \leq u$, again using (3). Thus $x \otimes (y \vee z) = \sup\{x \otimes y, x \otimes z\} = x \otimes y \vee x \otimes z$.
(b) From (5) we get $x \otimes (y \wedge z) \leq x \otimes y, x \otimes z$, whence $x \otimes (y \wedge z) \leq (x \otimes y) \wedge (x \otimes z)$. Let l be another lower bound of $x \otimes y, x \otimes z$; then from (4) $l \leq x, y, z$ follows. So $l \leq y \wedge z$, which, again using (5), implies that $x \otimes (y \wedge z)$ is the greatest lower bound.

(7): (a) Recall that $x \oplus (y \vee z) = x' \to (y \vee z)$ from (14.7) and that $x' \to y, x' \to z \leq x' \to (y \vee z)$ by Prop. 14.36(5) since $y, z \leq y \vee z$. Thus we get $(x' \to y) \vee (x' \to z) \leq x' \to (y \vee z)$, i.e. $(x \oplus y) \vee (x \oplus z) \leq x \oplus (y \vee z)$. Let $u \in L$ be such that $x \oplus y, x \oplus z \leq u$; then, by (4), $x, y, z \leq u$. Therefore, $y \vee z \leq u$, and $x \oplus (y \vee z) \leq x \oplus u$ by (5). So $x \oplus (y \vee z)$ is the least upper bound of $x \oplus y, x \oplus z$, i.e. $x \oplus (y \vee z) = (x \oplus y) \vee (x \oplus z)$.
(b) From (5) we get $x \oplus (y \wedge z) \leq x \oplus y, x \oplus z$, whence $x \oplus (y \wedge z) \leq (x \oplus y) \wedge (x \oplus z)$. The rest is analogous to 6(b). □

Corollary 14.39. *Let* $(L, \wedge, \vee, ', \to, 0, 1)$ *be a lattice implication algebra. Then* $(L, \wedge, \vee, \otimes, \to, 0, 1)$ *is a residuated, and in case L is complete, a Brouwerian lattice.*

Proposition 14.40. ([22, Proposition 1]) *Let L be a lattice implication algebra, then for any* $x, y, z \in L$*:*

(1) $x' \vee y \leq x \to y$.
(2) $(x \otimes y)' = x' \oplus y'$,
$(x \oplus y)' = x' \otimes y'$.
(3) $0 \otimes x = 0, \quad 1 \otimes x = x, \quad x \otimes x' = 0,$
$0 \oplus x = x, \quad 1 \oplus x = 1, \quad x \oplus x' = 1.$
(4) $x \otimes y = 0$ *implies* $x \leq y'$.

Proof. (1): $(x' \vee y) \to (x \to y) = (x' \to (x \to y)) \wedge (y \to (x \to y)) = (x' \to (y' \to x')) \wedge (x \to (y \to y)) = y' \to (x' \to x') = 1$, using in that order Def. 14.33(6), (3) and (1), as well as Def. 14.33(2) and Lemma 14.35(3). The assertion now follows by Lemma 14.34(1).

(2): $(x \otimes y)' = ((x \to y')')' = x \to y' = x'' \to y' = x' \oplus y'$.

(3): $0 \otimes x = (0 \to x')' = 1' = 0$, using Lemma 14.35(3) and Lemma 14.34(2).

(5): $x \otimes y = (x \to y')' = 0$ implies $(x \to y') = 0' = 1$, so $x \leq y'$. □

14.3.3.2 *LI-Relation Algebras*

The material of this section is taken from [22]. We need the following

Lemma 14.41. ([33, Theorem 2.1.12]) *If $(L, \vee, \wedge, ', \to, 0, 1)$ is a lattice implication algebra, then $(L, \vee, \wedge)$ is a distributive lattice.*

Proof.

$$\begin{aligned} x \vee (y \wedge z) &= ((y \wedge z) \to x) \to x && \text{commutativity of } \vee, \text{Prop. 14.37(2)} \\ &= ((y \to x) \vee (z \to x)) \to x && \text{Def. 14.33(7)} \\ &= ((y \to x) \to x) \wedge ((z \to x) \to x) && \text{Def. 14.33(6)} \\ &= (y \vee x) \wedge (z \vee x) && \text{Prop. 14.37(2)} \end{aligned}$$

Together with Prop. 1.46 distributivity follows. □

Definition 14.42. ([22, Definition 4])
*A LI-***relation algebra** *(LIRA) is a structure $\mathfrak{L} = (L, \vee, \wedge, ', \to, 0, 1, ;, \breve{}, \mathbb{1})$ of type $\langle 2, 2, 1, 2, 0, 0, 2, 1, 0 \rangle$ such that*

(1) $(L, \vee, \wedge, ', \to, 0, 1)$ is a complete lattice implication algebra,
(2) $(L, ;, \mathbb{1})$ is a monoid,
(3) $(x \vee y); z = (x; z) \vee (y; z)$,
(4) $(x \to y)\breve{} = x\breve{} \to y\breve{}$,
(5) $x\breve{}\breve{} = x$,
(6) $\mathbb{1}' \vee \mathbb{1} = 1$,
(7) $(x; y)\breve{} = y\breve{}; x\breve{}$,
(8) $\left(x\breve{}; (x; y)'\right) \vee y' = y'$, i.e. $x\breve{}; (x; y)' \leq y'$,
(9) $(a \otimes x); (b \otimes y) \leq (a; b) \otimes (x; y)$,

for all $a, b, x, y, z \in L$, where $x \otimes y = (x \to y')'$.

Every relation algebra $\mathfrak{R} = (A, +, \cdot, \bar{}, 0, 1, ;, \breve{}, \mathbb{1})$ (see Def. 14.2) is a LIRA if we define $x \to y = x' \vee y$ for any $x, y \in A$. The motivating example, however, for the above definition is the following: Let X be any non-empty set and $(L, \wedge, \vee, ', \to, 0, 1)$ a lattice implication algebra; recall that $\mathcal{F}_L^X$ denotes the set of all L-fuzzy relations on X. For any $R, S \in \mathcal{F}_L^X$ and any $x, y \in X$ we define operations as follows:

(L1) $(R \vee S)(x, y) = R(x, y) \vee S(x, y)$,
(L2) $(R \wedge S)(x, y) = R(x, y) \wedge S(x, y)$,
(L3) $(R \to S)(x, y) = R(x, y) \to S(x, y)$,
(L4) $R'(x, y) = (R(x, y))'$, $R\breve{}(x, y) = R(y, x)$,
(L5) $R; S(x, y) = \bigvee_{z \in X} \left(R(x, z) \otimes S(z, y)\right)$, and

(L6) $I(x,y) = \delta_{xy}$.

Then one can show by routine calculations that $(\mathcal{F}_L^X, \vee, \wedge, ', \rightarrow, \emptyset, U_X)$ is a lattice implication algebra. Associativity for composition: $(R;S);T = R;(S;T)$, and $R;I = I;R = R$ hold by Th. 9.33(1),(3), i.e. $(\mathcal{F}_L^X, ;, I)$ is a monoid. So we have shown Def. 14.42(1) and (2). (5) is obvious, (6) follows from $I \vee I' = U_X$ due to Lemma 14.34(2), and (4) we get as follows: $(R \rightarrow S)^{\smile}(x,y) = (R \rightarrow S)(y,x) = R(y,x) \rightarrow S(y,x) = R^{\smile}(x,y) \rightarrow S^{\smile}(x,y)$. Concerning (3) we have

$$\begin{aligned}
[(R \vee S);T](x,y) &= \textstyle\bigvee_{z \in X} [R(x,z) \vee S(x,z)] \otimes T(z,y) \\
&= \textstyle\bigvee_{z \in X} \left[\left(R(x,z) \otimes T(z,y) \right) \vee \left(S(x,z) \otimes T(z,y) \right) \right] \\
&= \textstyle\bigvee_{z \in X} \left[\left(R(x,z) \otimes T(z,y) \right) \right] \vee \bigvee_{z \in X} \left[\left(S(x,z) \otimes T(z,y) \right) \right] \\
&= R;T(x,y) \vee S;T(x,y).
\end{aligned}$$

Similar straightforward calculations can be done to show (7) and (9). The proof of (8) is more sophisticated and maybe not easy to follow in [22, p. 414f]; so we include it below in a hopefully somewhat easier to grasp manner.

First note that it suffices to prove one of $R^{\smile};(R;S)' \leq S'$ or the equivalent $S \leq \left(R^{\smile};(R;S)'\right)'$ (see Prop. 14.36(6)). We will show the second inequality. Further recall from Prop. 14.37 and Prop. 14.40(2) that the De Morgan laws are valid in L for both $\vee, \wedge$ and $\oplus, \otimes$. This will be used for the transformation from the first to the second line below: $(\bigvee a \otimes b')' = \bigwedge a' \oplus b$. And by Lemma 14.41 L is distributive. Then, for any $z \in L$ we have

$$\begin{aligned}
\left(R^{\smile};(R;S)'\right)'(x,y) &= \left(\textstyle\bigvee_{z \in X} \left[R^{\smile}(x,z) \otimes \left(\bigvee_{w \in X} R(z,w) \otimes S(w,y) \right)' \right] \right)' \\
&= \textstyle\bigwedge_{z \in X} \left[(R(z,x))' \oplus \bigvee_{w \in X} \left(R(z,w) \otimes S(w,y) \right) \right] \\
&= \textstyle\bigwedge_{z \in X} \bigvee_{w \in X} \left[(R(z,x))' \oplus \left(R(z,w) \otimes S(w,y) \right) \right].
\end{aligned}$$

We use the following abbreviation: $RRS := \bigvee_{w \in X, w \neq x} \left[(R(z,x))' \oplus \left(R(z,w) \otimes S(w,y) \right) \right]$ for notational convenience and get

$$\begin{aligned}
&\textstyle\bigvee_{w \in X} \left[(R(z,x))' \oplus \left(R(z,w) \otimes S(w,y) \right) \right] \\
&\quad = RRS \vee \left[(R(z,x))' \oplus \left(R(z,x) \otimes S(x,y) \right) \right] \\
&\quad = RRS \vee \left[R(z,x) \rightarrow \left(R(z,x) \rightarrow S(x,y)' \right)' \right] && \text{by definition of } \oplus \text{ and } \otimes \\
&\quad = RRS \vee \left[\left(R(z,x) \rightarrow S(x,y)' \right) \rightarrow R(z,x)' \right] && \text{by Def. 14.33(3)} \\
&\quad = RRS \vee \left[\left(S(x,y) \rightarrow R(z,x)' \right) \rightarrow R(z,x)' \right] && \text{by Def. 14.33(3)} \\
&\quad = RRS \vee \left[S(x,y) \vee R(z,x)' \right] && \text{by Prop. 14.37(2)} \\
&\quad \geq S(x,y).
\end{aligned}$$

Summarizing the above, we have succeeded to prove the following result for L-fuzzy relations:

Corollary 14.43. *$(\mathcal{F}_L^X, \vee, \wedge, ', \rightarrow, \emptyset, U_X, ;, {}^{\smile}, I)$ is an LI-relation algebra.*

What as always remains is to check if this concept has the properties that we would expect taking Section 14.2.1 as reference:

Proposition 14.44. ([22, Proposition 2]) *Let $\mathfrak{L} = (L, \vee, \wedge, ', \rightarrow, 0, 1, ;, \breve{}, \mathbb{1})$ be a LI-relation algebra and $x, y, z \in L$. Then the following hold:*

(1) $(x \vee y)\breve{} = x\breve{} \vee y\breve{}, \quad (x \wedge y)\breve{} = x\breve{} \wedge y\breve{}$.
(2) $1\breve{} = 1, \quad \mathbb{1}\breve{} = \mathbb{1}, \quad 0\breve{} = 0$.
(3) $(x')\breve{} = (x\breve{})'$.
(4) $(x \oplus y)\breve{} = x\breve{} \oplus y\breve{}, \quad (x \otimes y)\breve{} = x\breve{} \otimes y\breve{}$.
(5) $x = y \Leftrightarrow x\breve{} = y\breve{}$.
(6) $x; (y \vee z) = (x; y) \vee (x; z)$.
(7) $x \le y$ *implies* $x; z \le y; z$.
(8) $y \le z$ *implies* $x; y \le x; z$.
(9) $y' = y' \vee \big((y; x\breve{})'; x\big)$, *i.e.* $(y; x\breve{})'; x \le y'$.
(10) $x; y \le z' \;\Leftrightarrow\; x\breve{}; z \le y' \;\Leftrightarrow\; z; y\breve{} \le x'$.
(11) $x\breve{} \le y\breve{} \;\Leftrightarrow\; x \le y$.
(12) $x; 0 = 0; x = 0$.

Proof. (1): Using Def. 14.42(4) and Prop. 14.37(2) we get $(x \vee y)\breve{} = ((x \rightarrow y) \rightarrow y)\breve{} = (x\breve{} \rightarrow y\breve{}) \rightarrow y\breve{} = x\breve{} \vee y\breve{}$. The second part can be proved using Prop. 14.37(3).

(2): By (1) $1 = 1\breve{} \vee 1 = 1\breve{} \vee 1\breve{}\breve{} = (1\breve{} \vee 1)\breve{} = 1\breve{}$. Moreover, $0 \rightarrow 0\breve{} = 1$, whence $(0 \rightarrow 0\breve{})\breve{} = 0\breve{} \rightarrow 0\breve{}\breve{} = 0\breve{} \rightarrow 0 = 1\breve{} = 1$; and $\mathbb{1}\breve{} = \mathbb{1}; \mathbb{1}\breve{} = \mathbb{1}\breve{}\breve{}; \mathbb{1}\breve{} = (\mathbb{1}; \mathbb{1}\breve{})\breve{} = (\mathbb{1}\breve{})\breve{} = \mathbb{1}$.

(3): Using Lemma 14.35(2) twice, Def. 14.42(4), and (2) we get $(x')\breve{} = (x \rightarrow 0)\breve{} = x\breve{} \rightarrow 0\breve{} = x\breve{} \rightarrow 0 = (x\breve{})'$.

(4): With (3) we get $(x \oplus y)\breve{} = (x' \rightarrow y)\breve{} = (x')\breve{} \rightarrow y\breve{} = (x\breve{})' \rightarrow y\breve{} = x\breve{} \oplus y\breve{}$.

(5): $x = y$ clearly implies $x\breve{} = y\breve{}$. Applying this to $x\breve{} = y\breve{}$ gives $x\breve{}\breve{} = y\breve{}\breve{}$.

(6): Using (1) and Def. 14.42(3),(7) we get $(x; (y \vee z))\breve{} = (y \vee z)\breve{}; x\breve{} = (y\breve{} \vee z\breve{}); x\breve{} = (y\breve{}; x\breve{}) \vee (z\breve{}; x\breve{}) = ((x; y) \vee (x; z))\breve{}$. Now use (5) to conclude $x; (y \vee z) = (x; y) \vee (x; z)$.

(7): $x \le y$ implies $y; z = (x \vee y); z = (x; z) \vee (y; z)$, whence $x; z \le y; z$.

(9): By Def. 14.42(8) $(y\breve{})' = (y\breve{})' \vee (x\breve{}; (x; y\breve{})')$, whence $((y\breve{})')\breve{} = \big[(y\breve{})' \vee (x\breve{}; (x; y\breve{})')\big]\breve{}$. Applying (1) and (3) proves the assertion, since $((y\breve{})')\breve{} = (y\breve{}\breve{})' = y'$ and $(x\breve{}; (x; y\breve{})')\breve{} = (x; y\breve{})')\breve{}; x\breve{}\breve{} = ((x; y\breve{})\breve{})'; x = (y; x\breve{})'; x$.

(10): Let $x; y \le z'$; then $z \le (x; y)'$, whence $x\breve{}; z \le x\breve{}; (x; y)' \le y'$ by (8) and Def. 14.42(8). $x\breve{}; z \le y'$ implies $y \le (x\breve{}; z)'$, so $x; y \le x\breve{}\breve{}; (x\breve{}; z)' \le z'$, again applying Def. 14.42(8). So the first equivalence is proved. Similarly: $x; y \le z' \Rightarrow z \le (x; y)' \Rightarrow z; y\breve{} \le (x; y)'; y\breve{} \le x'$, using (9) for the last inequality (with $x \leftarrow y, y \leftarrow x\breve{}$). Conversely, $z; y\breve{} \le x' \Rightarrow x \le (z; y\breve{})' \Rightarrow x; y \le (z; y\breve{})'; y \le z'$.

(11): Using (1) yields $x \le y \Leftrightarrow x \vee y = y \Leftrightarrow (x \vee y)\breve{} = y\breve{} \Leftrightarrow x\breve{} \vee y\breve{} = y\breve{}$, so $x\breve{} \le y\breve{}$. Applying this argument to $x\breve{} \le y\breve{}$ shows the other direction.

(12): By Prop. 14.40(3),(4) we have $(x\breve{}; 1) \otimes 0 = 0$, so $(x\breve{}; 1) \le 1$. And the first equivalence of (10) yields $(x\breve{}; 1) \le 1 \Leftrightarrow x; 0 \le 0$. □

Compared to classical relation algebras and MV-relation algebras, relatively few typical properties can be shown for LI-relation algebras; in particular, analogues to the

De Morgan and Tarski equivalences are missing (see page 447). This should not come as a surprise when noticing the difference between the implication Prop. 14.40(4): if $x \otimes y = 0$ then $x \leq y'$, and the equivalence Prop. 14.19: $x \leq y$ iff $x \odot \neg y = 0$.

14.3.4 *Other Non-Classical Relation Algebras*

To complete the picture we only mention three more variants of relation algebras to be found in literature, since too many technical details would be required even for a short description. So we confine our information to relevant links for each of them: Concerning *Rough Relation Algebras,* [6, 7] and [8] together provide the basic concepts and results. *Quantum Relation Algebras* are discussed in [32], and *Stone Relation Algebras* in [13].

References

[1] Chang, C.C.: Algebraic analysis of many valued logics, Trans. of the American Math. Soc. **88**, 467-490 (1958).

[2] Cignoli, R.L.O., D'Ottaviano, I.M.L., Mundici, D.: *Algebraic Foundations of Many-Valued Reasoning,* Springer Science+Business Media Dordrecht (2000).

[3] De Morgan, A.: On the syllogism no. IV, and on the logic of relations, Trans. Cambridge Philos. Soc. **10**, 331-358 (1864).

[4] Demri, S.P., Orłowska, E.S.: *Incomplete Information: Structure, Inference, Complexity,* Springer-Verlag Berlin Heidelberg (2002).

[5] Di Nola, A., Russo, C.: The semiring-theoretic approach to MV-algebras: A survey, Fuzzy Sets and Systems **281**, 134-154 (2015).

[6] Düntsch, I.: Rough relation algebras, Fundamenta Informaticae **21**, 321–331 (1994).

[7] Düntsch, I.: Rough sets and algebras of relations, in: Orłowska, E. (Ed.), *Incomplete Information: Rough Set Analysis,* Springer-Verlag Berlin Heidelberg, pp. 95-108 (1998).

[8] Düntsch, I., Winter, M.: Rough relation algebras revisited, Fundamenta Informaticae **74**, 283-300 (2006). (Also TR # CS-05-11, Brock University, Department of Computer Science, St. Catharines, Ontario (2005).)

[9] Furusawa, H.: Algebraic Formalisations of Fuzzy Relations and Their Representation Theorems, PhD thesis, Kyushu Univ., Dpt. of Informatics, Fukuoka, Japan (1998).

[10] Givant, S.: *Introduction to Relation Algebras-Relation Algebras, Volume 1,* Springer International Publishing AG (2017).

[11] Gottwald, S.: *A Treatise on Many-Valued Logics,* Research Studies Press Ltd., Baldock, Hertfordshire, England (2001).

[12] Grattan-Guinness, I.: *The Search for Mathematical Roots, 1870-1940,* Princeton University Press, Princeton, New Jersey (2000).

[13] Guttmann, W.: Stone relation algebras, in: Höfner, P. et al. (Eds.), *RAMiCS 2017,* LNCS 10226, pp. 127-143 (2017).

[14] Huntington, E.V.: New sets of independent postulates for the algebra of logic, with special reference to Whitehead and Russell's Principia Mathematica, Trans. of the American Math. Soc. **35**, 274-304 (1933).

[15] Jónsson, B., Tarski, A.: Representation problems for relation algebras, Bull. of the American Math. Soc. **54**, 79-80 (1948).

[16] Kawahara, Y., Furusawa, H.: An algebraic formalization of fuzzy relations, Fuzzy Sets and Systems **101**, 125-135 (1999).

[17] Kondo,M., Dudek, W.A.: Characterization theorem of lattice implication algebras, Far East J. Math. Sci. **13**, 325-342 (2004).

[18] Maddux, R.D.: Some varieties containing relation algebras, Trans. of the American Math. Soc. **272**, 501-526 (1982).

[19] Maddux, R.D.: The Origin of Relation Algebras in the Development and Axiomatization of the Calculus of Relations, Studia Logica **50**, 421-455 (1991).

[20] Maddux, R.D.: Relation-algebraic semantics, Theoretical Computer Science **160**, 1-85 (1996).

[21] Maddux, R.D.: *Relation Algebras*, Elsevier Science B.V., Amsterdam (2006).

[22] Pan, X., Xu, Y.: On the algebraic structure of binary lattice-valued fuzzy relations, Soft Computing **17**, 411-420 (2013).

[23] Peirce, C.S.: Description of a notation for the logic of relatives, resulting from an amplification of the conceptions of Boole's calculus of logic, Memoirs of the American Academy of Arts and Sciences, New Series **9**, 317-378 (1873). (Reprinted in *Writings of Charles S. Peirce, A Chronological Edition, Volume 2, 1867-1871*, Indiana University Press, Bloomington and Indianapolis, pp. 359-429 (1984).)

[24] Popescu; A.: Many-valued relation algebras, Algebra universalis **53**, 73-108 (2005).

[25] Popescu; A.: Some algebraic theory for many-valued relation algebras, Algebra universalis **56**, 211-235 (2007).

[26] Schmidt, G., Ströhlein, T.: *Relations and Graphs*, Springer-Verlag Berlin Heidelberg (1993).

[27] Schröder, E.: *Vorlesungen über die Algebra der Logik (Exakte Logik). Dritter Band: Algebra und Logik der Relative*, B.G. Teubner, Leipzig (1895). (Reprinted by Chelsea Publishing Company, New York (1966).)

[28] Stillwell, J.: *Mathematics and Its History*, 3rd ed., Springer Science+Business Media (2010).

[29] Tarski, A.: On the Calculus of Relations, J. Symbolic Logic **6**, 73-89 (1941).

[30] Tarski, A., Givant, S.: *A Formalization of Set Theory without Variables*, American Mathematical Society (1987).

[31] Toth, H.: *Relations: Concrete, Abstract and Applied - An Introduction*, World Scientific Publishing, Singapore (2020).

[32] Winter, M.: Towards Quantum Relation Algebras, TR # CS-12-05, Brock University, Department of Computer Science, St. Catharines, Ontario (2012).

[33] Xu, Y., Ruan, D., Qin, K., Liu, J.: *Lattice-Valued Logic*, Springer-Verlag Berlin Heidelberg (2003).

[34] Yon, Y.H.: On quasi-lattice implication algebras, J. Appl. Math. & Informatics **33**, 739-748 (2015).

[35] Zadeh, L.A.: Fuzzy Sets, Inform. and Control **8**, 338-353 (1965).

CHAPTER 15

FUZZY RELATIONS AND CATEGORIES

Like a *category* can be viewed as the first abstraction step away from sets and functions between them, with **Set** as the prototypical example, so does the concept of *allegory* with respect to sets and relations. And not really surprising, **Set** turns out to be an important subcategory of **Rel**, the motivating standard example for allegories.

As you will see from the references mentioned, the results obtained under a category theoretical viewpoint on fuzzy relations are the work of few people. Our exposition in this chapter mainly rests on the book of Winter [13], who may be called the leading figure, since it comprises also the general case of L-fuzzy relations. The classical reference on allegories is Freyd and Scedrov [5]; there are also chapters on relations and allegories in [3] and [8].

15.1 Categorical Prerequisites

In this section, we provide the most basic definitions and properties around categories in the shortest possible way. For a more detailed introduction you may have a look at e.g. [11, Chapter 11] or one of the references mentioned there like [2, 10].

The following nice description of category theory is given on p. 42 of [4]: "Category theory … provides a bag of concepts which, through abstraction, describes many different structures and formalisms from the various branches of mathematics in a uniform way. … Categories are essentially collections of objects (abstract version of a mathematical structure) and morphisms (abstract version of structure preserving functions) between them."

This could easily lead to the impression of category theory being another highly formal, but otherwise simple set of tools generalizing some well known mathematical concepts. However, the first few sentences from the *Introduction* of Leinster [10] give a wonderful characterization of the category theoretical reality, but at the same time also a kind of warning for the innocent reader, indicating where he will have to climb to: "Category theory takes a bird's eye view of mathematics. From high in the sky, details become invisible, but we can spot patterns that were impossible to detect from ground level. ... seeing patterns in mathematics that you may never have seen before."

Definition 15.1. *A* **category** $\mathcal{C}$ *consists of*

(a) *a collection* $\mathsf{obj}_{\mathcal{C}}$ *of* **objects** *of* $\mathcal{C}$*;*
(b) *a collection* $\mathsf{mor}_{\mathcal{C}}$ *of* **morphisms** *(or* **arrows***);*
(c) *operations assigning to each arrow* f *an object* $\mathsf{dom}\, f$*, called the* **domain** *(or* **source***) of* f*, and an object* $\mathsf{cod}\, f$*, called the* **codomain** *(or* **target***) of* f*. The collection of all arrows with domain* A *and codomain* B *is denoted by* $\mathsf{mor}_{\mathcal{C}}[A,B]$ *and called the* **hom-set** *between* A *and* B*;*
(d) *a* **composition operator** *assigning to each pair of arrows* f *and* g *with* $\mathsf{cod}\, f = \mathsf{dom}\, g$*, an arrow* $g \cdot f \in \mathsf{mor}_{\mathcal{C}}[\mathsf{dom}\, f, \mathsf{cod}\, g]$ *that satisfies the following associative law: for all arrows* $f : A \to B$*,* $g : B \to C$*, and* $h : C \to D$
$$h \cdot (g \cdot f) = (h \cdot g) \cdot f.$$
(e) *for each* $B \in \mathsf{obj}_{\mathcal{C}}$ *a morphism* $id_B \in \mathsf{mor}_{\mathcal{C}}[B,B]$*, called the* **identity on** B*, such that for* $f \in \mathsf{mor}_{\mathcal{C}}[A,B]$ *and* $g \in \mathsf{mor}_{\mathcal{C}}[B,C]$ *we have*
$$id_B \cdot f = f \text{ and } g \cdot id_B = g.$$

Common alternative notations are: $A \in \mathcal{A}$ for $A \in \mathsf{obj}_{\mathcal{A}}$, and $f : A \longrightarrow B$ or $A \xrightarrow{f} B$ for $f \in \mathsf{mor}_{\mathcal{C}}[A,B]$; and simple concatenation gf for $g \cdot f$.

Given objects and morphisms, one often uses diagrams to depict the relationships between them: Roughly speaking, a **diagram** in a category is a collection of objects together with a collection of arrows between these objects. A diagram as a whole is called a **commuting diagram** if whenever there are two paths *containing at least two arrows* from an object A to an object B, the morphism from A to B obtained by composing along one path is equal to the morphism obtained by composing along the other.

From Def. 15.1 we immediately get the following examples for commuting diagrams: the left one commutes by definition of composition, and the right diagram represents the identity laws.

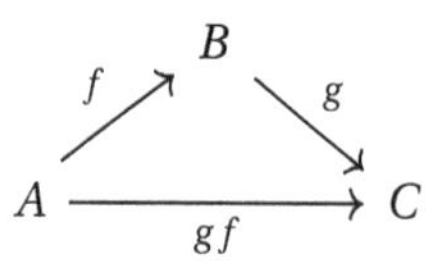

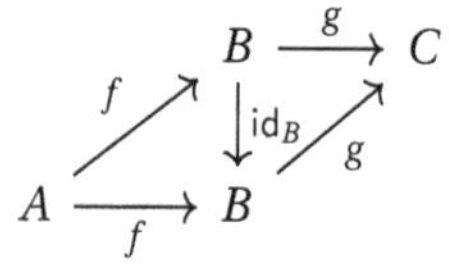

The prototype and motivating example for categories is **Set** which has sets as its objects and total functions between them as arrows. The composition of arrows is set-theoretic function composition, and the identity arrows are the identity functions. From Cor. 2.38(2) we know that totality is preserved by composition. Obviously, for each set A, id_A is total and the associativity and the identity laws hold.

The category **Poset** has as objects all partially ordered sets and as morphisms all order-preserving total functions. This category is one of the most elementary cases where the objects bear some structure that has to be preserved by the morphisms.

Other examples are **Mon** with monoids as objects and monoid homomorphisms as morphisms, or **Top**, the category of topological spaces and continuous maps.

Arrows may have some special properties. In order to give you a feeling about how the abstraction process in category theory works, the most simple case will be motivated by considering analogies for sets and functions, and will be spelled out in some detail. Recall that a function $f: X \longrightarrow Y$ is said to be

- *injective* (or *one-to-one*) when no two distinct inputs give the same output, i.e. for all $x, y \in X$, if $f(x) = f(y)$ then $x = y$; and
- *surjective* (or *onto*) if the codomain B is the range of f, i.e. for each $y \in Y$ there is some $x \in X$ such that $f(x) = y$.

Let $x \in A$ and $f \cdot g(x) = f \cdot h(x)$, i.e. $f(g(x)) = f(h(x))$. If f is injective, we get $g(x) = h(x)$, whence g and h, yielding the same output for every argument, are the same function. So we have shown that an injective function f is left cancellable:

$$\text{whenever } f \cdot g(x) = f \cdot h(x), \text{ then } g = h.$$

Generalizing this to the categorical setting results in

Definition 15.2.

(a) *An arrow $f: B \longrightarrow C$ in a category $\mathcal{C}$ is a* **monomorphism** *(or "is* **monic***") if, for any pair of arrows $g, h: A \longrightarrow B$, the equality $f \cdot g = f \cdot h$ implies that $g = h$.*

$$A \overset{g}{\underset{h}{\rightrightarrows}} B \xrightarrow{f} C$$

(b) *An arrow $f: C \longrightarrow B$ in a category $\mathcal{C}$ is an* **epimorphism** *(or "is* **epic***") if, for any pair of arrows $g, h: B \longrightarrow A$, the equality $g \cdot f = h \cdot f$ implies that $g = h$.*

$$A \overset{g}{\underset{h}{\leftleftarrows}} B \xleftarrow{f} C$$

Proposition 15.3. *In* **Set***, the monomorphisms are just the injective functions.*

Proof. Let $f: B \longrightarrow C$ be an injective function, and let $g, h: A \longrightarrow B$ be such that $f \cdot g = f \cdot h$ but $g \neq h$. Then $g(x) \neq h(x)$ for some $x \in A$. But since f is injective, $f(g(x)) \neq f(h(x))$, a contradiction, whence $g = h$ and f is a monomorphism. Conversely, let $f: B \longrightarrow C$ be a monomorphism. If f is not injective, then there are distinct elements $y, y' \in B$ with $f(y) = f(y')$. Let $A = \{x\}$, and let $g: A \longrightarrow B$, $x \mapsto y$, and $h: A \longrightarrow B$, $x \mapsto y'$. Then $f(g(x)) = f(h(x))$ and $g \neq h$, a contradiction to the assumption that f is a monomorphism. $\square$

15.2 Allegories: The Basics

An allegory is a category enriched with three operators which are inspired by the category of relations **Rel**: One can compare relations using a partial order $\subseteq$, take the intersection of two relations with $\cap$, and transform a relation to its converse using an unary operator $\breve{}$. This results in the following definition.

(*Important remark:* Note that the order of factors in composition now is again the one of usual relational composition as in [13]. This makes sense since allegories are much more directly tailored towards relations from the outset than the approach via regular categories. Moreover, using this kind of notation avoids the cumbersome and error-prone mirroring between categorial and relational formulations. But be careful: In e.g. [5, 2.14] notation for morphism composition is used.)

Definition 15.4. ([13, Definition 4.4]) *A category $\mathcal{C}$ is called an* **allegory** *if the following holds (where composition binds stronger than meet):*

(1) *For all $A, B \in \mathsf{obj}_{\mathcal{C}}$ the hom-set $\mathsf{mor}_{\mathcal{C}}[A, B]$ is a meet semilattice, where meet and the induced ordering are denoted by $\cap$ and $\subseteq$, respectively.*
(2) *There is a monotone operation $\breve{}$ assigning to each arrow $r : A \longrightarrow B$ a* **converse arrow** *$r\breve{} : B \longrightarrow A$ such that for all $s : B \longrightarrow C$ and $t : A \longrightarrow C$*

$$r\breve{}\breve{} = r,$$
$$(r \cdot s)\breve{} = s\breve{} \cdot r\breve{},$$
$$r \cdot s \cap t \subseteq r \cdot (s \cap r\breve{} \cdot t). \quad \textit{(modular law)}$$

(3) *For all arrows $r : A \longrightarrow B$ and $s, t : B \longrightarrow C$ we have subdistributivity of composition over meets:*

$$r \cdot (s \cap t) \subseteq r \cdot s \cap r \cdot t.$$

No wonder that the prototypical example of an allegory is the category **Rel** which has sets as objects and binary relations as arrows (i.e. an arrow $\rho : A \longrightarrow B$ is a subset $\rho \subseteq A \times B$), with subset inclusion of relations as its partial ordering on arrows, intersection as the meet of two arrows, and relational conversion as conversion of arrows. The identity arrow on a set A is I_A, and since composition of relations is associative, **Rel** is indeed a category.

The structure imposed on allegories mirrors the property that the collection of binary relations on a fixed set is a meet semilattice, and that composition is subdistributive over meets. So, allegories certainly are a suitable kind of category to start with when considering binary relations. Again, a kind of additional check from the reverse point of view is to consider which properties known to hold for binary relations are derivable for allegories from their definition. Some of them are given in

Proposition 15.5. ([13, Lemma 4.5]) *Let $\mathcal{C}$ be an allegory, $A, B, C \in \mathsf{obj}_{\mathcal{C}}$, and $r, s \in \mathsf{mor}_{\mathcal{C}}[A, B]$, $t \in \mathsf{mor}_{\mathcal{C}}[B, C]$ and $u \in \mathsf{mor}_{\mathcal{C}}[A, C]$. Then*

(1) $id_A\breve{} = id_A$.

(2) $(r \cap s)^{\smile} = r^{\smile} \cap s^{\smile}$.

(3) $(r \cap s) \cdot t \subseteq r \cdot t \cap s \cdot t$.

(4) $r \subseteq s \to r \cdot t \subseteq s \cdot t$ *and* *(monotonicity of* $\cdot$*)*
$t \subseteq u \to r \cdot t \subseteq r \cdot u$.

(5) $r \subseteq r \cdot r^{\smile} \cdot r$.

(6) $r \cdot s \cap t \subseteq (r \cap t \cdot s^{\smile}) \cdot s$. *(modular law, dual variant)*

(7) $r \cdot s \cap t \subseteq (r \cap t \cdot s^{\smile}) \cdot (s \cap r^{\smile} \cdot t)$. *(Dedekind rule)*

(8) *The two modular rules together are equivalent to the Dedekind rule.*

Proof. (1): $id_A^{\smile} = id_A^{\smile} \cdot id_A = id_A^{\smile} \cdot (id_A^{\smile})^{\smile} = (id_A^{\smile} \cdot id_A)^{\smile} = (id_A^{\smile})^{\smile} = id_A$.

(2): Using monotonicity of $^{\smile}$ and $r^{\smile\smile} = r$ we get

$$\begin{aligned} x \subseteq r^{\smile} \cap s^{\smile} &\leftrightarrow x \subseteq r^{\smile} \wedge x \subseteq s^{\smile} \\ &\leftrightarrow x^{\smile} \subseteq r \wedge x^{\smile} \subseteq s \\ &\leftrightarrow x^{\smile} \subseteq r \cap s \\ &\leftrightarrow x \subseteq (r \cap s)^{\smile} \end{aligned}$$

(3): Using (2) and Def. 15.4(2) we get

$$\begin{aligned} (r \cap s) \cdot t &= [(r \cap s) \cdot t]^{\smile\smile} \\ &= [t^{\smile} \cdot (r^{\smile} \cap s^{\smile})]^{\smile} \\ &\subseteq [t^{\smile} \cdot r^{\smile} \cap t^{\smile} \cdot s^{\smile}]^{\smile} \\ &= [(r \cdot t)^{\smile} \cap (s \cdot t)^{\smile}]^{\smile} \\ &= [(r \cdot t \cap s \cdot t)^{\smile}]^{\smile} \\ &= r \cdot t \cap s \cdot t \end{aligned}$$

(4): From (3) we get $(r \cap s) \cdot t \subseteq r \cdot t \cap s \cdot t$; assume $r \subseteq s$, then $r \cdot t = (r \cap s) \cdot t \subseteq r \cdot t \cap s \cdot t \subseteq r \cdot t$. Thus $r \cdot t \cap s \cdot t = r \cdot t$, whence $r \cdot t \subseteq s \cdot t$.

(5): Using the modular law we have

$$r = r \cdot id_B \cap r \subseteq r \cdot (id_B \cap r^{\smile} \cdot r) \subseteq r \cap r \cdot r^{\smile} \cdot r \subseteq r \cdot r^{\smile} \cdot r.$$

(6): By applying conversion to the modular law of Def. 15.4(2) we get $s^{\smile} \cdot r^{\smile} \cap t^{\smile} \subseteq (s \cap r^{\smile} \cdot t)^{\smile} \cdot r^{\smile} = (s^{\smile} \cap t^{\smile} \cdot r) \cdot r^{\smile}$. Now the assertion follows by appropriate renaming: $s^{\smile} := r, r^{\smile} := s, t^{\smile} := t$.

(7):
$$\begin{aligned} r \cdot s \cap t &= (r \cdot s \cap t) \cap t \\ &\subseteq (r \cap t \cdot s^{\smile}) \cdot s \cap t && \text{by (6)} \\ &\subseteq (r \cap t \cdot s^{\smile}) \cdot (s \cap (r \cap t \cdot s^{\smile})^{\smile} \cdot t) && \text{by the modular law} \\ &\subseteq (r \cap t \cdot s^{\smile}) \cdot (s \cap r^{\smile} \cdot t) && \text{by monotonicity of } ^{\smile} \text{ and } \cdot \end{aligned}$$

(8): From the monotonicity of composition it is straightforward that the Dedekind rule implies the modular laws. Conversely, assume that the modular laws (6) and Def. 15.4(2) hold. Then

$$\begin{aligned} r \cdot s \cap t &= (r \cdot s \cap t) \cap t \\ &\subseteq u \cdot s \cap t && \text{using (6), with } u := r \cap t \cdot s^{\smile} \\ &\subseteq u \cdot (s \cap u^{\smile} \cdot t) && \text{using Def. 15.4(2)} \\ &\subseteq u \cdot (s \cap r^{\smile} \cdot t) && u \subseteq r, \text{ so } u^{\smile} \subseteq r^{\smile}\text{; monotonicity} \end{aligned}$$

□

15.3 Special Properties of Arrows

Some basic facts on relations that we have learned about in Prop. 2.18 should not come as a surprise: In **Rel** these definitions correspond to the set-theoretic ones.

Definition 15.6. ([13, Definition 4.6]) *Let $\mathcal{C}$ be an allegory, $A, B \in \mathsf{obj}_{\mathcal{C}}$, and $r \in \mathsf{mor}_{\mathcal{C}}[A, B]$. Then r is called*

(a) **univalent** *iff* $r^{\smile} \cdot r \subseteq id_B$.
(b) **total** *iff* $id_A \subseteq r \cdot r^{\smile}$.
(c) *a* **map** *iff it is univalent and total.*

Proposition 15.7. *Let $\mathcal{C}$ be an allegory, $A, B, C \in \mathsf{obj}_{\mathcal{C}}$, $r \in \mathsf{mor}_{\mathcal{C}}[A, B]$, and $s \in \mathsf{mor}_{\mathcal{C}}[B, C]$. Then the following holds:*

(1) *The class of univalent relations is closed under composition.*
(2) *The class of total relations is closed under composition.*
(3) *The class of maps is closed under composition.*
(4) *If $r \cdot s$ is total, so is r.*

Proof. (1) $(r \cdot s)^{\smile} \cdot r \cdot s = s^{\smile} \cdot r^{\smile} \cdot r \cdot s \subseteq s^{\smile} \cdot s \subseteq id_C$.

(2) $r \cdot s \cdot (r \cdot s)^{\smile} = r \cdot s \cdot s^{\smile} \cdot r^{\smile} \supseteq r \cdot r^{\smile} \supseteq id_A$.

(3) Follows from (1) and (2).

(4): $id_A = id_A \cap r \cdot s \cdot (r \cdot s)^{\smile}$	$r \cdot s$ total
$= r \cdot s \cdot s^{\smile} \cdot r^{\smile} \cap id_A$	
$\subseteq (r \cdot s \cdot s^{\smile} \cap id_A \cdot r) \cdot r^{\smile}$	Prop. 15.5(6), monotonicity
$\subseteq r \cdot r^{\smile}$.	

□

For univalent and total arrows the modular laws and subdistributivity of composition over meets can be strengthened as follows (see Def. 15.4, and also Cor. 2.39):

Proposition 15.8. *Let $\mathcal{C}$ be an allegory, and $A, B, C \in \mathsf{obj}_{\mathcal{C}}$. Then the following holds:*

(1) *For $r : A \longrightarrow B$ univalent and $s, t : B \longrightarrow C$: $r \cdot (s \cap t) = r \cdot s \cap r \cdot t$.*
(2) *For $r, s : A \longrightarrow B$ and $t : B \longrightarrow C$ total: $(r \cap s) \cdot t = r \cdot t \cap s \cdot t$.*
(3) *For $r : A \longrightarrow B$, $s : B \longrightarrow C$ univalent, and $t : A \longrightarrow C$: $r \cdot s \cap t = (r \cap t \cdot s^{\smile}) \cdot s$.*
(4) *For $r : A \longrightarrow B$ total, $s : B \longrightarrow C$ and $t : A \longrightarrow C$: $r \cdot s \cap t = r \cdot (s \cap r^{\smile} \cdot t)$.*

Proof. (1): Due to Def. 15.4(3). we have to show one inequality only and get $r \cdot s \cap r \cdot t \subseteq r \cdot (s \cap r^{\smile} \cdot r \cdot t) \subseteq r \cdot (s \cap t)$ using the modular law first, and then that r is assumed univalent, i.e. $r^{\smile} \cdot r \subseteq id_B$.

(2): Due to Prop. 15.5(3) we have to show one inequality only and get $r \cdot t \cap s \cdot t \subseteq (r \cap s \cdot t \cdot t^{\smile}) \cdot t \subseteq (r \cap s) \cdot t$ using Prop. 15.5(6) first, and then that t is assumed total.

(3): We only have to show one inequality and get $(r \cap t \cdot s^{\smile}) \cdot s \subseteq r \cdot s \cap t \cdot s^{\smile} \cdot s \subseteq r \cdot s \cap t$ using Prop. 15.5(4) first, and then that s is univalent.The other direction is Prop. 15.5(6). □

Lemma 15.9. ([13, Lemma 4.8]) *Let C be an allegory, $r : A \longrightarrow B$, $s : A \longrightarrow C$, and $t : D \longrightarrow B$ be relations; and let $f : B \longrightarrow C$ and $g : A \longrightarrow D$ be maps. Then*

(1) $r \cdot f \subseteq s$ *iff* $r \subseteq s \cdot f^{\smile}$;
(2) $g^{\smile} \cdot r \subseteq t$ *iff* $r \subseteq g \cdot t$.

Proof. (1): From $r \cdot f \subseteq s$ we get $r \subseteq r \cdot f \cdot f^{\smile} \subseteq s \cdot f^{\smile}$ since f is total. Conversely, $r \subseteq s \cdot f^{\smile}$ implies $r \cdot f \subseteq s \cdot f^{\smile} \cdot f \subseteq s$ since f is univalent. □

Recall from Prop. 2.18 that two-valued relations ρ such that $\rho \subseteq I$ have been called coreflexive or a subidentity. Quite correspondingly, in allegories all $r \in \mathsf{mor}_C[A, A]$ with $r \subseteq id_A$ are called **subidentities** or **partial identities.** In Prop. 2.24(5)-(7) the following properties have been shown for coreflexive relations ρ, σ: ρ is symmetric, transitive, and idempotent; $\rho\sigma = \rho \cap \sigma = \sigma\rho$. The following proposition states the analogous results for allegories.

Proposition 15.10. ([6, Proposition 6.1.1], [12, Lemma 11]) *Let C be an allegory, $A \in \mathsf{obj}_C$, and $r, s \in \mathsf{mor}_C[A, A]$ be partial identities. Then*

(1) $r^{\smile} = r = r \cdot r$;
(2) $r \cdot s = r \cap s$.
(3) $r \cdot s = s \cdot r$.

Proof. (1): Using Prop. 15.5(5) we get $r \subseteq r \cdot r^{\smile} \cdot r \subseteq id_A \cdot r^{\smile} \cdot id_A \subseteq r^{\smile}$. Since $^{\smile}$ is monotone and involutive, we have $r \subseteq r^{\smile} \rightarrow r^{\smile} \subseteq r^{\smile\smile} = r$. Transitivity is implied by monotonicity: $r \subseteq id_A \rightarrow r \cdot r \subseteq id_A \cdot r = r$. And for idempotency we use the modular law and monotonicity again: $r = r \cdot r \cap \top \subseteq r \cdot (r \cap r^{\smile} \cdot \top) \subseteq r \cdot r$.

(2): We have $r \cdot s \subseteq r \cdot id_A = r$ and $r \cdot s \subseteq id_A \cdot s = s$, whence $r \cdot s \subseteq r \cap s$. Noting that $r \cap s$ is a partial identity, by (1) we get $r \cap s = (r \cap s) \cdot (r \cap s) \subseteq r \cdot s$, .

(3): Follows immediately from (2): $r \cdot s = r \cap s = s \cap r = s \cdot r$. □

15.4 Special Types of Allegories

We have already learned that the set of all binary relations among others exhibits the structure of e.g. a distributive lattice with a least element. Following [13], this gives rise to a whole hierarchy of allegory specializations which will be described below. We will also have a short look on some characteristic properties of these variants together with some example proofs .

15.4.1 *Distributive Allegories*

Definition 15.11. ([13, Definition 4.11]) *A* **distributive allegory** *is an allegory $\mathcal{C}$ which additionally satisfies the following conditions:*

(1) *The classes* $\mathsf{mor}_{\mathcal{C}}[A,B]$ *are distributive lattices with union* $\cup$ *and least element* $\perp_{AB}$.
(2) *For all relations* $r : A \longrightarrow B$ *we have* $r \cdot \perp_{BC} = \perp_{AC}$.
(3) *For all relations* $r : A \longrightarrow B$ *and* $s,t : B \longrightarrow C$ *we have* $r\cdot(s\cup t) = r\cdot s\cup r\cdot t$.

As already mentioned on page 268, $L = \mathbf{2}$ and $L = [0,1]$ are special cases of the general concepts of *L-fuzzy relations*, $(L, \wedge, \vee, \otimes, \frown, 0, 1)$ being a complete residuated lattice. Accordingly, **Rel** and [0,1]-**Rel** are special cases of L-**Rel** which has sets as objects and L-fuzzy relations as arrows. So it will suffice to state assertions for the latter one.

Corollary 15.12. L-**Rel** *is a distributive allegory.*

Proof. See Th. 9.33, with $0_{X\times Y}$, defined by $0_{X\times Y}(x,y) = 0$ for all $x \in X, y \in Y$, as least element. □

Proposition 15.13. ([13, Lemma 4.12]) *Let $\mathcal{C}$ be a distributive allegory, $A,B,C \in \mathsf{obj}_{\mathcal{C}}$, and let $r,s \in \mathsf{mor}_{\mathcal{C}}[A,B]$, $t \in \mathsf{mor}_{\mathcal{C}}[B,C]$. Then*

(1) $\perp_{AB}^{\smallsmile} = \perp_{BA}$,
(2) $\perp_{CA}\cdot r = \perp_{CB}$,
(3) $(r\cup s)^{\smallsmile} = r^{\smallsmile}\cup s^{\smallsmile}$,
(4) $(r\cup s)\cdot t = r\cdot t\cup s\cdot t$.

Proof. (1): For any $x \in \mathsf{mor}_{\mathcal{C}}[B,A]$, $\perp_{AB} \subseteq x^{\smallsmile}$, whence $\perp_{AB}^{\smallsmile} \subseteq x$.

(2): Using (1) and Def. 15.11(2) we get $\perp_{CA}\cdot r = (r^{\smallsmile}\cdot\perp_{CA}^{\smallsmile})^{\smallsmile} = (r^{\smallsmile}\cdot\perp_{AC})^{\smallsmile} = \perp_{BC}^{\smallsmile} = \perp_{CB}$.

(3): $r^{\smallsmile}\cup s^{\smallsmile} \subseteq (r\cup s)^{\smallsmile}$ is straightforward since $^{\smallsmile}$ is monotone by definition of an allegory. The converse inclusion follows from

$$\begin{aligned}(r\cup s)^{\smallsmile}\subseteq r^{\smallsmile}\cup s^{\smallsmile} &\leftrightarrow r\cup s\subseteq (r^{\smallsmile}\cup s^{\smallsmile})^{\smallsmile}\\ &\leftrightarrow r\subseteq (r^{\smallsmile}\cup s^{\smallsmile})^{\smallsmile} \ \&\ s\subseteq (r^{\smallsmile}\cup s^{\smallsmile})^{\smallsmile}\\ &\leftrightarrow r^{\smallsmile}\subseteq r^{\smallsmile}\cup s^{\smallsmile} \ \&\ s^{\smallsmile}\subseteq r^{\smallsmile}\cup s^{\smallsmile}.\end{aligned}$$

(4):
$$\begin{aligned}(r\cup s)\cdot t &= (t^{\smallsmile}\cdot(r\cup s)^{\smallsmile})^{\smallsmile}\\ &= (t^{\smallsmile}\cdot(r^{\smallsmile}\cup s^{\smallsmile}))^{\smallsmile} && \text{by (3)}\\ &= (t^{\smallsmile}\cdot r^{\smallsmile}\cup t^{\smallsmile}\cdot s^{\smallsmile})^{\smallsmile} && \text{by distributivity Def. 15.11(3)}\\ &= ((r\cdot t)^{\smallsmile}\cup(s\cdot t)^{\smallsmile})^{\smallsmile}\\ &= r\cdot t\cup s\cdot t && \text{by (3).}\end{aligned}$$

□

15.4.2 *Division Allegories*

The next step in the hierarchy of allegories are (variants of) division allegories which are characterized by the fact that $\cdot$ is a lower adjoint.

Definition 15.14.

(a) *A* **division allegory** $\mathcal{R}$ *is a distributive allegory such that for all relations* $r : B \longrightarrow C$ *and* $s : A \longrightarrow C$ *there is a relation* $s \backslash r : A \longrightarrow B$, *called the* **left residual** *of* s *and* r, *with* $t \cdot r \subseteq s \leftrightarrow t \subseteq s \backslash r$ *for all* $t : A \longrightarrow B$.

(b) *A* **Dedekind category** $\mathcal{R}$ *is a division allegory such that the classes* $\mathsf{mor}_{\mathcal{R}}[A, B]$ *are complete Brouwerian lattices (i.e. complete Heyting algebras) with a greatest element* $\top_{AB}$. *(They are also called* locally complete division allegories *in* [5].*)*

(c) *A* **Schröder category** $\mathcal{R}$ *is a Dedekind category in which the* $\mathsf{mor}_{\mathcal{R}}[A, B]$ *are (complete) Boolean algebras.*

The corollary below shows that L-**Rel** is also an example for a division and Dedekind category:

Corollary 15.15.

(1) L-**Rel** *is a division allegory.*

(2) L-**Rel** *is a Dedekind category.*

Proof. (1): By Th. 12.5(2) we have $\rho\sigma \subseteq \tau$ iff $\rho \subseteq (\sigma @ \tau^{\smile})^{\smile}$.

(2): By (1) L-**Rel** is a division allegory. The greatest element is the L-fuzzy relation 1_{XY}, with $1_{XY}(x, y) = 1$ for all $x \in X, y \in Y$. So every L-**Rel**$[A, B]$ is a complete Brouwerian lattice by Prop. 4.30(1). □

The Dedekind category L-**Rel** is a Schröder category iff L is a Boolean algebra. But the substructure of crisp relations constitutes a Schröder category since $\overline{0} = 1$ and $\overline{1} = 0$ in any complete residuated lattice; see Prop. 1.33(6).

We continue with some properties of division allegories which we have shown to hold for binary relations in Section 2.8.3:

Corollary 15.16. ([13, Section 4.4]) *Let* $\mathcal{C}$ *be a division allegory,* $A, B, C \in \mathsf{obj}_{\mathcal{C}}$, *and let* $r, r' \in \mathsf{mor}_{\mathcal{C}}[A, B]$, $s, s' \in \mathsf{mor}_{\mathcal{C}}[B, C]$, *and* $t, t' \in \mathsf{mor}_{\mathcal{C}}[A, C]$. *Then we have*

(1) $\mathcal{C}$ *has right residuals.*

(2) $r \subseteq (r \cdot s) \backslash s$ *and* $s \subseteq r / (r \cdot s)$.

(3) $(s \backslash r) \cdot r \subseteq s$ *and* $r \cdot (t / r) \subseteq t$.

(4) *If* $r' \subseteq r$ *and* $s \subseteq s'$ *then* $s \backslash r \subseteq s' \backslash r'$.

(5) *If* $r' \subseteq r$ *and* $t \subseteq t'$ *then* $r / t \subseteq r' / t'$.

(6) $r \backslash id_B = r$ *and* $id_A / r = r$.

Proof. (1): Using Def. 15.4(2), we have

$$r \cdot s \subseteq t \leftrightarrow s^{\smile} \cdot r^{\smile} \subseteq t^{\smile} \leftrightarrow s^{\smile} \subseteq t^{\smile} \backslash r^{\smile} \leftrightarrow s \subseteq (t^{\smile} \backslash r^{\smile})^{\smile}.$$

So, $(t^{\smile} \backslash r^{\smile})^{\smile}$ is the **right residual** of t and r, denoted by t / r.

(2): Obvious from $r \cdot s \subseteq r \cdot s$.

(3): By definition we have $t \cdot r \subseteq s \leftrightarrow t \subseteq s \backslash r$, so $(s \backslash r) \cdot r \subseteq s$.

(4): Using (3) we get $(s\backslash r)\cdot r' \subseteq (s\backslash r)\cdot r \subseteq s \subseteq s'$, so $s\backslash r \subseteq s'\backslash r'$.

(6): By (2) we have $r \subseteq (r\cdot id_B)\backslash id_B = r\backslash id_B$; and from (3) $r\backslash id_B = (r\backslash id_B)\cdot id_B \subseteq r$. □

15.4.2.1 *Dedekind Categories*

By moving on to Dedekind categories we enter the area of complete lattices as underlying structure. Cor. 15.15(2) indicates that Dedekind categories are a suitable structure that embraces the general concept of L-fuzzy relations.

Proposition 15.17. ([13, Lemma 4.21]) *Let $\mathcal{D}$ be a Dedekind category. Then the following holds:*

(1) $\top\breve{}_{AB} = \top_{BA}$.
(2) $\top_{AA}\cdot\top_{AB} = \top_{AB}\cdot\top_{BB} = \top_{AB}$.
(3) $\top_{AB} = \top_{AB}\cdot\top_{BA}\cdot\top_{AB}$.

Proof. (1): For $t \in \mathrm{mor}_{\mathcal{D}}[B,A]$ we have $t\breve{} \subseteq \top_{AB}$, whence $t \subseteq \top\breve{}_{AB}$.

(2): $\top_{AB} = \top_{AB}\cdot id_B \subseteq \top_{AB}\cdot\top_{BB}$, and analogously $\top_{AB} = id_A\cdot\top_{AB} \subseteq \top_{AA}\cdot\top_{AB}$.

(3): Follows directly from (1) and Prop. 15.5(5). □

Proposition 15.18. ([13, Corollary 4.22]) *Let $\mathcal{D}$ be a Dedekind category, and let $r, r_i \in \mathrm{mor}_{\mathcal{D}}[A,B]$ and $s, s_i \in \mathrm{mor}_{\mathcal{D}}[B,C]$. Then*

(1) $\left(\bigcup_i r_i\right)\cdot s = \bigcup_i r_i\cdot s$ *and* $r\cdot\left(\bigcup_i s_i\right) = \bigcup_i r\cdot s_i$.
(2) $\left(\bigcap_i r_i\right)\cdot s \subseteq \bigcap_i r_i\cdot s$ *and* $r\cdot\left(\bigcap_i s_i\right) \subseteq \bigcap_i r\cdot s_i$.

Now we can conclude that L-fuzzy relations and Dedekind categories do indeed fit together quite well. However, it has been shown in [12] that the latter are too weak to express some standard properties of L-fuzzy relations, e.g. that a given L-fuzzy relation $\rho \in \mathcal{F}^L_{X\times Y}$ is a two-valued Boolean relation, also called 0-1-*crisp*, i.e. $\rho(x,y) \in \{0,1\}$, for all $x \in X, y \in Y$. A detailed exposition of the technical prerequisites, involving special formal languages together with appropriate semantics, and two more kinds of categories, are beyond our scope. The topic is studied in detail in [13, Section 4.10 and Chapter 5].

In the last part of this chapter we will give a cursory overview on the two additional kinds of categories, called *arrow* and *Schröder categories*. For the former we need the notion of a *scalar* in a Dedekind category, the definition of which can be motivated as follows: Let us call an L-fuzzy relation a **scalar** on A (induced by $l \in L$) iff $\delta_l(x,y) = \begin{cases} l & \text{iff } x = y \\ 0 & \text{iff } x \neq y \end{cases}$. This obviously constitutes an isomorphism between L and the lattice of scalars, whence one can identify the elements of L and the scalar relations. Moreover, the set of scalars is closed under arbitrary intersections and unions. Clearly, both $\emptyset$ and id_A are scalar relations. Scalar L-fuzzy relations $\delta_l \neq \emptyset$ are strictly l-reflexive in the sense of Section 9.3.2 and can be characterized algebraically as follows:

Proposition 15.19. ([9, Proposition 3.2]) *ρ is a scalar L-fuzzy relation on X iff $\rho \subseteq I$ and $\rho U = U\rho$.*

Proof. For $\rho \subseteq I$ we have $\rho U(x,y) = \bigvee_{z\in X}(\rho(x,z) \wedge U(z,y)) = \rho(x,x) \wedge 1 = \rho(x,x)$ for all $x,y \in X$, and similarly $U\rho(x,y) = \rho(y,y)$. So, assuming ρ to be a scalar, we trivially get $\rho \subseteq I$, and, by what we have just shown, $\rho U(x,y) = \rho(x,x) = \rho(y,y) = U\rho(x,y)$. Conversely assume that $\rho \subseteq I$ and $\rho U = U\rho$. Thus, $\rho(x,y) = 0$ for $x \neq y$. Moreover, $\rho(x,x) = \rho U(x,y) = U\rho(x,y) = \rho(y,y)$, whence ρ is a scalar L-fuzzy relation. □

Definition 15.20. *Let $\mathcal{D}$ be a Dedekind category; a relation $s: A \longrightarrow A$ is called a* **scalar** *on A iff $s \subseteq id_A$ and $\top_{AA} \cdot s = s \cdot \top_{AA}$.*

Lemma 15.21. *Let $\mathcal{D}$ be a Dedekind category, s be a scalar on A, and $r : A \longrightarrow A$ a relation. Then the following holds (we let $\top := \top_{AA}$):*

(1) *$\perp_{AA}$ and id_A are scalars on A.*
(2) *$s \cdot r = r \cap s \cdot \top$ and $r \cdot s = r \cap \top \cdot s$, in particular, $s = id_A \cap s \cdot \top$.*
(3) *$s \cdot r = r \cdot s$.*

Proof. (1): Obviously, $\perp_{AA}$ and id_A are partial identities on A, and $\top \cdot id_A = \top = id_A \cdot \top$ and $\top \cdot \perp_{AA} = \perp_{AA} = \perp_{AA} \cdot \top$, the latter one following from Prop. 15.13(2).

(2): From $s \subseteq id_A$ we get $s \cdot r \subseteq r$, and $r \subseteq \top$ implies $s \cdot r \subseteq s \cdot \top$, so $s \cdot r \subseteq s \cdot \top \cap r$. By the modular law Def. 15.4(2) and Prop. 15.10 we get $s \cdot \top \cap r \subseteq s \cdot (\top \cap s^{\smile} \cdot r) = s \cdot (s^{\smile} \cdot r) = s \cdot r$. The second assertion can be proved in a similar way to the above using Prop. 15.5(6). For the last assertion take $r = id_A$.

(3): Since s is a scalar, by (2) we have $s \cdot r = r \cap s \cdot \top = r \cap \top \cdot s = r \cdot s$. □

15.4.2.2 *Arrow Categories*

We continue with building the category hierarchy by enriching Dedekind categories with two operations that map every relation to its support and core, respectively (see Def. 9.3). In other terms, these operations map a relation to the greatest 0-1 crisp relation it contains and to the least 0-1 crisp relation it is included in.

Definition 15.22. ([13, Definition 5.2]) *An* **arrow category** *$\mathcal{A}$ is a Dedekind category with $\top_{AB} \neq \perp_{AB}$ for all objects A and B together with two operations $^{\uparrow}$ and $^{\downarrow}$ such that the following holds:*

(1) $r^{\uparrow}, r^{\downarrow} : A \longrightarrow B$ for all $r : A \longrightarrow B$.
(2) $r^{\uparrow} \subseteq s \leftrightarrow r \subseteq s^{\downarrow}$ for all $r : B \longrightarrow A$ and $s : B \longrightarrow C$.
(3) $(r^{\smile} \cdot s^{\downarrow})^{\uparrow} = r^{\uparrow\smile} \cdot s^{\downarrow}$ for all $r : B \longrightarrow A$ and $s : B \longrightarrow C$.
(4) $(q \cap r^{\downarrow})^{\uparrow} = q^{\uparrow} \cap r^{\downarrow}$ for all $q, r : A \longrightarrow B$.
(5) If $d \neq \perp_{AA}$ is a nonzero scalar, then $d^{\uparrow} = id_A$.

As indicated above, the interpretation of $^{\uparrow}$ and $^{\downarrow}$ as support and core of an L-fuzzy relation gives the standard model as shown in

Theorem 15.23. ([13, (Lemma 3.3) Theorem 5.3]) *Let L be a complete Brouwerian lattice with $0 \neq 1$. Then L-**Rel** together with $^{\uparrow}$ and $^{\downarrow}$ is an arrow category.*

Proof. Note that for all nonempty sets $\top_{AB} \neq \bot_{AB}$ since $0 \neq 1$.
(1): Obvious.

(2): Suppose $\rho^{\uparrow} \subseteq \sigma$ and $\rho^{\uparrow}(x,y) \neq 0$. Then $1 = \rho^{\uparrow}(x,y) \leq \sigma(x,y)$, and thus $\rho(x,y) \leq 1 = \sigma^{\downarrow}(x,y)$. An analogous proof shows the reverse implication.

(3): We define two operations $^{\uparrow},^{\downarrow} : L \longrightarrow L$ by $x^{\uparrow} = 1$ if $x > 0$ and $x^{\uparrow} = 0$ otherwise; and by $x^{\downarrow} = 1$ if $x = 1$ and $x^{\downarrow} = 0$ otherwise. Then the following holds:

(a) $\rho^{\uparrow}(x,y) = \rho(x,y)^{\uparrow}$,
(b) $\rho^{\downarrow}(x,y) = \rho(x,y)^{\downarrow}$,
(c) $(x \wedge y^{\downarrow})^{\uparrow} = x^{\uparrow} \wedge y^{\downarrow}$, for all $x, y \in L$,
(d) $(\bigvee M)^{\uparrow} = \bigvee_{m \in M} x^{\uparrow}$, forall $M \subseteq L$.

Concerning (c) we have $1 = (x \wedge y^{\downarrow})^{\uparrow}$ iff $x \wedge y^{\downarrow} > 0$ iff $x > 0$ & $y = 1$ iff $x^{\uparrow} = 1$ & $y = 1$ iff $x^{\uparrow} \wedge y^{\downarrow} = 1$, and similar for $0 = (x \wedge y^{\downarrow})^{\uparrow}$. Then the assertion is shown as follows:

$$\begin{aligned}
(\rho^{\smile}\sigma^{\downarrow})^{\uparrow}(x,y) &= (\rho^{\smile}\sigma^{\downarrow})(x,y)^{\uparrow} && \text{by (a)} \\
&= \big(\textstyle\bigvee_z(\rho^{\smile}(x,z) \wedge \sigma^{\downarrow}(x,y))\big)^{\uparrow} \\
&= \big(\textstyle\bigvee_z(\rho(z,x) \wedge \sigma(x,y)^{\downarrow})\big)^{\uparrow} && \text{by (b)} \\
&= \textstyle\bigvee_z\big((\rho(z,x) \wedge \sigma(x,y)^{\downarrow})\big)^{\uparrow} && \text{by (d)} \\
&= \textstyle\bigvee_z\big(\rho(z,x)^{\uparrow} \wedge \sigma(x,y)^{\downarrow}\big) && \text{by (c)} \\
&= \textstyle\bigvee_z\big(\rho^{\uparrow}(z,x) \wedge \sigma^{\downarrow}(x,y)\big) && \text{by (a), (b)} \\
&= \textstyle\bigvee_z\big(\rho^{\uparrow\smile}(x,z) \wedge \sigma^{\downarrow}(z,y)\big) \\
&= (\rho^{\uparrow\smile}\sigma^{\downarrow})(x,y).
\end{aligned}$$

(4): Using the operations $^{\uparrow},^{\downarrow}$ defined above in (3) we get

$$\begin{aligned}
(\rho \cap \sigma^{\downarrow})^{\uparrow}(x,y) &= (\rho \cap \sigma^{\downarrow})(x,y)^{\uparrow} && \text{by (a)} \\
&= (\rho(x,y) \wedge \sigma^{\downarrow}(x,y))^{\uparrow} \\
&= (\rho(x,y) \wedge \sigma(x,y)^{\downarrow})^{\uparrow} && \text{by (b)} \\
&= \rho(x,y)^{\uparrow} \wedge \sigma(x,y)^{\downarrow} && \text{by (c)} \\
&= \rho^{\uparrow}(x,y) \wedge \sigma^{\downarrow}(x,y) && \text{by (a),(b)} \\
&= (q^{\uparrow} \cap r^{\downarrow})(x,y).
\end{aligned}$$

(5): Let δ_l be a scalar L-fuzzy relation, with $\delta_l(x,y) = l > 0$. Then

$$\delta_l^{\uparrow}(x,y) = (\delta_l(x,y))^{\uparrow} = \left\{\begin{matrix} 1 \text{ iff } x = y \\ 0 \text{ iff } x \neq y \end{matrix}\right\} = I(x,y).$$

□

15.4.2.3 *Schröder Categories*

The Dedekind category L-**Rel** is a Schröder category iff L is a Boolean algebra; see Def. 15.14. But the substructure of crisp relations constitutes a Schröder category since $\overline{0} = 1$ and $\overline{1} = 0$ in any complete residuated lattice; see Prop. 4.33(3).

Theorem 15.24. ([13, Theorem 4.49]) (Schröder equivalences) *Let $\mathcal{S}$ be a Schröder category, and $q : A \longrightarrow B$, $r : B \longrightarrow C$, and $s : A \longrightarrow C$ be relations in $\mathcal{S}$. Then*

$$q \cdot r \subseteq s \quad \text{iff} \quad q^{\smile} \cdot \overline{s} \subseteq \overline{r} \quad \text{iff} \quad \overline{s} \cdot r^{\smile} \subseteq \overline{q}.$$

Proof. Assume $q \cdot r \subseteq s$; this is equivalent to $q \cdot r \cap \overline{s} = \bot_{AC}$ since $\mathsf{mor}_{\mathcal{S}}[A, B]$ are (complete) Boolean algebras by Def. 15.14(c). Using the modular law Def. 15.4(2) and Def. 15.11(2) we get $q^{\smile} \cdot \overline{s} \cap r \subseteq q^{\smile} \cdot (\overline{s} \cap q \cdot r) = q^{\smile} \cdot \bot_{AC} = \bot_{BC}$. Hence, $q^{\smile} \cdot \overline{s} \subseteq \overline{r}$. The converse direction can be shown in a similar way: $q^{\smile} \cdot \overline{s} \subseteq \overline{r}$ iff $q^{\smile} \cdot \overline{s} \cap r = \bot_{BC}$. As above we have $q \cdot r \cap \overline{s} \subseteq q \cdot (q^{\smile} \cdot \overline{s} \cap r) = q \cdot \bot_{BC} = \bot_{AC}$, whence $q \cdot r \subseteq s$. By analogous arguments one can prove the second equivalence. □

Corollary 15.25. ([13, Lemma 4.50]) *Let $\mathcal{S}$ be a Schröder category, and $q : A \longrightarrow B$, $r : B \longrightarrow C$, and $s : A \longrightarrow C$ be relations in $\mathcal{S}$. Then we have for the*

(1) *right residual:* $q / s = \overline{q^{\smile} \cdot \overline{s}}$;
(2) *left residual:* $s \backslash r = \overline{\overline{s} \cdot r^{\smile}}$.

Proof. (1): Note that, being a division allegory, $\mathcal{S}$ has left and right residuals by definition and Cor. 15.16(1). The assertion follows from

$$x \subseteq q / s \iff q \cdot x \subseteq s \iff q^{\smile} \cdot \overline{s} \subseteq \overline{x} \iff x \subseteq \overline{q^{\smile} \cdot \overline{s}},$$

where we use the definition of a residual, Th. 15.24, and Boolean algebra in that order to show the above equivalences. □

15.5 Crispness in Allegories

An important concept in fuzzy set theory is 0-1-crispness, and is so not only from a formal syntactical, but also, and perhaps mainly, from a pragmatic semantical point of view. Especially in applications, where the concept of fuzzy sets gets in touch with reality, crispness becomes essential: Evaluating the rule base of a fuzzy controller for a certain set of input values results in a fuzzy set as output, in many cases a fuzzy number. But you cannot tune a physical system by feeding in a fuzzy set. Therefore, so-called defuzzification algorithms are applied to transform the fuzzy output into some definite crisp value. As another example take fuzzy decision theory: here one has to select a specific element from a fuzzy set of alternatives, possibly by using a cut operation. It is thus interesting to investigate how crispness can be handled in a category theoretical context.

We begin our discussion with the definition of 0-1-crispness of L-fuzzy relations: An L-fuzzy relation ρ over $X \times Y$ is called **0-1-crisp** (or **crisp** for short), if $x \rho y \in \{0, 1\}$ for all $\langle x, y \rangle \in X \times Y$. So the empty $\emptyset = 0_{X \times Y}$, the universal $U_{X \times Y}$ and the identity relation I_X are all crisp.

Definition 15.26. *A relation $r : A \longrightarrow B$ in a Dedekind category is called s-**crisp** iff $s \cdot q \subseteq r \rightarrow q \subseteq r$ for all nonzero scalars s and all relations $q : A \longrightarrow B$.*

Concerning L-fuzzy relations the following connection between 0-1 crispness and s-crispness holds:

Proposition 15.27. ([9, Proposition 5.4]), [13, Lemma 4.44]) *In L-**Rel** all s-crisp relations are 0-1 crisp.*

Proof. Suppose $\rho : A \longrightarrow B$ is s-crisp with $\rho(x', y') = l > 0$. Consider the scalar $d_l : A \longrightarrow A$ induced by l (see page 468) and $\sigma : A \longrightarrow B$, with $\sigma(x, y) := \rho(x, y) \frown l$, the relative pseudocomplement of $\rho(x, y)$ in l, i.e. the greatest element in $\{c \in L : l \wedge c \le \rho(x, y)\}$. Then, using Lemma 15.21(2), we get

$$(d_l \cdot \sigma)(x, y) = (d_l \cdot \top_{AB} \cap \sigma)(x, y) = l \wedge (\rho(x, y) \frown l) \le \rho(x, y).$$

Since ρ is s-crisp we have $\sigma \sqsubseteq \rho$, whence $1 = l \frown l = \rho(x', y') \frown l = \sigma(x', y') \le \rho(x', y')$. Thus ρ is 0-1 crisp. □

The converse is, in general, not true, i.e. the notion of s-crispness coincides with 0-1 crispness only if an additional assumption holds in L as shown in the next proposition. For this we need the following notion: An element x of a lower semilattice L with least element 0 is called **linear** if $x \wedge y = 0 \rightarrow y = 0$, for all $y \in L$. (Note that in a linear poset every nonzero element has this property. The converse implication does not hold; consider the poset $\{0 \le a \le b, a \le c\}$.)

Proposition 15.28. ([9, PROPOSITION 5.5]), [13, Lemma 4.45]) *In L-**Rel** the following properties are equivalent:*

(1) All 0-1 crisp relations are s-crisp.
(2) All nonzero elements of L are linear.

Proof. Assume (1), and let $0 < k, l \in L$ such that $k \wedge l = 0$. Then d_k is nonzero and by Prop. 15.10(2) we have $d_k \cdot d_l = d_k \cap d_l = 0_{AA}$, where $0_{AA}(x, y) = 0$ for all $x, y \in A$ clearly is the least L-fuzzy relation on A. Since 0_{AA} is 0-1 crisp it is also s-crisp by assumption, whence $d_l \sqsubseteq 0_{AA}$ and $l = 0$.

Conversely, assume (2) and suppose that $d_k\sigma \sqsubseteq \rho$ for $0 < k \in L$, $\sigma : A \longrightarrow B$ and that $\rho : A \longrightarrow B$ is 0-1 crisp. To prove that ρ is s-crisp we have to show $\sigma \sqsubseteq \rho$. Since $\rho(x, y) \in \{0, 1\}$ it suffices to prove $\rho(x, y) = 0 \rightarrow \sigma(x, y) = 0$. If $\rho(x, y) = 0$ we get $0 = \rho(x, y) \le d_k\sigma(x, y) = k \wedge \sigma(x, y)$. Bus since k is linear by assumption we must have $\sigma(x, y) = 0$. □

Thus, under the assumption that all nonzero elements of L are linear, the notion of s-crispness coincides with 0-1 crispness for L-fuzzy relations. In particular, this holds for the classical case of $[0, 1]$-fuzzy relations.

Recall that by Th. 15.23 L-**Rel** is an arrow category, if L is a complete Brouwerian lattice with $0 \ne 1$. Obviously, $\rho \in L$-**Rel** is 0-1 crisp iff $\rho^{\uparrow} = \rho$ iff $\rho^{\downarrow} = \rho$.

Additional Reading

We close this chapter with a few remarks:

- For classical relations there exists another approach which uses concepts and constructions different from the one presented in the current chapter (like e.g. pullback, epi-mono factorization structure, regular categories, 2-categories etc.). Interested readers may want to have a look at [11, Chapter 11] for an introductory overview, where the relationships between regular categories and allegories are described in Section 11.4.3.
- Harding et al. [7] discuss a category **FRel** with fuzzy sets as objects, i.e. $\mathsf{obj}_{\mathbf{FRel}} = \{A | A : X \longrightarrow [0,1]\}$ for some set X, or simply $\mathsf{obj}_{\mathbf{FRel}} = \{(X, A)\}$; and with morphisms $f : (X, A) \longrightarrow (Y, B)$ the relations $R \subseteq X \times Y$ such that $A(x) \leq B(y)$ for all $\langle x, y\rangle \in R$.
- In the recent survey [1] the authors investigate the properties of monoidal categories with multivalued relations as morphisms: According to how the underlying sets and relations are taken as either crisp or $\mathbb{L}$-fuzzy one clearly gets four possibilities: (crisp, crisp), (crisp, $\mathbb{L}$-fuzzy), ($\mathbb{L}$-fuzzy, crisp), ($\mathbb{L}$-fuzzy, $\mathbb{L}$-fuzzy).

References

[1] Alcântara, M.S.S., Dias, T. et. al: A survey of categorical properties of $\mathbb{L}$-fuzzy relations, Fuzzy Sets and Systems **425**, 62-82 (2021).

[2] Awodey, S.: *Category Theory*, 2nd ed., Oxford University Press, New York (2010).

[3] Bird, R., de Moor, O.: *Algebra of Programming*, Prentice Hall Europe (1997).

[4] Chowdhury, A.W.: An Abstract Algebraic Theory of L-Fuzzy Relations for Relational Databases, Master Thesis,, Faculty of Mathematics and Science, Brock University, St. Catharines, Ontario, Canada (2015).

[5] Freyd, P.J., Scedrov, A.: *Categories, Allegories*, North-Holland, Amsterdam (1990).

[6] Furusawa, H.: Algebraic Formalisations of Fuzzy Relations and Their Representation Theorems, PhD thesis, Dpt. of Informatics, Kyushu Univ., Fukuoka, Japan (1998).

[7] Harding, J., Walker, C., Walker, E.: Categories with fuzzy sets and relations, Fuzzy Sets and Systems **256**, 149-165 (2014).

[8] Johnstone, P.T.: *Sketches of an Elephant - A Topos Theory Compendium*, Oxford University Press (2003).

[9] Kawahara, Y., Furusawa, H.: Crispness in Dedekind categories, Bull. of Informatics and Cybernetics **33**, 1-18 (2001).

[10] Leinster, T.: *Basic Category Theory*, Cambridge University Press, Cambridge, UK (2014), available at `https://arxiv.org/abs/1612.09375`, and released under the Creative Commons Attribution-NonCommercial-ShareAlike 4.0 International licence (CC BY-NC-SA 4.0).

[11] Toth, H.: *Relations: Concrete, Abstract and Applied - An Introduction*, World Scientific Publishing, Singapore (2020).

[12] Winter, M.: A new algebraic approach to L-fuzzy relations convenient to study crispness, Information Sciences **139**, 233-252 (2001).

[13] Winter, M.: *Goguen Categories - A Categorical Approach to L-fuzzy Relations*, Springer Science & Business Media (2007).

INDEX

www.ingramcontent.com/pod-product-compliance
Lightning Source LLC
LaVergne TN
LVHW081935200926
848484LV00009B/2434